D0856174

Safety and Accident Prevention in Chemical Operations

Second Edition

SAFETY AND ACCIDENT PREVENTION IN CHEMICAL OPERATIONS

Second Edition

Howard H. Fawcett and
William S. Wood

Released from
Samford University Library

A Wiley-Interscience Publication
JOHN WILEY & SONS
New York   Chichester   Brisbane   Toronto   Singapore

Samford University Library

Copyright © 1982 by John Wiley & Sons, Inc.

All rights reserved. Published simultaneously in Canada.

Reproduction or translation of any part of this work
beyond that permitted by Section 107 or 108 of the
1976 United States Copyright Act without the permission
of the copyright owner is unlawful. Requests for
permission or further information should be addressed to
the Permissions Department, John Wiley & Sons, Inc.

Library of Congress Cataloging in Publication Data:

Main entry under title:
    Safety and accident prevention in chemical operations.

    Bibliography: p.
    Includes index.
    1. Chemical industries—Safety measures.
I. Fawcett, Howard H.   II. Wood, William Samuel,
1913–
TP149.S197   1982   363.1'1966   82-2623
ISBN 0-471-02435-X   AACR2

Printed in the United States of America

10 9 8 7 6 5 4 3 2 1

Stanford University Library

Annex
363.1
2nd ed.

TP
149
.S197
1982

Contributors

GEORGE T. AUSTIN, Department of Chemical Engineering, Washington State University, Pullman, Washington

ROBERT D. COFFEE, Eastman Kodak Company, Rochester, New York

ARTHUR D. CRAVEN, Consultant, West Yorkshire, England

HOWARD H. FAWCETT, Consultant, Wheaton, Maryland

D. J. FORBES, Exxon Research and Engineering, Florham Park, New Jersey

GEORGE L. GORBELL, Consultant, Louisiana, Missouri

ARTHUR B. GUISE, Fire Consultant, Marinette, Wisconsin

J. K. KIELMAN, Consultant, Washington, D.C.

D. JACK KILIAN, School of Public Health, University of Texas, Houston, Texas

TREVOR A. KLETZ, Imperial Chemical Industries, Ltd., Wilton Middlesborough, Cleveland, England

F. OWEN KUBIAS, SCM Glidden Coatings and Resins, Cleveland, Ohio

D. M. LISTON, King Research, Inc., Rockville, Maryland

ELLIOTT MACDERMOD, deceased

CLYDE MCKINLEY, Air Chemicals & Products, Inc., Allentown, Pennsylvania

D. N. MELDRUM, National Foam Systems, Lionville, Pennsylvania

A. J. MURPHY, Nursing Consultant, Greensboro, North Carolina

JOSEPH NICHOLS, Consultant, Richmond, Virginia

GAIL P. NORSTROM II, Industrial Risk Insurers, Chicago, Illinois

RICHARD W. PRUGH, Engineering Department, E. I. duPont de Nemours & Co., Wilmington, Delaware

DONALD RICHMOND, Anheuser-Busch, Inc., St. Louis, Missouri

RICHARD D. ROSS, Chemical Waste Management, Inc., Norristown, Pennsylvania

v

84-00911

IRVING R. TABERSHAW, Rockville, Maryland

KATHLEEN J. TIERNEY, Department of Sociology, University of California, Los Angeles, California

DAVID.T. SMITH, Tucson, Arizona

JACK S. SNYDER, Merck & Company, Rahway, New Jersey

FLOYD A. VAN ATTA, Consultant, Portland, Oregon

WILLIAM S. WOOD, Consultant, West Chester, Pennsylvania

EDMUND D. ZERATSKY, The Ansul Company, Marinette, Wisconsin

Foreword

The benefits which chemicals have brought modern society are well documented and continue to increase in potential. The concept of "taking apart and reassembling molecules" has been the basis of the search for "better living through chemistry." However, as chemicals have progressed from the glassware of the research scientist to the everyday components of our lives, certain problems involving their hazard properties have become obvious.

In one sense, chemicals are packaged energy and will react with whatever is available and receptive. There has been a natural reluctance to widely publicize the risks or hazards associated with chemical production, use, transportation, and disposal, because such information is not understood by persons with limited training in chemistry. However, society has been made aware that with the benefits have come risks. The media, including newspapers, magazines, radio, and television, have, at times, given coverage to events in a way that distorts the true story and leaves little room for a full airing of the facts. Corporate legal staffs often feel compelled to recommend that industry refrain from presenting its sincere views. This defensive stance has impeded communication of toxicological and other hazard information.

When the first edition of this book was being prepared in the early 1960s, much was already known and in place. Our documentation of the control measures and procedures then in common practice clearly reflected the concern that was manifest in many circles. Within a decade, however, the socio-political climate became charged with information that had not been scientifically validated. In addition, cases of admittedly lax control of chemical hazards were documented. Subsequently a series of laws (which are reviewed in Chapters 1, 2, 10, 11, 28, 31, and 37) were promulgated and enforced by the federal government and by many states.

Awareness that certain chemical materials may produce in humans carcinogenic, mutagenic, teratogenic, and other adverse effects many years after exposure, and the widespread publicity to each new utterance in this field regardless of its validity, eroded much of the public confidence that the "chemical age" with its benefits was really under control.

This book is dedicated to the proposition that all chemicals, biologicals, and radiation sources can be processed, handled, used, and disposed of safely *if* precautions, based on specific knowledge and careful planning, are observed. Where there are gaps in application of knowledge, lapses of attention, or failure of mechanical controls, harm probably will occur.

We have not applied all our available knowledge about man's oldest and more unruly servant—fire. Even in the chemical industry, fires and explosions account for too much human suffering and economic loss. It is small wonder, then, that only in recent times have we become concerned about the less obvious effects, such as chronic toxicity, in the four million known chemical compounds, or even of the 100,000 in actual daily commerce.

The authors of chapters were chosen for their expertise in specific areas. It is unrealistic to expect that any person can anticipate all the permutations and combinations of chemical utilization. For that reason, we suggest that the material in this book, while accurate within its context, be used with a measure of discrimination.

Preface

The wide range of subjects covered in this book may be considered a reflection of the highly specialized nature of chemical health and safety. In selecting the contributions made by many dedicated persons, we are hopeful that an interdisciplinary exchange will be encouraged.

The "safety inspectors" of yesteryear who preached rubber gloves and goggles have been succeeded by professionals in such specialties as human factors, cost/benefit analysis, systems safety analyses, industrial hygiene, fire prevention engineering, stability and reactivity evaluation, toxicology, health physics, occupational medicine, and industrial nursing. We are hopeful that this second edition will encourage more interplay and teamwork among professionals toward the goal of more complete control of chemicals.

It is perhaps significant that no one has heretofore collected under one cover a treatise on safety in the chemical industry. The diversity of problems has caused most of the outstanding men in the field of chemical safety to specialize to a considerable degree. Thus few candidates would be qualified to write a book covering the whole subject. We, the editors, likewise felt our incompetence, and for this reason we have tried to collect some of the best thinking of leaders in their respective specialties. We feel that herein is reflected the distilled wisdom and judgment of practical men and women who have made some phase of accident prevention their life work, and who have demonstrated over and over in their daily endeavors that accidents can be prevented and controlled.

Since effective control of accidental injury and property damage in chemical operations is based on such a broad spectrum of knowledge, the scope of this book is necessarily quite wide. Chemical safety requires the study of man in his industrial environment, considers the properties of the materials involved, requires exhaustive analysis of the chemical reaction or operation, and also covers the final packaging and shipping of the product in an acceptable manner. At any place in this chain, loss of control may lead to injury—from mechanical forces, from fires, from explosives, or from toxicity. These accidents, in addition to injuring the worker, possibly causing permanent injury or even death, also constitute a financial loss. They delay the process and reduce the contribution that the industry makes to society. Losses due to accidents can be materially reduced by application of the principles set forth in this book. Each chapter discusses a phase of chemical safety with direct application in many operations. Included are numerous literature references which will enable the reader to make a more thorough study than is possible here.

Even if this book were perfect, we concede that it would still not prevent all accidents, for, as we have tried to emphasize, we do not completely understand the human being who works surrounded by the bewildering array of tanks, pipes, valves, vessels, and pumps used in chemical operations. The more nearly we approach, biologically and psychologically, an adequate understanding of man, the more nearly we will find success in accident control.

Many whose names do not appear in print have supported and encouraged us in the preparation of this edition. To these behind-the-scenes friends, we extend our sincere appreciation and gratitude.

To the supervisors in the chemical industry, to the designers of chemical equipment, to the safety personnel involved with chemical operations, to the management of chemical and related industries, we submit this edition in the sincere hope that it will be a useful tool in helping to prevent injuries and losses due to accidents in chemical operations.

HOWARD H. FAWCETT
WILLIAM S. WOOD

Wheaton, Maryland
West Chester, Pennsylvania
August 1982

Contents

Safety and Accident Prevention in Chemical Operations

Second Edition

1

Chemical Health and Safety: Is Our Future Secure?

Howard H. Fawcett

"The past is prologue" has been often cited as an out for anyone who wishes to shun his or her concern for chemical health and safety. Not too long ago, safety and health professionals were given little more than lip service, as both industry and academe largely ignored the subject of health and safety. During the past two decades, and especially since 1970, legal requirements were established that the lawyers could cite to the courts as reasons why or why not certain health and safety activities or programs should be undertaken or enforced. To do something because it is a law is to admit that one has not really analyzed the problem; much of the opposition to chemical health and safety comes from an inability to present a comprehensive view that human life, including the environment in which humans live, are interesting and rewarding subjects for more intensive study. The awareness of "society" to the potential hazards of chemicals, when mishandled, is now established.

In this chapter, we present three somewhat different viewpoints on the same theme: *to know is to survive* and *to ignore fundamentals is to court disaster.*

Section 1.1 relates to the stake which society has in chemical health and safety.

Section 1.2 relates to a redefinition of chemical safety in terms to which the professional chemist and engineer can relate.

Section 1.3 suggests the changing nature of chemical hazards, real or potential, and the emergencies they may be expected to cause if lessons are not learned.

1

We recommend that this chapter be read in its entirety, in order that the larger concept may be understood.

1.1.   SOCIETY'S STAKE IN CHEMICAL HEALTH AND SAFETY

Concern for human values—health, safety, economic independence, meaningful employment, decent housing, adequate and nutritious food, a livable environment, and ultimately, death with dignity from natural causes at an advanced age—has not always been highly regarded on the scale of values of society or its institutions. Within the past two decades, however, an awareness has sprouted and begun to grow that perhaps additional attention must be paid to these concerns.

R. H. Tawney[1] expresses his feeling that:

> A reasonable estimate of economic organization must allow for the fact that, unless industry is to be paralyzed by the recurrent revolts on the part of outraged human nature, it must satisfy criteria, which are not purely economic. [We might extend this suggestion to all decision-makers at all levels.]

We are realists; we fully recognize there are no simple or one-shot cures to the many complex problems which are posed by the human condition and aspirations. The world is *not* really populated by four billion *people*—rather, it is Mother Earth for four billion *individuals*, each one a physically and mentally unique chemical system, each with diverse and continually changing aspirations, hopes, and desires. In the real world, absolutes or zero-risks are *not* realistic health and safety goals. Where the four and one-half million chemicals react with four billion individuals, each a complete chemical system within himself or herself, and with the changing environment, errors, both of commission and of omission are almost certainly to be expected.[2]

William Rowe has accurately stated the case:[3]

> The only certainty in life is death; uncertainty lies in when and how death occurs, and whether it is final. Man strives to delay its onset and extend the quality of life in the interim. Threats to these objectives involve risks, some natural, some man-made, some beyond our control, and some controllable.

Risk is defined by Rowe as the "potential for realization of unwanted, negative consequences of an event." Expressed in other terms, risk is part of the price one pays for the benefit expected.[4a] In the widely publicized incident of loss of cooling water at the Three Mile Island nuclear power plant in March 1979, the potential risk to the population from uncontrolled release of radioactive isotopes to the air and to the water was seen as a highly undesired cost of the operation. Until the three-year studies of population at especial risk, namely, pregnant women (of whom about 8000 were estimated in the Three Mile Island area), and the more detailed action for complete draining and shutdown of the plant, with subsequent removal of the waste to some remote area for ultimate disposal or reprocessing, the total cost of this incident will not be taken as a complete picture.[4b,c,d,e,f] On the other hand, the benefits of nuclear power are great, and the importance of atomic versus fossil fuel energy has been well documented.

When considering chemical health and safety, the trade-off is usually noted as "risk" versus "benefit," between so-called "safe" and "unsafe," between "healthy" and "unhealthful"—all vague terms which must be related to the conditions, circumstances, human conduct, and the alternatives. The key question remains: "What is an acceptable risk?" To answer such a question for a particular situation, we must determine acceptance to whom (the individual, the environment, or society at large) within the framework of the risk/benefit analysis. Neither term in this ratio is simple, constant, or absolute, for if the benefits are sufficiently large, most humans will elect to exposure to high risks or great odds. Gambling casinos are witnesses to the human feeling that "I can beat the odds."

Risk/benefit analysis is generally viewed as a special form of cost/benefit analysis. In recent years, it has become one of the considerations of legislation in matters affecting human and environmental health and safety.

In a forum on risk/benefit analysis cosponsored by the American Association for the Advancement of Science and the science subcommittees from the House and Senate, a consensus was projected:

1. Zero-risk situations are not realistic, and may stifle innovation.
2. Risk/benefit analysis is here to stay, and policy makers have no choice but to deal with it.
3. The methodology for risk/benefit analysis is nascent, inadequate and imprecise because neither risks or benefits can be identified, let alone quantified, independently, from the subjective judgmental perception of the assessor and of society.

The conclusion was that risk/benefit analysis is useful to Congress only if it is not taken too seriously. It is an uncertain guide for public policy, but it can stimulate thoughts and provide a framework for discussion, and there are ways to improve its reliability and usefulness. A caveat, however, is that the power in Congress is so decentralized that consensus is never reached, and risk/benefit analysis, as imperfect as it is, will be used by opposing parties to support their interests.[4d]

Our standards of values are rapidly changing, and "value vertigo," which Rowe defines as the loss of fixed standards or points of reference, has replaced established norms in many situations. Even the question of the dollar value of a human life is answered by a series of legal judgments, each of which has within itself several considerations. Dean Chauncy Starr has advanced the theory that as long as the individual has a choice to opt for or to decline the risk, he can decide on its acceptance to him. When the risk is either not known, or its magnitude too uncertain for quantification, the decision is often made on emotional or economic grounds alone. Starr feels that the risk of disease is an average level of risk which society considers acceptable in judging other risks.[4a]

The above discussion should in no way detract from an awareness and appreciation for the major contributions which chemicals and other products of our developing technology have made and are continuing to make to our standard of living, to our food supply and to food quality, to our clothing, to our shelter, and to transportation and communications. There is, however, growing awareness that one should *examine* or *test before deciding* that a substance or system is or is not sufficiently "safe," using whatever criteria may be appropriate. To continue to achieve acceptable benefit from chemicals and other technology, we must *control* the *risks* by *understanding* them, keeping in mind that controls which are passed through several levels of command may differ in implementation from the controls originally proposed and assumed. Evaluating "safety" thus becomes the process of detecting, analyzing, and measuring risk, and by applying personal and social judgment as to the acceptability of that risk to us, as opposed to the benefits derived.

William Lowrance has reduced the process of risk analysis to:[5]

1. Define the conditions of exposure.
2. Identify the adverse effects.
3. Relate exposure with effect.
4. Estimate overall risk.

In considering risk/benefit trade-offs, it is essential to remember that for every benefit we usually incur some risk or cost, however small or inconspicuous it may be. We have maintained for years that every chemical and chemically related substance, if misused, or misapplied, or improperly disposed, can produce some degree of risk. This is true throughout the whole cycle of synthesis, production, handling, transportation, use, and ultimate disposal. Due partly to the patchwork of legal considerations, attention seems to shift from one emphasis to another. In sequence, the Occupational Safety and Health Administration (OSHA) attempted to develop and impose standards for chemicals in the workplace, the Environmental Protection Agency (EPA) under the Clean Air, Clean Water, Toxic Substances Control, and the Solid Waste Act, including the Resources Conservation and Recovery Act of 1976, in turn, attempted to define toxic and hazardous substances control and disposal. Currently, a major effort is underway to define and control the hazards to the public from thousands of questionable disposal sites where toxic and other hazardous waste products have been disposed without full appreciation of the potential hazard, although the disposal may have been perfectly legal at the time. In one state alone, 210 miles of roads were used as a disposal area for transformer oil containing polychlorinated biphenyl (PCB).[6] (See also Chap. 28.)

We maintain that it is *not* the chemical substance itself which is hazardous (since chemicals are only functional groups designed to react under certain conditions and can hardly be expected to act differently in vivo than in vitro), but the consequences of its misapplication and improper disposal. When the inherent properties of a substance are not properly appreciated at all stages in the cycle, or the substance is improperly handled or disposed, the risks or unwanted effects, such as injury from toxicity, or damage from fire or explosion, may overwhelm the expected benefits. A net loss then may result, measured both in human life and in economic terms. If the material is produced, shipped, and used, and then ultimately disposed properly, the benefits should be high and the risks small. It is to this end—the intelligent and adequate controls at each step in this system—that this book is dedicated. A formal system analysis, resulting in a better understanding of the interrelationship of all elements, should be undertaken, such as the management control system called Chemical Hazard Assessment and Inventory Network (CHAIN).[7]

"Dangerous articles" and "extra-hazardous commodities," legally defined under 49 CFR Section 72.5 (as revised) are well identified and usually well recognized for their hazardous properties, based on prior history of unfortunate incidents, and hence these substances pose *relatively low risk* as long as they are given the *knowledge, respect, and care* in handling and disposal to which they are entitled.[8a,b] In addition, the labeling or other identification of these substances with hazardous properties encourages their ultimate substitution by less hazardous materials. For example, nitroglycerin, used for a century since Nobel invented dynamite, is being replaced by blasting agents using ammonium nitrate water gel explosives which are far safer since they are less sensitive to accidental detonation. Vaporizing liquid fire extinguishers employing carbon tetrachloride mixtures or chlorobromomethane, once widely used, have been gradually phased out and replaced by less toxic, more efficient fire-extinguishing agents.[8c]

As agencies and groups have different mandates to consider in evaluating hazardous substances (with particular emphasis on their own area of concern), we have noted an increasingly large number of lists compiled with the objective of classifying substances as hazardous or nonhazardous. For example, the U.S. Coast Guard (which, before the formation of the Department of Transportation was part of the Treasury Department) did not feel that any rating system then in use adequately reflected the various properties of substances in the context of safe water transportation. The National Academy of Sciences–National Research Council Committee on Hazardous Materials was formed and funded during 1964–1975 for the purpose of giving technical assistance to the Coast Guard, including a hazard classification system. The system which was evolved consisted of 10 numbers, on a scale of 0 to 4, for the 10 parameters which were considered necessary to adequately describe the hazards of the chemical during water transportation. These were: fire, vapor or gas irritation, liquid or solid irritation, poisons, human toxicity, aquatic toxicity, aesthetic effects (including odor, color, adverse effects on beaches or vessels in contact with the spill), reaction with other chemicals (cargo compatability), reaction with water, and reaction with self (polymerization). This rating system was used to classify chemicals for regulatory purposes to encourage increased safety in water transport.[9]

As other systems were developed for specific concerns, the need for some integration of systems into a hazard index of general application has become apparent. Often this has included various numbers reflecting the various hazardous properties, weighted for the relative population exposed, pounds of production, or other considerations. In an attempt to assess the present state of the hazard assessment science (perhaps it is both an art and a science at present), C. J. Jones has noted the shortcomings of existing models, and the difficulties of developing a consistent, generally applicable model. Jones concludes that both experimental and theoretical work are necessary if a satisfactory model is to be developed for evaluating waste-management options.[10] This point is extremely timely in view of the attention to improper disposal sites or landfills in the United States and elsewhere, where frequently the identity of the substances to be considered is not known.

Current uneasiness exists both in the public mind and in the regulatory arena over several substances which may or may not pose serious problems to humans and wildlife throughout the environment, and for which the data base is still not firm. For example, the accidental release of recombinant DNA research products has been given much attention in the media as well as scientific circles.[11a,b,c] The waste product from nuclear power plants, as well as from defense operations, has been studied extensively for nearly 30 years, by highly competent experts and committees, but a clear national policy for handling and ultimate disposal is still not recognized.[12a,b,c] Polychlorinated biphenyl[13] and polybrominated biphenyl[14] (PBB) have been shown to persist in water systems and soils for long periods, affecting the animal and human food chains. For five years, a major river was closed to fishing because a residue of chemicals persist from a now-abandoned operation.[15] Herbicides have a significant economic contribution to control of vegetation, but questions have arisen about their adequate control, such as the possible health effects of 2,4,5-T relating to the risk criteria in 40 CFR 162.11 (a)(3) relating to oncologic and teratogenic effects, and/or fetotoxic activity in mammalian test species.[16] We have noted with much interest the emergence of an environmental ethic in our society and its impact in such areas as energy, the economy, and in human health. Even scientists of the highest credibility may differ on their assessment of the risk/benefit ratio of many issues. With new information, it is hoped that agreement can be reached before such issues as noted above become after-the-fact disasters, which legal judgments and financial settlements can do little to really restore.

Much greater emphasis than previously is being placed by industry, by government agencies, by unions, and by society in general towards a more complete understanding and control of chronic (long-term) health effects of chemicals and related materials. Such concerns are long overdue, and even now many aspects of control are poorly understood and existing knowledge is not completely utilized. We are acutely aware that, in the long term, exposures to excessive concentrations of certain materials may produce insults beyond the ability of certain persons to tolerate them, and may result in cause or aggravation of lung and other cancers, pulmonary edema, congestive heart failure, angiosarcoma and other liver diseases, nervous disorientation, mutagenic damage, teratogenic and other reproductory effects, and in other modes not always immediately obvious. Only recently has a national toxicology program been established to coordinate, or at least make known the work, plans, and results of various agencies and institutes responsible for research and development in the field of human toxicology.[17]

Reflecting the impact of the new laws and regulations, as well as the general awareness of society, a new cognizance and sense of responsibility has arisen among professional personnel, including chemists, chemical engineers, environmental engineers, industrial hygienists, occupational health nurses and physicians, and safety professionals at all levels. Much of this new awareness is reflected in the new images of long-established organizations. For example, the Manufacturing Chemists Association has changed both its name (now Chemical Manufacturers Association) and its approach to health and safety problems. The National Safety Council (a private nonprofit organization chartered by the Congress, established in 1915) has done and is doing exceptionally fine and progressive work, especially in its Industrial Department, through such groups as the Chemical Section and the Research and Development Section. The American Society of Safety Engineers has sponsored the concept of certified safety professionals. The American Chemical Society established a Council Committee on Chemical Safety in 1962, and a Division of Chemical Health and Safety in 1976, which was granted permanent status September 12, 1979 since the consensus was that it filled a real need. The American Institute of Chemical Engineers established an Occupational Safety and Health Committee in 1976, and in 1979 it evolved into a Division. The American Institute of Chemists formed a Safety and Health Committee in 1977. Alpha Chi Sigma, a professional chemical fraternity, established a safety committee in 1940, which is still active.

Professional interest is not limited to the United States. The Royal Institute of Chemistry organized an effective symposium on the Social and Professional Responsibilities of the Chemist for Health and Safety in 1975.[18] In Sweden, graduates of chemistry and engineering at the Royal Institute have been studied in terms of mortality and cancer, a study which is still incomplete at this writing. The Royal Australian Chemical Society has recently drafted recommendations for the safe use of environmentally dangerous chemicals.

Officials of the Institute of Chemistry of the People's Republic of China have expressed interest in the health and safety educational activities. The work of the World Health Organization in Geneva, Switzerland, through its various publications, such as the Chemical Information Service (CIS) abstract service continues to be a positive force in international understanding.

One of the fundamental motivations for additional attention to health and safety in the workplace, beyond the legal and economic, is the renewed concern for the effect of workplace exposure on women. While hazards have been recognized for several years, as witnessed in 1956 by Mary Louise Brown: "Women may be employed at jobs which involve the use of toxic materials if the exposure is kept below harmful limits," even today we are not clear what is a harmful limit. Ms. Brown mentions that no woman who is expecting a child should be exposed to aniline, benzol (benzene), chlorinated hydrocarbons, heavy metals, or radioactive substances, to name only a few.[19] More recent studies have begun to add additional materials to the list of substances which may be harmful, and a guideline on pregnancy and work has been prepared.[20a] A conference on women and the workplace has been convened to study the many factors not previously widely connected with occupations.[20b,c] From such bases of new understanding, new insights may be expected.

In the first edition of this book, we placed considerable emphasis on the cost of accidents as a factor in assigning proper priority for health and safety programs. It is our opinion that the true value of health and safety still has not been recognized. Within the decade, however, the costs have greatly increased for a variety of reasons including inflation and the increased public concern for injuries.[21] Product liability and malpractice, even among consultants and university professors and public school teachers, have required new understanding both by the professionals and the legal community.[22a,22b] A few examples, selected from relatively recent history, may serve to illustrate our concerns

Awards totaling $717.5 million were granted to next-of-kin and injured (29 were killed and 56 injured) following an explosion at a pyrotechnic plant in 1971, in which, according to the legal testimony, the plant operator had previously classified the ingredients and products as "flammable" instead of "explosive," and the Federal Court admitted that the Federal Government had been guilty of negligence and liable under Georgia law for the entire damage. Another contractor now operates the plant.[23]

A chemical company was fined $13.2 million for illegally dumping toxic chemicals into a city waste-treatment system. The city itself had previously been fined $10,000 for permitting discharge of the pollution into the river. In a separate action, two managers of a now-bankrupt plant, ordered closed and dismantled, were fined. The contamination extended down river into a bay, and has resulted in a major disruption of both commercial and sport fishing through 1980. One estimate is that if the entire 65-mile stretch of the river is to be dredged, it would require several years and up to $200 million. In addition, no effective method of treatment is known for the former employees who were affected by the chemical, and the long-term aspects of the exposures are not known.[24,25]

A major manufacturer was fined $4 million, in addition to agreement to conduct a research program on environmental effects of PCB, to assist in partial clean-up of the upper Hudson River, because of contamination over many years of the PCB used in electric capacitors and transformers. The present level of control is such that the plant now discharges less than 1 g/day into the river (according to plant management). The cost of freeing the 35.7-mile stretch of the river above Troy, New York, has been estimated at $150 million. Several towns and communities draw their water supply from this river.[13]

Forty men died during repairs to a supposedly empty liquefied natural gas (LNG) container on Staten Island, New York in 1973, probably due to ignition of flammable vapors by unsafe procedures and equipment. The resulting settlements for liability equal approximately $15 million.[26]

Silicon tetrachloride leaked for several days from a large storage tank in the Calumet, Illinois, harbor area during April 1974. At times, the cloud of hydrolyzed vapors measured from 5 to 10 miles long as it moved across the city, affecting several thousand persons. No dollar loss estimate is available, but the incident was a major hazard and survival problem for a large number of people, including nine government agencies and departments.[27]

Major problems, which eventually equate to both economic loss and disruption of public confidence, have no geographical borders. On July 10, 1976, in Seveso, a town of 17,000 population in Northern Italy, the safety-valve stack (pressure relief) of a reactor for making trichlorophenol (TCP), released, forming a cloud which spread over the town and surroundings. The fall out contained the TCP with an impurity of dioxin (dioxin is more toxic than TCP). Evacuation of the area, followed by decontamination and treatment of 500 persons followed. No recognized treatment is available for dioxin poisoning. A compensation fund of $11.5 million was established for compensation; $48 million emergency funds were also established by the Italian government (see Chapters 6 and 7).[28a,b,c,d]

A temporary bypass intended to keep operations continuing during repairs, ruptured, causing a cyclohexane vapor cloud to be released into the air, which then ignited; 28 were killed and 36 injured inside the work site, and 53 recorded casualties occurred outside the works. The loss is estimated at approximately $100 million.[29]

The United States Government has agreed to pay $5.7 million to a group of 445 asbestos workers, as part of a settlement in a case in which the workers sued on grounds that public and private negligence resulted in their exposure to a type of asbestos called amosite, which apparently can cause lung cancer in humans. Although the U.S. Public Health Service surveyed the plant in 1964, and found the levels high, the information was held confidential, and the Public Health Service did not warn the workers.[30a,b]

The Legal Approach

The tendency of society to attempt a solution of technical problems by legislation, and the regulation of chemical health and safety issues by governmental agencies (local, state, regional, and Federal) has been especially evident in the past decade (1967–1978). As noted in Fig. 1.1, the number of laws which pertain to toxicology and safety-related matters has increased from three to thirteen, not including the laws establishing the National Fire Prevention and Control Administration (1974), the Department of Transportation (1968), The Transportation Safety Act of 1974 establishing the independence of the National Transportation Safety Board, and the Department of Energy Act of 1977, all of which have considerable interests in health and safety. Every law has had sufficient justification to be passed by Congress after much review and voluminous legislative history that often extended over several years of hearings, and, at the time of signing by the President, was hailed as a historic landmark panacea to the problem it addressed. Unfortunately, when laws are interpreted by the courts, translated

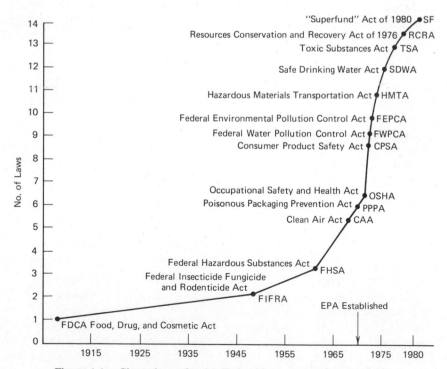

Figure 1.1. Chronology of major Federal laws concerning chemicals.

into regulations, and then enforced by persons far removed from the spirit of the original intent, often with little or no background in the problem, the interests of society may be less adequately served than originally visualized, or may actually be less than before the law was passed. The evolution of PL 91-596 (Occupational Safety and Health Act of 1970), certainly laudable in its prime objective but not implemented properly for a variety of reasons, is an example of the difficulty society has with legal remedies. The study by Nicholas Ashford explores many of these defects in detail.[31] It is hoped that the work of the Interagency Task Force on Workplace Safety and Health will be fruitful.[32] An undated reference from the *New York Times* summarizes the basic problem:

> Safety comes from man's mastery of his environment and of himself. It is won by individual effort and group cooperation. It can be achieved only by informed, alert, skillful people who respect themselves, and have a regard for the welfare of others.

The changing nature of the legal viewpoint, to place emphasis on the individual, rather than to penalize the corporation, is highly interesting. We know personally of one safety director who, along with two other corporate officials, is awaiting trial on charges of manslaughter as a result of an explosion in a facility for which they were responsible. The thought is being explored that jail sentences imposed on high officials will have a more direct effect on the industrial community management understanding the importance of health and safety. Whether or not this will prove to be true remains for the future. We are hopeful that awareness will occur and action be taken before the injuries occur.[33] Perhaps education of the proper type will be the ultimate answer, as more effective than legal actions.[34,35]

1.2. SAFETY: WHAT DO WE REALLY MEAN?

Professional chemists and engineers are beginning to appreciate that society is now critically evaluating the products of their labors—in research, production, engineering, marketing, and disposal—and that society no longer automatically assumes that chemistry and engineering produce unmixed blessings. In spite of the major contributions which chemistry and the other physical sciences have made and continue to make to our civilization, much attention has recently been directed to the less desirable byproducts. To our dismay, television and the news media continue to bring into our living room almost nightly new or recycled stories of persons and places which have allegedly suffered from the misuse and improper disposal of chemicals. The coverage is often emotional, and without balance. In another paper, we have reviewed acute chemical emergencies from a historical perspective, noting that society is becoming aware of the new potential hazards as related to the interface of man/ecology/economics.[36]

This section suggests thought processes that professionals might consider in order to catalyze the improvement in our collective image, and our professional relations, so society will again appreciate and have faith in the positive contributions of chemistry and engineering as positive constructive forces in the national interest.

Recognize and Admit That All Chemicals Are Potentially Hazardous If Mishandled or Improperly Disposed

Our observation over the years has convinced us that professionals often deny that the materials with which they work, synthesize, handle, or discard are in any way hazardous, either to themselves, to their colleagues, or to society. The suggestion is often made that "the other fellow" in the adjoining laboratory or plant creates hazards, but the speaker is totally safe, so personnel concerned with safety hazards should look elsewhere. The fact is that chemical molecules play no favor, nor do they respect security fences or the degree status of individuals who may be exposed. In commenting on the two worlds of professionals, J.Y. Oldshue recently noted that the world of nature is where we harness laws for processes, and that nature is easier to handle than the world of society where one has to encounter and deal with the rules of people, countries, politics, religion, and customs.[37] We must be concerned with people; we cannot ignore the human element. A decade or two ago, perhaps indifference to safety and health might be excused or condoned, since data was not readily available in a really retrievable form; however, today such data is available, often in-house, or by a phone call. The American Chemical Society (ACS) Division of Chemical Health and Safety and the Council Committee on Chemical Safety constitute valuable resources available to ACS members, as does the ACS Safety and Health Reference Service.[38] Other professional organizations, including the American Institute of Chemical Engineers Division of Occupational Safety and Health, the American Institute of Chemists Safety Committee, the Royal Institute of Chemists, and the Royal Australian Chemical Institute now have active health and safety programs. Professionals should know how to obtain credible information and how to apply it wisely in their activities to protect both themselves personally, their companies, and also society. If the in-house sources, such as the medical department, the occupational health nurse, the safety engineer, the fire department, or the environmental services engineer cannot help, the services of an outside consultant should be considered.

Learn at Least the Basic Aspects about Chemical Regulations no Matter How Far Removed You Consider Yourself from the Legal Staff

Regulations now constitute a body of knowledge and law which affect the professional chemist and engineer, both personally as well as professionally.

At the Federal level, OSHA, EPA, National Institute for Occupational Health and Safety (NIOSH), Federal Emergency Management Agency (FEMA), Department of Transportation (DOT), Department of Energy (DOE), Department of Defense (DOD), Food and Drug Administration (FDA), and the Consumer Product Safety Commission (CPSC) have already had significant influence on our understanding and control of chemicals and related materials. Whether we agree or not with the regulations, they constitute the law of the land, and until repealed or overturned by courts, they must be obeyed. Ignorance of the law has never been accepted as adequate defense. The collective message is clear—society expects and legally charges the professional to assume responsibility for his or her actions and the consequences of the activity. Two recent documents which reflect this concern are: (1) the OSHA policy document on chemical carcinogens, effective April 21, 1980, which will still apply in spite of the recent decision by the Supreme Court in the benzene standard case, and (2) the EPA manifest system and control pathway to track hazardous wastes from "cradle to grave," effective August 26, 1980.[33,39,40]

As Fig. 1.1 illustrates, many Federal regulations which affect chemists and engineers are less than a decade old.

Apply the Same Professional Standards to Cost/Benefit Or Risk/Benefit Analyses Before Accepting or Rejecting Them

Risk/benefit analysis is a popular buzz term for any attempt to compare the benefits from an activity with the costs. Most chemists and engineers have been trained to think in terms of benefit to the project on which they are engaged. These benefits may be entirely different in both kind and magnitude from the costs. Costs, usually expressed in hard financial terms, are then compared to the benefits in terms which are familiar to the sociologist or political scientist.[3,6,41] Even today, the value of human life is not absolute: the courts and society have complex formulae for calculating which is often foreign to the physical scientist. Frequently, the effects may be long removed in time and space from the immediate producer or user. The long induction period of chemical carcinogens is one example. Another is the waste chemical disposal, an area which may have been secure for decades[42] suddenly becomes indited for breach of containment, by pollution of the air, the soil, or the water. Emissions from power plants burning high sulfur coal in the midwest United States may affect life in lakes in the northeast United States and Canada, from the increased carbon dioxide and "acid" rain.[43]

Insure That Your Own Personal "Data Base" Records Are Up-to-Date and Available for Future Analysis

Most of the discussion and uncertainty over the effects of chemical exposures could be reduced to scientifically defensible recommendations and scientific facts if we each had complete medical files and exposure data immediately at hand. From such records, the epidemiologist can then construct retrospective analyses, which permit factual cause/effect conclusions. Unfortunately, the data base on humans is often very limited, and animals are not a true substitute for humans. We suggest that each person maintain his or her own personal medical and exposure record, perhaps in a diary form, just as he or she records time charges or research efforts. Copies of medical records and examinations should be considered personal records, just as are income tax files, and kept so they are easily and quickly retrieved. This is especially important to anyone who has more than one employer during his lifetime, and should start with undergraduate laboratory exposures. The fact is that each of us is a chemical indicator, if we would only take time to read and interpret the information in a professional manner. While there will always be controversy about confidential records, we feel strongly that personal human data, and exposure records are vital to each human, and should be so treated.[44]

If You Discover or Observe New or Unreported Health and Safety Knowledge, Share It with Your Fellow Professionals

Every science or profession is built on the accumulated wisdom or data from many sources. Our knowledge of chemical health and safety is gradually developing and evolving from our collected data base. Not all knowledge is applied as quickly as it should. In 1949 we pointed out the need for integration of chemical safety in academic courses; even today the question of safety education is not resolved. In 1950 we published a warning about the potential hazards of domestic refrigerators when used to store volatile flammable solvents, and even incorporated this into a motion picture.[45] In 1979, two domestic refrigerators used for chemical storage exploded in a large university. On the other hand, the increasing interest in chemical incompatability, in the laboratory, pilot plant, production facility, or in hazardous waste disposal, has been a welcome sign that reactions are being noted and reported that were previously ignored or unreported. As a recent example, we note a contribution from a major company, noting that mixing potassium permanganate with dimethylformide (DMF) may be hazardous under some conditions. Although it has been previously recorded that DMF will react with oxidizing agents such as bromine and chromium trioxide, specific reference to $KMnO_4$ was not given. It appeared in the Letters to the Editor section of *Chemical and Engineering News*. (See *Chemical and Engineering News*, page 65, Sept. 22, 1980.) Unfortunately, not all publications abstract the Letters to the Editor, hence, the information may not be made part of the permanent literature by that route. Contributions of this nature should also be reported to the Chemical Section of the National Safety Council, 444 N. Michigan Avenue, Chicago, IL. 60611, and to the NFPA

491-M Committee on Hazardous Chemical Reactions, National Fire Protection Association, Batterymarch Square, Quincy, MA. as well as to the *Journal of Hazardous Materials*, published in Amsterdam by Elsevier. In that manner, the information will appear in future editions of permanent references, such as the *Handbook of Hazardous Chemical Reactions*, by Leslie Bretherick, published by Butterworth, London.

It is obvious that voids exist in our safety and health data transfer even today, and every incident or unusual reaction, whether physical or physiological, deserves to be included in our health and safety data base.

1.3. THE CHANGING NATURE OF CHEMICAL EMERGENCIES AND RELATED POTENTIALS

The chemical and allied industries have excellent safety records, compared to all industry. According to the National Safety Council reports the chemical industry is one of the safest industries for workers, using the OSHA formula based on 100 full-time equivalent workers.[46]

The chemical industry has historically been a leader in the safe and proper manufacture, use, and disposal of chemicals. However, each chemical substance, of which more than 4.5 million are recognized by *Chemical Abstracts*, may be considered units of energy, each possessing unique properties which make it a useful, profitable, and hence desirable member of our society. Properly made, used, and disposed of, chemicals are essential building blocks of the high standard of our life. Improper handling and disposal may be unforgiving, and most chronic as well as acute chemical emergencies occur beyond the plant gate.

In discussing emergencies arising from misuse or improper handling of chemicals, a historical prospective will put into focus the need for more complete understanding of control of chemical materials, both on the plant site and in the consumer domain.

Fire

Perhaps the first acute chemical emergency resulted from an uncontrolled fire at the hands of ancient beings.[47] Fire, chemical oxidation, is both a servant and a foe of humans. Simple fires first used for food preparation were supplemented by controlled fires for heating, for smelting, and working of metals (such as iron, copper, and lead), for land-clearing, and for other constructive purposes.

Open fire sources have caused many emergency situations. In 1653, for example, a conflagration occurred in Boston, resulting in deaths and devastation of much of the new city.[48] The cause: ignition of a straw roof from chimney embers. Wooden roofing material is still a major factor in fire spread where it is used, especially in the West.

Candles and lamps fueled by whale oil, and later, kerosene, were a prime source of illumination until 1882, when Edison first developed electrical-lumination and power in New York. Electricity, which replaced many fire sources, is a source of ignition itself, unless it is properly controlled. The June 23, 1980 fire on the twentieth floor of a building in Manhattan, which apparently originated in computer-related equipment, burned for $3\frac{1}{2}$ hours and sent 100 firemen to emergency rooms, in spite of the availability of self-contained breathing apparatus for protection against smoke inhalation.[49]

Somewhat similar situations have been reported in the numerous night-club fires of the past, and, more recently, in the fire on the tenth floor of a major clinical center, May 23, 1979, where furniture, carpets, drapes, wall coverings, and other accessories ignited and spread rapidly. Most commonly used polymeric materials burn or decompose under fire conditions, and produce decomposition products which may include carbon monoxide, carbon dioxide, hydrogen chloride, and other gases with significant smoke. The comprehensive study of this subject by the National Research Council National Materials Advisory Board deserves careful reading in engineering, scientific, and emergency-control circles.[50]

BLEVE

BLEVE (boiling liquid expanding vapor explosion) may result from a fire situation if the boiling liquid has insufficient pressure relief. The inadequate attention which has been given to pressure relief over the years has resulted in many poorly engineered systems, with inadequate relief systems. Pressure relief was studied in depth by a National Research Council (NRC) committee which analyzed over 50 different formulae proposed or used. Although the NRC study was funded by the Coast Guard, the principles of calculation apply to all large tanks and containers, either fixed or mobile, which may become overpressured, as in fire situations, or from chemical reactions within the vessel.[51] (See Chap. 7.)

BLEVEs have occurred sufficiently often that they must be recognized, especially as pressures and temperatures are increased to achieve higher production efficiency. In the 20 years prior to 1970, for example, 18 BLEVEs occurred when LP Gas or propane tanks were exposed to fire; between 1970 and 1975, 12 BLEVEs were reported. Substances which are reported to have been involved in BLEVEs include vinyl chloride, butane, butadiene, and propylene. Propylene, in an incident in Spain in 1978, was released from a tank truck near a camping area, and resulted in over 200 deaths and many injuries. In 1962, a propane tank truck, out of control on a narrow road, overturned and released cargo in Berlin, New York, with significant damage and loss of life.[53] These experiences, and others, suggest that fire and other emergency personnel should use extreme care in approaching flammable gas or flammable liquid fires, or even fires in liquids of higher-flash points, since a ruptured tank or vessel may fragment and travel hundreds of feet horizontally and many feet vertically, even from stationary, as well as from mobile tanks on trucks, rails, or barges.[54]

Explosion and Explosives

Explosion may be crudely defined as release of energy in a rapid and uncontrolled manner. The potential for explosion, deflagration, or detonation, is far greater than usually appreciated. For thousands of years, mankind has been making and using explosive materials; the Chinese are credited with invention of black powder. Even today, it may be encountered in fireworks and other pyrotechnic devices. The potential for creation of a sudden force by chemical means has long been recognized and used as a constructive force in construction, mining, and earth-moving. The first powder plant in America was on the Brandywine River, and began operations in 1802. Later, Nobel invented nitroglycerine, and added dynamite, which is nitroglycerine absorbed on an inert base, to the workhorses. Unfortunately, black powder and nitroglycerine are shock and heat sensitive, and numerous accidental explosions have been recorded, in manufacture, handling, and storage. A detailed study of the effects of these explosions, both in the United States and abroad, resulted in the development of an American Table of Distances, widely recognized as the basic reference for safe storage of various quantities of explosives materials with respect to railroads, highways, and inhabited dwellings.[55] (Available from the Institute of Makers of Explosives, 1575 I Street, N.W., Suite 550, Washington, D.C. 20005.) In recent times, the use of water-base gelled slurries has largely replaced the previously used blasting agents.

While similar in its effects to the BLEVE discussed above, UVCEs (unconfined vapor cloud explosions) are another combustion process creating major emergency problems, but only recently recognized as a threat of total devastation to process plants and surroundings. The delayed ignition of a large vapor cloud in air can create pressure effects which may be highly damaging long distances from the incident. In a recently published book, Gugan discusses the UVCE process in detail, noting that even today much is not understood or agreed upon.[56] A compilation of 100 incidents prior to February 1977, for which technical data is available, is included. We especially noted that the Ludwigshaven, Germany UVCE on July 28, 1948, is mentioned. In this incident, which is poorly documented in American literature, over 200 were killed and 3800 injured when 33 metric tons of dimethyl ether was released from a tank car, probably due to hydrostatic overfilling and thermal expansion on a very warm day. Chemicals as diverse as hydrogen, ethylene, butane, LPG, cyclohexane, ethylene oxide, ethyl chloride, isopropyl alcohol, acrolein, and vinyl chloride, as well as crude oil, have caused serious UVCEs. It is hoped that more attention will be given to this process so more complete understanding and effective protective measures may be developed.

Uncontrolled Release

When a container or containment system fails, the resulting spill or release may have effects which may be dramatic or undetected. Chemical materials have a wide range of properties, producing effects of diverse nature. It is difficult to separate "chronic" from "acute" emergencies in this context, since at some point, the "chronic" low level release may reach a point or concentration that at least for some persons at high risk, is very "acute." As more handicapped employees enter the workplace, additional attention must be given to the effects of any unexpected or unscheduled release.

Containment failures may take many forms. On January 15, 1919, a large tank of molasses in Boston ruptured, producing a flood of perhaps two million gallons of viscous material, which killed 19 persons and caused heavy property loss. One interesting factor in this release is that the tank was constructed of steel plate sufficiently strong for water containment ($62\frac{1}{2}$ lb/ft^2) but not for the molasses (90 lb/ft^3).[57] Even well-designed and constructed vessels can break with amazing speed, as witness the release of anhydrous ammonia from a tank car on a siding in Crete, Nebraska, February 18, 1969. Half of the tank car was intact; the other half was broken into several fragments of various sizes as though the heavy steel were an eggshell. The car had been struck by another train.[58]

Water-reactive chemicals deserve special attention, since the release almost always results in water contact with the material. In the Somerville, Massachusetts tank car rupture of April 3, 1980, phosphorous trichloride leaked from the car into a ditch. One observer reported that the responding fire company deliberately applied water to hasten the hydrolysis, and hence increased the acidity and opacity of the cloud. In any event, 23,000 persons were reported evacuated, 120 persons reported to the area hospitals for treatment, and corrosion damage alone from the acid gases formed was estimated at $500,000.[59]

Information on hazardous material spills has not been readily available until the reporting requirements of the Federal Water Pollution Control Act Amendments of 1972, and the operation of the National Spill Contingency Plan by the U.S. Environmental Protection Agency and the U.S. Coast Guard. One estimate places the annual number of hazardous material spills at approximately 3,000. Reported spills range in size from a few pints to hundreds of gallons. In most cases, such spills hardly constitute an acute chemical emergency, but where the substance enters the water supply of an individual, a community, or a city, serious "acute" problems can develop. Since 89 of the 100 largest cities in the U.S. derive most, if not all, of their domestic water from navigable waterways or tributaries, the potential is serious. Substances which were not previously recognized as hazardous, but now known to be, may have been discharged or leached into waterways for years. As an example, PCBs have been widely used as a dielectric fluid in electrical capacitors and transformers, and the release of even gram quantities is now seen as a cause for great concern. Our level of knowledge and understanding is often a factor in chemical emergencies.[60]

On a basis of the study of over 15,000 spill reports, Buckley and Wiener[61] rated primary causes as follows:

	Hazard Potential Probability	
	High	Low to Moderate
Tank rupture or puncture	0.23	0.77
Tank overflow, and other leakage	0.19	0.81
Hose, transfer system failure	0.08	0.92
Nontank container rupture or puncture	0.03	0.97

The five materials most frequently reported spilled are:

Sulfuric acid
Ammonium nitrate fertilizer
Sodium hydroxide
Hydrochloric acid
Ethyl parathion

One major factor in any spill or emergency is the prompt obtaining of technical data and information useful to personnel at the scene, no matter how remote. In 1969, the National Academy of Sciences Committee on Hazardous Materials reviewed the status of information resources, and recommended that systems be developed for immediate guidance of on-scene personnel. From this recommendation has come three systems presently available: CHEMTREC, a telephone 24-h response service of the Chemical Manufacturers Association; CHRIS, a response information system primarily available through the Coast Guard, and EPA-OHM-TADS, or technical assistance data, available from regional EPA offices. All three of these systems have performed well in emergencies. CHEMTREC has responded to 21,700 transportation emergencies during its first decade since its activation in September 1971. Unfortunately, not all the information in all three is correct or documented, and the total number of chemicals covered by the three systems (eliminating duplication of names) is perhaps less than desired. However, the existence of these systems makes a major step forward in control of chemical and related emergencies. We hope the requirement of the UN identification number will not set back the information systems.[62]

The four digit number is now required on tank cars and trucks. As an example, gasoline is 1203; sulfuric acid is 1830.[62b]

It may be of interest that major progress has been reported by the Federal Railroad Administration, with respect to hazardous material releases in two recent years.[63]

	1978	1979
Cars damaged	1205	1057
Cars releasing HM	338	165
People evacuated	25,981	16,093
Killed	24	0
Injured	221	15

Uncontrolled Chemical Reactions (Chemical Boobytraps)

In 1940, the late George Jones, a chemist in the Bureau of Mines, began to compile a list of hazardous chemical reactions which had not been previously recognized. In 1950, a committee on hazardous chemical reactions, now known as NFPA 491, was formed to review and expand that listing. Several editions of the work have since been published, and new items are being added regularly.[64] Meanwhile, Leslie Bretherick, a research chemist with BP in England, became interested, and has produced a book, now in its second edition, with 7000 entries.[65] It should not be necessary to rediscover hazardous reactions by an emergency, and emergency personnel would be well advised to become familiar with these and related publications.

Time-Bomb of Toxic or Explosive Materials from Improper Disposal

Hazardous-waste management is not a new problem, but one whose effects, while still largely unappreciated, are rapidly becoming "acute" emergencies. Movements in the early 1970s to deal with air and water pollution, as evidenced by the Clean Waters and Clean Air Acts, forced the closing of many incinerators and other disposal facilities, with the result that land disposal of hazardous wastes today represents a major problem. Pollutants previously discharged into waterways and the atmosphere often accumulate, even when diluted, and become part of the biological chain. The fact is that over 3000 hazardous waste disposal sites have been identified, including perhaps 500 to 1000 potential time bombs of the Love Canal magnitude.[66]

Hazardous wastes may contain toxic chemicals, pesticides, acids, caustics, infectious, radioactive, carcinogenic, flammable, or explosive materials.

According to the EPA, ten states, Texas, Ohio, Pennsylvania, Louisiana, Michigan, Indiana, Illinois, Tennessee, West Virginia, and California generate 65% of all hazardous wastes. To these must be added Massachusetts,

New Jersey, New York, North and South Carolina, Kentucky, Alabama, Arizona, Washington and Oregon, for the second ten, or about 25%. Only 10–15% of wastes are believed "hazardous," but the sheer quantity is such that 57 million metric tons may be classified as hazardous.

The scope of this paper does not include a detailed discussion of hazardous wastes. However, the potential time-bomb aspect must be recognized. As this is written, the much-discussed "superfund" legislation, to provide significant funds for clean-up of known sites injurious to health and property, is still not fully implemented. Acute chemical emergencies may occur at any time in the hundreds of hazardous waste sites which have been identified, and perhaps many more.[67-69] (See also Chaps. 28 and 29.)

In summary, the acute chemical emergency has changed from the obvious situations, where fires and explosions, regardless of their origin, were tangible and of relatively short duration, to less obvious "chronic" potential problems, which manifest themselves as "acute" with little or no warning.

Biohazards

The recent ruling by the Supreme Court that new biological systems are patentable will doubtlessly bring a significant increase in biological research. While guidelines for recombinant DNA research have been specified, specific guidelines for biohazards are still incomplete. If we can place any credibility in reports that over 1000 persons died from an explosion in an anthrax-production facility, the potential effects of biohazards, if not controlled, may be significant.[70,71] The present questions regarding "yellow rain" as a biological–chemical warfare agent, in which mycotoxins, such as T-2, may have caused serious and fatal effects, in Yemen, Laos, Afghanistan, and Cambodia, will doubtlessly be the subject of much future study.[72]

REFERENCES

1. R. H. Tawney, *Religion and the Rise of Capitalism*, Peter Smith, New York, 1976.

2. R. E. Scully, "Normal Reference Laboratory Values," *N. Engl. J. Med.* **298**(1), 34–45 (January 5, 1978).

3. W. D. Rowe, *An Anatomy of Risk*, Wiley-Interscience, New York, 1977.

4. a. C. Starr, Benefit-Cost Studies in Sociotechnical Systems, *Perspectives on Benefit-Risk Decision Making*, National Academy of Engineering, Washington, D.C., 1972, pp. 17–42.

   b. E. Teller, "I Was the Only Victim of Three-Mile Island" (advertisement). *Chicago Tribune* (and other papers) October 18, 1979, pp. 10, 11.

   c. "Report of the President's Commission on the Accident at Three-Mile Island," J. Kemeny, Chairman, The White House, Washington, D.C. October 1979.

   d. "Washington Alert," Department of Chemistry and Public Affairs, American Chemical Society, Washington, D.C., August 10, 1979, referring to the House Commerce Committee and the Oversight and Investigations and Consumer Protection Committee Hearings, July 1979.

    e. J. R. Emshwiller, "Many Nuclear-Plant Perils Remain Three Years After Three Mile Island," *Wall Street Journal*, **33**, (Feb. 26, 1982).

    f. Anon, "A Nuclear Power Primer: Issues for Citizens," League of Women Voters, Washington, D.C. 20036, 1982.

5. W. W. Lowrance, *Of Acceptable Risk: Science and the Determination of Safety*, W. Kaufmann, Inc., Los Altos, Calif., 1976. See also Interview, *Chemical and Engineering News*, 13–20 (July 6, 1981).

6. Personal correspondence, November 14, 1980, Deputy Attorney General of North Carolina to H. H. Fawcett; see also "Waste Disposal Site Survey," Committee Print 96-IFC 33, Committee on Interstate and Foreign Commerce, U.S. Govt. Printing Office, October 1979.

7. Chemical Hazard Assessment and Identification Network (CHAIN) is a trademarked system of management of chemical hazards.

8. a. L. Benner, "Risk Concepts in Dangerous Goods Transportation Regulations," U.S. National Transportation Safety Board, Washington, D.C. 1971.

    b. R. A. Scott, Jr., Technical Paper No. 10, "Methodology for Chemical Hazard Prediction, "Department of Defense Explosives Safety Board, Washington, D.C. 1975.

    c. H. H. Fawcett, "Carbon Tetrachloride Mixtures in Fire Fighting," *AMA Arch. Ind. Hyg. Occup. Med.*, **6**, 435–440 (1952).

9. National Research Council, *Evaluation of the Hazard of Bulk Water Transportation of Industrial Chemicals, A Tentative Guide*, a Report Prepared by the Evaluation Panel of the Committe on Hazardous Materials, National Academy of Sciences, Washington, D.C., January 1974. See also R. F. Griffiths, *Dealing with Risk*, Interscience/Wiley 198.

10. C. J. Jones, The Ranking of Hazardous Materials by Means of Hazard Indices, *J. Hazard. Mater.*, **2**(4), 363–390 (1978).

11. a. *Recombinant DNA Research*, Vol. 1, DHEW Pub. No. (NIH) 76-1138, National Institutes of Health, August 1976.

    b. V. Cohn, "NIH sees a Small Risk" in Gene Work," *Washington Post*, Nov. 18, 1977; "Restricting DNA Experiments" (Editorial), *Washington Post*, December 19, 1977.

    c. "Program to Assess the Risks of Recombinant DNA Research"; Final Plan, National Institutes of Health, *Fed. Reg.*, Part IV (September 13, 1979); also "Guidelines for Research Involving Recombinant DNA Molecules," *Fed. Reg.* (July 20, 1979).

12. a. Symposium on Waste Chemistry of the Nuclear Fuel Cycle as it Relates to Health and Safety, September 10, 1979, co-sponsored by Division of Chemical Health and Safety and Division of Nuclear Chemistry and Technology, American Chemical Society, Washington, D.C.

    b. "The Geological Criteria for Suitable Sites of High-Level Radioactive Waste Depositories," Panel on Geological Site Criteria, Committee on Radioactive Waste Management, National Research Council, Washington, D.C. August 3, 1978.

    c. "Alternatives for Long-Term Management of Defense High Level Radioactive Wastes, Hanford Reservation, Richland, Washington, ERDA 77-44, September 1977.

13. R. Severo, "Cost Estimates Soar for Taking PCBs from Hudson River," *New York Times*," October 19, 1977, p. 1; W. Claiborne, "PCBs Ruin Housatonic, River of Writers," *Washington Post*, July 11, 1977, p. 1; "PCBs and the Environment," Department of HEW, COM-72-10419, March 20, 1972, NTIS, Springfield, Va.

14. E. Chen, *PBB, An American Tragedy*, Prentice-Hall, Englewood Cliffs, N.J., 1979.

15. "Kepone/Mirex/Hexachlorocyclopentadiene: An Environmental Assessment (Scientific and Technical Assessments of Environmental Pollutants)," Commission on Natural Resources, National Research Council, Washington, D.C., 1978.

16. "The Biologic and Economic Assessment of 2,4,5-T; A Report to the USDA - States - EPA, 2,4,5-T RPAR Assessment Team," February 15, 1979.

17. "National Toxicology Program, Annual Plan, Fiscal Year, 1981, NTP-80-62," Chairman, David P. Rall, Department of HHS, Washington, D.C., Dec. 1980.

18. *Health and Safety: The Social and Professional Responsibilities of the Chemist*, Proceedings, RIC Symposium, York, April 10, 11, 1975, Royal Institute of Chemistry, London.

19. M. L. Brown, *Occupational Health Nursing*, Springer, New York, 1956; see also M. L. Brown, *Occupational Health Nursing, Principles and Practices*, Springer, New York, 1980.

20. a. American College of Obstetricians and Gynecologists, "Guidelines on Pregnancy and Work," prepared for National Institute for Occupational Safety and Health, September 1977.
    b. A. Hricko, and M. Brunt, "Working for your life: A Women's Guide to Job Health Hazards," Labor-Occupational Health Program, University of California, Berkeley, 1976.
    c. *Proceedings, Conference on Women in the Workplace*, June 17–19, 1976, Society for Occupational and Environmental Health, Washington, D.C. 1977.

21. S. Crapnell, "Accident Prevention: Your Key to Controlling Surging Workers' Compensation Costs," *Occup. Hazards*, 35–38 (November 1979).

22. a. M. M. Nemec, "The High Cost of Product Liability," *Occup. Hazards*, 47–50 (November 1979).
    b. "Setting Your Defenses Against Product Liability Claims," *Occup. Hazards*, 51 (November 1979).

23. Federal District Court, Savannah, Georgia, October 1, 1977.

24. "The Costs of Kepone" (Editorial), *Washington Post*, October 9, 1976, p. A-16.

25. "Chemical Plants Leave Unexpected Legacy for Two Virginia Rivers (Mercury and Kepone)," *Science, **198**, 1015–1019 (December 9, 1977).

26. Committee on Interstate and Foreign Commerce, "Staten Island Explosions: Safety Issues Concerning LNG Storage Facilities," House of Representatives, July 10–12, 1973, Serial 93–42, Washington, D.C., 1973; see also *Conference Proceedings on LNG Importation and Terminal Safety*, Boston, Mass., June 13–14, 1972, Committee on Hazardous Materials, National Academy of Sciences-National Research Council, Washington, D.C.; see also "Liquefied Natural Gas, Views and Practices," Policy and Safety, CG-478, Department of Transportation, U.S. Coast Guard, Washington, D.C., February 1, 1976.

27. W. C. Hoyle, "Bulk Terminals: A Toxic Substance Leak in Retrospect," Division of Chemical Health and Safety, New Orleans National meeting, American Chemical Society, March 22, 1977.

28. a. J. G. Fuller, *The Poison That Fell from the Sky*, Random House, New York, 1977.
    b. D. B. Richardson, "The Continuing Agony of a Town Buried in Industrial Poison," *U.S. News and World Report*, August 1, 1977, pp. 44–45.
    c. *Chem. Week*, August 4, 1977.
    d. P. Cooper, "Dioxin: A New Biological Probe?" *Food Cosmet. Toxicol.* 15(5), 481–483 (October 1977).

29. Report of the Court of Inquiry, *The Flixborough Disaster*, June 1, 1974, Department of Employment, H.M. Stationery Office, London, 1974.

30. a. B. Richards, "U.S. Will Pay $5 millions in Asbestos Suit, Asbestos Settlement Bid Unprecedented," *Washington Post*, December 9, 1977, p. 1.
    b. D. Burnham, "Asbestos Workers Illness—and Their Suit—May Change Health Standards," *New York Times*, December 20, 1977, p. 30.

31. N. A. Ashford, *Crisis in the Workplace—Occupational Disease and Injury*, MIT Press, Cambridge, Mass., 1976.

32. Interagency Task Force on Workplace Safety and Health, "First Recommendations," Rosslyn, Va., August 1, 1978.

33. a.   M. Mintz, "Jail Terms Sought for Business Health, Environmental Violators,"
          Washington Post, Nov. 25, 1979, p. 1; "Support for a Crackdown on White Collar
          Crime by Corporate Officials Whose Decisions Create Products or Situations That
          Harm the Environment and Public Health," *Toxic Mater. News*, 6(48), 377 (November
          28, 1979).
     b.   J. M. Brown and L. A. Schiller, "A Study of Federal Authority to Mandate Property
          Insurance Coverages," *Insur. Counsel J.*, 222–276 (April 1979).

34. "Part of the Human Condition, Health and Safety Hazards in the Workplace," DHEW
     (NIOSH) Publication No. 78-137, National Institute for Occupational Safety and Health,
     Cincinnati, Ohio, 1978.

35. "A Toxics Primer," League of Women Voters Education Fund, League of Women Voters
     of the United States, Washington, D.C., Pub. No. 545, 1979.

36. H. H. Fawcett, "The Changing Nature of Acute Chemical Hazards (A Historical Perspec-
     tive)," Second Chemical Congress of North America, Las Vegas, Nev., August 29, 1980.

37. J. Y. Oldshue, "The Two Worlds of the Chemical Engineer," *Chem. Eng. Progr.*, 30
     (February 1980).
     R. Fleming, "How Industry Views Itself," *Chem. Eng. Progr.*, (May 1980).
     A. L. Conn, "Effective Interaction in an Adversary Climate," *Chem. Eng. Progr.*, 15–18
     (July 1980).

38. American Chemical Society Health and Safety Information Service. [Contact Ms. Barbara
     Gallagher, (202) 872-4511.]

39. F. L. Cross, "What the New Source Performance Standards (Clean Air Act Amendments
     of 1977) Mean to You," *Pollut. Eng.*, 12(2), 35–37 (February 1980).
     "Congress Eyes Criminal Penalties for Concealing Hazards," *Occup. Hazards*, 41–43
     (March 1980).
     Note: Hazardous waste generators, handlers, or transporters who failed to notify EPA by
     August 18, 1980 were subject to fines up to $25,000 a day after August 18. Treatment, storage
     and disposal facilities operators must have reported before November 18, 1980.

40. Federal Regulations are published ultimately as the Code of Federal Regulations. They are
     issued in the Federal Register. 29 CFR 1990 is published as *Fed. Regist.*, **45,** No. 15, Book 2
     of 2, Part VII, Department of Labor, Occupational Safety and Health Administration,
     *Identification, Classification and Regulation of Potential Occupational Carcinogens,
     Final Rules*, January 22, 1980, pp. 5001–5296.
     "Hazardous Waste Management, Final Rules," published as *Fed. Regist.* 12722–12744
     (February 26, 1980), and 33287–33588 (May 19, 1980).

41. Discussions of risk/benefit will be found in:
     "Cost/Benefit Studies for American Industrial Hygiene Association," conducted by Jack
     Bailey. [Available from Edward Howard & Co., 1021 Euclid Ave., Cleveland, Ohio
     44115.]
     B. R. Putnam, "Cost-Benefit Analysis for Regulatory Policy Making," *Chem. Eng.
     Progr.*, 20–23 (February 1980).
     National Research Council, *Regulating Pesticides*, National Academy of Sciences,
     Washington, D.C. 1980.
     D. P. Rall, *First Annual Report on Carcinogens*, July 1978. [Available, in two volumes,
     from National Toxicology Program, U.S. Public Health Service, P.O. Box 12233, Re-
     search Triangle Park, NC 27709.]
     "Estrogen, Not Cost Effective, Still May Be Worth Taking," *Washington Post*, August 7,
     1980, p. A-7.
     "The Cost-Benefits Argument—Is the Emphasis Shifting?", *Occup. Hazards*, 55–59
     (February 1980).
     B. Crickmer, "Regulation: How Much is Enough?", *Nation's Bus.*, 26–33 (March 1980).
     N. A. Ashford, "The Regulation of Chemicals and the Role of the Professional Chemist,"

presented at the Symposium on the Interface of Government Safety Regulations and the Chemical Professional, American Chemical Society Meeting, Chicago, Ill., August 26, 1975.

"Union Calendar No. 848, Chemical Dangers in the Workplace," Thirty-Fourth Report by the Committee on Government Operations, Sep. 27, 1976, U.S. Government Printing Office, Washington, D.C.

R. Murray, "The Emotional Impact of Chemicals," *J. Occup. Med.*, **20**(4), 267, 268 (April 1978).

N. Root and J. McCaffrey, "Targeting Worker Safety Programs: Weighting Incident Against Expense," Monthly Labor Review, U.S. Dept. of Labor, January 1980, pp. 3–8.

a. J. S. Turner, "How Safe is Safe? A Consumer's Viewpoint," *Global Aspects of Chemistry, Toxicology and Technology as Applied to the Environment*, Vol. 3 in Series EQS Environmental Quality and Safety, Georg Thieme Verlag Stuttgart, Academic Press, New York, pp. 1–6.

b. O. H. Lowry, "How Safe is Safe? A Scientific Viewpoint," *Global Aspects of Chemistry, Toxicology and Technology as Applied to the Environment*, Vol. 3 in Series EQS Environmental Quality and Safety, Georg Thieme Verlag Stuttgart, Academic Press, New York, 1974, pp. 7–10.

c. W. C. Wescoe, "How Safe is Safe? A Producer's Viewpoint," *Global Aspects of Chemistry, Toxicology and Technology as Applied to the Environment*, Vol. 3 in Series EQS Environmental Quality and Safety, Georg Thieme Verlag Stuttgart, Academic Press, New York, 1974, pp. 11–16.

d. P. B. Hutt, "How Safe is Safe? A Regulator's Viewpoint," *Global Aspects of Chemistry, Toxicology and Technology as Applied to the Environment*, Vol. 3 in Series EQS Environmental Quality and Safety, Georg Thieme Verlag Stuttgart, Academic Press, New York, 1974, pp. 187–195.

G. H. Collings, Jr., "Medical Confidentiality in the Work Environment," *J. Occup. Med.* 461–468 (July 1978).

"Urinary Cytologic Examinations of Laboratory Workers" *J. Occup. Med.* 482–487 (July 1978).

R. E. Eckardt and R. A. Scala, "Toxicology: Accessing the Hazard," *J. Occup. Med.*, 490–493 (July 1978).

W. E. Morton, "The Responsibility to Report Occupational Health Risks," *J. Occup. Med.*, 258–260 (April 1977).

EPA, "Test Methods for Evaluating Solid Waste, Physical/Chemical Methods," SW-646, Office of Water and Waste Management, U.S. EPA, Washington, D.C., May 1980. (Refers to 40 CFR Part 261, Identification and Listing of Hazardous Waste.)

"Toxics on Tap: Chemical Contamination of Long Island Drinking Water Supplies." [Available from New York Public Interest Research Group, Room 1000, 5 Beekman St., New York, NY 10030.]

42. L. R. Ember, "Uncertain Science Pushes Love Canal Solutions to Political, Legal Arenas," *Chem. Engl. News*, 22–29 (August 11, 1980).

43. G. E. Likens, R. F. Wright, J. N. Galloway, and T. J. Butler, "Acid Rain," *Sci. Am.*, **241**(4), 43 et seq.

44. OSHA, "Access to Employee Exposure and Medical Records," *Fed. Regist.* 35212–35303 (May 23, 1980).

"Damon Clinical Colliquium, Update," *Medicine and the Workplace*, **2**, No. 1 (1979). [Available from Damon Medical Laboratories, 115 Fourth Ave., Needham Heights, MA 02194.]

Diamond Shamrock Corp., "Influencing the 'Need to Know'," *Occup. Health Safety*, **49**(1), 22–24. See also "The Industrial Physician's Role as Data Manager," *Occup. Health Safety*, **49**(1), 25–27.

R. C., Atherley, *Occupational Health and Safety Concepts, Chemical and Processing Hazards*, Applied Science Publishers, London 1978.

"Environmental Surveillance Specifications," 3rd ed. [Available from Occupational Health Association, P.O. Box 12854, Pittsburgh, PA 15241 (lists over 120 OSHA medical and environmental monitoring specifications).]

M. M. Key, ed., "Occupational Diseases, A Guide to Their Recognition," DHEW (NIOSH) Publ. No. 77–181, June 1977.

R. R. Jones, ed. "Readers Reveal the Dangerous Lines of R&D Scientists," *Ind. Res. Dev.*, 127–130 (July 1980).

45. "Chemical Boobytraps," 16-mm sound motion picture, 10 min. 1959. [Available from General Electric Co., Audio-Visual Dept., Scotia, NY 12301.]

46. "Accident Facts," National Safety Council, Chicago, IL, revised annually.

47. Genesis III: 24 "and a flaming sword"; Exodus III: 2 "burning bush was not consumed."

48. Paul R. Lyons, *Fire in America*. National Fire Protection Association, Quincy, MA, 1976.

49. a. "Towering Inferno," *New York Daily News*, June 24, 1980, p. 1.
    b. R. F. Mendes, *Fighting High-Rise Building Fires, Tactics and Logistics*, National Fire Protection Association, Boston, Mass. 1975.

50. Fire Safety Aspects of Polymeric Materials, National Research Council, National Materials Advisory Board, in 10 volumes, Technomic Publishing Co., Westport, Conn. 1979–81.

51. "Pressure-Relieving Systems for Marine Cargo Bulk Liquid Containers," National Research Council, Washington, D.C., 1973.

52. Paul R. Lyons, "The Terrible Blast of a BLEVE," *Fire in America,* pp. 226, 227, National Fire Protection Association, Quincy, MA, 1976.

53. George H. Proper, Jr., Personal correspondence. The accident involved release of 12,500 kg LPG, which formed a cloud 122–183 m in diameter, 24 m high, which exploded.

54. W. E. Isman and G. P. Carson, *Hazardous Materials*, Glencoe Publishing Co., Encino, Calif., pp. 114–115, 1980.

55. R. Assheton, *History of Explosions*, Institute of Makers of Explosives, 1930.

56. K. Gugan, *Unconfined Vapor Cloud Explosions*, Gulf Publishing Co., Houston, 1979.

57. "Collapse of Molasses Tank, Boston, Mass., January 15, 1919," *NFPA Q.* 12(4), 371–373 (April 1919).

58. "Report on Crete, Nebraska Rail Accident," National Transportation Safety Board, Washington, D.C., 1973.

59. "120 Hospitalized by Toxic Gases after Rail Crash," *Washington Post*, April 4, 1980, p. A-6. See also NTSB-HZM-81-1, Somerville, Mass. Derailment, National Transportation Safety Board, Wash., D.C., 1981.

60. E. J. Shields, "Prevention and Control of Chemical Spill Incidents," *Pollut. Eng.*, 52–55 (April 1980); see also A. J. Smith, Jr., *Managing Hazardous Substances Accidents*, McGraw-Hill, NY, 1981.

    "350 square miles Louisiana Marshes Closed to Fishing from PCP Spill," *Wall Street Journal*, August 5, 1980, p. 1.

61. J. L. Buckley and S. A. Wiener, "Hazardous Material Spills, A Documentation and Analysis of Historical Data," Industrial Environmental Research Laboratory, Office of Research and Development, U.S. EPA, Cincinnati, Ohio (PB 281 090).

62. a. "Transport of Hazardous Wastes and Hazardous Substances," *Fed. Regist.*, Part II, 34560–34705 (May 22, 1980).
    b. *Emergency Response Guidebook* 1980, DOT P5800.2, U.S. Department of Transportation, Washington, D.C. 20590.

63. "Advance Accident/Incident Report," Federal Railroad Administration, Washington, D.C., December 1979.

64. NFPA No. 491-M, *"Hazardous Chemical Reactions,"* National Fire Protection Association, Quincy, MA.

65. L. Bretherick, *Handbook of Reactive Chemical Hazards*, 2nd ed., Butterworth, London, 1979.

66. "Waste Disposal Site Survey," Report by Subcommittee on Oversight and Investigations, House of Representatives, 96th Congress, Oct. 1979; see also L. R. Ember, "Uncertain Science Pushes Love Canal Solutions to Political, Legal Arenas," *Chem. Eng. News*, pp. 22–29 (August 11, 1980).

67. "Everybody's Problem: Hazardous Waste," SW-826, U.S. EPA; see also: "Hazardous Waste—Fifteen Years and Still Counting," EPA 98/0, June 1980.

68. a. M. H. Brown, *Laying Waste: The Poisoning of America by Toxic Chemicals,* Pantheon Books, New York, 1980.

    b. M. Clark, et al., "Fleeing the Love Canal," *Newsweek*, pp. 56–57 (June 2, 1980).

    c. J. O. Mang, "Stirrings of Life Return to Love Canal," *Washington Post,* A-12 (Nov. 9, 1981).

69. "Hazardous Waste and Consolidated Permit Regulations," *Fed. Regist.* 33063–33286 (May 19, 1980); see also S. M. Senkan and N. W. Stauffer, "What to Do with Hazardous Waste," *Technology Review* (M.I.T.), **84**(2), 34–49 (Nov./Dec. 1981).

70. "Eyewitness Account Claims 1,000 Died after Germ-Plant Blast in Soviet City, *Baltimore Sun,* July 12, 1980.

71. "Introduction to Biohazards Control," slide-audio cassette program. Audio-Visual Section, National Safety Council, 444 N. Michigan Ave., Chicago, IL 60611.

72. B. Wain and J. Leach, "Yellow Rain: The UN's 'Pitiful Farce' and the Choices Now Facing the U.S." *Wall Street Journal*, December 30, 1981, p. 6; see also W. Kucrwicz, "Asian Refugees: Death in the Night," March 1, 1982, p. 22.

    R. H. Myers, "The Chromosone Connection," *The Sciences*, **20**, No. 5, 18–20 (May/June 1980).

    "Guidelines for Research Involving Recombinant DNA Molecules," *Fed. Regist.* **45**(20), 6724–6749 (January 29, 1980).

    "The Miracles of Spliced Genes," *Newsweek*, pp. 62–72 (March 17, 1980.)

    E. Milewski and S. Barban, eds., "Recombinant DNA Technical Bulletin," **2**, No. 4, (April 1980). [Available from Office of Recombinant DNA Activities, Bldg. 31, Room 4A52, National Institutes of Health, Bethesda, MD 20205.]

    M. Slade and T. Ferrell, "Cell Factories Produce Antibodies," *New York Times*, August 3, 1980, p. 20E.

    N. Myers, "Corn Genes and Big Dollars," *Technology Review*, **84**(2), 8, 9, 64 (Nov./Dec. 1981).

BIBLIOGRAPHY

Bresco, F., "Spread and Evaporation of Liquids," *Prog. Energy/Comb. Sci.* **6** (1980).

Brown, James M., "Probing the Law and Beyond: A Quest for Public Protection from Hazardous Product Catastrophes," Staff Discussion Paper 402, Program of Policy Studies in Science and Technology, The George Washington University, Washington, D.C., July 1969.

Cave, L., "Risk Assessment Methods for Vapour Cloud Explosions," *Prog. Energy/Comb. Sci.*, **6** (1980).

Cox, R. A., "Dispersion of Vapour," *Prog. Energy/Comb. Sci.*, **6** (1980).

DeReamer, R., *Modern Safety and Health Technology*, 2nd ed., John Wiley & Sons, New York, 1980.

Guide to Occupational Safety Literature, 1977, National Safety Council, Chicago, IL 60611 (or most recent issue).

Haas, Thomas J., "Toxic Hazard Awareness in Response to Hazardous Materials Spills," *Proceedings of the Marine Safety Council*, **37**(6), 119–123 (June/July 1980), CG 129, U.S. Department of Transportation.

Hilado, C. J., *Fire and Flammability Series: Bedding and Furniture Materials*, Technomic Publishing, Westport, Conn. 1976.

Hilado, C. J., *Fire and Flammability Series: Floor and Floor Covering Materials*, Technomic Publishing, Westport, Conn. 1976.

Lefebvre, A. H., "Air Blast Atomization," *Prog. Energy/Comb. Sci.*, **6** (1980).

Lewis, David J., "Unconfined Vapour-Cloud Explosions: Definition of Source of Fuel," *Prog. Energy/Comb. Sci.*, **6** (1980).

Lewis, David J., "Unconfined Vapour-Cloud Explosions—Historical Perspective and Predictive Method Based on Incident Records," *Prog. Energy/Comb. Sci.*, **6** (1980).

Nicholson, W. J., "Management of Assessed Risk for Carcinogens," *Ann. N.Y. Acad. Sci.*, **363** (1981).

Tierney, K. J., *A Primer for Preparedness for Acute Chemical Emergencies*, The Disaster Research Center Book and Monograph Series, The Ohio State University, Columbus, Ohio, 1980. (See Chap. 34.)

Urbanek, G. L. and Barber, E. J., "Development of Criteria to Designate Routes for Transporting Hazardous Materials—Final Report," DOT-FH-11-9595, Federal Highway Administration, U.S. Department of Transportation, Report No. FHWA/RD-80-105, Washington, D.C., September 1980.

Van de Putte, T., "Purpose and Framework of a Safety Study in the Process Industry," *J. Haz. Mater.* **4**(3), 225–234 (January 1981).

Kennedy, D., "The Politics of Preventive Health," *Technology Review*, **84**(2), 58–60 (Nov./Dec. 1981).

DOT P5800.2, *Emergency Response Guidebook, Hazardous Materials 1980*, U.S. Department of Transportation, Washington, D.C. 20590.

Katzenstein, A. W., *An Updated Perspective on Acid Rain*, Edison Electric Institute, Washington, D.C. 20036, Nov. 1981.

2

Federal Standards on Occupational Safety and Health

Floyd A. Van Atta

With the passage of PL 91-596, the Williams-Steiger Occupational Safety and Health Act and now generally referred to as the OSH Act, the primary responsibility for regulation of occupational safety and health conditions was transferred to the Federal government from the various state governments.[1] This act, which was signed into law on December 29, 1970 and took effect on April 28, 1971, led to the first uniform standards for occupational safety and health across the nation. It established the new position of Assistant Secretary for Occupational Safety and Health and the Occupational Safety and Health Administration in the U. S. Department of Labor and authorized the Secretary of Labor to promulgate Standards for Occupational Safety and Health which would have the force and effect of law with regard to any employer of one or more employees whose business affected interstate commerce. This language was a considerable broadening of the traditional wording of Federal laws based on the commerce clause. They generally apply only to businesses which introduce a product into interstate commerce. In effect, it says that the Occupational Safety and Health Act and its regulations apply to every business, large or small, in the United States, its possessions, and its territories except the United States government and the governments of states and subdivisions of states. However, in Executive Order 12196, the President has asked Federal agencies to comply with OSHA.

The Act specifies two duties for employers: (1) to provide to every employee employment and a place of employment free from recognized hazards which are causing or are likely to cause significant physical harm or death to the employee even if the exposure is throughout the working lifetime of the employee, and (2) to comply with the occupational safety and health standards promulgated under the act. The one duty specified for employees is to comply with the occupational safety and health standards which are applicable to his or her own actions. The major difference concerning these duties is that penalties may be assessed against the employer for non-compliance. There is no provision for penalties against employees.

In order to provide a body of regulations for starting the operation of the first comprehensive occupational safety and health program in the history of the United States the act provided that for two years following the effective date of the law the Secretary could promulgate, as regulations under the act, any existing Federal regulations or any voluntary national consensus standards without any formal proceedings. This authorization expired on April 28, 1973 but it provided for the assembling of all of the original OSHA standards. They consisted of the standards which had been issued previously under the Walsh-Healey Public Contracts Act, the McNamara-O'Hara Service Contracts Act, the Longshoremen's and Harbor Workers' Compensation Act, the Vocational Rehabilitation Act, the National Foundation for the Arts and Humanities Act, the Contract Work Hours and Safety Standards Act (Construction Safety Act), and certain voluntary standards which had either been adopted by reference in previous Federal regulations or which were adopted by reference for this purpose.

Since the components of this set of standards came from a variety of sources and had to be adopted without modification it was a pretty motley assortment. Some sections were vertical standards with reference to a specific industry or trade, some were horizontal standards with application to any occupation where the subject matter occurred, most were specifications with precise descriptions of what must be done to attain compliance. For instance: "the general slope of grain in side rails (of wood ladders) shall not be steeper than 1 in 12, except that for ladders under 10 feet in length and having flat steps for treads, the general slope of grain shall not be steeper than 1 in 10" [Paragraph 1910.25(b)(2)]. A few were performance codes, stating only the required result with no indication as to how it should be attained, for instance Subpart Z, Tables Z-1, Z-2, and Z-3. Some fell between these descriptions. The result was a volume with slightly more words than the King James version of the Bible and at least as confusing to the uninitiated as that venerable testament.

From the vantage point of the Department of Labor there is a considerable advantage to having detailed specification standards. All that the local inspector is required to do is read the book. The process being inspected is either according to specifications in every detail or it is not in compliance with the law. The same argument has an irresistible appeal to the Solicitor's Office and, presumably, to the courts. The requirements are spelled out in precise detail which makes it very easy to prepare and win a case of non-compliance.

To the person trying to run a business the advantages are not quite so obvious. There is one, and only one, method to operate in compliance with the law. That method may, or may not, be compatible with the particular process being used. The particular process being used may be one which was not even dreamed of at the time that the standard was written, or simply one which the standard writing committee did not consider. Whatever the reason, it is so obvious as to be indisputable that the existence of this kind of standard makes the introduction of any innovation measurably more difficult or, in the worst case, impossible.

There are a number of methods outlined in PL 91-596 for getting standards amended or revoked, or new standards introduced. By far the most common method for starting such a process is for the National Institute for Occupational Safety and Health (NIOSH) to present a criteria document to the Secretary of Labor for consideration. NIOSH is designated in PL 91-596 as the research and development agency for OSHA and has the primary responsibility for such developments. Any citizen or group of citizens has the right to petition the Secretary to establish a standard or the Secretary may, upon his or her own initiative, appoint an ad hoc advisory committee to inquire into the need for and the feasibility of a standard on any subject. Such ad hoc committees are specifically provided for in PL 91-596. None of these actions are binding upon the Secretary except to the extent that he or she is required to state for the record and publish in the Federal Register his decision and the reasons for it.

If the decision is to proceed with the development of a standard, or the modification of an existing standard, a project officer will be assigned to it by the Office of Standards Development in OSHA and the Office of the Solicitor will assign an attorney. It is the responsibility of these two to shepherd the standard through all of the steps to its final promulgation. They do such chores as writing the draft standards, reviewing the pertinent literature, and getting the necessary environmental impact and inflationary impact statements. When a draft which has the approval of the Solicitor's Office and the Office of Standards Development has been produced, it is published in the *Federal Register* as a proposal and an appropriate time is allowed for written public comments. Anyone who wishes to, during the comment period, can request a public hearing on the proposal. Such a request may not be denied and a place and time for one, or more, public hearings, depending upon the apparent demand, will be designated shortly after the close of the time for written comments. These hearings are informal in the sense that they are not conducted in accordance with judicial rules of evidence. An administrative law judge presides but does not render a decision. It is the function of the judge to see that the hearing proceeds in a reasonably orderly manner and to certify to the Secretary at the close of the hearing that the record is complete and accurate. All of the written comments become part of the hearing record and a stenographic record of oral proceedings is maintained and certified.

After the hearing record is closed and certified it is the function of the Project Officer and Attorney to review the record and rewrite the standard in the light of the evidence presented. The final standard so produced is published in the *Federal Register* and, in the absence of a court challenge, becomes a mandatory standard. This whole rather involved process is shown in graphic form in Fig. 2.1.

This process is designed to permit the parties most directly involved to have an influence, which could be a decisive influence, on the nature of the regulations under which they will have to operate. One reason that it regularly fails to have that effect is that so very few people bother to understand the regulatory process or to take part in it when they have the opportunity. An excellent example of this is what is happening to paragraph 1910.1000, "Air Contaminants." It is well known that this section was adopted almost in toto from the American Conference of Governmental Industrial Hygienists 1968 Threshold Limit Values table which represented in a general way the best estimates at that time of the maximum concentrations of a variety of contaminants which most people could inhale for a long time without overt signs of poisoning. One of the first major projects which NIOSH proposed was what, as good and dedicated bureaucrats, they entitled the "Standards Completion Project."[2] The argument goes that the people subject to the regulation should know, on the face of the standard, precisely what they must do to comply. So it is necessary to add to each entry in the table requirements for monitoring the airborne concentration at specified intervals and for keeping records of the results, for medical examinations of employees and preserving the records, for personal protective equipment for employees, for posting of warning signs, and for employee observation of monitoring and access to the records. In all of this bureaucratic foofaraw there was no suggestion to address the question of the appropriateness of the 1968 TLVs in 1982, or even in 1975 when the project was started, nor was any question raised as to whether an academic teaching laboratory necessarily needs the same type and quantity of monitoring as does an industrial control laboratory, both certainly are subject to the act. There are those who might maintain that that is the major similarity between them. This is a situation which cries out for correction. Not only for this particular set of standards but, more importantly, for the general philosophy of the people who are responsible for producing the standards. This particular set could be substantially simplified by writing one standard to set out the general principles which will govern the necessity for and the frequency of monitoring. There is certainly neither necessity for nor advantage in repeating this information in every standard having to do with air contaminants.[2] There is also no real advantage and very real disadvantage in specific rules concerning the frequency and method of monitoring for any specific contaminant. There is a good reason for specifying a referee method of sampling and analysis which will be used in case of argument but not necessarily as

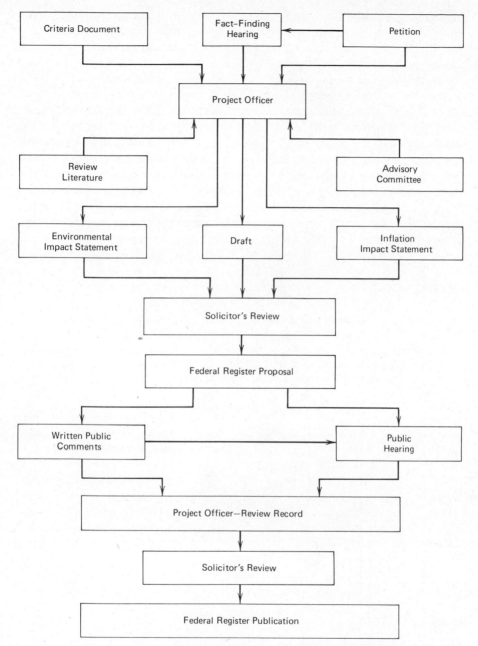

Figure 2.1. Development and promulgation of standards by OSHA.

the required method for routine use. Precisely the same kind of arguments apply to the requirements for employee observation of monitoring and access to records. These are areas which can and should be governed by principles of general applicability.[3]

This should not be interpreted as a plea for less regulation or for more regulation.[4] In the dolorous state of our reporting of occupational disease and even of occupational injuries it is not possible to determine what is really happening. What we measure is being reduced and if what we measure is significant that is encouraging, but the effect of regulations on what we measure is moot. However, there seems to be a consensus that we need regulations, a consensus which is reinforced by what we know about the history of occupational injuries. This is simply a plea that, since we are going to have regulations let us have them framed in a rational fashion. In such a fashion, in short, that whatever effect they are going to have will be attained with the minimum expenditure of time and energy.

REFERENCES

1. Public Law 91-596, Williams-Steiger Occupational Safety and Health Act, December 29, 1970.
2. NIOSH/OSHA Occupational Health Guidelines for Chemical Hazards (320 chemicals), Edited by Frank W. Mackison, DHHS(NIOSH) Publication 81-123, January 1981. For sale by Superintendent of Documents, U.S. Governments Printing Office, Washington, D.C. 20402.
3. Miller, D. E., *Occupational Safety, Health and Fire Index*, Marcel Dekker, New York, 1976.
4. Garmet, S., "Visting the Front in the Battle Against Regulation," *Wall Street Journal*, December 4, 1981, p. 34.

3

Safety Aspects of Site Selection, Plant Layout, and Unit Plot Planning

Donald M. Liston

3.1. INTRODUCTION

In this chapter, we are concerned with the problems of relative physical locations of facilities. There is the problem of locating the plant relative to its surroundings and the problem of locating components of the plant relative to each other. If we consider these problems, starting with the most general aspects and working toward specific details, we will pass through four recognized phases: location, site selection, plant layout, and unit plot planning.

First, let us consider location. Given one proposed chemical plant and one world, the problem is where to locate said plant on said world. This is what we refer to as the location study. Obviously, the major considerations in this phase are economic ones. Thoughts about safety have little to contribute here. Before the threat of atomic war, we might say there were no safety considerations in this phase. Today we might deliberate at some length on steering clear of possible atomic blast or fallout. However, such deliberations will not be subject matter for this chapter. Of course, we cannot simply forget such factors as high wind velocities, earthquakes, snow loads, and frequency of electrical storms. But, as far as location of the plant is concerned, these factors, too, become submerged in the economics. This is because safety factors against these hazards can be designed into the plant at a given cost.

For these reasons, this chapter will commence with the second phase, site selection. The location study may have told us to locate our plant on the east coast of the United States in the vicinity of Atlantic City, or in the northern midwest where efficient transportation by water as well as rail is available. Now it is time for site selection—the selection of a specific plot of ground on which to build. The concern is still one of location of a whole plant relative to its surroundings.

The third step toward detail is plant layout. We are beyond studying surroundings and can concentrate on the plant itself. The blocks to shuffle around here are functional operating units of the plant. We will be setting up the relative locations of things like process units, boiler plants, water treating facilities, maintenance shops, administrative facilities, etc.

Finally, there is unit plot planning. Our playthings now become individual pieces of equipment like pumps, exchangers, drums, columns, pipes, etc. The problem is the proper location of these things within a given operating unit. In these latter three phases, what can we do to enhance safety? This is the content of this chapter.

3.2.   THE HAZARDS AND THE LINES OF DEFENSE

In site selection, plant layout, and unit plot planning, it is possible to make allowances for only certain types of hazards. To facilitate discussion of these hazards, let us break them down into first degree hazards and second degree hazards.

First degree hazards are those which provide the *potential* for trouble. Under normal conditions, these hazards do not cause damage to either persons or property, but they set the stage for accidents that can cause injury, fires, or explosions. Typical of first degree hazards are:

1.   The presence of flammable or combustible materials.
2.   The presence of heat.
3.   The existence of ignition sources.
4.   The presence of oxygen.
5.   The presence of compressed materials.
6.   The presence of toxic materials.
7.   The possibility of human error.
8.   The possibility of mechanical failure.
9.   The movement (normal or emergency) of people and equipment through the plant.
10.   Reduced visibility from vapor clouds, etc.

When first degree hazards get out of hand, the result can be second degree hazards which are capable of directly inflicting damage to life, limb, and property. These second degree hazards are:

1. Fire.
2. Explosion.
3. Release of free toxic materials.
4. Stumbling.
5. Falling.
6. Collision.

In combating all of these hazards, we draw three lines of defense. The first line of defense consists of trying to deal with first degree hazards in such a fashion as to prevent the occurrence of second degree hazards. Success here depends mostly on careful engineering of the equipment we use. However, there are things to watch for during plant layout and unit plot planning to assist this first line of defense. For example:

1. Locating ignition sources upwind (based on prevailing wind direction) from points of possible release of flammable materials.
2. Providing adequate accessways for the movement of equipment and people.

Despite all such efforts, there are occasions when second degree hazards, such as fires, do occur. Now the second line of defense comes into play. In the event of a second degree hazard, how can we minimize the extent of damage to life, limb, and property? Here, again, there are steps we can take in selecting the site and arranging the plant to help limit such damage. Examples of these are:

1. Separating the most hazardous areas from the areas most often occupied by people.
2. Strategically locating fire fighting equipment.

But some damage will result from these hazards. People do get hurt in modern industrial plants no matter how carefully we plan against accidents. The third line of defense is to provide efficient first aid and hospital facilities to care for those who do get hurt. The essence of this last line of defense is the rapid repair of damages we have not been able to prevent.

The specific safety problems in Sections 3.4, 3.5, and 3.6, are all applications of these three lines of defense against first and second degree hazards.

3.3.　THE TOOLS AND TECHNIQUES OF DEFENSE

Now let us consider the tools and techniques that are available to us to help implement these lines of defense. Nature provides some of these tools and others must be supplied by man. Let us take the natural ones first.

Topography is one factor which we can use in planning for safety. Just as liquids flow downhill, so will many of the flammable or toxic gases which may be released in an operating plant. Proper utilization of this feature can make it work for us as a safety tool is disposing of these hazardous gases.

Sources of large volumes of water are exceedingly valuable when it becomes necessary to fight fires. Adequate water supply can spell the difference between success and failure.

Another important natural element is the direction of the prevailing wind. The local weather bureau in almost any area can supply a wind rose which will tell the percent of the time the wind blows from the direction of each point on the compass. Prevailing wind will help in preventing flammables from drifting toward ignition sources. It will also help prevent vapor clouds or toxic gases from drifting through highly populated areas or across roads. Lest it should appear, later on, that we are overrating the value of this tool, let us emphasize that utilizing prevailing wind is a matter of playing the odds. Obviously the wind does not always blow in the prevailing direction, and everything in the plant cannot be either upwind or downwind as we seem to suggest. However, it is always best to be on the right side of the odds as often as physically possible.

Going beyond nature, the intelligence of man supplies some of the elements to enhance safety. Separation by distance is one such element. We can use distance to separate one hazard from another (such as a furnace from a relief valve which may discharge to the atmosphere). We can also use distance to separate hazards from people (such as high pressure vessels from a control house). A similar tool is separation by physical barrier. A typical example here is the confinement of liquid spills by walls, such as the dikes around storage tanks.

Two tools which work hand-in-hand are the *concentration* of hazards and the *identification* of hazards. Consider the location of pressure storage vessels (spheres and spheroids). It is best to isolate this type of vessel into one particular area of the plant. And the fact that hazards are thus concentrated facilitates the demarcation of the area as being a particularly hazardous area. This helps in two ways. First, it becomes much more practical to keep people out of these hazardous areas except those who have specific duties there. Second, those people who must work in or pass through these areas can be made fully aware of the hazardous conditions which prevail. Hazards that are dispersed throughout the plant become all the more dangerous by their inconspicuousness. We must recognize that along with these advantages comes the possibility that a fire or explosion on one vessel may result in a larger total loss through the involvement of adjacent vessels. But the consensus is that the reduced likelihood of trouble with concentrated hazards which are properly looked after does result in a safer installation.

The final tool is the ability to design and build physical facilities with which to combat hazards. Fire water systems, safety showers, and first aid stations are all examples of applying this tool.

These are the tools we have. We must apply them as we select sites, lay out plants, and plot plan individual operating units. In this problem of application, we have found two techniques to be most helpful.

Enlisting the services of a consulting safety committee is extremely valuable. The members of the safety committee provide a strong bias toward a high level of safety. This helps balance the bias of the project manager who will normally lean quite heavily toward minimum costs. Actually, most good designers have an understanding of safety problems, and the function of the safety committee becomes one of *reviewing* and *recommending* arrangements to enhance the level of safety. It will be wise to include a representative of the operations side of the organization as a member of the safety committee—one who will represent the people who must safely operate the plant after the project people have finished with it and moved along to the next project.

As a typical requirement, the safety committee should conduct a thorough review of the proceedings at these times:

1. Before the final selection of a building site.
2. When a rough, block-type layout of the plant is available—one that shows relative locations of operating units and administrative and service facilities. This layout should also indicate the major features of the surrounding community.
3. Before finalizing the plant layout when major pipe alleys, roadways, tank dikes, and similar physical features may be inspected.
4. Early in the plot planning of units before any considerable amount of detailed engineering is under way.
5. In the final stages of design when it is possible to settle the exact locations of safety showers, hose cart stations, fire hydrants, and so on.

Strongly supporting the efforts of the safety committee is the technique of scale modeling. Particularly during the plot planning of units, the scale model has much to contribute. The ability to see three dimensions at once adds much to the speed and effectiveness of searching out and solving safety problems.

In the early plot planning, we can make use of a model made from simple rectangular and cylindrical blocks to represent the location and general size and shape of equipment. Placing these blocks on a cross-sectioned baseboard allows us to quickly observe the distances involved. For this purpose, a scale of $\frac{1}{10}$ inch to the foot or smaller will suffice. Figure 3.1 shows an example of this type of scale model.

Reviews of the final design stages will require models of much more detail. Usually built to a scale of $\frac{1}{4}$ or $\frac{3}{8}$ of an inch to the foot, these models show piping, valves, instruments, platforms, ladders, and all such details of

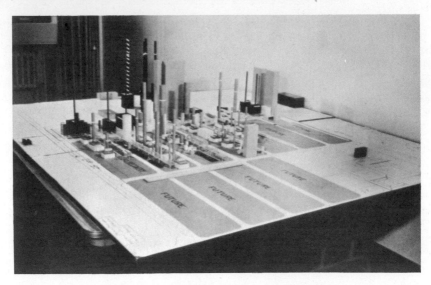

Figure 3.1. A preliminary plot-planning model.

the actual plant. Figure 3.2 shows a detailed model of the same plant that appears in Fig. 3.1. The progression from preliminary model to detailed model is quite well portrayed in these two photographs.

Timing! Here is the big problem with models: models must be available soon enough for reviews, and the resulting changes to occur before much engineering is done that may be wasted. The greater the waste, the harder it is to justify the changes, be they for purposes of safety or whatever. Many companies, and contractors too, are adopting new design methods to accommodate scale models. Moving directly from detailed flow charts to models completes the model before any appreciable engineering occurs on the drafting board. This method, of course, completely solves the timing problem. It does, however, require the services of people with combined designer-model building capabilities.

3.4. SAFETY PROBLEMS IN SITE SELECTION

We have been speaking in generalities up to this point. The discussion has covered the general principles, tools, and techniques to apply in achieving a good level of safety. Now, let us take the three areas of this chapter, one at a time, and discuss the specific problems involved.

The first problem is site selection. Remember that the plant to be built may actually be hazardous to the community in which it will be placed. Toxic gases may drift out of the plant into residential areas or into other areas containing a concentration of people. Flammable gases may drift out of the plant to ignition sources such as incinerators of other industrial plants.

Figure 3.2.   A detailed engineering model.

Fog from cooling towers may drift across high-speed highways or roads with high traffic density.

There are ways to combat these problems. Separation by distance is one. This can be accomplished by selecting an isolated site. If other considerations make this solution impractical, we may have to rely on the prevailing wind. Unfortunately, the wind does not always blow in the prevailing direction. Placing the plant downwind from the community simply improves odds against trouble occurring from the above hazards.

The possibility of collapse of high structures is another potential hazard to the community. In many cities, building codes require that tall structures or vessels be located a certain distance within property lines. This is to prevent such structures from falling on pedestrians, motorists, or neighboring facilities. The effective area of the site may thus be significantly reduced. Do not forget to check on similar local restrictions which may apply to the site under consideration.

The plant will have waste streams of which to dispose. We should be sure that the intended means of disposal will not foul the drinking water for the community. Toxic effect on marine life can be a serious problem especially where people depend on fiishing for a livelihood. Here again, be aware of local, state, or federal regulations. Try to avoid routing sewers, which may contain explosive mixtures, across public or private property (see Chaps. 19, 28, 29, and 30).

Be careful about the main point of entry to and egress from the plant. The sudden surge of traffic at starting or quitting time, in and out of the plant, may cause a serious highway hazard if not properly located or dispersed. High-speed highways adjacent to the plant present the danger of vehicles leaving the road and crashing into the plant.

Now turn about for a few moments. It is also entirely possible that the community may present some definite hazards to the plant and to the people who run it. Or if not presenting actual hazards the community may lack some of the facilities that would enhance the safety of the plant. Neighboring plants may emit toxic gases or flammable gases which could drift *into* the plant to sicken employees or ignite due to sparks or hot surfaces. In such cases, it would be better for the plant to be located upwind or again, separated by distance if such is feasible.

What can the community offer in the way of fire-fighting assistance? There may come a time when such assistance could mean the difference between just a fire or a real disaster. In a similar vein, we may ask what can the community offer in the way of first aid and hospital facilities? The balance of life or death for injured persons may hinge on the existence of good medical facilities (see Chaps. 9 and 11).

These latter two instances are examples where separation of the plant from a community by distance can be a disadvantage in safety terms. This serves well to illustrate that no pat formula insures maximum safety in site selection. The problem is simply to try to obtain the most favorable balance of forces for any specific situation.

An adequate source of water will enhance the ability to fight major fires. An adequate source can best be assured by having a stream or lake adjacent to the plant so that fire water does not have to be pumped from the ground. Have a look at the local city water system as a possible source of supplemental source of fire water.

Topography is also a factor to consider. Surely everyone involved in designing the plant will agree that a nice flat and level site is the thing to have. For safety, we would like to avoid low areas within the site in which pools of toxic or flammable vapors or liquids can form. Relative to the surrounding area, it would be better for the site to be on high ground rather than in a basin.

Remember to check each proposed site for existing rights-of-way for pipelines, roads, railroads, and power lines. We must assess the existing or potential hazards to the plant which may develop as a result of such rights-of-way.

3.5.   SAFETY PROBLEMS OF PLANT LAYOUT

Now we move along to the layout of the plant itself. The center of attention here always seems to be the processing units—the real backbone of the plant—so we will discuss them first.

The processing units are probably the most hazardous areas in the plant. Here we purposely bring together in close proximity many of the first degree hazards. We find toxic or flammable materials, high temperatures, high pressures, and ignition sources. These areas are full of mechanical equipment, all of which is subject to failure. Things can happen fast in the process unit, enhancing the possibility for human failure. Perhaps the only redeeming factor is that normally there are few people in the process unit.

Processing units should be removed from the boundaries of the plant and should be consolidated rather than scattered. This latter point will improve their identification as hazardous areas and help reduce the amount of transient traffic passing through. Watch for the major ignition sources in the plant and the major concentrations of people. The processing units should be downwind from both of these since the release of flammable or toxic materials is a possibility. It is well to maintain a considerable separation of the process areas from the major tankage areas. These two are mutually hazardous to each other.

While we have said that process areas should be consolidated, we must be careful the area is not too consolidated. Some separation between units is necessary because here again they are mutually hazardous to each other. This is particularly true where the units are not integrated process-wise. In such a case, one unit may be in full operation while its neighbor is shut down for major maintenance, presenting an increased potential for trouble. Ignition sources, heavy activity, movement of machinery, and a concentration of people in a hazardous area are all signposts of which to beware.

To date, in the chemical industry, the spacing of process units, one from the other, remains a matter of good judgment for the most part. The major factors which come to bear on these judgments are:

1. Operating temperatures.
2. Operating pressures.
3. Types of materials present in the units.
4. Quantities of materials present in the units.
5. Types of structures in the units.
6. Relative values (investment) of the units.
7. Space required for fire fighting or other emergency operations.

A good written source of material to assist in these judgments appears in a book on safety in petroleum and related industries by George Armistead, Jr.[1] Armistead's recommended minimum desirable distances, while applying specifically to petroleum refineries, will be found analogous to many similar situations in chemical plants. Another applicable reference in the handling of explosives is the American Table of Distances in the *Ordnance Safety Manual*.[2]

The integration of processes is a subject still wanting adequate treatment with regard to requirements for separating facilities by distance. We critically need some careful consideration of the effect on physical separation of facilities by these aspects of integration:

1. Process integration (elimination of intermediate tankage, etc.).
2. Mechanical integration (combined or multipurpose items of equipment).
3. Integrated or interacting control systems.

Every plant, of course, will need some administrative facilities. Safety considerations dictate that the main offices should be located on the periphery of the plant and as isolated as possible from the hazardous components of the plant. There are several reasons for this.

First, sales people, suppliers, and others who must conduct business with the plant personnel may do so, for the most part, without having to enter the plant proper. It is well to keep such visitors out of the plant whenever possible since they are not acquainted with the nature and locations of the hazards in the plant, and their normal habits may make them prone to inadvertent smoking in hazardous areas without thinking.

Second, isolated office buildings will permit smoking in the offices without endangering the plant. This also permits a very clear demarcation between the areas where smoking is permitted and those where it is not. Third, the offices will house probably the largest concentration of people in the whole plant. Separation of these people from the hazards improves the odds in favor of safety.

Fortunately, these motives for peripheral location of the offices are compatible with most of the other considerations in plant layout. There is one factor, however, that may be an opposing force on this point. Certain people, including operating foremen and maintenace supervisors, must split their time between their offices and duties in the plant. They, of course, would like their offices adjacent to their in-plant duties. An ideal solution to this problem is not easy. Careful consideration of their needs, and provisions for efficient transportation may coax them into the main offices. If they insist on in-plant offices, do not forget the safety considerations we discussed during the plant layout.

Laboratories will normally be situated adjacent to the other administrative facilities from a functional standpoint. Fortunately, the laboratory is compatible with these facilities safety-wise with two glaring exceptions. Smoking is one. There are small quantities of flammables in laboratories which may be inadvertently released into the building. Toxics, too, may be present. For these reasons, a direct connection between the laboratory building and the other administrative buildings may be inadvisable.

Boiler plants are major ignition sources normally requiring an upwind location. Maintenance shops, too, represent a major ignition source as well as a concentration of personnel. Thus, the shops, too, should be upwind and separated by distance as much as practical. Warehouses will normally be adjacent to the shops and both should be accessible by railroad spurs. Good plant layout will avoid routing railroad spurs through the plant, thus avoiding the concomitant hazards. Thus, peripheral location of these items is desirable.[3]

Similar considerations prevail for tank car and tank truck loading and unloading facilities. Traffic to these areas must not pass through the plant. Spills of toxics or flammables may occur at loading racks suggesting a downwind location as desirable. Peripheral location, also, seems best, but care should be taken not to locate tank car loading racks too close to the main line of the railroad. Locomotives may be ignition sources, and during possible derailments, could cause extensive damage to loading facilities. Regardless of these conditions, some railroad spurs will probably have to be routed into the interior of the plant. Ignition and derailment hazards must be accounted for with regard to these spurs.

Waste-water treating facilities may turn out to be the ultimate point of collection of toxics or flammables spilled anywhere in the plant and thus a downwind, as well as remote, location is indicated.

Cooling towers have the similar characteristic of collecting toxics or flammables which may leak into the water side of coolers in the operating units. This, in itself, would indicate a downwind location, a decision reinforced by the problem with fog. Cooling tower fog can block visibility on roadways, units, and in elevated structures frequently traveled by operating personnel. Thus, cooling towers should be downwind from roadways, units, and elevated structures.

Elevated flares or burning pots at grade present a contrary problem. As sources of ignition, we would tend to locate them upwind. But there is another consideration. Severe operating upsets can cause flares or burning pots to belch out considerable quantities of flammables which may fall, burning or not, on people or facilities. This, of course, would call for a downwind location. The only solution here is a location at the side (relative to the prevailing winds) of the facilities in question. We will call this a sidewind location. A sidewind location is a useful compromise in cases where combinations of location problems in plant layout prevent the ideal upwind or downwind location of any single facility. Separation by distance is also very advisable in the case of flares or burning pots.

Storage vessels, such as tanks, spheres, and spheroids, are items to be handled with caution. Each such vessel is a tremendous storehouse of energy or of toxic materials, as the case may be. Obviously, it is wise to keep people, operating units, and tankage as far apart as possible. These vessels are capable of releasing tremendous spills of toxics or flammables and, therefore, by all means should occupy a downwind location. We have already mentioned that tankage should be relegated to its own private area in the plant to enhance the identification as a hazardous area and minimize all extraneous traffic through the area.

Three problems are inherent in the location and allocation of space for tankage:

1. Separation of tanks from each other.
2. Separation of tanks from other facilities.
3. Area required to provide adequate dikes.

Two major hazards with tanks have great effect on the above three problems. First is the possible rupture of a tank shell releasing the entire contents in a very short time. Second is the boil-over tendency of some viscous stocks when tanks containing a water layer are heated above the boiling point of the water. As in the case of processing units, little has been compiled for the chemical industry in the way of actual dimensional recommendations for the three problems above. But here again, Armistead's book[1] contains recommendations which will be helpful guides.

Plant layout includes the problem of adequate roadways which are very important to safety. Each process unit should be completely surrounded by roadways. Every tank should be accessible on at least one side, preferably two opposite sides, from a roadway. Try for a layout which will make every point in the plant accessible by road from two directions. Along this same line of thought, it is good to have an alternate main entrance to the plant to be used when the normal entrance is blocked by traffic or road maintenance. This may be very important in emergencies. The roads which are heavily traveled should be arranged to minimize the chances of damage to hazardous equipment from vehicles leaving the road. The layout should also attempt to minimize the road crossings by ground-level pipe alleys.

The major transfer pipe alleys also come under surveillance at this point. One very influential factor in pipe alley layout is the need for looped piping arrangements in certain services. A looped system is one so arranged and valved that a failure at any point in the system can be isolated by closing valves, thus maintaining service to the rest of the system. To accomplish this, the layout must supply these services to the critical points in the plant from at least two directions. To enhance safety, especially during emergencies, the piping for these services should be looped:

1. Fire water.
2. Steam for power or heat.

Consider also the locations of the sources of these services. The boiler plants and pumping stations should occupy positions as well protected as possible from damage by fires or explosions occurring in other facilities in the plant. Try to minimize the number of road crossings. Pay particular attention where alleys cross above roadways. Overhead clearance must allow easy passage of heavy equipment such as cranes with minimum danger of collision. And, finally, pipe alleys must not pass through diked areas. A fire in a diked area can rupture pipes, adding to the fuel for the fire or interrupting the supply of important services like steam. Also, pipes passing through dikes present potential leaks in the dikes. Avoid a layout which requires using trenches for pipe alleys. Trenches transport flammables or toxics (liquid or vapor) from one area to another.

Electrical power, though carried in wires rather than pipes, is another service to treat in a fashion similar to the water and steam systems discussed above. It is most desirable to keep electrical power lines below ground inside the plant. If economics will not permit this, take care to prevent electrical power lines from falling on critical facilities. And, turning about, locate guy wires and elevated structures so they cannot fall across the power lines.

The plant may also involve docks for the loading or unloading of water-going vessels. The basic hazards here are similar to those of the other loading facilities, except that the quantities of materials involved compare more to a storage tank than to a tank truck or tank car. For docks, an isolated downwind location is certainly desirable.

It is time now to look carefully at the locations for safety facilities. Particularly, the site for the first aid station is a problem for plant layout. The first aid station must be remote enough from the hazardous areas to remain unscathed in times of emergency. And yet, it is from these hazardous areas that most people will come who urgently need first aid. So it cannot be too remote. The time required to reach the first aid station may someday mean life or death to an injured person. Obviously, a compromise is the answer in this case (see Chap. 11). A similar compromise will be needed in the case of the fire station (see Chap. 12).

We should take a preliminary look at the general pattern of locations for safety showers, hose carts, and fire hydrants. However, the exact locations for these will carry over into the plot planning phase.

One final note on topography before we leave the subject, plant layout. It is, of course, not always possible to acquire the smooth, flat site that is so desirable. We may, in the final analysis, be stuck with a hilly or sloping site on which to lay out the plant. Several precautions will be in order. Sources of flammables, liquid or vapor, should not be uphill from points of ignition. Sources of toxics or flammable liquid should not be uphill from concentrations of people. The site may be subject to flooding. In this case, it is wise to locate boiler houses, electrical substations, and pumping stations on high ground. Continued operation of these services becomes most imperative during emergency situations, such as will surely exist during periods of flooding. Tanks, too, are vulnerable to flooding. An empty tank will float on surprisingly little water. This can rupture lines connected to the tanks, perhaps resulting in very large spills, to further aggravate the emergency. We may even want to consider including a system of barriers to prevent the flow of liquids or the spread of fires from one area of the plant to another.

3.6. SAFETY PROBLEMS IN UNIT PLOT PLANNING

Proceeding into the final area of this chapter, we face the safety problems which arise in the plot planning of the process units. This involves the physical locations of individual pieces of equipment relative to each other within the battery limits of each processing unit. It is not easy to develop a plot plan which will minimize the construction and operating costs, and at the same time provide an adequate level of safety. In general, the more compact the unit, the lower will be the cost of piping, pumping, and real estate. On the other hand, safety considerations would dictate that the unit be well spread out for the separation of hazards and the provision of adequate space for fire fighting or other emergency operations.

A compromise is necessary. Fortunately, safety has a couple of allies which help to justify a more open plot plan with plenty of elbow room. Overcrowding will be very detrimental to the efficiency of the construction as well as the maintenance of the unit, adding to the initial and continuing costs.

Experience over the years in trying to balance these various factors has led to a method of plot planning pretty well accepted by many people in the processing industries. The key element in this method is a long, straight, "in-line" arrangement of most of the towers, drums, exchangers, pumps, and the main pipe alley of the unit. Figure 3.3 will assist in discussing the "in-line" arrangement. This is a cross section showing the main features involved.

Starting from the left side of Fig. 3.3, notice the main features are:

1. A roadway.
2. A gantry-way, which is an area adjacent to the roadway along which travels a gantry crane on rails for handling the channels and tube bundles of the heat exchangers.
3. Cooling water headers beneath the gantry-way.
4. A line of fractionating towers, heat exchangers, accumulators, reflux

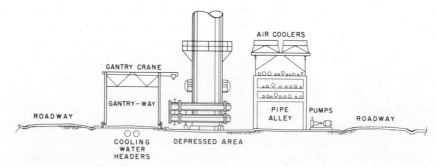

Figure 3.3. A cross section of the in-line arrangement of processing equipment.

drums, etc. It is usual for the grade below this line of equipment to be depressed about 8 in.

5. The main pipe alley of the unit. If the unit includes fan-type air coolers, these can be mounted very nicely atop the pipe alley.
6. A row of pumps.
7. A roadway.

In addition to serving well the operating and maintenance requirements of the unit, the "in-line" arrangement makes notable contributions to safety:

1. The two roadways flanking the in-line arrangement provide excellent access to both sides of a considerable portion of the processing equipment in the unit; ideal for fire fighting or other emergencies. The same two roadways plus the gantry-way serve as fire breaks, separating the in-line equipment from the other blocks of equipment in the unit.

2. The gantry crane is specifically designed for its job. Thus, it provides safe handling of the heavy exchanger components.

3. The area required for the gantry-way provides space for the tube bundles from exchangers to be worked on at the unit or stored until picked up and taken to the shops. All of this is possible while maintaining full traffic right-of-way on the roadway next to the gantry-way. In some of the smaller units, the gantry crane itself may be difficult to justify. Even so, the provision of some open area between the roadway and the line of equipment will help serve these purposes.

4. The indicated location of the cooling water headers makes practically the entire cooling water system accessible for repairs without digging around and under the equipment.

5. The gantry-way provides an open area beside every fractionating tower. Tray segments and other tower internals may be lowered into this area or, in many cases, directly onto a truck bed for transportation to a working area. This minimizes chances of dropping these parts on the other equipment or on the men who may be working on the equipment.

6. The depressed area below the tower, drum, and exchanger line-up will retain spills of flammables or toxics from spreading into other areas until the sewers carry the material away. Firewalls 8 in. high crossing the depressed area from side to side break it up into segments. This prevents the spreading of spills along the length of the depressed area. Rather than the depressed area, some people prefer the use of low firewalls all the way around this line-up of equipment. The theory here is that the walls will prevent the spread of spills *into*

Samford University Library

as well as out of the area beneath the equipment. On the other hand, these walls may present somewhat more of a tripping hazard than the depressed area. Maintenance people seem to prefer the depressed area to the walls.

7. The strategic location of the pipe alley permits a clean and efficient piping layout. It virtually eliminates any runs of pipe with low overhead clearance or just above grade. Operating and maintenance personnel will find very few head-bumpers or tripping hazards in this type of arrangement. With this arrangement, there is no need for pipe trenches which are good carriers of hazardous liquids or vapors.

8. Notice that the row of pumps is immediately adjacent to a roadway and is completely free of overhead obstructions. This will permit the removal of pumps or drivers for maintenance, even while the unit is operating, with minimum chance for damage to any other equipment. Furthermore, the pumps may be safely handled by almost anything from a fork-lift truck to a large crane, whichever may be handy at the moment.

9. Fan-type air coolers are perplexing items in plot planning. Located at grade, air coolers require considerable space, and are hazardous to personnel. This is especially true in configurations which place the fans below the tube banks. Placing the air coolers above the pipe alleys relieves both of these situations without adding noticeably to the cost of structural support.

10. The arrangement provides clear areas for walking traffic along the length of the rows of equipment. The openness enables people to retreat rapidly from fires, explosions, or spills of toxic materials which may occur. It also allows rapid access to safety showers, fire hydrants, or monitors.

With the in-line equipment serving as the backbone of the unit plot plan, the other components of the unit (control house, compressors, reactors, knockout drums, surge drums, flash drums, furnaces, etc.) may be located along either side of the in-line grouping. This approach will usually result in a plot plan with battery limit dimensions approximating a square.

The in-line principle is also applicable in the case of integrated process units (integrated both process-wise and physically). In one specific instance, such a plot plan developed into a series of parallel, side-by-side, in-line groupings. The other components of the units occupied an area adjacent to these groupings extending along the ends of the lines of equipment. Separating these two basic areas was the ''master'' pipe alley for the entire process area. And, of course, branching from the master pipe alley were the individual pipe alleys for each in-line grouping. Figure 3.4 illustrates this concept graphically.

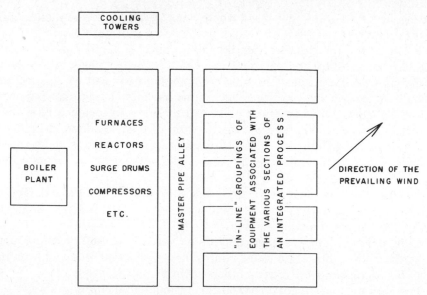

Figure 3.4. An example of applying the in-line arrangement to integrated processing units.

We shall turn now to discussing those components of the unit which are not part of the in-line group. The control house, the nerve center of the unit, will be our first topic. Operating considerations, by themselves, would probably put the control house in the center of the unit area. This would certainly make for the shortest routine tours of the operating check points. However, the glaring hazards thus created usually persuade us to locate the control house on the perimeter of the unit area. This will facilitate escape in the event of disaster. The control house should have doors to permit this escape in a direction away from the unit. The strategic placement of the control house for the best view of the total unit will also help. In units handling toxics, the control house should occupy an upwind location. Finally, keep the control house well separated from high-temperature or high-pressure vessels, or from vessels normally containing considerable amounts of flammable or toxic liquids.

Furnaces present two basic problems. As obvious ignition sources, they should be upwind from the rest of the unit. This, however, may cause a problem with flue gas drifting through elevated platforms on towers or other structures. In some cases, the best solution may be to compromise on a sidewind location. Make every effort to maintain at least a soft separation between the furnaces and other blocks of hazardous equipment.

Compressors may act violently, and, as such, should be respected with a reasonable separation from other hazardous equipment. They are also prone to leak gas, and, therefore, call for a downwind location. (Modern practice seldom puts compressors or pumps into houses because of this tendency toward leakage. In the few cases where weather protection is a must, it is best to stick to a pavilion-type building, having a roof but no sides.)

For reactors, the main consideration is to provide ample space and gear for safe handling of internals and the catalysts which may be involved. Sometimes we find reactors running hot enough to be dealt with as ignition sources. Occasionally, piping configurations outside the unit, causing pockets in the blowdown lines, will make it necessary to locate a blowdown drum at the unit. In such cases, the location of the blowdown drum must permit the piping from the relief valves to the drum to be as short and as straight as possible without any pockets ahead of the drum. Minimum pressure drop is essential.

Electrical power must enter the processing unit below ground. Try to arrange the point of entry so that there need not be any electrical manholes within the battery limits of the unit.

If the unit will have an emergency dump valve or a snuffing steam manifold, those should be close to the control house and well removed from the most likely locations for fires or other hazards.

Fire hydrants or monitors must be close enough to the hazardous spots to be effective and yet not so close as to be inaccessible in emergencies. Watch for obstructions which might prevent the stream of water from reaching the critical points. See that there is a pathway for a quick retreat if necessary. Similar strategic locations are needed for hose carts, safety showers, etc.

Where it is absolutely necessary to bring railroad spurs into a unit area, try to provide adequate clearances for possible derailments. Also avoid locating equipment opposite the end of the spur in case railroad cars may overshoot, tearing out the bumper and colliding with the equipment.

Before finalizing the plot plan, give a good hard look at the possibilities of adding equipment in the future. A unit, beautifully arranged when built, can be completely fouled up by squeezing in additional equipment where inadequate space exists. Using the in-line arrangement, we usually find enough room in the pump row for some additional pumps and perhaps one or two additional drums. But consider carefully where to install future exchangers, towers, furnaces, reactors, etc.

There may be cases where it will be necessary to house a part or the whole of a processing unit indoors. Requirements for very precise temperature control or for constant operator attention are examples which might bring about such a necessity. In dealing with indoor facilities, two of the tools discussed in Section 3.3 become ineffective. There will, of course, be no prevailing wind to consider indoors. Separation by distance will most likely cause too great a financial burden because of the size of the buildings required. Even if this were not true, the effectiveness of distance is reduced indoors because released toxic or flammable vapors will be contained in the building and not dispersed as they would be outdoors.

However, we can still make good use of some of the other tools. For indoor installations, we will lean most heavily, perhaps, on separation by physical barrier. If ignition sources and sources of flammables both must be indoors, it will be well to allocate them to separate compartments of the building. The walls separating the two must have the absolute minimum of doors or other openings which would permit vapors or liquids reaching the sources of ignition. Facilities especially prone to fires, explosions, or the release of toxics (such as high-temperature, high-pressure, or large-volume containers) should be isolated from areas most often occupied by people, like the control center. Walls for this purpose must also have a minimum number of openings. But in addition, these walls should be designed for strength and fire resistance. Compartments subject to explosion may have one or more intentionally weak walls to help direct explosive forces in a direction away from people or other facilities.

A building of multiple floors or of varying floor levels will have a topography all its own. Sources of flammable or toxic liquids should not be uphill from ignition sources or from people. Where vapors are involved, the location of people and ignition sources will depend on whether the vapors are heavier or lighter than air.

As before, the concentration of hazards will be helpful in the demarcation of particularly hazardous areas. And, in addition, concentration of hazards will improve the practicability of providing special facilities for safety. Examples of such facilities are:

1. High-capacity ventilating systems to help keep air-vapor or air-dust mixtures below the explosive limits.
2. High-capacity drainage systems for very quick removal of spilled liquids.
3. Remotely operated handling devices.
4. Automatic fire-fighting devices such as water sprays, steam blanketing, and foam or inert gas systems.

Despite these possibilities, when indoors, we must definitely limit the ultimate size of the possible combined effect of several hazardous pieces of equipment concentrated in one area or compartment because of the inherent proximity to people or other facilities.

As compared to outdoor facilities, the paths of retreat from housed facilities will be severely limited at best. It is difficult to overstress the need for a carefully planned system of platforms, ladders, stairs, walkways, doors, and escape chutes.

Permanent platforms should provide for access to all operating points in the unit which are not accessible from grade. From all operating platforms (with the possible exception of tower platforms) there should be two means of descent to help prevent trapping operating personnel during emergencies. Insofar as possible, all ladders, stairways, or slidepoles leading down from elevated platforms should land at grade at points which are least likely to be

TABLE 3.1. Checklist of Safety Aspects, Site Selection, Plant Layout, and Plot Planning

Block or Unit Facilities under Consideration	Safety Aspects
Site selection Entire chemical plant	1. Downwind from populated areas 2. Downwind from outside ignition sources 3. Downwind from major highways 4. Avoid stream and air pollution problems—check the legal restrictions 5. Traffic problems at main entrance 6. Isolated site—separation by distance from populated areas, ignition sources, or highways 7. Topography of the site—flatter the better 8. Sources of toxics or flammables not uphill from populated areas or outside ignition sources 9. Rights-of-way across the site
Community surrounding the chemical plant	1. Sources of flammables or toxics downwind from the plant 2. Ignition sources upwind from the plant 3. Sources of flammable or toxics not uphill from the plant 4. Proximity of fire stations to the plant 5. Proximity of hospitals or first-aid stations to the plant 6. Proximity of water sources to the plant
Plant Layout Processing Units	1. Separation from the boundaries of the plant 2. Consolidated—not scattered 3. Downwind from ignition sources 4. Downwind from concentrations of people 5. Separation from tankage[1] 6. Separation from each other[1,2]—effect of integration 7. Not uphill from ignition sources or concentrations of people

TABLE 3.1. *(Continued)*

Block or Unit Facilities under Consideration	Safety Aspects
Administrative buildings	1. Peripheral location 2. Separation from hazardous areas 3. Upwind from sources of flammables or toxics
Boiler plants	1. Upwind from sources of flammables 2. Separation from hazardous areas 3. High ground if site is subject to flooding
Maintenance shops	1. Upwind from sources of flammables 2. Peripheral location because of railroad spurs and truck traffic
Loading and unloading facilities for tank cars and trucks	1. Peripheral location because of railroad spurs and truck traffic 2. Downwind from ignition sources 3. Separation from main line of railroad 4. Not uphill from ignition sources
Waste-water treating facilities	1. Downwind from ignition sources 2. Peripheral location 3. Downwind from concentrations of people
Water-cooling towers	1. Downwind from ignition sources 2. Downwind from concentrations of people 3. Downwind from roadways, operating units, and elevated structures
Flares or burning pots	1. Sidewind from ignition sources, operating units, or tankage 2. Downwind from concentrations of people 3. Separation from people and hazardous areas
Storage tanks	1. Separation of tanks from each other[1] 2. Separation of tanks from other facilities[1] 3. Size and arrangement of dikes[1] 4. Downwind location 5. Not uphill from ignition sources, process units, or concentrations of people

TABLE 3.1. *(Continued)*

Block or Unit Facilities under Consideration	Safety Aspects
Roadways	1. All four sides of process units 2. One or two sides of each tank 3. Two-way access to every point in the plant, and two entrances 4. Minimum crossings of pipe alleys
Pipe alleys	1. Looped arrangement for fire water, cooling water, boiler feed water, and steam 2. Minimum crossings of roadways 3. Clearances over roadways 4. Avoid passing through diked areas
Electrical power lines	1. Below ground if possible 2. Avoid passing over critical facilities 3. Separation from elevated structures which may fall across power lines 4. Looped arrangement to critical users
Electrical switch gear	1. Upwind from sources of flammables 2. High ground if site is subject to flooding 3. Not downhill from sources of flammables
Docks	1. Separation from other hazardous areas 2. Downwind from ignition sources
Unit plot planning	
Fractionating towers	1. See in-line arrangement Section 3.6
Drums	1. See in-line arrangement Section 3.6
Heat exchangers	1. See in-line arrangement Section 3.6
Pumps	1. See in-line arrangement Section 3.6
Pipe alleys	1. See in-line arrangement Section 3.6
Roadways	1. See in-line arrangement Section 3.6
Control houses	1. Locate on perimeter of the unit 2. Doors facing away from unit as well as toward the unit 3. Locate for good view of unit 4. Upwind from sources of toxics

TABLE 3.1. *(Continued)*

Block or Unit Facilities under Consideration	Safety Aspects
	5. Separation from high temperature or high-pressure vessels
	6. Separation from vessels containing flammable or toxic liquids
Furnaces	1. Upward or sidewind from the rest of the unit
	2. Separation by 50 ft from other hazardous equipment
Compressors	1. Downwind from rest of unit
	2. Separation from other hazardous equipment
	3. Not in closed building
Reactors	1. Space and gear for safe mechanical handling
	2. Hot reactors may be ignition sources
Blow-down drums	1. Locate for best piping arrangement with minimum pressure drop
Electrical power lines	1. Below grade
	2. No electrical manholes inside the unit
Emergency dump valves and snuffing steam manifolds	1. Close to control house
	2. Remote from hazardous areas
Fire hydrants or monitors, safety showers, hose carts, etc.	1. Close but not too close to the hazardous areas
	2. No obstructions for monitors
	3. Good path for retreat
Railroad spurs	1. Clearance for derailments
	2. No equipment opposite the end of the spur
Indoor facilities	1. See Section 3.6
Elevated platforms	1. Permanent platforms for all operating points
	2. Two routes for descent
	3. Landings at grade in safe locations
Piping	1. Watch for head bumpers, shin splitters, and tripping hazards
	2. Personnel protection from hot lines
	3. Safe steam trap discharges

subject to fires or toxic spills. Also, these landing points should be close to pathways affording good retreat from the unit.

The platforms at manholes on towers should provide enough working space for the safe handling of tower internals. The direction in which the manholes swing open should be away from the ladder leading down from the platform. Poorly placed valve stems can be head-bumpers, shin-splitters, and tripping hazards especially to personnel in a hurry.

Steam-trap discharges which are aimed across walkways, as well as uninsulated hot pipes may cause serious burns to the skin. Hot piping, where it passes within reach of operating people in their routine duties, must be insulated or otherwise protected.

3.7.   SUMMARY

This discussion has been oriented around the planning of new plants. The same problems, of course, apply to existing plants. While existing plants cannot be made over immediately to solve the safety problems built into them, the changes and revamps which occur over the years should follow a master plan which progresses toward a satisfactory level of safety. The ideas we have discussed here will help in developing such a master plan.

Sections 3.4, 3.5, and 3.6 have dealt most specifically with plants designed for the continuous processing of fluids. Nevertheless, many of the ideas in these three sections, plus the material in Sections 3.2 and 3.3 will find application in plants or units involving batch-type operations or the handling of solid materials.

The discussions in Sections 3.4, 3.5 and 3.6 are summarized in a checklist to make them more useable. This check list appears in Table 3.1. It is important to keep these items in proper perspective. Remember that these are factors related to safety only. They will not be compatible in every instance with the other factors which come to bear on site selection, plant layout, and plot planning. Compromises may be necessary. And, finally, do not accept this as the final, all inclusive, checklist complete for all time. By applying the general principles discussed in Sections 3.2 and 3.3 of this chapter, we may make additions.

REFERENCES

1.   G. Armistead, Jr., *Safety in Petroleum Refining and Related Industries,* 2nd ed., Simmons, New York, 1959, pp. 69–82.
2.   *Ordnance Safety Manual,* ORD 7-224, with Supplements 1-6, U.S. Government Printing Office, Washington, D.C., 1951.
3.   D. V. Gagliardi and N. R. Pratt, "Guidelines to Follow When Designing for Plant Safety," *Plant Eng.,* **17,** 108–111 (June 1963).

BIBLIOGRAPHY

Banister, R. M., "Feasibility Study: Fuel Cell Cogeneratlion at the Anheuser-Busch Los Angeles Brewery," HN 1540, Holmes and Narver, Inc., Orange, Calif., 1980.

Boeri, G. "Methodology for the Analysis of Environmental Parameters on a Nationwide Scale," UCRL—TRANS 11607, NTIS PC A03/MF A01, 1980.

Delile, G., "Implementing Data from Environmental Research Conducted for Nuclear Plants," *Rev. Gen. Nucl.,* No. 1, 51–61 (1980).

Douglas, L. A., "Use of Soils in Estimating the Time of Last Movement of Faults," *Soil Sci.,* **129**(6), 345–352.

Eagles, T. M., "Modeling Plant Location Patterns: Applications." EPRI-EA-1375, NTIS, PC A09/MF A01, February 1980.

Gee, J., "Nuclear Power: The French Do It Their Way," *Electr. Rev. Int.,* **206**, No. 14, 12, 13 (1980).

Kotin, A. D., "Relationship of Eligible Areas to Projected Electrical Demand," *Solar Power Satellite Program Review*, Lincoln, Neb., 22 April 1980, NTIS PC A99/MF A01, 1980, pp. 521–524.

Madsen, C., "Utilities and Site Selection," *Energy Plan. Net.* No. 12, 2–4 (Spring 1980).

McCoy, H. A. and Singer, J. F., "Socio-Economic Impacts of Large Power Plants: An Input for the Siting Process," *Int. J. Environ. Stud.*, **15**(2), 95–108 (1980).

Moss, J., "Soviet Nuclear Programme in Site of Difficulty," *Electr. Rev. (London),* **206**(2), 16, 17 (1980).

Pasternak, A. D. "Regional and State Roles in Nuclear Siting, Acceptable Future Nuclear Energy System," ORAU/IEA—80—3 (P), DEP. NTIS PC A11/MF A01, 1980, p. 61–65.

Raynor, G. S. and Hayes, J. V., "Transport and Diffusion Climatology of the U.S. Atlantic and Gulf Coasts," Second Conference on Coastal Meterology, Los Angeles, Calif., January 30, 1980, American Meterology Society, Boston, Mass. 1980, pp. 30–33.

Raynor, G. S., "Recommended Changes in Meteorological Measurements and Prediction Methods for Coastal Sites," BNL 26840, Second Conference on Coastal Meteorology, Los Angeles, Calif., January 30, 1980, Dep. NTIS PC A02/MF A01, 1980.

Renne, D. S. and Sandusky, W. F., "DOE Candidate Site Meteorological Measurement Program," PNL-SA-7840, NTIS PC A02/MF A01, 1980.

Site Selection and Evaluation for Nuclear Power Plants with Respect to Population Distribution: A Safety Guide, International Atomic Energy Agency, Vienna, Austria, 1980.

Talmage, S. S. and Coutant, C. C., "Thermal Effects," *J. Water Pollut. Control Fed.* **52**(6), 1575–1616 (June 1980).

Tokarz, F. J., "Earthquake Engineering Programs at the Lawrence Livermore Laboratory," UCRL 82234 (Rev. 1), NTIS PC A02/MFA01, 1980.

Wang, F. C. "Energy Analysis for EIS." *J. Water Resour. Plan. Manage. Div., ASCE* **106**(WR2), 451–465 (1980).

4

Hazards of Commercial Chemical Reactions

George T. Austin

Commercial chemical production of useful products results from the conversion of one substance into another. Some conversion reactions are simple and subject to very little hazard, but some are most complex and hard to control or understand. When the exact chemical nature of the reactants and products is known, it would appear that it should be possible, by applying known laws of chemistry, to assess the hazards of commercial production of the substance. In practice, this has not proved to be true. The exact paths taken by substances in changing from one substance to another are often devious. The alternate and sometimes dangerous paths which the substances may take under only slight variations in temperature, pressure, or composition, and the kinetics of the various steps are rarely known and frequently most difficult to foresee.

Temperature is the most frequent cause of divergence from expected reaction rates. Where no data are available, the assumption that organic reaction rates will double with a 10°C rise in temperature is often surprisingly good. This approximation assumes that the temperature of the mix is uniform throughout. This is rarely true in commercial reactors, so reaction rates may vary considerably within a unit, particularly if a considerable radial temperature gradient exists.

The Arrhenius equation is more exact than the rule of thumb and expresses the effect of temperature on the reaction rate in the form:

$$k = Ae^{-E/RT} \tag{4.1}$$

where k = rate constant for the reaction; A = a constant; E = energy of activation for the reaction; R = gas constant; T = absolute temperature; and

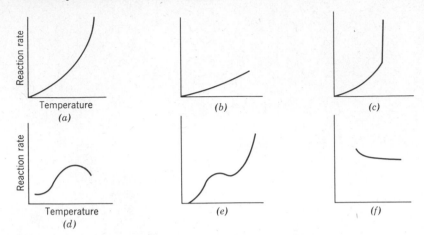

Figure 4.1.   Effect of temperature on reaction rate.[1]

e = natural base for logarithms. A reaction whose rate changes with temperature in accordance with Equation (4.1) is generally thought of as a normal reaction. Figure 4.1a shows this type of behavior graphically. Also Fig. 4.1 shows several other types of rate change with temperature curves frequently observed in practice. It is interesting to note that if k doubles for a 10°C rise, this corresponds to an energy of activation of approximately 13,000 cal/mole.

It is evident from Fig. 4.1 that many other types of reaction rate curves occur besides the simple Arrhenius (logarithmic) one. Walas[1] has tabulated the reasons for the existence of the types shown in Fig. 4.1 as follows:

    a.   Normal behavior; rapid increase in rate with increased temperature.

    b.   The behavior of certain heterogeneous reactions dominated by resistance to diffusion between phases; a slow increase in rate with increased temperature.

    c.   Typical of explosions, where the rapid rise takes place at the ignition point.

    d.   Catalytic reactions controlled by the rate of adsorption (in which the amount of adsorption decreases at elevated temperatures) and enzyme reactions where high temperatures destroy the enzyme.

    e.   Some reactions, combustion of carbon for example, complicated by side reactions which become significant as the temperature is increased.

    f.   Diminishing rate with increased temperature, for example, the reaction between oxygen and nitric oxide, where the equilibrium conversion is favored by lower temperatures and the rate appears to depend on the displacement from equilibrium. Only Fig. 4.1a is a simple

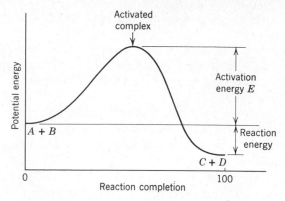

Figure 4.2.   Potential energy change for a reaction.

reaction; all the others result from complex reactions or are controlled by exterior physical rate processes.

Before a chemical reaction can occur, there must be a collision (or collisions) between the reacting molecules. It is a rare case indeed that follows precisely the path indicated by the usual chemical stoichiometric equations, which merely show the weight and energy relationships between the reactants and the products. When a collision occurs, a reaction will occur only if the participants possess sufficient energy to penetrate each other's electron fields to such a degree that rearrangement occurs. The amount of energy required to cause a chemical change to occur as the result of a collision is called the activation energy. The rate of a chemical reaction depends upon the number of collisions occurring per unit of time and the fraction which possesses sufficient energy to make the collisions effective. The larger the energy of activation, the slower the reaction at a given temperature, for only a few pairs of colliding molecules will have sufficient energy to react. Conversely, fast reactions will have low activation energies.

Consider the case of two molecules A and B which react to form two other molecules C and D. When A collides with B, there is a period in which they are essentially a unit, commonly called an activated complex. If the energy content of the AB pair is low, they then simply fly apart again. If energy is large, there may be exchanges of energy and matter between A and B, resulting, on their separation, in the emergence of C and D, rather than the original substances. A potential energy diagram for this system is shown in Fig. 4.2. The horizontal axis shows the percent of the total reaction which has occurred (0 = none, 100 = all finished). The distance E shows the amount of energy which must be acquired by A and B before they can react to form C and D. Since C and D have a lower potential energy than A and B, the reaction is exothermic, $A + B \rightarrow C + D +$ heat. The effect of a catalyst on a reaction is to make the activation energy much less. End products are the same, but reaction is faster at a given temperature because the number of successful collisions is increased since the energy required for success is lower.

Reaction rate theory is complicated in the extreme because, in general, several choices of the ultimate product may be open to the activated complex. The reader interested in pursuing this in greater detail is referred to standard works on kinetics.[1-3]

Commercial reactions must be examined carefully for possible side reactions, and the kinetics of these side reactions should also be observed, if possible. The existence of reactions of the types shown in c and e of Fig. 4.1 are the usual cause of difficulty.

Most observed chemical reactions proceed by complex paths and many steps, each having different rates, may occur between reactant and final product. Physical adsorption or absorption, diffusion, and other processes in addition to chemical reaction may be involved. In some cases, all the intermediate processes with the exception of one will be fast, yielding a rate-determining step. Where several are of similar magnitude, complex rate-temperature curves can be expected. King[4] has examined several cases with extraordinary clarity.

Hazards associated with flammability, toxicity, or instability of the primary product are usually discovered in the laboratory. Those which arrive in the form of unexpected explosions are not. The fact that laboratory scale tests have been completed without incident is no guarantee of freedom from hazard in the pilot plant or the production plant. Difficulties arise in the plant because impure chemicals replace pure ones, radial temperature gradients lead to maximum temperatures well above the averages indicated by thermometers, processing times are often much longer, by-products accumulate and may have catalytic effects, pressures are often higher, and the use of metal instead of glass apparatus may exert profound effects on the course of the reaction. Safe practices require that tests be run on the reactants at conditions substantially worse than the averages expected. Laboratory screening tests have been suggested[5,6] and add greatly to the safety of pilot plant work. The National Fire Protection Association (NFPA) has issued a Manual of Hazardous Chemical Reactions[7] which can help greatly in avoiding known problems.

A particularly difficult kind of reaction to control occurs in the manufacture of certain types of explosives. Partial decomposition of the compound being formed occurs at hot spots in the reactor and the products of the decomposition are catalysts for the decomposition. This, of course, leads to a hotter hot spot, a faster rate, and a runaway reaction which, if explosion does not occur, is commonly called a "fume off." A better term for this type of runaway is an autocatalytic reaction. Control requires elimination of "hot spots," removal of the source of catalysis, or destruction of the catalytic agent as rapidly as it is formed.

Runaway reactions which give some warning have been controlled by dumping the charge into large quantities of water, rapid dilution or limiting of the quantity of reactable material in process. With the advent of more rapidly responsive sensors it has become possible[8] to suppress explosions which have already begun by the use of explosion suppressors. These devices act with great rapidity on quick rise of temperature or pressure and suppress the explosion by any of several techniques such as venting, isolation, automatic shutdown, and sudden quenching.

Despite much study, the values of many reaction rates are unknown or unpublished. Each reaction must be studied in the laboratory to obtain kinetic data; these data must then be used to estimate the constants for plant size reactors. Concentration of reactants, pressure, temperature, and the presence of reaction-promoting or inhibiting substances are the fundamental variables involved in determining reaction rates.

It is rarely possible to carry out enough experiments to determine the effect of all the variables under all conditions which may be encountered, for such experiments are expensive. Rate constants for all the reactions, side reactions, and by-reactions may all be determined without ever determining the actual mechanism by which the reaction occurs.[3] The mechanism of a reaction is the exact series of steps by which the reaction takes place. Such exact mechanisms are known for only a few reactions and the experimental determinations of them are most difficult. Most design engineers feel that a satisfactory rate equation is one that may be used with confidence to design commercial scale equipment for carrying out the reaction. In speaking thus of a "reaction," the principal reaction is usually meant and this may be insufficient knowledge from a safety standpoint, since another reaction may become predominant at slightly changed conditions. Design kinetic equations are universally empirical.

Where rate is controlled principally by concentrations, the reactions are called homogeneous—and these are classified by the "order" of the reaction. The reaction rate, r, of a constant volume homogeneous reaction can be expressed by

$$r = k_c C_A^a C_B^b \tag{4.2}$$

where r is the reaction rate, mol/h-volume; k_c the reaction rate constant, concentration units; C_A, C_B the concentration of A and B mol/volume; and a, b, constants.

The order of the equation is $a + b$, the number of molecules which combine in the actual rate-determining step—not the molecular proportions shown in the usual equation as frequently erroneously used. All reaction orders are empirical, and the methods for obtaining them are summarized by Corrigan.[9] An approach is being made to predict reaction rates from absolute physical properties of systems, but this approach is many years away from commercial usefulness. Computer use is sharply reducing the number of laboratory experiments which must be performed before reliable kinetic data are possible (see Chapters 16 and 17).

Rate equations, once determined with a fair degree of accuracy by laboratory observation, can be applied to the design of optimum reactors, but there are complicating factors which may obscure kinetic effects as limiting factors in reactor design. Stevens[10] lists the following limiting factors which must be carefully considered, in addition to the validity of all the mathematical assumptions made: longitudinal diffusion, radial gradient, wall effects, impurity accumulation, and catalyst life. There are others. Churchill[11] notes that theory has a very limited role in the description, measurement or derivation of rates, but is helpful in interpreting and organizing data.

Reactions used industrially have been classified by Shreve[12] into 32 categories, or unit processes, which offer a systematic approach to the study of chemical reactions. These have been expanded and studied in detail in several major works concerned with industrial chemical reactions.[13-15] Using a somewhat less comprehensive list than Shreve's, each of these types of unit processes is briefly considered, looking for some hazards characteristic of the reactions.

4.1. COMBUSTION

Solid, liquid, and gaseous fuels oxidized for the creation of heat represent such common reactions that they are rarely considered to be chemical. Normal combustion reactions are fast, but controllable. The ranges over which controls are possible are known for many substances in both air and oxygen. Tables of explosive and flammability limits are available in the literature[15] and should be observed with great care when setting up furnaces. Rates are usually held under control by regulating temperature, availability of oxidizing substances, or availability of fuel. Under most conditions, ignition of mixtures is necessary, but reactive substances, for example oxygen + tung oil, may ignite spontaneously.

4.2. OXIDATION

Oxidation differs from combustion only in the fact that the decomposition reaction is stopped enroute to CO_2 and H_2O. A combustion hazard always exists if sufficient oxidant and fuel are present. Oxidation reactions are all highly exothermic. Equilibrium is nearly always in favor of the complete reaction and steps must be taken to limit the extent of the oxidation to prevent loss of product. Marek, in reference 13, lists 10 types of oxidative reactions which represent the principal types encountered industrially. Vigorous oxidizing agents are frequently used, and the following should be used with extreme caution: salts of permanganic acid, hypochlorous acid and salts, sodium chlorite and chlorine dioxide, all chlorates, all peroxides, nitric acid and nitrogen tetroxide, and ozone. Safety is usually represented by low concentrations of oxidizing agents, low concentrations of fuel, low temperatures, or better yet—all three. Talmage[16] has written a fine summary of precautions to use in vapor phase oxidations.

4.3. NEUTRALIZATION

Aside from thermal effects resulting from the too rapid addition of reactants, these reactions are relatively free from hazard. Low concentrations allow better control.

4.4. ELECTROLYSIS

Reaction hazards are almost nonexistent. The usual hazards present where large amperages are used, poisoning hazards from the use of cyanides,[17] possible explosion hazards due to the presence of combustible gases, and products in high oxidation states (persalts, etc.) are present, but not reaction problems.

4.5. DOUBLE DECOMPOSITION

These reactions are usually of the equilibrium type with small driving forces and relatively low heats of reaction. Hazards are low.

4.6. CALCINATION

As endothermic reactions, these are readily controllable and rarely troublesome.

4.7. NITRATION

All nitration reactions are potentially hazardous, not only because of the frequently explosive nature of the end products, but also because most nitrating agents are also strong oxidizing agents, especially when diluted. This dual nature of nitrating agents makes many by- and co-products possible and some of these reactions are rapid and uncontrollable. Both the nitration reaction itself and the oxidation reactions are highly exothermic.[13] Temperature control must be exceptionally good if runaway reactions or explosions are to be avoided. Maximum temperatures permissible during nitrations can be estimated from temperature sensitivities of the final products, which may be found in Kirk and Othmer,[15] Groggins,[13] Davis,[18] and others. Sensitivity to temperature is increased by the presence of impurities, particularly oxides of nitrogen in liquid phase nitrations, for these act as catalysts for the further oxidation. Rapid, autocatalytic decompositions sometimes occur without explosion. These "fume offs" may be quite violent. Heat evolved is rapid and end products include N_2, nitrogen oxides, and free carbon.

The heat of nitration of benzene is 423 cal/g, but the heat of reaction in commercial apparatus is greater (498 cal/g), for the water formed in the reaction causes a large heat of dilution and the heat capacity of the mixed acid used in nitrating is not large. Temperature control of liquid phase nitrations is, therefore, both vital and difficult.

Continuous processes for nitration are most attractive because they limit the amount of material in process and hence greatly reduce the explosion potential.

In vapor phase nitrations, undesired oxidation always accompanies the nitration. Control consists of careful regulation of the concentration of the critical reactant and overdesign of the heat exchange apparatus.

4.8.   ESTERIFICATION

Both organic and inorganic esterification reactions are generally slow. Catalysts are usually necessary. Hazards are usually small except when the esterifying material is a powerfully reacting material (esters of nitric or perchloric acid, for example) or one of the reacting materials or products is unstable.

4.9.   REDUCTION

Hazards of the reaction are negligible and difficulties are associated almost solely with handling reactive reducing agents.

4.10.   AMINATION BY AMMONOLYSIS

The aminating agent is usually ammonia. Reactions, which are frequently second order, usually appear to be first order because of the necessity of using a large excess of ammonia. Gas phase reactions are usually run under pressure. Most common reactions are exothermic, but not strongly so.

4.11.   HALOGENATION

Chlorine, fluorine, bromine, and iodine represent the industrially important halogens (in that order of importance). Chlorine derivatives are most important because of their much lower cost. All heats of reaction are highly exothermic, with fluorine being extremely high. Chain reactions occur in liquid and gaseous addition and substitution reactions, making detonations possible over considerable concentration ranges. Corrosion effects are extraordinarily difficult to solve in halogen systems. Metals may serve as active catalysts.

Chlorination

Addition and substitution reactions are common with a wide variety of chlorinating agents and systems: chlorine gas, hydrochloric acid, sodium hypochlorite, phosgene, thionyl chloride ($SOCl_2$), sulfuryl chloride (SO_2Cl_2), phosphorus chlorides. Liquid and gas phase chlorinations are common and the extremely diverse paths which these reactions run make prediction without experimentation impossible. All reactions are potentially hazardous.

Fluorination

Fluorine is the most reactive element and its reactions are, therefore, the most difficult to control. Direct reactions with hydrocarbons are violent and frequently explosive. Unwanted C—C bond cleavage is frequent. The new bonds formed between fluorine and other substances are so strong and the heats of reaction liberated are so great that extreme precautions are necessary to keep the reaction under control. Gas phase reactions are usually controlled by dilution with an inert gas. Extreme care is essential.

Bromination and Iodination

Essentially the remarks under chlorination apply, but all reaction conditions are far less critical.

4.12. SULFONATION

Most commercial sulfonations use sulfuric acid as the sulfonating agent. The reactions require strong concentrations to give good driving forces and usually high temperatures. Reactions are mildly exothermic, but usually relatively easy to control.

4.13. HYDROLYSIS

Hydrolysis means decomposition with water, but only a few reactions (usually inorganic) use water alone and unaided to effect hydrolysis. The inorganic reactions—lime slaking and sulfuric and phosphoric acid formation—have problems not common to the usual organic reactions. While the inorganic hydrolyses have control problems, they are all concerned with heat removal and contacting and none is sufficiently unique to warrant discussion.

Organic hydrolyses are run in both liquid and vapor phases and utilize the following reagents: (1) water alone, (2) acidic solutions, (3) alkaline solutions, (4) alkali fusion (usually anhydrous), and (5) enzymes. Most such reactions are comparatively slow and only slightly exothermic. The most important problem is how to speed up the reaction. High temperatures and/or pressures sometimes help.

4.14.   HYDROGENATION AND HYDROGENOLYSIS

The use of high pressure hydrogen in a plant is always hazardous, but, aside from this problem, hydrogenation is not excessively troublesome. Hydrogenation is usually exothermic and the reactions frequently require the use of a catalyst in order to proceed at a reasonable rate. Since the heat released by the reaction is on the surface of the catalyst, local catalyst temperatures may be extremely high—perhaps enough to cause sintering. When such high temperatures are allowed to occur, cracking and various side reactions may take place which cause waste of reactants. Many hydrogenation reactions are operated at high pressure, but conditions of uncontrollability are rare.

4.15.   ALKYLATION

Carbon alkylations are generally mildly exothermic reactions which proceed slowly, even under high pressures and temperatures and in the presence of a catalyst. Competing reactions such as polymerization, isomerization, hydrogen transfer, and destructive alkylation occur when catalytically alkylating isoparaffins and aromatics with olefins. Thermal alkylations require high temperatures and pressures, but are less generally used than catalytic alkylations. Hazards exist from the use of powerful corrosives (HF, dimethyl sulfate, etc.) as alkylating or catalytic agents, but the reactions themselves are very trouble free.

4.16.   CONDENSATION

The energy of activation of polymerized substances lies between 15,000 and 30,000 cal/mole. Condensations are equilibrium reactions and the split-off condensation product (usually water) must be removed in order for the reaction to proceed to high molecular weight. As molecular weights become high and the reaction mass becomes quite viscous, agitation and heat removal become extremely difficult and burning can occur, since these reactions are exothermic. All condensations are step-wise and are usually easy to manage.

4.17. POLYMERIZATION

Polymerization differs from condensation since as the materials combine, nothing is split off. The intermediates are usually short-lived active radicals or ions and the step-wise process is rarely observable. The polymer chain is generally formed in a single reaction and in a fraction of a second.[13] Chain reactions proceed quickly following slow initiation by some readily decomposable substance such as tertiarybutyl-hydroperoxide or benzoyl peroxide.

REFERENCES

1. S. M. Walas, *Reaction Kinetics for Chemical Engineers,* McGraw-Hill, New York, 1959.
2. O. Levenspiel, *Chemical Reaction Engineering*, 2nd ed., Wiley, New York, 1972.
3. J. M. Smith, *Chemical Engineering Kinetics*, 2nd ed., McGraw-Hill, New York, 1970.
4. E. L. King, *How Chemical Reactions Occur*, W. A. Benjamin, New York, 1964.
5. R. H. Albisser and L. H. Silver, "Safety Evaluation of New Processes," *Ind. Eng. Chem.*, **52**(11), 77a (1960).
6. J. C. Rapean, D. L. Pearson, and H. Sello, "A Test for Hazardous Chemical Decomposition," *Ind. Eng. Chem.*, **51**(2), 77a (1959).
7. *Manual of Hazardous Chemical Reactions*, NFPA 491M-1975; see also *National Fire Codes*, Section NFP49-1975. [These are revised periodically and are available from the National Fire Protective Association, Batterymarch Square, Quincy, Mass. 02269.]
8. C. B. Hammond, "Explosion Suppression: New Safety Tool." *Chem. Eng.*, **68**(26), 85 (1961).
9. T. E. Corrigan, "Introduction to Reaction Kinetics," *Chem. Eng.*, **61**(7), 230 (1954); "Kinetics of Homogeneous Reactions, No. I," **61**(8), 820 (1954); "Kinetics of Homogeneous Reactions, No. II," **61**(9), 210 (1954); "Kinetics of Homogeneous Reactions, No. III," **61**(10), 210 (1954).
10. W. F. Stevens, "Chemical Engineering Kinetics," *Ind. Eng. Chem.*, **50**(4), 591 (1958).
11. S. W. Churchill, *The Interpretation and Use of Rate Data*, McGraw-Hill, New York, 1974.
12. R. N. Shreve, *The Chemical Process Industries*, 4th ed., McGraw-Hill, New York, 1977.
13. P. H. Groggins, *Unit Processes in Organic Synthesis*, 5th ed., McGraw-Hill, New York, 1958.
14. J. A. Kent, *Riegel's Industrial Chemistry*, Reinhold, New York, 1962.
15. R. E. Kirk and D. F. Othmer, *Encyclopedia of Chemical Technology*, Interscience, New York, 1950.
16. W. P. Talmage, "Safe Vapor Phase Oxidation," *Chemtech*, **1**(2), 117 (1971).
17. C. L. Mantell, *Electrochemical Engineering*, 4th ed., McGraw-Hill, New York, 1960.
18. T. L. Davis, *Chemistry of Powder and Explosives*, Wiley, New York, 1943.
19. L. Bretherick, *Handbook of Hazardous Reactions*, 2nd ed., Butterworth, London, 1979.

5

Hazards of Commercial Chemical Operations

George T. Austin

The "unit operations" utilized in the manufacture of chemicals are the individual physical processes of manufacture. The number of unit operations existing lies somewhere between 6 and 100, depending on the exclusiveness of the divisions selected. Only the more common ones are discussed in this chapter. The following subdivisions are used: (1) heat transfer and reactors; (2) size reduction; (3) mixing; (4) materials handling; (5) mass transfer; simultaneous heat and mass transfer; (6) humidification; (7) drying; (8) evaporation and crystallization; (9) pumps, compressors and agitators; (10) phase separations based on fluid mechanics; and (11) instruments and control systems.

Hazards existing within manufacturing systems are not solely the function of the physical properties of the materials being handled. In addition to hazards such as toxicity, corrosivity, explosivity and other health and safety problems associated with the process in general, other hazards exist which are connected with the unit processes themselves. Each unit operation should be considered as a safety problem and safety details should be considered along with selection of size of the unit to be used. It is not sufficient to consider the condition of normal operation only. Operations during the abnormal conditions of start-up, shutdown, extreme emergency, and extraordinary levels of heat and cold must also be planned for.

Codes of various types (Federal, state and local) must be complied with in selecting equipment of all types. The National Fire Codes[1,2] published by the NFPA are extremely useful. OSHA has occupational safety and health standards[3] for a wide variety of situations. Supplementary standards are constantly being issued.

Reactions will not always function in the direct manner visualized by laboratory workers. Designs are rarely prepared with provision for deviation from projected norms. An astonishing number of chemical reactor accidents can be blamed on elementary design inaccuracies.[4] One useful concept, the *severest credible incident*, commonly used in the nuclear field warrants general adaptation to design procedures. Under this concept, the engineer considers the effect of malfunction of each item in a design. The effect of several simultaneous malfunctions is then considered. The worst that can conceivably happen is thus arrived at. Reasonable plans do not require design to the severest possible combination, but the possible scope needs examination. Instrument failure is to be expected and planned for. Instruments must "fail safe" or be backed up by safety devices and limit controls. A process fails safe if, on loss of control by any means, the unit assumes a nonhazardous condition. This usually means a shutdown, sharp temperature reduction, stopped flow of reactants, or perhaps emergency venting. The possibility of fire must be considered and inert gas or steam may be automatically provided.

When designing for new operations, storage areas should be planned to prevent, control, and contain fires. Emergency pump connections, dikes, and sprinklers both internal and external should be provided. Use is infrequent, but one small incident can repay all costs. Pump-out connections may salvage expensive materials as well as limiting fire size by removing combustibles. Foam can be spread far more effectively in diked areas. Drains must be channeled through pollution control devices. Jennett[5,6] has discussed components for pressure relieving systems.

5.1. HEAT TRANSFER

Almost every chemical conversion requires the transfer of heat. The equipment is well standardized, but highly diverse in character. Where streams exchanging heat can react with one another, special test equipment (such as conductivity meters) should be installed to detect the first trace of leakage and sound an alarm. (See Chap. 4).

Fouling of heat-transfer surfaces slowly reduces the effectiveness of heat transfer. Where heat transfer is rapid and occurs at high temperature, e.g., radiant tubes in a furnace, fouling leads to hot spots on the tube, softening, and ultimately to the loss of the tube. Fouled jackets on pressure vessels and interior or exterior heating or cooling coils lead to reduced capacity of the apparatus and occasionally to loss of control of the reaction. Temperature and pressure measurement at the inlet and outlet connections generally give adequate warning of reduced transfer capacity. The use of visual or audible alarms when dangerous limits are reached is desirable. More critical installations should include devices to automatically shut off fuel supply, start auxiliary ventilating systems, call the fire department, introduce blanketing steam, or other protective measures.[7]

Problems arise in scale-up because the volume of a reactor goes up as the cube of its size while the area for heat transfer of a jacket goes up only as the square of the size. Larger reactors must have coils or other systems providing surface extended beyond that possible with a jacket.

Unless reactor agitation is very effective, the material being handled is not very viscous and the temperature drop across the cooling surface is small, there may exist a substantial radial temperature gradient within the reactor. If the temperature drop from the wall to the center of the reactor is as much as 10°C, the reaction rate at the center would differ from that at the walls by a factor of about 2. Thermometers are usually installed at fixed points in the vessel, so the existence of radial gradients may not be observed or allowed for. Indicated temperatures are frequently not even good averages.

Heat exchange is usually provided to control or balance a reactive process and for stable operation the removal of heat must exactly balance its rate of generation. The rate of generation of an exothermic reaction is commonly exponential in temperature (see Chap. 4), but heat transfer varies linearly with the difference in temperature between the jacket coolant and the reactants. This can create a runaway condition. The several conditions leading to reactor instability are discussed by Perlmutter.[8]

When a reaction is sufficiently exothermic to make possible an adiabatic temperature rise of about 200°C if all the reagent should react without loss of heat to the surroundings, reactor design can be critical. Reactors for such reactions are extremely sensitive to small changes in reaction or jacket water temperatures and are uncontrollable unless the temperature at which they run is near ambient.[9] The design of such "parametrically sensitive" reactors has been examined by several groups.[8,10,11] The following example is taken from Boynton, Nichols and Spurlin.[9] Consider a simple exothermic reaction which follows Arrhenius' law and whose rate doubles for a 10°C rise in temperature (very common, see Chap. 4). Operation is at 120°C with an activation energy E of 23,000 cal/mol. The rate of heat production is

$$\frac{dQ}{dt} = Ae^{-E/RT} \tag{5.1}$$

If reaction is in a jacketed kettle, the rate of heat removal is

$$\frac{dQ'}{dt} = UA(T - T_0) \tag{5.2}$$

where Q is the heat of reaction, cal/mol; Q' the heat transferred, cal/mol; t the time; A a constant; E the activation energy, cal/mol; R the gas constant, cal/mol-K; T the temperature of reacting mixture, K; UA a constant including the heat transfer coefficient U and the area for heat transfer A, cal/sec-K-mol; and T_0 the cooling water temperature, K. Figure 5.1 shows plots of Equations (5.1) and (5.2). At a given water rate, line A shows the heat removal as a function of mixture temperature. This line intersects the reaction heat generation curve at 123°C and 143°C. The intersection at 123°C is

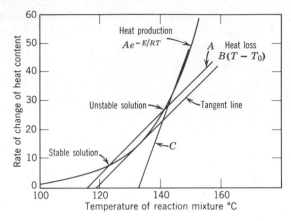

Figure 5.1.   Rates of heat production (curve) and heat loss (straight lines).

stable because the temperature will tend to return to this point if displaced slightly above or below it. The 143°C intersection is unstable. If displaced just below it, the contents will drop to 123°C. Once exceeded, the temperature will rise until the reagents are exhausted by reaction, explosion, fire, or volatilization. This is the familiar phenomenon of ignition. Operation with a 7°C differential (jacket temperature = 116°C) was used in making line 7, so represents stable operation if care is utilized.

When the differential temperature is allowed to rise to 14°C, a curve results which is tangent to the heat production curve at 130°C. This is the maximum stable temperature of operation, and this offers no operational leeway.

By increasing the heat-transfer surface at constant reactor volume, line C results. The point of tangency is 134°C, a higher permissible temperature, but the maximum Δt for stable operation is still 14°C.

The result is general for reactions doubling in rate with a 10°C temperature rise. For the more general case, Equations (5.1) and (5.2) can be differentiated and equated to determine the maximum stable temperature differential

$$(T - T_0)_{\text{critical}} = \frac{RT^2}{E} \tag{5.3}$$

If the activation energy is replaced by the known temperature T_d required to double the reaction date

$$(T - T_0)_{\text{critical}} = \frac{2T^2}{E} = 2(\ln 2)T_d = 1.4T_d \tag{5.4}$$

Reaction rates requiring over 20°C to double are uncommon and operation near the critical temperature difference is hazardous, so unknown exothermic reactions should not be operated, even in the laboratory, at rates demanding Δt over 10°C unless emergency quenching and cooling apparatus are at hand. Several methods of controlling exothermic reactions may be possible: (1) use dilute solutions or suspensions, (2) feed one component slowly enough to prevent heat rise to a dangerous value, (3) add around 10% of a volatile solvent to soak up heat by volatilization if temperature rises suddenly, (4) provide automatic controls to shift operative conditions if rise commences, and, (5) redesign to permit sufficient transfer with a safely low Δt.

Hot material stored in bulk can cause problems. Powders placed in containers at reasonable temperatures may absorb oxygen from air and rise sharply in temperature because the insulating value of the powders prevents heat escape. Coal piles heat badly and finely powdered zinc stearate loaded into 18 in. $\times$ 18 in. $\times$ 2 ft cartons well below the melting point has reached the melting point and lumped within two days.

Flow-induced vibrations in shell and tube heat exchangers can be both noisy and destructive. Barrington[12] discusses control means.

Air-cooled heat exchangers are becoming increasingly utilized. These units have unusual problems when used in cold climates. These problems are described by Shipes[13] with suggestions for safe, satisfactory operation.

5.2. SIZE REDUCTION

Fire and explosion hazards exist where it is necessary to grind combustibles. Stray iron should be removed by magnetic separators and/or screens just ahead of any high-speed grinding machinery to insure the absence of sparks. All machines should be carefully grounded with low electrical resistance in the grounds. Ducts must also be grounded if dust is conveyed. Safety relief vents should be provided to prevent pressure build-up in case of minor explosions. High humidity and low concentration of oxygen in the gas streams used for conveying and cooling the substances being ground reduces the hazard. Stern[14] reports that a sulfur grinding unit has operated for years without fires by using cooled flue-gas atmospheres containing half or less of the oxygen in air as the grinding atmosphere.

Explosive mixtures of air and solids are similar to air–vapor mixtures in having high and low explosive limits. Within a single unit, it is possible to be both above and below explosive limits. Volume 5 of the *National Fire Codes*[1] has much data concerning dusts in industry. Explosion-proof lights and nonsparking fans and mill parts reduce the hazard in grinding areas. Dust should be rigorously controlled to protect worker health (see Chapter 25).

5.3. MIXING

Industrial processes require mixing of all combinations of gas, liquid, and solid.

Mixing Gas with Gas

Simple contact is usually sufficient and no machinery is required. Explosive mixtures require the usual precautions.

Gas–Solid Mixing

Solid fuel burners are common devices of this sort. Long-term experience has made these units remarkably free of problems. Volumes 5 and 7 of the *National Fire Codes*[1] outline special procedures to be observed. Where gases and solids are mixed intimately to bring about chemical reaction, as for example in fluid-bed processing, the very high weight of solid compared to the weight of gas present means that sudden rises in gas temperature are readily controlled by the flywheel effect of the large heat capacity of the solid. All containers should be well grounded.

Gas–Liquid Mixing

When mixing gas with a liquid without reaction, the explosion temperature of the air–vapor mixture produced is within 1°C of the Pensky–Martens flash point. Possible sources of ignition should be eliminated and all apparatus well grounded. Combustible foams often form. These can be troublesome because they may bury material under a dangerous flammable layer and may spread up and out to possible sources of ignition.

When gas is being dispersed into liquid, many agitator designs will overload if the gas supply is interrupted. Interlocks between gas and motor are desirable, overload protection essential. If it is necessary that the agitator continue to run without gas flow, a two-speed motor can permit operation without overload.

Liquid–Liquid Mixing

With liquids of low viscosity, motionless mixers or some type of high-speed mixer are usual. There are no special hazards above the usual ones connected with rotating machinery. Side-entering propeller mixers should not be used in large tanks if operation with the propeller partially submerged is essential. Excessive vibration, destruction of shafts, shaft seals, and motors frequently occur when operating partially submerged. When very viscous fluids are processed in kneaders or rolls, extraneous solids must be excluded or they become hazardous to workers and machines.

When processing explosive mixtures, remote operation with weak walls, barricades and some type of snuffing device is advisable.

Liquid–Solid Mixing

Low-viscosity mixes are generally done with high-speed propellers or turbines. The problems are essentially the same as with liquid–liquid mixing. Material properties present the primary hazard.

Solid–Solid Mixing

Dust explosions represent the principal hazard. Use of inert atmospheres is desirable. High humidity and careful grounding of units reduces the hazards. Removal of tramp iron or other spark-producing material is desirable.

5.4. MATERIALS HANDLING

This section is restricted to techniques involving moving solids, since pumps and blowers are considered separately.

Materials are moved by an enormous variety of devices: conveyors, cranes, bucket elevators, lift trucks, etc. Most of the hazards connected with the use of these devices arise from the increasing use of automation and the possibility of injury from contacting moving machinery. These hazards are not special or characteristic of the chemical industry.

Chemicals require special attention to reduce corrosion of mechanical devices. Problems generally first appear in the boots and unloading shoes of bucket elevators and in charging and discharge chutes. Where possible, inert atmospheres or controlled humidities in conveying systems offer good control safeguards.

5.5. MASS TRANSFER

Most mass-transfer operations such as gas absorption and solvent extraction have fire hazard problems, but few other problems characteristic of the operations. Columns operating at high temperatures, especially those utilizing steam in processing should have special provisions to prevent the injection of liquid water into the system. Such water may collect in an idle steam line. One of the most spectacular accidents which the author ever witnessed was caused by injecting a few gallons of liquid water into the bottom of a fractionating tower distilling hot asphalt. The flipped-off top and ensuing rain of bubble caps which covered almost 10 acres testified to the vigor of the flash evaporation that ensued.

Low-pressure towers should be equipped not only with pressure, but with vacuum-relieving devices. Wash-out or flooding connections, opened at inappropriate moments, can cause disasters. The installation of line blinds instead of valves can provide some degree of safety. A lead-lined still which had not been properly equipped with a vacuum valve was collapsed when a spray of cold water was injected into it by opening the wrong valve. The still was full of steam, which condensed quickly, forming a high vacuum inside. Atmospheric pressure then crushed the still.

Simultaneous Heat and Mass Transfer

5.6. HUMIDIFICATION

Humidification in itself is generally a most innocuous operation. The blower and exhaust systems, however, may constitute a fire hazard in themselves, and the ducts may introduce hazards contributing to the spread of fires. Ducts should not pierce fire walls, floors, or make connections between hazardous areas. Where it is essential that this be done, provision must be made for automatic shut-off in case of fire or explosion.

Fume scrubbers designed to reduce the amount of combustible or objectionably odored material released into the atmosphere may accumulate large quantities of dangerous combustibles at points within their systems unless they are specifically designed to hold up a minimum of combustible liquid or solid.

5.7. DRYING

Drying equipment comes in a wide variety of types which differ sharply in the hazards of their operations. Those which possess moving parts are subject to the usual difficulties found around moving machinery. Where the material being dried may become dusty, release a combustible substance, or is simply combustible, it is advisable to provide some type of emergency over-range protection to insure that overheating of the dryer does not result in a fire or explosion. Steam or inert gas injection offer good protection and sprinklers may sometimes be used. The design of continuous dryers should eliminate the possibility of the deposition of combustibles on electric heaters or the undue accumulation of dusts at points within the apparatus itself. Spray dryers are subject to the same possibilities of dust explosions that one might find handling any other combustible material in an atmosphere containing oxygen. Dust or partially dried material accumulating on temperature controller bulbs can cause loss of control. Using flue gas directly for drying reduces fire hazard where the composition does not cause reactions with the material being dried. Superheated steam or gases can sometimes be used. Flash dryers should be treated in the same fashion as grinding equipment, and their ducts should be provided with explosion vents if there is a possibility of an explosive mixture being formed. Drum dryers should have their feed screened, for large particles can be violently thrown out from between the rolls, endangering operators.

5.8. EVAPORATION AND CRYSTALLIZATION

Because most evaporation is conducted within a closed system, there is rarely a major difficulty with this type of equipment. The bodies should be equipped with pressure and vacuum relief valves and with devices to protect the equipment in case of fire or accident within the unit area. Like heat exchangers, evaporators can become fouled, thus greatly reducing the heat-transfer ability. Scale accumulations are particularly troublesome when the scale is soft, because soft scale is a very poor conductor of heat. Scales should be removed at regular intervals by means of standard commercial devices.

Crystallizers are generally simple agitated tanks or vessels designed to form crystals of controlled size from saturated or slightly supersaturated solutions. These units possess almost no hazard other than those associated with the materials being handled. Provide protective equipment such as shear pins to prevent strain on motors or gears in case of an interruption or freeze-up.

Microwave evaporators are coming into use and require very specialized design to prevent leakage of radiation. Regularly scheduled tests for leakage should be programmed.

Submerged combustion evaporators show little hazard unless the fire is extinguished. Gas interlocks should be used to interrupt for restarting.

5.9. MOMENTUM TRANSFER: PUMPS, COMPRESSORS, AND AGITATORS

Pumps and compressors are among the most thoroughly standardized, widely used, and trouble-free devices found within the plant. Most difficulties with new pumps and blowers arise from corrosion, cavitation, water hammer, misalignment, or misuse. Plant start-up groups know that if a pump can get through the first hour of operation that it may expect a reasonable life. This simply confirms the fact that almost all difficulties with pumps occur during their start-up. J. R. Caddell[15] has an excellent discussion concerning the operation and maintenance of fluid-flow equipment which is particularly valuable in its description of the hazards of starting centrifugal pumps. J. E. Troyan[16-18] has provided an excellent comprehensive report on what we should actually do to prepare for a plant start-up including training operators, preparing manuals, and debugging.

Pumps and connecting valves, installed where normal working pressures are reasonable, fail because of ram effects caused by sudden shutting of valves. This is commonly known as "water hammer." If water is flowing in a pipe and is suddenly stopped, its kinetic energy is suddenly released. The head h (in feet of water) thus suddenly developed is:

$$h = \frac{cV}{g}$$

(5.5)

where c is the velocity of pressure wave disturbance up the pipe, ft/sec; v the reduction in velocity ft/sec; and g the gravitational constant ft/sec². The velocity of wave propagation, c, may be estimated by Jonkowsky's formula

$$c = 4660/1 + KB \qquad (5.6)$$

where K is the ratio of elastic modulus of water to pipe shell material (0.01 for steel); and B is the ratio of pipe diameter to pipe thickness. These equations are both rather approximate, since they assume instantaneous closing, but the results are conservative.

Water-hammer effects can be large and destructive, particularly to concrete or similar fragile pipe and to large cast-iron pumps. Prevention includes providing air cushions within the system or by making it impossible to close valves fast.

If a pump operates in such a fashion that areas within the stream of fluid around the impeller fall to pressures below the vapor pressure of the fluid being pumped, cavities form in the flowing fluid. When these cavities move into an area of higher pressure, they collapse very suddenly. When the bubbles collapse, fluids meet at high velocity and pressures at the point of collapse become quite high. If the point of collapse occurs on the surface of the metal, it may be stressed beyond its elastic limit resulting in rapid failure. Cavitation can be avoided by provision of adequate net positive suction head.

There are basic differences in the handling of centrifugal and positive displacement pumps which are particularly troublesome during start-up and shutdown. Centrifugal pumps use minimum power (electrical current) when their outlet valves are closed. They require maximum power when there is an open output—no valve and no pressure to pump against. Positive displacement pumps react in exactly the reverse fashion. They require maximum power when the outlet valve is closed; minimum power with the outlet wide open. Centrifugal pumps often have their motors destroyed by attempts to test them on the job with no outlet line attached (thinking that this is a no-load condition). Positive displacement pumps operate with close clearances, requiring strainers for carefree operation. It is desirable to provide pressure relief on the outlet side in case the discharge is inadvertently shut off. No pump should be run dry for periods beyond a few seconds or damage, even fire, may result.

Check for strains placed on the pump by installed piping. If misalignment is appreciable, it will often be detectable by the fact that the pump shaft cannot be moved by hand power. Piping stresses lead to rapid wear of the seals on the pump and, if sufficiently large, may cause rubbing of the impeller or case cracking. See that piping is installed so that thermal expansion (or contraction) does not set up intolerable stresses. Pump and prime-mover bearings should be checked thoroughly for proper lubrication. Suction and discharge pressure gauges should be installed with each pump, as this will permit an immediate check on the operating conditions of the pump.

When starting a centrifugal pump, it is desirable to close the discharge valve to reduce the load on the driver, and this valve should not be opened until the motor has come up to speed. If this valve is then left closed for any substantial length of time, the work done on the small volume of liquid contained within the casing may result in its boiling, and trouble may arise from volatilization (such as loss of prime or destruction of the packing). Packing glands on pumps must be kept adjusted so that there is a continuous drippage from them. This drippage serves as a lubricant and coolant between the packing and the shaft. Where the material dripping out is dangerous, provide an internal seal with a second fluid. The secondary fluid must not react with or cause problems with the fluid being pumped, since a small amount will enter the pump. When repacking, always repack both sides of the sealing fluid inlet.

After a shutdown, the centrifugal pump should not be started again until the discharge valve is shut off or the pump motor may burn up while coming up to speed. Packing is usually selected by trial. Where corrosive fluids are handled, shaft seals usually greatly reduce the leakage to the outside and make operation far safer, but seals in chemical plants are often special and require more than normal care in selection and maintenance.

Starting up reciprocating or other types of positive displacement pumps requires as much care in seeing that the valves are open as is required in seeing that the valves are closed in the case of centrifugals. Most difficulties with reciprocating pumps arise from lack of attention to their automatic lubricators. Do not exceed the speeds recommended by the manufacturers and watch carefully for vibration in pipe supports which may lead to failures in the pipelines.

It is possible to move solids in pipelines, and this operation has become increasingly economical. The maximum allowable particle diameter of the solid to be handled should not exceed one-half of the smallest canal dimension of the pump, and the maximum permissible size of the particle to prevent blockage of the lines must be such that the particle diameter is not greater than $\frac{1}{3}$ that of the pipe. Stable operation of this type of unit is essential, and startup and shutdown can be most troublesome. Data on designing and operating this type of unit have been presented in the literature.[19-21]

5.10. MOMENTUM TRANSFER-PHASE SEPARATIONS BASED ON FLUID MECHANICS

Filtration, centrifugation, screening, elutriation, jigging, classification, and sedimentation all show only the difficulties which we would expect due to the handling of difficult materials. Avoid fires and excessive exposure of personnel to toxic materials. Centrifuges require heavy supports of highest quality. Pressure filters should be installed to assure that personnel are not liable to be sprayed with injurious or offensive solutions, but the precautions are no more elaborate than those taken in installing all types of pressured equipment.

5.11. INSTRUMENTS AND CONTROL SYSTEMS

The selection and installation of control systems is a complex subject requiring specialized training. The selection of instruments for apparently simple applications is frequently made by unskilled persons and can lead to poor or even disasterous results. The subject is complex but a few frequently encountered problems may alert installers to potential difficulties.

The addition of a control to a previously designed process is rarely effective, for the controller and the process must function as a unit. Quality control can be obtained only when all parts of the unit are balanced. For example, if a tank is being heated by steam condensing inside a coil, the size of the coil will have much more effect on the ability to control concisely at a selected point than the sensitivity of the instrument.

Advanced controls commonly possess three adjustments, for proportional band, reset, and rate. The assumption that such a control will perform in a fashion superior to an on–off control is frequently false. For example, the introduction of proportional control may give very consistent control, but not at the point selected if the overall load changes.

Much difficulty has been encountered in using central digital controls because of too-frequent down time, but this has been constantly improving. The entire instrument field is in a state of flux and good results require the assistance of one skilled in the art of selection and installation. Even more important, the process equipment must be selected to be controllable. See Chapter 6.

REFERENCES

1. *National Fire Codes* (15 vol.), National Fire Protection Association, Quincy, Mass., 1975 (reissued periodically).
2. *Manual of Hazardous Chemical Reactions*, NFPA 491M-1975, National Fire Protection Association, Quincy, Mass. (1975) (revised periodically).
3. *Fed. Regist.* **37,** Part 2 (Oct. 18, 1972).
4. H. Popper, "Last Year's Major Explosions Prod This Year's Safety Push," *Chem. Eng.,* **70**(1), 91 (1963).
5. E. Jennett, "Components of Pressure-Relieving Systems, Part II," *Chem. Eng.,* **70**(17), 151 (1963).
6. E. Jennett, "Design Considerations for Pressure-Relieving Systems," *Chem. Eng.,* **70**(14), 125 (1963).
7. D. M. Considine, *Process Instruments and Controls Handbook*, McGraw-Hill, New York, 1957.
8. D. D. Perlmutter, *Stability of Chemical Reactors*, Prentice-Hall, Englewood Cliffs, N.J., 1972.
9. D. E. Boynton, W. B. Nichols, and H. M. Spurlin, "How to Tame Dangerous Chemical Reactions," *Ind. Eng. Chem.*, **51**(4), 489–94 (1959).
10. O. Bilous and M. R. Amundson, "Chemical Reactor Stability and Sensitivity," *AIChE J.*, **1**(4), 513 (1955).

11.   O. Bilous and M. R. Amundson, "Chemical Reactor Stability and Sensitivity II, Effect of Parameters on Sensitivity of Empty Tubular Reactors," *AIChE J.*, **2**(1), 117 (1956).

12.   E. A. Barrington, "Acoustic Vibrations in Tubular Exchangers," *Chem. Eng. Prog.*, **69**(7), 62 (1973).

13.   K. V. Shipes, "Air Cooled Heat Exchangers in Cold Climates," *Chem. Eng. Prog.*, **70**(7), 53 (1974).

14.   A. B. Stern, "A Guide to Crushing and Grinding Practice," *Chem. Eng.*, **69**(25), 129 (1962).

15.   J. R. Caddell, *Fluid Flow in Practice*, Chapman and Hall, London, 1956.

16.   J. E. Troyan, "How to Prepare for Plant Start-ups in the Chemical Industries," *Chem. Eng.*, **67**(18), 107 (1960).

17.   J. E. Troyan, "Trouble-Shooting New Processes, Hints for Plant Start-up: Part I," *Chem. Eng.*, **67**(23), 223 (1960).

18.   J. E. Troyan, "Pumps, Compressors, and Agitators, Hints for Plant Start-up, Part III," *Chem. Eng.*, **68**(9), 91 (1961).

19.   E. Condolios and E. E. Chapus, "Transporting Solid Materials in Pipelines," *Chem. Eng.*, **70**(13), 93 (1963).

20.   E. Condolios and E. E. Chapus, "Designing Solids Pipelines," *Chem. Eng.*, **70**(14), 131 (1963).

21.   E. Condolios and E. E. Chapus, "Operating Solids Pipelines," *Chem. Eng.*, **70**(15), 145 (1963).

6

Instrumentation for Safe Operation

Donald Richmond

Automatic control of chemical processing equipment has made it possible for a chemical operator to assume the responsibility of additional units of operation. This is because controllers are performing tasks which were formerly handled by people. This simple concept of greater productivity applies to batch as well as to continuous operation. This chapter presents some of the safety considerations which are engineered into process design, where safety engineering begins. While this chapter covers only a few of the most common safety devices generally associated with instrumentation, it illustrates the type of thinking which is required of designers in order to maintain a high degree of safety in present day plants.

For the purposes of this discussion, the term *instrumentation* means the use and application of industrial instruments of measurements and control. This definition includes indicators, recorders, controllers, and transmitters for measurements such as temperature, pressure, flow, liquid level, and analysis. Instrumentation also embraces the fields of data reduction, alarms, and interlocks.

At the present time 5% of the capital investment in large chemical plants may be in instrumentation. The rate of growth of instrumentation cost in one plant over the past five years has been approximately four times as fast as the plant average. The total invested capital per employee in this company has increased approximately fourfold in the past 10 years. In a large oil refinery the investment per employee may be as high as $500,000 or more. These statistics provide clues as to the long range effect of the increasing demands of greater productivity upon instrumentation.

Process design engineers are just becoming generally aware of the safety concepts of instrumentation, including some of its inherent dangers. Actually, safety and instrumentation go hand in hand because good engineering includes safety in its design. The types of manufacturing equipment which we will find in the plants of tomorrow will dictate special safety requirements in engineering. The hazards that we find associated with special products and special materials which have to be handled, will generally dictate a certain amount of safety instrumentation. Efficient control implies safety. Everyone is interested in safety from the standpoint of protection of personnel and equipment. The operating department, however, has an extra interest in safety because of the desire to prevent production losses such as spills. The maintenance department likewise has a special interest in safety because of the desire to easily maintain the operating facilities. For example, instead of installing a control valve near a hazardous area where an employee might easily be injured, it might be better to locate the valve at a safe location where the maintenance crew can easily repair it. A few of the common safety devices which are commonly used in chemical plants will be discussed.

6.1. SELF-ACTING PRESSURE REGULATORS

Two basic types of pressure regulators are shown in Fig. 6.1. One is the type with the integral downstream pressure tap, and the other has the external downstream pressure tap. The first valve is a single-seated valve which normally is used either with small flows and small valve sizes or on those applications involving small pressure drops. The other valve is double seated generally because of large flows or large pressure drops. Usually double seating begins at the 1½-in. size. In both cases, however, the action of the spring is opposed to the pressure that is placed upon the diaphragm. The

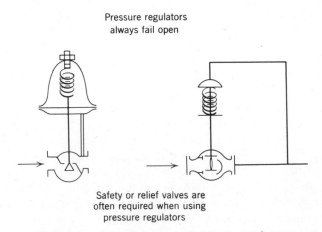

Pressure regulators
always fail open

Safety or relief valves are
often required when using
pressure regulators

Figure 6.1. Two basic types of pressure regulators.

purpose of the regulator is to reduce a fluid pressure to a smaller value. Rupture of the diaphragm is the commonest cause of failure in these self-acting regulators. Should the diaphragm fail for any reason, the action of the spring in both valves will be to return the valve to the wide open position. Because of its design, therefore, pressure regulators fail in the unsafe position, such as open. Upon such a failure full upstream pressure will exist on the downstream side of the valve. Normal leakage through the valve can also raise the downstream pressure at no load.

This brings up the first rule about pressure regulators—they always fail in the unsafe position. Because of this inherent weakness, it is customary to find pressure relieving devices or safety disks installed downstream of self-acting pressure regulators to safeguard equipment from over pressure.

6.2.   SELF-ACTING TEMPERATURE REGULATORS

In Fig. 6.2, the diaphragm motor is replaced by a bellows motor in which the vapor of a fluid in a temperature sensitive bulb exerts its pressure. The bulb contains a liquid such as ether or alcohol. The temperature sensitive bulb is

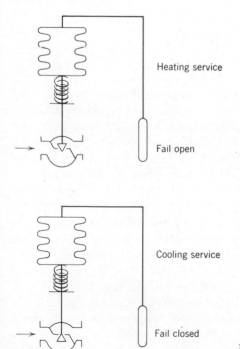

Heating service

Fail open

Cooling service

Fail closed

Figure 6.2.   Temperature regulators.

located in a tank or vessel which is to be temperature controlled. The heating medium for the vessel passes through the automatic valve as the fluid pressure inside of the bulb increases because of an increased temperature. The pressure in the bellows motor will gradually close off the valve. This action reduces the rate of heat input to the system and in that way the temperature is regulated. As in the case of the pressure regulator a failure of the thermal system will cause the valve to fail in the wide open position because of the action of the spring. These regulators with valve plugs on top of the valve seats are used on heating service. This points up the similar rule—all self-acting temperature regulators on heating service fail in the open position which is the unsafe position.

For cooling service the same instrument is used by reversing the action of the valve. The plug is on the underneath side of the seat instead of the top side. A failure of the thermal system will cause the spring to close the valve, in which case the coolant is shut off. Therefore, self-acting temperature regulators for cooling service fail in the unsafe position. Because of the inherent weakness of self-acting temperature regulators, some external temperature switch is required in order to protect vessels from overheating or overcooling. There are self-acting temperature regulators on the market which feature "fail-safe" design. This is because the fluid chosen for the bulb operates in the vacuum range. This is unsatisfactory because of the limited available power for the movement of the valve stem. One other disadvantage of the self-acting temperature regulator is that the temperature is not controlled very closely.

6.3.   THE PNEUMATIC CONTROLLER

The need for closer control of temperature usually justifies the installation of an instrument such as in Fig. 6.3. This is a temperature controller wherein the internal bourdon tube is connected to a temperature-sensitive bulb, and becomes a pressure controller when the internal bourdon tube is connected directly to a pressure vessel or pipeline. In this particular instrument the sensitivity, proportional band or throttling range (these terms are synonymous), can be easily adjusted to suit the process. Furthermore, the action of the controller can be easily changed so as to use normally open or normally closed diaphragm motor valves as shown in the figure. Many believe this arrangement constitutes a fail-safe system because the control valve can be chosen to fail open or closed, but it is not. The control valve fails safe only in the event of an air supply failure. However, a failure of the thermal element will result in the pointer falling to the lower end of the temperature scale. This makes the instrument think the temperature is too low, so the action of the controller will then be able to cause the valve to open wide regardless of the action of the valve. Therefore, the normal closed-tube temperature systems likewise never fail safe. However, it is now possible to buy completely

Temperature and pressure controllers fail open or closed

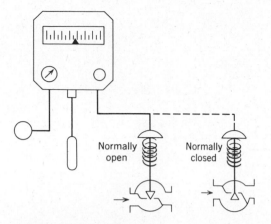

Closed–tube systems never fail safe.
Relieving devices or temperature switches often required.

Figure 6.3.   Temperature and pressure controllers fail open or closed.

fail-safe temperature controllers at an additional cost. A piece of invar inside of the thermal bulb causes the fluid pressure to drop on a rise in temperature, which is the reverse action of the conventional system just described. Under this condition, when the tube system fails the pointer moves up scale. In this way, the instrument thinks the system being controlled is too hot and the heating control valve will be automatically closed. This particular controller is advantageous on small process vessels which are easily overheated; that is, small in relation to the rate of heat input.

6.4.   THE POTENTIOMETER CONTROLLER

A temperature controller of the electric potentiometer or wheatstone bridge type in Fig. 6.4 is more commonly used in process control. They are generally considered the safest because on a failure of the primary element the pen moves up scale, down scale, or stationary, as desired. Upon an electrical failure the pen stops moving. If the pen happens to be above the control point at the time of electrical failure, the controller fails safe. If the pen is below the control point at the time of electrical failure, the controller fails in the unsafe position. This is especially true on those controllers with automatic reset. Again, the engineer has the choice of using a normally open or normally closed valve depending upon which is the safest way for the valve to fail in the event of a failure of the air supply. Thus, the rule for electrically operated temperature controllers is the control valves can be chosen to fail safe upon air failure but not upon electric failure.

Electrical temperature controllers
sometimes fail safe

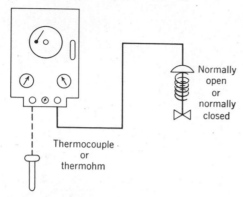

Normally
open
or
normally
closed

Thermocouple
or
thermohm

On failure of primary element
Pen goes upscale

Figure 6.4. Electrical temperature controllers sometimes fail safe.

6.5. FLOAT SWITCHES

Two types of float switches are shown in Fig. 6.5. The first one is the simple float-operated switch which is generally installed at the top of a tank. As the float rises because of a rise in level a magnet which is connected to the float rises into the magnetic field of the external magnet. This action attracts and tilts the mercury bottle, and in that way switch action is obtained. In the

High–level float switches

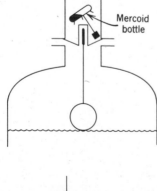

Mercoid
bottle

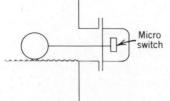

Micro
switch

Figure 6.5. High level float switches.

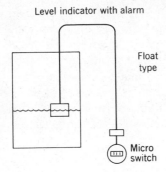

Figure 6.6. Level indicator with alarm.

lower picture of Fig. 6.5, the side entrance float switch is shown. In this case the same size of ball float is used to operate a micro switch through a packless flexure tube. This type of switch is applicable to liquids under any pressure or vacuum where the wetted parts are compatible with the fluids. This type of switch is not suited to applications where deposits on the float, or its guide rod interfere with float motion, or applications where sublimation or corrosion can interfere with float motion.

Figure 6.6 illustrates another version of a float-operated level gauge. This type of level gauge is accurate and quite dependable. For this reason, it is generally advisable to attach the level switch alarm to it.

6.6. BUBBLER-TYPE LEVEL GAUGES

Figure 6.7 shows the bubbler-type level gauge. The back pressure in the dip pipe which motivates the U tube manometer for liquid level measurement can also be used for actuating an ordinary pressure switch. A contactor

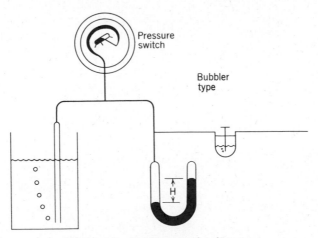

Figure 6.7. Bubbler-type level gauge.

manometer shown in the lower part of Fig. 6.7 is used to indicate liquid level and to provide switch action. Tungsten probes are adjusted down into the mercury. The motion of the mercury will either close or open the electrical circuit which will actuate other devices as required. The mercury U tube contactor manometer, while inexpensive, cannot be made explosion-proof. Upon loss of air to the bubbler or serious tube leaks, this device is undependable. If liquid rests on the left side of the manometer, or the indicating fluid is lost, the device becomes unreliable. In modern plants, the U tube is replaced with a *D/P* cell.

6.7. ALARMS CONNECTED TO TRANSMITTERS

It is customary to add a switch to a process variable transmitter for alarm purposes. An illustration of this principle is shown in Fig. 6.8 where a displacer is used as a level transmitter. Other transmitters for pressure, flow, or temperature, for example, may be used in conjunction with switches attached to the transmitted signal lines for alarm or interlock services.

The displacer of Fig. 6.8 is based on Archimedes' buoyancy principle. The change in weight of the displacer varies in accordance with the weight of fluid displaced by the displacer at various liquid levels. The displacer moves vertically only a fraction of an inch upon level changes. The transmitter senses the change in weight and thereby transmits an output signal proportional to level.

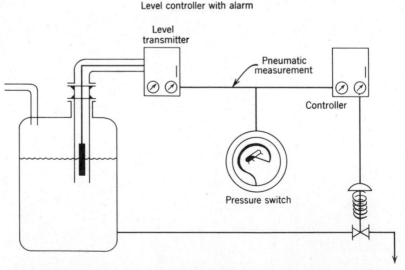

Pressure switch must be connected to the pneumatic measurement, not the controller.

Figure 6.8. Level controller with alarm.

In using pneumatic transmitters, the usually accepted output signals are 3 psig representing 0% to 15 psig representing 100%. Since the pressure switch depends on a dependable supply of air, a failure of air supply to any pneumatic transmitter can result in a corresponding failure of a high signal alarm. If the switch is designed for actuation upon a preset low signal, the same air supply failure will initiate a low signal alarm.

6.8. ALARMS ON RECORDING CONTROLLERS

Figure 6.9 is a picture of any controller scale. The shaded portion represents the normal control band and the heavy black marks indicate the low and high alarm. If the controller can be depended upon to hold the measurement within the limits of the shaded portion, then alarm contacts would serve a useful purpose if their settings were made just outside the normal control band. This method is generally accepted in practice among the industry today, because it usually provides adequate advance warning of system failure.

6.9. ALARMS ON MULTIPOINT RECORDERS

Figure 6.10 illustrates a conventional multiple point recorder. These are most widely used for temperature measurements where the thermocouple or resistance bulb becomes the primary element. It is possible to have selected temperature alarms even on a multiple point recorder. The circuit consists of a terminal board, selector switch, print switch, standardizing switch, and a temperature switch which is shown arranged for one temperature only.

The standardizing switch is normally closed and opens only when the instrument goes into standardization, either automatic or manual. When the

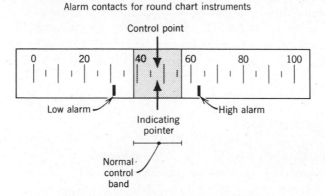

Figure 6.9. Alarm on recording controllers.

Alarm contacts for multipoint recorders

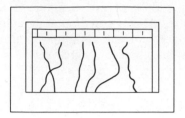

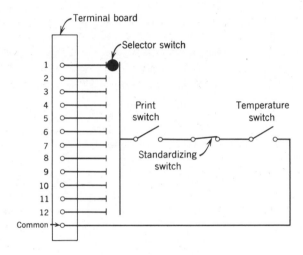

One temperature only

Figure 6.10.   Alarm on multipoint recorder.

print wheel comes down to print, the print switch makes a momentary contact. Should the temperature of the selected point be in excess of what it was set for, the temperature switch will close and the entire circuit which is normally open will be momentarily closed when the print switch closes. It is this momentary signal which can be used for alarm and other shutdown purposes.

6.10. SOLENOID VALVES

Some applications of solenoid valves are shown in Fig. 6.11. It is assumed that on air failure the control valve is to be held in the last position before the air failure. A two-way solenoid valve connected to a pressure switch in the air line will accomplish this purpose.

Applications of solenoid valves

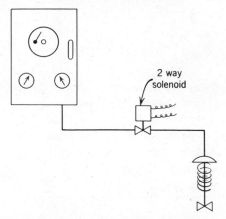

On failure, valve is stuck in last position before failure.

Figure 6.11.   Solenoid valve application.

Figure 6.12 illustrates the use of three-way solenoid valves. Of course, three-way solenoid valves can easily be made into a two-way solenoid valve merely by plugging off the third port. In this particular application the control valve can either be vented upon failure, or by using a pressure regulator the control valve can be moved to any preset position between wide open and fully closed merely by setting the pressure regulator at the desired pressure. Solenoid valves can be actuated by other process controls, master

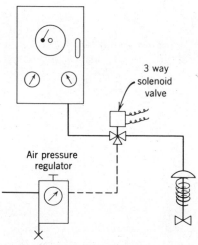

On failure, valve is vented, moved to a preset position or controlled by another source.   **Figure 6.12.** 3-way solenoid valve.

shutdown devices, and other devices, upon loss of agitation, pump failures, or even fire alarms.

6.11.   ANNUNCIATORS

When an instrument engineer considers alarms for shutdown devices he or she usually thinks of three basic design premises:

1.   All the *critical* points of operation should be protected by alarms and/or shutdown devices.
2.   The device should fail safe.
3.   Field testing should be simple.

An annunciator is a device which calls the operator's attention to the fact that something has gone wrong. This is usually accomplished by an audible and visual alarm.

Annunciator with Manual "Stop Horn" Button and Automatic Reset

The basic elements of the most modern annunciators on the market today include:

1.   A horn or audible device.
2.   A light which attracts attention to the particular event which initiated the alarm.
3.   An alarm acknowledge pushbutton.
4.   One or more relays to maintain the alarm and satisfy several other choice options.

The basic alarm of Fig. 6.13 illustrates these points. While modern annunciators have undergone several improvements in design to improve reliability, such as encapsulation of the relays in a nitrogen environment and the transition to solid state relays, a working knowledge of the annunciator is integrated into designs for safe operation.

6.12.   INTERLOCKS

High-Level Interlocks

Interlocks are really controls for safe operation. In the event of a serious process upset the interlock does something about it in addition to warning the operator of danger. Figure 6.14 shows an interlock for high level. A float switch on top of a tank can be easily interlocked with a motor starter on a

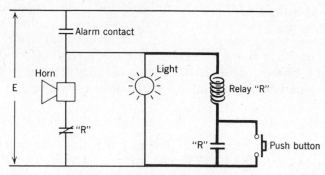

Figure 6.13. Annunciator with manual "stophorn" button and automatic reset.

Some common interlocks

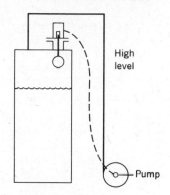

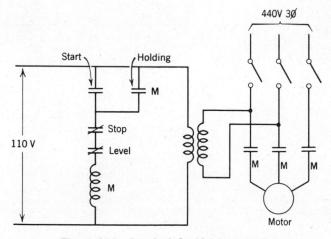

Figure 6.14. Interlock for high level control.

pump which pumps the material into the tank in order to prevent spills or overflows of that tank. The way in which it is accomplished is shown in the lower section of Fig. 6.14. Notice that the control voltage is limited to 110 V for safety reasons. Since most motor voltages are 440 V, three phase, it is merely necessary to install a small 4:1 step-down transformer inside the magnetic starter box in order to obtain the 110 V required for the system. The normally closed contact of the level controller or level switch is connected in series with the stop circuit of the magnetic starter.

Prevention of Operating Errors

Figure 6.15 shows one versatile method of preventing operating errors. Let us assume that this process calls for the sulfonation of an organic material in

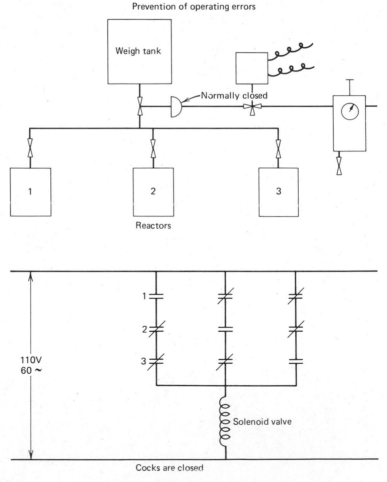

Figure 6.15. Prevention of operating errors.

three batch sulfonators. Oleum is measured out in a weigh tank for each batch. In operation the valves are lined up to the proper sulfonator. Full cooling is applied to the sulfonator. The temperature is controlled either manually or automatically by the rate of oleum addition until the oleum weigh tank is empty. It is not unusual to make an operating error by leaving a valve open on another sulfonator. In many cases such an error is hazardous as well as costly in wasting valuable chemicals. To prevent this kind of error, it is necessary first to mount a lever switch on each of the three cocks. On each switch will be mounted three mercury bottles. In practice there will be the same number of bottles mounted on the shaft on each lever switch as there are cocks which are included in the interlock system. These 4-A bottles are arranged as shown in the electrical drawing, and the bottles are all shown when the cocks are in the closed position. Assuming it is desired to sulfonate #1 sulfonator the cock to it is opened and when it is open, current will automatically flow through the first circuit through the solenoid valve. However, if any one of the other cocks are opened, the flow of current to the solenoid valve will be interrupted. This causes the oleum flow to be shut off by venting the diaphragm motor of the normally closed control valve. This technique is being used at the present time with very good results with negligible maintenance.

Agitation Failure

Another common interlock is known as agitation failure protection. It is designed to prevent accidents which can occur in many chemical reactions when agitation ceases. Since the agitator shaft or the gears can break off, a measure of the current flow to the motor can be used to detect loss of agitation. A current sensitive relay or current switch is connected in series with one wire which is connected to the motor. On large motors a current step-down transformer is inserted here and the relay is connected to the secondary of the transformer. The switch is set when the liquid level is about one-third of normal. At higher levels, more than enough current will keep the relay energized. A drop in current below the set point will deenergize the relay. A typical wiring diagram is shown in Fig. 6.16 using a three-way solenoid valve in the control air line to the diaphragm motor valve. Upon loss of agitation the current switch deenergizes the solenoid valve which in turn vents the diaphragm of a normally closed control valve. This shuts off the flow of chemicals to the reactor to prevent an explosion. The switch action can also be used to actuate an annunciator system and perform other jobs as required by safety considerations.

Miscellaneous Interlocks

Other interlocks include temperature and temperature differential, pressure and pressure differential, liquid level, flow, pH, conductivity, density and analyzers. All of these interlocks or contacts can be made to shut down

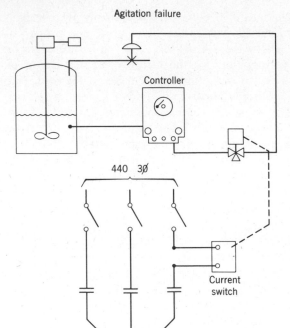

Figure 6.16.   Agitator failure protection.

pumps, blowers, vacuum jets, or other parts of processes. Interlocks can be used to shut off steam, natural gas, cooling water, or fluid flow or to vent vessels.

In summary, remember to use normally closed actuating contacts whenever practicable. Of course this cannot be done in the case of multipoint temperature recorders because of the cyclical nature of a multipoint instrument. Select valves which fail safe, employ normally energized circuits, provide for easy start-up; use timers if necessary, protect all the critical points of operation. Do not over instrument processes. Avoid highcontrol voltages in alarm systems and do not mix up voltages as this practice may cause electrical difficulties. Avoid temporary wiring and exposed terminal boards. Limit the use of plant air on alarms; never use sprinkler air. Avoid bypasses whenever possible or practicable. Do not locate control valves in dangerous areas, and do not forget to put drains in the low sections of piping in order that the entire line may be properly drained when parts of the piping or control valves must be taken down.

Electronic Instrumentation

Industrial plants are increasingly turning towards electronic instrumentation in place of pneumatic measurements and controls for a variety of reasons.

There is no doubt about the present-day reliability now that solid state transistors can be economically produced to perform under wide temperature limits and can withstand a certain amount of vibration or mechanical shock. The practical applications of electronic control systems have, of necessity, required attention to maintenance of signal integrity through grounding and shielding. We will not pursue this subject here.

The safety aspects of electronic instrumentation can be identified in the following areas of concern:

1. Applications in hazardous areas containing flammable gasses, vapors and combustible dusts.
2. Applications in wet areas.
3. Hazards to maintenance personnel, such as electrical shock.
4. Design problems involved in the fail safe concept.

Comments on the four categories follow:

1. *Hazardous Areas.* The concern in this area is the possibility of an electrical device setting off a spark capable of igniting an explosive mixture, with obvious disastrous results. This subject is covered by the National Electric Code, together with ISA (Instrument Society of America) Recommended Practices No. 12.1, 12.2, and 12.4. Both standards set forth the minimum requirements of installation practices to comply with most national, state and local electrical codes. The reader is referred to these sources because of the highly technical nature of the subject.

2. *Wet Areas.* The area of concern from a practical installation viewpoint deals with the water damage to improperly protected electrical instruments. In addition, there is always a possibility of personnel electrical shock which is enhanced by wet flooring. Further, the electrical instrument in a wet area may contain short circuits and dangerous voltages, a double hazard to unsuspecting personnel who may come in direct physical contact with this type of hazard.

3. *Electrical Shock.* There is a continual danger whenever voltages of 110 V or more are employed in congested panels, particularly if there are exposed terminals. Service personnel who must come in direct contact inside of enclosures such as panels to perform their work are subject to electrical shock. In addition, it is just as easy to damage electrical equipment by accidental grounding or shorting terminals with the tools that are carried by these personnel. Any efforts to reduce congestion will reduce this hazard. It is generally unwise to mix 440 V in 110-V control wiring. It is doubly unwise because the ordinary service-person does not suspect higher than 110 V in 110-V circuits. It is not uncommon to find limit switches of all kinds directly wired to 440-V circuits to keep installation costs down. This bad practice requires continual policing vigilance to insure that high voltages be kept out of low voltage areas. Whenever higher than 110 V are to be used in external circuits such as limit switches, it should be a plant practice to tag the device to that effect.

4. *"Fail Safe" Concept.* It is not uncommon to supply electrical power to remote instruments such as transmitters from a source separate from the controlling instruments. In this case, a transmitter power failure will cut off the generated signal. The controller in a control loop will send out control signals in accord with a zero signal. On cooling applications, the coolant will be shut off. On pH applications, the alkali control valve will be wide open. To correct problems like these, it may be possible to:

A.   Furnish the power supply to all parts of the control loop from one source.

B.   Interlock the air supply to the control valve with the field power supply so that upon a field power failure, the valve will be closed, or through the action of a three-way solenoid, a preselected signal may be directed to the control valve.

In reviewing an electrical or electronic control loop, it is recommended that each device be reviewed to insure fail safe action in the event of a power failure. Of course, the same procedure should be used in the event of air failure or utility failure on solenoids, switches, controllers and control valves.

One final word about control valves would be appropriate when considering fail safe concepts. Depending upon the actuator employed to power a control valve, a designer often has three choices of failure positions, depending upon the application:

1.   Fail closed.
2.   Fail open.
3.   Fail in last position.

The most reliable actuators are the spring-diaphragm type due to the simplicity of construction. The wound-up power in a spring is capable of positively stroking a valve to either the closed or open position, without using any outside power source. Generally, the next most reliable actuator for strictly on–off service is the spring return piston and cylinder actuator which is controlled by a large capacity three-way solenoid valve which is mounted on or near the control valve. The "fail in last position" valve can be handled by:

1.   A spring diaphragm actuator equipped with a tight shut off two-way solenoid in the pneumatic signal line connected to the diaphragm.
2.   A piston-cylinder operator of the double acting or double solenoid type.

7

Pressure Relief for Chemical Processes

Clyde McKinley

7.1. INTRODUCTION

Chemical processing equipment and related storage tanks should be constructed with pressure relief systems capable of venting the contained fluids to prevent pressure rise from exceeding an acceptable maximum. Should the pressure rise to some intermediate level between normal operating pressure and upper safe pressure, the relief system should release the quantity of fluid necessary to lower the pressure effectively to the predetermined safe value. This control is necessary to avoid rupture of the processing equipment and to limit the release of the product.

The safety of personnel, those directly engaged in the operation and those who may be in the neighborhood, is the primary concern in selection and location of relief devices. After this primary concern is satisfied one again examines the processing system to see if any additional pressure relief devices may be needed for equipment protection.

Overpressure may be caused by abnormal heat absorption into the processing equipment (as in an external fire), by abnormal exothermic chemical reactions, by improper operation of the equipment, or by mechanical failure. Such abnormal pressure rises may be dealt with through identifying the maximum credible pressure rise rate for each of the causes and providing appropriately sized relief devices.

7.2. OVERPRESSURE PROBLEMS AND RELATED RELIEF SYSTEMS

Overpressure Caused by Abnormal Heat Absorption

Rapid pressure rise in chemical processing equipment may result from fire exposure or from loss of insulation on equipment containing fluids below the ambient temperature. Both situations may be objectively appraised. Bare (uninsulated) process equipment and liquid storage tanks may absorb heat into the liquid-wetted surfaces at rates approximating 34,500 Btu/h·ft² when exposed to free-burning fires.[1] Loss of insulation from lines and vessels containing cryogenic fluids may result in high heat absorption rates, but generally lower than those arising from fire exposure. The attendant generation of vapor may be estimated from knowledge of the thermodynamic properties of the fluid and selection of a credible area of exposure.

Overpressure Caused by Chemical Reaction

The selection of a relief device for chemical processing equipment containing reactant systems with the potential to undergo rapid chemical reaction is difficult because of the unsteady-state nature of the vapor generation. Polymerizations may "run away" with very rapid increases in vapor generation rate. Relief devices, properly sized for the vapor load, may be ineffective because of the problem of disengagement of the vapor from the liquid and the resulting two-phase flow through the relief opening. The overpressure hazard associated with an undesired chemical reaction may also be addressed through avoidance of reaction initiation (use of inhibitors, inert atmospheres, insulation to prevent heat influx from fire or high ambient temperatures).

Overpressure Caused by Improper Operation

Equipment failure and human error, alone or in combination, can lead to overpressurization of processing equipment. Overfilling is a common cause of overpressure. Another is the entrapment of liquid in portions of equipment containing no relief device. Generally, each process equipment section which may be isolated from all other sections by valve closures should contain a relief valve. The relief capacity and relief locations required for individual systems can be predicted for practically all operational difficulties following a careful analysis of credible modes of equipment failure and likely operational errors.

7.3. ESTIMATION OF NEEDED RELIEF CAPACITY IN PRESSURE-RELIEVING SYSTEMS VENTING VAPOR.

The Problem in Summary

In order to estimate or design an adequate pressure-relieving system, one must estimate the volume rate of excess fluid that must be allowed to escape from the system to provide pressure relief and one must size the valve (relief device opening) and the related venting system.

Needed Capacity in Case of Fire

Fire is frequently the most critical threat to a system. A good understanding of the characteristics of a potential fire and of the system's behavior during fire exposure is needed in order to determine the required pressure-relieving capacity of the safety system. Many formulae have been used over the past 50 years for calculation of the vapor flow capacity required in a system exposed to fire. (Fifty-two are summarized in Appendix B of reference 1.) Some of these formulae relate the area of the system to the rate of heat transferred into the system or to the equivalent amount of vapor generated. If the formula develops an estimated heat absorption rate, the thermodynamic properties of the fluid in the system must be used to translate from heat absorption to vapor generation.

A simple fundamental equation may be used to evaluate the heat absorption potential of a system

$$Q = qFEA \qquad (7.1)$$

where Q is the rate of heat absorption by the fluid in the system, Btu/h; q the heat flux from the fire to the system, Btu/(h·ft^2); F an environmental factor to account for conditions exterior to the system, dimensionless; E a fire exposure factor—simply, the fraction of total surface exposed to the fire, dimensionless; and A an area corresponding to the wetted surface in the system, ft^2.

The heat absorption rate equation, $Q = qFEA$, provides a framework for the analysis and evaluation of the risk. Each element of the equation may be examined separately and a credible value assigned.

The heat flux from the fire to the system, q, is a function of the kind of fuel, the geometry of the fuel source relative to the exposed system, and the radiation characteristics of the flame. Experimentally determined heat flux values for free convection vary from about 13,000 to 47,000 Btu/(h·ft^2). A good approximation, used in many equations, is the value 34,500 Btu/(h·ft^2). A "clean" burning fuel such as methanol or acetone will yield a maximum flux of about half this value; a "sooty" flame will have high radiant heat-transfer capability and may yield a flux of as much as 47,000 Btu/(h·ft^2). Impinging flames result in much higher local heat fluxes; therefore, special protection is advised where impinging flames are possible because of the higher probability of local failure at the site of flame impingement.

The environmental factor, F, combines the effects of any environmental conditions which may reduce the heat flux. Examples are insulation and external water spray. To take credit for such environmental factors they must be deemed reliable, F may range from 1.0 for a system with no environmental protection to less than 0.10 for a completely insulated system.

The fire exposure factor, E, is at best a conservative estimate of the fraction of the total system surface which may be exposed to the fire. It is to be borne in mind that the system under consideration is that portion open to the relief device. If the exposure of that entire system to a fire is not credible an estimate of the portion potentially exposed must be made. Small systems may be completely exposed. E equals 1.0. Large systems may be partially exposed with E becoming a small fraction. Several time-honored formulae derate the area of a system by equating the total heat flux to the total area of the system raised to the 0.82 power. This method provides a mathematically consistent way to express the probability that the larger the system the smaller the portion of it that may be engulfed in a fire.

The total *wetted* surface area of the system, A, is determined from knowledge of the liquid-filled portion of the system. Protection of the system is provided by evaporation of liquid in the liquid-filled portion of the system. The relief device must be sized to allow the generated vapor to escape. It is important that the fire burn out or be brought under control before the unwetted surfaces reach a softening temperature. The relief devices are intended to provide the needed grace period for fire fighting. The product EA represents the fraction of the total wetted area in the system in actual contact with the fire.

Effect of Liquid–Vapor Mixtures on Capacity

Calculation of the effective discharge area of a pressure-relief device for a liquid-filled system exposed to fire is based on the rate of vapor generation caused by the heat absorbed from the fire. The rated capacity of the relief device should be greater than the rate of vapor generation, otherwise the pressure will build up to an unsafe level. When adiabatic and reversible flow may be assumed, the capacity of the relief device for gas and vapor can be calculated (see Section 7.4). The capacity of the relief device becomes uncertain when it encounters two-phase flow, which could be the case if the vapor contained entrained liquid. If the relief device, designed for vapor, is exposed to saturated liquid its volumetric capacity will generally be reduced. It will not be able to release enough liquid from the system to provide protection. Sylvander and Katz[2] examine this question, liquid versus vapor relief, for saturated liquid propane at 138°F relieving to the atmosphere. A relief valve venting flashing liquid propane (138°F, 301.4 psia) will discharge 1.87 times the mass of propane that it would discharge of vapor propane at this same inlet conditions (138°F, 301.4 psia). For equal pressure relief the relief valve, however, must pass 8.9 times as much propane if the valve receives liquid rather than vapor, the specific volume of gas at 138° and 301.4 psia being 8.9 times greater than liquid propane at the same temperature and pressure. The valve, at these conditions, is effectively derated from 100 to 21% when it "sees" liquid rather than vapor at the inlet.

Pressure-relief should be based on venting of vapor, if at all possible, with precautious taken to avoid two-phase or liquid flow to the valve.

Needed Capacity Other Than in Case of Fire

The fire hazard frequently determines the required capacity for a pressure-relief system. After the fire hazard risk is addressed other conditions requiring pressure relief should be methodically examined. Candidate conditions include misoperation and exothermic chemical reaction such as polymerizations and decompositions.

Systems containing chemicals which may undergo exothermic chemical reactions are subject to a unique hazard. A reaction, a polymerization or a decomposition, may "run away." It may liberate heat at a rate greater than can be dissipated resulting in higher temperatures, higher reaction rates, higher heat-liberation rates, higher temperatures. . . . a runaway. In this case the reaction can be quenched if the pressure rise can be halted, thus putting a ceiling on the temperature (temperature and pressure roughly correspond to the equilibrium vapor pressure of the system). The pressure relief system should be designed to quench a runaway if such a hazard exists. Styrene, vinyl chloride, butadiene, acetylene, ethylene oxide, and molten ammonium nitrate are a few examples of systems which have runaway potential.

Experimentation with "worse case" conditions may be desirable to establish relief device adequacy for chemical systems having behavior uncertainty.

7.4. SIZING OF PRESSURE-RELIEF DEVICES

By assuming adiabatic and reversible flow and by assuming also that the fluid follows the ideal gas law, the maximum mass flow rate of a safety or relief valve may be determined by application of the ASME (1968) Code:

$$W = CKAp\left(\frac{M}{T}\right)^{\frac{1}{2}} \tag{7.2}$$

In this equation, W is the mass flow rate of gas or vapor, lb/h; C a constant (with dimension) for gas or vapor, a function of the specific heats ratio, $k = c_p/c_v$; K the coefficient of discharge, dimensionless; A the actual discharge area, orifice or nozzle throat area, in.^{2}; p the upstream pressure—the set pressure multiplied by 1.10 or 1.20 (depending on the permissible pressure), plus the atmospheric pressure, psia; M the molecular weight of gas or vapor, lb/lb mole; and T the temperature of the gas at the relieving conditions, degrees R. Equation (7.2) assumes that the absolute upstream and downstream pressures permit critical flow, which is the necessary condition for the maximum flow rate. For real gases, an empirical correction can be made in Equation (7.2) by using the equation of state in the form

$$pv = ZRT \tag{7.3}$$

where v is the gas specific volume, ft³/lb; R the gas constant; and Z the compressibility factor, dimensionless. Consequently, Equation (7.2) becomes

$$W = CKAp\left(\frac{M}{ZT}\right)^{\frac{1}{2}} \tag{7.4}$$

Values of C for different k are given in Fig. 7.1.

For liquid filled systems, the vapor flow rate through the pressure-relief valve is given by

$$W = \frac{Q}{L} \tag{7.5}$$

where Q is the amount of heat absorbed by the liquid, given by Equation (7.1)

$$Q = qFEA, \qquad \text{Btu/h}$$

L is the latent heat of vaporization at relieving conditions, in Btu/lb. Assume that a given relief valve is tested with dry air at 60°F (520°R); the pressure differential is sufficient for critical flow. For this case, Equation (7.4) can be written as

$$W_a = C_a KAp\left(\frac{M_a}{Z_a T_a}\right)^{\frac{1}{2}} \tag{7.6}$$

solving for KAp,

$$KAp = \frac{W_a}{C_a}\left(\frac{Z_a T_a}{M_a}\right)^{\frac{1}{2}} \tag{7.7}$$

where $C_a = 356$ (for air, $k = 1.4$); $M_a = 28.97$; $Z_a = 1.0$; and $T_a = 520°R$. Substituting these values in Equation (7.7), one obtains

$$KAp = 0.0119 W_a \tag{7.8}$$

When this relief valve is designed for use in vapor relief on a liquid containing system, the air-rated capacity of the valve can be determined by substituting Equations (7.5) and (7.8) in Equation (7.4), as

$$W_a = \frac{Q}{0.0119LC}\left(\frac{ZT}{M}\right)^{\frac{1}{2}} \tag{7.9}$$

where W_a is the air-rated equivalent capacity, converted to pounds of air per hour at 60°F inlet temperature. This equation assumes either that K in Equa-

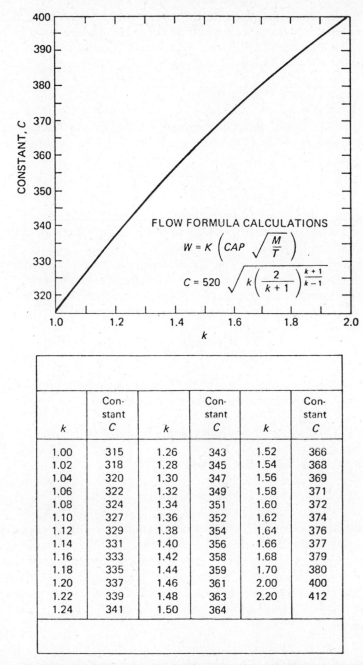

Figure 7.1. Constant C for gas or vapor related to ratio of specific heats ($k = c_p/c_v$ (Source: Figure UA230, ASME Boiler and Pressure Vessel Code, Section VIII, Unfired Pressure Vessels.)

The graph shows:

CONSTANT, C (y-axis from 320 to 400)
k (x-axis from 1.0 to 2.0)

FLOW FORMULA CALCULATIONS

$$W = K \left(CAP \sqrt{\frac{M}{T}} \right)$$

$$C = 520 \sqrt{k \left(\frac{2}{k+1} \right)^{\frac{k+1}{k-1}}}$$

k	Constant C	k	Constant C	k	Constant C
1.00	315	1.26	343	1.52	366
1.02	318	1.28	345	1.54	368
1.04	320	1.30	347	1.56	369
1.06	322	1.32	349	1.58	371
1.08	324	1.34	351	1.60	372
1.10	327	1.36	352	1.62	374
1.12	329	1.38	354	1.64	376
1.14	331	1.40	356	1.66	377
1.16	333	1.42	358	1.68	379
1.18	335	1.44	359	1.70	380
1.20	337	1.46	361	2.00	400
1.22	339	1.48	363	2.20	412
1.24	341	1.50	364		

tion (7.8) is independent of pressure p or that the air-rated capacity is determined under the pressure conditions of the equivalent vapor flow.

Air equivalent rate of discharge in cubic feet of air per minute at standard conditions (60°F and 14.7 psia) is given by

$$(cfm)_a = W_a v_0/60, \tag{7.10}$$

where $v_0 = 13.1$ ft³/lb, specific volume of air at 60°F, 14.7 psia.

Substituting Equation (7.9) in Equation (7.10)

$$(cfm)_a = \frac{18.34Q}{LC}\left(\frac{ZT}{M}\right)^{\frac{1}{2}} \tag{7.11}$$

When the amount of heat to be absorbed by the liquid in the system, Q, is known, air equivalent capacity of the pressure relief valve can be calculated from Equation (7.11).

If the specific heats ratio k or the isentropic expansion coefficient is not known, it is suggested that C be set at 315, the smallest value in Fig. 7.1. This may result in somewhat larger relief valve size, which is not objectionable. In addition, if $Z = 1$ is used, Equation (7.4) becomes

$$W = 315KAp\left(\frac{M}{T}\right)^{\frac{1}{2}} \tag{7.12}$$

where A is the discharge area, in.² and p is the upstream pressure, psia.

For saturated steam at atmospheric pressure,

$$T = 672°R \qquad M = 18.016$$

Substituting these values in Equation (7.12),

$$W_s = 51.5KAp \tag{7.13}$$

This equation is the ASME (1968) Code formula for official capacity rating of steam pressure relief valves. The rated capacity W_s is given in pounds of steam per hour. At atmospheric conditions, the compressibility factor of saturated steam is slightly less than unity. The coefficient C, on the other side, is 350 ($k = 1.329$ for low-pressure steam), rather than the 315 used in Equation (7.12). This is not objectionable, since it allows about 10% more discharge area.

7.5.   RELIEF VENT LOCATION AND VENT SYSTEM DESIGN

Injection Into Air

When pressure relief is obtained by venting fluid, the discharge system should be so engineered that the gases and vapors are dissipated to harmless concentrations as rapidly as possible. This is especially important in the case of substances of highly flammable or toxic nature. The venting system should lead the vapors well away from personnel and from nonventilated areas which could collect a "pocket" of the vapors being discharged.

Another consideration in the design of nozzles or discharge piping systems is the avoidance of any possible impingement upon other parts of the system being protected, or any adjacent system. Such impingement with flammable vapors may lead to an impinging flame and attendant very high local heat fluxes several times higher than in normal free-burning fires. Such impinging flame heat fluxes may lead to film boiling inside the system and local blowouts due to metal softening.

The purpose of emergency pressure relief is to prevent, or at least minimize, the possibility of catastrophic system failure. The emergency pressure relief results in small releases which can be handled safely, being planned for; the catastrophic failure would greatly increase the exposure and risk.

Burning of Vented Vapors

Burning of nontoxic flammable vapors is attractive in some environment protection situations. The vent system must be designed and operated to avoid flammable vapor—air mixtures *inside* the relief system. Such mixtures may result in flashback and system rupture. Manifolding of pressure relief discharges into a common header must be carefully designed to avoid cross contamination and to ensure compatibility.

Security of the Discharge Piping System

The pressure relief device must operate as designed to pass the rated volume of vapor from the system. That vapor must now be securely transported inside the discharge piping system to its properly planned exit. Unfortunately, many accidents happen because an improperly designed and secured discharge piping system failed. If the discharge piping is not securely mounted it may break loose and "whip" under the high thrust of the venting gases thus releasing vapors at unplanned and hazardous locations.

Poisonous and Toxic Materials

When especially toxic substances are involved, release of even small amounts may lead to serious consequences. Burning with attendant clean-up of the combustion products, chemical scrubbing, adsorption, and total containment for subsequent disposal are possible approaches for handling of pressure-relief vent discharges.

7.6. INSPECTION, TESTING, MAINTENANCE, AND REDUNDANCY

Pressure-relief devices must be highly reliable. Life and investment depend upon their proper functioning. Plant operations must include a planned program combining inspection, testing, and preventive maintenance to ensure reliable pressure-relief device performance. In critical systems it will be found that redundancy in relief-device installation is needed to achieve the desired certainty of protection.

7.7. SAFETY CHECKLIST SAMPLE QUESTIONS

Pressure relief devices may be depended upon to offer protection to personnel and to equipment if these devices are properly sized, selected, and maintained *and* if their utilization is accompanied by a complete plant safety program. A checklist of safety related questions provides a relatively straight forward and fast way to audit a plant safety program. A safety checklist should be prepared tailored to each specific plant (the checklist questions will thus all be relevant and meaningful). Some sample questions follow, in three categories: facility, process fluids and solids, and people.

Facility

1. Is the site well situated with regard to topography and adequate drainage?
2. Will the climate and related factors materially affect operations? (earthquake, floods, fog, hurricane, lightning, smog, snow, tornados, excessive heat, and very low temperatures)
3. Will toxic fumes from fire, explosion, or other accidents affect the surrounding community?
4. Are major highways, airports, or congested areas located near the site? Is there unusual risk from low-flying aircraft during pressure relief venting? Can emergency equipment get through traffic at all times?
5. Are utilities dependable? (water, gas, electricity, compressed air, etc.)
6. Does the community provide adequate fire-fighting personnel and equipment? Is emergency personnel well trained?
7. Does the community provide adequate ambulance, hospital, and police protection? Do plant or local nurses and physicians understand the specific medical treatment for exposure to substances likely to be encountered?
8. Can air, river, landfill, and sewage systems be used for waste disposal without violating local or Federal ordinances or the health and welfare of the surrounding community?
9. Do nearby facilities present fire, explosion, or toxic hazards?

10. Are there any nearby sources of toxic or flammable vapors?

11. Is the site enclosed by adequate fences and gates?

12. Are process areas separated from utilities, storage, office, and laboratory areas?

13. Do all buildings conform to the National Building Code? To local code?

14. Are structural (steel) members and supports insulated so as to be fire resistive?

15. Are buildings that are exposed to explosion hazards vented for relief according to standards?

16. Are all buildings properly ventilated to limit exposure of personnel to toxic substances and to reduce hazard from flammable substances?

17. Are there sufficient and clearly marked exits, stairwells, or escape chutes in all buildings?

18. Are drainage facilities in buildings adequate? Where do they ultimately discharge?

19. Are chemical reactions possible in the drains—for example, acids plus cyanides or sulfides?

20. Are hazardous units separated from all critical areas, such as control room or process computer installations?

21. Does spacing of equipment consider the nature of the product, the quantity, the operating conditions, the sensitivity of the equipment, the need to combat fires, and the concentration of valuables?

22. Are waste-disposal systems downwind from concentrations of personnel?

Process Fluids and Solids

1. Have the quantities of materials in process in all stages of handling, transport, and storage and in all physical states been considered in relation to the hazards of fire, explosion, toxicity, pollution, and corrosion?

2. Have the pertinent physical properties of each process chemical been determined? (melting point, boiling point, vapor pressure, particle size, etc.)

3. Have the chemical properties of each process chemical been classified? (especially chemical reactivity with other products.)

4. Have highly hazardous materials been identified and their location in the plant determined?

5. Are the process chemicals toxic? Have threshold limit values and acute short-term limits been established?

6. Is the material corrosive? (skin, metal, plastics, and materials of construction.)

7. Have the effects of impurities been taken into account as related to fire and explosion, toxicity, corrosivity, and stability of the product?

8. Are overpressure protection devices adequate in size and construction?

9. Have the overpressure protection devices been properly set and sealed? By whom? When?

10. Have the potential hazards of all process chemicals been evaluated?

11. Are precautionary measures taken to guard against accidental release of flammable or toxic liquids, gases, or combustible dusts?

12. Are unstable chemicals handled in a way to minimize exposure to heat, pressure, shock, or friction?

13. Have all heat-transfer operations been properly evaluated for hazards?

14. Are waste disposal and air pollution problems handled in accordance with current regulations?

People

1. Has an adequate "Standard Operating Procedure" manual been prepared and understood in detail by all personnel? Is it reviewed periodically and reviewed when process changes are made? Is a copy at hand?

2. Are there adequate personnel job training programs? Do they cover both supervisory and operating personnel?

3. Have adequate start-up and shut-down programs been initiated?

4. Do operations include a "permit" system for hazardous jobs? Is it enforced?

5. Are qualified technical personnel available to monitor or checkout work in confined spaces, gas-freeing or inerting, and related entry procedures?

6. Are personnel trained to recognize potential equipment or facilities malfunctions?

7. Are employees trained to handle emergency situations?

8. Are operators trained in the utilization as well as limitations of protective equipment (face, head, eye, and respiratory protection)? Is full suit available if needed?

9. Are emergency procedures understood and all situations covered?

10. Are personnel actually able to use emergency equipment without hesitation?

11. Are facilities for control of spills adequate? On all shifts?

12. Are spills promptly reported to both on-site and off-site authorities? Are procedures for reporting plainly displayed and understood by all personnel?

REFERENCES

1. *Pressure-Relieving Systems for Marine Bulk Liquid Containers.* National Academy of Sciences, Washington, D.C., 1973.
2. N. E. Sylvander and D. L. Katz. "The Design and Construction of Pressure Relieving Systems," *Engineering Research Bulletin* No. 31, University of Michigan, Ann Arbor, 1948.

8

Design and Inspection of Pressure Vessels

Elliott MacDermod

8.1. DEVELOPMENT OF THE BOILER AND PRESSURE VESSEL CODES

The need for a boiler and pressure-vessel code[1,2] was demonstrated during the latter part of the nineteenth century and the first decade of the twentieth century when some 10,000 boiler explosions killed an equal number of people and seriously injured 15,000 others.

Massachusetts issued the first rules for the construction and installation of boilers in 1907 following disastrous explosions in 1905 and 1906. Other states and cities followed the example of Massachusetts and enacted laws or ordinances for the construction, installation, and inspection of steam boilers with resulting chaos. The lack of uniformity in requirements of the different states did not allow a manufacturer to build stock boilers that would be acceptable in other states, second-hand boilers could not be shipped across state lines, and a qualified boiler inspector in one state was not recognized in others.

The American Society of Mechanical Engineers (ASME) was requested to formulate standard specifications for the construction of steam boilers and pressure vessels, and in 1911 a committee, later known as the Boiler Code Committee, was formed.

Editors' note:

Despite the unfortunate demise of Elliott MacDermod, the editors have included his chapter from the first edition. Some effort has been made to update the text and references but the tables are as he assembled them in the early sixties.

The ASME Unfired Pressure Vessel Code gives no definition of a pressure vessel. A pressure vessel is usually considered to be a closed container of a fluid under pressure used to perform some process function such as a storage tank, heat exchanger, evaporator, or reactor. Code jurisdiction over piping external to the vessel terminates at the first circumferential seam or joint for welding end connections; the face of the first flange for bolted connections; and at the first threated joint in that type of connection.

Paragraph U-1 of the Unfired Pressure Vessel Code lists the exceptions from code jurisdiction, however the "Synopsis of the Boiler and Pressure Vessel Rules and Regulations" and the local authority having jurisdiction should be consulted for further details as to requirements and exemptions which may vary widely from place to place.

A noncode pressure vessel may be defined as a vessel not meeting the minimum construction requirements of the code for design, fabrication, inspection, and certification. It could not be stamped with the code symbol and could not be installed in a jurisdiction that had adopted the ASME Code unless some special ruling was in effect.

8.2.   THE ASME BOILER AND PRESSURE VESSEL CODE AND THE NATIONAL BOARD OF BOILER AND PRESSURE VESSEL INSPECTORS

Those in the chemical industry associated with the procurement and operation of pressure vessels should have a thorough understanding of the functions and relationship of the ASME Boiler Code Committee and the National Board of Boiler and Pressure Vessel Inspectors. For that reason, excerpts from the ASME Code Foreword and the NBBPVI Code of Practice are reproduced here, as being the least known or most frequently overlooked or misunderstood functions of the two organizations.

Excerpts from the ASME Code Foreword

The Boiler and Pressure Vessel Committee's function[3] is to establish rules of safety governing the design, the fabrication; and the inspection during construction of boilers and unfired pressure vessels, and to interpret these rules when questions arise regarding their intent. In formulating the rules, the committee considers the needs of users, manufacturers, and inspectors of pressure vessels. The objective of the rules is to afford reasonably certain protection of life and property and to provide a margin for deterioration in service so as to give a reasonably long, safe period of usefulness. Advancements in design and material and the evidence of experience have been recognized.

The Boiler and Pressure Vessel Committee meets regularly to consider requests for interpretations and revisions of the rules. Inquiries must be in writing and must give full particulars in order to receive consideration.

Interpretations of general interest are published in *Mechanical Engineering* as "Code Cases," and inquirers are advised of the action taken. Code revisions approved by the committee are published in *Mechanical Engineering* as proposed addenda to the code to invite comments from all interested persons. After final approval by the committee and adoption by the ASME Council, they are printed in the addenda supplements to the code.

Code cases (interpretations) may be used in the construction of vessels to be stamped with the ASME Code symbol beginning with the date of their approval by the ASME Council.

Manufacturers and users of pressure vessels are cautioned against making use of revisions and cases that are less restrictive than former requirements without having assurance that they have been accepted by the proper authorities in the jurisdiction where the vessel is to be installed.

After code revisions are approved by council they may be used beginning with the date of issuance shown on the pink-sheet addenda. Revisions become mandatory as minimum requirements six months after such date of issuance, except for boilers or pressure vessels contracted for prior to the end of the six-month period.

The National Board of Boiler and Pressure Vessel Inspectors is composed of chief inspectors of states and municipalities in the United States and of provinces in the Dominion of Canada that have adopted the Boiler and Pressure Vessel Code. This board, since its organization in 1919, has functioned to uniformly administer and enforce the rules of the Boiler and Pressure Vessel Code. The cooperation of that organization with the Boiler and Pressure Vessel Committee has been extremely helpful. Its function is clearly recognized and, as a result, inquiries received which bear on the administration or application of the rules are referred directly to the National Board. Such handling of this type of inquiry not only simplifies the work of the Boiler and Pressure Vessel Committee, but action on the problem for the inquirer is thereby expedited. Where an inquiry is not clearly an interpretation of the rules, or a problem of application or administration, it may be considered both by the Boiler and Pressure Vessel Committee and the National Board.

It should be pointed out that the state or municipality where the Boiler and Pressure Vessel Code has been made effective has definite jurisdiction over any particular installation. Inquiries dealing with problems of local character should be directed to the proper authority of such state or municipality. Such authority may, if there is any question or doubt as to the proper interpretation, refer the question to the Boiler and Pressure Vessel Committee.

The specifications for materials given in Section II of the code are identical with or similar to those of the American Society for Testing Materials

(ASTM) as indicated, except in those cases where that organization has no corresponding specification.

Excerpts from NBBPVI Code of Practice

Protection of life and property is the first duty of any government. Many governmental subdivisions in the United States and Canada, recognizing this duty, have adopted mandatory laws requiring that the design, fabrication, and inspection of boilers and unfired pressure vessels comply with the rules of The American Society of Mechanical Engineers' Boiler and Pressure Vessel Code[4] as administered by The National Board of Boiler and Pressure Vessel Inspectors (National Board).

The development of rules and codes and the construction of safe boilers and unfired pressure vessels, confirmed by competent inspection, is a three-pronged endeavor:

1. Voluntary cooperation of the ASME Boiler and Pressure Vessel Committee which writes the rules;
2. Boiler and pressure-vessel manufacturers who comply with the rules; and
3. The National Board which issues commissions to inspectors who diligently enforce the rules.

Boilers and unfired pressure vessels which meet these requirements in every respect proudly bear the symbols of approval in the form of the ASME and National Board stamps.

For a manufacturer to become a fabricator of code vessels he must first apply for the official symbol of the ASME and authorization for its use for a specified period. Form letters for this purpose are included in the Codes as is an illustration of the Certificate of Authorization. After obtaining the authorization, the manufacturer must make arrangements for shop inspections, either with an insurance company or with the local authority.

The ASME Boiler and Pressure Vessel Committee—a voluntary group of engineers organized in 1911—is composed of designers, manufacturers, users, and insurance and inspection authorities, all of whom are interested in developing and maintaining a safety code for boilers and unfired pressure vessels.

The National Board of Boiler and Pressure Vessel Inspectors is a voluntary organization of officials who are charged with the enforcement of boiler and pressure-vessel inspection regulations of any political subdivision of the United States and Canada that has adopted one or more sections of the ASME Boiler and Pressure Vessel Code. The National Board was organized in 1919 for the purpose of promoting greater safety to life and property by securing concerted action among the states, cities, and provinces; and uniform inspection practices and consistency in the construction, installation, and inspection of safe boilers and pressure vessels. National Board inspectors' commissions have been issued to over 3000 individuals who have met and maintained the board's high standards of qualification.

In order to retain the confidence necessary to maintain the integrity of the codes and the inspection authorities, complete cooperation must be exercised by all concerned. Any deviation from the codes would tend to discredit them as a safety measure.

8.3. SECTIONS OF THE ASME BOILER AND PRESSURE VESSEL CODE

Section I	Power Boilers
Section II	Material Specifications
Section III	Boilers of Locomotives (discontinued) Now Nuclear Vessels
Section IV	Heating Boilers
Section V	Nondestructive Examination
Section VI	Heating Boilers
Section VII	Power Boilers
Section VIII	Pressure Vessels
Section IX	Welding and Brazing Qualifications
Section X	Fiberglass Reinforced Vessels
Section XI	Inspection of Nuclear Power Plant Components

8.4. STATUS OF THE ASME CODE IN THE UNITED STATES AND CANADA

While much effort has been expended to obtain uniformity in Boiler and Pressure Vessel Laws and Regulations, there still exist wide differences in practice.[5] A chemical industry contemplating construction of a plant in another state may obtain information concerning that state's rules and regulations in the *Synopsis of Boiler and Pressure Vessel Laws, Rules, and Regulations,* published by the National Bureau of Casualty Underwriters. Tables 8.1, 8.2, and 8.3 are based on the synopsis which gives additional information concerning:

Department having jurisdiction and address
Date of law
Rules of construction and stamping
Objects subject to rules

TABLE 8.1. Status of ASME Sections I and VIII in the United States

State	Section I	Section VIII	State	Section I	Section VIII
Alabama		L	Nebraska	L	L
Alaska		L	Nevada	L	L
Arizona	L	L	New Hampshire	L	NJ
Arkansas	L	A	New Jersey		L
California			New Mexico		
Colorado			New York	L	NJ
Connecticut	L	NJ	North Carolina	L	L
Delaware	L		North Dakota	O	NJ
District of Columbia	L	L	Ohio (SI)		O
Florida			Oklahoma	L	NJ
Georgia			Oregon	L	L
Hawaii	L	A	Panama Canal Zone	LX	LX
Idaho			Pennsylvania (SI)	L	L
Illinois	L	NJ	Puerto Rico	L	L
Indiana	L	L	Rhode Island	L	A
Iowa	L	A	South Carolina		
Kansas	L	NJ	South Dakota		
Kentucky			Tennessee	L	A
Louisiana	L	NJ	Texas	L	A
Maine	L	A	Utah	L	L
Maryland	L	NJ	Vermont	L	L
Massachusetts	L	L*	Virginia	L	L
Michigan	L	A	Washington	L	L
Minnesota	L	L	West Virginia	L	A
Mississippi	NJ	L**	Wisconsin	A	NJ
Missouri			Wyoming		
Montana					

L — Legal requirement
A — Acceptable
NJ — No jurisdiction
* — Compressed air tanks only

X — Under Coast Guard jurisdiction if on floating equipment
** — LPG containers and equipment
O — Ohio Rules
SI — Shop Inspectors must have state commission in addition to National Board

TABLE 8.2.   Status of ASME Code Sections I and VIII in Canada

	Section	
	I	*VIII*
Alberta	L	L
British Columbia	L	L
Labrador		
Manitoba	L	L
New Brunswick	A	L
Newfoundland	L	L
Northwest Territories	L	L
Nova Scotia	L	L
Ontario	L	L
Prince Edward Island		
Quebec	L	L
Saskatchewan	L	L
Yukon Territories	L	L

Objects subject to field inspection

Inspection required

Insurance company requirements

Certificate of inspection

State inspection fees

8.5.   BOILER AND PRESSURE VESSEL REQUIREMENTS IN FOREIGN COUNTRIES

Many chemical industries have or may contemplate foreign manufacturing facilities.[6] Table 8.4. lists the legal standing and administration of pressure-vessel codes in various countries.

8.6.   THE PRESSURE-VESSEL DESIGNER

The ASME Unfired Pressure Vessel Code is not a design manual and does not contain rules to cover all details of design and construction. The code covers only the minimum requirements for design, fabrication, and inspection of unfired pressure vessels. Whether these minimum requirements are suitable for a vessel's service conditions or whether more exacting criteria is indicated is the responsibility of the pressure-vessel designer. The importance of design and the views of some pressure vessel designers are contained in references 7 and 8.

TABLE 8.3. Status of ASME Code Sections I and VIII in Cities and Counties of the United States

	Section I	Section VIII		Section I	Section VIII
Albuquerque, N. Mex.	L		Niagara Falls, N. Y.	L	NJ
Barton, Fla.	I and I	I and I	Niles, Mich.	L	NJ
Buffalo, N. Y.	L	L	North Miami Beach, Fla.	L	NJ
Chicago, Ill.	L	L	Oklahoma City, Okla.	L	NJ
Dade County, Fla.	L	L	Omaha, Neb.	L	L
Dearborn, Mich.	L	L	Oswego, N. Y.	L	NJ
Denver, Col.	L	NJ	Peoria, Ill.		
Des Moines, Iowa	L		Phoenix, Ariz.	L	L
Detroit, Mich.	L	L	Pueblo, Col.		
East St. Louis, Ill.	L	L	Richmond, Va.	L	L
Evanston, Ill.	L	L	St. Joseph, Mo.	L	L
Greensboro, N. C.	L	L	St. Louis, Mo.	L	L
Jefferson Parrish, La.	L	L	St. Louis County, Mo.	L	L
Kansas City, Mo.	L	L	San Francisco, Cal.	L	L
Los Angeles, Cal.	L	L	Seattle, Wash.	L	L
Memphis, Tenn.	L	L	Spokane, Wash.	L	L
Miami, Fla.	L	L	Tacoma, Wash.	L	L*
Miami Beach, Fla.	I and I	I and I	Tampa, Fla.	L	L
Milwaukee, Wis.	L	L	Tulsa, Okla.	L	L
Mt. Vernon, N. Y.	L	L	University City, Fla.	L	NJ
New Orleans, La.	L	L	Yonkers, N. Y.	L	NJ
New York, N. Y.	L	NJ			

I and I — Covers field Inspection and insurance
L — Legal requirement

* — Air tanks
NJ — No jurisdiction

126

The design of the many shapes, sizes, and types of pressure vessels used in chemical plants requires the skills of an experienced engineer specializing in this field. Vessel design requires a thorough knowledge of the code requirements, the properties of the fluid being contained, the properties of suitable materials of construction, and the problems involved in the fabrication of these materials. Design factors may be so varied that the engineer can only rely on his or her engineering judgment and experience with similar problems to produce a safe economical design that will last the intended life of the equipment.

Due to space limitations, no attempt will be made here to discuss or interpret the requirements of the pressure-vessel code in detail. Figure 8.1 illustrates the various features of pressure-vessel construction referenced to the applicable code paragraphs. The table of references is arranged by subject for those whose interest may require more detailed information.

8.7. BASIS FOR DESIGN

Pressure and Temperature

The design pressure is customarily set at 15 psi or 10%, whichever is the greater, over the operating pressure for vessels without cyclic swings. For the latter the design pressure is usually 5 to 10% above the highest pressure anticipated. The design temperature is usually 50°F above the maximum temperature that will not result in a decrease of the code-allowable stress which would cause an increase in thickness.

For example, a carbon steel vessel operating at 350°F should have a design temperature of 650°F, not 400°F, as the allowable stress is the same for either temperature. If the vessel is code stamped at the lower temperature, and at some future date it is desirable to raise the operating temperature, complications arise in having the code stamp changed to the desired increased temperature. For the same reason the vessel should be stamped with the maximum allowable working pressure based on the nominal thicknesses of the parts rather than the design pressure used to calculate the minimum thicknesses.

Minimum Plate Thickness

The plate thickness of large-diameter low-pressure vessels is often based on arbitrary plate thicknesses rather than on calculated thickness based on design pressure. This results in a calculated maximum allowable working pressure that may be much greater than the design pressure. The reason for using the heavier plate is that large-diameter thin-plate shells are difficult to handle in the shop without spiders or rings to hold the shell sections cylindrical. The purchaser benefits by paying for and obtaining a heavy tank rather

TABLE 8.4. Legal Standing and Administration of Pressure-Vessel Codes in Various Countries

Country	Laws Relating to Construction of Boilers and Pressure Vessels	National Code — Title	National Code — Scope	Legal Standing of Code	Code Writing Body	Administration — Organization Responsible for Code Administration	Administration — Submission of Drawings	Acceptability of Imported Vessels
Australia	The Factory and Machinery Acts of the various states (e.g. the Factories & Shops Act 1912–1957 of New South Wales) under which boiler and pressure-vessel regulations are issued	Standards Association of Australia Boiler Code I'ts. I to V	Boilers and unfired pressure vessels	Code has the force of a state regulation in most states and is acceptable, with modifications, in the remainder	Standards Association of Australia	State inspecting authority	Drawings must be approved by state inspecting authority before starting construction	Vessels fabricated outside Australia may be accepted at the discretion of the State Chief Inspector
Austria	Art. 48 Verwaltungsentlastungsgesetz (VEG) Bundesgesetzblatt No. 277/1925 as amended by BGBl No. 55/1948	Dampfkesselverordnung (DKV) BGBl No. 83/1948 and Werkstoff und Bauvorschriften (WBV) BGBl No. 264/1949	Boilers and unfired pressure vessels	Codes have the force of a state regulation	Dundesministerium fur Handel und Wideraufbau	Government or government-licensed inspectors	Drawing approvals recommended but not mandatory	Vessels must meet the requirements of the Austrian Code

Country	Code	Scope	Equipment	Legal status	Issuing body	Inspecting authority	Remarks
Belgium	La Reglement General pour la Protection du Travail. La Reglementation des Appareils a vapeur	None. A code is currently being written by N.B.N. Most major national codes are acceptable in Belgium at the present time				Ministere du Travail, Inspection technique	Vessels built to major national codes are acceptable
Canada	The Boiler and Pressure Vessel Acts of the various Provinces	C.S.A. Standard B51-1957. This standard specifies the use of the ASME Code, with some amendments	Boilers and unfired pressure vessels	C.S.A. Standard B51-1957 (and hence the ASME Code) has been adopted by all Provinces. It has the force of a Provincial regulation	Canadian Standards Association and American Society of Mechanical Engineers	Provincial inspecting authority	Vessels must be inspected by an inspector holding the Canadian National Board Commission. Drawings must be approved by the Chief Inspector of the Province before starting construction
Finland	Finlands For-fattnings-samling No. 573–575, which constitutes the pressure-vessel code	Dimensioning, Materials and Welding of steel pressure-vessels	Boilers and pressure vessels	Code is a state regulation			

TABLE 8.4. (Continued)

Country	Laws Relating to Construction of Boilers and Pressure Vessels	National Code		Legal Standing of Code	Code Writing Body	Administration		Acceptability of Imported Vessels
		Title	Scope			Organization Responsible for Code Administration	Submission of Drawings	
France	The laws, decrees, and circulars collected and published as Reglementation des Appareils a Vapeur and Reglementation des Appareils a Pression de Gaz	S.N.C.T. No. 1 (published by Syndicat National de Chaudronnerie et Tolerie)	Unfired pressure vessels (There is no national code for boilers)	Only legal requirements are those collected in the two documents given in column 2. These do not constitute a code as normally understood. The S.N.C.T. Code has no direct legal recognition	S.N.C.T.	Service des Mines is responsible for verifying that boilers and pressure vessels are in accordance with legal requirements	Not normally required	Vessels must conform with French law and in particular Article 4 of the decree of April 2, 1926, which specifies the tests and certificates which are required
Germany	Various decrees of the Ministry of Labour	Dampfkessel-Bestimmungen (currently replacing Werkstoff und Bauvorschriften fur Dampfkessel) (W.U.B.) AD-Merkblätter	Boilers Unfired pressure vessels	Dampfkessel-Bestimmungen, those parts of W.U.B. still applicable, and the AD-Merkblatter have the force of law	Technische Überwachungs Vereine (T.U.V.)	Technische Überwachungs Vereine (T.U.V.)	Drawings must be submitted to T.U.V. before construction is started	Vessels, including those manufactured outside Germany, must be inspected by T.U.V., A.S.M.E. Code generally accepted for operating temperature below 650°F.

Country	Regulation	Code	Scope	Code status	Board	Authority	Drawings	Additional requirements
Holland	Stoombesluit dated 22.12. 1953 (boilers) and Outwerp Drukhouder-besluit dated 15.3.1960 (unfired pressure vessels)	Grondslagen waarop de beo-ordeling van de constructie en het materiaal van stoomtoe-steelen, damp-toestellen en drukhouders-berust	Boilers and unfired pressure vessels	The Code "Grondslagen . . ." has the force of a state regulation	Dienst voor het Stoomwezen	Dienst voor het Stoomwezen	Drawings must be submitted to Stoomwezen before starting construction	Imported vessels must meet requirements of Stoomwezen
India	Indian Boilers Act 1923	Indian Boiler Regulations, 1950	Boilers	Code is a regulation under the Indian Boilers Act, 1923	Central Boilers Board	Chief Inspector of Boilers, India	Submission of drawings may be required by approved inspecting authority before starting construction. Drawings of completed boiler must be submitted to Chief Inspector of Boilers, India and approved before boiler is registered and used	Design must be approved by Chief Inspector of Boilers. Inspection during manufacture is delegated to an approved inspecting authority, e.g. Lloyds, which must submit inspection certificate to Chief Inspector of Boilers, India, before vessel can be approved, registered, and used

TABLE 8.4. *(Continued)*

Country	Laws Relating to Construction of Boilers and Pressure Vessels	National Code		Legal Standing of Code	Code Writing Body	Administration		Acceptability of Imported Vessels
		Title	Scope			Organization Responsible for Code Administration	Submission of Drawings	
Israel	Factories Ordinance Sections 31, 32, and 33. This law is almost identical with the relevant sections of the British Factories Act	None. A.S.M.E., British, Austrian and German Codes are commonly used and all are accepted				No national inspection authority responsible for the acceptance of new construction. Periodic inspection of boilers, however, is carried out by inspectors authorized by the Chief Inspector of Labour	Not required	Vessels constructed to American, British, Austrian, and German Codes are acceptable
Italy	Various royal and governmental decrees dating from September 9, 1926 which, together with a number of A.N.C.C. circulars, form the A.N.C.C. Code	A.N.C.C.	Boilers and unfired pressure vessels	The A.N.C.C. Code constitutes the law relating to boilers and pressure vessels. The new proposed Code published as "Proposta di Nuova Regolamentazione della verifiche	A.N.C.C.	A.N.C.C.	Not required	Imported vessels must meet the A.N.C.C. requirements

				"di Stabilita degli Apparecchi a Pressione" has not yet received legal sanction and is applicable only to certain classes of vessel not covered by the original code, e.g. refinery vessels			
New Zealand	Boilers, Lifts and Cranes Act, 1960	Boilers and unfired pressure vessels	New Zealand Boiler Code. New Zealand Fusion Welded Pressure Vessels Code	The code has the force of a governmental regulation	Marine Department, New Zealand	Marine Department, New Zealand	Drawings must be submitted and approved prior to use, preferably prior to construction	Design must be approved by Marine Department. Inspection during manufacture not required. Final inspection prior to certification and use by Engineer Surveyor of Marine Department. Certificate of Workmanship must be submitted by inspecting authority (e.g. Lloyd's) for vessels manufactured outside New Zealand

TABLE 8.4. (Continued)

Country	Laws Relating to Construction of Boilers and Pressure Vessels	National Code Title	Scope	Legal Standing of Code	Code Writing Body	Administration Organization Responsible for Code Administration	Submission of Drawings	Acceptability of Imported Vessels
Norway	Kongelig resolusjon av 29 September 1925, Kjelsforskrifter (Boiler Laws)	Norwegian Code under revision. Norway uses Swedish Tryck-karlsnormer, German and A.S.M.E. Codes	Boilers and unfired pressure vessels	When current the Norwegian Code has the force of law	Arbeidstilsynets Kjelkontroll, in cooperation with technical committee appointed by Ministry of Labour	Arbeidstilsynets Kjelkontroll.	Drawings must be submitted and approved before starting construction	Imported vessels must be acceptable to Kjelkontroll.
Pakistan	Pakistan Boilers Act, 1923	Pakistan Boiler Regulations, 1951, are currently under revision				Chief Inspectors of Boilers of East Pakistan, West Pakistan, and federal capital		Until revision of regulations is complete, boilers constructed to other codes will be acceptable subject to modification of construction and working pressure at the discretion of the Chief Inspector of Boilers

Country	Legislation	Code	Scope	Status	Authority	Requirements
South Africa	Mines and Works Act, No. 27 of 1956 and the Factories, Machinery and Building Works Act, 1941	None. South Africa uses B.S.1500, the A.S.M.E. Code and the German Codes			Department of Mines is broadly responsible for boiler and pressure-vessel construction. Detailed supervision carried out by inspection authorities such as Lloyd's	Governmental authorization or inspection of imported vessels is not normally required
Sweden	Arbetarskyddslagen: Svensk Forfattningssamling No. 1/1949 (Law for the Protection of Workers) and Arbetarskyddskungorelsen: Svensk Forfattningssamling No. 476/1956 (Notice for the Protection of Workers)	Tryckkarlsnormer	Boilers and unfired pressure vessels	Code has no legal status but has been accepted by Workers Protection Board and is in practice mandatory	Pressure Vessel Commission of the Swedish Academy of Engineering Sciences	Drawings of welded vessels must be approved by welding supervisor authorized by W.P.B.
		Angpanneformer	Boilers only		National Board of Workers Protection (W.P.B.)	Vessels must meet the requirements of the Swedish Code
		Pannsvetsnormer	Boiler Welding Code (applicable also to pressure vessels)			

TABLE 8.4. (Continued)

| Country | Laws Relating to Construction of Boilers and Pressure Vessels | National Code | | Legal Standing of Code | Code Writing Body | Administration | | Acceptability of Imported Vessels |
		Title	Scope			Organization Responsible for Code Administration	Submission of Drawings	
Switzerland	Federal Regulations dated April 9, 1925 and March 19, 1938	Regulations of Swiss Association of Boiler Proprietors	Boilers and unfired pressure vessels	Code has the force of a federal regulation	Swiss Association of Boiler Proprietors	Swiss Association of Boiler Proprietors	Drawings must be submitted before construction is started	Vessels manufactured outside Switzerland must conform to Swiss regulations
Turkey		Draft Code based on German rules has been prepared but is not yet in force	Unfired pressure vessels		Turkish Institution of Mechanical Engineers			

Country	Laws / Regulations	Codes / Standards	Scope	Legal status	Issuing body	Inspection authority	Drawings	Imported vessels
United Kingdom of Great Britain	Sections 29, 30, and 31 of Factories Act, 1937	Lloyd's Rules	Boilers and unfired pressure vessels	None. Safety is the responsibility of the user who must ensure that steam boilers are of "good construction, sound material, adequate strength, and free from patent defect" (Section 29 of the Factories Act)	Lloyd's Register of Shipping	Lloyd's Register of Shipping	Drawings must be submitted and approved before construction is started	Regulations as to the acceptability of imported vessels are lacking
		Rules of the Associated Offices Technical Committee (A.O.T.C. Rules)	Boilers and unfired pressure vessels		A.O.T.C.	A.O.T.C.		
		B.S.1113 B.S.1500	Boilers Welded pressure vessels		British Standards Institution	None	Drawings approved by purchaser	
United States of America	Various states and municipal regulations	A.S.M.E. Code	Boilers and unfired pressure vessels	In those states and municipalities where it has been adopted, the code has the force of law	American Society of Mechanical Engineers	Inspection authorities of states and municipalities adopting code	Not required by state or municipality. Authorized Code Inspector examines drawings	Imported vessels subject to approval by individual local authorities and are normally acceptable if they meet A.S.M.E. Code

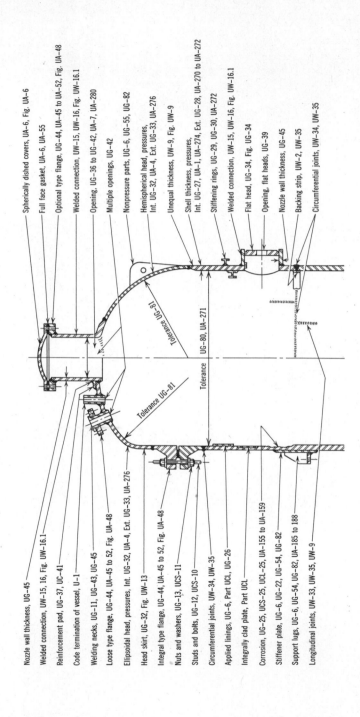

Spherically dished covers, UA–6, Fig. UA–6
Full face gasket, UA–6, UA–55
Optional type flange, UG–44, UA–45 to UA–52, Fig. UA–48
Welded connection, UW–15, UW–16, Fig. UW–16.1
Opening, UG–36 to UG–42, UA–7, UA–280
Multiple openings, UG–42
Nonpressure parts, UG–6, UG–55, UG–82
Hemispherical head, pressures, Int. UG–32, UA–4, Ext. UG–33, UA–276
Unequal thickness, UW–9, Fig. UW–9
Shell thickness, pressures, Int. UG–27, UA–1, UA–274, Ext. UG–28, UA–270 to UA–272
Stiffening rings, UG–29, UG–30, UA–272
Welded connection, UW–15, UW–16, Fig. UW–16.1
Flat head, UG–34, Fig. UG–34
Opening, flat heads, UG–39
Nozzle wall thickness, UG–45
Backing strip, UW–2, UW–35
Circumferential joints, UW–34, UW–35

Tolerance UG–81
UG–80, UA–271
Tolerance
Tolerance UG–81

Nozzle wall thickness, UG–45
Welded connection, UW–15, 16, Fig. UW–16.1
Reinforcement pad, UG–37, UC–41
Code termination of vessel, U–1
Welding necks, UG–11, UG–43, UG–45
Loose type flange, UG–44, UA–45 to 52, Fig. UA–48
Ellipsoidal head, pressures, Int. UG–32, UA–4, Ext. UG–33, UA–276
Head skirt, UG–32, Fig. UW–13
Integral type flange, UG–44, UA–45 to 52, Fig. UA–48
Nuts and washers, UG–13, UCS–11
Studs and bolts, UG–12, UCS–10
Circumferential joints, UW–34, UW–35
Applied linings, UG–6, Part UCL, UG–26
Integrally clad plate, Part UCL
Corrosion, UG–25, UCS–25, UCL–25, UA–155 to UA–159
Stiffener plate, UG–6, UG–22, UG–54, UG–82
Support lugs, UG–6, UG–54, UG–82, UA–185 to 188
Longitudinal joints, UW–33, UW–35, UW–9

138

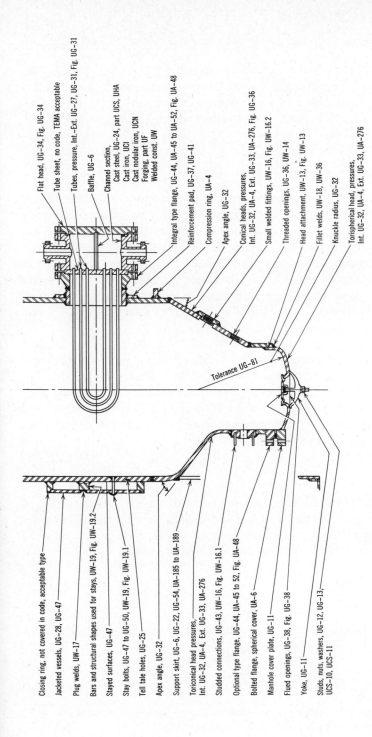

Closing ring, not covered in code, acceptable type
Jacketed vessels, UG–28, UG–47
Plug welds, UW–17
Bars and structural shapes used for stays, UW–19, Fig. UW–19.2
Stayed surfaces, UG–47
Stay bolts, UG–47 to UG–50, UW–19, Fig. UW–19.1
Tell tale holes, UG–25
Apex angle, UG–32
Support skirt, UG–6, UG–22, UG–54, UA–185 to UA–189
Toriconical head pressures,
  Int. UG–32, UA–4, Ext. UG–33, UA–276
Studded connections, UG–43, UW–16, Fig. UW–16.1
Optional type flange, UG–44, UA–45 to 52, Fig. UA–48
Bolted flange, spherical cover, UA–6
Manhole cover plate, UG–11
Flued openings, UG–38, Fig. UG–38
Yoke, UG–11
Studs, nuts, washers, UG–12, UG–13,
  UCS–10, UCS–11

Flat head, UG–34, Fig. UG–34
Tube sheet, no code, TEMA acceptable
Tubes, pressure, Int.–Ext. UG–27, UG–31, Fig. UG–31
Baffle, UG–6
Channel section,
  Cast steel, UG–24, part UCS, UHA
  Cast iron, UCI
  Cast nodular iron, UCN
  Forging, part UF
  Welded const. UW
Integral type flange, UG–44, UA–45 to UA–52, Fig. UA–48
Reinforcement pad, UG–37, UG–41
Compression ring, UA–4
Apex angle, UG–32
Conical heads, pressures,
  Int. UG–32, UA–4, Ext. UG–33, UA–276, Fig. UG–36
Small welded fittings, UW–16, Fig. UW–16.2
Threaded openings, UG–36, UW–14
Head attachment, UW–13, Fig. UW–13
Fillet welds, UW–18, UW–36
Knuckle radius, UG–32
Torispherical head, pressures,
  Int. UG–32, UA–4, Ext. UG–33, UA–276

Tolerance UG–81

Figure 8.1.   Reference guide to ASME unfired pressure vessel code.

139

than paying for jigs, fixtures, or structural rings. The thicker plate also allows for better fit up.

The API-ASME Code specifies that the plate thickness should not be less than $(D\text{-}100)/1000$, where D is the nominal diameter in inches. The minimum plate thickness specified for welded construction by many organizations is $\frac{3}{16}$ or $\frac{1}{4}$ in. The code minimum for riveted construction is $\frac{3}{16}$ and $\frac{3}{32}$ in. for brazing or welding of carbon steel.

External Pressure or Vacuum

Many process vessels operate under external pressure or vacuum or may be accidentally subjected to these conditions. The code specifies that a code vessel occasionally subject to an external pressure of 15 psi or less need not meet the code requirements for construction for external pressure. Consideration should be given to many consequences before taking advantage of this waiver, such as personnel hazard in event the tank collapsed, cost of replacing vessel, property damage, and hazards, due to release of contents, and value of lost contents. If at all possible, the use of a vacuum breaker or a seal leg is much cheaper than designing for vacuum.

Designing for external pressure is a trial and error affair that has resulted in many unnecessarily expensive vessels. Many times the plate thickness required for internal pressure has been used and stiffener rings added as necessary, when it would have been more economical to increase the shell thickness and eliminate some or all of the stiffeners. Unless the designer has reliable fabrication costs it would be worthwhile to submit several alternates to a fabrication shop for pricing.

Selection of Material

The selection of construction materials is in most instances a matter of experience. The material that has proved satisfactory in the past under similar conditions is likely to be used again, unless there is a very good reason for changing. When experience is lacking, a number of factors must be considered.

STRESS LIMITATIONS

The material's allowable stress should not require excessive thicknesses for the design pressure and temperature.

TEMPERATURE LIMITATIONS

Some materials such as copper and aluminum, and their alloys, and cast iron have specific limitations on temperature (400–450°F).

MATERIALS RESISTANCE TO CORROSION BY PROCESS FLUID

The resistance to corrosion that may be expected of a construction material used in a process vessel may be approximated by field tests conducted under identical conditions as those of the vessel's operation. The actual operating corrosion rate and the simulated operating corrosion rate obtained by laboratory test are often at great variance because of the many unforeseen operational factors. An excellent reference for corrosion rates is *Corrosion Data Survey*.[9] It furnishes information at least as reliable as that obtained by laboratory testing. For new processes and with no previous experience under similar conditions, the services of an experienced corrosion consultant would be desirable.

PERMISSIBLE CONTAMINATION OF PROCESS FLUID BY CORROSION PRODUCTS.

While high purity requirements may limit the selection of material it does not necessarily mean that the most expensive material is the best. Where pressure and temperature conditions permit, cladding, glass, rubber, lead, gunnite, or plastic linings may be used rather than expensive alloys.

RELATIVE FABRICATED COST OF VESSEL FOR VARIOUS MATERIALS CONSIDERED.

It is often desirable and profitable to request alternate proposals from fabricators such as stainless steel versus aluminum and clad material versus solid alloy. Clad material does not always effect the savings anticipated on thicknesses $\frac{1}{2}$ in. and less. Although the cost of the clad material may be less than that of the solid alloy the cost of fabrication may be approximately the same for either, with the overall cost remaining about the same.

DESIGN OF EQUIPMENT FOR PERIODIC REPLACEMENT RATHER THAN FOR PLANT LIFE.

For other than large, complicated vessels difficult to remove, it may be economically feasible to design equipment using less expensive metals and replace the vessel after a few years than design the vessel for a 10- or 15-year plant life using an expensive or super alloy. A cost comparison of a vessel fabricated of carbon steel will range to about $1\frac{1}{2}$ for a low-alloy steel, to around 4 for stainless, and to 10 or more for the more expensive alloys presently in use.

In the past a corrosion allowance was customarily employed in the design of carbon steel vessels. This was usually an arbitrary value added to the calculated minimum thickness, and it was applied indiscriminately to all vessels in a unit process regardless of the vessel's importance. Present day techniques in determining corrosion rates by inserting corrosion racks in vessels and pipe lines should result in more realistic corrosion allowances for similar process conditions.[10] As the corrosion rates can be predicted more accurately, so may the vessel's operating life expectancy.

In specifying material for code-construction vessels either the ASME SA number[3] should be used or the complete ASTM A number[11] including the year and any suffix such as ASTM-A987-79T. The ASTM may reissue the specification with a new year and without the tentative (T) with the result that the shop inspector may reject the material as not conforming to ASME requirements.

ASME and ASTM Specifications include grades for plate, pipe, tubing, forgings, and castings with similar chemical and physical properties to allow for compatibility or matching of materials.

Loadings Other Than Pressure

The vessel and its supports must be designed to accommodate loadings resulting from:

Weight of vessel and contents
Weight of internals—trays, baffles, coils, etc.
Weight of externals—equipment, agitators, exchangers, drums, etc.
Weight of externals—structures, ladders, platforms, piping
Weight of externals—dead and live loads
Weight of insulation and fireproofing
Wind and earthquake loads

In addition to the foregoing, investigations may be necessary as to the effect of reactions due to supporting lugs, ring stiffeners, piping, and thermal gradients that may cause excessive localized stresses.

Estimating weights can be tedious and time consuming unless the vessel designer has manufacturers catalogue and reference material, records or tabulated data from previous jobs, and assorted handbooks.

The weights of structural shapes may be obtained from *Steel Construction*. Lukens Steel Company catalogues *Spun Heads* or *Clad Steel Heads* furnish volumes and weights of different types of heads.

The weights of flanges, fittings, piping, or valves, can be obtained from any of the suppliers catalogues such as Ladish, Tube Turns, Midwest, Crane, or Taylor. The weights of trays, agitators, exchangers, drums, or other equipment should be obtained from the manufacturer unless the designer has sufficient experience to make an educated guess as to probable weights.

For estimating the weight of caged ladders and platforms, 25 lb/lineal ft and 50 lb/ft² are fair averages for the weight of steel. Platforms should be designed for a live load of 100 lb/ft² if a walkway only, 150 lb/ft² if a working platform, or 150% of the weight of any equipment the platform may support during turnarounds.

The minimum wind load should not be less than that specified by the AISC or 20 lb/ft² on the vertical projection of the finished structure. Lacking any local legal requirements the recommended wind pressure may be obtained from the American Standard Building Code ASA A58.1, or if the maximum sustained gust velocity is known it may be determined by an empirical formula developed by the U.S. Weather Bureau.[12]

The design of the vessel should be such as to be self-supporting under the combined conditions of wind with vessel full, operating, or during erection. The combination of erection weight and wind is usually the governing condition.

In addition to designing the vessel so that the stresses are within those permitted by the code, it is desirable that the deflection of the tower be checked as excessive deflection may upset operation of the top trays, and it is also disconcerting for workmen.

Many areas require that consideration be given to earthquake hazard in design. The seismic probability in various localities of the United States has been outlined by the U.S. Coast and Geodetic Survey (see Fig. 8.2) according to the amount of damage caused by earthquakes in that locality. The areas are designated Zone 0 (no damage), to Zone 3 (major damage). The seismic coefficients used in design range from 0.02 to 0.20 depending upon the zone and the vibration period of the structure.

Data and detailed design procedure for vertical vessels subject to applied forces, vibration, earthquake, and their supports and foundations may be found in references 12–19.

To prevent localized stresses due to lugs, brackets, and supports for platforms, equipment, and piping, circumferential rings are often used to distribute the load. These rings may cause excessive secondary stresses in the shell immediately adjacent to the ring unless the rings are properly proportioned. The design of these rings and resulting shell stresses are discussed in references 20 and 21.

Until recently the effect of piping reactions, moments at vessel nozzles, and the magnitude of the resulting secondary stresses at the shell openings had no satisfactory solution in general usage. The designer usually applied a heavy reinforcing pad around the nozzle with a number of gussets to distribute the loading around as large an area of the shell or head as deemed practical. Satisfactory methods have been developed within the past few years for computing the stresses in cylindrical and spherical shells resulting from pipe loads.[22–25] Reference 26 presents the data and formulas of 22 through 25 in readily useable form. This reference[26] also gives basic formula for thermal stress calculations at nozzles not readily available in standard tests.

Vessels are insulated for any or all of three reasons, which are to conserve

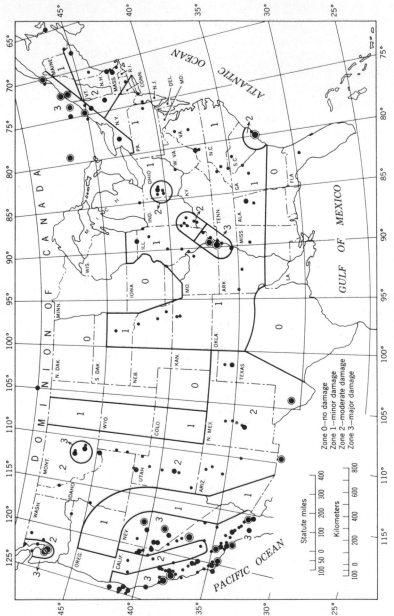

Figure 8.2. Seismic map (U.S. Coast and Geodetic Survey).

Zone 0—no damage
Zone 1—minor damage
Zone 2—moderate damage
Zone 3—major damage

Statute miles

Kilometers

heat, to prevent temperature variation affecting the process, and personnel protection, 150°F max. The diameter over the insulation should be used in determining the area on which the wind acts, and the weight of the insulation should be used in weight calculations.

The supports, skirts, saddles, legs, and lugs, of vessels containing flammable fluids are customarily fireproofed with 2–4 in. of gunnite or 1–2 courses of brick. Vessels over about 4 ft in diameter will be fireproofed both inside and outside of the skirt. As the skirt and baseplate usually support the fireproofing and give the vessel greater stability under wind conditions, the weights are included in the calculations.

The effects of wind pressure and earthquake are seldom figured in design as acting simultaneously. Each is figured separately, and the worse condition governs the design. (See Figs. 8.3–8.10.)

Computation sheets such as those illustrated may be of use but too much dependence should not be placed on them, especially where widely different conditions exist in the design of vessels for a plant. The computation sheets illustrated were prepared for us in the design of vessels in one plant where pressures and temperatures were relatively low and not cyclic. Earthquake was not a factor and the minimum thickness specified by the client was $\frac{1}{4}$ in. Unless the form sheets are used judiciously the tendency is to plug in values in the formulas and grind out the answers without due consideration of all the factors that may be involved.

Supports

The design of supports for other than small tanks, either horizontal or vertical, may be quite complex due to secondary stresses, moments, and shears caused by support attachments.

Vertical vessels may be supported by legs, lugs, or brackets, either with or without a ring girder, or a skirt may be used. Large, heavy vessels supported by legs or lugs should have the support reactions at the shell carefully investigated. Methods for analyzing the reactions are given in reference 27. Skirt attachment has been accomplished in a number of ways, but the consensus seems to be to have the skirt and shell outside diameter the same, with the skirt butted to head knuckle and welded. This arrangement should not be used for other than ellipsoidal or spherical heads; for flanged and dished heads, the skirt should be fitted to the outside diameter of the bottom head flange and fillet welded. The effects of securing heads by various methods are discussed in references 28–30.

Large, horizontal tanks frequently have been supported on three or more saddle supports. This is a hand-down from riveted construction when it was usual practice to provide a saddle adjacent to each riveted girth seam in an effort to prevent leakage at the riveted seam. Preferred practice today is not more than two supports. The saddles are designed and located so as to utilize the strength of the heads to maintain roundness of the shell, or stiffening

SKETCH OUTLINE OF EQUIPMENT SHOWING PRINCIPAL DIMENSIONS, NOZZLES & SIZES, ETC.

NAME OF COMPANY_____

VESSEL   EQUIP.   No.
OR TANK_____ No____ REQ'D_____

J. O. No_____
Sheet No_____of_____
Date_____
Comp. By_____C'k'd By_____

Shell

Mat'l_____
_____C.A_____
Contents_____
_____Sp. Gr._____
Operating
Pressure_____psig @_____°F
Vacuum or
Ext. Press_____psig @_____°F

Jacket

Mat'l_____
_____C.A_____
Contents_____
_____Sp. Gr._____
Operating
Pressure_____psig @_____°F
Vacuum_____@_____°F

Heater coils or lances

Mat'l_____
_____C.A_____
Contents_____
_____Sp. Gr._____
Type_____

Dia_____Spacing_____Sq. ft_____
Operating
Pressure_____psig @_____°F

Trays

Material_____
_____C.A_____
Type_____

No.
Caps_____Size_____Spacing_____

Special conditions

_____
_____
_____
_____
_____
_____
_____

Figure 8.3. Sketch outline of equipment showing principal dimensions, nozzles and sizes, etc.

COMPUTATION SHEET

NAME OF COMPANY_____

J. O. No._____

Sheet No._____of_____

SUBJECT_____ EQUIP. No._____

Date_____

Comp. By_____C'k'd By_____

Diameter_____(OD)(ID) T.L.–T.L._____(Hor.)(Vert.) Support_____

Operating Press._____psig @_____°F Static Press. = .433 × _____ = _____psig

Design Press. (use larger value of)
- operating press. + 15   = _____ + 15   = _____psig
- operating press. × 1.15   = _____ × 1.15   = _____psig
- operating press. + static press. = _____ + _____ = _____psig

Min. Des. Temp. (larger value) = 1.10 × operating temp. = ___°F or operating temp. + 50 = ___°F

Base Mat'l ASME SA_____Gr._____Qual._____Inter. Case_____

Clad Mat'l ASME SA_____Gr._____Type_____Th'k _____

XR____SR___Shell $E =$____Hd. $E =$ ____ C.A. $= c =$ ___min. $t = .094''$ or_____''

Atm. Temp. Shell and Head $S_a =$ _____psi $S_aE =$ _____psi

Des. Temp. Shell and Head $S_d =$ _____psi $S_dE =$ _____psi

t = calc. thickness without corr. allow. $t + c$ = calc. thickness including corr. allow.

$t_n - c$ = nom. thickness without corr. allow. t_n = nom. thickness including corr. allow.

	I.D. (D) Shell	O.D. (D_o) Shell	2:1 Ellipt. Head	Code F&D Head
$t + c$	$\dfrac{P(D + 2c)}{2S_dE - 1.2P} + c$	$\dfrac{PD_o}{2S_dE + .8P} + c$	$\dfrac{P(D + 2c)}{2S_dE - .2P} + c$	$\dfrac{.885P(L + c)}{S_dE - .1P} + c$
MAWP	$\dfrac{2S_dE(t_n - c)}{(D - 2c) + 1.2(t_n - c)}$	$\dfrac{2S_dE(tn - c)}{D_o - .8(t_n - c)}$	$\dfrac{2S_dE(t_n - c)}{(D + 2c) + .2(t_n - c)}$	$\dfrac{S_dE(t_n - c)}{.885(L + c) + .1(t_n - c)}$
MAP	$\dfrac{2S_aEt_n}{D + 1.2t_n}$	$\dfrac{2S_aEt_n}{D_o - .8t_n}$	$\dfrac{2S_aEt_n}{D + .2t_n}$	$\dfrac{S_aEt_n}{.885L + .1t_n}$

S.R. req'd it at any welded seam $t_n > \dfrac{D + 50}{120} = \dfrac{}{120} =$

Press. – Temp. Rating of_____ASA Flanges = _____psig @_____°F

Max. Test Pressure on Flanges = _____psig

Figure 8.4. Computation sheet 1.

COMPUTATION SHEET

J. O. No._____

Name of Company_____   Sheet No._____of_____

Equip.   Date_____

Subject_____ No._____   Comp. By_____C'k'd By___

Shell

Diameter_____(OD)(ID) T.L.–T.L._____Corr. Allow. $= c =$ _____"

Mat'l ASME SA_____Gr._____Type or Qual._____Y.S. $=$ _____

If jacketed, jacket design press. $=$ _____psi @_____°F

Ext. press. due to full or partial vacuum $=$ _____psi @_____°F

Design External Pressure (sum of above) $= P =$ _____psi @_____°F

Min. Des. Temp. − greater of (max. operating temp. $\times$ 1.10) or
(max. operating temp. $+$ 50)

   $t =$ nom. req'd th'k without corr. allow. $(t + c) =$ nom. req'd th'k including corr. allow.

$D_o =$ outside dia. $P =$ ext. design press. $P_a =$ allowable external press.

$L =$ design length of vessel section (See UG-28 and 29)

Use chart (Fig. #_____) Appendix V

Check for P_a (or MAP_a) with given $D_o =$ _____t or $(t + c) =$ _____$L =$ _____

$\dfrac{D_o}{t}$ or $\dfrac{D_o}{(t+c)}$	$\dfrac{L}{D_o}$	B (from Chart)	P_a or $MAP_a = \dfrac{Bt}{D_o}$ or $\dfrac{B(t+c)}{D_o}$

Check for L with given $P =$ _____ and $D_o =$ _____ with variable t

t	$\dfrac{D_o}{t}$	$B = \dfrac{PD_o}{t}$	$L - D_o = X$ (from Chart)	$L = D_o X$	$t + c$

Stiffening Rings

$E =$ mod. of elast. for max. temp. under ext. press. conditions $=$ _____

$I =$ mom. of inertia of ring $\geq I_s =$ req'd mom. of inertia

$A_s =$ cross-sect. area of ring.   A and B = Chart factors—Appendix V

$D_o =$ _____ $D_o{}^2 =$ _____ $\dfrac{D_o{}^2}{14} =$ _____ $D_o{}^3 =$ _____

Trial $I_s = \left(\dfrac{.14 D_o{}^3 P}{E}\right) L = \left(\underline{\hspace{2cm}}\right)\underline{\hspace{0.5cm}} =$ _____ Use_____

                                                   $W =$ _____$A_s =$ _____I_____

t	L	$\dfrac{A_s}{L}$	$t + \dfrac{A_s}{L}$	PD_o	$B = \dfrac{PD_o}{t + \dfrac{A_s}{L}}$	A	$\left(\dfrac{D_o{}^2}{14}\right)L$	$I_s = \left[\left(\dfrac{D_o{}^2}{14}\right)L\right]\left[t + \dfrac{A_s}{L}\right]A$

Figure 8.5. Computation sheet 2.

COMPUTATION SHEET

J. O. No._____

NAME OF COMPANY_____ Sheet No._____of_____

EQUIP. Date_____

SUBJECT_____ No._____ Comp. By_____C'k'd By____

Diameter_____(OD)(ID) Type_____Corr. Allow. = c = _____

Mat'l ASME SA_____Gr._____Type or Qual._____Y.S._____

If jacketed, or intermediate head, design press. on convex side_____psi @_____°F

Ext. press. due to full or partial vacuum on concave side_____psi @_____°F

P = Design ext. press. = sum of above = _____psi @_____°F

Min. Des. Temp. = greater of (max. operating temp. $\times$ 1.10) or (max. operating temp. + 50)

Atm. Temp. S_a = _____psi Design Temp. S_d = _____psi $E = 1.0$

t_h = Nom. th'k without C.A. $(t_h + c)$ = Nom. th'k with C.A.

D_o = OD for F&D = _____ or ID + $2(t_h + c)$ for Ellipt. = _____

L_I = (Rad. of Dish + c) for F&D = _____ or $K_I D_o$ for Ellipt. = _____

B = Factor from Chart (Figure #_____) Appendix V

Max. Allow. Ext. Working Pressure

t_h	$\dfrac{L_I}{100 t_h}$	$\dfrac{L_I}{t_h}$	B	$P_a = \dfrac{B}{L_I/t_h} \geq P$

Use larger value of t_h above, or $t = \dfrac{1.67 P(\text{ID} + 2c)}{2 S_d - .334 P}$ (Ellipt.) or

$$\dfrac{1.475 P L_I}{S_d - .167 P} \text{ (F\&D)}$$

Max. Allow. External Pressure

$(t_h + c)$	$\dfrac{L_I - c}{100(t_h + c)}$	$\dfrac{L_I - c}{(t_h + c)}$	B	$\text{MAP}_a = \dfrac{B}{(L_I - c)/(t_h + c)} \geq$ jacket MAP

Use larger value of $(t_h + c)$ above, or $(t + c) = \dfrac{1.67(\text{MAP})(\text{ID})}{2 S_a - .334(\text{MAP})}$ (Ellipt.) or

$$\dfrac{1.475(\text{MAP})(L_I - c)}{S_a - .167(\text{MAP})} \text{ (F\&D)}$$

Figure 8.6. Computation sheet 3.

COMPUTATION SHEET

NAME OF COMPANY_____

SUBJECT_____

J. O. No._____
Sheet No._____of_____
Date_____
Comp. By_____C'k'd By_____

Estimated Weights		*Estimated*	*Corrected*
Top head			
Shell			
Manholes			
Nozzles			
Tray support rings			
	Weight *A*		
Bottom head			
Sump shell			
Sump head			
	Weight *B*		
Skirt			
Base plate			
	Weight *C*		
1. Insulation			
2. Fireproofing			
3. Ladders and Platforms			
4. Trays			
5. Liquid on trays			
6. Liquid holdup: shell = head =			
7. Total liquid: shell = heads =			
8. Others: piping, reboilers, etc.			

Figure 8.7. Computation sheet 4.

COMPUTATION SHEET

J. O. No.________________

Name of Company________________________ Sheet No.________of________

Date________________

Subject________________________ Comp. By______C'k'd By______

Shipping Weight (D)	*A* = *B* = *C* = 3 = 4 = 8 =
Operating & Test Weight @ Bot. T.L. (E)	*A* = 1 = 3 = 4 = 5 = 8 =
Test Weight on Skirt (F)	*A* = *B* = 1 = 3 = 4 = 7 = 8 =
Operating Weight @ Base (H)	*A* = *B* = *C* = 1 = 2 = 3 = 4 = 5 = 6 = 8 =
Test Weight @ Base (K)	*A* = *B* = *C* = 1 = 2 = 3 = 4 = 7 = 8 =

Figure 8.8. Computation sheet 5.

COMPUTATION SHEET

J. O. No._____

Name of Company_____   Sheet No._____of_____

Equip.   Date_____

Subject_____ No._____   Comp. By_____C'k'd By____

Point under consideration (Base) (Bot. Tang. Line) other_____

P_w = Wind pressure per sq. ft. of projected area_____

D_o = OD shell section plus 2 × insulation thickness_____

L = Length of shell section under consideration_____

H = Moment arm of shell section under consideration_____

W_8 = Wt. of any eccentric load_____ e = moment arm of W_8 = _____

M_w = Wind moment = D_oLP_wH = _____

M_e = Eccentric moment = W_8e = _____

M = Summation of M_w and M_8 about point under consideration_____

P = $MAWP$ = _____ or $P = P_a$ = _____ S_a = _____ S_d = _____ E = _____

d = Nom. shell dia. in._____ D = Nom. shell dia. ft._____

$Z = 9.42D^2$ = _____ $C = 37.7D$ = _____

L_t = Tension, lbs. per linear in. L_c = Comp. lbs. per linear in.

AT BOTTOM TANG. LINE OR OTHER POINT_____

$\pm \dfrac{M}{Z}$ = ——— = + $\quad$ —   Use larger value of t or t_h

$-\dfrac{W_A}{C}$ = ——— = $\quad$ xxxxxx $\quad$ — $t = L_t/S_aE$ = _____

L_i = I.R. = _____$L_i/100t_h$ = _____

$L_t(+)$ or $L_c(-)$ $\quad$ + $\quad$ — B = _____ $\geqq (L_c/t_h$ = _____) $\leqq S_a$

$\pm \dfrac{P_d}{4}$ = ——— = + $\quad$ —   Use larger value of t or t_h

$\pm \dfrac{M}{Z}$ = ——— = + $\quad$ — $t = L_t/S_dE$ = _____

$-\dfrac{W_E}{C}$ = ——— = - $\quad$ — L_i = I.R. = _____$L_i/100t_h$ = _____

B = _____ $\geqq (L_c/t_h$ = _____) $\leqq S_d$

$L_t(+)$ or $L_c(-)$ $\quad \pm \quad$ —

$\pm \dfrac{M}{Z}$ = ——— = + $\quad$ —   Use larger value of t or t_h

$t = 2L_t/S_a$ = _____

$-\dfrac{W_F}{C}$ = ——— = + $\quad$ xxxxxx $\quad$ — L_i = I.R. = _____$L_i/100t_h$ = _____

B = _____ $\geqq (L_c/t_h$ = _____) $\leqq S_a$

$L_t(+)$ or $L_c(-)$ $\quad$ + $\quad$ —

AT BASE LINE OR OTHER POINT_____

$\pm \dfrac{M}{Z}$ = ——— = + $\quad$ —   Use larger value of t or t_h

$t = 1.54L_t/S_a$ = _____

$-\dfrac{W_d}{C}$ = ——— = $\quad$ xxxxxx $\quad$ — L_i = I.R. = _____$L_i/100t_h$ = _____

B = _____ $\geqq (L_c/t_h$ = _____) $\leqq S_a$

$L_t(+)$ or $L_c(-)$ $\quad$ + $\quad$ —

$\pm \dfrac{M}{Z}$ = ——— = + $\quad$ —   Use larger value of t or t_h

$t = 1.54L_t/S_a$ = _____

$-\dfrac{W_k}{C}$ = ——— = $\quad$ xxxxxx $\quad$ — L_i = I.R. = _____$L_i/100t_h$ = _____

B = _____ $\geqq (L_c/t_h$ = _____) $\leqq S_a$

$L_t(+)$ or $L_c(-)$ $\quad$ + $\quad$ —

Figure 8.9. Computation sheet 6.

J. O. No._____

Name of Company_____   Sheet No._____of_____

Date_____

Subject_____   Comp. By_____C'k'd By____

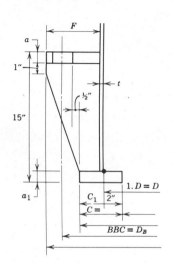

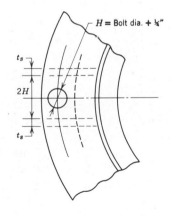

L_c @ Base Line =   _____

$C = \dfrac{L_c}{500} = \dfrac{}{500} = $ ″

$a_1 = .335C_1 = .335x$ =   _____ ″

Wt. "D" =   _____lbs.

M @ Base Line =   _____ft. lbs.

N = No. of anchor bolts =   _____

Dia. Bolt Circle = D_B = _____

$T_M = \dfrac{4M}{ND_B} = \dfrac{}{} =$

$C_W = \dfrac{\text{Wt. } D}{N} = \dfrac{}{} =$

_____

$T_B = T_M - C_W$ =

$A_B = \dfrac{T_B}{15000} = \dfrac{}{} =$

Bolt size (d_b) = _____A_a = _____

$T = 15000A_a = 15000x$ ____ = _____

$4H = 4x$_____ =

$1.5d_b = 1.5x$_____ = _____

$S =$ = $4H - 1.5d_b$ =

$M_a = \dfrac{TS}{8} = \dfrac{x}{8} =$ _____

$a = \sqrt{\dfrac{M}{3333(F - 1.5d_b)}}$

$= \sqrt{\dfrac{}{3333x}} =$ _____

$t_s' = \dfrac{A_a}{2(C_1 - t_s)}$ = _____ = ____

$t_s \geq \dfrac{15 - (a + a_1)}{21} =$ _____ = ____

Figure 8.10. Computation sheet 7.

rings may be employed. Tanks over 10 ft in diameter and 50 ft long are supported in this manner. See references 31–34 Tables 8.5 and 8.6, and Figs. 8.11–8.13. Leg supports for horizontal tanks may be designed using the data in Fig. 8.14.

Heads

Pressure-vessel heads may be any of the following which are listed in order of increasing thickness for a given pressure, diameter, and material. (The approximate shell thickness $t = 0.06$ in./ft diam/100 psi for an allowable stress of 10,000 psi.)

1. Hemispherical   ($t = \frac{1}{2}$ shell thickness—approx.)
2. Ellipsoidal   ($t =$ shell thickness—approx.)
3. Conical   ($t =$ shell thickness/cos of half apex angle)
4. Torispherical   ($t = 1.77$ shell thickness—approx.)
   (F and D)
5. Flat   no simple relationship

Hemispherical heads have the best stress distribution of any of the head types, with less material required to contain a given volume. Relatively few fabricators have the forming equipment necessary for producing these heads, and therefore they are not as commonly used as are ellipsoidal heads and flanged and dished heads.

Ellipsoidal heads having a major to minor axis ratio of 2:1 have a better stress distribution than torispherical, and for pressures over 150–175 psi and for diameters over about 5 ft they are usually more economical than the F and D head. Ellipsoidal heads are usually sized by their inside diameter rather than by their outside diameter as are F and D heads.

Conical heads are usually formed by pressing, not by rolling or spinning, and they are expensive to fabricate. They are used for digestors, rendering tanks, or where it is desirable to drain off solids or heavy viscous materials. A truncated conical head is commonly used as a transition piece connecting two different diameters of a vessel. Conical heads without a transition knuckle are permitted by the code under certain conditions, but they should be used only for low-pressure, low-temperature conditions.

The Lukens Steel Company catalogue, *Lukens Spun Heads* lists three types of flanged and dished heads:

1. Flanged and dished heads—Standard
2. Flanged and dished heads—ASME
3. Flanged and shallow dished heads

TABLE 8.5. Stress Due to Saddle Supports

Longitudinal Bending Stress

$$S_1 = \pm \frac{3K_1QL}{\pi r^2 t} \le S_a - \frac{Pd}{4tE} \le \frac{S_y}{2} \le \left(\frac{E}{29}\right)\left(\frac{t}{r}\right)\left(2 - \frac{66.6t}{r}\right)$$

$$\left|\leftarrow \text{ neglect if } \frac{t}{r} \ge .005 \rightarrow\right|$$

Tangential Shear Stress

$$S_2 = \frac{K_2Q}{rt}\left(\frac{L - 2A - H}{L + H}\right) \le .8S_a \text{ (saddles away from head } A \le .25L)$$

$$S_2 = \frac{K_2Q}{rt} \le .8S_a \text{ (stress in shell—saddles at head)}$$

$$S_2 = \frac{K_2Q}{rt_n} \le .8S_a \text{ (stress in head—saddles at head)}$$

Circumferential Stress at Horn of Saddle

$$S_3 = -\frac{Q}{4t(b + 10t)} - \frac{3K_3Q}{2t^2} \le 1.25S_a \text{ when } S_c = S_t \ (L \ge 8R)$$

$$S_3 = -\frac{Q}{4t(b + 10t)} - \frac{12K_3QR}{Lt^2} \le 1.25S_a \text{ when } S_c = S_t \ (L < 8R)$$

Additional Stress in Head used as Stiffener

$$S_4 = \frac{K_4Q}{rt_h} \le 1.25S_a - S_h \text{ (saddles at head)}$$

Ring Compression in Shell over Saddles—Wear Plates

$$S_5 = \frac{K_5Q}{t(b + 10t)} \le \frac{S_y}{2} \ (t = t_s + t_{wp})$$

Design of Ring Stiffeners

$$S_6 = -\frac{K_7Q}{na} \pm \frac{K_6Qr}{nZ} \le \frac{S_y}{2} \text{ or } S_a$$

Design of Saddles

$$F = K_8Q \text{ acting } \le .33r \text{ below shell } S \le .66S_a$$

NOMENCLATURE

Q = load on one saddle (lb)
L = tangent length of vessel (ft)
A = distance from ₵ saddle to T.L. (ft)
H = depth of head (ft)
R = radius of shell (ft)
r = radius of shell (in.)
t = thickness of shell (in.)
t_h = thickness of head (in.)
b = width of saddle (in.)
F = force across bottom of saddle (lb)
$S_1S_2S_3$ = calc. stresses psi
$K_1K_2K_3$ = constants

S_a = allowable stress psi
S_y = yield stress psi
S_h = head stress due to int. press. psi
E = mod. of elasticity psi
Z = section modulus in.^{3}
n = no. of stiff. each saddle
a = cross. sect. area each stiff. sq in.
θ = contact angle of saddle with shell
t_{wp} = thickness of wear plate (in.)

TABLE 8.6. Values of Coefficients in Formulas for Various Support Conditions

Longitudinal Bending Stress

$$S_1 = \pm \frac{3K_1QL}{\pi r^2 t} \leq S_a - \frac{Pd}{4tE} \leq \frac{S_y}{2} \leq \left(\frac{E}{29}\right)\left(\frac{t}{r}\right)\left(2 - \frac{66.6t}{r}\right)$$

$$\left|\leftarrow \text{ neglect if } \frac{t}{r} \geq .005 \rightarrow\right|$$

Tangential Shear Stress

$$S_2 = \frac{K_2Q}{rt}\left(\frac{L - 2A - H}{L + H}\right) \leq .8S_a \text{ (saddles away from head } A \leq .25L)$$

$$S_2 = \frac{K_2Q}{rt} \leq .8S_a \text{ (stress in shell—saddles at head)}$$

$$S_2 = \frac{K_2Q}{rt_n} \leq .8S_a \text{ (stress in head—saddles at head)}$$

Circumferential Stress at Horn of Saddle

$$S_3 = -\frac{Q}{4t(b + 10t)} - \frac{3K_3Q}{2t^2} \leq 1.25S_a \text{ when } S_c = S_t (L \geq 8R)$$

$$S_3 = -\frac{Q}{4t(b + 10t)} - \frac{12K_3QR}{Lt^2} \leq 1.25S_a \text{ when } S_c = S_t (L < 8R)$$

Additional Stress in Head used as Stiffener

$$S_4 = \frac{K_4Q}{rt_h} \leq 1.25S_a - S_h \text{ (saddles at head)}$$

Ring Compression in Shell over Saddles—Wear Plates

$$S_5 = \frac{K_5Q}{t(b + 10t)} \leq \frac{S_y}{2} (t = t_s + t_{wp})$$

Design of Ring Stiffeners

$$S_6 = -\frac{K_7Q}{na} \pm \frac{K_6Qr}{nZ} \leq \frac{S_y}{2} \text{ or } S_a$$

Design of Saddles

$$F = K_8Q \text{ acting } \leq .33r \text{ below shell } S \leq .66S_a$$

NOMENCLATURE

Q = load on one saddle (lb)
L = tangent length of vessel (ft)
A = distance from ₵ saddle to T.L. (ft)
H = depth of head (ft)
R = radius of shell (ft)
r = radius of shell (in.)
t = thickness of shell (in.)
t_h = thickness of head (in.)
b = width of saddle (in.)
F = force across bottom of saddle (lb)
$S_1 S_2 S_3$ = calc. stresses psi
$K_1 K_2 K_3$ = constants

S_a = allowable stress psi
S_y = yield stress psi
S_h = head stress due to int. press. psi
E = mod. of elasticity psi
Z = section modulus in.³
n = no. of stiff. each saddle
a = cross. sect. area each stiff. sq in.
θ = contact angle of saddle with shell
t_{wp} = thickness of wear plate (in.)

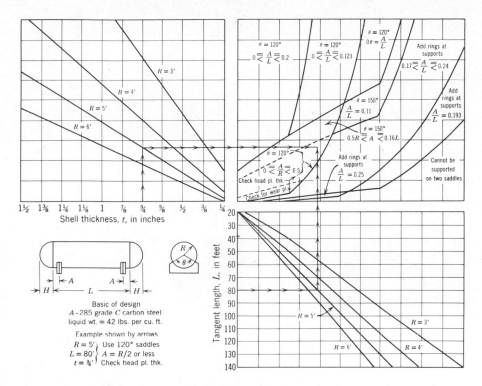

Figure 8.11. Location and type of support for horizontal pressure vessels on two supports.

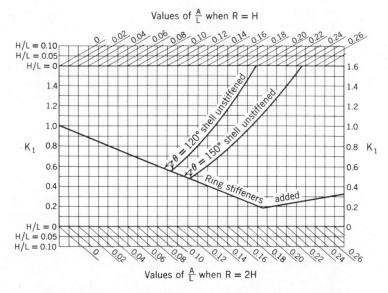

Figure 8.12. Plot of longitudinal bending-moment constant K_1.

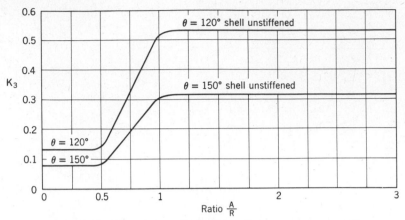

Figure 8.13. Plot of circumferential bending-moment constant K_3.

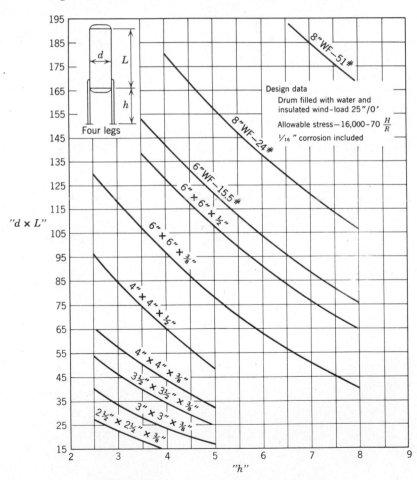

Figure 8.14. Leg supports for vertical vessels.

Standard flanged and dished heads have a dished crown radius equal to, or less than, the outside diameter of the head skirt or straight flange, but the transition knuckle between the crown and the skirt has a radius of only three times the metal thickness. About 30 years ago the code refused further acceptance of heads having these knuckle proportions following the discovery of many cracked heads, and after several disastrous explosions that were traced to cracks that had developed at the knuckle as a result of the stress intensification at the short radius. These heads are used for atmospheric tanks and should not be used for pressure vessels.

The ASME flanged and dished heads have a dished radius equal to or less than the outside diameter of the head skirt or straight flange, and a knuckle radius not less than 6% of the outside diameter of the head skirt but not less than three times the metal thickness. The latter provision increases the knuckle radius for thicker heads. When flanged and dished heads are used ASME F and D heads should be specified.

Flanged and shallow dished heads have a dished crown radius greater than twice the head diameter, and they more nearly approach a flat or flanged only head.

The use of welded flat heads, except for small, low-pressure closures, are seldom justified as, in addition to the extreme thicknesses required, severe discontinuity stresses are introduced in the cylindrical shell. A more satisfactory closure is obtained by a pipe cap or a code head. Bolted on blind flanges are used as manhole or access openings but above 24 in. in diameter, and for high pressures and temperatures it is often possible to save money and effect easier handling by using a formed head attached to a pipe flange. Illustrative of this is that a 6 ft diameter vessel carrying 350 psi required a flat head 12 in. thick weighing 13,800 lb, whereas a 2 to 1 ellipsoidal head $\frac{7}{8}$ in. thick weighing 1700 lb could have been used (the equivalent of the bolting portions of the closure not included for either type head).

Literature on the stresses occurring in pressure vessel heads is extensive. Some references are 30 and 35–37.

8.8. FABRICATION

Pressure vessels may be fabricated by bolting, riveting, forging, casting, brazing, welding, or a combination of any of these. Other than for flanges, bolting has never been used extensively for pressure-vessel construction, although for years the box header for one make of water-tube boiler had been bolted to the steam and water drum.

Riveted construction is almost a thing of the past, and it is doubtful if more than a handful of shops in the country today still have equipment for bull riveting a boiler or pressure vessel.

Forgings, other than forged fittings, may be used for high-pressure drums where the ratio of thickness to diameter is high, and it would be difficult to form a cylindrical shell by rolling or pressing. The code has no provision for forging by hammer welding, although this method was used for years in the fabrication of large paper-mill digestors.

Castings may be used for unusual shapes. Cast iron has definite pressure-temperature limitations established by the code. Cast steel has no such limitations. Brazing has been used extensively for production of small air tanks for service station use. The code limits the temperature to 406°F and the thickness to 1 in. maximum. Many nonferrous materials are brazed.

With the exception of a few special cases, practically all industrial pressure vessels are of welded construction. Welding processes are many and consist essentially of brazing, forge welding, gas welding, thermit welding, arc welding, induction welding, resistance welding, and flow welding. Pressure-vessel welding is usually either arc welding or gas welding and may be manual or automatic. Gas welding may be air–acetylene, oxygen–acetylene, or oxygen–hydrogen. Arc welding covers a wider range as follows:

Carbon	Metal Electrode
Shielded-carbon arc	Consumable electrode
Inert-gas carbon arc	Nonconsumable electrode (tungsten)
Unshielded-carbon arc	Shielded electrode (coated)
Unshielded twin-carbon arc	Submerged arc
	Inert-gas metal arc
	Atomic hydrogen
	Impregnated tape

The most common arc-welding processes used in pressure-vessel fabrication are shielded or coated electric, submerged arc, and inert-gas metallic arc. For details of these and other processes consult references 38–40.

For fabricating pressure vessels by welding or brazing the code requires that the manufacturer qualify the welding or brazing procedure to be employed and the welding or brazing operator that will use that procedure in the fabrication of the vessel.

Section 9, "Welding Qualifications," of the ASME Boiler and Pressure Vessel Code details the requirements for qualification of procedure and operators. Although a procedure or procedures may have been qualified for certain welding positions, materials, and filler rod, it may be necessary to requalify if any changes are made. Unless there is a very good reason the designer should not specify any welding details, groove dimensions, or filler metal, that will require the fabricator to change from the qualified procedure and necessitate a new procedure and operator qualification. Qualifying a welding procedure and welding operators is expensive.

8.9. INSPECTION AND TEST

While the code has requirements for forming, fit-up, and tolerances it is well for the purchaser to have his own inspector in the shop to supplement the code inspector. The code inspector is only concerned that the code requirements are met, and not with location, orientation, and projection of nozzles, installation of internals and other attachments, cleanliness, or finish. What the code inspector sees he will no doubt report, but vessels have been shipped with certain idiosyncrasies such as a column shipped to the field with all five manhole davits, including the hinge, welded only to their respective coverplates. Tolerances that are generally available are shown in Fig. 8.15. Closer tolerances may be specified and obtained but usually at additional cost.

For critical service it is often desirable to have more stringent requirements for testing than those presently required by the code, such as ultrasonic testing (now covered by a Code Interpretation Case for plate only), magnetic particle testing, fluid penetrant examination, and helium leak testing. Much time and effort may be saved by extracting procedures and acceptability standards from MIL-STD-271 which covers these test methods in detail.[41] Similarly, MIL-C-19874 is a convenient reference for cleanliness requirements.[42]

8.10. CERTIFICATION

Upon completion of fabrication and testing of the vessel it is stamped as required by the code in the presence of the code inspector who will also sign the proper ASME form:

1. U-1 Mfg. Data Report covering shop and/or field assembly inspection
2. U-1A for single chamber shop fabricated vessel
3. U-2 partial data report for a part fabricated by one manufacturer for another manufacturer
4. U-3 data report for vessels that owing to small size are not shop inspected and are certified by the manufacturer and stamped UM.

The purchaser usually specifies the required number of copies to be finished of:

1. Data forms
2. Rubbings or facsimiles (photographs are better) of the stamping on the vessel
3. Stress relief or heat treatments, time-temperature recordings
4. Pressure recordings

Tolerances given herein shall apply for any pressure vessel, whether vertical or horizontal, using the Reference Line or Working Point shown on the drawings.

(1) Height from Reference Line to face of top nozzle ± 1½", or + ⅟₆₄" per foot of height (whichever is greater).

(2) Far side of towers to weir plate — ± ³⁄₈"

(3) Height of weir above tray support ring — ± ⅛"

(4) Bottom of down flow plate above tray support ring — ± ⅛"

Manway & nozzle drilling:
Bolt circle — ± ⅟₁₆"
Bolt hole spacing — ± ⅟₃₂"
Maximum eccentricity of bolt circle with respect to bore — ⅟₃₂"

(12) Location of top nozzle flange face from head seam shall be ± ¼"

(13) Maximum deviation of straightness along sides for every 10'-0" in length — ⅛"

(14) Top of tray supports out of level over any diameter

Tower Diameter	Tolerance
4'-0" and under	⅛"
Over 4'-0" to 7'-0"	³⁄₁₆"
Over 7'-0" to 13'-0"	¼"
Over 13'-0" to 16'-0"	⁵⁄₁₆"
Over 16'-0"	³⁄₈"

(15) Top of weir plate out of level over any diameter

Tower Diameter	Tolerance
7'-0" and under	⅛"
Over 7'-0" to 13'-0"	³⁄₁₆"
Over 13'-0" to 16'-0"	¼"
Over 16'-0" to 24'-0"	⁵⁄₁₆"
Over 24'-0"	³⁄₈"

(16) Tolerance between adjacent tray supports — ± ⅛"

(17) Location of tray supports from Reference Line — ± ¼"

162

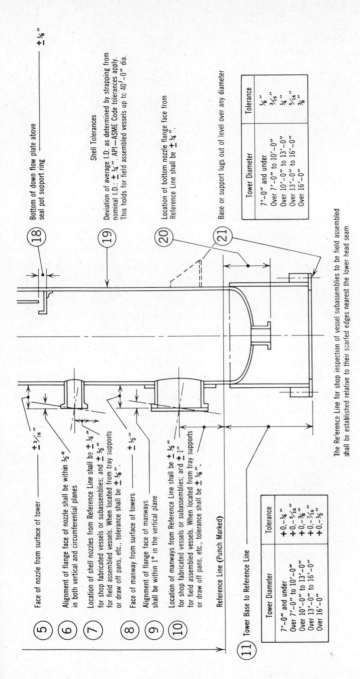

5   Face of nozzle from surface of tower ±3/16"

6   Alignment of flange face of nozzle shall be within ½° in both vertical and circumferential planes

7   Location of shell nozzles from Reference Line shall be ±¼" for shop fabricated vessels or subassemblies; and ±½" for field assembled vessels. When located from tray supports or draw off pans, etc., tolerance shall be ±⅛".

8   Face of manway from surface of towers ±½"

9   Alignment of flange face of manways shall be within 1° in the vertical plane

10   Location of manways from Reference Line shall be ±½" for shop fabricated vessels or subassemblies; and ±1" for field assembled vessels. When located from tray supports or draw off pans, etc., tolerance shall be ±⅛".

Reference Line (Punch Marked)

11   Tower Base to Reference Line

Tower Diameter	Tolerance
7'-0" and under	+0,−¼"
Over 7'-0" to 10'-0"	+0,−5/16"
Over 10'-0" to 13'-0"	+0,−⅜"
Over 13'-0" to 16'-0"	+0,−7/16"
Over 16'-0"	+0,−½"

±⅛"

18   Bottom of down flow plate above seal pot support ring

Shell Tolerances

19   Deviation of average I.D. as determined by strapping from nominal I.D.: ±½". API–ASME Code tolerances apply. This holds for field assembled vessels up to 40'-0" dia.

20   Location of bottom nozzle flange face from Reference Line shall be ±¼".

21   Base or support lugs out of level over any diameter

Tower Diameter	Tolerance
7'-0" and under	⅛"
Over 7'-0" to 10'-0"	3/16"
Over 10'-0" to 13'-0"	¼"
Over 13'-0" to 16'-0"	5/16"
Over 16'-0"	⅜"

The Reference Line for shop inspection of vessel subassemblies to be field assembled shall be established relative to their scarfed edges nearest the tower head seam.

Figure 8.15. Vessel tolerances.

163

5. Mill test certificates
6. Other records pertinent to the vessels fabrication

Similar documents are furnished to the state or other jurisdiction where the vessel will be installed. Future possible difficulties in relocating in another state may be avoided if National Board inspection has been specified and documents sent to their headquarters.

8.11. INSTALLATION

When a new vessel has been installed with its necessary piping and appurtenances, such as relief valve, vents, drains, gauge glass, and controls, local requirements may require state or local inspection prior to putting equipment in service, or inspection may be made by an insurance inspector in lieu of the state inspection. If the installation is satisfactory a certificate of inspection is issued that is valid for 1 year or 2 years depending on the jurisdiction. Prior to the expiration of the certificate arrangements must be made for removing the vessel from service to permit internal inspection.

It is extremely important that the user of pressure vessels have a thorough knowledge of the local laws concerning the installation and operation of pressure-vessel equipment, otherwise he may experience considerable inconvenience and expense. If through ignorance (or contempt) of requirements, equipment is installed and operated without authorization, the user may be subject to penalties which include shutting down the equipment. This could mean a plant shutdown depending on the nature of the installation.

After some costly experience a nice degree of cooperation usually exists between the user and the inspection agency whether it be an insurance company, a municipal agency, or a state agency.

The inspector is kept informed as to new equipment installation with inspection and certification of a matter of routine. Subsequent inspections are made without disrupting the plant's operation by notifying the inspector when scheduled downtime occurs, and inspection may be made then rather than to wait for the inspection certificate to expire which would require a shut down when it may be least convenient.

8.12. SECOND-HAND EQUIPMENT

No statistics are available as to the money wasted on the purchase of second-hand boilers, pressure vessels, heat exchangers, and refrigeration vessels, that are rejected as being noncode or otherwise unacceptable for operation after reinstallation at a new location.

Any user contemplating the purchase of second-hand equipment should have the equipment inspected by an authorized code inspector and obtain a written report that the equipment meets the requirements of the jurisdiction where it is to be reinstalled. The report should also furnish sufficient information concerning the condition of the equipment for the user's decision in purchasing. The code inspector will report conditions and make recommendations for repairs, but will not make a recommendation that purchase be made, his function is inspecting, not appraisal. Arrangements for inspection of second-hand equipment can usually be made with the user's insurance company or through his insurance agent. The cost is a per diem charge for the inspector's services plus expenses.

8.13. REPAIRS

Nonmandatory Appendix X of the Unfired Pressure Vessel Code covers recommended practices for inspection, repair, and allowable pressure for vessels in service. Further information concerning inspection and repair is given in the National Board Inspection Code[4] which will guide a code inspector in code jurisdictions.

The recommended procedure is for a competent pressure vessel designer to prescribe the nature and extent of repairs to be undertaken together with correct welding procedures, heat treatment, and nondestructive testing, that may be required depending on the original construction of the vessel and its present or future intended service. The proposed repair should then be discussed in detail with a code authorized inspector, regardless of whether or not the vessel is under any code jurisdiction. If the vessel is not insured or not under any legal jurisdiction the services of a code inspector may be obtained as outlined in Section 8.12. After the inspector has approved the proposed repair, the work should be done by a reputable fabricator, preferably one listed as authorized under the National Board.[4] After witnessing the satisfactory completion of the repair and testing the inspector will sign the manufacturers Record of Welded Repair to be furnished the governmental unit having jurisdiction.

This may appear as an unduly rigorous approach to a seemingly simple problem, but consider, an unauthorized repair in a code jurisdiction may result in an immediate condemnation of the vessel when the inspector learns of it. At best the inspector will demand removing the vessel from service immediately in order that the repair be inspected and its acceptability judged. This requires an equipment and possibly a plant shutdown. Should the worst happen and a failure causing loss of life occur with resultant public hearings it would be exceedingly difficult to justify the repair.

8.14. HIGH PRESSURE VESSELS

An erroneous conclusion has frequently been drawn that because no specific rules were laid down for vessels carrying over 3000 psi, that such vessels were exempt from any and all code requirements. A proposed revision appearing in *Mechanical Engineering*, April, 1962 reads as follows:

UNFIRED PRESSURE VESSELS, 1959

Paragraph U-1(d),

The rules of this Division have been formulated on the basis of design principles and construction practices applicable to vessels designed for pressures not exceeding 3000 psi 20,670 kPa. For pressures above 3000 psi 20,670 kPa, deviations from and additions to these rules usually are necessary to meet the requirements of design principles and construction practices for these higher pressures. Only in the event that after having applied these additional design principles and construction practices the vessel still complies with all of the requirements of the Division may it be stamped with the Code symbol.

The design of a high-pressure vessel is, as it should be, usually done by an organization or individual with long experience in this highly specialized field. Pressure, temperature, nature of process, and whether hydrogen embrittlement must be considered will influence the material selection and may determine which of the various theories is to be used in computing thicknesses. The theories that may be considered in high-pressure vessel design are:

Maximum principal stresses theory (Lamé)
Maximum strain theory (Clavarino or ICC)
Maximum strain energy theory
Maximum shear theory

The maximum principal stress theory (Lamé) is ordinarily used for pressures around 3000 to 5000 psi and the maximum strain energy theory for higher pressures.

The vessel may be made of two half shells hot formed and welded together, a turned and bored forging, or a number of concentric, thin shells making a composite heavy wall. The heads may be formed hemispherical, ellipsoidal, or flanged.

In order that material thicknesses may be reduced, it is desirable that the inner material of the shell be in compression, which is accomplished by prestressing. This is done in multilayer construction by each successive shell compressing its inner shell. Solid shells are prestressed by overstraining hydrostatically.

Extensive literature is available on the design of high-pressure vessels, see references 43–51.

REFERENCES

1. *History of The Boiler Code,* American Society of Mechanical Engineers, New York, 1955.
2. *The ASME Pressure Vessel Code—A Joint Effort for Safe Construction,* E. O. Bergman, C. F. Braun Co., Los Angeles, 1954.
3. "Section I. Power Boilers," "Section II. Material Specifications," "Section VIII. Unfired Pressure Vessels," "Section IX. Welding Qualifications," "Code Case Interpretations," *The ASME Boiler & Pressure Vessel Code,* American Society of Mechanical Engineers, New York, 1980.
4. *Code of Practice for the Boiler and Pressure Vessel Industry,* National Board of Boiler and Pressure Vessel Inspectors, Columbus, Ohio; 1961.
   National Board Inspection Code, 1956.
   Rules for Repairs of Power Boilers and Unfired Pressure Vessels, 1956.
   National Board Requirements for Inspections and Stamping, and List of Manufacturers Authorized under the National Board, 1963.
5. *Synopsis of Boiler and Pressure Vessel Laws, Rules, and Regulations,* 1960; National Bureau of Casualty Underwriters, New York.
6. J. F. Lancaster, "A Comparison of United States, European, and British Commonwealth Codes for the Construction of Welded Boilers and Pressure Vessels," ASME Paper 61-SA-40 (1961) *Engineer* (London), **211,** 5479, 122–125 (January 27, 1961).
7. Walter Samans, "Importance of Design to Tanks and Pressure Vessels," *Weld. J.,* **29,** 7 (January 1950).
8. J. J. Murphy, C. R. Soderberg, Jr., and D. B. Rossheim, "Considerations Affecting Future Pressure Vessel Codes," ASME Paper 56-MET-4 (1956); Similar article in API Proceedings, **33,** Sect. 3, 258–279 (1955); Discussion in *Weld. R. Coun. Bull. Ser.* **27,** 10–31 (May 1956).
9. G. A. Nelson, *Corrosion Data Survey* Shell Development Corp., Emoryville, Calif. 1954.
10. H. H. Uhlig, *Corrosion Handbook* Wiley, New York, 1948, pp. 1052–1058.
11. *Material Specifications Part I–Ferrous Metals* and *Material Specifications Part II–Non-Ferrous Metals,* American Society for Testing Materials, Philadelphia, 1980.
12. S. M. Jorgensen, "Pressure Vessel Design Calculation," *Pet. Refiner,* **24,** 109–113 (October 1945).
13. E. O. Bergman, "The Design of Vertical Pressure Vessels Subject to External Forces," ASME Paper 54-A-104 (1954).
14. Raymond C. Baird, "Aerodynamic Vibration of Tall Cylindrical Columns," ASME Paper 58-PET-4 (1958).
15. C. E. Freese, "Vibration of Vertical Pressure Vessels," ASME Paper 58-PET-13 (1958); ASME Trans., *J. Eng. Ind.* **81,** Ser. B, No. 1, 77–86 (February 1959).
16. V. O. Marshall, "The Design of Foundations for Stacks and Towers," *Pet. Refiner,* **22,** 8, 251 (August 1943).
17. *Minimum Design Loads in Buildings and Other Structures,* ANSI A58-1, American Standards Association, New York, 1972.
18. *Uniform Building Code,* Pacific Coast Building Official Conference, Los Angeles, Calif. 1952.
19. "Lateral Forces of Earthquake and Wind," ASCE Proc. **77,** 66 (April 1951). *ASCE* **117,** (1952.) Discussion in ASCE Proc., **78,** D-66 (May 1952).
20. Raymond J. Roask, *Formulas for Stress and Strain,* 3rd ed., McGraw-Hill, New York, 1954, Chap. 12.

21. F. W. Catudal and R. W. Schneider, "Stresses in a Pressure Vessel with Circumferential Ring Stiffeners," *Welding J.,* (London), **36** (12), 550S-552S (December 1957).

22. P. P. Bijlaard, "Computation of Stresses from Local Loads in Spherical Pressure Vessels or Pressure Vessel Heads," *Weld. Res. Coun. Bull.,* Ser. No. 34 (March 1957).

23. P. P. Bijlaard, "Stresses in a Spherical Vessel from Radial Loads Acting on a Pipe," 1–130, *Weld. Res. Counc. Bull.,* Ser. No. 49 (April 1959).
     P. P. Bijlaard, "Stresses in a Spherical Vessel from External Moments Acting on a Pipe," 31–62, *Weld. Res. Coun. Bull.,* Ser. No. 49 (April 1959).
     P. P. Bijlaard, "Influence of a Reinforcing Pad on the Stresses in a Spherical Vessel under Local Loading," 63–73, *Weld. Res. Coun. Bull.,* Ser. No. 49 (April 1959).

24. P. P. Bijlaard, "Stresses in Spherical Vessels from Local Loads Transferred by a Pipe;" "Additional Data on Stresses in Cylindrical Shells under Local Loading;" *Weld. Res. Coun. Bull.,* Ser. 50 10–50 (May 1959).

25. E. O. Waters, "Theoretical Stresses near a Circular Opening in a Flat Plate Reinforced with Cylindrical Outlet," *Weld. Res. Coun. Bull.,* Ser. 51 (June 1959) also *J. Eng. Power,* **81,** Ser. A., 189–200 (April 1959).
     D. W. Hardenbergh, "Stresses in Contoured Openings of Pressure Vessels," *Weld. Res. Coun. Bull.,* Ser. 51, 13–24 (June 1959).
     C. E. Taylor, N. C. Lind, and J. W. Schweiker, "A Three-dimensional Photoelastic Study of Stresses around Reinforced Outlets in Pressure Vessels," *Weld. Res. Coun. Bull.,* Ser. 51, 26–40 (June 1959).
     F. S. G. Williams and E. P. Auler, "Unreinforced Openings in a Pressure Vessel," *Weld. Res. Coun. Bull.,* Ser. 51 42–46 (June 1959).

26. *Tentative Structural Design Basis for Reactor Pressure Vessels and Directly Associated Components,* U.S. Dept. of Commerce, Office of Technical Services PB 151987 (December 1958).

27. F. E. Woloscewich, "Supports for Vertical Pressure Vessels, Part I," *Pet. Ref.,* **30,** 7, 137–140 (July 1951).
     F. E. Woloscewich, "Supports for Vertical Pressure Vessels, Part II," *Pet. Ref.,* **30,** 8, 101–108 (August 1951).
     F. E. Woloscewich, "Supports for Vertical Pressure Vessels, Part III," *Pet. Ref.,* **30,** 10, 143–145 (October 1951).
     F. E. Woloscewich, "Supports for Vertical Pressure Vessels, Part IV," *Pet. Ref.,* **30,** 12, 151–153 (December 1951).

28. N. A. Weil and J. J. Murphy, "Design and Analysis of Welded Pressure Vessel Skirt Supports," ASME 58-A-153 (1958), *J. Eng. Ind.,* **82,** Ser. B, 1 pp. 1–14 (February 1960).

29. J. T. McKean and G. P. Eschenbrenner, "Thermal Analysis and Design of Intermediate Heads of Pressure Vessels," ASME 58-PET-32 (1958).

30. R. G. Sturm and H. L. Obrien et al., "Stresses in Head to Shell Junction of Pressure Vessels," *Weld. J.,* **29,** 285–292 (June 1950).

31. Herman Schorer, "Design of Large Pipe Lines," *ASCE Trans.,* **97,** 101–119 (1933); Discussion (paper no. 1829), 120–191 (1933).

32. A. C. Barton, "Design of Ring Girders for Horizontal Tanks," *Pet. Ref.,* **33,** 6, 207–218 (June 1944).

33. O. L. Garretson and Harry R. Ziegler, "Design of Concrete Piers for Horizontal Storage Tanks," *Pet. Ref.,* **33,** 6 207–218 (June 1944).

34. L. P. Zick, "Stresses in Large Horizontal Cylindrical Pressure Vessels on Two Saddle Supports," *Weld. J.,* **30,** 9, 435S–445C (September 1951).

35. "Report on the Design of Pressure Vessel Heads," *Weld. Res. Suppl.* (Jan. 1953); *Weld. J.,* **32,** 1, 31S–41S (1953); Appendix 41S–52S (January 1953).

36. G. W. Watts and H. A. Lang, "The Stresses in a Pressure Vessel with a Flat Head Closure," *ASME Trans.*, **74**, 6, 1083–1090 (1952); Discussion, 1090–1091 (August 1952).

37. G. D. Galletly, "Torispherical Shells—A Caution to Designers," *J. Eng. Ind.*, **81**, Ser. B, 1, 51–62 (February 1959).
     G. D. Galletly, ASME 59-A-163, 1959, *Weld. Res. Coun. Bull.*, Ser. 54, 9 (October 1959).

38. *Welding Handbook,* American Welding Society, 1953, Section II.

39. R. D. Stout and W. D. Doty, *Weldability of Steels,* American Welding Society, 1953.

40. O. H. Henry and G. E. Claussen, *Welding Metallurgy,* 2nd ed., revised by G. E. Linnert, 1949.

41. *Non-destructive Test Requirements for Metals*, MIL-STD-271, Bureau of Ships, 1977.

42. Cleaning Requirements for Special Purpose Equipment, MIL-STD-767B, 1973.

43. S. J. Jorgensen, "Overstrain and Bursting Strength of Thick Walled Cylinders," ASME 57-PET-4 (1957), *Pet. Ref.*, **37**, 2, 163–169 (February 1958).

44. S. M. Jorgensen, "Overstrain Tests on Thick Walled Cylinders," ASME 59-PET-1 (1959), *J. Eng. Ind.*, **82**, (Series B) No. 2, 103–121 (May 1959).

45. S. M. Jorgensen, "Reduce Thickness by Overstrain," *Pet. Ref.*, **37**, 2 163 (February 1958).

46. R. W. Schneider, "Chart Compares Vessel Design Theories," *Chem. Eng.*, **63**, 1, 218 (January 1956).

47. R. R. Maccary and R. F. Fry, "Design of Thick Wall Pressure Vessel Shells," *Chem. Eng.*, **56**, 8, 124–127 (August 1949).

48. "High Pressure," *Chem. Eng.*, **56**, 8 171 (August 1949).

49. E. D. Narduzzi and George Welter, "High Pressure Vessels Subject to Static & Dynamic Loads," *Weld. Res. Suppl.* (May 1954).

50. C. E. Freese, "Mechanical Design Problems Connected with Ammonia Synthesis," ASME 56-PET-1 (1956), *Pet Ref.*, **36**, 1 193–197 (January 1957).

51. E. Jennett, "Design Considerations for Pressure-Relieving Systems," *Chem. Eng.*, **70**, 125–130 (July 8, 1963).

BIBLIOGRAPHY

J. Barford, "What is the Role of Materials in the Safety and Reliability Aspects of LWR's," International Congress on Material Aspects of World Energy Needs, Reston, Va., March 26, 1979, NTIS, PC A25/MF A01, 1980.

A. W. Barsell, "HTGR Safety Research Program," IAEA Specialists Meeting on Gas Cooled Reactor Safety and Licensing Aspects, Lausanne, Switzerland, September 1, 1980, NTIS, PC A02/MF A01.

G. Becker, "Concepts for the Closures of Large-Diameter Cavities," Annual Meeting on Nuclear Technology," Berlin, March 25, 1980.

R. Blondeau, "Study of the Different Aspects of Stress Relief Treatment in the Production of Pressure Vessels," *Rev. Metall. (Paris),* **77**(1), 61–81 (1980).

R. Blundell and P. B. Bamforth, "Special Concretes for Special Problems," *Nucl. Engr. Int.,* **25**(294), 43–46 (1980).

A. J. Boland, et al., "Development of Ultrasonic Tomography for Residual Stress Mapping," Final Report, EPRI-NP-1407, NTIS PC A05/MF A01, 1980.

M. K. Booker, "Fatigue of Weldments in Nuclear Pressure Vessels and Piping," Report No. NUREG/CR 1351, Oak Ridge National Laboratory, 1980.

D. M. Boyd, "Holographic and Acoustic Emission Evaluation of Pressure Vessels, March 5, 1980, UCRL 84046, NTIS PC A02/MF A01.

D. M. Boyd and W. W. Wilcox, "Weld Evaluation on Spherical Pressure Vessels Using Holographic Interferometry," UCRL 84079, NTIS PC A02/MF A01, 1980.

J. M. Brear and B. L. King, "Assessment of the Embrittling Effects of Certain Residual Elements in Two Nuclear Pressure Vessel Steels (A533B and A508)," *Philos. Trans. R. Soc. London, Ser. A.,* **295**(1413), 291 (1980).

K. B. Broughton, "Wire Winding System for Stressed Concrete Pressure Vessels," *Nucl. Eng.,* **21**(1), 84–86 (1980).

R. A. Buchanan, "Analysis of the Ultrasonic Examinations of PVRC Weld Specimens 155, 202, and 203 by Standard and Two-Point Coincidence Methods," *Weld. Res. Counc. Bull.,* No. 257 (February 1980).

D. A. Canonico, "Material Considerations in Assessing Safety and Reliability of Light-Water Reactor Pressure Vessels," International Congress on Materials Aspects of World Energy Needs, Reston, Va., March 26, 1979, NTIS, PC A25/MF A01, 1980.

A. B. Christensen, et al., "Preliminary Safety Evaluation of a Commercial-Scale Krypton-85 Encapsulation Facility," NTIS PC A02/MF A01, Paper Delivered at 16 DOE Nuclear Air Cleaning Conference, San Diego, Ca., October 20, 1980.

J. Ciprian, "Selected chapters from National and International Catalogues of Technical Rules on the Problem of Pressure Vessel Design, Pt. 2: Cylindrical Shells under External Pressure," *Verfahrenstechnik (Mainz),* **14**(2), 100–106 (1980).

J. Clements, "Organic Materials Division Research Bulletin," Lawrence Livermore Laboratory, May 1980, UCRL 52941, NTIS, PC A02/MF A01, May 1980.

M. P. Gomez, "J-Integral Plastic Fracture Mechanics Evaluation of Stability of Cracks in RPV," NTIS, PC A99/MF A01, pp. 802–810, 1980.

M. P. Gomez, "J-Integral Elastic Plastic Fracture Mechanics Evaluation of the Stability of Cracks in Nuclear Reactor Pressure Vessels, SAND- 80–0688, NTIS, PC A05/MF A01, 1980.

G. E. Grotke, "Survey of Welding Processes for Field Fabrication of 2.25 CR-1 Mo Steel Pressure Vessels," DOE/ET/13511-T1, NTIS PC A06/MF A01, April 1980.

M. A. Hamstad, "Acoustic Emission for Equality Control of Kevlar 49 Filament-Wound Composites," NTIS, PC A02/MF A01, UCRL 83783, 1980.

C. E. Harris, "Program for Addressing the Fracture Toughness Requirements of Appendix G to 10 CFR 50 (Thermal Reactor Safety), NTIS, PC A99/MF A01, 1980.

G. Henjes, "Systematic Measures for the Prevension of Damage to Pressure Vessels, Steam Boilers and Pipelines," *Schweissen Schneiden,* **32**(3), 108–112 (March 1980).

P. Hoekler, "Identification of Crack-Like Flaws and Determination of Crack Parameters in Ultrasonic Non-Destructive Evaluation," *Materialpruefung,* **22**(1), 32–39 (1980).

A. Holt and J. Brophy, "Defect Characterization by Acoustic Holography, Vol. 1, *Imaging in Field Environments,* Final Report, NTIS, PC A15/MF A01, September 1980.

I. W. Hornby, "Strain Behavior of a Pre-Stressed Concrete Reactor Vessel After Twelve Years Operation," *Int. J. Pressure Vessels Piping,* **8**(1), 25–40 (1980).

P. H. Huppertz and A. Retter, "Selection of Materials for Pressure Vessels and Chemical Plants," *Z. Wertstofftech.,* **11**(4), 124–133 (1980).

L. D. Kenworthy and C. D. Tether, "Users Manual Data Base MATSURV.: Reactor Pressure Vessel Material Surveillance Data Management System," Report No. ALO-63, Dep. NTIS PC A06/MF A01, 1980.

K. Klein, "Determination of the Optimum Prestressing of Cylindrical Pressure Vessels by Plastic Deformation," *Verfahrenstechnik (Mainz),* **14**(2), 107–110 (1980).

A. Kuhlmann, "Safety in Conventional Power Stations," *Brennst.-Waerme-Kraft,* **32**(6), 241–247 (1980).

L. Laska and A. Schumacher, "New Steam Boiler Regulation," *Brennst.-Waerme-Kraft,* **32**(6), 231–241 (1980).

H. J. Manthey, Use of a system for the detection of minimum-size leaks in the reactor pressure vessel of the NS Otto Hahn, pages 799–802, Annual Meeting on Nuclear Technology, Berlin, March 25, 1980 (in German).

D. E. McCabe and J. D. Landes, "Design Properties of Steels for Coal Conversion Vessels: Report on Long-term Testing, NTIS, PC A 03/MF A01, September 1980.

R. J. Morgan, "Failure Modes and Durability of Kevlar/Epoxy Composites," URCL 83875, NTIS, PC A02/MF A01, 1980.

Oak Ridge National Laboratory, "Gas-Cooled Reactor Programs: High-Temperature Gas-Cooled Reactor Base-Technology Program: Annual Progress Report for Period Ending December 31, 1979," ORNL 5643, NTIS PC A13/MF A01 1980.

G. Oestberg, "Role of Materials in the Safety and Reliability of Light-Water Reactors," International Congress on Materials Aspects of World Energy Needs, Reston, Va., March 26, 1979, NTIS, PC A25/MF A01, 1980.

V. Provenzano et al., Fractographic and Microstructural Analysis of Stress Corrosion Cracking of A533 Grade B Class 1 Plate and A508 Class 2 Forging in Pressurized Reactor-Grade Water at 93 Degrees C," NTIS, PC A02/MF A01, 1980.

D. T. Ramani et al., "Dynamic Analysis of a Primary Reactor Coloant System Subjected to Seismic and Local Loads, Including Shield Building Interaction," *Int. J. Pressure Vessels Piping,* **8**(1), 1–13 (1980).

G. Rump, "Failures in Steam Boilers—Repair and Testing, Part 2," *Tech. Ueberwach,* **21**(1), 36–39 (1980).

W. Schmitt and R. Wellein, "Methods of Determining the Influence of Quality Assurance on the Reliability of Primary Components of a PWR," *Int. J. Pressure Vessels Piping,* **8**,(3), 187–195 (1980).

W. Schmitt et al., "Linear Elastic Stress Intensity Factors for Cracks in Nuclear Pressure Vessel Nozzles Under Pressure and Temperature Loading," *Int. J. Pressure Vessels Piping,* **8**,(1), 41–68 (1980).

Schneider, Jr., "HP-Series of AD-Specifications and On-Site Fabrication of Pressure Vessels," *Schweissen Schneiden,* **32**(4), 129–133 (April 1980).

H. D. Schulze, and H. Fuhlrott, "Stable Crack Growth and Variation of Crack Opening Displacement of Pre-Cracked Specimens under Sustained Load," *Int. J. Pressure Vessels Piping,* **8**(2), 131–142 (1980).

T. E. Scott, "Pressure Vessels for Coal Liquefaction: An Overview," NTIS, PC A03/MF A01, 1980.

D. A. Shockey, "Computational Modeling Microstructural Fracture Processes in A533B Pressure Vessel Steel," Final Report, EPRI-NP-1398, NTIS, PC A06/MF A01, May 1980.

C. O. Smith, "Design stresses in Probabilistic Form for Ellipsoidal and Toroidal Pressure Vessels, *Int. J. Pressure Vessels Piping,* **8**(2), 143–153 (1980).

R. Stahlberg, "Comparison of Methods to Determine JSUB(IC) Values for a Reactor Pressure Vessel Steel," *Schweissen schneiden,* **32**(3), 115–116 (1980).

D. Sturm, M. Loss, and W. H. Cullen, "Autoclaves for Fatigue Crack Growth Tests with Unirradiated and Irradiated Specimens, *Materialpruefung,* **22**,(1), 5–10, (1980).

H. Tada and P. C. Paris, "Further Results on the Subject of Tearing Instability," Report No. NUREG/CR 1220, Vol. 2, NTIS, 1980.

"Trends in Steel Technology," *Met. Prog.,* **117**(1), 24–32 (1980).

L. R. Tripp, "High Pressure Gas Metering Project," NTIS, PC A03/MF A01, UCID 18643, 1980.

W. C. Turner and J. F. Malloy, *Handbook of Thermal Insulation Design Economics for Pipes and Equipment,* Robert E. Krieger Publishing Co., Huntington, N.Y., 1980.

M. Tvrdy, "Fracture Toughness and other mechanical metallurgy characteristics of pressure vessel steels," *Int. J. Pressure Vessels Piping,* **8**,(2), 91–103 (1980).

J. Vazquez and P. C. Paris, "Application of the Plastic Zone Instability Criterion to Pressure Vessel Failure, CSNI Specialists Meeting on Plastic Tearing Instability, Nuclear Regulatory Commission, CSNI 39, Conf. 790963, 1979.

M. E. Verkade and T. E. Scott, "Bibliography of Hydrogen Attack of Steel and Relevant Ancillary Topics," IS 4747, NTIS, PC A12/MF A01, 1980.

J. J. Wagschal et al., "LWR-PV Damage Estimate Methodology," ANS Topical meeting on 1980 Advances in Reactor Physics and Shielding, Sun Valley, Idaho, September, 14 1980, NTIS, PC A02/MF A01, 1980.

J. Wiberg, "Investigation of 'Aging' of Pressure Vessel Steel during the Growth of a Fatigue Crack," *Int. J. Pressure Vessels Piping,* **8**,(2)79–90 (1980).

T. Wilkie, "Cracks in French Pressure Vessels Pose No Danger," *Nucl. Eng. Int.,* **25**(294), 27–29 (1980).

M. L. Wilkins and J. E. Reaugh, "Plasticity Under Combined Stress Loading," UCRL 83322, NTIS PC A02/MF A01.

Witschakowski, W., "Material Properties Influence the Geometry of Pressure Vessels," *Maschinenmarkt.,* **86**(31), 624–627 (1980).

9

The Relationship Between Safety Management and Occupational Health Programs

D. Jack Kilian, M.D.

The goals of the occupational physician cannot be clearly separated from the goals of the safety manager or loss-control specialist, even though it may have seemed, in the past, that the work of these two professionals was within quite different areas.

The compensation-oriented specialist of earlier times, both in safety and medicine, was largely influenced by the need for attention focused on an appalling rate of death and disability associated with machinery and equipment.[1] Because of this, the early safety practitioner concentrated almost exclusively on traumatic injury prevention, while the early industrial physician or "company doctor" concentrated on treatment of acute injuries or poisoning, the return of the worker to full employment, or certification of a claim for compensation. The early industrial physician was often someone working part-time, without specialized training in occupational medicine, and without sufficient knowledge of just what went on in the work situation.

Too often, there was a complete lack of understanding and communication between safety and medical directors as to causes of and responsibility for specific types of work-related morbidity; the same lack of understanding, on both sides, would often characterize the degree and expected duration of disability. The physician, trained to see patients as individuals, not as members of a larger work force, would tend to be unsympathetic to the tendency of some safety engineers to see illness and injury only in terms of lost-time records and compensation cost. The safety engineers, on the other hand, could sometimes feel that the physician did not really know what the industrial environment involved or that the company doctor did not appreciate either the tremendous costs associated with high injury/illness rates and the amount of human suffering that the safety practitioner has been able to prevent.

To some extent, much of these attitudes and misunderstandings can be expected to remain as handicapping factors in safety manager/occupational physician interaction, now and in the future. There will still be safety managers who only see worker health in terms of lost-time statistics and who openly or covertly discourage workers from seeking medical help for on-the-job problems. There will still be physicians who do not know, or even ask about, the actual work environment that the patient must confront day after day, year after year, or, who are too readily amenable to the patient's suggestions that his problems are work-related when they are due, in actuality, to his personal habits and lifestyle.

However, recognition of the fact that the goals of occupational health and industrial safety are still far from being met may help professionals in both fields move toward these goals in a cooperative manner. Recognition of the conceptual and practical barriers that have divided the safety manager and the occupational physician in the past may help to resolve these differences and to find common areas of concern in the future.

Without doubt, the occupational health specialist must recognize that the injury-oriented approach of the early safety specialist resulted in a drastic curtailment of worker death and disability rates. Both specialists, now, must come to grips with the need for a different emphasis upon new, multifactorial problems that continue to emerge from the worker equipment, material/ environment components of occupational systems in general and the chemical industry in particular. Both specialists must learn to work, cooperatively, within greatly expanded definitions and concepts of their roles.

More than 25 years ago, the World Health Organization defined occupational health as

> . . . the promotion and maintenance of the highest degree of physical, mental, and social well-being of workers . . . the adaptation of work to man and of each man to his job.

Within recent years, the meaning of the term *physical harm*, within the context of occupational/industrial accident prevention, has come to include both tramatic injury and disease as well as adverse mental, neurological, or systemic effects resulting from work-place exposure. In addition, the very concept of the *accident* itself has broadened to that of a *contact* with a source of energy (electrical, chemical, kinetic, thermal, ionizing, etc.) above the threshold limit of the body or with a substance that interferes with normal bodily processes.[2]

Occupational medicine, meanwhile, has also broadened its horizons by considering that workers (who are, statistically, one of the healthiest segments of society) can best be served by an emphasis on health rather than disease. From this point of view, the goals of an occupational health program should be the preservation and, if possible, the improvement of the health of the entire workforce, from the chief executive officer to the newest trainee.[3] Such a program would embrace four basic elements: *prevention*, with renewed emphasis on education; *diagnosis/treatment/care* of the acutely sick or injured; *rehabilitation,* with emphasis on proper placement of the handicapped individual; and *counseling* of the troubled employee.

The scope of each program will depend on the size and stability of the business, its geographic location, the availability of trained health-care professionals, the nature of the hazards to be reasonably expected, and, perhaps most important of all, the philosophy of management and labor. In some instances, many of the functions of an occupational health program will be mandated by legislation requiring particular types of examinations, at given time intervals, of workers exposed to specific substances. The successful occupational health program, however, will shape itself to the specific needs of its own organization and work force, which are often far different from legislatively imposed requirements.

The occupational health program of the author's organization operates within a conceptual framework of "Product Stewardship," a voluntarily issued set of internal guidelines to implement a total commitment to appropriate protection of employee and public health. Without a commitment of this type on the part of management, successful operation of occupational health and safety programs can only be a goal and not a reality. In addition, it is necessary to have intensive, informed, cooperative interaction on the part of many individuals. In this light and since the goals of both safety and health workers overlap in so many ways, cooperation between the occupational physician and the safety manager is crucial.

9.1. ACUTE ILLNESSES AND INJURIES

The function that people associate most often with an industrial medical department is that of diagnosis/treatment/care of individuals who have suffered a disabling injury or who have been overcome by some substance in the work place. It goes without saying that a medical department should be prepared—with personnel, supplies, and equipment—to deal with these injuries and illnesses as they occur. In addition, the medical department should have well-thought-out contingency plans for the prompt transportation of persons who cannot be treated adequately within the department or the community and for disastrous situations in which large numbers of people must be cared for. In regard to disasters, coordination of planning with safety professionals is essential. Planning needs must be individualized:

we on the Gulf Coast must think of hurricanes and rattlesnakes and do not give much thought to the prevention and treatment of frostbite; other plants in other places would have different concerns.

The personnel of the occupational health department of a chemical company must know (or be able to find out quickly) the hazardous properties of the chemicals to which the workers are actually or potentially exposed and the appropriate treatment for injury caused by these substances. There is often difficulty in getting the necessary information in the shortest possible time, and this problem becomes more complex the larger the company and the greater the number of compounds that are manufactured, handled, and transported. Knowledgeable operating and safety personnel are invaluable sources of this data. The information itself can be stored on cards, in notebooks and files, or in a computer. From the point of view of the medical and nursing personnel, the important thing is that the information is current, complete, and rapidly accessible.

The well-run department of occupational medicine will maintain good working relationships with community physicians, health associations, and hospitals; it will be equipped with up-to-date knowledge of specialized facilities, such as a burn center, and skills, such as a retinal surgery service, to which patients can be referred without delay when needed. In some situations, it is useful to make provision for a visiting nurse, usually an R.N. trained in occupational health, to follow patients laid up at home or confined to a hospital for a period of time. The patient is reassured at not being forgotten by his company and its medical people, and the medical department can make plans on the basis of regular reports of the patient's progress and attitude. The disabled person usually has a good idea of just how he was injured, and his painfully acquired insight is sometimes an invaluable tool in prevention strategies.

In some companies, it is a matter of policy that *all* injuries, no matter how minor, are to be reported to and treated by the medical department. In other organizations, selected groups of personnel are trained in first aid and given basic supplies for on-the-spot treatment of small cuts, bruises, and the like. Unfortunately, in some small plants, there are no formal provisions for even first-aid procedures.

Assuming a medical department, there are several factors that will influence the extent to which it is utilized. The first and foremost of these is the attitude of management, from the top on down to the level of the supervising foreman. It is within the power of management to steadfastly maintain a policy that encourages early treatment of all injuries or to communicate a subtle, but pervasive, message of disapproval throughout the organization if such treatment is not given. In a growing number of companies, enlightened self-interest has shown top-level management that early and prompt treatment of minor injuries is the best way to avoid some of the cost of major disabilities. In regard to the implementation of management policy, the philosophy of the safety program can be a decisive factor, particularly if safety performance indices are maintained on the basis of the actual or potential severity of the incident (as now required in many circumstances by the Occupational Safety and Health Administration), rather than on merely a numerical count of visits to the medical department.

Utilization of an occupational health program will also depend on how conveniently the medical facility is located in relation to the work sites. A worker who does not consider himself or herself seriously injured may decide it is not worth the trouble if he or she must take an inordinate amount of time or effort to even show up at the medical facility's front door. Crowded, unpleasant conditions, long periods of waiting, and the need for even a minimum of paperwork can discourage many people.

The attitude of the medical department and its personnel will also affect the worker's attitudes. Friendly, caring physicians, nurses, technicians, and clerical personnel who treat all employees with dignity and respect have powerful voices when it comes to safety and health. Employees tend to trust the advice of those who have actually cared for the injured and sick. This trust and creditability are, in fact, the occupational health team's most valuable currency.

Accident prevention is not something to be preached by a health worker after an injury has occurred, but a matter of day-to-day alertness to the hazards that the worker may encounter and the means available to avoid them. Noticing that an employee has had a series of minor burns, for example, might be a signal that this individual is working in unsafe conditions or that he or she is being careless. In either case, recognition that a pattern exists will allow an investigation to be made and positive steps to be taken—by both health and safety personnel—to prevent a more serious burn.

Members of the occupational health team can also be excellent models of healthful and safe behavior in the chemical environment, particularly in matters of life style. Industrial physicians and nurses are often observed to see how they handle such things as the appropriate use of hard hats and safety glasses, the use of automobile seat belts, the use of tobacco and the abuse of alcohol, the control of cardiac risk factors, and the careful practice of safety rules within the medical department's treatment rooms and laboratories. Occupational health workers (and safety managers) who practice what they preach can be greatly influential in persuading others to lead healthier and safer lives.

The liberal use of photographic slides, tape recordings, film strips, and videotapes can be of great value in the direct education of workers-at-risk and health and safety professionals. Figure 9.1 illustrates a slide from a presentation, prepared by the author, on chemical burns of the eye.[23]

It is unfortunate that many smaller plants cannot afford or justify the services of a full-time physician. Except in the largest of companies, there are relatively few employer-maintained departments of occupational health that provide round-the-clock facilities and the services of fully qualified

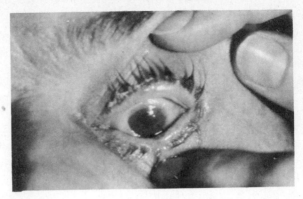

Figure 9.1. Illustration of the long-term effects of a chemical burn to the eye. While there was no visual impairment, permanent loss of many of the eyelashes of the lower lid can be seen (Medical Graphic Arts).

occupational physicians, nurses, technicians, and other health personnel. Many smaller or marginal operations find it completely impractical to think in terms of medical and nursing services and are finding it a struggle to provide the minimum of legally required medical monitoring. Many worker populations are poorly educated and/or highly transitory, making it difficult to implement the best-intentioned schemes of health and safety intervention. Outside of the most highly developed industrial nations of the world, effective health and safety programs are few and far between.

9.2. THE PREVENTION OF CHRONIC ILLNESS

It is now recognized that exposure to relatively low levels (i.e., not enough to cause acute injury or immediate symptoms) of many industrial chemicals can produce, in some persons, serious chronic disease* leading to significant disability or premature death. The detection and, when possible, the prevention of even the earliest possible signs of this insidious type of exposure has been the objective of occupational health professionals in government and private industry for many years.[5] Intense public concern in these matters led to passage of the Occupational Safety and Health Act of 1970, and since then, there has been increasing emphasis of governmental regulation of permissible levels of exposure to various substances and worker-health examination requirements.

Unfortunately, chronic diseases are often characterized by long latency

---

*Since the turn of the century, average life expectancy in the United States has lengthened, and the average age of the population has increased. At the same time, chronic conditions (such as heart disease, cancer, and stroke), which are characteristic of older populations, have replaced infectious diseases (such as pneumonia, diphtheria, and influenza) as the leading cases of death.[4]

periods, of years or decades, between exposure to the causative agent and development of the overt disease process. These long periods of latency can make it difficult, or impossible, to determine if exposure to a particular chemical is indeed the sole or contributing cause of a specific disease process in an individual, particularly if the disease itself is a common one, such as heart disease in American males, breast cancer in American females, high blood pressure in blacks, and skin cancer in whites of both sexes. Only when the disease is rare, such as mesothelioma in the general population, and is found to occur with increased frequency in a particular occupational groups, as was the case with asbestos workers, can a strong cause-and-effect relationship be assumed. Even in the case of asbestos, other factors can complicate the picture: we know today, for instance, that the male asbestos worker of years past is now at increased risk of lung cancer if he has been a cigarette smoker. Some people seem to be less or more susceptible to occupationally induced disease; perhaps they have inherited differing immunity patterns or some factors in their life styles makes the difference.[6]

It is of interest to note that one of the most widely publicized occupational disease discoveries in recent years—that of angiosarcoma of the liver in vinyl chloride workers exposed years ago to high concentrations of the monomer—was made and communicated by a physician associated with one of the affected industries.[7,8] As in other instances of unsuspected hazards, this work by an alert occupational physician led to the rapid development of industry-wide standards and engineering controls to prevent unsafe levels of exposure in the future.

Increasingly sensitive instrumentation now makes it possible to monitor the work-place environment and the worker's breathing zone with great precision. In addition, increasingly sophisticated methods and instruments are being developed for evaluation of subtle bodily changes and laboratory analysis of body fluids and tissues. However, for the reasons mentioned above and since we are all susceptible, in one way or another, to disease processes that start "all on their own," it is still very difficult to draw conclusions and to make predictions on the basis of environmental and personal monitoring. Even with sophisticated examination methods, it remains almost impossible to pick up signs of many occupationally induced diseases at a point where intervention is still feasible. Except in the most general way (and except in the case of a few uncommon enzyme deficiency states, such as $alpha_1$-antitrypsin and glucose-6-phosphate dehydrogenase deficiencies), it remains impossible to predict who will develop a disease and who will not.

9.3. ANIMAL TESTING

The safety of chemicals in the industrial environment and in general use is most easily evaluated by means of animal tests. This is particularly true in regard to those chemicals that can cause acute injury. In trying to evaluate chemicals for long-term chronic toxicity, however, it must be remembered that data derived solely from animal work has inherent limitations.[9] Only human beings, for example, can evaluate odor or report on such subjective symptoms as irritation and nausea. The type of sensitivity reactions that so often plague chemical workers is difficult to induce in animals, and direct human experience is necessary if this effect is to be detected. There are genetic defects (such as G-6-PD deficiency) found only in human beings. Effects upon the human nervous system, often seen as behavioral changes, cannot be adequately detected in animal models, and reactions mediated by human metabolic pathways (which are themselves influenced by age, sex, heredity, and other factors) are sometimes difficult to extrapolate with confidence from animal findings.

9.4. PREEMPLOYMENT/PREPLACEMENT EXAMINATIONS

The preemployment, or preplacement, examination is of value to both the company and the employee. The company benefits from having prospective workers screened for conditions that would seriously hamper on-the-job performance. The objective is not to screen out the less-than-perfect speci-mens that most of us are, but to be reasonably sure that workers will be physically able to carry out their responsibilities without undue stress or demands upon their physical or mental well-being. New employees also benefit by not being expected to do more than their actual physical status will permit. The preplacement examination, however, is of even greater value to both workers and the company when it is used to document the history and baseline measurements of standard physical parameters against which any future change can be measured.

9.5. MEDICAL SURVEILLANCE

The preemployment examination can be the cornerstone of an effective, comprehensive program of health surveillance for all workers exposed to toxic chemicals. As outlined in Table 9.1, a medical monitoring, or health surveillance, program ideally starts with the recording of a complete history of the person's medical, personal, family, and occupational background. The wording of the questions to be asked in obtaining this history and the specific types of information that are desired should be carefully evaluated in ad-vance. For the purposes of comparability and later analyses, the same ques-tions should be asked of each person and the answers recorded in a standardized format, preferably one that can be easily transferred to com-puter storage. In some cases, it is useful to ask the individual to fill out a health-inventory questionnaire at home. The questions and answers can then

TABLE 9.1 Comprehensive Surveillance Program for Workers Exposed to Toxic Chemicals

A. Preplacement examination
  1. History: medical, personal, family, occupational
  2. Physical examination, standard protocol
  3. Laboratory work-up
    a. Standard protocol
    b. Tests specific for job
    c. Mutagenic baseline
  4. Reevaluation upon job transfer
B. Periodic evaluation
  1. Update history
  2. Repeat physical examination
  3. Laboratory tests
    a. Standard and specific
    b. Routine exposures
    c. Overexposures
    d. Mutagenic evaluation
  4. Pre- and post-retirement evaluation
C. Epidemiology
  1. Case-control, prospective, historical-prospective
  2. Reproductive and other special studies
  3. Chronic disease, death diagnosis registries

be reviewed with the individual for consistency and completeness at the time of the health evaluation appointment. If a questionnaire is used, it is always desirable to pretest its design and format with a few trial runs.

Getting a complete and accurate history is one of the medical department's most important jobs. It is a particularly useful tool in accident prevention, since it is usually within the medical department that the train of events leading up to the accident is first and fully put together. In almost every case, the history obtained by the medical personnel will prove to be the most accurate since later attempts to reconstruct the incident can be easily influenced by faulty recall and work place pressures.

Ideally, the preemployment and periodic examinations should be carried out according to a standardized protocol, so that the findings of various physicians and the measurements recorded by technicians upon examination of many different people can be compared and evaluated in relation to one another. A standard battery of laboratory tests should also be used for the basic evaluation of each individual. The development of automated laboratory equipment has made it relatively simple to perform a standard set of laboratory procedures in a uniform manner. If the prospective job requires exposure to a chemical known to be hazardous to a particular body system, as, for instance, overexposure to benzene is known to be potentially injurious to the blood and blood-forming organs, special tests can be done to rule out the presence of a predisposing condition. The format of the preplacement examination should not be reserved for new employees—it is equally useful when a worker is being considered for transfer from one job area to another.

9.6.   SPECIAL PROBLEMS: MUTAGENICITY

In recent years, it has become known that many chemicals and physical states will produce mutations in plants, bacteria, and animals.[10] Although it is difficult to prove that specific agents are mutagenic in the human being, the evidence of mutagenicity in lower life forms and the prospect that similar processes might occur in the human species have become matters of serious concern for many chemical industry occupational health specialists. The question of possible chemical mutagenicity is so troublesome, first, because of the risk of damage to the unique genetic heritage that each of us has received from our ancestors and that we can pass on to our children and children's children; since the human being is a finely honed product of many eons of evolution, any change in our genetic constitution is most likely to be expressed in a disadvantageous way. The possibility of mutagenic response to certain chemicals is also a matter of grave concern to many researchers because it has been claimed that substances that can cause mutation can also cause cancer.[11]

At the present time, there are a number of tests available to check for possible mutagenicity in various substances.[12,13] Most tests rely on experimental changes induced in plants, bacteria, and animals. Many investigators consider any extrapolation from these observations (many of which involve the mutation back to normal of a mutated form) to the complex metabolism of the human being to be of doubtful validity.

One method of evaluating for mutagenicity, however, is based upon the direct observation of genetic material from persons exposed to possible mutagens and, thus, may have more than experimental relevance to the concerns of those involved in chemical operations. This method is human cytogenetic evaluation[14,15]; in essence, it is the observation of chromosomes (Fig. 9.2) for visible abnormalities of structure or number. Cytogenetic evaluation, then, is the analysis of cellular sensitivity to the many different agents (such as radiation chemicals, viruses, and many medicines) that can produce loss, breakage, or rearrangement of the chromosomal material.

Most of us have a number of chromosomal aberrations present in the cells of the body from time to time; some people, as a result of exposure to a chromosome-breaking agent like ionizing radiation, have a great many aberrations that persist for years. While in almost all cases, the body manages to repair chromosome aberrations, in some instances, viable abnormal cell lines may be established. It has been found that these abnormal cell lines are sometimes statistically associated with increased incidence of serious dis-

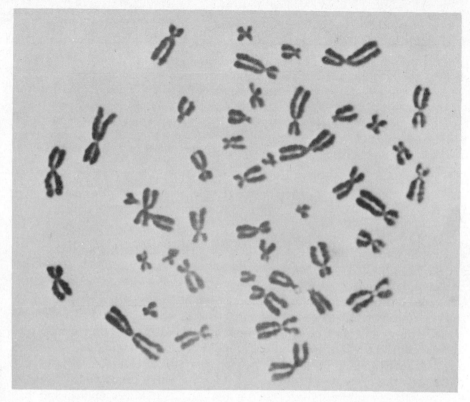

Figure 9.2. Normal chromosome complement in blood cell of human male.

case or sometimes passed along to subsequent generations, being expressed in increased rates of fetal wastage or birth defects.

Except in a few cases, evaluation of the chromosomes of an *individual* will not allow prediction of future health status for that person or his/her children. On the other hand, evaluation of the frequency of chromosome aberrations in a *group* of persons has sometimes revealed an apparent association between exposure of the group to a chromosome-breaking agent and an increased frequency of chromosome abnormalities. This association warrants concern if comparison with a similar group—that has not been exposed to the substance in question—shows a significant difference in the aberration rates of the two groups. The concern is warranted because it is known that some groups of people with a greater than usual average number of chromosome aberrations also have, as a group, a statistically increased risk of developing cancer. This association has been seen in A-bomb survivors, some worker groups, and in patients treated for various conditions with chromosome-breaking drugs and radiation (Table 9.2); the association is also found in some diseases states (Table 9.3), most of which are relatively rare.

TABLE 9.2   Neoplastic Conditions That Also Display an Increased Frequency of Chromosome Aberrations

1. *Leukemia and thyroid cancer* in A-bomb survivors
2. *Leukemia* in patients given x-ray treatment for ankylosing spondylitis
3. *Lymphoma and leukemia* after treatment with a chromosome-breaking drug for rheumatoid arthritis
4. *Lung cancer* in uranium miners (risk is more greatly increased in those who smoke tobacco)
5. *Liver cancer* in persons diagnostically exposed to thorium dioxide and in vinyl chloride workers
6. *Osteogenic cancer* in luminous watch dial painters
7. *Visceral cancers* after treatment with a chromosome-breaking drug for psoriasis

9.7. SPECIAL PROBLEMS: REPRODUCTIVE DISORDERS

Because of the great number of factors that can influence the conception and development of a normal, healthy human infant,[16] it is extremely difficult to make definitive judgments about the possible effects of industrial chemicals, drugs, and diet on the reproductive cells of the human male and female and on the intrauterine environment of the developing fetus.[17,18] We are, again, hampered in our efforts at prevention of adverse effects by the fact that "there are as yet no precise rules by which teratogenic effects in humans may be inferred from experimental results in lower animals and that . . . observations on birth defects cannot be extrapolated uncritically from one species to another.[19]

A discussion of the problems involved in the formulation of public and private policy regarding the prevention of reproductive disorders (for review, see reference 20) is beyond the scope of this chapter. It can be said, however, that in a world increasingly dependent on chemicals to produce the

TABLE 9.3   Conditions (Neoplastic or with Neoplastic Association) Displaying Chromosome Anomalies

Chronic Myelocytic Leukemia	Ataxia-Telangiectasia
Acute Myelocytic Leukemia	Fanconi's Anemia
Chronic Lymphatic Leukemia	Bloom's Syndrome
Polycythemia Vera	Glutathione Reductase Deficiency
Myeloid Metaplasia	Down's Syndrome
Erythroleukemia	Turner's Syndrome
Sideroblastic Anemia	Klinefelter's Syndrome
Meningioma	Wilms' Tumor-Aniridia Syndrome
Burkitt's Lymphoma	Kostmann's Infantile
Retinoblastoma in D-Deletion	Agranulocytosis
Syndrome	

necessities of life for expanding populations, the study of the causes of birth defects and of reproductive impairment deserve greatly increased attention by the public and private sectors of the scientific community and that the reproductive status of the individual workers should be considered, in the same manner as other physical characteristics, when the occupational physician is called upon to give advice and counsel and to make recommendations about the work place environment.

9.8. FUTURE CHALLENGES

As shown in Table 9.4, the emphasis in occupational health and safety programs has changed considerably, and even greater changes are to be expected in the future. The challenges before us include the extension of effective medical surveillance to the 90% of all businesses in the United States that employ fewer than 25 persons and the development of practical and economical methods of long-term health surveillance for our increasingly mobile worker populations. We need methods to accurately measure the impact of industrial operations on surrounding community populations and greatly increased numbers of people trained in all aspects of occupational health and safety.

There will be a greater and greater need for compatible information systems within companies that operate in different localities and between companies that deal with similar products. Closely entwined with this will be the problem of proper safeguards for confidential medical records.

Because there is a risk that increased automation of testing procedures and increased governmental reporting requirements will result, not in useful information, but in mountains of unrelated, unanalyzable data, there will be a tremendous need for the development of innovative record-management and record-linkage systems. In addition, since many occupational health statistics are based only on death records of traceable, employed persons, company-, state-, and nation-wide registries of cause-of-death diagnoses for the total population will almost certainly need to be established and maintained.[21] Much greater emphasis will need to be placed on comparative toxicology and on the development of accurate screening tests to detect those individuals susceptible to various types of disease and to detect signs of incipient disease while there is still time for effective intervention.

It is also vitally important that we find and utilize the most effective methods of health and safety education. In recent years, it has become more and more clear that elements of life style—diet, smoking, drinking, driving, and general hygiene—have a profound influence on health status and life expectancy.[22] While any program of occupational health will continue to have its traditional responsibility to prevent adverse effects resulting from worker/work place, person/material interactions, it must assume an even greater responsibility for educational efforts that will lead to the workers'

TABLE 9.4   An Occupational Health and Safety Program

Historically

1. Treatment of traumatic injury, acute illness
   Evaluation and certification of Disability
   Emphasis on lost-time costs, statistics
2. Preemployment examinations
3. Primary reliance on animal toxicology

Current Trends

1. Prevention of acute and chronic injury/disease
   Education and counseling
   Periodic examinations
   Environmental/personal monitoring
2. Preplacement evaluation
   Establishment of baseline values
   Counseling of troubled employees
   Rehabilitation, placement of handicapped
3. Biomedical research

Future challenges

1. Human-oriented research
   Epidemiology
   Metabolic studies
   Reproductive and other special studies
   Interindustry research
2. Effective screening tests
   Chronic disease
   Individual susceptibilities
3. Job-exposure-health record linkage
   General population data banks
4. Education
   Lifestyle factors
   Emergency procedures

elimination, on their own initiative, of those risk factors that are within their own control and from which no government regulation or industry standards will protect them.

ACKNOWLEDGEMENT

I thank Trudy Barna Lloyd, M.P.H., for her help in preparation of this manuscript.

REFERENCES

1. Kenneth H. Rohner, "Occupational Safety in Industry," In: *Occupational Medicine. Principles and Practical Applications*, Carl Zenz, Ed., Year Book Medical Publishers, Chicago, 1975, pp. 21–40.

2. Frank E. Bird, Jr., "Safety." In: *The Industrial Environment—Its Evaluation and Control*, NIOSH, Rockville, MD., 20857, 1973, pp. 681–691.

3. Jan L. Kanzen, "Design and Operation of an Occupational Health Program," *The Industrial Environment—Its Evaluation and Control*, NIOSH, Rockville, M.D., 20857, 1973, pp. 693–701.

4. President's Science Advisory Committee, *Chemicals and Health: Report of the Panel on Chemicals and Health of the President's Science Advisory Committee*, Science and Technology Office, National Science Foundation. Washington, D.C., 1973.

5. Alice Hamilton and Harriet L. Hardy, *Industrial Toxicology*, 3rd ed., Publishing Sciences Group, Acton, Mass., 1974.

6. J. J. Mulvihill, "Congenital and Genetic Diseases." In: *Persons at High Risk of Cancer. An Approach to Cancer Etiology and Control*, F. J. Fraumeni, Jr., Ed., Academic Press, New York, 1975.

7. J. L. Creech Jr. and M. N. Johnson, "Angiosarcoma of the Liver in the Manufacture of Polyvinyl Chloride," *J. Occup. Med.*, **16**, 150–151, (1974).

8. "How They Found Vinyl-Cancer Link," *Chem. Week*, 30, 32, 56 (July 17, 1974).

9. B. D. Dinman, "Principles and Use of Standards of Quality for the Work Environment, *The Industrial Environment—Its Evaluation and Control*, NIOSH, Rockville, MD., 20857, 1973, pp. 75–93.

10. M. W. Shaw, "Human Chromosome Damage by Chemical Agents," *Ann. Rev. Med.*, **21**, 409–432 (1970).

11. Bruce N. Ames, William E. Dustron, Edith Yamasaki, and Frank D. Lee, "Carcinogens Are Mutagens: A Simple Test System Combining Liver Homogenates for Activation and Bacteria for Detection," *Proc. Nat. Acad. Sci.*, **70**, 2281–2285 (1973).

12. Committee 17, "Environmental Mutagenic Hazards," *Science* **187**, 503–514, (1975).

13. A. Hollaender, Ed., *Chemical Mutagens: Principles and Methods for Their Detection*, Vol. 4, Plenum Press, New York, 1976.

14. D. J. Kilian, D. J. Picciano, and C. B. Jacobson, "Industrial Monitoring: A Cytogenetic Approach," *Ann. N.Y. Acad. Sci.*, **269**, 4–11 (1975).

15. D. Jack Kilian, and Dante Picciano, "Cytogenetic Surveillance of Industrial Populations," In: *Chemical Mutagens: Principles and Methods for Their Detection*, Vol. 4, Plenum Press, New York, 1976, pp. 321–339.

16. James Wilson, *Environment and Birth Defects*, Academic Press, New York, 1973.

17. R. W. Smithells, "Epidemiology of Malformations: Inspiration and Perspiration," (Editorial) *Teratology* **10**, 217–220 (1974).

18. Thomas H. Shepard, James R. Miller, and Maurice Marois, Eds., *Methods for Detection of Environmental Agents that Produce Congenital Defects.* Proceedings of the Guadeloupe Conference Sponsored by l'Institute de la Vie, American Elsevier, New York, 1975.

19. Teratology Society, Teratogens and the Delaney clause, (Resolution), *Teratology,* **10**(2), 1 (1974).

20. Joan I. Samuelson, "Employment Rights of Women in the Toxic Workplace," *Calif. Law Review,* **65**, 1113–1142 (1977).

21. Brian MacMahon, Ed., "Methods and Resources for Estimating Disease Risk in Humans, *Human Health and the Environment: Some Research Needs,* Report of the Second Task Force for Research Planning in Environmental Health Science. National Institute of Environmental Health Sciences, DHEW (NIEHS) Publication No. 77-1277, Washington, D.C., 1977.

22. Ernst L. Wynder, and Gio B. Gori, "Contribution of the Environment to Cancer Incidence: An Epidemiological Exercise" (Editorial), *J. Natl. Cancer Inst.,* **58,** 825–832 (1977).

23. Medical Graphic Arts, "Chemical Burns to the Eye," "Chemical Burns to the Skin," "Prevention of Low-Back Pain," 35-mm slide-tapes with written commentary by D. J. Kilian, M.D., P.O. Box 34, Lake Jackson, Texas 77566.

BIBLIOGRAPHY OF SUGGESTED READING

H. Bartsch, "Predictive Value of Mutagenicity Tests in Chemical Carcinogenesis," *Mutat. Res.,* **38,** 177–190 (1973).

Charles H. Brigham, "The Influence of Occupations Upon Health," Third Annual Report of Michigan Board of Health, Reprinted in *J. Occup. Med.,* **18,** 235–241 (1975).

B. A. Bridges, "The Three-Tier Approach to Mutagenicity Screening and the Concept of Radiation-Equivalent Dose," *Mutat. Res.,* **26,** 335–340 (1974).

Geoffrey Ffrench, *Occupational Health,* Medical and Technical Publishing Co., Lancaster, U.K., 1973.

Marilyn K. Hutchinson, *A Guide to the Work-Relatedness of Disease*, National Institute for Occupational Safety and Health, NIOSH Publication No. 77-123, Rockville, MD., 20857, 1976.

S. Moeschlin, Outstanding Symptoms of Poisoning, pp. 644–678 in *Poisoning: Diagnosis and Treatment,* Grune & Stratton, New York, 1965.

Herbert Moskowitz, "Drug Effects in Relation to Industrial Safety," *Behavioral Toxicology: Early Detection of Occupational Hazards,* Charles Zinteras, Barry L. Johnson, and Ida de Groot, Eds., National Institute for Occupational Safety and Health, NIOSH Publication No. 74-126, Rockville, MD., 20857, 1974.

M. Gerald Ott, Benjamin B. Holder, and Ralph R. Langner, "Determinants of Mortality in an Industrial Population," *J. Occup. Med.,* **18,** 171–177 (1976).

H. A. Waldron, *Lecture Notes on Occupational Medicine.* Blackwell Scientific Publications, Second Edition, Oxford, 1979.

Proceedings of the Conference on Stress, Strain, Heart Disease and the Law, Jan. 26–28, 1978, Boston, Mass., sponsored by American Heart Asso., American Society of Law and Medicine, and President's Committee on Employment of the Handicapped, White House, Washington, D.C. 20006.

10

The Impact of TSCA on the Practice of Occupational Medicine

Irving R. Tabershaw, M.D.

In the last decade Congress has enacted landmark legislation, including the Toxic Substances Control Act (TSCA) which is affecting the practice of occupational medicine. In the past, some health hazards in industry were either unrecognized, overlooked, treated indifferently, or simply neglected. TSCA's aim is to help correct these deficiencies. If the law is administered wisely with the understanding of the limitations inherent in regulating scientific and medical endeavors, TSCA combined with these other legislative acts will force the occupational physicians to reorganize their skills, and to integrate them with the other industrial health disciplines. In so doing, the occupational physician will become effective in assisting to remedy the deficiencies still extant in protecting worker health. On the other hand, if equivocal, preliminary, or merely suggestive findings must be reported and then either misinterpreted or exaggerated by TSCA, the law may inhibit needed scientific investigations and lead to frustration and confusion of the health professionals and industry with consequent negation of the benefits that our society expects.

The practice of occupational medicine had undergone changes over time according to the laws and the social demands of that particular era. Occupational medicine was originally identified primarily as industrial surgery because the workers' compensation laws which were enacted subsequent to 1910 demanded that the employer provide medical care for injured employees. The practice of occupational medicine still embraces this provision of care after injury and in a given plant may still represent the entire practice

Table 10.1   Chemicals or Industrial Processes Associated with Cancer Induction in Humans Based Upon Study of Exposed People (Epidemiological Evidence)

Chemical or industrial process	Target organ
Aflatoxins	Liver
4-Aminobiphenyl	Bladder
Arsenic compounds	Skin, lung, liver
Asbestos	Lung, pleural cavity, gastrointestinal tract
Auramine (manufacture of)	Bladder
Benzene	Hemopoietic system
Benzidine	Bladder
Bis (chloromethyl) ether	Lung
Cadmium-using industries (possibly cadmium oxide)	Prostate, lung
Chloramphenicol	Hemopoietic system
Chloromethyl methyl ether (possibly associated with bis (chloromethyl) ether	Lung
Chromium (chromate-producing industries)	Lung, nasal cavities
Cyclophosphamide	Bladder
Diethylstilbestrol	Uterus, vagina
Hematite mining (? radon)	Lung
Isopropyl oils	Nasal cavity, larynx
Melphalan	Hemopoietic system
Mustard gas	Lung, larynx
2-Naphthylamine	Bladder
Nickel (nickel refining)	Nasal cavity, lung
N,N-Bis(2-chloroethyl)-2-naphthylamine	Bladder
Oxmetholone	Liver
Phenacetin	Kidney
Phenytoin	Lymphoreticular tissues
Soot, tars, and oils	Lung, skin (scrotum)
Vinyl chloride	Liver, brain, lung

Source:.   Tomalis et al., "Evaluation of the Carcinogenicity of Chemicals: A Review of the Monograph Program of the International Agency for Research on Cancer (1971 to 1977)," *Cancer Res.,* April 1978.

of occupational medicine. In World War II because there was a need to place workers properly in their jobs because of insufficient worker power, medical examinations to determine physical abilities and limitations were introduced and the practice of occupational medicine enlarged its dimensions to include medical examinations before employment, periodically, upon retirement, upon transfer, etc. These two activities, care of the injured and physical examinations, were the keystones of medical occupational practice until this last decade.

The Occupational Safety and Health Act with its mandate to promulgate health standards has turned attention specifically to the work environment and the consequences of exposure to noxious agents. This is now becoming the dominant feature of current occupational medical practice. TSCA focuses even further on an aspect that has long been neglected, namely, early recognition of the potential toxicity of the chemical agents with which the worker is in contact. TSCA's interest forces several disciplines which have previously been called upon irregularly and infrequently to function in a mutual effort to close this loophole in the prevention of workers' diseases.

The burden on industrial health specialists under the premanufacturing regulation in TSCA is significant. The potential hazards related to the manufacture of a *new* chemical must be defined by extensive toxicologic studies in experimental animals prior to manufacture. This burden falls most heavily on toxicologists and their associates, the pathologists, the geneticists, and the metabolic chemists. These are the disciplines most concerned with the potential of harm from any agent used in industry *before* it is introduced into active use. However, every chemical which is foreign to normal physiologic function may have, under some set of circumstances, adverse effects on the health of the experimental animal or man. Therefore, while the results of toxicological studies may demonstrate an element of relative safety, man must necessarily be monitored, after the chemical is approved for routine production, if only as a safeguard against untoward effects arising from species differences. The same skills of toxicology, pathology, oncology, and physiologic chemistry in addition to other disciplines, notably clinical medicine, epidemiology, and industrial hygiene are needed for a reasoned judgment on the potential cause and health effect relationship.

In addition to new compounds, the industrial health specialists must also be concerned with these 100,00+ compounds which have been compiled in NIOSH's *Registry of Toxic Effects of Chemical Substances,* Volumes I and II, revised annually since 1971. Presumably, experience has demonstrated these compounds to be relatively safe under specified conditions of use and production. Again the safety of a compound foreign to the human body is recognized to be relative and TSCA is systemizing its monitoring mechanism to ensure continued safety in the work environment.

In a number of cases, specific potential health risks have been recognized for identified molecular entities and for which medical monitoring programs have been devised. During the implementation of such programs, if untoward effects are demonstrated, i.e., never before reported—under Section 8(e) TSCA, an obligation exists to report the possible cause-and-effect relationship between the specific chemical and the adverse health effect observed. Where a potential toxin has been identified and surveillance carried out, untoward health effects are more easily recognized. However, the occupational physician now has an additional obligation. Under Section 4(c) of

TSCA, the physician is required to report suspect adverse health effects based on an unusual incidence of abnormalities which may be found in a selected population, even though the causative principle may not be immediately recognized. In most cases, these findings would be a chance observation in routine periodic examinations, in review of absenteeism for medical reasons, or in repetitive workers' compensation cases.

It should be pointed out that physicians, in their undergraduate education, are trained in the diagnosis of pathological conditions, that is, disease patterns which need correction, and in the treatment of these conditions to restore physiological balance. In occupational medicine, the focus is also on clinical diagnosis, but the physician's responsibility is not primarily treatment, but prevention. If a worker is overtly ill, then he or she is usually referred to the appropriate specialist. But, the occupational physician's basic responsibility is to prevent recurrence of the disease in that individual and to prevent its appearance in the individual's co-workers. Some general preventive measures are applicable to several classes of noxious agents but effective prevention demands that the causal agent be identified. Identification of the specific toxin in industry is necessary for at least two reasons:

1. To demonstrate to plant management that control of the specific agent will solve the problem as general preventive measures may be expensive. However, if a specific agent is identified, some measure of control may be exercised within reasonable financial limits, e.g., by segregation, by local exhaust ventilation, etc.

2. There is a likelihood of a workers' compensation or negligence lawsuit. The causal relationship between an illness and exposure in the work environment is the focal issue in these cases.

In order to carry out effectively this goal of identifying and controlling noxious agents, in addition to the clinical diagnosis, the occupational physician must be cognizant of a number of scientific disciplines, to which the medical student is minimally or not exposed at all. These skills are:

1. Toxicology—which determines the potential and the mechanism of a noxious biological effect from a specific chemical.

2. Industrial hygiene—which defines the degree of a hazardous exposure to a chemical agent in a plant, and its control.

3. Epidemiology—which studies whether the incidence and prevalence and patterns of distribution of illness within the plant population is consistent with the known exposures.

Chiefly through OSHA and TSCA's activities, it is increasingly being recognized that exposure to some chemicals may have serious and far-reaching health effects even though no obvious or subtle acute effect has occurred.

The latency period in cancers initiated by industrial chemicals is of continuing and growing concern to all health professionals in the field. There is another serious health effect which is causing concern, namely, the potential of damage to reproduction, including such effects, as teratogenesis, mutagenesis, fetal toxicity, etc.

Companies and particularly chemical companies, are regrouping internally the professional health skills resident in their various corporate headquarters, divisions and plants. Some companies are creating an executive position—a corporate health executive—to be responsible for all health matters.

In any event, industry's current attempts to respond to the demands of TSCA is severely limited by the number of physicians, industrial hygienists, toxicologists, and epidemiologists, who are knowledgeable and available to carry out the studies necessary to monitor the exposed workers, and to make a competent judgement on whether or not a given health condition is, or is not, caused by a plant exposure. Eleven or more so-called educational resource centers in various universities are currently being funded by the government to educate doctors, industrial hygienists, nurses, and safety personnel to fill this gap. In the next few years, these occupational health practitioners will be numerous enough, and distributed widely enough, to respond to TSCA's demands. Finding an incipient or cryptic physiological abnormality which may be caused by an industry agent demands that the worker be repeatedly examined, that baselines and trends in the biochemical parameters of each individual and of each group be established, that longitudinal studies to ascertain the long-term outcome of these changes be carried out, and that information on biochemical and clinical data of various groups of workers in different work environments be made available for comparative purposes. TSCA's demand that unexplained, or adventitious health findings be examined for its potential relationship to a worker's exposure, creates a new and uncharted field of endeavor for the medical occupational practitioner and co-workers in the allied industrial health professions. The success of TSCA is dependent in great measure on how well these health professionals, with industry's cooperation, perform these tasks. The scientific problems are great and are compounded by economic, social, legal, labor relations, and other factors. The potential of success is small in the near term, but it should increase as we learn to cope with the issues and with the administrative demands of TSCA. While it is hard to estimate the magnitude of TSCA's impact on occupational medical practice, there is no doubt that it exists and its continuing influence will depend a great deal on the wisdom and understanding of those who administer the act.

11

The Role of an Occupational Health Nurse in a Chemical Surveillance Program

A. J. Murphy, R.N.

11.1. PUBLIC AWARENESS

Chemical manufacturers, users, and disposers have indicated their concern and demonstrated interest in making the public, the private sector, and the industry itself more knowledgeable regarding the chemicals produced, the various properties of these chemicals, and their effects on humans and the environment. In addition to the information presently made available by the chemical manufacturers and users, they are also attempting, through various media, to educate the public and make them more aware of potential hazards that may exist if the substances are misused, and what responsibilities they must assume if such situations arise.

11.2. INDUSTRY'S TEAM APPROACH

To accomplish this, various avenues of communication may be addressed. In industry, where various types of chemicals are used, the team approach is most effective to promote, educate, and develop a safety consciousness among the employees in dealing with whatever chemical substance may be present.

The team may be made up of an environmentalist, a hygienist, an industrial physician, an occupational health nurse, a safety engineer, and a toxicologist. Of the team members, the occupational health nurse is the one most directly responsible for maintaining the wellness of the employee, from age 18 to 65+ years, during their employment and beyond. In occupational health we strive to maintain a healthy work place populated by a healthy work force of informed employees who accept the responsibility for maintaining their own health and safety. To accomplish this goal, the occupational health nurse must become involved in many different facets.

11.3. PREPLACEMENT HEALTH HISTORY

At the time of the placement health interview and assessment, the nurse carefully records past chemical exposure data supplied by the prospective employee. It is important to record previous places of employment and to obtain data as to chemicals handled and processes with which the applicant was involved, whether or not protective devices (personal protective equipment) actually worn and whether or not medical monitoring was maintained. If it appears to be advisable to obtain medical records from previous employers, the applicant should be asked to sign a release of information request which should state specifically the information desired.

11.4. NEW EMPLOYEE ORIENTATION

During the interview, the nurse informs the prospective employee, as required by OSHA, of the chemical exposures in the plant operation to which he or she may be subjected. It should be emphasized that an employee's health and safety is a shared responsibility between the employee and the company, and that he or she must comply with all safety regulations and wear the personal protective devices required for the job.

11.5. THE OCCUPATIONAL HEALTH NURSE'S RESPONSIBILITY

Since approximately 80% of occupational health nurses work alone, under the medical directives and indirect supervision of a plant physician, and often without a safety director, the maintaining of a safe, healthful environment is primarily the responsibility of the occupational health nurse. Specialists such as an industrial hygienist should be available for consultation. Small companies usually obtain such services from their insurance carrier, from a government agency, or from a private consulting group.

In large corporations, the Safety and Medical Departments work as a team, and the occupational health nurse works as a collaborative member with the hygienists, toxicologists, and safety personnel.

In order to evaluate the work environment of present employees, the placement of new employees, or employees being transferred, chemical exposures must be constantly evaluated. The occupational health nurse must know the various manufacturing processes, the chemicals involved with each, and their characteristics, as well as other types of problems which may be encountered, for example, that an asthmatic should not be exposed to a respiratory irritant or some one with a history of skin problems to a substance likely to cause dermatitis.

This data is essential for the nurse to make informed decisions and assist management in the proper placement of new employees as well as assist in keeping present employees well and productive. He or she needs to be informed by the procurement personnel of any new chemicals being ordered for introduction into the plant, and by management of any change in the manufacturing processes; the properties of these new chemicals and procedures must then be investigated in order to determine what reactions may be encountered (Fig. 11.1).

11.6. CONTINUING EDUCATION FOR NURSES—AN ESSENTIAL

To develop and maintain the expertise to make judgments as suggested above is an ongoing educational process for the occupational health nurse,

Figure 11.1.   Frequent visits to operating areas alert occupational health nurse to changes in employees and chemicals.

supported by the team approach, whether in the plant or available on a consultant basis to the nurse working alone, who is responsible for the entire employee health program including the chemical safety of the employees. Information available from chemical associations, manufacturers, the National Safety Council, reference books, NIOSH, OSHA, and other sources should be utilized. The nurse should avail himself or herself to short courses in continuing education offered by professional associations and societies, such as the American Association of Occupational Health Nurses, the American Industrial Hygiene Association, the American Conference of Governmental Industrial Hygienists, the National Safety Council, the American Society of Safety Engineers, the American Chemical Society, the American Institute of Chemical Engineers, government agencies such as NIOSH Educational Resource Centers, and State industrial commissions, as well as college and university short courses. There is an abundance of educational resources available. The better informed the nurse, the greater the contribution to the management team and to the employees—the ultimate benefactors.

Many industries which are not chemical companies, per se, use thousands of chemicals. For instance, the paper, furniture, textile, and automobile manufacturing industries use a wide spectrum of chemicals. With better medical surveillance and thorough research, we are realizing that many chemicals that in the past had been considered relatively harmless are now known to be harmful if not used properly. This is particularly true in the area of carcinogens, mutagens, and teratogens. The full extent of this danger is not known at this time. As research continues, the list grows longer and the problem becomes even more serious—another obvious reason for the need for continuing education.

A reference library of materials, periodicals, texts, and three volumes of the revised edition of *Patty's Industrial Hygiene and Toxicology*, OSHA and NIOSH bulletins, reports and alerts, are recommended. The American Industrial Hygiene Association's *Basic Industrial Hygiene—A Training Manual,* and the *Industrial Hygiene Manual* of the National Safety Council are among a list of references which should be available for immediate reference. It is also a requirement for the occupational health nurse to obtain and maintain for reference the OSHA final rules from the Federal Register which describe the compliance mandated for chemical surveillance in the work place. Copies of OSHA, NIOSH, and state regulations and recommendations should be made available to the occupational health nurse. The nurse should be on the mailing list of such agencies in order to remain current of the fast-growing list of regulations regarding the harmful effects of various chemicals.

With the references mentioned above, the medical data in files, and the chemical surveillance techniques in place, the occupational health nurse is a valuable data base and information source to the occupational health and safety team and to plant management.

The American Chemical Society Chemical Abstract Service has indexed over four million chemical substances. Perhaps 70,000 are actually in industry, but the list is continually changing and growing. The NIOSH *Registry of Toxic Effects of Chemical Substances* contains in the 1980 register data on 45,000 substances, of which 2,500 are suspected carcinogens, 2,700 suspected mutagens, and 370 suspected teratogens. The cumulative supplement to EPA's Toxic Substances Control Act Chemical Substances Inventory, contains 55,103 chemical substances, including 686 generic names generated for chemicals with confidential identities, as of publication in July 1980. From these numbers, it is clear that a major task is involved by anyone who must keep aware of the passing parade of chemicals.

The Occupational Safety and Health Act of 1970 (PL 91-596) has, in spite of its shortcomings in implementation and understanding, altered both the employees and the public to the risk to which many workers may be exposed in manufacturing processes, and by identification of harmful chemicals.

11.7. IDENTIFICATION OF CHEMICALS AND THEIR PROPERTIES AND CONTROL

Identification is important. To control or monitor occupational exposures to chemicals, the first task is to identify all of the chemicals used within the plant. In smaller plants this may be the responsibility of the nurse, in cooperation with the supervisor and the purchasing agent. A toxicity manual, including safety data sheets, should be prepared on each chemical with both the trade and generic name, to what extent, and how and where the chemical is used. The information obtained may be separated by departments, and names of chemicals listed. It is also important to write a description of *every* job and list *every* chemical used in performing that job, being mindful that a combination of chemicals can be dangerous, while one alone may be harmless. (See Chaps. 16, 22, and 24.) Information obtained through the purchasing department from the chemical supplier for the OSHA safety data sheets as required from the Department of Labor and other sources, for each chemical ordered, is helpful resource information.

For each chemical used, the chemical, generic, and trade name, the hazards involved from overexposure, symptoms of acute and chronic reactions, body organ affected, mode of entry, and the emergency treatment described in event of overexposure, should be included in the manual. The name and phone number (both day and night) of the medical director or consultant of the manufacturer should also be included, along with the local poison control center number.

Copies of the safety data sheet with important data underlined should be sent to management for their information, with the date noted on the sheets. By preparing a manual, data sheets, and job description listing chemicals involved, a ready reference is quickly available when it is necessary to check the properties and descriptions of chemicals in the event of a chemical or medical emergency. Emphasis should be on prevention.

11.8   RESOURCES FOR INFORMATION

The supplier company or manufacturer's emergency telephone number (both day and night) should be noted at the top of each safety data sheet, and is a valuable reference in the event of an exposure, leak, or spill. In the event of a spill of a chemical, the employees will be able to take correct action immediately, thus preventing or avoiding harmful exposures, providing they have had previous instruction, practice drills, and have guidelines to follow on the best approved method of clean-up.

Employees and line supervisors must be made aware of any potential chemical hazards. To accomplish this, the nurse, working as the chemical coordinator and advising management concerning distribution of safety data sheets, contacts line supervision, with whom chemical data are reviewed. Close communication and team effort is extremely important, and mutual trust must be established.

11.9.   EMPLOYEE EDUCATION

At a mutually convenient time, the occupational health nurse, after determining the area and employees involved, should go out to the department and review the chemical data sheet personally with the employees, either in a group or individually. He or she should cooperate with management in planning and carrying out an information program for the employees. At this time the occupational health nurse determines at what level of understanding the employees and supervisory personnel are operating, and then the terminology and information are explained at an appropriate level. At this time questions may be raised and satisfactorily answered. If the nurse does not know the answer, a major effort to find out and report back to the inquirer at a later date will be instituted. The employees are assured that knowledge of the chemical products, compliance with rules, and use of personal protective equipment will be a safeguard in their working situation, and no threat to their job security. They are advised that literature and other information on any new chemical products being considered for use in the facility will be made available to them, and that they will be informed of the proper procedure to use in the event of overexposure. At this time the nurse outlines the emergency procedures the employees can use prior to the arrival and treatment by the medical team. Adequate training in emergency care should be provided on a regular basis. If emergency equipment, such as positive pressure self-contained breathing apparatus, is required and available, the responsibility for its use and care should be delegated and clearly understood.

In small plants it is recommended that the occupational health nurse be a first-aid and cardiopulmonary resuscitation (CPR) instructor, and that employees should be trained in first aid and CPR. In this way an adequate support team can assist the nurse in case of an emergency as well as to care for minor cases when the nurse is not available. It is even better if the nurse and other members of the team avail themselves of emergency medical technicians training (EMT) which is taught in many community colleges.

The more training one receives the more safety conscious they become, which in turn contributes to a safer work place. Fewer accidents are likely to occur, and their severity lessened. There are also many psychological advantages to having the employees feel a part of this safety and preventive effort. They become less apprehensive toward management and more cooperative. As a guideline, it is recommended that there be at least two first-aid trained employees available on each shift in each department. If departments are large, even more first-aid-trained personnel should be available.

In the corporate medical departments, the occupational health nurse can review the OSHA safety data sheets with the safety director, industrial hygienist, and allied personnel, review and ascertain procedures to be developed in alerting and educating employees as to the properties of chemicals with which they work. Documentation of this type of employee education and surveillance is at this time recorded individually in the employee's file and also documented in the training program records which the occupational health nurse maintains of the occupational health involvement in employee protection. Employees should be made aware of their responsibilities to their family and community. By such protective measures as removing contaminated clothing, showering, and personal hygiene, the wearing of protective clothing and personal protective equipment, and general safe work practices they protect others as well as themselves.

11.10. PLANT TOURS BY NURSE

In order to become more knowledgeable of plant processes, the occupational health nurse must tour the plant and observe the working processes and the environment to which the employee is exposed (Fig. 11.2). With this type of information, the nurse is able to evaluate symptoms presented by the employee when he or she visits the medical department. As a result of a coordination of this data, the nurse is in a position to reference data to the plant physician and to the corporate team member involved, for further checks. While touring the various departments he or she should be aware of any environmental hazards, poor housekeeping or inadequate ventilation, unsafe work practices, as well as check the use and fit of personal protective equipment or devices if required.

Figure 11.2.   Plant tours present opportunity to observe real-world conditions.

11.11.   SURVEILLANCE

Since the occupational health nurse can recognize symptoms and the processes involved, he or she can make an invaluable contribution to maintaining healthy employees and a safe environment. As an example, if he or she knows that a certain chemical (such as mercury) may affect the nervous system, he or she would be alert to any early neurological symptoms long before the supervision or fellow employees would suspect a work exposure. The nurse should always be alert to the possibility of a "lone" or single adverse reaction among a group of employees exposed.

With the tremendous increase in the number of new chemicals and processes, continuous education of the employees is necessary, both on a "one-to-one" basis and in groups. The better informed, the safer the employee will be. The nurse, through observation, investigation and treatment, may be able to alert management to the need for a change in process or substitution of a different chemical or material. A rule of thumb should be to use the chemical with the lowest toxicity possible that will perform in the process (assuming that flammability, reactivity, and other considerations are equal). Management should advise the nurse of any potentially dangerous experimental processes. Data should be referred to the medical department so that a protocol can be established and an educational program in toxicology can be developed, including a section on how to prevent hazardous exposures. An adequate ventilation system must be maintained. Respiratory or other special personal protective devices must be provided, if needed, and employees must be properly fitted and trained in their use, care, and maintenance. To accomplish all this, it is of utmost importance to keep communications open and the occupational health nurse is in a key position, as the liaison, to insure communication in both directions. Training programs must be developed and adequate educational processes maintained at all levels.

11.12. WASTE DISPOSAL

Wherever chemicals are used in the manufacturing process, wastes result from the process. The occupational health nurse must be vigilant as to how these wastes are disposed, not only to protect the employee's health and safety, but that of the community as well. Full knowledge of the various acts such as the Toxic Substances Control Act, the Resources Conservation and Recovery Act (RCRA), and related laws, both Federal and state, is highly desirable if the nurse is to achieve the higher objectives of the occupational health nurse. There is also a responsibility to assist management in maintaining compliance and avoiding harmful, even catastrophic situations (see Chaps. 28 and 29).

In some plants, a materials chemical recognition system has been instituted as a retrospective and prospective approach. In reference to the retrospective, consultants have been engaged to evaluate and screen new chemicals, as well as the more traditional materials, for mutagenic and carcinogenic effects. The question of teratology and effects of chemicals on both the male and female reproductive systems is a very open one at present, and much more data will develop in the near future. Since 52% of the work force is female, the effects of chemical exposures on the unborn fetus must be considered seriously. The prospective approach deals with the new substance, how it travels through the plant process, how it could contact the employee, to what degree they might be exposed, and what resultant protection is required. As a consequence, the occupational health nurse is better able to ascertain, with the team, the degree of protection required for the particular operation.

11.13. PERSONAL PROTECTION

Each employee has been advised during the preplacement health interview, of the health and safety protection required for each task. The type of equipment required, if any, has already been evaluated, described, and recommended (for each chemical used in the plant) on the data sheet which has also been discussed with the worker.

In the advanced orientation process, both the supervisory personnel and the nurse describe in detail the equipment required, its use, proper fit, cleaning, and storage; recommendations regarding impervious clothing (previously tested for the particular exposure expected) to protect the skin are made and added safeguards such as respirators against inhalation of chemicals are discussed. Periodic physical examinations provide an ongoing record of screening against any health defects. Any additional laboratory or other tests required will depend upon the specific chemical exposure.

Whether the exposure may come from a liquid, a vapor, a dust, or a fume (aerosol), the occupational health nurse is aware that employees are protected by a combination of personal protective equipment and engineering controls against chemical exposures. If feasible, engineering controls should be installed and properly maintained. If protective devices are necessary, the nurse, when consulting with an employee, constantly checks to determine if the proper protective device is being worn correctly and maintained in proper condition. In order to monitor this program, reference to the toxicology manual is advantageous, particularly if a chemical product might be substituted, from a solid to a liquid form. The nurse should have readily available a copy of the latest TLVs from the ACGIH and OSHA.

If respirators are required, the nurse works with the management and the employees to maintain compliance with government regulations and company policy. Periodic inspection is necessary to assure proper fit of correct type respirator for the specific exposure. Documentation in employee's records of the type of respirator worn in specific jobs is required. Education regarding respiratory protection can be on a one-to-one or in groups by the occupational health nurse, utilizing films, pamphlets, data sheets, materials supplied by Federal and state agencies, including NIOSH and the manufacturers of respirators.

To insure the correct fit, the employee is involved in a qualitative as well as a quantitative fit. The occupational health nurse can develop procedures for the "buddy system," for entry into closed spaces, vessels, or other confined areas. Any problems in wearing respirators should be referred to the plant physician for evaluation, as some employees may hyperventilate, or have a psychological difficulty so that they cannot wear respiratory protection. Others may need an air-supplied respirator regardless of the type of exposure, due to existing chronic obstructive lung disease or a chronic heart condition. See Chapter 25.

11.14. MEDICAL MONITORING

Periodic evaluation with lung function tests by the occupational health nurse establishes data in the medical files. The nurse fully explains to the employee the results of any medical monitoring and counsels according to need. An informed employee is less apprehensive, more ccoperative, and a safer worker. To be most effective, the nurse must be well informed and must carry out a comprehensive envirosurveillance health monitoring and employee education program.

Another objective involving surveillance in the employee education program is to create an awareness of the chemical products delivered to the plant by stressing the importance of labels on containers, drums, and barrels. The employee must read all labels and become aware of the hazardous warnings on the drum, and what the manufacturer recommends for compliance in the event of overexposure. Proper storage, handling, dangers that may be present through mixing of various chemicals, excessive heat or moisture, and the importance of proper disposal should be stressed.

Careful attention to eye and body chemical splashes, and immediate flushing of the eyes with water, and drenching of the body with water is mandatory. It is the nurses' responsibility to see that adequate facilities are available and operating properly and that employees are trained in how to use them in the event of an accident. See Chapters 25, 26, 27, 32, 33 and 34.

11.15. RECORDS

Recent Federal laws mandate the maintenance of medical records for the duration of exposure of the employee in the work environment, and for 30 years thereafter. Therefore, the occupational health nurse is much involved in a chronological report of data, surveillance, monitoring, and compliance for the employee. Symptomatology combined with epidemiology provide a relevant historical pattern. Employees now have a right to know what is in their medical records, and must be so informed.

As a part of the total team approach, the occupational health nurse has an important and vital role in chemical surveillance in the occupational setting, with the objective of safeguarding the employee and the work site, as well as the community in which the operation is located. In the 1980s and 1990s, the occupational health nurse will be increasingly at the forntier of "real world" occupational health and safety.

ACKNOWLEDGMENT

The author acknowledges with deep appreciation the assistance of Mary Seaver, R.N. COHN, in reviewing an earlier version of this chapter.

SUGGESTED REFERENCES

The following publications available from NIOSH are typical of those available on occupational related nursing subjects from the National Institute for

Occupational Safety and Health, Division of Technical Services, Cincinnati, Ohio 45226.

"A Guide to Industrial Respiratory Protection" (NIOSH 76-189).

"A Guide to the Work-Relatedness of Diseases" (NIOSH 77-123).

"A Nationwide Survey of the Occupational Safety and Health Work Force (NIOSH 78-164).

"Behavioral Toxicology—Early Detection of Occupational Hazards (NIOSH 74-126).

"Community Health Nursing for Working People" (NIOSH 70-10253).

"Compendium of Materials for Noise Control" (NIOSH 80-116).

"Comprehensive Bibliography on Pregnancy and Work" (NIOSH 78-132).

"Guidelines on Pregnancy and Work" (NIOSH 78-118).

"Health Consequences of Shift Work" (NIOSH 78-154).

"Hospital Occupational Health Services Study; Vol. II: Employee Health and Safety Statistics and Records" (NIOSH 74-102).

"Human Variability and Respirator Sizing" (NIOSH 76-146).

"Information Profiles on Potential Occupational Hazards" (CONTR 210-77-0120).

"Licensed Practical Nurses in Occupational Health—An Initial Survey (NIOSH 74-162).

"Occupational Medical Symposia" (NIOSH 75-189).

"Pocket Guide to Chemical Hazards" (OSHA/NIOSH 78-210).

"Pre-Employment Strength Testing" (NIOSH 77-163).

"Respiratory Protection—A Guide for the Employee" (NIOSH 78-193B).

"Respiratory Protection—An Employer's Manual (NIOSH 78-193A).

"The New Nurse in Industry" (NIOSH 78-143).

OTHER SUGGESTED REFERENCES

P. B. Beeson, and W. McDermott, *Textbook of Medicine,* 14th ed. Saunders, Philadelphia, 1975.

M. L. Brown, *Occupational Health Nursing*, Springer, New York, 1956.

Emergency Care and Transportation of the Sick and Injured, 2nd ed., American Academy of Orthopedic Surgeons, George Banta Co., Menasha, Wisc., 1977.

Excerpta Medica, International Medical Abstracting Service, Princeton, N.J.

Fundamentals of Industrial Hygiene, J. Olishifski and E. McIlroy, National Safety Council, Chicago, Ill. 60611.

W. M. Grant, Toxicology of the Eye, 2nd. ed., C. C. Thomas, Springfield, 1978.

Handbook of Statistical Tests for Evaluating Employee Exposure to Air Contaminants, NIOSH-75-147, April 1975.

Health Hazard Evaluation Summaries, National Institute for Occupational Safety and Health, Cincinnati, 1980.

D. Hirschfelder, Better Industrial Vision Testing and Safety, *Occup. Health Safety*, 49 30 et seq. (November/December, 1980).

G., Hommel, *Handbuch der gefahrlichen Guter*, Springer-Verlag, Berlin and Heidelberg, 1980.

D. Hunter, *Diseases of Occupations*, 6th ed, Little, Fort Lee, N.J., 1978.

Index Medicus, National Library of Medicine, Bethesda, Md.

R. T. Johnstone and S. W. Miller, *Occupational Diseases and Industrial Medicine*, Saunders, Philadelphia, 1960.

A. Q. Maisel, *The Health of People Who Work*, National Health Council, New York, 1960.

Merck Index, Merck & Co., Inc., Rahway, N.J. 1980.

NIOSH Registry of Toxic Effects of Chemical Substances, (revised annually). [Available on microfiche on quarterly basis from the National Institute for Occupational Safety and Health, Rockville, Md. 20857.]

"Occupational Diseases, A Guide to Their Recognition," DHEW (NIOSH) Publication 77-181, 1977. [Available from Government Printing Office, Washington, D.C. 20402.]

Occupational Diseases: A Syllabus of Signs and Symptoms, E.R. Plunkett, Barrett Book Co., Stamford, Conn., 1977.

Occupational Health Manual, Naval Medical Training Institute, National Naval Medical Center Bethesda, 1972.

OSHA Compliance Manual, Petersen, Don, McGraw-Hill, New York 1975.

Patty's Industrial Hygiene and Toxicology, 3rd. revised ed., edited by Clayton and Clayton, Wiley, Interscience, New York, 1978–1980.

M. B. Schmidt, and C. Silberstein, Assessing Complaints About Itching Eyes, *Occup. Health Safety*, 42–49 (November/December, 1980).

H. B. Selleck, *Occupational Health in America*, Wayne State University Press, Detroit, 1962.

C. Zenz, *Occupational Medicine: Principles and Practical Application*, Year Book Medical Publishers, Inc., Chicago, Ill. 1975.

H. A. Waldron, *Lecture Notes on Occupational Medicine*, 2nd ed., Blackwell, Oxford, 1979.

R. S. Rowley, "Chemical and Thermal Burns of the Eye," *Occu. Health Nursing*, 28–30, 37, June 1981.

NIOSH/OSHA Occupational Health Guidelines for Chemical Hazards, 320 OSHA Regulated Chemicals, Publications Dissemination, NIOSH, 4676 Columbia Parkway, Cincinnati, OH 45226 (1981).

NIOSH Certified Equipment List, Revised Annually, NIOSH, Rockville, MD. 20857.

M. M. Nemec, "Special Report, Eye, head, and face protection at DuPont," *Occu. Hazards*, 44–47 (March 1979).

"Chemical Hazards to Human Reproduction," Council on Environmental Quality, Jan. 1981, U.S. Government Printing Office, Washington, D.C.

NIOSH *Current Awareness Series*, Issued occasionally, NIOSH, Rockville, MD. 20857.

BIBLIOGRAPHY

Albert, R. E. "Chronic Central Nervous System Damage by Environmental Agents," CIBA Found. Study Group, **35**;73–75.

American Medical Association, *Guide to the Significance of Occupational Exposure Limits*, Council on Occupational Health, Chicago, 1966. *Epidemiology in Occupational Disease and Injury*, Council on Occupational Health, Chicago, 1967.

Ashford, N. A., *Crisis in the Workplace: Occupational Diseases and Injury*. MIT Press, Cambridge, Mass., 1974.

Bahn, A. K. and Friedlander, B. R., "Put on Your Epidemiologist's Hat," *Occup. Health & Safety*, **46** (January/February 1977).

Beck, L., "The Occupational Health Nurse Should Take an Active Role in Hazardous Materials Surveillance," *Occup. Health & Safety*, **46**(6);36(1977).

Berger, J. and Riskin, G., "Economic and Technological Feasibility of Regulating Toxic Substance Under the Occupational Safety and Health Act," *Ecol. Law Quarter.* **7**;285–360 (1979).

Bodnar, E. M., "Management of Toxic Substance in the Workplace: The Role of the Occupational Health Nurse," *Occup. Health Nurse*, 27(3), 7–12 (March 1979).

Brown, M. L., "Occupational Health Nursing Emerging Keystone of Plant Medical Surveillance," *Occup. Health Safety*, 14–18 (March/April 1976).

Brown, M. L., *Occupational Health Nursing Principles and Practices* Springer Publishing, New York, 1981.

Burkeen, O., "The Nurses and Industrial Hygiene," *Occup. Health Nurs.* 24(4); 7–10 (April 1976).

Cohen, R., "Industrial Toxicology: A Practical Approach," *Occup. Health Nurs.*, 11, 9(June 1980).

Collen, M. F., "Dollar Cost per Positive Test for Automated Multiphasic Screening," *N. Engl. J. Med.* 283; 459(1970).

Cordasco, E. M., "The Health Effects of Halogens in the Air," *Occup. Health Safety* 47, 36–38. (January/February 1977).

Cralley, L. V., ed., *Industrial Environmental Health: The Worker and the Community*, Academic Press, New York, 1972.

Cralley, L.V. and Clayton, G. H., *Industrial Hygiene Highlights*, Vol. I, Industrial Hygiene Foundation, Pittsburgh, 1968.

Deichmann, W. F. and Gerarde, H. W., *Toxicology of Drugs and Chemicals*, Academic Press, New York, 1969.

Dollberg, D. D. and Verstuyft, A. W., *Analytical Techniques in Occupational Health Chemistry,* ACS Symposium Series 120, American Chemical Society, Washington, D.C. 20036, 1980.

"Dow Chemical's Using Computers to Promote Health and Safety," *Occup. Health Safety*, 13c–13d, 17 (May/June 1979).

Drug Data For Nurses, American Journal of Nursing, New York, 1979.

Early Detection of Health Impairment in Occupational Exposures to Health Hazards, WHO Technical Report Series 571, Geneva, 1975.

Eckardt, R. E., *Industrial Carcinogens*, Grune & Stratton, New York, 1959.

Enterline, P. E., "Epidemiology: Nothing More Than Common Sense?" *Occup. Health Safety*, 48(1), 45–48 (1979).

Fawcett, H. H., "Exposures of Personnel to Laboratory Hazards," *Am. Ind. Hyg. Assoc. J.*, pp. 559–567 (August 1972).

Friedman, C. and Hazuka, B. T., "Basic Epidemiology for Occupational Health Nurses, *Occu. Health Nursing* (7 Feb. 1981).

Guidelines for a Chemical Plant Safety Program and Audit, Chemical Manufacturers Asso., Publications Service, Washington, D.C., 1978.

Hamilton, A. and Hardy, H. L., *Industrial Toxicology*, 3rd ed., Publishing Sciences Group, Inc., Acton, Mass., 1974.

Hutchinson, M. K., *A Guide to the Work-Relatedness of Disease.*, NIOSH Publication No. 77-123, U.S. Government Printing Office, Washington, D.C., 1979. (Revised Edition Publication No. 79-116).

Hygiene Guide Series, American Industrial Hygiene Association, Akron, 1960–1981.

Industrial Data Sheets and Guide to Data Sheet Development, National Safety Council, Chicago, 1981.

Johnston, R. and Miller S., *Occupational Disease and Industrial Medicine*, W. B. Saunders, Philadelphia, 1960.

Keane, W. T., "Federal Courts Have Ruled Worker Must be Warned," *Occup. Health Safety*, 47(2), 22–23 (1980).

Key, M. M., *Occupational Disease: A Guide to Their Recognition*, NIOSH, Cincinnati, 1977.

Knox, E. G., "Multiphasic Screening," *The Lancet*, 14–34 (December 1974).

Kornbacher, G. The Nurses' Role in Toxicology," *Occup. Health Nurs.* **24** (August 1976).

Lemen R. and Dement, J. M., eds., *Dust and Disease*, Pathatox Publishers, Park Forest South, Ill., 1979.

Linch, A. L., *Biological Monitoring for Industrial Chemical Exposure Control.*, CRC Press, Cleveland, 1974.

Loomis, T. A., *Essentials of Toxicology*, 3rd ed., Lea & Febiger, Philadelphia, 1978.

Major Drugs: Their Use and Effects. Public Relations Department, Playboy Enterprises, Chicago, 1980.

Montgomery, R. R. and Reinhardt, C. F., "A Capsule Dose of Toxicology," *Occup. Health Nurs.*, **24** (5), 7 (May 1976).

S. Moeschlin, Outstanding Symptoms of Poisoning, pp. 644–678, in *Poisoning: Diagnosis and Treatment*, Grune & Stratton, New York, 1965.

Morse, J. M., "Epidemiological Research Design: Utility for the Occupational Health Nurse," *Occup. Health Nursing,* 18–21 (April 1982).

Nelson, E., "The Occupational Health Nurse is Often the Worker's 'First Line of Defense' Against Workplace Hazards," *Occup. Health Safety*, 36–39 (November/December 1976).

National Institute of Occupational Safety and Health, *Occupational Diseases: A Guide to the Recognition*, Pub. no. 77-181, U. S. Government Printing Office, Washington, D.C. (Rev. ed. June 1977).

The Industrial Environment—Its Evaluation and Control. US Public Health Service, U.S. Government Printing Office, Washington, D.C. 1973. (GPO S/N 017-001-00396-4.) *Publications on industrial toxicology.* (Available from state departments on labor and industry.)

NIOSH Certified Equipment (revised annually) NIOSH, Morgantown.

NIOSH/OSHA, "Pocket Guide to Chemical Hazards," September 1978.

Occupational Health Hazards: Their Evaluation and Control (Bulletin 198) U.S. Department of Labor, Washington, D.C., 1968.

Olishifski, J. B. and Elroy, F. E., *Fundamentals of Industrial Hygiene*, Occupational Health Series National Safety Council, Chicago, 1976.

Oynett, H. P., "The Nurse—Valuable New Resource in Hazards Monitoring," *Occup. Health Safety*, 18–22 (November/December 1976).

J. R. Schroeder, "Silica Classification for the Workplace Environment," *ASTM Standardization News*, **9**(11), 9, 10–14 (Nov. 1981).

Guenther, R., "Employers Try In-House Fitness Centers to Lift Morale, Cut Cost of Health Claims," *Wall Street Journal*, p. 29 (Nov. 10, 1981).

J. S. Lublin, "OSHA's Review of Cotton-Dust Standard Broadened; Step May Aid Drive to Ease it," *Wall Street Journal*, p. 20 (Feb. 10, 1982).

J. R. Brobeck, *Physiological Basis of Medical Practice*, 9th ed., Williams & Wilkins, Baltimore, MD, 1973.

S. Sodeman and J. Sodeman, *Pathologic Physiology: Mechanisms of Disease*, 5th ed., W. B. Saunders Co., Philadelphia, PA, 1981.

Patty, F. A., *Theory and Rationale of Industrial Hygiene*, Vol. III Interscience, Wiley, New York, 1979.

Pearson, J. C., "Devising Your Own Surveillance Program," *Occup. Health Safety*, **48** 26–27 (May/June 1979).

Physicians Desk Reference, 35th ed. (published annually), Medical Economics Company, Oradell, N.J., 1981.

Proctor, N. H. and Hughes, J. P., *Chemical Hazards of the Workplace*, J. B. Lippincott, Philadelphia, 1975.

Raniere, T. M., "Chemical Hazard Identification—Our Need to Work Together," *Occup. Health Nurs.*, **15**, 19–21 (September 1978).

Sackett, D. L., "Can Screening for Serious Diseases Really Improve Health?" *Science Forum*, **15**, 9 (1970).

Sax, N. I., *Dangerous Properties of Industrial Materials*, 5th ed., Van Nostrand Reinhold, New York, 1979.

Schering, N. J., *State Association of Industrial Nurses Symposium*, Industrial Toxicology for Nurses, 1973.

Schilling, R. S. F., ed., *Occupational Health Practice*, Butterworths, London, 1973.

Sprout, W. L., "How do you Inform Labor, Management and Other Professionals About Toxicity Data?" *Occup. Health Safety*, 20–21 (March/April 1978).

Stallones, R., "Close Disease Surveillance, Environmental Awareness are Epidemiological Keys," *Occup. Health Safety*, 32–33 (March/April 1976).

Standards, Interpretation and Audit Criteria for Performance of Occupational Programs, American Industrial Hygiene Association, Akron, 1979.

Stokinger H., and Schul, L., "Hypersusceptibility and Genetic Problems in Occupational Health Medicine—A Consensus Report," *J. Occup. Med.*, **15**, 564–573 (July 1973).

Strasser, A. L., "Workplace Hazards: We Have the Skills to Reduce Them—If We Work Together," *Occup. Health Safety*, 33–34 (March/April 1978).

"The Carcinogen Question: How Valid The Date?" *Occup. Hazards*, 47–48, 51 (November 1973).

The Industrial Environment—Its Evaluation and Control, Superintendent of Documents, U.S. Government Printing Office, Washington, D.C., 1973.

TLV Booklet, American Conference of Governmental Industrial Hygienists (updated annually), Cincinnati, Ohio.

U.S. Department of Health, Education and Welfare, *A Guide to the Work-Relatedness of Disease*, NIOSH Publication No. 79-116, Washington, D.C., 1979.

Registry of Toxic Effects of Chemical Substances, two volumes (NIOSH publication, revised annually and quarterly).

Effects of Perchlorethylene/Drug Interaction on Behavior and Neurological Functions, NIOSH Publication No. 77-191, 1977.

Approaches to Determining the Mutagenic Properties of Chemicals: Risk to Future Generations, NIOSH, 1977.

Reliability and Utilization of Occupational Disease Data, NIOSH Publication No. 77-189, 1977.

Pilot Study for Development of an Occupational Disease Surveillance Method, NIOSH Publication No. 75-162, Rockville, Md., 20857, 1975.

Secretary's Committee to study Extended Roles for Nurses. Extending the scope of Nursing Practice. DHEW Publication No. (HSM) 73-2037. U.S. Government Printing Office, Washington, D.C., 1971.

Simmons, R. S., "Is your company under surveillance? It should be!" *Occu. Health and Safety*, 50–54 (April 1982).

Wilson, J. M. G. and Jungner, G., *Principles and Practice of Screening for Disease*, WHO Public Health Paper 34, 1968.

Zeng, C., ed., *Occupational Medicine*, Yearbook Publishers, Chicago, 1975.

D. Kennedy, "The Politics of Preventive Health," *Technology Review (M.I.T.)*, **84**(2), 58–60, (Nov./Dec. 1981).

B. MacMahon, "In Pursuit of the Public Health," *Technology Review (M.I.T.)*, **84**(2), 51–57, (Nov./Dec. 1981).

12
Services and Facilities

David T. Smith

The services and facilities covered here are those directly used by personnel in the chemical plant or laboratory, such as electricity, water, air, service gases, steam heating, waste disposal, locker and shower facilities, eating facilities, traffic, parking, walking and climbing facilities, etc. These items may be not only of tremendous importance to the health and safety of personnel by the adequacy of their design and arrangement, but failure to maintain them properly may be a source of injury. Key points that injury experience has shown need close attention will be discussed rather than details of design and installation because of space limitations.

12.1. ELECTRIC LIGHTING AND POWER

The National Electrical Code, NFPA 70-, ANSI C1-, is accepted as the bible for all industry and by regulation is incorporated in OSHA Safety and Health Standards in subpart S, 1910.308. The provisions of the code should be considered the minimum necessary for safety in all installations. In addition, each plant or laboratory should adopt definite rules and procedures for electrical work to insure that provisions of the code are met. OSHA records show that "Electrical Standards" is the second highest category in percentage of violations found by their inspectors.

Underwriters Laboratories, Inc. approval is the accepted norm for approval of electrical equipment, appliances, and materials of construction both for hazardous locations and most general applications. Check for approval before purchasing electrical items.

It is good safety practice as well as good economy in most chemical manufacturing operations to install all electrical wiring, equipment, and lighting to meet requirements of Class I, Group D, Division II, Hazardous Locations. This protects against the temporary presence of explosive mixtures of most classes from leaks or spills of flammable liquids or gases. Regardless of the flammability of operations when installed, temporary changes in process may occur without anyone realizing the electrical installation is unsuitable. The Class I, Group D, Division II requirements include vapor-proof lighting fixtures, totally enclosed motors, oil-immersed or properly enclosed switches, and complete absence of open sparking devices. (Details may be found in the National Electrical Code.)

It is generally more economical to prevent explosive atmospheres in working areas than it is to try to provide the special explosion-proof electrical equipment necessary for such conditions. Personnel should never be allowed to work in an explosive atmosphere. Where the creation of such atmospheres cannot be avoided through containment or control of flammable liquids, gases, and dusts, the space involved should be limited and segregated by hoods or special ventilation. All electrical equipment in the exposed segregated area should conform to the requirements for Class I, Group D, Division I, Hazardous Locations, or whatever class and group the specific situation falls in.

Electrical equipment on open, outdoor structures more than 20 ft above ground usually are considered to be free from exposure to more than temporary, local explosive mixtures near leaks.

All building steel and outdoor structures, all tanks, drums, pipelines, open-end hoses, tank cars, trucks, and chemical equipment handling or using flammable liquids or gases, or in areas where handled, should be grounded to dissipate static electricity in accordance with the National Electrical Code. Flexible grounds should be connected to large water pipes preferably, or to driven grounds or plates, but never to electrical conduit, branch sprinkler lines, gas, steam, or process piping. These grounds should be properly maintained, and periodically inspected for damage and the electrical resistance to ground measured. (Grounding conductors which will give lightning protection require much larger capacity than those for dissipation of static electricity.)

To avoid mistakes in use controls and equipment, the voltage, name and number of equipment controlled should be clearly marked on all switch boxes, compensators, and starters. The same applies to the equipment controlled. Pins and chains should be provided on all butterfly switches.

Extension cords should all be three-wire, and limited to 25 ft in length.

Only flashlights approved by the Underwriters Laboratories for Class I, Group D, locations should be permitted in the plant. Unapproved flashlights have a tendency to find their way into hazardous locations from nonhazardous locations.

Some hearing aids worn by employees are theoretically capable of producing a spark of sufficient energy to ignite flammable vapors. Thus the switch of a hearing aid should not be operated during exposure of the wearer to a spill or leak which might result in an explosive mixture. A more important source of sparks is the ordinary telephone which makes a high energy spark when dialed, punched, or rung. Explosion-proof telephones are available and should be used where applicable. Radio paging systems and intercom equipment may also present an electrical hazard when exposed to flammable vapor situations. They should be checked as to their electrical classification before purchase and installation.

Due to the continued development of new equipment, there is limited UL approved electrical instrumentation available. Much of it can furnish an ignition source if faults develop. However with care in selection and special installation, it is not difficult to avoid ignition sources from operation of the electrical instruments. Sometimes the instruments can be enclosed and continuously purged by a very slow flow of clean air through a small air line. Often the instrument can be located outside the danger zone. Care should be given to the design and installation of each instrument from the electrical safety standpoint.

If current-carrying parts or conductors must be exposed, they should be guarded by elevating them at least 8 ft above walking surfaces, or providing enclosures. Avoid using metal rules, metal tapes, metal hard hats, metal ladders, or other metal objects near energized exposed electrical conductors or equipment.

12.2. ILLUMINATION

Inability to see effectively contributes to many errors and injuries. Seeing depends not only on lighting intensity as measured in footcandles, but on contrasts in color, light–dark contrast, and in reflectivity of surfaces. Accenting hazard points, controls, and key points with contrasting color is an important aid to better seeing. Reducing or removing glare is necessary, as is eliminating too great a contrast in the light intensity of the field of sight. If the general level of illumination is high, we cannot see into an exceptionally dark area. Similarly, in an area of low illumination, a very bright spot such as an unshielded light tends to blind us. However, medium contrast within the limits shown in Table 12.1 helps us to see better. Tables 12.1 and 12.2 from the Illuminating Engineering Society should be followed.

12.3. WATER

Hot water is used by many plants for washing down floors and equipment, by operating or janitorial personnel. Steam cleaning should only be performed by specially trained persons under carefully controlled conditions. Do not connect steam and water lines directly to make hot water for cleaning except through an adequately designed and maintained water–steam mixer

TABLE 12.1. General Recommended Values of Illumination

Tasks	Footcandles
Most Difficult Seeing	200–1000[a]
Finest precision work involving finest detail, poor contrasts, over long periods of time.	
Examples: extra fine assembly, precision grading, extra fine finishing.	
Very Difficult Seeing	100
Precision work involving fine detail, fair contrasts, long periods of time.	
Examples: fine assembly, high-speed work, fine finishing.	
Difficult and Critical Seeing	50
Prolonged work involving fine detail, moderate contrasts, long periods of time.	
Examples: ordinary bench work and assembly, machine shop work, finishing of medium-to-fine parts, office work.	
Ordinary Seeing	30
Involving moderately fine detail, normal contrasts, intermittent periods of time.	
Examples: automatic machine operation, rough grinding, switchboards, continuous processes, conference and file rooms, packing and shipping.	
Casual Seeing	10
Examples: Active pedestrian walkways, entrances, vital exterior structures or locations.	
Building Exteriors	5
Examples: Active pedestrian walkways, entrances, vital exterior structures or locations.	
Building Surroundings	1

[a]Obtained with a combination of general lighting plus specialized supplementary lighting. Care should be taken to keep within the recommended general brightness ratios (Table 12.2) and to avoid glare conditions when lightcolored materials are involved.

Source: Illuminating Engineering Society.

that will shut off completely when the cold water supply is interrupted. The maximum possible hot water temperature should be known and posted, and all lines identified by name at the mixer and at all outlets. Water hotter than 135°F (57°C) coming in contact with the skin usually results in at least first degree burns. If water temperatures greater than this are used, appropriate training and controls should be utilized.

Never make hot water for bathing or washing the skin by direct mixing of steam and water; always use a water heater involving heat exchange, which will give accurately controlled temperature water. Water at shower heads or wash basins must not be hotter than 135°F (57°C), regardless of the operation of shower or faucet mixing valves under control of the users.

TABLE 12.2. Recommended Permissible Brightness Ratios[a]

Areas Involved	Environmental Classification		
	A[b]	B[c]	C[d]
Between task and adjacent darker surroundings	3 to 1	3 to 1	5 to 1
Between tasks and adjacent light surroundings	1 to 3	1 to 3	1 to 3
Between tasks and more remote dark surfaces	10 to 1	20 to 1	e
Between tasks and more remote light surfaces	1 to 10	1 to 10	e
Between luminaires, windows skylights, etc., and surfaces adjacent to them	20 to 1	e	e
Anywhere within normal field of view	40 to 1	e	e

[a]From the normal viewpoint, brightness ratios of apprecsize in industrial areas should not exceed those given in the table.

[b]Interior areas where reflectances of entire space can be controlled in line with recommendations for optimum seeing conditions.

[c]Areas where reflectances of immediate work area can be controlled, but control of remote surrounding is limited.

[d]Areas (indoor and outdoor) where it is completely impractical to control reflectances and difficult to alter environmental conditions.

[e]Brightness Ratio control not practical.

Source: Illuminating Engineering Society; ANSA Std. All. 1.

Hot water heaters should have not only a pressure relief valve for overpressure but also a fusible plug for over temperature relief.

Safety showers should be installed where handling corrosive or toxic materials. Even where employees wear flame-retardant clothing, safety showers may be of great value in extinguishing fire or flames of flammable liquids on the clothing. It is important that a large volume, low velocity discharge from directly overhead be used, so that the employee will be easily drenched, completely and continuously. A satisfactory minimum flow is 50 gal/min. The high velocity spray from the ordinary domestic type shower head is unsuitable.

Water to outside showers may be heated by chasing the water line with a low temperature electric ground heating cable having a maximum temperature of 80°F. (27°C).

The valves of all safety showers should be at the same height and the same relative position to the shower head. They should operate in the same direction and be exactly alike. These requirements are very important because the employee needing the protection of a shower may be unable to see, and may be groping.

Frequent, scheduled testing of each shower is essential. Training and retraining in the location and use of the safety shower is important to enable each employee to use one effectively in an emergency.

The safety of potability of water supply is important both from the standpoint of securing a supply that is adequate in purity for the use required, and from the standpoint of prevention of contamination of the source of supply.

Municipal and other public sources for sanitary and drinking use are usually satisfactory. Potable water must meet the U.S. Public Health Service Drinking Water Standards, 42 CFR part 72. Private wells and other sources should be carefully controlled and inspected, with tests daily.

Extreme care must be taken with certain types of flush toilets to prevent sewage contamination to water supplies in case of loss of water supply pressure. Antisiphon devices are required.

In many plants, water which is not potable is used for cooling or other process use. In some cases it is used for washing floors and equipment. It is important that all water outlets where there is nonpotable water supplied be identified as to type of water, and precautionary rules posted to warn employees that the water is unsafe, and is not to be used for drinking, washing of the person, cooking or eating utensils, washing of food preparation or processing premises, or personal service rooms or for washing clothes. Nonpotable water may be used for washing other premises only if it does not contain chemicals, fecal coliform, or other substances which could create unsanitary conditions or be harmful to employees. This is covered in OSHA Safety and Health Standards 1910.141.

Any connections made to water supplies or systems for process water or process cooling water, or other systems must be made in such a manner that there can be no back flow of possible contaminated water to the municipal, public, or plant system that is used for drinking, bathing, washing, or other potable uses as defined above. We cannot rely on check valves. There must be a positive mechanism to prevent siphoning back of water in case of loss of pressure in the supply line. An effective arrangement is to have the municipal supply, or other potable supply, empty by free fall into a tank from which we take water to a pump inlet for process purposes. The tank must be supplied with an overflow outlet beneath the inlet from the municipal supply.

It is important to check that all installations are in conformance with local and state plumbing codes. Permits issued only after official inspections are required in many cases for original installations, alterations, and periodically thereafter.

Portable drinking water dispensers must be kept closed, equipped with a tap, and maintained in a sanitary condition. Ice for portable drinking water dispensers must be made of potable water. Common drinking cups are prohibited. Where single service cups are supplied, both a sanitary container for the unused cups and a receptacle for disposing of the used cups shall be provided.

12.4. BREATHING AIR SUPPLY

Air is generally compressed and piped to various locations for process and mechanical use, for instrumentation, but is sometimes used by humans for breathing with air masks. Because of the difference in quality required, separate systems should be provided for any breathing use. Air for human consumption must be free of contaminants such as carbon monoxide, oil vapors, flakes of rust, or other foreign material, and the ratio of oxygen to nitrogen and carbon dioxide must be correct.

The ordinary oil-lubricated air compressor cannot be relied on to deliver the needed quality of air during sustained operation, primarily because of the possibility of high temperatures breaking the oil down, unless specially equipped.

In selecting a compressor, it is important that the seller thoroughly understand that it is to be used to supply breathing air for humans, and that written assurance be given that it will meet this requirement when operated in accord with the manufacturer's instructions. It is equally important to follow these instructions.

Special equipment required for the oil-lubricated compressor includes a carbon monoxide alarm, oil vapor removal device, odor removal device, a carbon monoxide removal device which catalytically converts the carbon monoxide to carbon dioxide, and dust screens to remove particulate matter. Even then, frequent testing for carbon monoxide level is desirable.

Special air compressors designed for producing breathing-quality air are a much better choice. Compressors may be internally lubricated by water, water with the addition of a trace of soap, or by natural mineral oil. So-called "unlubricated" compressors do employ lubrication for the bearings and working parts of the compressor, but the compressor chamber or cylinder is not lubricated because of the use of low friction seals on the piston, eliminating the need for lubrication of the cylinder walls. The diaphragm type compressor likewise is considered unlubricated.

After producing air of breathing quality, it must not be contaminated. The only way to insure this is to have no interconnections for instrument, process or other usage. Check valves have been known to fail (see Chapter 25).

Cylinder air for human consumption may be purchased in many areas, but should be clearly marked and guaranteed as being of "Breathing Air" quality. There are advantages in reliability from freedom of contamination in the use of cylinder-supplied air over distribution through piping systems.

When using piped air distribution systems for air masks, filters and traps should be installed near the outlets to intercept and remove foreign particles and liquids, such as condensed moisture. Piping down stream from the filters should be nonferrous, preferably copper, brass or "Delrin." It is not necessary to add or remove moisture from breathing air as the body can tolerate extremely wide variations in humidity without discomfort or harm.

All breathing air outlets should be plainly identified and personnel trained never to use an unidentified outlet.

For discussion of emergency breathing apparatus per se, see Chapter 25, "Respiratory Hazards and Protection."

Ventilation

Satisfactory indoor general ventilation requires air to be of comfortable temperature and humidity, and free from harmful or disagreeable contaminants. Humidity alone is not a major factor, humans can enjoy air in widely varying percentages of humidity if the temperature is adjusted accordingly. Extremes of heat and cold may be health and safety problems.

Natural ventilation through windows and roof ventilators may suffice if adequate to prevent overheating in the summer. However, careful checks should be made to assure that concentration of air contaminants does not build up above maximum allowable limits in the winter when the building is closed up. Where ventilation is used to control potential exposure to workers, it must be adequate to reduce the concentration of the air contaminant to the degree that a hazard to the worker does not exist. One problem often overlooked is exhausting contaminated air to locations where it may be drawn in again with the fresh air intake. For details of methods of ventilation, see ANSI Code Z9.2-, "Fundamentals Governing the Design and Operation of Local Exhaust Systems." Experienced engineering by specialists in ventilation work is very desirable; almost every system requires individual design.

12.4a.   HEAT AND COLD

The stress of hot environment and cold environment on the body is sometimes overlooked. Identifying the extent of exposure and taking control methods will permit a work climate of thermal comfort. The heat stress for the individual worker depends on:

1. the bodily heat produced by the work exertion;
2. the heat produced by the equipment and work area;
3. the number and duration of the exposures;
4. the heat exchanges to the body by radiation, convection, and evaporation;

5. the thermal conditions of the rest area;
6. the clothing worn.

12.5 SERVICE GASES

Nitrogen, carbon dioxide, and other inerting gases are sometimes referred to as "service gases" rather than process gases. Usually installed with schedule 40 piping for use at 150 psi or less, the principal safety problem has been one of identification. Fatalities have occurred when inerting gases have mistakenly been used for breathing air during work inside tanks or confined spaces. While color coding is helpful, all outlets and valves of service gases (as well as all other materials) should be identified by having the name of the material affixed. Pressure sensitive tapes, metal and plastic tags or painted signs are all suitable.

12.6. STEAM

While high-pressure steam is required for process and power use, only low pressure steam, 15 psi or less, is suitable for most room-heating equipment. Identification of steam lines by pressure is important from a safety standpoint if there are steam services of more than one pressure. Discharge of steam traps, blowdowns, and such should always be carried to points where unpredicted discharges will not expose personnel. An outside stone-filled dry well can be used for a small steam trap condensate discharge. If steam condensate discharges into a sewer line, provision must be made to prevent build-up of pressure in case of trap malfunction and escaping steam.

Severe thermal burns can result from persons contacting unprotected steam lines or radiators. All lines, fittings, and radiators located less than 7 ft above the floor should be insulated or guarded, or at other locations where personnel may come in contact with them.

12.7. WASTE DISPOSAL

Air pollution, a subject of much concern from a public relations viewpoint as well as safety of personnel, demands that we do not liberate gases in harmful concentrations to the air, liquids in harmful concentrations outside of our own plants, dusts and particulate matter to the air, even visible smokes and vapors. Each case is an individual engineering problem, involving recovery or removal with consideration of equipment of varying degrees of sophistication, or perhaps revisions in process to eliminate or reduce the cause of the emission. Consultation with specialists in this field is desirable.

Limits for allowable concentrations of noxious and toxic gases and particulates liberated from vents or stacks into the atmosphere are usually set by federal, state, and local regulations. The same is true for disposal of liquid or solid wastes. In most areas, waste disposal contractors are available who can remove undesirable and toxic liquid and solid wastes and handle them away from the site for safe and legal disposal through specialized treatment, burial, etc. (see Chapters 28 and 29).

The disposal of liquids containing chemical wastes to public sewers, streams, or ground water requires that we have sufficient neutralization and treatment to insure that they have been completely removed or are nonobjectionable and harmless in the concentration present in the carrier stream where they leave our property. Skimmers and traps must be used to remove any oils or other nonwater soluble materials. Settling basins can be effective in removing entrained solids.

If flammable volatile materials are handled, transportation of wastes by underground piping systems may present the problem of explosive mixtures in the air space in the pipe. It is important to prevent any flammable volatile waste materials from entering closed underground sewer lines. There should also be enough openings for clean-out and maintenance in the sewer pipe. Sometimes it is necessary to install open ditches for waste liquid sewers and equip them with adequate fire stops. The open ditch, lined with or constructed of material resistant to the chemical exposure, has the added advantage of accessibility for clean-out and ease of inspection. Properly treated, laminated wood-box ditches have proved very successful. Dilution with sufficient quantities of water before entering the waste stream disposal system helps with many waste disposal problems.

We are a nation of litterers; possibly we are inherently lazy. To prevent employees from discarding waste paper, gloves, gaskets and a hundred other miscellaneous objects on our grounds, walks and even in out of the way corners of buildings, we must make it easy for them to dispose of waste materials quickly and with a minimum of physical effort. This means we must provide commodious, attractive trash collection containers at strategically placed locations, frequently spaced. In addition, we must have an efficient system for keeping them collected and emptied. Separate containers for receiving metals and for glass are usually justified. If used, each should be so labeled and their correct use enforced.

Employees should be responsible for the cleanliness of their own work location, and to keep it clean. If they know that they are going to have to clean it up themselves, they are much more careful not to create a housekeeping problem. The example set by supervision usually determines the overall effectiveness of the good housekeeping program. If the supervisor or foreman ignores an item of discarded trash, or contributes one, then the area will be trashy.

Efficient methods of collection of unused maintenance and repair materials can not only prevent poor housekeeping but result in major savings in their return to reuse. "Tote" boxes for collection and return of small unused items are useful.

12.8   WALKING AND CLIMBING FACILITIES

Falls constitute a surprisingly large percentage of serious injuries in the chemical industry. The poor condition of access routes such as aisles, platforms, stairs and ladders accounts for many injuries; their misuse accounts also for a large number.

Walking Tips

Walking surfaces such as floors, walks, etc. should have nonslip surfaces. They should be free of tripping hazards, obstacles and debris, and items which may roll or slide if stepped on, including paper clips, rubber bands, bolts, nails, etc.

Look where you are stepping.

Walk only on recognized pathways unless required to walk elsewhere.

Do not step on railroad rails or other possibly slippery surfaces unnecessarily. Do not step into puddles, on wet or icy spots, or on oily spots.

Run only in case of extreme emergency.

Climb and descend stairs one step at a time, with one hand on or near rail. Never use stairs while both hands are occupied in carrying objects.

Platforms

Platforms 48 in. or higher from the ground should have standard guard railings. A standard railing consists of a top rail, an intermediate rail, and posts. It is 42 in. from the floor level to the top of the top rail; the intermediate rail is halfway between. Toe boards 4-in. high are required where height and traffic present the hazard of objects falling off the platform on persons.

Where material is being moved from such a platform the rail may be removed temporarily if the edge is clearly marked with a broad yellow stripe and all persons warned of the falling hazard. On platforms under 48 in. in height, the guard rail may be omitted, but if there is any liklihood of the edge not being clearly observed, stripe painting of the edge is desirable.

When stairs are cut into the edges of a platform, they should always be guarded by hand railings; hand railings are always desirable even if the steps are exterior to the platform. See section on "Stairs" for stair hand railing heights.

The importance of adequate illumination on platforms cannot be overemphasized. It is difficult to predict when some person will use or cross a platform.

Ramps

Ramps should have nonslip surfaces, and have a gradient of not greater than 1-in. rise per 10 in. of horizontal run. Ramps are preferable to stairs for movement of groups of people at one time. They must be provided with hand railings as if stairs. When used for vehicles they should also have curbs.

Stairs

All stairs should meet the requirements of the Building Exits Code (NFPA No. 101, ANSI A-9.1, OSHA 1910.23). One point often overlooked is that stair railings must be not more than 34 in. nor less than 30 in. from the top rail to the stair tread in line with the face of the riser at the forward edge of the tread.

Maintenance of stairs in safe condition is extremely important. Periodic scheduled inspections should be made. A checklist used by one chemical firm is as follows:

1. Are there cracks, loose treads or other repair work indicated?
2. Are the nosings excessively rounded, worn, or slippery?
3. Are the handrails free of rough or sharp points?
4. How nonskid are the tread surfaces. Are they kept dry, clear of ice, snow, and debris?
5. Are there any tripping hazards present?
6. Is the landing adequate in size, and arranged so that there is no change in level as one passes through a door?

Ladders

From a safety standpoint, ladders are a second choice to stairs. Permanently installed ladders should be of steel, not wood. Detailed standards for installation and use of fixed and portable ladders are included in OSHA Safety and Health Standards 1910.25-1910.27 and should be consulted. The following are points that experience has shown are frequently overlooked:

Cage guards should be installed on all fixed ladders which extend more than 20 ft above the ground or floor, and when installed should start at the 7-ft. level Straight fixed ladder runs over 30 ft should be broken with platforms every 20 ft or less.

Fixed ladders providing access to roofs, etc. where installed at a wall should have the rungs a minimum of 7 in. from the wall. Where the ladder reaches the access level, the side rails (and cage if applicable) should be carried 42 in. above the landing level. Such point of entry should always be protected by a chain or gravity swinging bar gate to protect a person from falling from the elevated surface.

Check to see that portable ladders, either straight or step, are in accordance with ANSI standard A-14.1 and OSHA before purchasing. Heavy-duty ladders give an extra margin of safety. Wood stepladders should have a metal tie rod beneath each step, and be free from racking and twisting.

Avoid the installation of ship's ladders in chemical plants. They present a fascinating temptation to employees to be used as a stair, descending facing away from the ladder, and to some the allure of a sliding chute. Use either standard stairs or standard ladders—nothing in between.

All portable straight ladders should be equipped with nonskid feet, and with permanently affixed tie ropes 9 ft in length, spliced to the second rung from the top. Extension ladders should be equipped with box-type enclosed dogs to prevent fingers from being crushed.

Safe use of ladders is covered in Chapter 13.

12.9   ELEVATORS

ANSI Code A-17.1 covers the requirements for installation of elevators, hoists, dumb waiters, and escalators, and should also be used as a guide for inspection and maintenance. It is not incorporated in OSHA Safety and Health Standards yet.

Man hoists are potentially dangerous and should not be used. Key points for safety checks are:

All elevators should have mechanically and electrically interlocked car gates and hoistway gates. Do they operate to stop the car if either gate is open?

Safety dogs to check free fall required. Is their an adequate annual test of these?

Biparting-type doors should be equipped with safety astragals to prevent crushing of a hand caught during closing. What is their condition?

Monthly testing and inspection of interlocks by qualified personnel should be not only made but adequate records should be kept.

12.10.   EXITS

It is important that exits of buildings be adequate to accommodate all the employees in case of fire or other emergency. The Building Exits Code (ANSI A-9.1 and NFPA 101) cover requirements of construction and should be consulted when installations are being planned. They are incorporated in OSHA Safety and Health Standards as Subpart E 1910.35-1910.40. Key points are:

Exits must have free and unobstructed egress.

Every exit shall be clearly visible or the route to it shall be conspicuously indicated.

In every building or structure equipped with artificial illumination, adequate illumination shall be provided for all exit facilities.

In general, every building must have at least two means of escape from each floor, in some cases from each room. Where quick egress from hazardous operation is vital, quick opening emergency doors and chutes should be provided. Each employee should be trained to see each day that the exits are not obstructed.

12.11.   AISLES AND ACCESS TO WORK LOCATIONS

Each employee should be able to reach his or her work area and carry out his or her duties without being exposed to bumping, tripping, or slipping hazards. This requires clean, adequately maintained walking surfaces, freedom from protruding objects, and 7 ft of head room. If employees must rod materials inside bins, there should be a heavy grating or other safe surface for the employees to stand on. Pits must be provided with metal covers, or guard rails, fences, etc. to protect nearby employees.

Where obstructions are necessary, they should be padded if possible, covered with bright plastic sleeves, or marked with yellow paint.

Minimum aisle widths for one-way foot traffic should be 3 ft, for two-way traffic, 5 ft. Greater width is preferable.

Plastic tape 3-in. or paint stripes should be used at blind corners to designate center lines of halls and aisles, and personnel trained to "keep to the right".

12.12.   PERSONAL HYGIENE FACILITIES

Exposures to harmful substances in the work place have three routes of entry into the body of the employee:

Mouth to gastrointestinal tract
Inhalation to the respiratory system
Skin contact, i.e., absorption through the intact epidermis

Good personal hygiene entails:

Washing hands before eating
No food or tobacco allowed in the work area, or allowed to be exposed to harmful substances
Shower and clothing changes

Facilities to permit these defense mechanisms of personal hygiene must be provided in almost every chemical manufacturing and laboratory installation. All employees must be encouraged, in fact required, to adhere to high standards of personal bodily cleanliness. In many operations, this justifies mandatory washing of the hands and face before smoking, drinking, or eating. If exposed to chemicals, the hands should also be washed before touching the more sensitive parts of the body.

Similarly mandatory daily showering and daily change to clean work clothes may be required, or at least encouraged. Providing separate lockers for street clothing and work clothing is often advisable. The opinion of a qualified industrial physician and occupational health nurse should be sought as to adequacy of facilities and personal health requirements for each exposure. (See Chaps. 9, 10, 11, 14 and 15.)

Shower and Locker Rooms

To reduce slipping hazards in shower rooms, perforated rubber mats may be used if cleaned and dried regularly. A nonabsorbent floor surface with roughened surface is generally much more desirable. Grab rails installed at waist height on the walls at the showers have been found helpful to prevent slips and falls.

The round group wash basin of nonabsorbent cement is excellent from the standpoint of service and ease of maintenance.

Liquid or powdered soap recommended by the industrial physician is preferable to bar soap. Soap scraps frequently remain on the floor as slipping hazards.

As mentioned in the section on water service, all water to showers and wash basins must be delivered to the outlet at a temperature not in excess of 135°F. Water hotter than this coming in contact with the skin will usually result in first degree burns. Showers should be fitted with a dial-type mixing valve which requires the user to go through the cold water phase before reaching the hot water. Direct mixing of steam and water to make hot water for use on the body is forbidden. Hot water must be manufactured by use of a water heater involving heat transfer.

Adequate ventilation, control of humidity, and efficient janitor service are a must for expecting cooperation from employees in maintaining the needed cleanliness in shower, lockers, and toilet facilities.

In spite of advertising to the contrary, foot baths of various liquid antiseptic solutions present little assistance in controlling or preventing athlete's foot, a fungus disease. Complete drying of the feet after bathing, keeping them dry, and use of antiseptic powders recommended by a capable industrial physician are far more effective measures.

Lockers with sloping tops discourage employees from leaving articles on top. Periodic inspections in the presence of the occupant insures good housekeeping and cleanliness in lockers.

Toilet Facilities

The number of toilet facilities, in toilet rooms separate for each sex, based on the number of employees, is now specified by OSHA Safety and Health Standards 1910.141 as follows:

Number of Employees	Minimum No. of Water Closets
1–15	1
16–35	2
36–55	3
56–80	4
81–110	5
111–150	6
Over 150	1 additional per 40 additional employees

Where toilet facilities will not be used by women, urinals may be substituted for water closets, except that the number of water closets shall not be reduced to less than ⅔ of the minimum specified. Where toilet rooms will not be occupied by more than one person at a time, can be locked from the outside, and contain at least one water closet, separate toilet rooms for each sex need not be provided. Other OSHA regulations cover construction, equipment and accessories, etc., required for toilet facilities.

12.13.   EATING FACILITIES AND FOOD DISPENSING

If food or drink is supplied employees, either by personnel of the employer, or by a contractor, or by a dispensing mechanism, a whole new area of food safety and industrial hygiene must be considered. Food dispensed must be wholesome and free from spoilage and contamination. Dispensing personnel and all facilities for eating and food dispensing must be under adequate hygiene and medical supervision and continued inspection and records kept of the supervision and inspection.

Provide refrigeration for safe storage of perishable foods that the employees bring to work. A household refrigerator, if of sufficient size and frequently cleaned, is satisfactory. Both the refrigerator and lunch-box storage facilities should be convenient and free from the likelihood of chemical contamination. This means that they must be away from places where chemicals are handled. Chemicals should not be stored in domestic refrigerators, but specially modified refrigerators should be used.

Provide a clean, well-lighted location for eating, drinking, and smoking away from chemical operating or laboratory work. In many cases it is necessary to prohibit the carrying of food, smoking, or chewing materials in work areas to prevent possible toxic contamination.

12.14. TRAFFIC AND PARKING

We are nation on wheels. Outside the plant or laboratory, we ride. Inside we may walk, but not far. This means parking lots for employees' cars close to their place of work, and adequate walking facilities from the parking location to the job. Snow and ice removal from the parking lot and walks is usually a management responsibility. For buses, loading and unloading facilities should be located for off-the-street access.

The configuration of the parking layout in the lot depends largely on the proportions and size of the parking lot. For a two-way traffic driveway, a 20-ft minimum width is needed, 30 ft preferred. Each parking stall should be, for standard size passenger cars, 9 × 20 ft. A comfortable backing space is required, and will vary with the parking angle. A suggested table of backing lengths as used by a large company is as follows:

Parking Angle	*Backing length (width of driveway)*	
	Minimum	Preferred
45°	13 ft	20 ft
60°	18 ft	25 ft
90°	24 ft	30 ft

Guidelines should be white, 4–6 in. wide. In some arrangements a drive-through arrangement can be used, but this uses more total space per car usually. Employees exiting after a work shift or period tend to leave at the same time; employees simultaneously backing out of parking spaces into drives often have minor collisions. To avoid this, some employers have found that having employees back into the parking spaces so that they can drive out facing the drive is very effective at preventing collisions.

Concrete or wood parking guides to insure uniform and minimum space per car can be used, though painted lines with some training generally is effective. Fixed parking guides are an obstruction during snow removal and may present a tripping hazard.

If the parking lot is used at night, 5 foot candle/illumination at a height of 36 in. above ground on walking surfaces is desirable (see Table 12-1).

Separate pathways in parking lots appear to be diminishing in use. Since these are not provided in shopping center parking lots, people have learned to do without them.

Walkways

Whether separate walkways should be provided inside the plant or the employees requested to use plant roadways for walking depends primarily on the vehicular traffic intensity, secondarily on the adequacy of the road surface, width of road, and lighting. If roadways are used for walking, it is better to paint walkway boundaries on them and give the pedestrian the right-of-way inside the walk lines than to depend on employees facing traffic.

A 5-ft width of walk appears to be adequate for the usual walk involving movement of several people at the same time.

Lighting of plant roads and walkways for pedestrian illumination is required if there is night travel on them. Illumination levels for paths, and roads used for walking should be 5 ft candles at a height of 36 in. above the walking surface (see Table 12.1). Lights should be elevated and shielded so they do not tend to blind pedestrians or drivers approaching them. Great contrast in illumination levels on the same path should be avoided.

Blind intersections between paths and vehicle routes, either automotive or railroad should be avoided. Guard rails and gates are useful solutions.

BIBLIOGRAPHY

American Conference of Governmental Industrial Hygienists, Committee on Industrial Ventilation
"Industrial Ventilation—A Manual of Recommended Practice"

American Industrial Hygiene Association
"Heating and Cooling for Man in Industry"
"Ergonomics Guide to Assessment of Metabolic and Cardiac Costs of Physical Work" (J32:560-1971)

American National Standards Institute (ANSI)
"Building Exits Code" (A-9.1)
"Practice for Industrial Lighting" (A-11.1)
"Safety Requirements for Floor and Wall Openings, Railings and Toe Boards" (A-12.1)
"Safety Code for Portable Wood Ladders" (A-14.1)
"Safety Code for Portable Metal Ladders" (A14.2)
"Safety Code for Fixed Ladders" (A-14.3)
"Safety Code for Elevators, Dumbwaiters, Escalators, and Moving Walks" (A-17.1)
"American Standard Practice for the Inspection of "Elevators—Inspectors Manual" (A-17.2)
"National Electrical Code" (C-1)
"Requirements for Fixed Industrial Stairs" (A-64.1)
"Standard Practice for Office Lighting" (A-132.1)
"Standard for Emergency Eyewash and Shower Equipment (Z358.1-198X, 1980)
"Minimum Requirements for Sanitation in Places of Employment" (Z-4.1)
"Specifications for Drinking Fountains" (Z-4.2)
"Fundamentals Covering the Design and Operation of Local Exhaust Systems" (Z-9.2)

A.M. Best Co., Oldwick, N.J. 08858
"Best's Safety Directory"

L. Brouha
"Physiology in Industry"—Pergamon Press, N.Y. (1967)

Compressed Gas Association
"American National Standard Method of Marking Portable Compressed Gas Containers to Identify the Material Contained (C-4)
"Compressed Air for Human Respiration" (G-7)
"Commodity Specification for Air" (G-7.1)
"Safe Handling of Compressed Gases" (P-1)

Illuminating Engineering Society
"Lighting Handbook"

National Bureau of Standards, U.S. Department of Commerce.
"Safety Rules for the Installation and Maintenance of Electric Utilization Equipment" (Handbook H-33)

National Fire Protection Association
"National Electrical Code" (70)
"Life Safety Code" (101)

National Safety Council
"Accident Prevention Manual for Industrial Operations"
"Fundamentals of Industrial Hygiene"

Underwriters' Laboratories, Inc.
"Electrical Appliance and Utilization Equipment Lists"
"Electrical Construction Materials List"
"Hazardous Location Equipment List"

U.S. Department of Labor, Occupational Safety and Health Administration
"OSHA Safety and Health Standards" (29CFR/Part 1910)
U.S. Public Health Service, U.S. Dept. of Health, Education and Welfare
"Drinking Water Standards" (Pub. 956)
"Manual of Individual Water Supply Systems" (Pub. 24)
Manual of Recommended Water - Sanitation Practice" (Pub. 525)
"Food Service Sanitation Manual" (Pub. 934)

U.S. Public Health Service, U.S. Dept. of Health, Education and Welfare
"The Industrial Environment—Its Evaluation and Control" (Center for Disease Control—NIOSH 1973)

13
Maintenance of
Chemical Equipment

David T. Smith

Maintenance in the chemical industry differs from that in other industries because of the nature of the materials, processes, and types of equipment used. Since much chemical work involves the movement of fluids, gases, and powdered solids from one piece of equipment to another, many pipelines, conveyors, fork-lift trucks, and other material-handling devices are used. Containers are more likely to be tanks, drums, or some form of closed container than in other industries.

Much of the chemical-reaction equipment involves mixing, "cooking," cooling, and stirring under widely varying levels of temperature and pressure in sizes from one-quart laboratory autoclaves to huge columns that loom against the sky like grain elevators. Disassembling and reassembling in itself requires special rigging and millwrighting techniques. Specialized problems such as high corrosion rates, flammable liquids, and toxic or noxious materials are typical of this industry. Therefore a high quality of work is more necessary for safety of operation than in most other industries. Leaking pipes, equipment improperly reassembled, and so on expose personnel to hazards.

The cases where the process line was connected into the stair-pipe railing, where steam was connected into the toilet, and where the electronic recording instrument began to play radio programs are part of the folklore of chemical maintenance. If they did not happen, they could have, because they are so typical of the many "boners" that are possible as a result of inadequate identification of pipes, wiring, and equipment.

Records show many serious injuries and even deaths from mistakes such as connecting inert or toxic gases into air lines supplying air for humans to breathe or connecting piping so that wrong materials were added to reactions with resulting fires, explosions, and emission of toxic gas. Connections have been left open which allowed the escape of dangerous materials to expose personnel. Relief devices have been left inoperative or removed allowing vessels and systems to rupture from overpressure.

A second source of hazard to personnel and property from improper maintenance is the failure of equipment during either normal or abnormal operations. For example, leaking acid pipes can drip on persons, escape of flammable gas may result in fire or explosion, and loss of sulfide liquids to an acid stream will result in formation of deadly hydrogen sulfide gas. Even broken steps and stair rails take their toll of injuries every year.

In summary, in no other industry is high quality of maintenance, workmanship, and proper identification as important to safety as in the chemical industry.

13.1   PREVENTIVE MAINTENANCE

By inspecting, repairing, and replacing equipment on an intelligently planned schedule, we can prevent many failures. There are tremendous cost advantages to this type program as well as safety advantages. Breakdowns during manufacturing operations result in loss of production, inefficient use of worker power, and increased repair cost because of overtime and other emergency penalties. These factors cost far more than the investment in a planned preventive maintenance program. The advantages to safety are obvious because of the reduction of exposure to emergency conditions and the ability to plan and provide safe procedures and equipment. The advantages to quality and quantity of production are not to be overlooked.

Preventive maintenance essentially is the system of determining the probable frequency of failure of a device or system, and inspecting, repairing, or replacing it before it fails. We can draw a curve showing the probability of failure of a given device against time, and it will appear as the ''normal probability'' curve (see Fig. 13.1). There is a very low probability that it may fail very early, then there is a rapid rise of probability of failure to an average value, then the curve drops again. There is a very low probability that failure will be deferred until long after the average time. The shape of the curve will

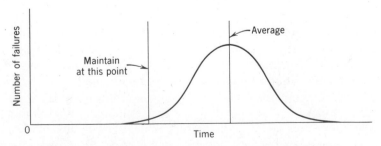

Figure 13.1.   Selection of periods for preventive maintenance.

appear as a bell, with the peak as the average. Our job is to schedule maintenance so that we are confident that we take action before the likelihood of failure is 2%, or some other chosen figure. In other words, we would be confident that most of the time we will have applied maintenance before a failure occurs.

It is obvious that this requires a knowledge of the average length of time before breakdown of the particular device under the specific conditions of operation. Often this is not easy to determine. Past experience in similar operations, studies of rate of deterioration, and other means of analysis can be used. Knowledge of how the device or system can fail, and of the most likely ways for it to fail can help to determine the frequency of failure.

Needless to say, the ability to match optimum maintenance dates and time with availability of equipment (release by operations for shutdown), availability of materials, maintenance facilities and manpower are required. To accomplish this, planning ahead, using schedule sheets, and keeping everybody informed, appear to be key points. It is important to control *all* equipment from a preventive maintenance standpoint by making schedules for every piece, even if the frequency is as long as once every 10 years.

An example of the working of preventive maintenance in a typical plant follows.

PUMPS P-16–P-22 IN BUILDING A310

STEP 1.   Area-maintenance engineer and operating-department supervisor agree on frequency of inspection and removal of pumps to pump shop for overhaul. Spare pumps will be installed. Frequency is annual and one pump is scheduled to be replaced each month, January through July.

STEP 2.   Shutdown dates and times are entered in production control schedule to coincide with down time for other mechanical work.

STEP 3.   Maintenance engineering group enters dates and times on work-order control-system cards.

STEP 4.   Monthly maintenance work-order schedule shows dates and times.

STEP 5.   Work order is issued one week in advance by maintenance engineering group. Time and date are agreed on again by operating supervision. Instructions are issued to operating personnel on readying equipment.

STEP 6.   Work procedures are reviewed with maintenance foreman and men. Workers, equipment, and materials are lined up.

STEP 7.   Work done.

STEP 8.   Results are recorded on equipment maintenance record of the pump, and next overhaul date is scheduled.

Proper maintenance to prevent accidents means more than achieving freedom of failure of the equipment and buildings maintained. It means protection of people while doing maintenance. This can be considered in three phases:

First, protection of operating personnel from temporary environmental hazards caused by maintenance work.   For example, roping off temporary openings, keeping areas cleared below overhead work to protect against falling objects, keeping discarded parts and tools from being a stumbling hazard on floors, shielding of welding arcs, etc.

Second, protection of maintenance personnel against hazards of the operating conditions.   This requires clearance with operating supervision before doing any mechanical work, coordination of operating activities with mechanical work, and use of permits where necessary to formalize this coordination. For example, a mechanic could step back from a pump repair job into an aisle in front of a moving fork-lift truck carrying a pallet of drums, unless operating personnel is familiar with his or her assignment, has posted warning signs to protect the work location, and is familiar with the movements of the operating personnel in their work.

Third, protection of maintenance personnel against hazards of their own work.   Methods of doing work should be standardized where possible and written procedures prepared covering steps and key points. In this way we can avoid overlooking the lessons learned from previous mistakes and injuries. Too often we say, "every maintenance job is different, we cannot freeze them into a neat procedure." What we really mean is that we are too lazy to work out "agreed on" methods for repetitive jobs, and to put them down on paper.

13.3   SHUTDOWN AND START-UP

Clearcut responsibility for each step is the keynote of planning and execution of shutdown and start-up. Shutdown of operating equipment should be the responsibility of operations, not maintenance or other service groups. Auxiliary service groups should handle their own specialties, such as a power group handling the outside steam, electric, air, gas, and other services to buildings. Close coordination is required. Notification of shutdown to interested groups should be made as far in advance as possible.

Operating supervision must be held responsible for emptying, washing out, steaming out, purging with inert gas, and other steps to make equipment safe before turning it over to the maintenance group for working on it. Maintenance supervision has the responsibility to check as far as feasible to see that this has been accomplished before accepting the equipment. Key points are proper blanking off of pipelines connected to the equipment, proper ventilation, and testing for flammable vapors. Vessels that have been purged with nitrogen or other asphyxiant should be identified and posted.

During actual shutdown, and even more likely during start-up, unusual thermal stress may be added to the normal mechanical stresses. Exposure of personnel to leaks and ruptures of equipment must be guarded against. No connections should be tightened during warm-up periods—we must wait until temperatures have leveled off.

During shut downs and start ups special care must be taken to see that all "lock-out" procedures are followed closely for electrical work, for moving machinery, and valves.

Written procedures for emergency shutdowns as well as for normal shutdowns must be prepared, rehearsed, kept up to date, and kept available to people that have to use them.

Provision must be made for emptying safely all lines that are to be opened, for accounting for conditions in all equipment, and for knowing the condition of all vents and relieving devices.

Personnel must not be permitted to break into any pipeline without full protection against it being "loaded" regardless of draining and other procedures. Too often pipes have been plugged or partially plugged, trapping dangerous materials in them. Complete protection to the body means "acid-suit" protection or better against corrosive materials, and supplied air of breathing quality for respiratory protection against noxious fumes and gases (see Chapters 25, 26 and 27).

13.4 ENTERING VESSELS AND CONFINED SPACES

Entering vessels presents special problems because of the inability of the workers to get out of the vessel without outside help in case of emergency and because of difficulty of communication. For these reasons, plus the possibility of emergency chemical exposures, rigidly enforced special precautions must be taken. These are required by statute or governmental regulation in many states. In general these include:

1. The vessel must be thoroughly cleaned by operations.
2. All connecting pipelines must be disconnected and blanked off.
3. All power driven devices (such as agitators) must be locked out at positive disconnect switches.
4. Air samples must be taken to prove absence of flammable vapors, and also in some cases, of toxic or noxious materials.
5. Air samples must be taken to prove presence of a normal amount of oxygen.
6. A tank entry permit must be signed by operating and maintenance supervision verifying that above steps have been satisfactorily complied with, and posted at the vessel site.
7. Workers to enter vessel and watcher must be equipped with life belts and ropes. (Belts should be the type fitting high under armpits for ease in lifting, not around waist. Handcuff types are preferred as alternates by some, opposed by others.)

8. In most cases, an air mask and fresh-air supply must be available for each worker entering the tank and for the watcher. A chemical protective suit of impervious material completely enclosing the body, and provided with fresh breathing air and air conditioning for comfort not only serves that purpose but provides a safe comfortable working environment. It is equipped with an air-supply hose which also acts as a life line (see Chapter 25).

9. There must be one watcher at the vessel entrance who can keep in touch at all times with the worker or workers inside. In addition there must be at least one other worker within call of the watcher to help in case of emergency. One worker can do little alone. Devices for signaling for additional help such as a "freon" actuated horn are acceptable.

10. Rope or chain ladders with rigid rungs of wood or metal may be used for vessel entry where straight ladders cannot be used. However, lowering of a worker into a vessel without facilities for exiting by the individual alone should be avoided.

11. A worker should not be allowed to go into a vessel through an opening which requires "squeezing through." He or she cannot be removed quickly enough in emergency conditions. A 22-in. utility hole should be standard. Small workers can enter as small an opening as 18 in. Anything less is extremely doubtful.

Similar principles should be followed when placing workers in any confined space from which emergency exit may be hampered. Any pit or trench deeper than 5 ft, and any work on a roof or column where a worker may be trapped in case of liberation of toxic fumes, requires special control procedures of similar principles (see Chapter 36).

13.5. BURNING, WELDING, AND OTHER FLAME-PRODUCING WORK

Because the chemical industry handles so many flammable and volatile materials, and because of the rapidly changing technology which in turn causes frequent changes in materials handled, control of ignition sources to prevent fires and explosions is of prime importance. Therefore, special procedures to reduce the hazard of fire from heat and flame producing work are essential. In general they cover these key points:

1. No welding or burning is permitted outside of designated welding shops and specified welding areas, without first obtaining permission of the operating supervision responsible through the use of a written,

signed "burning permit" or "hot work permit." This should indicate the nature and location of the work, any particular restrictions required for safe performance, and the time during which the work is to be permitted. Since the issuance of a permit also entails the control of operations to prevent the creation of any hazardous conditions during the work period, it is important that the permit be made a responsibility of the operating-line organization, not a service or staff group such as the safety or fire-protection group. It may be necessary to assign additional personnel to act as fire watchers during the job, and they may be supplied by the safety or fire protection group, but the responsibility for authorizing the work in any given location should be put squarely on the supervision responsible for the operating conditions of that location.

2. Only qualified burners and welders should be given authority to perform this type of work, and mere mechanical skill is not enough. These workers must be trained in the responsibility to properly analyze and protect surrounding flammable materials, to extinguish fires, and handle emergencies. A permit system certifying qualified welders has been proved useful in many plants.

3. All burners and welders must have first-aid fire protection on the job with them. Water hoses are generally needed also. Cracks and flammable materials must be protected by incombustible covers or barriers. Overhead work must be analyzed to provide sufficient protection from sparks falling below. No flammable gas or oxygen cylinders should be permitted inside of operating, laboratory, or storage buildings, except under most unusual and carefully controlled circumstances (see Chapter 19).

13.6. PRESSURE TESTING, CORROSION INSPECTION

While various instruments are now available to permit accurate measurement of thickness of vessel walls at selected areas, they do not guarantee that a weakness has not developed at some location not measured. Therefore, they do not take the place of periodic hydrostatic pressure testing to determine the ability of the vessel to withstand pressures in excess of those of normal operation. The generally accepted ratio for periodic testing during usage of a vessel is one and onehalf times the normal working pressure; the frequency of hydrostatic test may be determined by analysis of the results on instrument studies of rate of thinning of the container walls, by measurement of container wall thickness by drilling or metallurgical study of plugs, and by experience. Once set, the hydrostatic testing schedule should be rigidly followed. Plans should be made in advance with operations so that production schedules can be arranged to fit.

Hydrostatic testing is normally done with water, but if water is incompatible with the use of the vessel, other nonhazardous inert liquids may be used—never air or other gas. The only time that gases may be used for pressure testing is when use of liquids is *not* feasible, and then only when the vessel is barricaded to protect personnel and property from missiles and blast force in case of rupture. The gas pressure in a vessel acts as an energy spring which will propel missiles at high velocities. Liquid pressure in a vessel drops instantly to zero on failure of the container.

There is no need to hydrostatically test pipe and certain equipment used for noncritical services from the safety viewpoint, unless failure can expose personnel to injury. No hydrostatic testing program is better than its control. Adequate card files on each piece of equipment, adequate scheduling and communications with all parties concerned are needed to make it work (see Chapter 8).

13.7. LOCKING AND TAGGING FOR SAFEGUARDING PERSONNEL

While details vary from plant to plant, safe maintenance requires that no one may work on (or where exposed to) power-driven equipment without locking out or physically disconnecting the source of power beforehand.

This may be done either by (1) locking the electric switch on the power circuit (not the starting circuit) in the off position, (2) disconnecting the motor electrically or mechanically, (3) by removing the belt drive, or (4) locking feed valves to prime movers in the "off" position and blocking movement of pistons, crank arms, or flywheels. Any exceptions, such as adjusting glands and seals, must be approved by supervision.

Written procedures should be prepared and thorough training given in their use. In general these procedures will stress:

Operating supervision must first identify the equipment and equipment controls.

Operating supervision and each person who is to work on the equipment go to the controls and individually place their lock on the controls after they have been placed in the off position. Each person keeps the key to his or her own lock.

The equipment is rechecked then to be certain that it cannot be started. Work on it can then be commenced.

When the work is completed, the reverse procedure is completed before turning the equipment back to operating supervision.

When electric motors are disconnected, the switch box is still locked with the switch in the open position.

In large-scale mechanical jobs where numerous workers and crafts are concerned with the same equipment, most plants permit a "group lock-out" where only the mechanical supervisor's lock is placed on the switch, and each mechanic places his or her individual tag on the lock. Only the mechanic can remove his or her own tag, and the foreman will not remove the lock until all tags are accounted for.

Whenever electricians work on electrical circuits or equipment, the same procedures can be followed unless testing or other work involving energizing the circuit is required. In these cases, after proper identification of the equipment and circuits by operating supervision, the circuits and equipment and turned over to the electrician, who affixes his or her "blocking out" tags to the control points. No one but this electrician can operate any controls after that until he or she removes the tags. As an additional safeguard, the electrician will lock the switch box during periods when work will permit.

Similar principles should be followed in doing work on pipelines or piping systems, with the valves tagged and locked if possible. All lines should be disconnected and blanked if practicable.

Rotating Equipment

Rotating equipment such as turbocompressors and reciprocating compressors are subject to failure due to imbalance, bearing failure, unequal loading, misalignment, or foundation settling. Vibration measurements on a continuous or periodic basis is used to detect operational problems before damage results. Attention given to such early indicators of trouble has been called "predictive" maintenance. Usually a specialist in vibration analysis is needed to take the measurements and to interpret abnormalities.

13.8.   DISPOSAL OF CHEMICALLY CONTAMINATED EQUIPMENT

It is the responsibility of operating supervision to see that all equipment leaving the area is clean to a point where an employee, untrained in the hazards of corrosive, toxic, or flammable materials, can work on it safely. All equipment that is removed from service should be tagged immediately as to its state of chemical contamination, and the tag changed as conditions change. When it is clean it should be so tagged. Otherwise, there is always the danger someone may assume it is clean, and either move it or work on it. Obviously, it is impractical to decontaminate equipment before dismantling in many cases. In such cases cleaning and decontamination should be accomplished if possible before removal from the area. If it cannot be decontaminated at the area, it must be handled separately on a planned basis and taken to an adequate salvage area where it can be worked on. Each job may have to be individually studied and planned from a decontamination standpoint.

Among the details to remember is that it is important to give special attention to prevent pipe and valves from being discarded with materials trapped in them—they can become bombs. All valves should be cleaned, the bonnet should be loosened, and the gate opened before being discarded. Pipe may be heated at a burning ground. Flammable liquid wastes should not be allowed to reach drainage ditches or sewers. Glass should be handled separately from other waste materials. Other special problems will be apparent on detailed consideration. The important point is not to leave this phase of the maintenance work to chance. (See Chaps. 28 and 29).

13.9.   EMERGENCY MAINTENANCE

Emergency maintenance should more properly be termed emergency repairs, since maintenance implies preventing emergency repairs. However, some emergencies do occur, and we must be prepared to meet them. In general, day-to-day maintenance procedures will meet the situations arising, but some special problems require additional planning to meet. For example, what do we do if a pipe connection breaks off a chlorine tank? The time to plan to take care of it is in advance, not after it happens. Considering this case, special clamp fittings can be made in advance and kept in an emergency kit with supplementary equipment to meet just such an emergency. Suppose a sulfuric acid pipeline ruptures. Planning shutdown, wash down, and repair activities in advance will make it possible to take such mishaps in stride with minimum exposures to personnel and property. The key point is to "brainstorm" such incidents, decide in advance who should do what, with what, and line it up so that it can be accomplished if it happens.

How about rupture of an underground pipeline? How are we going to handle the shut down? Who will be available to excavate? How will we flush out the broken pipe? How can it be plugged to keep it clean while the excavation is being pumped out? Are emergency sleeves suitable? Are they available? How about pipe stock?

What do we do when an electric line fails? Do we have alternate routes of supply? Where are the switching facilities? What effect does loss of power have on emergency shut down of operations? These are all typical problems of emergency maintenance for which solutions can and should be planned in advance.

There is one principle to adhere to in emergency maintenance. If it is not safe to do it a certain way under normal maintenance, then it is doubly unsafe to try that way under emergency conditions. A reasonably safe method can be found if we really try. If it is not safe to do electrical hot work under normal conditions, then it should be even more strongly forbidden during emergencies.

13.10   EQUIPMENT INSPECTION AND MAINTENANCE

Hand Tools

Several books could be written on ways in which hand tools can deteriorate into an unsafe condition; suffice to say the pipe wrench with dulled jaws has injured more mechanics than the most spectacular explosions. Mushroomed chisel heads, cracked or bent wrenches, faulty hammer and screw-driver handles, and poor chisel and screwdriver blade condition are all too common. We can meet these hazards only by special effort to educate our mechanics to keep their tools up to standard, and by making sufficient tool inspections to know that they are kept that way. Doing it after the injury helps, but periodic checking is better. A check list is desirable.

Ropes, Slings, Chains, and Hoisting Equipment

Because the danger of failure presents unusually severe safety hazards, and because of the high wear factor, periodic inspection at a central point, with good record control, is a must for proper maintenance of ropes, slings, chains, and hoisting equipment. This includes lifelines and lifebelts. Inspection for defects seems the best procedure in most cases. Overloading for test purposes in itself may cause failure later on the job. However, tests up to rated capacity may uncover hidden defects. Because of wide variation of deteriorating exposures, no rules can be given for frequency of inspection and test, but it pays to err on the conservative side. There is a wealth of material available from suppliers on safe use and maintenance of this equipment which can be used to advantage.

For example, synthetic fiber ropes may be much less subject to chemical attack and therefore give longer life with a greater margin of safety—but the user should allow for their being smoother and hence they provide a lesser hand grip.

Ladders

Ladder failures during use usually result in falls and injuries—therefore they must be prevented. Standards for their purchase should be established. Use only the stronger industrial grade ladders, even if cheaper ones will pass American National Standards Institute standards. The stronger ones give greater margin of safety against abuse and wear. The length of ladders permitted should be spelled out in standards. Ladders should be carefully inspected by trained personnel, including hammer testing, and the inspection dates as well as ownership should be marked on each ladder. Strays have a way of developing among ladders. Many plants inspect each ladder annually, others more often.

13.11. SUPERVISION OF MAINTENANCE PERSONNEL

Supervision of maintenance workers presents special problems of its own from a safety standpoint, but these problems are capable of solution. Because of the changing work locations and assignments, mechanics are usually "on their own" more than chemical operating or laboratory people. It requires more effort on the part of maintenance supervision than in the case of other groups to know how their workers are performing, but it can and must be done. Sampling techniques applied to observation can be used because continuous observation is impractical. However, the setting of standards of performance in maintenance work and the training of men to achieve these standards is no more difficult than in other work groups. The same management principles apply. We have to know the capabilities and characteristics of each worker and give guidance where needed so that he or she can reach his or her highest potential. Management tools of communication commonly effective are daily short "tail gate" or "tool box" meetings scheduled individual contacts on safety subjects, scheduled inspection of work in various stages of completion to observe work techniques, condition of tools and equipment, and housekeeping.

In summary, good maintenance prevents accidents, by use of careful well-planned preventive maintenance, planning for emergencies, and day-to-day good techniques in management of men.

BIBLIOGRAPHY

American Iron and Steel Institute, Committee of Wire Rope Producers, 1000 Sixteenth St. NW, Washington, DC 20036
"Wire Rope Users Manual"

American National Standards Institute, 1430 Broadway, New York, NY, 10018
"Safety Requirements for Demolition," ANSI-A10.6
"Safety Code for Scaffolding," ANSI-A10.8
"Safety Requirements for Temporary and Portable Space Heating Devices and Equipment Used in the Construction Industry," ANSI-A10.10
"Safety Code for Steel Erection," ANSI-A10.13
"Safety Code for Floor and Wall Openings, Railings, and Toe Boards," ANSI-A12.1
"Safety Code for Pressure Piping," ANSI-A13.1
"Safety Code for Portable Wood Ladders," ANSI-A14.1
"Safety Code for Portable Metal Ladders," ANSI-A14.2
"Safety Code for Fixed Ladders," ANSI-A14.3
"Safety Code for Job-Made Ladders," ANSI-A14.4
"Safety Code for Elevators, Dumbwaiters, Escalators, and Moving Walks," ANSI-A17.1
"Practice for Inspection of Elevators," ANSI-A17.2
"Requirements for Fixed Industrial Stairs," ANSI-A64.1

"Safety Code for Abrasive Wheels," ANSI-B7.1
"National Electrical Code," ANSI-C1
"Protection Against Lightning," ANSI-C5.1
"Scheme for Industrial Accident Prevention Signs," ANSI-Z35.1
"Safety in Welding and Cutting," ANSI-Z49.1
"Safety Color Code for Marking Physical Hazards," ANSI-Z53.1

Associated General Contractors of America, 1957 "E" St. NW, Washington, DC 20006
"Manual of Accident Prevention in Construction"

Broderick and Bascom Rope Co., 10440 Trenton Ave., St. Louis, MO 63132
"Riggers Handbook"
"Wire Rope Handbook"

Crom, R. C. W., "Safeguarding Against Shock Hazards," *Chem. Eng.*, **84** (7), 90–96, (Mar 28, 1977).

Finley, R. W., "Incipient Failure Detection in Rotating Machinery," *Chem. Eng.*, **87** (14), 104–112 (July 14, 1980).

Jones, F. G. "Prevention of Catastrophic Failure in Reciprocating Compressors," *Loss Prev.*, **5**.

Lee. R. P., "Systematized Failure Analysis—Some Unusual Failure Modes," *Chem. Eng.*, **84** (1), 107 (January 3, 1977).

Lee, R. P., "How Poor Design Causes Equipment Failures," *Chem. Eng.*, **84** (3), 129 (Jan. 31, 1977).

National Fire Protection Association, Batterymarch Square, Quincy, MA 02269
"Flammable and Combustible Liquids Code," NFPA-30
"National Electrical Code," NFPA-70
"Life Safety Code," NFPA-101
"Cleaning or Safeguarding Small Tanks and Containers," NFPA-327

National Safety Council, 444 No. Michigan Boul., Chicago IL 60611
"Accident Prevention Manual for Industrial Operations"—Eighth Edition 1980
"Industrial Data Sheets"

NIOSH, "Criteria for a Recommended Standard—Working in Confined Spaces," DHEW (NIOSH) Publication No. 80-106, U.S. Government Printing Office, Washington, D.C., 1980.

Occupational Safety and Health Administration, United States Department of Labor, Washington, DC 20210
"Occupational Safety and Health Standards for General Industry—29 CFR Part 1910"

Piper, J., "Selecting a Maintenance Training Medium," *Plant Engineering,* **36** (8), 81–84 (April 15, 1982).

Posiseal Co., "Valve Meets Fire-Safe Tests," *Chem. Eng.*, **84** (21), 95–96 (October 10, 1977).

Roebuck, A. H., "Safe Chemical Cleaning—the Organic Way," *Chem. Eng.*, **85** (17), 107–110 (July 31, 1978).

Sack, Thomas F, *A Complete Guide to Building and Plant Maintenance*, McGraw Hill Book Co., New York, 1980.

Schumacher, W. J., "Wear and Galling Can Knock Out Equipment," *Chem. Eng.*, **84**, 57–59 (May 9, 1977).

Tustin, W. "Measurement and Analysis of Machinery Vibration," *Loss Prev.*, **5** (1968).

Wells, C. and Antus, C. H., "Understanding Electrical Properties of Hose," *Plant Engineering*, **36** (8), 87–89 (April 15, 1982).

Wong, S. M., Koehler, K., Husseiny, A. A., Sabri, Z. A., and Sprung, J. L., "Statistical Analysis of Reportable Events Due to Maintenance and Testing Activities in Nuclear Power Facilities," Reported No. SAND-80-0837C of ANS Thermal Reactor Safety Meeting, NTIS, PC A02/MF A01, 1980.

14
Toxicity Versus Hazard

Howard H. Fawcett

HOW TOXIC—HOW HAZARDOUS?

These two related thoughts are frequently combined into one compound question, yet in their practical aspects, they may require entirely different answers. This discussion will point out some of the factors which must be considered in evaluating the true meaning of the "hazard" of a substance suspected of being "toxic" (Fig. 14.1). To some people, the Toxic Substance Control Act (P.L. 94-469) was a new thought and an entirely new approach to the control of potentially hazardous materials. Some authorities feel it was the most important piece of legislation pertaining to the environment to be enacted by the Congress. The act, which became generally effective January 1, 1977, gave the EPA's administrator broad authority to compile inventories of existing substances; require chemical industry to conduct extensive testing of substances; delay manufacture and marketing of a new product if questions arise as to whether or not it is safe for intended use; bar or place restrictions on the marketing of existing new substances or of new applications for the substances; and, require the maintenance of such records as the administrator may reasonably require. However, this act, which placed the word "toxic" in a specific context, probably created as much confusion and legal dispute as any previous legislation. It was not the first attempt to control "toxic" substances by law (see Fig. 1.1 and Table 14.1). The word "toxic" has been loosely applied to many different effects, when actually it should be related to general, systemic effects of a substance in living animals or in human beings. Almost every substance will produce injurious effects to some degree in a living body, because even safe substances, or those that at various times have been officially classified as GRAS (generally regarded as

245

Acute Inhalation Toxicity for Rats in	PPM	for Hours
Acrolein	8	4
Allyl chloride	2,900	3
Thiophene	8,700	½
Nitroethane	30,000	1
DMF	Sat. Vapor	4
1, 1, 1–Trichloroethane	10,000	3

Toxicity Scale

Rating Term	Lethal Dose for Man
1. Extremely	Taste
2. Highly	Teaspoon
3. Moderate	Ounce
4. Slightly	Pint
5. Pract. Non–	Quart
6. Relat. Harmless	>Quart

Threshold Limit Values

Are
Time-Weighted Average
Day-After-Day Exposure
Required Number of Factors
    in Deciding on Hazard

Are not
Common Denominators
Sole Criteria for Disease
Air–Polution Values
Permanent (Reviewed Yearly)
"Toxicity" Data

A lead grinder forgetful named Peck
Parked his respirator south of his neck
'Twas a very poor notion
The lad lost all motion
For lead dust made him a wreck!

Figure 14.1. Various attempts to explain toxicity.

TABLE 14.1   Major laws controlling toxic substances

EPA

Toxic Substances Control Act of 1976: places heavy reporting burden on industry; EPA can demand premarket testing of some chemicals.

Safe Drinking Water Act of 1975: carcinogens and toxic substances in public drinking water supplies.

Resource Conservation and Recovery Act of 1976: disposal of toxic and other hazardous wastes in landfills, by incineration, and so forth.

Water Pollution Control Act Amendments of 1972 and 1977: discharges of hazardous effluents into the nation's waterways; and the ocean landward of the three-mile limit; cleanup of hazardous spills on land and in water.

Marine Protection, Research, and Sanctuaries Act of 1972; ocean dumping from three to 12 miles at sea (enactment of 200 mile limit in spring 1977 may extend this).

Clean Air Act Amendments of 1970: allows EPA to set national emission standards for hazardous air pollutants; standards now exist for beryllium, asbestos, mercury, and vinyl chloride.

Federal Insecticide, Fungicide, and Rodenticide Act of 1972: covers the sale and use of economic poisons, and foodstuff treated with such poisons; counterpart legislation exists for enforcement by USDA and FDA.

FDA

Federal Food, Drug, and Cosmetic Act of 1906, amended in 1938 and 1962; bars any detectable amounts of carcinogens in foods, cosmetics. Specifically exempt hair dyes from enforcement provisions.

Occupational Safety and Health Administration

Occupational Safety and Health Act of 1970: hazardous materials in the workplace, or in products bought for use in the workplace.

Consumer Product Safety Commission

Consumer Product Safety Act of 1972: Specifically excludes tobacco, foods, drugs, and cosmetics; created commission to enforce, among other things, earlier statutes.

Federal Hazardous Substances Act of 1927, amended in 1976; flammable, corrosive, allergenic, or toxic materials in consumer products.

Flammable Fabrics Act of 1953 as amended: clothing as well as fabrics used for other purposes in the home.

Poison Prevention Packaging Act: the law that produced childproof caps.

DOT

Oil Pollution Act of 1961: oil spills, chemical spills from ships.

Dangerous Cargo Act; Tank Vessel Act; Ports & Waterways Safety Act 1972; Pipeline Safety Act: various railroad and truck transportation safety laws.

safe) in excessive amounts or in certain dosages can produce injurious effects. Substances such as salt, for example, which is by any definition a moderately toxic substance; baking powder, commonly found in practically every baked product; and, sugar, which certainly in excessive amounts can cause very serious dysfunctions and lack of coordination in the body's system, are all potentially toxic materials, if by toxicity we mean an adverse effect on the human body. The key questions, of equal importance, are:

How much is needed to produce a toxic effect?
How likely is this amount of material to enter the body where it can actually produce this effect?

Substances differ widely in their relative toxicity; in their ability to enter the body; and, in the effects they produce (Fig. 14.2).

The word "toxicity" has many definitions. We prefer the discussion be limited to a very simple definition for this purpose, namely, that toxicity is the effect produced by excessive amount of a substance being incorporated into the body system beyond that which the body can eliminate or tolerate without injury. *Toxicity is not a specific physical constant determined by standardized devices,* such as used to determine specific gravity or melting point. While there are protocols for determining the toxic effects of materials, most of these protocols require administering large quantities in animals, since they require excessive doses of the substance to produce an obvious effect, and, since they are also subject to interpretation, the absolute standardization of materials with respect to toxicity probably is beyond our present ability. This in no way means we should ignore toxicity, because toxicity is a very serious effect produced by many materials, and not so seriously considered with respect to others. Our knowledge is less than complete, and with due respect to the wisdom of Congress in passing the Toxic Substance Control Act, we know that the ability to specifically identify, quantify and make proper use of toxicity information, is a long series of evolutions which probably will require many years, if in fact absolutes are ever obtained in the biological systems involved. It cannot be stressed too strongly that toxicity is not a property of the substance itself, but, rather the degree to which the substance effects living cells. We stress the word "effects" and the word "living." No meaningful definition of toxicity can ignore these fundamentals.

Even standardized batches of laboratory animals where hundreds or thousands of rats, mice, guinea pigs, hamsters, monkeys, or other animals are bred under laboratory-type conditions, will exhibit considerable difference in their characteristics. In one study at the University of Texas some years ago, white rats were studied with respect to differences in individual characteristics. One rat drank 15 times more alcohol than another, and,

"Concern for man himself and his fate must always form the chief interest of all technical endeavor; Never forget this in the midst of your diagrams and equations"

A. Einstein

"Check this for toxicity, will you, Ed?"

Skin Penetration Toxicity is Greater than by Mouth

Chemicals	Times
Fluoroallyl alcohol	43
Di-2-n-hexyl ether	30
5-Indanol	8
Formaldehyde	7
Tridecyl acrylate	7
Alkyl pyridines	6
Aliphatic amines	2-6
2-Ethylbutyric acid	4
Vinyl butyl ether	4
Substituted adipaldehydes	3
Butyl Cellosolve	3
1-Heptanel	7

	Vapor Pressure m.m. Hg. (°C)		T.L.V. p.p.m.
Acetylene Tetrabromide	1	(65)	1
DMF	4	(25)	10
Nitroethane	20	(25)	100
Thiophene	60	(20)	—
1, 1, 1-Trichloroethane	100	(20)	350
Acrolein	213	(20)	0.1
Allyl chloride	400	(27)	1

Figure 14.2. Four views of toxicity.

another rat travelled 6 miles while his less-active sibling was moving 150 ft. If rats, bred and carefully selected for uniformity, are such nonconformists, consider how much more human beings differ from each other and from the standard curve. The problem of individual variation has been demonstrated many times in the field of toxicology, as well as in other phases of life. The cliché, "one man's meat is another man's poison," is appropriate in summing up man's experience in this respect; just as there is no "standard animal," there is no standard human. Both animals and humans are subject to wide variations.

Toxicity cannot be measured at all or even satisfactorily quantified until a definite recognized change has occurred in an animal or a human. These changes may be very small, easily overlooked and quite subtle; such changes as impaired judgment and delayed reaction time may be involved at levels too low for the production of body damage.[1] Nevertheless, effects can exist, and depending on the length of the experiment, the number of animals and the sophistication of the pathologist who make the final judgment, the results can either be meaningful or not meaningful. It is not unusual, for example, in the studies which have been made over the years, to find a chemical listed as producing carcinogenic effects in one species of animal, but not in another species of animal. Laboratory animal specialists have pointed out repeatedly that there is a great variation even among various strains of rats and of mice, with respect to the susceptibility to various changes; i.e., acute toxicity or carcinogenic, mutagenic, teratogenic response to given doses. Until we fully appreciate that no animal experiment can approach the precision of computers, mathematical models, and other sophistication which has been introduced into the system, can we approximate the knowledge gained from human observation and human exposure.

Animals may react in a much different manner than humans to the same exposures. Mules, for example, do not develop silicosis, even while working besides miners who do develop that disease of the lungs. Crystalline penicillin-G is essentially nontoxic in animals, with the exception of guinea pigs. Guinea pigs are particularly sensitive to penicillin and to certain other antibiotics. Doses as small as 7000 units per kilogram, may produce serious effects and finally death within a few days. Doses as high as 9,800,000 units or 5.93 g/kg were tolerated in mice, while humans can tolerate tremendous doses. Humans can tolerate doses on a daily basis as high as 86,000,000 units for a 28-day period, if they are perfectly normal and well; but, those that are allergic will find that even small doses produce severe and, occasionally, fatal shock. This is why penicillin is no longer the "wonder" drug that it was widely hailed to be 15 or 20 years ago. Animals may survive a relatively large or acute dose, but die from smaller doses over a long period of time, because the ability of the body to tolerate by various modes of metabolism or elimination is not sufficiently understood.

To understand toxicity, as related to hazard, we must examine certain fundamental real-world areas.

14.1   COMPOSITION OF THE SUBSTANCE AS ACTUALLY USED OR HANDLED

It is futile to attempt any evaluation of hazard without specific and definite information on composition. To say a solvent mixture "contains mineral spirits" is to present inadequate information, since mineral spirits vary widely in composition, and also the percentage of mineral spirits in the mixture may be very small or quite high. Other constituents in the mixture may be far more hazardous, such as benzene or carbon tetrachloride. If a proprietary-brand solvent or mixture is involved, the maker will usually reveal the complete formula on a confidential basis to a responsible person for the use, if given assurance that the information will not be used against his best interests or passed on to anyone else. The subject of confidential information has received much attention, especially with reference to release of information to government agencies, such as EPA, FDA, NIOSH, OSHA, and CPSC. Since the matter is largely legal, rather than technical, the advice of a legal expert or counsel should be sought. It should be noted that the composition of "trade-name" materials may change from time to time, with little or no change in the name or label.

If the supplier will not cooperate, two alternatives are always suggested: (1) analyze the substance, or (2) locate a more cooperative supplier. Usually the supplier will cooperate if he realizes his business depends on cooperation, and this approach is usually faster, more accurate, and more economical for all concerned than analysis. With the wide use of techniques such as chromatography and infrared spectography, however, analyses are much easier to obtain than previously, and the complete analysis should be obtained if we are suspicious a hazard may be involved.

14.2   TOXICITY IN ANIMALS

Once we have learned what is involved chemically, we can turn to the literature in the hope that our substance has been investigated and the data published. At this point extreme care must be exercised, since toxicity values are by no means absolute; they can be considered only as yardsticks of activity. Spector[1] lists five conditions that influence the toxicity of any given substance:

Dose

Generally the larger the dose, the more rapid the action.

Rate of Absorption

The faster the rate of absorption, the quicker the action. Food or oils in the stomach slow this action. For skin absorption, the larger the area involved, and the longer it is in contact, the faster the effect.

Route of Administration

Toxicity is greatest by the route that carries the toxic substance to the bloodstream most rapidly. In decreasing order of speed, routes for most substances are:

Intravenous (into a vein)

Inhalation (breathing)

Intraperitoneal (into the abdominal cavity)

Intramuscular (into a muscle)

Subcutaneous (under the skin)

Oral (by mouth)

Cutaneous (on the skin)

Site of Injection

With subcutaneous injections, toxicity may be affected by the density of the subcutaneous tissue. In intravenous administration, the rate of injection, or the amount of toxic material injected per minute, will considerably influence the value of the toxic dose.

Other Influences

Disease, environmental temperature, habit and tolerance, idiosyncrasy, diet, season of the year all may affect toxicity. The toxicity of chemicals will also vary with the species of animals used and sometimes with different strains of the same species. Within the same strain, the toxicity may differ with age, weight, sex, and the general conditions of the animals. The time to produce death, or the period of time for which fatalities are counted, may also be a factor.

There are several units in which the toxicity dose is expressed. The most frequently used are: LD or lethal dose (the amount which kills an animal), the MLD or minimum lethal dose (the smallest of several doses which kills one of a group of test animals), LD_{50} or lethal dose for 50% (the amount which kills 50% of a group of test animals, usually 10 or more), and LD_{100} or lethal dose for 100% (the amount which kills 100% of a group of test animals, usually 10 or more). Sometimes D is replaced by C in the above symbols, and the work "concentration" used instead of dose, as LC = lethal concentration, when referring to vapor concentration in air.

The usual form in which lethal doses of solids and liquids are expressed is in milligrams of substance per kilogram of body weight of the animal, or abbreviated as mg/kg or g/kilo. Since a 150-lb adult weighs about 70 kg, it might be expected that animal data could be translated into human data by multiplying the mgm/kg dose by 70. This practice is filled with pitfalls, and it should be used only as a "degree of magnitude" rough calculation. For reasons already mentioned, plus the important fact that there is no laboratory animal (except possibly the higher apes) which reacts to chemicals like man, extreme care should be used in applying animal data to humans.

14.3   HUMAN EXPOSURES

The real value and ultimate test of toxicity data, of course, is what actually happens to humans. Here we see even clearer that toxicity/hazard is not a simple matter. Some substances highly hazardous to young children (up to age 4 years) are relatively nontoxic to adults, because no adult would knowingly eat or drink them except by the highly unusual accident or a suicide measure.

Two common materials which cause serious poisonings in children are kerosene and aspirin. By chewing on cribs, windowsills, and toys painted with lead-containing paint, children may be poisoned with lead. A highly unusual case, reported a few years ago, developed from a teenage boy chewing on a lead "sinker" which he used in fishing. Medications, especially aspirin and sleeping pills, are often eaten by young toddlers, occasionally with fatal results. The operation of over 500 poison-control centers in the United States and Canada to advise the physician on an emergency basis the composition and recommended treatment in poisoning cases has helped to make the public more aware of accidental poisoning cases and hence to prevent accidents involving drugs and other chemicals. In general, these are coordinated by State health offices, in larger hospitals. A national clearinghouse, operated by the Department of Health and Human Services coordinates the information base.[3]

Another practical problem encountered in evaluation of hazards is that much necessary data can only be obtained by experience. For example, some substances such as the isocyanates have a very low level of true toxicity, but in extremely small concentrations in air they can cause bronchial irritation from sensitization in sensitive poisons. The irritating aspects of materials are not adequately reflected by animals. A mixture of diphenyl and diphenyl ether is a widely used heat-transfer agent, having a very low toxicity rating, but when accidentally sprayed in a face it may cause serious, almost fatal, respiratory effects due to irritation. Hydrogen sulfide is an example of a gas which in higher concentrations soon paralyzes the nose so the odor cannot be used as even a rough estimate of concentration for this highly dangerous substance. Irritation varies with people—some people frankly state they like the odor of low concentrations of pyridine and of mercaptan, while others are irritated, annoyed, and affected by the same concentration.[4] Boric acid is an example of a chemical which has been used in the treatment of burns for many decades, but whose toxicity has been recently recognized as too high to justify its use in this application.

Beyond the gross dosage problems, as reflected in poisonings, there are the practical day-to-day exposures encountered in industry. This is the real practical test of toxicity—what does the substance, either alone or in combination with other substances, do when breathed or absorbed in other ways at rates which will probably vary over wide limits during the day, for several hours a day for a certain period of time such as a 40-year working lifetime. To guide the control of such exposures, Professor Warren Cook tabulated and published in 1945 recommendations of maximum concentrations permissible for many common substances. Since 1947, the American Conference of Governmental Industrial Hygienists has sponsored a committee which publishes an annual revision to the threshold limit values, formerly referred to as MAC or maximum acceptable concentration values. Nearly 600 substances are currently listed by the committee. This list is not an official "standard," as such, but many states have adopted it as the working limits for their labor or health codes, and the 1969 edition became the official OSHA standard for most values.

Before referring to this list, however, it is wise to carefully read the preamble, in order to understand exactly what the list is as well as what it is not. These values are not a measure of *relative* hazard. The early values were established largely on the basis of safe level to prevent damage from chronic exposures. As more data became available based on actual experience, many values were lowered to reflect irritation and other transitory or acute changes. Recently, subacute levels have been introduced to reflect comfort levels. In addition to the threshold limit values, when evaluating hazards, we must consider the vapor pressure of the material which will determine the potential of attaining the exposure level under given conditions of use such as ceiling values and time weighted values (TWA) as well as other important factors which can relate to actual conditions of use. In recent years, short-term exposure limits (STEL) have been listed for many substances. Physical agents, such as noise, light, ionizing and non-ionizing radiation, and lasers have been considered by the ACGIH Committees, in addition to chemical exposures, and threshold limits established for them as well.

The United States is not the only country which has attempted to develop standards for exposure levels. While several countries use the basic data of the United States,[5,6] the Soviet Union has developed standards which are often considerably different than American standards. Apparently the Russian values are based on behavioral toxicity studies, and hence reflect a data base different from ours.[7] A more complete understanding of what represents the earliest manifestation of injurious effect would be highly valuable.

Another source of information which is often helpful in evaluating the hazards of a substance is the manufacturer—a person who is truly interested in seeing that the substance is used without adverse effects. Both in published data sheets and in answers to specific inquiry by telephone or mail, manufacturers will usually give practical recommendations as to the precautions they believe necessary. The more specific the inquiry, the more helpful will be the reply. Regardless of the quantity (a few grams may represent more hazard than a million gallons) the fundamentals are the same, and manufacturers will usually supply the information if requested, and assured the inquiry is genuine.

Another guide to health safety information is open literature. Much is available for those who seek it. A two-part article in *Industrial and Engineering Chemistry*, points out several excellent sources of safety information not widely used.[8] The National Safety Council[9] publishes information on many chemicals—information which helps in evaluating potential hazards. Hygienic Guides, published by the American Industrial Hygiene Association,[10] now cover nearly 200 substances in considerable detail. The National Center for Toxicological Information[11] in Oak Ridge, Tennessee, is another national resource, as is the National Institute for Occupational Safety and Health.[12] Several companies have made their product data sheets available.

14.4. CLASS OF TOXICITY INCLUDING THE SUBSTANCE

Ultimately we must decide on the degree of hazard presented by a substance. Classification of hazards goes far beyond toxicity, as illustrated by the excellent work of a National Fire Protection Association Committee by the late James J. Duggan, which classified and labeled substances, especially larger amounts in barrels or storage tanks, for emergency control purposes. The NFPA Standard 704 M, which resulted from this work, includes toxicity (hazard to life), flammability (fire hazard), explosion hazard (probability of explosion), and chemical reactivity (possible reactions with other nearby substances, if spilled or ruptured), which, combined with other essential information would be most valuable to emergency personnel, and its wide application should be encouraged.[13]

Hodge and Sterner tabulate toxicity into six classes[14]:

	Probable Oral Lethal Dose (Human)		
Toxicity Rating or Class	Dose		For 70 kg person (150 lb)
6 Super toxic	less than 5	mg/kg	A taste (less than 7 drops)
5 Extremely toxic	5–50	mg/kg	Between 7 drops and 1 tsp
4 Very toxic	50–500	mg/kg	Between 1 tsp and 1 oz
3 Moderately toxic	0.5–5	g/kg	Between 1 oz and 1 pint (or 1 lb)
2 Slightly toxic	5–15	g/kg	Between 1 pint and 1 quart
1 Practically nontoxic	above 15	g/kg	More than 1 quart (2.2 lb)

14.5. LIKELIHOOD OF RECEIVING A HARMFUL CONCENTRATION

The old cliché of the sea states, "Not all the waters of the seven seas can sink a boat until it gets inside the boat." Toxicity is much the same. A chemical in a bottle, in a tank, or in a boat is harmless as long as it is fully contained. In evaluating the hazard, therefore, the basic consideration should be based on how much of the substance will be in the air, or where it will be so it may be eaten, or absorbed through the skin or the eyes. Once this is established, we can add the toxicity data, expressed as dose, or irritation, or maximum allowable concentration, or threshold limit value, and find some measure for the actual hazard.

About 20 years ago, a relatively new chlorinated hydrocarbon was introduced on the market, and widely promoted as a substitute for other solvents such as carbon tetrachloride. (Carbon tetrachloride had been recognized as too hazardous for "bucket" or "open container" operations, in spite of its relative safety insofar as flammability is concerned.) The new substance had many characteristics and properties of carbon tetrachloride—its vapor pressure, cleaning ability, and ability to dry without residue were similar to carbon tetrachloride. Extensive animal exposure established a firm basis for assigning this substance a threshold limit value of 500 ppm in contrast to 10 ppm then in effect for carbon tetrachloride. This limit has since been lowered, but still the concensus is that the potential hazard is a very wide margin over carbon tetrachloride. Although we agree this solvent is much safer than carbon tetrachloride, and we continue to recommend it for many applications, we must point out that at least three fatalities have resulted from its improper use. Such solvent fatalities usually involve a careless use in a confined unventilated space. In one incident, a worker, while working at the bottom of the outer shell of a vacuum furnace, 38 in. in diameter by 49 in. deep, was cleaning oil and grease from the interior surfaces with steel wool and rags saturated with the solvent. The tank which was located in an open room had one 15-in. and three 6-in. diameter ports 28 in. from the bottom. The solvent was being used from an open coffee can. Approximately a quart had been used when the technician was found dead, about 50 min. after the supervisor had last checked the work. Such cases do not indicate the solvent is not relatively safe—it simply underscores again that any solvent must be used with respect. *Use of any solvent in a confined space without adequate ventilation, especially by a person working alone or with only nominal, occasional supervision is extremely unwise.*[15] In this connection, an advertisement used at one time by some distributors that a solvent is "20 times safer than carbon tetrachloride" is highly misleading, and represents a use of threshold limit values in an entirely different manner than intended (see Chaps. 15 and 25).

14.6. EMOTIONAL APPROACH TO THE SUBSTANCE

A few substances have come to be associated with high hazard in the public mind. If we consider "poisons" as synonymous with "hazards," we probably think of cyanide, lead, silica, arsenic, and carbon monoxide. Snake venom, curare, benzene, and carbon tetrachloride might be included on second thought. The strange enigma of the lung disease related to beryllium and beryllium compounds and the reluctance of some persons to accept the facts about the substances, indicate that publishing data alone does not insure that everyone will be informed or will believe. Some of the older tonnage chemicals such as aniline, nitrobenzene, and hydrogen sulfide, are finally being recognized as hazardous, and there are many substances about which we know so little that it is impossible to even guess how safe or how hazardous they are in the intended applications. Criteria documents, as developed by NIOSH for consideration by OSHA in standard setting, make a sincere attempt to evaluate hazard, but in some cases limited data has lead to emphasis which may not coincide with practices and procedures based on years of real-world experience. A western university reported the death of three students who were working with several bicycloheptadiene derivatives, previously not considered hazardous. Animal investigations are underway to elucidate the hazard.[16] Magic methyl, or methylfluorosulfonate, caused a fatality in the Netherlands before its respiratory hazards were recognized.[17] The need for new data even on the more familiar substances must not be overlooked.

Extreme fear and anxiety about hazards may actually create accident situations, just as may apathy and ignorance. Personnel may be so fearful that they will become nervous and supersensitive—perhaps even allergic. The reverse condition, lack of adequate and proper respect for a hazard, may also contribute to accidents by encouraging carelessness and lack of protective measures. Exactly how to present the degree of hazard to personnel in their specific use of the toxic or corrosive material, so they will actually react with respect and confidence but not fear, is one of the challenges of supervision. In addition, the emotional state of the workers, their emotional stability and adequate adjustment to their jobs, their bosses, their company, and their home life, all may be far more important in evaluating actual on-the-job hazards than toxicity data alone. In all cases, it must be *both* the worker, and the chemical, which is the center of our attention. See Chapters 1, 9, 11, 22 and 33.

The increased attention which the news media, especially television and radio, have given chemical emergencies has produced an awareness of the disadvantages of chemicals when they are misused or carelessly handled. Kepone, polychloronated biphenyl, polybromobiphenyl, and dibromochloropropane are chemicals which have received unfavorable publicity due to careless exposures or accidents which dilute the real benefits from their safe and proper use. The interface between emotions, science and politics is clearly shown by Lois Gibbs at Love Canal[18] and by the novel *Fever*, concerning bizarre exposures to a 12-year-old daughter.[19] A chemical emergency is news; the day-by-day safe use of the substance is not.

The emotions play a significant role in influencing the impressions which society has of chemicals, the chemical industry, and chemists. The risk/

benefit factor which may minimize the hazard of a substance is irritation ability. If a gas or vapor affects the upper respiratory tract sufficiently to cause sneezing, coughing, or extreme discomfort to the eyes or throat the normal reaction is one of escape or repulsion, which tends to decrease the exposure. On the other hand, if the substance is not sufficiently irritating or objectionable, the practical danger is much greater, since the warning will be less or inadequate. This is one factor why carbon monoxide remains the serious hazard in both vehicle and industrial exposures—virtually no warning of serious exposure occurs before awareness is overcome.[20] On the other hand, formaldehyde, which is highly irritating in low concentrations, was banned from home insulation of the UF-type by the Consumer Products Safety Commission, largely because the irritation was easily demonstrated.

14.7. SUMMARY

1. Chemicals, per se, are not toxic or nontoxic (common usage meaning the ability of excessive amounts to produce damage to life).

2. Toxicity refers to the *effect* on *living cells* (usually animals or humans). Toxicity data must be carefully screened and examined before extrapolating into human experience.

Chemicals, even "extremely toxic," may or may not be hazardous, depending on their use. All chemicals can be handled and disposed of safely. Until they reach and produce effects inside the body, toxic substances are not harmful. Toxicity, then is a phenomenon of living organisms, and is *not* a fundamental property of the contained material.

REFERENCES

1. W. Spector, *Handbook of Toxicology,* in two volumes, Saunders, Philadelphia, 1956–1957.
2. Sir Michael Foster (1836–1907), J. Am. Med. Assoc., Editorial, **186,** 1167–1168. (December 28, 1963).
3. National Clearinghouse for Poison Control Centers, Rockville, MD. 20857.
4. "Threshold Limit Values" (updated annually). [Available from the American Conference of Governmental Industrial Hygienists, P. O. Box 1937, Cincinnati, Ohio 45201.]
5. Charles Levinson, I.C.F., "Work Hazard, Chemical Agents in the Workplace, Threshold Limit Values in the United States, Germany, and Sweden," International Occupational Health Conference, Geneva Switzerland, October 28–30, 1974; "Control of Toxic Substances in the Workplace," Hearings before a Subcommittee of the Committee on Government Operations, House of Representatives, 94th Congress, May 11–18, 1976, U. S. Government Printing Office, Washington, D.C.
6. M. A. Winell, "An International Companion of Hygiene Standards for Chemists in the Work Environment," *AMBIO* 4(1) 34–36 (1975).
7. G. J. Ekel and W. H. Teichner, "An Analysis and Critique of Behavioral Toxicology in the U.S.S.R.," Contract No. HSM-99-73-60, National Institute for Occupational Safety and Health, Cincinnati, Ohio December 1976.

8. H. H. Fawcett, "Who Knows What About Chemical Safety," *Ind. Eng. Chem.* **52**(8) 85A–88A (June 1960); **52**(8), 75A–76A (August 1960); The Literature of Chemical Health and Safety, 182nd National Meeting, American Chemical Society, New York City, Aug. 27, 1981.

9. National Safety Council, 444 N. Michigan Avenue, Chicago, Ill. 60611.

10. American Industrial Hygiene Association, 475 Wolf Ledges Parkway, Akron, Ohio 44311.

11. Toxicology Information Response Center, Oak Ridge National Laboratory, P. O. Box X, Building 7509, Oak Ridge, Tenn. 37830.

12. National Institute for Occupational Safety and Health, Rockville, Md. 20857.

13. "Identification of the Fire Hazards of Materials, NFPA, No. 704M. [Available from National Fire Protection Association, Battery March Square, Quincy, Mass. 02269.]

14. H. C. Hodge and J. H. Sterner, "Tabulation of Toxicity Classes," *Am. Ind. Hyg. Assoc. Q.*, **10**(4), 93 (December 1949). See also: Gosselin, Hodge, Smith, Gleason, *Clinical Toxicology of Commercial Products, Acute Poisoning*, 4th ed., 1976, Williams and Wilkins Co., Baltimore, 1976, inside cover.

15. "Safety Solvent Fatality, Case History No. 442," *Case Histories of Accidents in the Chemical Industry,* Vol. 1, Chemical Manufactureres Association, Washington, D.C., 1962.

16. S. Winstein, "Communications to the Editor—Bicycloheptadiene Dibromides," *J. Am. Chem. Soc.*, **83**, 1516–1517 (March 20, 1961).

17. "Vapors Fatal," *Chem. Eng. News* **54**(36), Letters to Editor, 5 (August 30, 1976).

18. L. M. Gibbs, *Love Canal, My Story*, State University of New York Press, Albany, N.Y. 12246, 1982.

19. R. Cook, *Fever*, Putnam, New York, 1982.

20. *EPA Journal (Toxics)* **4**(8) (September 1978), U.S. Environmental Protection Agency, Washington, D.C. 20460.

BIBLIOGRAPHY

Annual Plan, Fiscal Year 1982, National Toxicology Program, NTP81–94, March 1982, Public Information Office, National Toxicology Program, P.O. Box 12233, Research Triangle Park, N.C. 27709.

Atherley, G. R. C., *Occupational Health and Safety Concepts*, Applied Science Publishers, London, 1978.

Bandal, S. K., Goldberg, L., Marco, G., and Leng, M., *The Pesticide Chemist and Modern Toxicology*, ACS Symposium Series 160, American Chemical Society, Washington, D.C., 1981.

Branson, D. R. and Dickson, K. L., ASTM/STP737, Aquatic Toxicity and Hazard Assessment, Sept. 1981; ASTM TP657, Estimating the Hazard of Chemical Substances to Aquatic Life, American Society for Testing & Materials, Philadelphia, PA 19103, 1978.

Bunt, R. "The U.S. Fight Against Chemical War," *The Wall Street Journal*, January 4, 1982, p. 31; Review and Outlook, *On the Agenda: Yellow Rain, The Wall Street Journal*, March 24, 1982, p. 26.

Choudhary, G., *Chemical Hazards in the Workplace,* ACS Symposium Series 149, American Chemical Society, Washington, D.C., 1981.

Clayton and Clayton, *Patty's Industrial Hygiene and Toxicology*, 3rd ed., (in 3 volumes), Wiley Interscience, 1978–1980.

Documentation of the Threshold Limit Values for Substances in Workroom Air, 4th ed., American Conference of Governmental Industrial Hygienists, 1980, revised frequently. [Available from ACGIH, P. O. Box 1937, Cincinnati, Oh. 45201.]

"Hazardous Waste Options," 16-mm sound color motion picture, Stuart Finley, 3428 Mansfield Road, Falls Church, VA., 1981.

Hopke, P. K., "Multitechnique Screening of Chicago Municipal Sewage Sludge for Mutagenic Activity," *ES&T,* **16**(6), 140–147 (March 1982).

Jones, C. J., "The Ranking of Hazardous Materials by Means of Hazard Indices," *J. Hazard. Mater.,* **2**(4), 363–389 (November 1978).

Lowrence, W. W., ed., *Assessment of Health Effects at Chemical Disposal Sites,* Kaufmann, Los Altos, Calif., 1981.

Mayer, C. E., "Report Paints Grim Picture of Asbestos," *Washington Post,* December 30, 1981, pp. D8–10.

McGrady, Pat, Sr., *The Persecuted Drug. The Story of DMSO,* Charter Books, New York, 1979.

Miller, J. A., "Are Rats Relevant?," *Sci. News* **112,** 12–13 (July 2, 1977).

Moeschlin, S., "Outstanding Symptoms of Poisoning," pp. 644–678, *Diagnosis and Treatment,* Grune & Stratton, New York, 1965.

Plimmer, J. R., *Pesticide Residues and Exposure,* ACS Symposium Series 182, 1982, American Chemical Society, Washington, D.C. 20036.

"Registry of Toxic Effects of Chemical Substances," prepared for NIOSH. [Revised annually, also available on microcards (quarterly), National Institute for Occupational Safety and Health, Rockville, Md., 20857.

Review and Outlook, "Highest Priority," *The Wall Street Journal,* January 7, 1982, p. 20.

Ross, S. S., *Toxic Substances Sourcebook* (Environment) [Available from Information Center, 292 Madison Avenue, New York, N.Y. 10017.]

Scanlan, R. A., Tannenbaum, S. R., *N-Nitroso Compounds,* ACS Symposium Series 174, 1981, American Chemical Society, Washington, D.C. 20036.

Sittig, M., ed., *Priority Toxic Pollutants, Health Impacts, and Allowable Limits,* Noyes Data Co., Park Ridge, N.J., 1980.

Spiro, T. G., and Stigliani, W. M., *Environmental Issues in Chemical Perspective* 1980, State University of New York Press, Albany, N.Y. 12246.

Stahl, W. H., *Compilation of Odor and Taste Threshold Value Data, E-18 Committee on Sensory Evaluation of Materials and Products,* American Society for Testing and Materials, Philadelphia, 1973.

"WHO Consultation on the Implementation of WHA Resolution 30.47: Evaluation of the Effects of Chemicals on Health," Geneva, May 1–5, 1978. [Available from World Health Organization, 1211 Geneva, 27, Switzerland.]

Tabor, M., "Minimizing the Menace of OTC Drugs," *Occu. Health and Safety,* **51**(5), 14–19 (May 1982).

15

Toxicity–Chemicals React with Living Systems

Howard H. Fawcett

The wide interest in toxicity, especially as NIOSH, OSHA, FDA, and EPA have approached the subject, might suggest that some new aspect of chemicals had recently been discovered. In fact, the ability of chemical substances to affect human and animal life was known to the ancient man. Alchemists of the Dark Ages, who were the progenitors of modern chemical science, were both seeking the Philosopher's Stone (with which to bestow eternal life on mortals), as well as a method for converting base metals, such as lead, zinc, and mercury, into gold, the uncorruptible metal. Several important drugs were first isolated from roots, flowers, or other parts of plants used for centuries by Indian tribes. For example curare and digitalis, both powerful drugs in common use today, were first isolated from natural sources. An understanding of toxicity is an important step towards the safe and proper utilization of drugs and chemicals. (See Chaps. 9, 10, 11, 14, 22, and 25.)

The emphasis that has been placed on toxicity frequently overlooks that even today, our knowledge base of toxicity is limited and fragmented. Society, through consumer groups and various governmental agencies, especially the FDA, CPSA, NIOSH, OSHA, Department of Agriculture, DHHS, [which includes the National Cancer Institute, NCI], the National Clearinghouse of Poison Control Centers, and the EPA, under the Toxic Substances Control Act of 1976 (P.L. 94-1302) as well as local and state interests, has fragmented the research and control activity, and we face a long period before the emotional and economic aspects of toxic substances control are resolved. It would be unfair to neglect mention of the outstanding toxicity and safety evaluation work conducted and published in the open literature over the years by several major chemical companies, including Union Carbide, Dow, DuPont and Eastman Kodak, long before public concern or legal requirements were at their present level. The Chemical Industry Institute of Toxicology, which has sponsorship from major companies, was established in 1975 and has underway an ambitious program of toxicological research and testing of commercial chemicals.

Toxicity is not a strange unknown; rather, it is a measure of the action of a given amount of a chemical or other substance on a designated living specie or system (plant or animal), to quantify the level or dose at which the system is damaged beyond prompt recovery when that dose enters and is absorbed by the living system. Even an element which is absolutely essential to human life, namely, oxygen, is toxic under conditions of prolonged inhalation at elevated pressure, as in Scuba diving below 32 ft. Several elements essential to life in low concentrations, such as zinc, copper, magnesium, selenium, and arsenic, are toxic or lethal at higher concentrations. Hence, to simply label a substance "toxic" or "nontoxic" without specifying the details and limits of the data, is misleading. Note that until the dose has actually entered the body and reached critical organs by one or more routes, such as inhalation, ingestion, skin absorption, toxicity will not be manifest. See Chapters 11 and 14.

Unlike the many values pertaining to a chemical tabulated in handbooks and critical tables, toxicity is not a specific property or physical constant derived by simple physical measurement in the abstract. Because it depends on measurements relative to living systems, it is never absolute, since individual members of all living species and systems, especially humans and animals, vary among themselves and with other species in their resistance or susceptability to insults from toxic materials. Defense mechanisms, by which humans and other hosts adapt or resist a toxic agent, include adaption, response to stress, and complex biochemical mechanisms which act to minimize or prevent toxicity. When these homothesis actions are overburdened, toxic effects are always observed.

We recall one such response "escape" measure we observed in connection with exposures to oxides of nitrogen, a not infrequent emergency and well recognized by experienced operators for serious or fatal consequences. When a nitrator operator suddenly saw a cloud of dark brown or orange-colored fumes in his immediate area, he would instinctly bring his operation under control, and then escape—first, by holding his breath, then by "shallow breathing," that is by restricting his breathing to a few hundred milliliters of tidal volume. He can thereby keep his respiratory system operating, avoiding hypoxia (oxygen deficiency), but not inhale significantly the fumes into the lower respiratory tract, where damage to the alveoli could occur and produce serious or fatal pulmonary edema, often several hours later. The urge to "panic" and "run" was suppressed. Shallow breathing permitted survival in an otherwise toxic environment until escape, or allowing sufficient time to a respiratory protective device. The recently reported "low temperature" process for nitrocompounds, operated from a remote control room, has hopefully made exposures to oxides of nitrogen less frequent, but any person involved in operations the environ of which suddenly contains any highly toxic gases may profit from the survival technique of shallow breathing. (See L. Albright and C. Hanson, *Industrial and Laboratory Nitrations,* ACS Symposium Series 22, American Chemical Society, Washington, D. C., 1976.)

Humans and animal systems have several other mechanisms for dealing with insults from toxic substances, often called detoxification. At some point on the dose-effect (or dose-response) curve, however, acute or chronic injury can occur—which may or may not affect every member of the subject group, or be immediately obvious. The "injury" may be manifest in one or several ways, such as modified or irregular behavior (intoxication), central nervous system disorientation, impairment of body functions, allergic reactions, skin and eye irritation or damage, respiratory and cardiac collapse, pulmonary edema, or by less obvious delayed reactions which may require years before producing carcinogenic, mutagenic, or teratogenic effects in essential organs. Until the toxic dose actually enters the body through one or more routes of entry, such as inhalation, ingestion, or cutaneous (skin) absorption, it is probably harmless. The possible exception to the previous statement is the possibility that a one-hit injury may occur, that is, one molecule of the substance contacts and damages the essential DNA and RNA of a critical organ, and may induce carcinogenic activity. This theory of no threshold for exposure to chemical carcinogens is widely debated, and impressive arguments may be cited both for and against this theory. It is possible that our understanding of this process is so incomplete that a threshold exists, but has not yet been recognized and accepted. However, for most chemicals, the consensus is that a threshold value exists, based on protection for the majority of the workers who will be exposed, but the degree of protection it affords any one individual in a group is not clear at this writing.

Table 15.1 presents a comparison of hygienic standards for the workplace between U.S. OSHA (1974), East Germany, West Germany, Sweden, Czechoslovakia, and Russia (1972). It will be obvious that many Russian values are lower than corresponding American values. While we do not have the complete information on which to judge such differences, we do know that the Russian values are based on a different criteria than American values. The reader should note these may not necessarily be the current values.

Since the original presentation of a list of threshold limit values by Professor Warren Cook in 1945, many changes have occurred annually, as the list was revised by the American Conference of Governmental Industrial Hygienists. The 1969 Threshold Limit Values published by the ACGIH became the OSHA standards for most substances (see *Federal Register,* **36**, No. 105, May 29, 1971). Since then, the annual revisions have continued by the ACGIH and been expanded to over 500 substances, as well as to physical

TABLE 15.1 Work Environment Hygienic Standards in Different Countries

	USA—OSHA 1974 (ppm)	(mg/m³)	BRD 1974 (mg/m³)	DDR 1973 (mg/m³)	Sweden 1975 (mg/m³)	CSSR 1969 (mg/m³)	USSR 1972 (mg/m³)c
Acetaldehyde	200	360	360	100	90	—	5
Acetic acid	10	25	25	20	25	—	5
Acetone	1000	2400	2400	1000	1200	800	200
Acetonitrile	40	70	70	—	—	—	10
Acrolein	0.1	0.25	0.25	0.25	0.25	0.5	0.7
Aldrin	—	0.25	0.25	—	—	—	0.01
Allyl alcohol	2	5	5	5	5	3	2
Ammonia	50	35	35	25	18	40	20
Ammonium sulfamate	—	15	15	—	—	—	10
Amyl acetate	100	525	525	200	525	200	100
Aniline	5	19	19	10	19	5	0.1
p-Anisidine	0.1	0.5	0.5	0.5	—	—	1
Antimony & compounds (as Sb)	—	0.5	0.5	0.5	0.5	—	0.3–2
Arsenic & compounds (as As)	—	0.5	0	0.3	0.05	0.3	0.3
Arsine	0.05	0.2	0.2	0.2	0.005	0.2	0.3
Benzene	10	30	0	50	30	50	5
Benzoyl peroxide	—	5	5	—	—	—	5
Benzyl chloride	1	5	5	5	—	—	0.5
Beryllium	—	0.002	0	0.002	0.002	—	0.001
Boron oxide	—	15	15	—	—	—	10
Boron trifluoride	1c	3c	3	—	—	—	1
Bromoform	0.5	5	—	—	—	—	5
1,3-Butadiene	1000	2200	2200	500	—	500	100
2-Butanone	200	590	590	300	440	—	200
Butyl acetate	150	710	950	400	710	400	200

Butyl alcohol	100	300	300	200	150	100	10
Butylamine	5	15	15	—	—	—	10
Cadmium (metal dust and soluble salts)	—	0.2	—	0.1ᵃ	0.05	—	0.1
Cadmium oxide fume (as Cd)	—	0.1	0.1	0.1ᵃ	0.02	0.1	0.1
Camphor	2	12	2	—	—	—	3
Carbaryl (Sevin)	—	5	5	—	—	—	1
Carbon disulfide	20	60	60	50	30	30	10
Carbon monoxide	50	55	55	55	40	30	20
Carbon tetrachloride	10	65	65	50	65	50	20
Chlorine	1	3	1.5	1	3ᶜ	3	1
Chlorine dioxide	0.1	0.3	0.3	—	0.3ᶜ	—	0.1
Chlorobenzene	75	350	230	50	—	200	50
Chlorodiphenyl (42% chlorine)	—	1	1	1	0.5	1	1
Chlorodiphenyl (54% chlorine)	—	0.5	0.5	1	0.5	0.5	1
Chloroprene	25	90	90	10	90	50	2
Chromic acid and chromates (as Cr)	—	0.1ᶜ	0.1	0.1	0.05	0.05	0.01
Cobalt, metal fume & dust	—	0.1ᶜ	0.5	0.1	0.1	0.1	0.5
Copper, fume	—	0.1	0.1	0.2ᵇ	—	—	1
Copper, dusts and mists	—	1	1	—	—	—	1
Crotonaldehyde	2	6	6	—	—	—	0.5
Cumene	50	245	245	50	—	—	50
Cyclohexane	300	1050	1050	—	—	—	80
Cyclohexanone	50	200	200	—	—	—	10
Cyclopentadiene	75	200	200	—	—	—	5

TABLE 15.1 (Continued)

	USA—OSHA 1974 (ppm)	(mg/m³)	BRD 1974 (mg/m³)	DDR 1973 (mg/m³)	Sweden 1975 (mg/m³)	CSSR 1969 (mg/m³)	USSR 1972 (mg/m³)c
2,4-D	—	—	10	—	—	—	1
DDT	—	1	1	1	—	—	0.1
Dibutylphtalate	—	5	—	—	—	—	0.5
o-Dichlorobenzene	50c	300c	300	150	—	—	20
p-Dichlorobenzene	75	450	450	200	—	—	20
Dichlorvos (DDVP)	0.1	1	1	—	—	•	0.2
Dieldrin	—	0.25	0.25	—	—	—	0.01
Diethylamine	25	75	75	50	—	—	30
Diethylamino ethanol	10	50	50	—	—	—	5
Diisopropylamine	5	20	—	10	—	—	5
Dimethylamine	10	18	18	—	—	—	1
Dimethylaniline (N-dimethylaniline)	5	25	25	—	—	—	0.2
Dimethylformamide	10	30	60	30	30	30	10
Dinitrobenzene	0.15	1	1	1	—	1	1
Dinitro-o-cresol	—	0.2	0.2	0.2	—	—	0.05
Dinitrotoluene	—	1.5	1.5	1	—	—	1
Dioxane	100	360	360	200	90	—	10
Epichlorhydrin	5	19	18	5	—	—	1
Ethyl acetate	400	1400	1400	500	1100	400	200
Ethyl alcohol	1000	1900	1900	1000	1900	1000	1000
Ethyl amine	10	18	18	20	—	—	1
Ethyl bromide	200	890	890	500	—	—	5
Ethyl chloride	1000	2600	2600	2000	—	—	50
Ethyl ether	400	1200	1200	500	1200	300	300
Ethyl mercaptan	10c	25c	1	—	—	—	1
Ethylene chlorohydrin	5	16	16	—	—	—	0.5

Ethylene diamine	10	25	25	—	—	—	2
Ethylene imine	0.5	1	1	1	0	—	0.02
Ethylene oxide	50	90	90	20	36	1	1
Fluoride (as F)	—	2.5	2.5	—	2.5	1	1
Formaldehyde	2	3	1.2	2	3^c	2	0.5
Furfural	5	20	20	10	—	—	10
Heptachlor	—	0.5	0.5	—	—	—	0.01
Hydrazine	1	1.3	0.13	—	0.13	0.1	0.1
Hydrogen chloride	5^c	7^c	7	5	7^c	8	5
Hydrogen cyanide	10	11	11	5	11	3	0.3
Hydrogen fluoride	3	2	2	1	2^c	1	0.5
Hydrogen sulfide	20^c	30^c	15	15	15	10	10
Iodine	0.1^c	1^c	1	—	1^c	—	1
Isopropylamine	5	12	12	—	—	—	1
Lead, inorganic fumes and dusts	—	0.2	0.2	0.15	0.1	0.05	0.01
Lindane	—	0.5	0.5	0.2	—	—	0.05
Maleic anhydride	0.25	1	0.8	—	1	1	1
Manganese and compounds (as Mn)	—	5(c)	5	5	2.5	2	0.3
Mercury, metal	—	0.1^c	0.1	0.1	0.05	0.05	0.01
Mercury, alkyl	—	0.01	0.01	0.01	0.01^c	—	0.005
Methyl acetate	200	610	610	200	—	200	100
Methyl acrylate	10	35	35	20	—	100	20
Methyl alcohol	200	260	260	100	260	100	5
Methyl amine	10	12	12	—	—	—	1
Methyl bromide	20^c	80^c	80	50	—	—	1
Methyl chloride	100	210	105	100	—	100	5
Methyl chloroform	350	1900	1080	500	540	500	20
Methyl cyclohexane	500	2000	2000	—	—	—	50
Methyl isocyanate	0.02	0.05	0.05	—	—	—	0.05

TABLE 15.1 (Continued)

	USA—OSHA 1974 (ppm)	(mg/m³)	BRD 1974 (mg/m³)	DDR 1973 (mg/m³)	Sweden 1975 (mg/m³)	CSSR 1969 (mg/m³)	USSR 1972 (mg/m³)ᶜ
α-Methyl styrene	100ᶜ	480ᶜ	480	—	350	—	5
Methylene chloride	500	1740	1750	500	—	500	50
Molybdenum soluble compounds	—	5	5	—	—	—	4
Molybdenum, insoluble compounds	—	15	15	10	—	—	6
Morpholine	20	70	70	—	—	—	0.05
Naphta (coal tar)	100	400	—	—	—	200	100
Naphtalene	10	50	50	20	—	—	20
Nickel carbonyl	0.001	0.007	0.7	0.5	0.007	—	0.0005
Nickel, metal	—	1	0	—	0.01	—	0.5
p-Nitroaniline	1	6	6	—	—	—	0.1
Nitrobenzene	1	5	5	5	5	5	3
p-Nitrochlorobenzene	—	1	1	1	—	1	1
Nitroethane	100	310	310	—	—	—	30
Nitrogen dioxide	5	9	9	10	9ᶜ	10	5
Nitromethane	100	250	250	—	—	—	30
1-Nitropropane	25	90	90	50	—	—	30
2-Nitropropane	25	90	90	50	—	—	30
Ozone	0.1	0.2	0.2	0.2	0.2	0.1	0.1
Pentachlorophenol	—	0.5	0.5	0.5	0.5	—	0.1
2-Pentanone	200	700	700	—	—	—	200
Perchloroethylene	100	670	670	300	200	250	10
Phenol	5	19	19	20	19	20	5
Phosgene	0.1	0.4	0.4	0.5	0.2ᶜ	0.4	0.5
Phosphine	0.3	0.4	0.15	0.1	0.4	0.1	0.1
Phosphorus (yellow)	—	0.1	0.1	—	—	0.03	0.03

Phthalic anhydride	2	12	5	10	12	5	1
Propargyl alcohol	1	2	2	—	—	—	5
n-Propyl acetate	200	840	840	400	—	400	200
Propyl alcohol	200	500	—	—	—	500	10
Propylene dichloride (1,2-Dichloropropane)	75	350	350	50	—	—	10
Propylene oxide	100	240	240	—	—	—	1
Pyridine	5	15	15	10	15	5	5
Quinone	0.1	0.4	0.4	0.1	0.1	—	0.05
Selenium compounds	—	0.2	0.1	2	2[c]	—	0.1
Sodium hydroxide	—	2	2	—	—	—	0.5
Stoddard solvent	500	2950	—	200	600	—	300
Styrene	100	420	420	10	210	200	5
Sulfur dioxide	5	13	13	1	5	10	10
Sulfuric acid	—	1	1	—	1	1	1
Tellurium	—	0.1	0.1	—	—	—	0.01
1, 1, 2, 2-Tetrachloroethane	5	35	7	10	—	—	5
Tetraethyl lead (as Pb)	—	0.075	0.075	0.05	0.075	—	0.005
Tetrahydrofuran	200	590	590	200	—	—	100
Tetranitromethane	1	8	8	—	—	—	0.3
Thallium	—	0.1	0.1	—	—	—	0.01
Thiram (tetramethyl-thiuramdisulfide)	—	5	5	1	—	—	0.5
Toluene	200	750	750	200	375	200	50
Toluene-2,4-diisocyanate	0.02[c]	0.14[c]	0.14	0.1	0.07[c]	0.07	0.5
o-Toluidine	5	22	22	10	—	5	3
Trichloroethylene	100	535	260	250	160	250	10
1, 2, 3-Trichloropropane	50	300	300	—	—	—	2
Triethylamine	25	100	100	20	—	—	10
Trinitrotoluene	0.2	1.5	1.5	1.5	—	1	1
Triorthocresylphosphate	—	0.1	—	0.1	—	—	0.1

TABLE 15.1 (*Continued*)

	USA—OSHA 1974 (ppm)	USA—OSHA 1974 (mg/m³)	BRD 1974 (mg/m³)	DDR 1973 (mg/m³)	Sweden 1975 (mg/m³)	CSSR 1969 (mg/m³)	USSR 1972 (mg/m³)c
Turpentine	100	560	560	300	560	—	300
Uranium, soluble compounds (as U)	—	0.05	0.05	—	—	—	0.015
Uranium, insoluble compounds (as U)	—	0.25	0.25	—	—	—	0.075
Vanadium, V_2O_5 dust (as V)	—	0.5c	0.5	0.5	0.5	—	0.5
Vanadium, V_2O_5 fume (as V)	—	0.1c	0.1	0.1	0.05c	—	0.1
Vinyl chloride	1	3	—	500	3	—	30
Vinyl toluene	100	480	480	—	—	—	50
Xylene	100	435	870	200	435	200	50
Xylidine	5	25	25	10	—	5	3
Zinc oxide fume	—	5	5	5	5	5	6
Zirconium compounds (as Zr)	—	5	5	—	—	—	4–6

aAs CdO.
bAs CuO.
cCeiling value.

agents including light, noise, lasers, microwaves, ionizing radiation, and heat stress. As noted in the preface to the ACGIH compilation for airborne contaminants in the workplace, to which the reader should refer for the most current values, three categories of threshold limit values (TLV's) are specified:

1. *Threshold Limit Value–Time Weighted Average (TLV–TWA)*. The time-weighted average concentration for a normal 8-hour workday or 40-hour workweek, to which nearly all workers may be repeatedly exposed, day after day, without ill effect. To cite a few typical values as examples from the 1981 listing:

Acetone	1000 ppm or	2,400 mg/m³	(TLV–TWA)
Acrolein	0.1 ppm or	0.25 mg/m³	(TLV–TWA)
Ammonia	25 ppm or	18 mg/m³	(TLV–TWA)
Carbon monoxide	50 ppm or	55 mg/m³	(TLV–TWA)
Triethylamine	25 ppm or	100 mg/m³	(TLV–TWA)
Zinc oxide fume		5 mg/m³	(TLV–TWA)

2. *Threshold Limit Value—Short Term Exposure Limit (TLV–STEL)*. The maximal concentration to which workers can be exposed for a period of up to 15 min continuously without suffering from (1) intolerable irritation, (2) chronic or irreversible tissue change, or (3) narcosis of sufficient degree to increase accident proneness, provided that no more than four excursions per day are permitted, with at least 60 min between exposure periods, and providing the daily TLV–TWA also is not exceeded. The STEL should be considered a maximal allowable concentration, or ceiling, not to be exceeded at any time during the 15-min excersion period. Examples of TLV–STEL for the same substances:

Acetone	1250 ppm or	3,000 mg/m³	(TLV–STEL)
Acrolein	0.3 ppm or	0.8 mg/m³	(TLV–STEL)
Ammonia	35 ppm or	27 mg/m³	(TLV–STEL)
Carbon monoxide	400 ppm or	440 mg/m³	(TLV–STEL)
Triethylamine	40 ppm or	160 mg/m³	(TLV–STEL)
Zinc oxide fume		10 mg/m³	(TLV–STEL)

3. *Threshold Limit Value–Ceiling (TLV–C)*. The concentration that should not be exceeded even instantaneously. Examples of TLV-C 1981 values are:

Boron trifluoride,	ceiling value of 1 ppm or 3 mg/m³
Hydrogen chloride,	ceiling value of 5 ppm or 7 mg/m³
Manganese and compounds, as Mn,	ceiling value of 5 mg/m³

For some substances, such as irritant gases, only one category, the TLV–ceiling, may be relevant; for others the TLV–TWA and TLV–STEL may be appropriate. In the compilation of the ACGIH, substances which are known to be toxic through the intact skin are so designated by the addition of the word "skin" after the chemical name. As an example, aniline, one of several substances which can be absorbed through the intact skin as well as by inhalation is listed:

aniline - skin   TWA 2 ppm or 10 mg/m^3 STEL 5 ppm or 20 mg/m^3

From the discussion of threshold limit values, it is obvious that the subject of exposure limits is complex. Therefore, the warning quoted from the preface is clearly in order:

> These limits are intended for use in the practice of industrial hygiene and should be interpreted and applied only by a person trained in this discipline. They are not intended for use, or for modification for use, (1) as a relative index of hazard or toxicity, (2) in the evaluation or control of community air pollution nuisances, (3) in estimating the toxic potential of continuous, uninterrupted exposures or other extended work periods, (4) as proof or disproof of an existing disease or physical condition, or (5) for adoption by countries whose working conditions differ from those in the United States and where substances and processes differ.

In experiments on humans using radioactive tracers in pesticides, and using the forearm as the frame reference, it has been shown that the palm, of which the thick stratum corneum is allegedly almost impenetrable, allowed approximately the same penetration as the forearm. The abdomen and dorsum of the hand had twice the penetration as the forearm. The follicle-rich sites, including the scalp, angle of the jaw, postauricular area, and forehead had fourfold greater penetration. The intertrigenous axilla had a fourfold to sevenfold increase; the scrotum allowed almost total absorption. The overall impression is that all atatomic sites studied show significant potential for penetration of pesticides, and hence systemic intoxication. [See H. I. Maibach, et al., "Regional Variation in Percutaneous Penetration in Man," *Archives of Environmental Health,* **23**, 208–211 (September 1971)].

Dose.   Generally the larger the dose (or higher the concentration), the more rapid the action.

Rate of absorption.   The faster this rate, the quicker the action. With oral administration, the intoxicating and the lethal doses may be considerably influenced by the condition of the gastrointestinal tract, especially by the amount of food and other matter in the stomach and intestine. A vehicle, such as oil, also affects absorption of a skin exposure.

Route of Administration.   For the most part, toxicity is greatest by the route that carries the toxic substance to the blood stream most rapidly. In *decreasing* order of speed of action, routes for most other drugs and other substances are:

Intravenous

Inhalation

Intraperitoneal

Intramuscular

Subcutaneous

Oral

Cutaneous (skin)

See Chapters 9, 10, 11, and 14.

Food in the alimentary canal may delay or decrease toxic action; digestive enzymes may destroy or alter the compounds with resultant changes in the toxicity. Certain compounds are virtually harmless if taken orally, but lethal when introduced parenterally; in many cases the converse is true. The toxicity of the material may also vary considerably with the form in which it is administered, for example, solid, in suspension, or in solution. In solution, the toxicity may be influenced by the solvent and the concentration. Synergism (or combined effect) may be a very real factor in the action, and probably occurs far more frequently than recognized. In the area of "mixed exposures," knowledge is very uncertain. As an example, exposures to inhalation of hydrogen sulfide, or to trichloroethylene, followed by ingestion of alcohol, greatly increases the effects of both.

Site of Introduction.   With subcutaneous injections, toxicity may be affected by the density of the subcutaneous tissue. With intravenous administration whether the injection is made into the femoral or jugular vein may be of importance, but in any case the rate of injection, or the amount of material injected per unit time, will considerably influence the value of the toxic dose.

Other Influences.   Disease, environmental temperature, habits and tolerance (such as smoking, alcohol intake, other drugs, low-level exposure to carbon monoxide and other substances), idiosyncrasy or allergy, diet, and season of the year all may influence the toxicity of a substance. The toxicity of chemicals also vary with the species of animals used, and sometimes with different strains of the same species. Within the same strain significant differences between litter mates have been observed. This variation is not unique to laboratory animals, but can be observed in humans as well. As Kubias has pointed out in Chapter 33, there is no average person.

Since in the real world, people have a wide variety of exposures, from the moment of birth into the air of this unclean world, and the defense processes which each person possesses may change from time to time, it is seldom possible to attribute the toxic effects of any substance specifically without consideration of the other forces from other exposures which are operating over the years. In discussing the defense processes which are often the determining factor in human reactions to a chemical, G. R. C. Atherley, in *Occupational Health and Safety Concepts,* published by Applied Science Publishers, Ltd., London, 1978, notes respiratory filtration of aerosols and mists (which we consider in Chapter 25), cellular defense, inflammatory response, immune response (which alters sensitivity), homeostasis (in which the internal milieu of the body is maintained in a stabilized condition), stress resistance (linked to the hormone cortisol), thermoregulation of body temperature, and metabolic transformation whereby a toxic substance is rendered less toxic and can be excreted by the kidneys.

Because of these defenses, which are most effective in healthy adults, the epidemiologists, when studying a large number of persons of a given population, such as employment on similar occupational tasks for a comparable period of time, often note that they are usually the population with the highest resistance, due partly to pre-selection, to excellent medical attention, to adequate and balanced diet, and other factors which are in favor of good general health.

Although a direct relationship between an occupation (chimney sweeping) and cancer was established by Potts in 1775, the subject of potential carcinogenic activity of common chemicals did not receive widespread attention until fairly recent times. Several studies have suggested that between 50% and 80% of human cancer has origin in occupational or ambient (real-world) environment. Since 1970, when it became a stated policy of the government to develop the information necessary to eliminate cancer from its position as one of the leading causes of adult death in the United States, various approaches have been made to the problem. The National Cancer Institute (NCI), which is part of the National Institutes of Health, has been studying cancer both by epidemiology studies of human populations exposed, and from studies in animals, beginning with rats and mice, and going up the scale to nonhuman primates. During the past several years, starting in 1972, several hundred chemicals and chemically related substances have been tested, either by the NCI itself, or through grants and research contracts. Unfortunately, most rodent studies require two years for the experiment, and usually another year before the data is fully tabulated and reported in acceptable form. Information on the availability of information from testing may be obtained from the Office of Information, National Cancer Institute, Bethesda, Maryland 20205, attention, Information Specialist, or call (301) 496-6095 or (301) 496-2351.

Reviews of the older literature have shown that perhaps 2000 chemicals are suspected carcinogens, based to data currently available. (See Appendix 2.)

Recognizing that the dissemination of information to workers is an important factor in educating persons to avoid contact, or, if they prefer, to seek alternate employment, the Occupational Safety and Health Administration requested the Assembly of Life Sciences of the National Research Council to recommend how to inform workers. The committee concluded that when controlled experiments demonstrate that an agent produces cancer in animals, that agent should be regarded as possibly carcinogenic in humans. It recommended that a single national source determine if an agent is of sufficient hazard to warrant informing workers. This source, such as Department of Health and Human Services, should be one that is, to the extent possible, credible to both management and labor and one that is not involved in the regulatory process. The committee noted that apparently effective employee information programs conducted by large corporations and associated labor unions do exist. However, it expressed concern with the large number of workers who might be exposed to carcinogens in small plants. These employees would not have access to the health staffs and other benefits of larger plants, nor to the resources of a union. For that matter, the very existence of their work places would probably not be known to the government agencies that sought to inform them. At least 75% of the nonagricultural work force is not affiliated with labor unions. Plant size varies considerably. For example, in 1973, the construction industry employed almost 600,000 people who could have been exposed to asbestos in their working environment, distributed throughout about 70,500 establishments. Only 2% of these were employers of more than 50 workers each. Fifty-one percent employed fewer than four workers each.

The use of bacteria as indicator both of carcinogenic and mutagenic potential has received much attention in the past few years. The Ames test which was developed by Professor Bruce Ames at the University of California, uses Salmonella microsome, and certain other recognized test strains, as the test subject. The test can be run and reported in a few days at nominal cost. Professor Ames estimates that the cost for a compound would be in the order of $100–$300, and one person could do several compounds a day. The acceptance of Ames' approach has set the stage for development of new and ingenious techniques that, used together, may provide accurate information on the possible hazards of chemicals. Six short term tests for detecting carcinogenicity have been evaluated using 120 compounds, half of which were carcinogens and the remainder noncarcinogens. The results obtained indicate that the Ames test and a "cell transformation" assay are both sufficiently sensitive to carcinogenicity, or the lack of it, in the compounds studied to enable them to be employed for detecting potential carcinogens, if used under carefully controlled conditions.

At a conference held at Cold Spring Harbor, Long Island in September 1976, a large assembly of scientists from many disciplines considered the problem of carcinogenesis, and the best approach to resolving acknowledged unknowns or uncertain aspects of our present knowledge. They reaffirmed what was mentioned previously, namely, that humans *are* animals in this context, and that test data that suggests or shows that a certain chemical produces neoplasms in lower animals can be used as predictive for humans. Citing specific cases where well-documented animal data were available that proved applicable to humans, Dr. Rall of the National Institute of Environmental Health Sciences mentioned that in 1945 data showed diethylstilbestrol (DES) produced cancer in mice. Since the medical profession prescribed widespread use of this substance as a female hormone supplement in early pregnancy, it has been shown to have produced vaginal cancer in the daughters of some women. Likewise, 4-aminobiphenyl was studied in rats and dogs from 1952 to 1954, and shown to produce bladder cancer. Production in the United Kingdom never was attempted; in the United States, starting in 1935, the compound was produced for 20 years. Today it is one of the "OSHA carcinogens."

As Dr. Lorenzo Tomatis of the International Agency for Research on Cancer in Lyon, France has pointed out, the prevailing preoccupation in the research community is "not to extrapolate unduly for experimental data to man," yet very little emphasis, if any, is put on the mistakes made when results obtained in animals were not taken as indicative of a possible danger to man.

A concensus reported from the conference included:

1. The chain of events which converts normal healthy cells into irreversible carcinogenic neoplasms is still incompletely understood.
2. Really effective new cancer treatments are not on the horizon.
3. Prevention—the only long-range "cure"—is still not understood or applied to any real degree. (Cigarette smoking and auto exhausts containing known carcinogens, may be cited as knowns about which little is being done, even today.)

We can offer the personal opinion that, as public and social pressures increase, more aggressive action will be forthcoming. The Toxic Substances Control Act (signed by President Ford October 1976; effective date 1 January 1977) is an example of such concern reduced to law, which hopefully will be administered wisely and intelligently, in the national interest. (See Chapters 9, 10, 11, 14, 22 and 25.)

THE TOXIC SUBSTANCES CONTROL ACT OF 1976 (P.L. 94-469)

After five years of Congressional hearings, study, and debate, the Toxic Substances Act of 1976 is now law. Before outlining the law's major sections, the Findings of Congress and Policy of the United States, as presented in Section 2 of the law are pertinent:

(a) *Findings: The Congress finds that—*
   (1) human beings and the environment are being exposed each year to a large number of chemical substances and mixtures;

(2)   among the many chemical substances and mixtures which are constantly being developed and produced, there are some whose manufacture, processing, distribution in commerce, use, or disposal may present an unreasonable risk of injury to health or the environment; and

(3)   the effective regulation of interstate commerce in such chemical substances and mixtures also necessitates the regulation of intrastate commerce in such chemical substances and mixtures.

(b)   *Policy: It is the policy of the United States that—*

(1)   Adequate data should be developed with respect to the effect of chemical substances and mixtures on health and the environment and that the development of such data should be the responsibility of those who manufacture and those who process such chemical substances and mixtures;

(2)   adequate authority should exist to regulate chemical substances and mixtures which present an unreasonable risk of injury to health or the environment, and to take action with respect to chemical substances and mixtures which are imminent hazards; and

(3)   authority over chemical substances and mixtures should be exercised in such a manner as not to impede unduly or create unnecessary economic barriers to technological innovation while fulfilling the primary purpose of this Act to assure that such innovation and commerce in such chemical substances and mixtures do not present an unreasonable risk of injury to health or the environment.

The law requires the EPA to order animal testing of any chemical or mixture for which data and experience are insufficient, and which are relevant to a determination that the manufacture, distribution in commerce, processing, use, or disposal of each substance or mixture, or any combination of such activities does or does not present an unreasonable risk of injury to health or the environment. The standards to be prescribed for the testing of health and environmental effects will include carcinogenesis, mutagenesis, teratogenesis, behavioral disorders, cumulative or synergistic effects, and any other effect which may pose an unreasonable risk of injury to health or the environment. Among characteristics to be considered are persistence (how long the material will still be active before it is degraded or otherwise converted to a biologically inactive form), acute toxicity (high dose over short time), subacute toxicity (produces effects but not fatal) and chronic toxicity (low-level doses over long periods) (Fig. 15.1).

Essentially, the law applies to all new chemical substances, and to new uses of existing substances. Notification and test data must be submitted 90 days before the manufacture or significant new use is begun. If the data are acceptable to EPA, clearance will be given, otherwise EPA may request additional evidence, or prohibit or limit production and use.

The EPA will compile and keep current a list of chemical substances which EPA finds unacceptable.

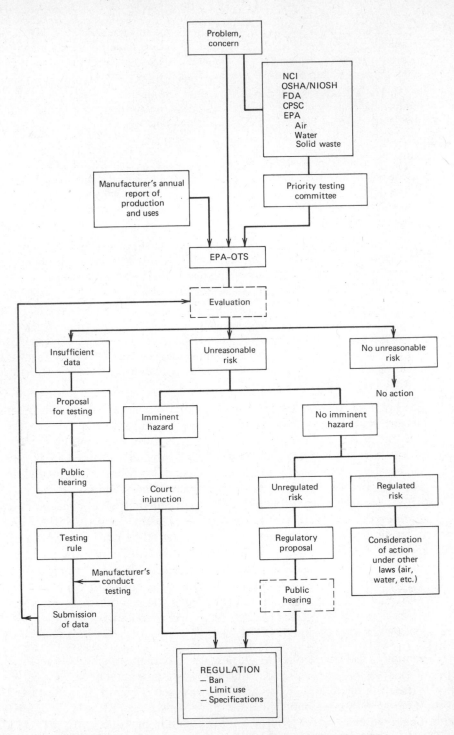

Figure 15.1. Regulatory scheme for existing product under TSCA.

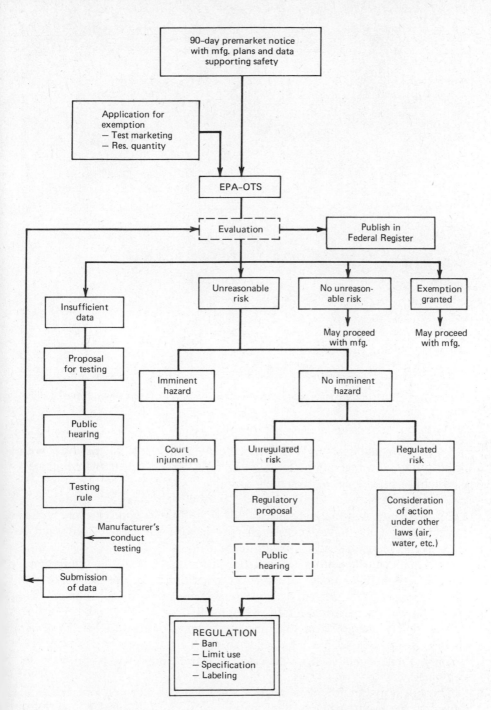

Figure 15.2. Regulatory scheme for new product under TSCA.

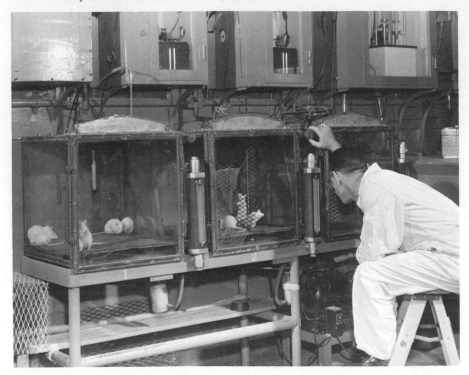

Figure 15.3. Inhalation chambers for study of inhalation toxicity in animals at Dow Chemical Co.

Polychlorinated biphenyls are specifically considered in the Act, with provision for new regulations for disposal of PCB, and "clear and adequate" instructions in the marking or labeling. In one year (January 1, 1978), totally enclosed systems were required for PCB, and after two years (January 1, 1979), no person may manufacture any PCB, nor possess or distribute it after $2\frac{1}{2}$ years (July 1, 1979), with certain provisions for exemptions. (Inasmuch as PCB is the main dielectric insulating fluid used in electrical distribution transformers and capacitors, and all the substitutes to date have limitations, such as flammability under electric arc conditions, the resolution of this substitution of PCB presents an interesting and important technical and social, as well as environmental development.)

Specifically exempt from the law are chemicals manufactured only in small quantities (as defined by rule) solely for purposes of scientific experimentation or analysis, or chemical research, analysis, or product development.

The bill provides for fines of up to $25,000 per day per violation, but considers the gravity and extent of the violation and ability to pay, as well as previous history of violations, in setting fines.

Criminal penalties may be imposed on persons who "knowingly or wilfully" violate any provision of the act which defines prohibited acts."

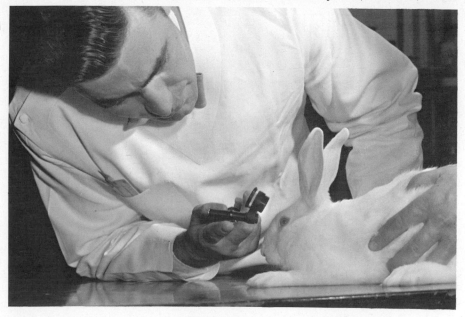

Figure 15.4.   Research scientist examines rabbit used in test method for chloracnegens at Dow.

Figure 15.2 outlines our understanding of control schemes for existing chemicals which EPA elects to review, while Fig. 15.2 outlines the procedure for new chemicals.

The TSCA Inventory Reporting Regulations were published in *Federal Register,* Vol. 42, Part VI, December 23, 1977. Reporting forms and the instruction manual can be obtained from all EPA Regional Offices and the EPA Headquarters' Industry Assistance Office, Washington, D.C. 20460.

While the legal aspects have received much attention in recent years, the long and sincere interest of major companies with the safe utilization of their products, both in manufacture and in the hands of the consumer, is often overlooked. The Dow Chemical Company established a toxicology laboratory in 1933. Much information was published in the open literature, including details of inhalation chambers useful for animal investigations (Fig. 15.3), which was described in an article by D.D. Irish and E. M. Adams, *Industrial Medicine,* Apparatus and Methods for Testing the Toxicity of Vapors, pages 1–5 (Industrial Hygiene Section) January 1940. Another example is the Fig. 15.4, which was published originally in E. M. Adams, D. D. Irish, H. C. Spencer, and V. K. Rowe, The Response of Rabbit Skin to Compounds Reported To Have Caused Acneform Dermatitis, *Industrial Medicine*, pages 1–4 (Industrial Hygiene Section), January 1941, in which the rabbit ear test was used to detect the presence of chloracnegens long before the "dioxin" controversy. In November 1955, Dow established a separate biochemical research facility to coordinate all toxicity and biological activities. Other major companies, especially DuPont and Union Carbide, have been active in studies of human as well as animal effects from chemicals, and the companies should be consulted for information when questions arise about safe handling and use of their products.

BIBLIOGRAPHY

American Mutual Insurance Alliance, *Handbook of Organic Industrial Solvents, Technical Guide No. 6*, 3rd ed. revised 1980. [Available from 20 N. Wacker Drive, Chicago, Ill., 60606.]

Borden, W. O. and Gibson, R. W., "Dangerous Materials and Related Problems," presented to California Fire Chiefs Association, University of California at Davis, March 18, 1969, Stanford Research Institute, Menlo Park, Calif.

"Yellow Rain: Gaining Speed, Review and Outlook," *Wall Street Journal*, 26 (March 11, 1982).

Brown, J. M., "Legal-Social-Economic Considerations Affecting Hazardous Materials Decision Processes," Preprint 38A, presented at the symposium on Legal, Social and Personal Implications of Materials, Part I, Materials Conference, Philadelphia, Pa., March 31–April 4, 1968, American Institute of Chemical Engineers, New York.

Burgess, William A., *Recognition of Health Hazards in Industry: A Review of Materials and Processes*, Wiley- Interscience, New York, 1981.

"Chen, E., *PBB: An American Tragedy*, Prentice-Hall, Englewood Cliffs, N.J., 1979.

Deaths Due to Toluene Poisoning," *Occup. Hazards*, **15**, 59–60 (January 1978).

deTreville, R. T. P., "The Role of Industrial Hygiene in the Proper Engineering of Materials," Preprint 43-E, presented at the Symposium on Legal, Social and Personal Implications of Materials, Part II, Materials Conference, Philadelphia, Pa., March 31–April 4, 1968, American Institute of Chemical Engineers, New York.

Katz, D. L., Woodworth, M., and Fawcett, H. H., "Factors Involved in Cargo Size Limitations," Report to the U. S. Coast Guard by the National Research Council Committee on Hazardous Materials, July 29, 1970, National Research Council, Washington, D. C.

Kraybill, H. F., and Mehlman, M. A., *Environmental Cancer, Advances in Modern Toxicology*, Vol. 3, Wiley, New York, 1977.

Murphy, A. J., "Prevention Requires Teamwork," Preprint 43C, presented at the Symposium on Legal, Social, and Personal Implications of Materials, Part II, Materials Conference, AIChE, Philadelphia, Pa., March 31–April 4, 1968.

"The New Multinational Health Hazards," International Chemical Federation Meeting, Geneva, Switzerland, October 28–30, 1974.

NTP Technical Bulletin, Issue No. 6, Jan. 1982, National Toxicology Program, Dept. of Health and Human Services, Public Health Service, P.O. Box 12233, Research Triangle Park, N.C. 27709. See also Annual Plan, Fiscal Year 1982, NTP-81-94 *National Toxicology Program*, March 1982, same address.

"PCB's in the United States—Industrial Use and Environmental Distribution, Task I," EPA 560/6-76-005, February 25, 1976, Final Report, Office of Toxic Substances, U. S. Environmental Protection Agency, Washington, D. C.

"PCB's and the Environment," Interdepartmental Task Force on PCB's Report No. ITF-PCB-72-1, March 20, 1972. [Available from National Technical Information Service, Springfield, Va., 22152.]

"Pipeline Accident Report, Mid America Pipeline System, Anhydrous Ammonia Leak, Conway, Kansas, December 6, 1973," National Transportation Safety Board, Washington, D. C., Report No. NTSB-PAR-74-6.

Safety Information SB 82-11/3365A, San Francisco, Aug. 25, 1981, issued 1982 by National Transportation Safety Board, Washington, D.C. 20594. (PCB in gas pipelines).

Squire, R. A., Ranking Animal Carcinogens: A Proposed Regulatory Approach, *Science,* **214** (4523) 877–880 (November 20, 1981).

Waldholz, M., "Potential of Costly New Superdrugs Leaves Doctors Excited but Wary," *Wall Street Journal*, 29, (March 11, 1982).

Occupational Exposures of Women

American Society of Anesthesiologists, "Effects of Trace Anesthetics on Health," NIOSH Contract HSM-99-73-003 10/16/72–10/15/78.

Anon, "Arsenic Standard Greatly Reduces Risk of Lung Cancer in Workers, OSHA Finds," *Wall Street Journal,* 14 (April 12, 1982).

Brown, W., NLRB Finds Safety Data Essential to Contract Talks, *Washington Post,* A21 (April 14, 1982).

Proceedings, Conference on Women and the Workplace, June 17–19, 1976, Washington, D.C., Society for Occupational and Environmental Health, Washington, D.C.

"The dilemma of Regulating Reproductive Risks," *Business Week,* (August 29, 1977).

Fishbein, L. *Potential Industrial Carcinogens and Mutagens,* EPA 560/5-77-005, Office of Toxic Substances, Environmental Protection Agency, Washington, D.C., May 5, 1977.

Friedan, B., *The Second Stage,* reviewed in *Wall Street Journal,* December 4, 1981, p. 34.

Hricko, A. with Brunt, M., *Working for Your Life; A Woman's Guide to Job Health Hazards,* Labor Occupational Health Program/Health Research Group, University of California, Berkeley, June 1976.

Hunt, V. R., *Occupational Health Problems of Pregnant Women,* April 30, 1975, A Report and Recommendations for the Office of the Secretary, Department of Health, Education and Welfare, April 30, 1975. [Order No. SA - 5304 - 75, The Pennsylvania State University, University Park, Pa.]

"Mutagens and Teratogens," Subfile of Registry of Toxic Effects of Chemical Substances, National Institute for Occupational Safety and Health, Rockville, Md. 20852.

Nichols. E. E., *Guidelines on Pregnancy and Work,* NIOSH Contract No. 210-76-0159, American College of Obstetricians and Gynecologists, Chicago, September 1977.

"Occupational Health, Part II, Occupational Health Hazards for Pregnant Women," Discussion Paper No. 14, December 1976. [Available from *Working Women's Centre,* 423 Little Collins St., Melbourne, Australia.]

"Women in Employment," Vol. II, *Occupational Health and Safety,* International Labour Organization, Geneva, 1971, pp. 1501–1504.

"The Workplace," *The Spokeswoman,* 7–11. (July 15, 1976.)

Schechter, D., "Untangling the Asbestos Mess," *Occu. Health and Safety,* **51**(2), 30–34 (Feb. 1982).

Waldron, H., *Lecture Notes on Occupational Medicine*, 2nd ed. Blackwell Scientific Publications, 1979, Oxford, U.K.

NIH Guidelines for the Laboratory Use of Chemical Carcinogens. NIH Pub. No. 81-2385, May 1982, available from Frederick Cancer Research Laboratory, Frederick Md. 21701.

Don Irish (Dow) left, and Hal Schoenk (Industrial Hygiene Foundation), discuss exhibit at meeting in 1951 two decades before OSHA—on role of toxicology research in occupational and consumer safety.

16

Testing Reactions and Materials for Safety

Jack S. Snyder

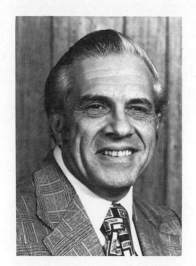

Norstrom, in Chapter 24, shows that runaway reactions and explosions are important causes of severe injury, property damage, and business interruption losses. Historically, catastrophies in the chemical industry have resulted from vapor cloud ignitions and from reaction vessel explosions.

The small, isolated pot-and-kettle plant of yesteryear has been replaced with complex automated processes. The reactions usually are fast because of the nature of the reactants, high temperatures, high pressures, and active catalysts. An intelligent appraisal of all new reactions, changes in conditions, or plant modifications must be made and presented to management with recommendations for minimizing the probability of serious accidents.

Chemists usually admit that in most organic reactions where compound A and compound B react to form compound C, three or four transitory compounds may be involved during the reaction. Not only the product is obtained, but varying amounts of other unwanted compounds as well. If the reaction is run every time exactly as the chemist describes, there will be routine operations, but, how close to or how far from an unstable situation the reaction may be, usually is not known. Ideally the parameters of safe operations for every step should be known, and procedures and equipment should be designed so that the limits will not be exceeded.

Knowledge as to what constitutes a potentially unstable condition, and why some compounds and mixtures behave in an undesired fashion, is relatively meager. Hence, the key to any determined effort to test for stability is the interest and backing of research management who must know it is important to find out how safe a reaction is as well as what is the yield. Most research chemists have not been trained or oriented in this direction.

No chemist would consider undertaking a new synthesis without a literature search; a good research chemist makes frequent use of the library. Unfortunately, the amount of safety information published in the chemical journals is very meager. Even standard textbooks on organic chemistry used in universities discuss dangerous reactions in a very routine manner. Hence, it is not difficult to understand why so many research chemists are not safety oriented. Perhaps another reason is that the consequences are minimal when instability is suddenly encountered in the laboratory. A reaction which could result in an explosion in a plant operation may merely cause one of the ground joints of the laboratory flask to be pushed apart or broken by materials frothing out over the bench. The chemist just does not equate the effect of the rapid gas evolution on a loosely connected glass system to the effect on tight steel equipment, containing large volumes, without the pressure-relieving capacity to take care of such a sudden surge.

Occasionally the feeling is expressed that it is an insult to professional ethics among chemists to warn that while a procedure described in the paper was being developed, variations in conditions resulted in misadventure, characterized by loss of reaction product to the ceiling or bench, or where flying glassware caused injuries. The research chemist is best qualified to recognize a potentially hazardous compound or reaction. No scale-up of laboratory to pilot plant or of pilot plant to factory should be permitted without adequate consideration being given to the stability of the compounds and reaction involved.

To undertake such a study, the following steps are suggested:

First, although little may be located in the classical literature, an ever-increasing body of safety literature can be helpful. Some research work has been done, and papers have been written on stability of materials. Second, there are ways to confirm or allay some suspicions with simple tests.

National Safety Council Industrial Safety Data Sheets[1] contain stability information on a limited number of common chemicals. NSC Data Sheet 486A, "Chemical Safety References," gives an indication of the data available on a number of compounds. "Case Histories of Accidents in the Chemical Industry," formerly published by the Chemical Manufacturers' Association,[2] contain an indexed compilation of some 2000 accidents, many of which illustrate the causes and results of unexpected instability. The Committee on Chemicals and Explosives of the National Fire Protection Association, Sectional Committee on Hazardous Chemical Reactions has published some 3500 reactions[3] that can cause serious problems if not controlled. Two of the American Insurance Association (formerly National Board of Fire Underwriters) Research Reports are "Nitroparaffins and their Hazards" and "Fire Explosion Hazards of Organic Peroxides".[4]

The United States Bureau of Mines Explosives Research Laboratory has contributed much of the chemical stability literature.[5] Much of the Bureau's research has been on explosives and includes several important tests for evaluating explosives. These same tests can also be used for evaluating nonexplosives. (See Chap. 17.)

One source of information frequently overlooked is the manufacturer. When confronted with a compound of suspicious configuration, it is possible the maker has literature on stability or his research staff has unpublished data he will furnish on request.

The Department of Defense has authorized considerable research on test methods for propellants, and these tests have application to commercial problems. One publication that contains information on both common and exotic materials is *The Handling and Storage of Liquid Propellants.*[6]

A search through this literature and others cited at the end of the chapter, will assist the chemist in the art of recognition, which is the heart of the matter, even though it may not yield specific reactants he or she plans to use. Certain types of compounds, mixtures, and reactions should always be suspected. Many may be familiar because they are difficult to handle, but the reactivity noted in the laboratory is amplified many fold in the closed pilot plant and factory equipment.

An explosion is a sudden release of pressure regardless of the source. For instance, a bursting steam boiler is truly an explosion even though steam is not an explosive.

An explosive is any substance that can produce in thousandths or even millionths of a second, high-pressure gas capable of destroying the surroundings. A detonation is the process in which an explosive undergoes chemical decomposition or oxidation—reduction, primarily within a high-pressure zone or shock wave that travels at greater than sonic velocity through the unreacted material. In other words, a true detonation is capable of almost instantaneous self-propagation once the reaction or decomposition is initiated.[6,7]

The reactive chemicals to be considered are not explosives even though they may be capable of self-reaction. As compared to explosives they react very slowly and seldom produce shock waves of any significance. Usually the reaction can be controlled after initiating it, by cutting off feed rates, or by lowering temperatures or pressures. With an explosive, this is impossible.

Steel and Duggan, in their paper "Safe Handling of Reactive Chemicals,"[8] characterize a reactive chemical as a material which will vigorously polymerize, decompose, condense, or otherwise react with itself in the pure state, or in the presence of a catalytic amount of some other material, or which will react violently with water. Classes of compounds which react with themselves include those containing a vinyl linkage, such as vinyl chloride, acrylates, and styrenes. Some of these require only high temperatures to catalyze polymerization. Others polymerize in the presence of light, peroxides, caustics, and strong acids. Most of the monomers have to be inhibited chemically or at least refrigerated to prevent polymerization. Another group involves the carbonyl radical contained in materials such as butyraldehyde and acetyldehyde. These can condense with extreme vigor in the presence of bases and sometimes strong acids.

Certain conjugated unsaturated compounds, such as butadiene, acrolein, and acrylonitrile can polymerize violently in the presence of suitable catalysts or when heated. Epoxy compounds are also self reactants. Ethylene oxide is the smallest and most reactive member of this group, and is capable of violent decomposition.

Strong oxidizers such as chlorine dioxide, chlorates, perchlorates, permanganates, peroxides, and chromic acid are all capable of uncontrolled reactions. Compounds of this type are capable of the instability which causes many chemical-plant and laboratory explosions.[9, 10] It is essential that the chemist determine the parameters of safety both for these compounds and for any reaction system that contains these compounds. The chemist should deliberately determine at what temperature the reaction accelerates dangerously, which common compounds can inadvertently or by design catalyze the reaction, and at what concentrations and temperatures this becomes difficult to control.

Compounds that are reactive with water are included in this study, because the results of this mixing can be violent and destructive. Among compounds requiring extreme care in this respect are sodium metal, aluminum alkyls, chlorosulfonic acid, phosphorous trichloride, and calcium carbide, as well as the concentrated mineral acids, including sulfuric and nitric acid and their mixtures.

Acetylene and acetylenic compounds are notoriously unstable. Metal acetylides are sensitive to heat, shock, and abrasion, and some can be classed as explosives.

Nitrogen-containing compounds fall into a class by themselves, and they can be very confusing. Nitrogen gas is quite inert, and it is usually added to a reaction mixture to minimize the possibility of an explosion. Gaseous ammonia, although it will form a flammable mixture in air, is not usually considered an unstable compound. The ammonium ion, however, increases the sensitivity and reactivity of some compounds. For example, compare the stability of ammonium nitrate with sodium nitrate.

Amines are not reactive in the above context. Trimethylamine is a prime example, although it makes up for its apparent innocence with an obnoxious odor. However, the fact that nitrogen is present in the compound does require a certain amount of caution, because when it comes into contact with strong oxidizers, vigorous and even explosive exotherms are experienced. In contact with mercury, explosive fulminates are formed and reactions with hydrochloric acid will produce diazo compounds with distinct thermal instabilities. Hydrazine, which is just two amine radicals with a nitrogen —nitrogen bond, is an extremely reactive compound and a propellant. Its vapors are shock sensitive and can explode with extreme violence.

The commercial diazo compounds with the —N=N— linkage are not generally shock sensitive, usually because the molecule is quite large, but practically all will decompose exothermically and even explosively on heating, and they are sensitized by a variety of compounds. All the hydrazines and hydrazones are reactive and can form explosive compounds, but unless intimate knowledge of the chemistry involved is applied, predictions are risky. For example, hexamethylene tetramine is used in making explosives, but it is not particularly sensitive by itself although it contains a high nitrogen percentage.

$N{=}N$ bonded compounds are not explosives, but if another nitrogen is added to form an azide, $N{-}N{\equiv}$ or $N{=}\overset{+}{N}{=}N$, the molecule becomes sensitive. The azides are thermally unstable, and all explode when heated. Most are shock sensitive even at room temperature.

The oximes $C{=}NOH$ are reactive but not really sensitive compounds, and neither are the nitriles with the $C{\equiv}N$ (cyanide) linkage. The carbon—nitrogen bond is much more stable than the nitrogen—nitrogen bond. The isocyanates $-N{=}C{=}O$ are not sensitive.

The nitrogen—oxygen bond is an interesting linkage. Nitro compounds with the $-O{-}N{=}O$ linkage are usually strong oxidizers and can be thermally unstable at relatively low temperatures. Compounds with more than one nitro group form our common explosives; the more nitro groups the more sensitive the compound. But if the molecule is small enough even one nitro group can be enough to make the compound shock sensitive. For example, a few years ago, two tank cars containing nitromethane exploded. This substance was manufactured for many years and shipping in tank-car quantities, until one day the right conditions occurred and the contents detonated. There are other materials in the same category that may not yet be identified as such (see Chaps. 16, 17, and 31).

Nitroethane has not exhibited the sensitivity of nitromethane. However, by test we have determined that if as little as 5% by weight of n-butylamine is added to nitroethane, the mixture is as shock sensitive as some explosives.

The ability to sensitize relatively stable compounds is not easily predicted. Not enough is known about the chemistry involved to be accurate. By sensitization is meant not only the increase in shock or friction sensitivity, but the ability of one material to catalytically decrease the thermal stability of a compound or mixture, either by lowering the decomposition temperature or by causing a rapid oxidation. Naturally this includes compounds which are reactive, and, therefore, capable of being affected more easily and by a surprisingly wide variety of materials.

Some of the references contain lists of incompatible compounds, such as the Department of Commerce Research Report, "Explosives, Propellants and Pyrotechnic Safety Covering Laboratory, Pilot Plant and Production Operations."[11] Some of the hazardous mixtures include ammonia with chlorine, mercury, or calcium hypochlorite; chlorine with acetylene, petroleum gases and finely divided metals, cyanides, and concentrated formaldehyde, and hydrazine and ferric oxide.

An unusual highly exothermic reaction with relatively common materials may be of interest. The contents of a receiver tank containing a chloroform–methanol mixture were being transferred to a drum that had originally contained tetrahydrofuran. The operator filling the drum heard a hissing noise and warned nearby workers to leave the building. While the building was being evacuated the drum exploded causing relatively minor building damage. A few windows were broken, a door was blown off its hinges, but no injuries resulted. Investigation revealed that sodium hydroxide was being used as a stabilizer for tetrahydrofuran and probably some caustic remained in the drum.

Laboratory studies proved that when chloroform is added to methanolic sodium hydroxide, a vigorous reaction occurs. When 1 g of sodium hydroxide was added to a mixture of 1 ml of methanol and 1 ml of chloroform, an exothermic reaction took place which, after 5 min, became vigorous enough to bubble out of a 6-in test tube. Aqueous sodium hydroxide also reacted in methanol–chloroform mixtures in the same way. Chloroform–tetrahydrofuran–sodium hydroxide mixtures did not react, thereby indicating that tetrahydrofuran was not responsible for the incident.

It should be noted that methanol plays the role of solubilizer by increasing the contact between base and chloroform. Other solvents which dissolve sodium hydroxide and are miscible with chloroform can be substituted as the solubilizer. The reaction products may be different, but the net effect will be the same. Potassium hydroxide and other alkalies may replace sodium hydroxide as a reactant in the solubilizer-chloroform mixture.

As a supplement to the recognition of reactive chemicals, the chemist should be alerted to certain types of reactions which utilize these chemicals.

All nitrations should be viewed critically. Many, if not most, of these reactions display thermal instability at certain temperatures and pressures especially if there is an excess of nitric acid. In many of the mixtures tested an excess of nitric acid will also cause the mixture to be shock sensitive.

Oxidation-reduction reactions using peroxides, chlorine, per acids, or per salts are notorious for explosions because of thermal instability or shock sensitivity. Included in this group would be the Wolff–Kishner reductions employing hydrazo compounds and even metal hydride or pure hydrogen reactions.

Condensations such as Friedel–Crafts, Claisen, and Cannizzaro types; polymerizations; Reppe chemistry reactions utilizing acetylenic compounds; and Grignards are examples. There are many more reactions and compounds which fit into these categories.

Simple tests can be run that are adequate for continuing laboratory work. There are more elaborate tests which should be used if the work is carried further into pilot plant or factory size equipment, or if the first test reveals a dangerous condition.[12]

The initial testing for thermal stability is based on the fact that practically all hazardous reactions involve rapid exotherms and that the rates of all

Figure 16.1. Thermal stability test apparatus.

reactions increase with temperature. Experts have determined that even for the majority of explosives, initiation is thermal in origin. With very little experience this test is remarkably reproducible, and it will reveal exotherms in oxidation and polymerization reactions involving reactants of low volatility. Hundreds of determinations may have been made without incident, but we should still insist on a rapid pretest to safeguard the tester. This pretest involves dropping a small quantity, less than one-tenth of a gram, on a hot plate at above 300°C. If a sound is heard like a pop, bang, snap, or crackel, the barricading is brought into the scene. If the sample decomposes, chars quietly, or sizzles first because of the evaporation of liquids, the test is continued. It is absolutely essential that the compound or mixture decompose for this pretest to be valid.

The simple thermal stability test makes use of a standard melting-point apparatus, as illustrated in Fig. 16.1.[13] The heating coil in the silicone oil-filled cyclic bath is regulated by a variable transformer. The material to be tested, about 1 g of active ingredient, is added to a 15-ml pyrex, tapered bottom, centrifuge tube. A fine wire thermocouple with a melting-point tube as a corrosion shield is inserted to a point near the bottom of the sample. The centrifuge tube and a second thermocouple are inserted through a cork into the melting-point apparatus. The thermocouples should be at the same level. The test is run in a hood with the hood door closed. The variable transformer and the temperature indicator are located outside the hood.

The stirrer is turned on and heat is applied. The rate of heating is kept at 6° to 10°C/min. The oil temperature is not permitted to exceed the sample temperature by more than 15°C preferably not more than 10°C. If it does, the rate of heating is slowed. For every 15°C increase in the temperature of the sample, the time, sample and oil temperatures, and transformer settings are recorded. Any change in the state of the sample is also recorded, as any evolution of gas, charring, melting, boiling, or fuming.

Particular note is paid to any rise in sample temperature above or approaching the oil temperature. If this occurs, readings are taken more rapidly. If the sample temperature even equals the oil temperature, the result of the test is considered to be positive. If an exotherm is not shown, heating is continued until the sample temperature reaches 315°C or until the sample is completely decomposed.

Table 16.1 is an example of one of the first tests. The exotherm experienced when this reaction got out of hand in large-scale equipment caused no injuries but considerable building damage. This explosion was the impetus that led to the development of the test. Note that the entire test took only 20 min to run. Slow bubbling was observed at 102°C. This bubbling increases with temperature and becomes quite vigorous. Note that at a bath temperature of 163°C the sample temperature equals the bath temperature and then begins to exceed it as the exotherm proceeds more and more rapidly. Incidentally, more precise tests that will be described, definitely showed that the exotherm begins at 90° to 95°C. Figure 16.2 shows the last 5 min of the test graphically.

Remember that this is a relatively crude test whose sole purpose is to uncover an exotherm and not to define precisely where it begins. The bubbling observed as a good indication, but not all exothermic reactions give off gases in the early stages. Whenever the sample temperature exceeds the bath temperature it is a sure indication that the reaction is exothermic, and in large-scale equipment this can be serious. To repeat, this test is most suitable for oxidation and polymerization reactions involving compounds of low volatility. Volatile compounds give erroneous results here but if the volatile is an inert solvent, not one of the reactants, the results can be meaningful to an experienced observer. Where two phases are involved agitation should be provided by bubbling nitrogen through the mixture, and it may also be desirable to use larger quantities here.

If an aging period is required to achieve a hazardous condition, such as in peroxide formation, this test will not discover the reaction.

Finally, since this test merely determines exotherms, we can expect to find a number of exotherms at very high temperatures which do not constitute a particular hazard. An experienced chemist, who knows the process involved and the meaning of this test, should be able to interpret the test results satisfactorily.

The reaction discussed previously was an oxidation by sulfuric acid. The mixture normally was heated to 90°C, held for ½ hr and then cooled. The reaction vessel was a 500-gal glass-lined reactor, rated 90 PSIG and insulated. This time it was decided to heat it to 90°C and let it cool by itself. Unfortunately, the vessel maintained excellent adiabatic conditions. The

TABLE 16.1.   Temperature Readings—Thermal Stability Test

Reaction 1

3 Ml. Open-Tube Test

Time From Start, Min.	Temperature, °C.		Transformer Setting	Comments
	Sample	Oil		
0	29	29	60	
2	38	54	60	
4	52	71	60	
5	66	79	60	
7	79	90.5	60	
8	93	102	60	
9	102	107	60	Slight gassing
10	107	113	60	Gassing
13	121	126.5	70	Gassing
15	135	139.5	70	Gassing rapidly
17	149	151.5	80	Gassing rapidly
18	163	163	90	Gassing rapidly
18½	171	168	100	Gassing rapidly
19	204.5	188	110	Gassing rapidly
19½	215.5	193	Off	

slow and easily controllable exotherm at 95°C became uncontrollable with this result.

The still buried 4 ft into the ground is shown in Fig. 16.3. It penetrated an acid brick-covered concrete floor. The still after removal is shown in Fig. 16.4. Note that it was not badly distorted. This was not a detonation, and there was not any fire after the explosion. The hole in the roof about 20 ft directly above the vessel is shown in Fig. 16.5. The head of the still embedded in concrete 250 ft away is in Fig. 16.6. It reached a height estimated at 170 ft.

An indication appeared at 102°C showing there was some bubbling occurring, some slight exotherm. At 163°C is was confirmed in the test. Even if it is assumed therre was no bubbling at 102°C, merely the reaction at 163°C, this should be enough indication that at some lower temperature, possibly 60°C to 70°C lower, an exotherm can occur, and if this reaction is to be carried out in anything larger than a small laboratory flask a more precise indication should be obtained of where the exotherm begins and how rapidly it progresses.

The next test employs a Dewar flask, and it must be run in a barricaded area. A diagrammatic sketch of the flask is shown in Fig. 16.7. It is useful for getting a more precise indication of where an exotherm begins. It approximates, but by no means approaches, the adiabatic conditions that are possible in an insulated, jacketed, factory reactor. Here 100 to 150 ml of the mixture is heated under carefully controlled, remotely operated conditions. Usually the agitated mixture is brought to a predetermined temperature level

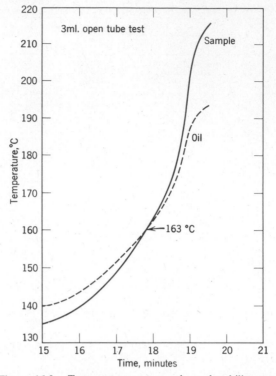

Figure 16.2. Temperature curves—thermal stability test.

Figure 16.3. Still buried 4 ft into the ground by uncontrolled reaction.

Figure 16.4. Still after removal.

Figure 16.5. Hole in the roof, 20 ft above vessel.

Figure 16.6. Head of still embedded in concrete 250 ft from still pot.

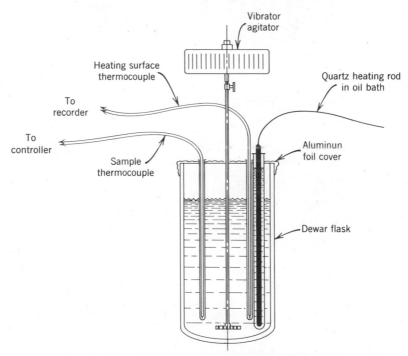

Figure 16.7. Dewar flask test apparatus.

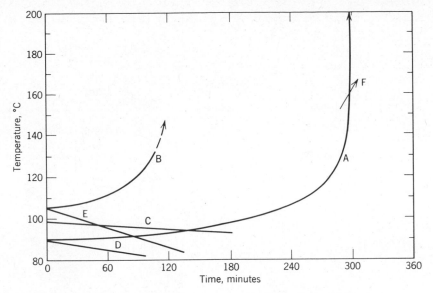

Figure 16.8. Time, temperature curves for Dewar flask stability test.

and the heat turned off. If nothing happens in 5 min heat is reapplied. The procedure is repeated to 300°C.

Let us return to the previous example and the Dewar flask determination in Fig. 16.8. Curve *A* shows the calculated time-temperature curve for a 500-gal jacketed and insulated reactor with only static air in the jacket. Curves *B*, *C*, and *D* show the curves for 125 ml of the reaction mixture in a 200-ml Dewar flask. Note how sharply the temperature affects the occurrence of an exotherm.

Curve *E* shows the Dewar flask curve for 125 ml of the reaction medium with the active ingredient removed. Curve *F* shows the rate of increase in temperature of 100 ml of reaction mixture at 160°C in an uninsulated 250-ml round-bottom flask. This rate is about 1°C/min as compared to about 40°C/min for Curve *A*, a factory reactor. A 10-ml sample in a test tube at 160°C will drop in temperature.

Incidentally, even though it could be demonstrated that the exotherm begins very slowly at 95°C, calculations indicate that with a 500-gal volume in this reactor and with full cooling water on, the reaction still could be controlled in the range of 130° to 140°C. With the Dewar flask test a combination of a rapid exotherm and substantial gas evolution is a sure indication that in factory equipment there would be an explosion.

Other tests are used in thermal stability studies. The pressure-tube test in Fig. 16.9 was developed by the Bureau of Mines.[12] The stainless-steel tube can be insulated as shown here or immersed in a heated oil bath. The heating-rate technique can be employed by observing the jacket versus sample temperatures, or the tube can be heated to a given temperature and the

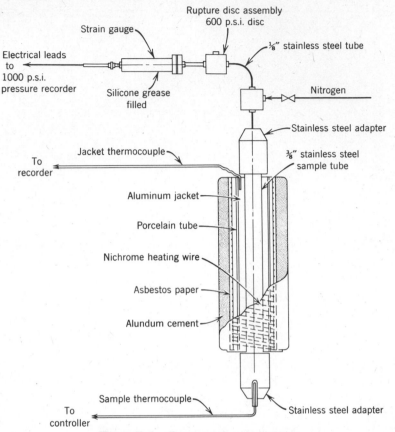

Figure 16.9. Pressure tube test apparatus.

temperature increase observed over a 5 to 15 min period. This is a good precision method where noncondensable gases are produced. It would not detect polymerization, and it can be rather difficult to interpret the data received. This bomb must be barricaded.

The improved thermal stability bomb in Fig. 16.10 developed by Dr. G. A. Mead of the Air Reduction Company is fully described in the publication *Liquid Propellant Test Methods, Recommended by the Joint Army-Navy-Air Force Panel on Liquid Propellant Test Methods.*[14] This small bomb is immersed in a bismuth–lead alloy bath. An air vibrator agitates both the bath and the sample bomb. This is an excellent and precise tool. Exotherms on the order of 2°C or 3°C can be detected. This entire apparatus must be barricaded, but it is possible to enclose it safely in an isolated laboratory.

Another interesting approach was developed by Dr. Joseph J. Martin when he was searching for the cause of the devastating 1952 explosion in a plant at Tonawanda, New York.[15] Tertiary butyl peracetate was the compound involved. It would not detonate when heated to dryness and would not detonate when subject to the shock from an exploding blasting cap.

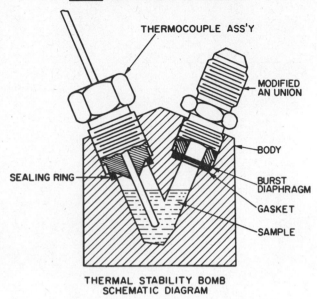

SCALE: ABOUT TWICE ACTUAL SIZE

THERMOCOUPLE ASS'Y

MODIFIED
AN UNION

BODY

SEALING RING

BURST
DIAPHRAGM

GASKET

SAMPLE

THERMAL STABILITY BOMB
SCHEMATIC DIAGRAM

Figure 16.10. Thermal stability bomb.

Dr. Martin discovered that if he inserted a 50-watt coil in the liquid and turned it on, the rapid heating of the sample caused it to detonate every time. Perhaps this phenomenon is due to compression of bubbles or to the localized high temperature which initiated the decomposition. The latter is very plausible because hot spots created when the heat could not be dissipated rapidly enough to the surrounding liquid does cause decomposition, and presents a yield problem in factory equipment even with relatively stable materials.

The thermal stability tests are recommended because they are reproducible. Many modifications of closed vessel tests are being used with success. Particularly valuable are the $\frac{1}{2}$ or 1-liter bombs used where the problem of high volatility is encountered.

The importance of thermal stability determinations cannot be overemphasized. If the various mixtures which can be reasonably expected are tested, the great majority of the possible exotherms will be revealed.

Where factory or pilot plant operations are contemplated, the more exact Dewar flask and closed vessel tests should be run on the most suspicious mixtures. If instability is observed at a relatively low temperature and if there is a possibility of mechanical instability—this means if a nitration is involved or peroxides, or diazo compounds—compounds and reactions where mechanical instability has been known, determinations should be made. There are organizations that will do this work, using apparatus such as the following:

In Fig. 16.11, an impact tester employs a fixed weight dropping at varying

Figure 16.11.   Drop-weight tester.

heights on a small amount of liquid in a precisely designed sample cup. The impact sensitivity is determined by observing a stainless-steel diaphragm. A hole in the diaphragm caused by a piston in contact with the diaphragm is positive. The mechanism of explosion of liquids in the drop-weight test is reported to be thermal ignition by compression of a bubble, followed by a normal deflagration process. A full description of this apparatus can be found in the previously mentioned *Liquid Propellant Test Methods*.[14] This tester will determine the ease of ignition through mechanical impact but not whether the ignition once initiated can be propagated. A further description of the meaning of this test and the next one (Fig. 16.12) can be found in the article "Evalating the Explosive Character of Chemicals," by R. Van Dolah.[12]

The drop-weight tester requires that it be operated remotely but no real barricading is necessary. The card-gap test shown in this figure requires a true explosion-resistant test cubicle and an isolated location in order to avoid the consequences of the shock wave and the noise. This is truly a rigorous test for any compound or mixture, and it will determine whether ignition can be propagated to full detonation. We noted previously that nitroethane which by itself was not sensitive to shock when tested by card gap, was sensitive when 5% by weight of butylamine was added. It is probably true

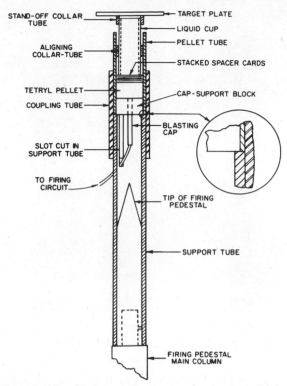

Figure 16.12. Card-gap test apparatus for shock sensitivity.

there is only a remote chance of encountering an initiator comparable to the 50 g of tetryl used in close proximity to the sample in this test. This test was run at room temperature with a small quantity of sensitizer. A process development chemist in the laboratory, or an operator's error can result in having a mixture at 50°C containing more sensitizer. Such a mixture could detonate from a much smaller shock, perhaps even dropping a flask on the floor. Knowing that the mixture is shock sensitive is essential in order to prevent these accidents.

In Fig. 16.13, the plate with the hole is similar to the hole produced by the nitroethane-butylamine mixture. This plate is just resting on the sample cup when the 40 cc of sample is shocked by a 50-g tetryl booster. The plate just cannot get out of the way fast enough, and the hole is blown through it.

It is most important to remember that a negative test does not mean a detonation is not possible. Low-order detonations of a type that can wreck a building may not show up here. The best procedure is to test at the highest temperature and under the worst conditions conceivable. Try to obtain a positive result. From there, usually enough inert diluents or desensitizers can be added, or the temperature lowered enough to pass below the threshold of detonation. Stability then is much more certain.

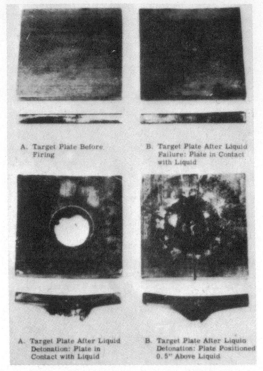

A. Target Plate Before Firing

B. Target Plate After Liquid Failure: Plate in Contact with Liquid

A. Target Plate After Liquid Detonation: Plate in Contact with Liquid

B. Target Plate After Liquid Detonation: Plate Positioned 0.5" Above Liquid

Figure 16.13. Card-gap test apparatus, typical test results.

One last facet of this stability problem should be noted. If thermal decomposition at a relatively low temperature, or a rapid oxidation, or shock sensitivity has been discovered, the reaction may still be run safely. In many cases an inert diluent or a densitizing agent is added.

For example, suppose an exotherm begins slowly at 100°C. A reaction in benzene is run without any problem. Since the benzene boils at 80°C, all the benzene must be vaporized before a higher temperature is reached. This could be a sufficiently good safety factor for a small scale supervised run with careful control of the reaction temperature, the quantity of reactants, the addition rates, and so on. Inert solvents also act as good desensitizing agents for shock sensitive systems. For instance, the shock sensitivity of nitromethane shows up very dramatically when tested in the card-gap apparatus. When 15% by weight of benzene is added to nitromethane, the mixture will not be shock sensitive by this test.

In conclusion: The tendency toward employing highly reactive compounds and mixtures must be recognized and dealt with.

The chemist's normal sources of information do not give proper attention to the possibility of instability, but there are other sources that could be of help.

The chemist should take the half hour necessary to use the melting point apparatus and from this simple screening test, determine if other tests should be run. The chemist is the one who has to prepare some of these potentially hazardous mixtures in order to run the test, and should start on a very small scale, in a hood, and keep protected at all times. Management should decide whether to run the other tests or have them run elsewhere. A final and most important word of caution on testing—there is no guarantee that an exploson will not occur. The art of testing is very new. Enough is not known of the various mechanisms involved in explosions to devise sufficient tests. The right contaminant or sensitizer or the right amount at the right temperature or the right pressure may not have been tested. But if testing is done intelligently reasonable assurances are obtained that the rumbling noise is just thunder or a jet plane.

REFERENCES

1. National Safety Council, 444 N. Michigan Ave., Chicago, Ill. 60611.
2. Chemical Manufacturers Association, 2501 M St., NW, Washington, D.C. 20037.
3. "Manual of Hazardous Chemical Reactions," National Fire Protection Association, Batterymarch Park, Quincy, Mass. 02269, 1980.
4. American Insurance Association, 85 John St., New York, N.Y. 10038.
5. Publications Distribution Section, U. S. Bureau of Mines, 4800 Forbes Street, Pittsburgh, Pa. 15213.
6. *The Handling and Storage of Liquid Propellants,* Office of the Director of Defense Research and Engineering, U. S. Dept. of Defense, U. S. Government Printing Office, Washington, D. C. (January 1963).
7. D. M. Tenenbaum, *Testing with Storables,* A. R. S. Paper 1263–60, American Institute of Aeronautics and Astronautics, New York, N.Y., 1963.
8. A. B. Steel and J. J. Duggan, "Safe Handling of Reactive Chemicals," *Chem. Eng.,* 66(8), 157–168 (April 20, 1959).
9. M. A. Cook, *The Science of High Explosives,* ACS Monograph No. 139, Reinhold, New York, 1958.
10. F. P. Bowden and A. D. Yoffe, *Initiation and Growth of Explosion in Liquids and Solids,* Cambridge University Press, 1952.
11. *Explosives, Propellants and Pyrotechnic Safety Covering Laboratory, Pilot Plant, and Production Operations,* U. S. Dept. of Commerce, Office of Technical Services, AD 272424, U. S. Government Printing Office, Washington, D. C.
12. R. W. Van Dolah "Evaluating the Explosive Character of Chemicals," *Ind. Eng. Chem.,* 53(7), 59A–62A (July 1961).
13. R. H. Albisser and L. H. Silver, "Safety Evaluation of New Processes," *Ind. Eng. Chem.,* 52(11), 77A (November 1960).
14. *Liquid Propellant Test Methods, Recommended by the Joint Army-Navy-Air Force Panel on Liquid Propellant Test Methods,* Chemical Propulsion Information Agency, Johns Hopkins Road, Laurel, Md. 20707.
15. J. J. Martin, "Test-Butyl Peracetate—An Explosive Compound," *Ind. Eng. Chem.,* 52(4), 65A (April 1960).

BIBLIOGRAPHY

American Insurance Association, NBFU Research Report No. 11, "Fire and Explosion Hazards of Organic Peroxides"; NBFU Research Report No. 12, "Nitroparaffins and Their Hazards," 1959.

ASTM E476-73, "Thermal Instability of Confined Condensed Phase Systems."

ASTM E487-74, "Test for Constant Temperature Stability of Chemical Materials."

ASTM E537-76, "Assessing the Thermal Stability of Chemicals by Methods of Differential Thermal Analysis."

Bellis, M.P., "Hazardous Chemical Reactions," *The Hexagon of Alpha Chi Sigma,* **40,** 40 (October 1949).

Benson, S.W., *Thermochemical Kinetics,* Wiley, New York, 1968.

Bowser, M. L. and Gibson, F.C., "Pulse-Forming Circuitry for Explosives Research," Bureau of Mines Report of Investigations 5985.

Case Histories of Accidents in the Chemical Industry, Vols. 1–4, Chemical Manufacturers' Association, Washington, DC., 1965–1976.

Douglas, J.R., and Thompson, C.E., "Some Studies in Chemical Fire Hazards," Oklahoma Engineering Experiment Station Publication, No. 73, November 1949.

Griffin, D.N., "The Intiation of Liquid Propellants and Explosives by Impact," presented at American Rocket Society Propellants, Combustion, and Liquid Rockets Conference, April 1961.

Kuchta, J.M., Martindill, G.H., Zabetakis, M.G., and Damon, G.H., "Flammability and Detonability Studies of Hydrogen Peroxide Systems Containing Organic Substances," Bureau of Mines Report of Investigations 5877, 1968.

"Manual of Sensitiveness Tests," published by Canadian Armament Research and Development Establishment on behalf of TTCP Panel 0-2, Working Groups of Sensitiveness.

Nawaisky, P., Ebersole, F., and Werner, J., "Explosive Reaction of Diazonium Compounds with Sulfides of Sodium," *Chem. Eng. News,* **23**(14) 1247 (July 25, 1945).

"Processing Under Extreme Conditions", *Ind. Eng. Chem.,* **48**, 826 (May 1956).

Robinson, C.S., *Explosions, Their Anatomy and Destructiveness,* McGraw-Hill, New York, 1944.

Siemens, A. M.E., "The Hazards of Organic Peroxides," *Br. Plast.,* **35** 357–360 (July 1962).

Stull, D.R., "Fundamentals of Fire and Explosion," AIChE Monograph Series 10, American Institute of Chemical Engineers, 1977.

17

Chemical Stability

Robert D. Coffee, Ph.D.

17.1. INTRODUCTION

In Chapter 19 it is shown that under suitable conditions most, but not all, organic materials will burn; that is, they are either flammable or combustible. Although not necessarily recognizing them as being organic, the average person is aware of this characteristic associated with many familiar substances: fuel oil, kerosene, propane, alcohol, sulfur, sugar, flour, etc. Most lay people have also probably heard that such things as nitroglycerine, TNT (trinitrotoluene), and the inorganic fertilizer ingredient ammonium nitrate may explode with disastrous consequences, but do not usually know why these materials behave differently nor what conditions contribute to adverse behavior. However, with the great proliferation of chemicals to which they are currently being exposed—at home, at work, and on the road—people are becoming greatly concerned. This concern contributed to the enactment of the Williams–Steiger Occupational Safety and Health Act of 1970 commonly known as OSHA.[1]

Fortunately, recent developments are enabling the safety professionals and the research chemists to more readily differentiate between those chemicals which just burn and those which are apt to explode. The differences are being characterized in terms of stability where chemical stability has been defined as:

> . . . the response of a chemical to external stimuli such as heat, light, friction or impact.

Those chemicals which remain essentially inert or unresponsive to the

305

stimuli (or merely burn when exposed to excessive heat) are considered stable whereas those which decompose or polymerize suddenly with the generation of considerable heat, light and/or gas are deemed unstable.

That it is imperative to be able to characterize the stability of all commercial chemicals is best illustrated by the following examples:

Peroxides (—O—O—)

On April 3, 1962, a trailer full of approximately 20 tons of organic peroxides was being unloaded at a warehouse in Norwich, Connecticut. Shortly after fire was discovered inside the trailer, the cargo exploded, killing four fire fighters and seriously injuring four other persons.

It is surmised that the impact from a bump or the shifting of the load in transit spilled organic peroxide which may have been ignited by friction between containers in the van.[2]

Nitromethane ($CH_3 NO_2$)

On January 22, 1958, a tremendous explosion occurred at a car switching yard of the Niagara Junction Railway, Niagara Falls, New York. This explosion was reported to have involved among other things, the detonation of a tank car containing 77,280 lb of nitromethane. [The bill of lading identified the cargo as "Lacquer Solvent—N.O.I. (not otherwise identified)—Nitroparaffin."] Although no one was killed, over 180 people were injured and the damage was estimated at over one million dollars. Six months later, June 1, 1958, another tank car of nitromethane detonated while undergoing switching operations near the city of Mt. Pulaski, Illinois. This time, however, two trainmen were killed, about forty people were injured, and losses were in the neighborhood of one million dollars.[3]

In both instances it has been surmised that adiabatic bubble compression initiated the detonation.

$$\text{Propargyl Bromide } (H—C\equiv C—\overset{\displaystyle H}{\underset{\displaystyle H}{\overset{|}{\underset{|}{C}}}}—Br)$$

On January 28, 1964, propargyl bromide (3-bromopropyne) was involved in a probable detonation which destroyed a large chemical plant in Linden, New Jersey. Three people were killed and five others were injured.[4]

Again, on December 22, 1966, propargyl bromide was involved in a detonation at a Freeport, Texas, plant which resulted in the loss of three lives.[5]

In the first case it is believed that sealed 55-gal drums of propargyl bromide were exposed to fire conditions. In the second case, pump cavitation was the probable initiating agent. The most tragic aspect is that the Russian literature had reported on the detonation of this compound.

These examples have been selected because they illustrate the dire consequences of handling chemicals whose instability either had not been adequately characterized at the time of the incident or for which available information had not been widely disseminated. The stimuli involved of course were: friction, heat, and impact.

Now that we have seen what can happen and has happened, let's find out how one can determine what materials are apt to be chemically unstable.

17.2 PREDICTING STABILITY

A definitive list of all unstable chemicals would be convenient but modern chemistry deals with too many materials to make such a list practical. It has been estimated that the chemists in this world are creating over one hundred thousand new chemicals every year. The total number compiled by *Chemical Abstracts* by the end of 1977 exceeded four million. But, since in some cases only a few grams or even just a fraction of a gram may be capable of causing considerable damage, some means is needed for screening these materials to determine those which may be potentially unstable.

Characterization by Analogy

Case Histories of Accidents in the Chemical Industry, Volumes 1–4, issued by the Manufacturing Chemists' Association,[6] contain an indexed compilation of over 2000 accidents. Many of these illustrate the causes and the consequences of unsuspected instability. The Committee on Chemicals and Explosives of the National Fire Protection Association[7] have been compiling data on hazardous chemicals for many years. The data is published in NFPA 49, "Hazardous Chemical Reactions." Another compendium, now in its fifth edition, is that of N. I. Sax, *Dangerous Properties of Industrial Materials*.[8]

Careful study of these and similar documents discloses that certain structural elements appear quite frequently. Thus the experienced chemist has learned to associate the structures shown in Table 17.1 with potential instability. Unfortunately the degree of hazard cannot be ascertained simply by noting that one or more of these groups may be present. The table does, however, assist in screening out those chemicals which should be investigated further before embarking upon large scale production.

Characterization via Thermodynamics

Since a study of the literature had shown that certain structural groups might be indicative of instability, and since physical chemists had been able to determine the energy associated with most of the structural groups commonly encountered, it seemed reasonable to surmise that thermodynamics should be helpful in characterizing stability. Thus the 1960s saw numerous individuals and groups exploring this premise.

TABLE 17.1.   Structural Groupings Indicative of Potential Instability

Acetylide	$-C{\equiv}C-METAL$
Amine Oxide	$\overset{\oplus}{\equiv}N-\overset{\ominus}{O}$
Azide	$-\overset{\oplus}{N}{=}N{=}\overset{\ominus}{N}$
Chlorate	$-ClO_3$
Diazo	$-N{=}N-$
Diazonium	$(-\overset{\oplus}{N}{\equiv}N)\ \overset{\ominus}{X}$
Fulminate	$-O-N{=}C^a$
N-Haloamine	$-N\overset{\displaystyle Cl}{\underset{\displaystyle X}{{<}}}$
Hydroperoxide	$-O-O-H$
Hypohalite	$-O-X$
Nitrate	$-O-NO_2$
Nitrite	$-O-NO$
Nitro	$-NO_2$
Nitroso	$-NO$
Ozonide	$-O\overset{\displaystyle}{\underset{\displaystyle O}{\diagdown\diagup}}O-$
Peracid	$-\overset{}{\underset{\displaystyle\overset{\|}{O}}{C}}-O-O-H$
Perchlorate	$-ClO_4$
Peroxide	$-O-O-$

*a*Structure uncertain.

   The laws of thermodynamics state that a reaction once initiated will proceed spontaneously from left to right (as chemical equations are normally written) *if the difference in free energy,* ΔG, between the reactants and the products is negative, that is:

$$\Delta G = (\Sigma H_{f_p} - \Sigma H_{f_r}) - T\,(\Sigma\,S_p - \Sigma S_r)$$

or

$$\Delta G = \Delta H - T\Delta S$$

where H_f is the enthalpy of formation, kJ; S the entropy, kJ/degrees Kelvin; T the absolute temperature, degrees Kelvin; p signifies products; and r signifies reactants. At ordinary temperatures the $T\Delta S$ term is often quite small since ΔS is commonly only a few tenths of a kilojoule per degree. On the other hand ΔH can be quite large in comparison frequently amounting to many kilojoules per mol. Thus only when the magnitude of ΔH is small does the effect of $T\Delta S$ become important. Being interested primarily in potentially hazardous reactions, attention is usually restricted to those cases where ΔH is negative; that is, the reaction under consideration is exothermic.

Because laboratory determinations of changes in free energy, ΔG, enthalpy, ΔH, and entropy, ΔS, are time consuming, they usually cannot be justified. However, by the mid 1960s it became apparent to numerous investigators that these changes could be predicted with considerable accuracy. Using the second-order additivity method developed by Dr. Sidney W. Benson, of the Stanford Research Institute,[9] ASTM Committee E-27 on Hazard Potential of Chemicals[10] developed the computer program CHETAH,[11] published in 1974. Other programs such as TIGER,[12] RUBY,[13] and NASA[14] had appeared earlier but were being used primarily by the military and the explosives industry. CHETAH, however, has become the tool of the chemical industry. Designed primarily to screen organic chemicals for potential instability, the program predicts from a knowledge of only the chemical formula and structure of a compound the maximum theoretical energy, ΔH_{max}, that may be released by the decomposition of that compound. This is the worst case which is seldom realized in real life because the CHETAH computation does not consider kinetics nor realistic equilibrium conditions. Nevertheless, the program is extremely helpful when used properly—as a screening tool.

For the thermodynamicist and the engineer, CHETAH provides excellent estimates of the enthalpy, entropy, and heat capacities from 300 to 1500K for a virtually unlimited number of organic and organometallic compounds. It also computes the net change in enthalpy, entropy, and free energy for balanced chemical equations.

By using CHETAH as a screening device, one is able to ascertain those chemicals and reactions which have little or no chance of explosive decomposition without ever running the reaction or having an experimental sample in hand. Thus one is alerted as to when special precautions are not warranted and is then able to devote available staff and laboratory time to the study of those chemicals and reactions which may be hazardous, that is, capable of the sudden release of energy.

By correlating the computations for approximately 200 compounds and mixtures which included known explosives, mixtures known to have caused explosions, and compounds considered to be nonhazardous, it has been found that a reaction with a maximum theoretical change in enthalpy, ΔH_{max} greater than 0.7 kcal/g can be potentially explosive. Thus CHETAH characterizes the potential hazard of a system as either *high*, *medium*, or *low* depending upon the value of the calculated heat of decomposition, ΔH_D, or reaction, ΔH_R; i.e.,

$$
\begin{aligned}
\text{high} \qquad & \Delta H_D > 0.7 \text{ kcal/g} \\
\text{medium:} \qquad & \Delta H_D < 0.7 - > 0.3 \text{ kcal/g} \\
\text{low:} \qquad & \Delta H_D < 0.3 \text{ kcal/g}
\end{aligned}
$$

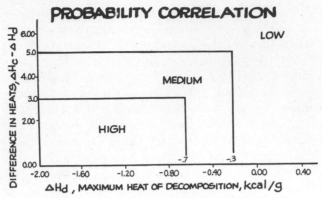

Figure 17.1. Probability correlation for CHETAH.

Recent work by Dr. D. R. Stull of Dow Chemical Company[15] seems to show that the potential hazard and possibly also the probability of this energy release may be indicated by correlating the difference between the heat of combustion in excess oxygen (computable by CHETAH) and the heat of decomposition, $\Delta H_C - \Delta H_D$, with the heat of decomposition, ΔH_D. Thus CHETAH also rates the potential hazard as *high, medium,* or *low* depending upon where the computed value of $\Delta H_C - \Delta H_D$ falls upon the diagram shown in Fig. 17.1.

When the computations for some 200+ bulk chemicals of interest to the U.S. Coast Guard are examined, it is found that almost all of the bulk compounds in water shipments and their binary mixtures may be considered *low* order. Some individual compounds and binary mixtures may produce heats of reaction capable of generating vapor pressures sufficient to rupture containers, but they do not seem to indicate a potential for explosive decomposition. The CHETAH program is not currently capable of further ordering the spectrum of hazard at this energy level. However, there are indications that other computational approaches may be successful in the future.

When and if the day ever arrives that one is able to predict kinetics with reasonable accuracy, then the engineer will be able to ascertain the actual hazard posed by any given set of circumstances. In the meantime, CHETAH can be used to determine which materials and reactions need further study. Fortunately CHETAH does not "cry wolf" too often.

17.3 TESTING FOR STABILITY

Having decided either by analogy or from thermodynamic computations that a chemical may be potentially unstable, one must then find a way to prove or disprove this conclusion. Currently this can only be done by experimentation.

To fully define the hazards posed by a potentially unstable compound under conditions of manufacture, use, storage, and transport requires extensive kinetic determinations. These are difficult, time consuming, and expensive and warranted only in relatively few instances. Thus for the majority of compounds minimization of the hazard is accomplished by avoiding conditions conducive to instability. Such conditions are fairly readily ascertained by a few relatively simple tests.

Thermal Stability

Since few chemicals exist naturally in a readily usable state, they must be made or processed under conditions quite different from the normal environment—usually at elevated temperatures. Therefore it is important to establish the stability of a chemical to both normal and abnormal process conditions—upsets do occur all too frequently!

Over the years many methods for testing for thermal stability have been developed using simple laboratory equipment. In essence, the tests depend on the facts that practically all hazardous reactions involve rapid exotherms and that the rates of all reactions increase with temperature. Accordingly, if one exposes a sample (single component or a mixture) to a progressively increasing temperature and explores the extremes of temperatures potentially available, one can feel reasonably sure that the sample is stable if no exothermic response is noted.

DIFFERENTIAL THERMAL ANALYSIS (DTA)

With the ready availability of modern thermal analyzers (e.g., Mettler, Du-Pont, Perkin-Elmer, etc.), the test has been standardized by ASTM Committee E-27. Utilizing a 5-mg sample and a heating rate between 10°C/min and 30°C/min the temperature range of interest may be rapidly explored. Accuracy is not critical since exothermic responses are rate dependent as is shown in Fig. 17.2. The objective of the test, ASTM E-537,[16] is to quickly determine whether or not the sample will produce an exotherm and to approximate the onset temperature of that exotherm. The magnitude (height) and sharpness (half-width) may be given some significance but only qualitatively. For besides shifting the onset of decomposition to higher and higher temperatures, the use of increasing heating rates also increases the sensitivity of response to the exothermic reaction. Table 17.2, taken from Maycock and Vernecker's studies of the decomposition of lead azide,[17] clearly shows both these phenomena.

That mass can also affect the response, possibly by changing the kinetics and/or reaction mechanism, is illustrated by Table 17.3 also taken from Maycock and Vernecker's studies. If necessary to differentiate between instability and a possible oxidation reaction, e.g., autoignition, the test should be rerun under an inert atmosphere.

Because of rate and mass dependency, the temperature corresponding to

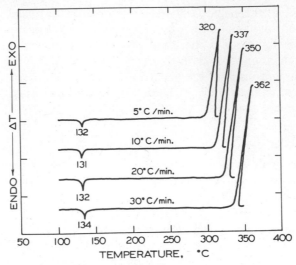

Figure 17.2.   Effect of heating rate on exothermic responses (Eastone Orange).

the initiation of the exotherm may be significantly higher than that which may be encountered in practice. Thus constant temperature stability (CTS) studies are recommended for those materials exhibiting exotherms. According to the standardized ASTM test, E-487[18], a small sample is held at some constant temperature below the temperature of initiation as approximated by the DTA curve from E-537. If an exotherm is found, the test is repeated at lower and lower temperatures until a temperature level is found which will not produce an exotherm in a 2-hr test period.

Further tests are normally not essential since it has been shown that if the temperature of onset is plotted versus delay time, the curve becomes asymptotic to a value that is closely approximated by the 2-hr test. Moreover, a plot of the logarithm of the delay time, seconds, versus the reciprocal of the absolute temperature, degrees K, should produce a straight line enabling one to estimate the maximum temperature to which the material should be subjected for any specified period of time.

TABLE 17.2.[17] Detonation of PbN_6 as a Function of Critical Heating Rate[a]

Heating Rate (°C/min)	Visible Criterion	Mass Loss (mg)	Sensitivity (height/ half width)	Inference Temperature of Reaction (°C)
10	Decomposition	0.50	10/3.5 = 2.9	<1515
15	Decomposition	0.55	15/2.0 = 7.5	<1515
25	Detonation	2.0	25/1.0 = 25.0	>1515

[a]Sample mass: 2.0 mg; Atmosphere: flowing He 10 liters/h.

TABLE 17.3.   Detonation as a Function of Critical Mass of PbN$_6^a$

Mass (mg)	Visual Criterion	Mass Loss (mg)	Inference Temperature of Reaction (°C)
1.0	Decomposition	0.25	<1515
1.6	Decomposition	0.35	<1515
2.0	Decomposition	0.50	<1515
3.0	Detonation	3.0	>1515
4.0	Violent detonation	4.0	>1515

^{a}Heating rate: 15°C/min; Atmosphere: flowing He 10 liter/h.

Should the DTA results indicate an endotherm closely followed by an exotherm as shown in Fig. 17.3*a*, then it is often helpful to repeat the test at a slower heating rate. Figure 17.3*b* shows what may be hidden by too high a heating rate. In Fig. 17.3*a* the sample melts, starts to decompose around 250°C but then starts to boil until finally the exothermic decomposition predominates.

In Fig. 17.3*b* the heating rate is slow enough so that closer attainment to thermal equilibrium permits the sample to decompose prior to attaining the boiling point. Similar results may be obtained by pressurizing the sample so that boiling is suppressed.

CONFINEMENT TEST

Just because DTA analysis may not indicate any exothermic reactions does not necessarily eliminate the possibility of a potential hazard. In the previous example, Fig. 17.3*a*, the exotherm was almost missed due to loss of sample.

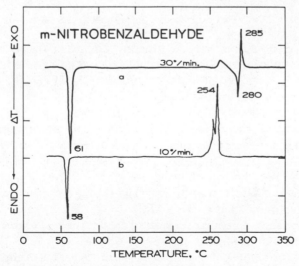

Figure 17.3.   Differential thermal analysis of m-nitrobenzaldehyde.

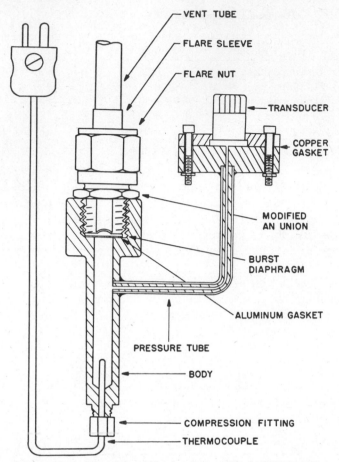

Figure 17.4.   Thermal stability bomb assembly (ASTM E-476).

This possibility should be suspected for low boiling materials in particular. Propargyl bromide, for example, boils at 84°C and no exotherms are noted. If heated while pressurized or under confinement, however, the material will be found to decompose violently around 190°C.

Furthermore, thermal analysis equipment normally gives no indication of the amount of gaseous products formed, their rate of formation, nor the potential pressure development in a confined space. In order to provide more quantitative data various so-called "confinement" tests have been developed. The best known of these are (1) the "pressure-tube test" developed by the U.S. Bureau of Mines[19] and (2) the "thermal stability test" (Test No. 6) recommended by JANAF.[20]

ASTM has issued a refined version of these two tests which should be simpler to use and easier to interpret, ASTM E476-73.[21] All the tests involve heating a sample in a confined space and measuring temperature and/or pressure responses as a function of bath temperature. Figure 17.4 illustrates the bomb assembly recommended by ASTM E-476.

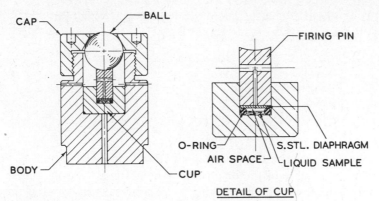

Figure 17.5. Olin drop-weight tester: sample holder (ASTM D-2540).

Shock Sensitivity

It has been found that if a material is thermally sensitive and capable of exothermic decomposition, it will also be sensitive to mechanical shock or friction provided that the stimulus is of sufficient magnitude. Because potentially unstable chemicals must often be pulverized, blended, pumped, and otherwise roughly treated during processing and transport, their reactions to these mechanical stimuli must be known. Thus many, many tests have been established for the determination of impact sensitivity. Only a few of these will be mentioned here.

HAMMER TEST

The simplest test for shock sensitivity is the hammer test. As the name implies, it is simply observing the response of a sample placed upon a suitable anvil to a sharp hammer blow. It is purely qualitative in nature and only positive results are dependable.

DROP-WEIGHT IMPACT

Next in sophistication is the drop-weight impact test. For liquid samples the tester most commonly used is the so-called Olin tester described in ASTM D2540-70.[22] The sample holder used in this test is shown in Figure 17.5. Typical results are given in Table 17.4.

For the testing of solid materials many devices have been built and used. Twenty-six of the thirty-eight sensitivity tests compiled by TTCP Panel 0-2, Working Group on Sensitiveness,[23] describe mechanisms for determining impact sensitivity. Probably no two of these give the same results although attempts to develop a normalizing factor look promising. They do however provide relative values of sensitivity which are quite valuable.

A modified version of the Bureau of Mines tester proposed by J. E. Guillet and M. F. Meyer[24] has been found to be quite helpful. The modification consists of the use of a gas buret to measure the amount of gas produced by

TABLE 17.4.   Liquid Propellant Drop Weight Test:
ASTM D2540

Sample	50% Level (2 kg weight)
Nitroglycerine	<1 cm
Propargyl Bromide	<2 cm
Iso-propyl Nitrate	3.5 cm
Nitromethane	44.5 cm

the decomposition. The combination of drop height, gas volume, and cup damage are indicative of the hazard. The height is always indicative of the sensitivity—the lower the height the greater the sensitivity—but the actual hazard is depicted by the other variables. Some possible combinations are (arranged in order of increasing potential hazard):

1. Little gas—no cup damage.
2. Lots of gas—no cup damage.
3. Little gas—much cup damage.
4. Lots of gas—much cup damage.

The amount of cup damage is dependent upon the rate of gas evolution and not upon the amount of gas evolved. In many cases, it will be found that shock-sensitive materials contain one of the structural groups depicted earlier as characterizing potentially explosive materials.

Card Gap Test

Impact levels greater than those available by dropweight are normally obtained by the use of explosives. The best known of these tests is the card gap test described in ASTM D2539.[25] More recent versions are instrumented to measure critical diameter and detonation velocity.[26] Although described for use with liquids, the tests require only minor changes for use with solid samples.

Trauzel Block Test

Since the drop-weight test has been designed to indicate sensitivity primarily, other tests have been devised to depict yield or the magnitude of the response to impact. The Trauzel or lead block test[27] measures the expansion of a cavity in a lead block from a sample initiated by a blasting cap. The greater the expansion of the cavity, the greater is the yield.

Lagoon Test

Another test which does not measure sensitivity but merely yield is the "lagoon test." Here, a sample (500 g) is encased in a plastic container together with 150 g of tetryl and a blasting cap initiator and immersed in a body of water. The yield is measured by the deflections produced in witness plates suspended at equal distances around the sample at a distance of 3–5 ft. In other versions, the pressure wave at some specified distance from the sample is measured. The results are correlated against the effects produced by known explosives and are usually expressed as TNT equivalents.

Under development by ASTM Committee E-27 is a test which may supplant both the Trauzel block test and the lagoon test. Tentatively being called the "heavy confinement test" it measures the height to which a steel block is thrown by a 4-g sample placed in a steel cavity and impacted by the explosive force of a #8 blasting cap.

17.4   CLASSIFICATION OF HAZARD

Simply because of the great variety of chemicals known today, it is to be expected that they run the gamut of hazard from completely innocuous materials to materials of extreme sensitivity capable of producing tremendous damage, e.g., water to nitroglycerine. Since it is impractical for the engineer to design for an infinite number of degrees of hazard, some means of classification is essential. Some means must be available to divide the spectrum of potential hazard into a manageable number of groups—each group to cover a range of hazards for which general design requirements can be established. Remember, in this instance only the potential hazard due to instability, i.e., the potential to release energy suddenly, is under consideration.

Using results from the three basic stability tests: computation, thermal analysis (DTA), and impact sensitivity (drop-weight), it can be shown that the spectrum may be conveniently divided into eight groups which upon further analysis may be reduced to three—representing low, medium, and high potential hazards.

Preliminary Groupings

Table 17.5 shows that using only positive or negative responses from the three different stability tests, eight combinations of results are possible. The table has been arranged so that the potential explosion hazard or the degree of instability increases as one proceeds down the table. How to use and interpret the data represented is best illustrated by presenting and discussing several typical examples.

COMBINATION NO. 1 $(-, -, -)$

Based on long experience, materials such as acetone, alcohol, benzene, sugar, starch etc., are considered stable. They show no exothermic activity, are not impact sensitive, and give values of $-\Delta H_d$ less than 0.7 kcal/g. Thus

TABLE 17.5.   Classification System

Combination Number	Heat of Reaction	Thermal Stability	Impact Sensitivity
1	−	−	−
2	−	−	+
3	−	+	−
4	+	−	−
5	+	+	−
6	+	−	+
7	−	+	+
8	+	+	+

Code

Heat of Reaction, $-\Delta H_d$:	− ,	<0.7 kcal/g
(CHETAH)	+ ,	>0.7 kcal/g
Thermal Stability:	− ,	no exotherm
(DTA)	+ ,	exotherm
Impact Sensitivity:	− ,	insensitive at 550 in.-lb (solids)
(Drop-Weight)		or 100 kg-cm (liquids)
	+ ,	sensitive at 550 in.-lb or
		100 kg-cm[a]

[a]Maximum values of impact for Guillet and Meyer solids tester[24] and Olin liquid tester (ASTM 2540).

this group heads the list and represents the least hazardous materials. Fortunately, the major portion of our common chemicals fall within this group.

However, being organic materials they are not without hazard. They will burn and are capable of potentially severe explosions when their vapors or dusts are admixed with air in suitable proportions.

COMBINATION No. 2 (−,−,+)

When computation indicates a low-order hazard and DTA shows no exothermic response, a positive impact test normally indicates an endothermic decomposition. Such a decomposition cannot propagate and is essentially innocuous. A typical example would be sodium bicarbonate which can be made to partially decompose under impact with the evolution of carbon dioxide.

In some instances, however, a positive impact test may indicate a reaction between the sample and the confining cup or sample holder.

COMBINATION No. 3 (−,+,−)

An exothermic response for a material exhibiting no impact sensitivity and having a low-order computer classification usually results from inadequacies within the computer program or from sample impurities. The combination occurs with sufficient frequency to warrant attention and in many cases serves as a warning flag for what may be an unsuspected hazard. Dihydrofuran falls in this group. Samples often indicate a sharp exotherm soon after the boiling endotherm at 72°C. Since the dihydrofuran should have all evaporated at 72°C, the exotherm has to be due to an unstable residue–in this case a peroxide. It has been found that this combination is frequently indicative of potential peroxide formers. Thus although such materials may be handled in relative safety, special precautions are needed during distillation or purification.

COMBINATION NO. 4 (+, −, −)

A high computer classification, accompanied by neither a positive DTA result nor a positive impact test does not occur very often. However, it must be considered real until proven otherwise. It is often indicative of instability at temperatures above the normal boiling point and/or at high impact levels. For example, 6-chloro-2,4-dinitroaniline exhibits no exothermic instability by routine DTA analysis, is not impact sensitive at 550 in.-lb, but does give a ΔH_d value in the high hazard area. Subsequent testing shows an exothermic response when heated under confinement and when tested by the "lagoon test" detonates with a yield equivalent to 50% that of an equal weight TNT.

Thus this combination warns that further and extensive testing is required before the material may be safely relegated to one of the stable groups.

COMBINATION NO. 5 (+, +,−)

A combination of +, +, − is most often associated with solid materials which may be classed as "high explosives." Such materials are thermally unstable, but often at relatively high temperatures, and are capable of releasing large amounts of energy instantaneously under the proper conditions. Fortunately, the necessary conditions are not normally encountered except intentionally.

These materials are impact sensitive but not to the standard drop-weight test. They require an impetus much greater than that to be expected from normal handling or processing conditions.

COMBINATION NO. 6 (+, −, +)

Usually it will be found that the sample is a liquid or a low melting solid when this combination is encountered. It does not mean that exothermic decomposition is not possible but only that it is not noticeable at atmospheric pressure.

The exotherm occurs at temperatures above the atmospheric boiling point. Again, further testing is indicated. Propargyl bromide is a typical example. When heated under confinement at about 150 lb psia and a temperature of 180°C, the liquid decomposes violently and may detonate.

COMBINATION NOS. 7 (−, +, +) AND 8 (+,+,+)

Whenever a material is found to be both thermally unstable and shock sensitive, it should be treated with utmost caution until larger scale tests have more clearly defined the degree of potential hazard. It will be found that azides, diazos, peroxides, perchlorates, nitro compounds and the like, fall into this group. The negative computer value may be due solely to a fault of the computer program, e.g., azides do not seem to indicate a hazard by CHETAH as currently written.

Final Classification

Further analysis of the groupings described in Preliminary Groupings shows that, in general, chemicals in groups 1, 2, and 3 are essentially stable and incapable of releasing energy suddenly; chemicals in groups 4 and 5 are capable of high energy release but are difficult to initiate; while groups 6, 7, and 8 represent chemicals which are sensitive both thermally and mechanically and which are capable of high energy release. Thus the spectrum of instability has been divided into three groups established by the results of three sensitivity tests and representing potentially low, moderate, or high hazards.

These three logical groupings are now amenable to the establishment of general design requirements. For specific design criteria additional tests are required to define hazards related to conditions of manufacture, storage and transport. Such tests include the determination of flash points, flammable limits, autoignition temperature, dust explosibility etc., as well as a comprehensive evaluation of toxicity and health hazards.

17.5   CONCLUSIONS

Because of the great number of chemicals in use or under development today, some means is needed to screen them for potential instability. This can be accomplished via thermodynamic computations, e.g., CHETAH. Then subsequent testing for thermal stability and impact sensitivity enables one to divide the spectrum of stability into either eight or three groups amenable to the establishment of general design criteria for safe manufacture, storage, or transport.

REFERENCES

1. *Code of Federal Regulations* Title 29, Chapter XVII, Part 1910.
2a. *NFPA Quarterly*, 62–70 (July 1962). [Available from NFPA, Batterymarch Park, Quincy, Mass. 02269.]
2b. V. K. Mohan, K. R. Becker, and J. E. Hay, "Hazard Evaluation of Organic Peroxides," *J. Hazard. Materials*, 5(3), 197–220 (Feb. 1982).

3. "Nitroparaffins and Their Hazards," NBFU Research Report No. 12 (1959), [Available from American Insurance Assoc., 85 John Street, New York, N.Y.].

4. *Newark Evening News,* January 29, 1964, p. 1.

5. *Houston Post,* December 23, 1966, p. 1.

6. Chemical Manufacturers Association, 2501 M St., N.W., Washington, D.C. 20037.

7. National Fire Protection Association, Batterymarch Park, Quincy, Mass. 02269.

8. N.I. Sax, *Dangerous Properties of Industrial Materials,* 5th ed., Van Nostrand Reinhold Co., New York, 1979.

9. S.W. Benson, *Thermochemical Kinetics,* Wiley, New York, 1968.

10. American Society for Testing and Materials, 1916 Race Street, Philadelphia, Pa. 19103.

11. *CHETAH-The ASTM Chemical Thermodynamic and Energy Release Evaluation Program,* DS51, American Society for Testing and Materials, 1974.

12. W.E. Wiebenson Jr., W.H. Zwisler, L.B. Seeby, and S.R. Brinkley Jr., "Tiger Computer Program Documentation," Final Report to Army Ballistics Research Laboratories, Aberdeen, Maryland, November 1968.

13. H. G. Levine, and R. E. Sharples, "Operator's Manual for Ruby," Lawrence Radiation Laboratory Report UCRL-6815 (1962), Livermore, CA 94550.

14. S. Gordon and B. J. McBride, "Computer Program for Calculation of Complex Equilibrium Compositions, Rocket Performance, Incident and Reflected Shocks, and Chapman-Jouquet Detonations," NASA SP-273, National Aeronautics and Space Administration, 1971.

15. D.R. Stull, "Identification of Reaction Hazards," *Loss Prev., 4* (1970).

16. ASTM E537-76, "Assessing the Thermal Stability of Chemicals by Methods of Differential Thermal Analysis."

17. J.N. Maycock and V.R.P. Vernecker, "Simultaneous Differential Thermal Analysis-Thermogravimetric Analysis Technique to Characterize the Explosivity of Lead Azide," *Anal. Chem., 40* (8), 1325–1329 (1968).

18. ASTM E487-79, "Test for Constant-Temperature Stability of Chemical Materials."

19. R.W. Van Dolah, "Evaluating the Explosive Character of Chemicals," *Ind. Eng. Chem., 53*(7), 59A–62A (July 1961).

20. "Liquid Propellant Test Methods, Recommended by the Joint Army-Navy-Air Force Panel on Liquid Propellant Test Methods," the Liquid Propellant Information Agency, Johns Hopkins University, Silver Spring, Md.

21. ASTM E476-73, "Test for Thermal Instability of Confined Condensed Phase Systems."

22. ASTM D2540-70, "Test for Drop Weight Sensitivity of Liquid Monopropellants."

23. "Manual of Sensitiveness Tests," published by Canadian Armament Research and Development Establishment on behalf of TTCP Panel 0-2, Working Group on Sensitiveness, 1966.

24. J.E. Guillet and M.F. Meyers, "Determining Shock Sensitivity of Liquid Organic Peroxides," *I & EC Prod. Res. Dev.,* (1) 226–30 (December 1962).

25. ASTM D2539, "Test for Shock Sensitivity of Liquid Monopropellants by the Card Gap Test."

26. ASTM D2541, "Test for Critical Diameter and Detonation Velocity of Liquid Monopropellants."

18

Cool Flames

Robert D. Coffee, Ph.D.

In this chapter on the hazards accompanying the handling of organic compounds we shall discuss "cool flames." However, in order to comprehend cool flames we must also understand "autoignition." The two terms are definitely related and in reality are merely two different manifestations of the same process: Oxidation.

Both cool flames and autoignitions are the "visible" phenomena accompanying the vapor phase oxidation of organic compounds in air or oxygen.

Autoignition may be defined as "the spontaneous ignition of a vapor–air mixture in the absence of an obvious source of ignition such as a spark or a flame." The autoignition temperature, or AIT, is the lowest temperature at which such an ignition will occur. As we shall see later, it is dependent on composition, pressure, time, and reaction vessel volume.

In general (and most noticeable for homologous series) the higher the molecular weight of a compound the lower will be its autoignition temperature under similar conditions (see Table 18.1).

18.1. COOL FLAMES

What is a cool flame? A cool flame is the "visual" phenomenon associated with the *low* temperature oxidation of an organic material. It is accompanied by very small temperature and pressure changes compared to those accompanying normal, or hot flame, ignitions and is often referred to as a partial or intermediate oxidation reaction (see Table 18.2).

For any given system ΔT_{cf} is greatest at high pressures but low starting temperatures (T_o).

TABLE 18.1. AIT versus Molecular Weight

Compound	Molecular Weight	Autoignition Temperature[a]	
Benzene	78.11	594°C	1100°F
Toluene	92.13	568°C	1055°F
o-Xylene	106.16	493°C	920°F
Hexane	86.17	267°C	513°F
Heptane	100.20	227°C	440°F
Methanol	32.04	446°C	835°F
Ethanol	46.07	435°C	815°F
Propanol	60.10	413°C	775°F
Butanol	74.12	362°C	684°F
Dodecanol	186.33	249°C	480°F

[a]ASTM D-2155.

It should be particularly noted that a cool flame

1. May occur at temperatures several hundred degrees lower than the AIT for the same compound. Typical values are given in Table 18.3.
2. Can not be seen in a lighted room.
3. Gives the appearance of a pale-blue luminescence when viewed in the dark with good dark vision adaptation. The luminescence is presumably caused by the formation of excited formaldehyde molecules resulting from the explosive decomposition of hydroperoxides.)

History[1]

Cool flames were first discovered by Humphrey Davy[2] in 1812 and observed by several other workers during the nineteenth century. Perkin[3] found that cool flames could be obtained from a wide variety of hydrocarbons, alcohols, aldehydes, acids, oils, and waxes.

The phenomena was most clearly demonstrated with diethyl ether and acetaldehyde. Cool flames could be initiated by holding a metal ball heated to 100–300°C in the vapor of the substance (in air). The flames were quite cool and did not char paper. They emitted a pale blue light visible only in a darkened room and produced quantities of partial oxidation products including aldehydes and acids.

TABLE 18.2. Flame Characteristics

	Cool Flame	Autoignition
ΔT	10–200°C	800–2000°C
P_f/P_0	low (<2)	6–10 (in confined spaces)
Products	CH_2O; CO	H_2O; CO_2

Little further work was carried out until the 1930s when the phenomenon was rediscovered by Prettre[4] and Townend.[5] Much work has been done since but unfortunately primarily to establish theory.[1,6]

Significance

Why are *we* interested in these phenomena? Our interest concerns the assessment of the safety of high-temperature chemical processes where unexpected fires and explosions may be quite hazardous. Earlier investigations were concerned with controlling engine knock in internal combustion engines.[6] More recently the Naval Research Laboratories became quite involved when the wing tanks of high speed jets started blowing up.

It seems that at speeds around mach 2, the leading edges of the wing tanks become quite hot—hot enough to initiate a cool flame reaction with some fuels. Then upon rapid descent to lower altitudes and higher pressures there frequently was a transition from cool flame to normal ignition resulting in an explosion and the loss of the tank.

Currently only a few laboratories are seriously studying the problem from a safety viewpoint: The Naval Research Laboratories, Bureau of Mines, the Technical Safety Research Lab (EK Co.), and perhaps one or two others. Thus the literature is essentially devoid of any practical data. Yet many previously unexplained industrial fires and explosions may be attributed to the formation of a cool flame with subsequent transition to hot ignition.

18.2. PREDICTIONS

First of all, when we realize that cool flames and autoignitions are associated with vapor-phase oxidation reactions and when we learn that all such oxidation reactions are exothermic in nature, we find that we should be able to make some general predictions.

Because the reactions are exothermic, the associated phenomena must be dependent upon heat transfer to the surroundings. Thus one should suspect that these phenomena should:

1. Be a function of the vapor/air ratio or composition.
2. Occur at lower temperatures for larger volumes due to the insulating effects of large masses (i.e., heat transfer is a function of thickness).
3. Initiate at lower temperatures at higher pressures because higher pressures mean more reactants and thus more heat generation per unit volume.

18.3. THEORY

But, before we take a look at some experimental data to find out how good our predictions are, let us review a few more *theoretical* concepts.

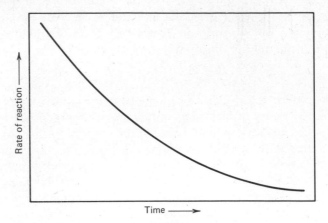

Figure 18.1. First-order reaction under isothermal conditions.

In the basic theory of thermal explosion, heating is the sole cause of the progressive increase in the rate of a chemical reaction and thus the rate of heat evolution. From the point of view of this theory, chemical transformations in themselves are only inhibiting factors since the rate of a chemical reaction is proportional to the concentrations of reactants and therefore must *tend* to decrease as the starting materials are consumed. However, there apparently is a large class of chemical reactions, including all oxidation reactions of hydrocarbons with air or with oxygen, whose rates increase up to a certain limit as the reaction products accumulate. Such a reaction is called autocatalytic.

Thus under isothermal conditions a first-order reaction will slow down with time as the starting materials are consumed (see Fig. 18.1) while an autocatalytic reaction will speed up with time and become more difficult to

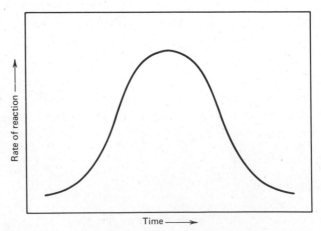

Figure 18.2. Autocatalytic reaction under isothermal conditions.

maintain at isothermal conditions. At some point in time, however, the rate will have to decrease due to the consumption of materials (see Fig. 18.2). In many cases this point of inflection occurs when about 50% of the starting ingredients have reacted.

Theory also states that for "first-order reactions," the rates are proportional to pressure and to temperature.

To summarize then, it seems that for the phenomena we are discussing, reaction rates should increase not only with increases in temperature and pressure but also with time (autocatalytic). Under adiabatic conditions at a sufficiently high initial reaction rate this should result in a runaway thermal explosion. Experience has shown that it often does!

18.4. EXPERIMENTAL

Cool Flames

Experimental studies have shown that at low temperatures the vapor-phase oxidation of hydrocarbons proceeds by the formation of hydroperoxides. At some concentration of peroxide a cool flame initiates and there is a jump in temperature but a reduction in rate since the hydroperoxide is now consumed faster than it is formed. Under suitable heat-transfer conditions, if the temperature drops, multiple cool flames may be noted. If the temperature continues to rise due to heat generation, the rate will reverse itself again and ultimately result in ignition (see Fig. 18.3). Cool flames seem only to occur for those reactions which possess a negative temperature coefficient at some point in time.

These studies have shown that formaldehyde is produced by the decom-

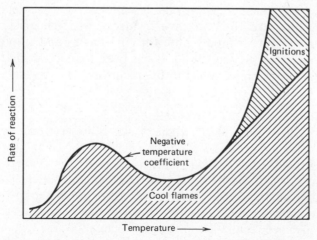

Figure 18.3. Typical reaction rate versus temperature diagram.

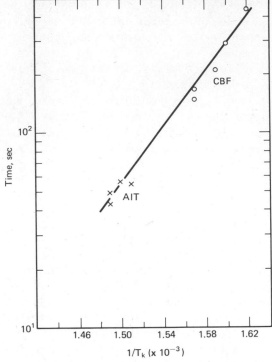

Figure 18.4. Delay time versus ignition temperature for ethylene glycol (5000-ml flask).

position of the peroxide radical. It is these excited formaldehyde molecules that are responsible for the chemiluminescence of the cool flames.

Since the cool flame does not initiate until a suitable concentration of hydroperoxides is reached and since the reaction rate is proportional to temperature, one would conclude that the time to the appearance of the cool flame (delay time) would be an inverse function of temperature. This has been confirmed experimentally both for cool flames and autoignition temperatures (see Figs. 18.4 and 18.5). It has been found that the logarithm of the delay time (τ) is proportional to the reciprocal of the absolute temperature; i.e.,

$$\log \tau = \frac{A}{T} + B, \text{ where A and B are constants}$$

Since the ignition delay vs. temperature curve passes through the experimental points for both the hot- and cool-flame ignition zones, it suggests that the cool-flame and hot-flame ignitions are part of a "two-stage" process and that the cool flame is the controlling factor in low temperature ignition delays. It has also been found[6] that at constant temperature, T_o, the delay time,

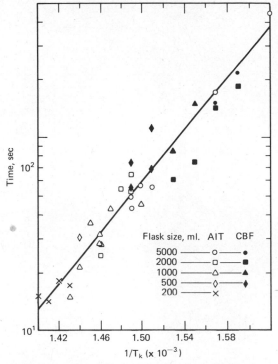

Figure 18.5.   Influence of flask size on delay times for ignition of ethylene glycol.

τ_{cf}, becomes shorter as the pressure, P_o, is increased, obeying the equation

$$\tau_{cf} = KP_o^n + c$$

where n is negative and K and c are constants. This equation holds both at low and at high pressures.

Composition

It does not take much experimental work to learn that both cool flames and autoignitions are quite dependent upon composition. The standard test methods—ASTM D-286, D-2155, and D-2883—all indicate that the composition must be varied to obtain the minimum temperature for response. One also soon discovers that autoignitions cannot occur unless the composition lies within the flammable range as defined by other tests (see Fig. 18.6). However, both cool flames and hot flames can occur at concentrations above the upper flammable limit as determined by conventional spark or flame ignition. This is shown by Figure 18.7, which illustrates the influence of pressure on the flammable ranges of hexane-air mixtures.[7]

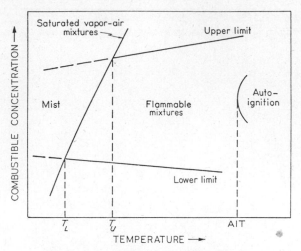

Figure 18.6. Typical flammability diagram.

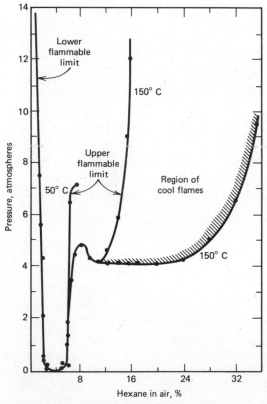

Figure 18.7. Effect of temperature and pressure on the flammability of hexane in air.

Normal flames were obtained up to 4 atmospheres, but at 4.1 atmospheres and 150°C a "cool flame" appeared with mixtures from 11 to 22 percent of hexane, the normal flame being limited to mixtures from 1.2 to 7.0 percent. At pressures above 4.8 atmospheres the two ranges met; for example, at 6.5 atmospheres mixtures of 1 to 14.5 percent of hexane propagated normal flames and 14.5 to 32 percent propagated "cool" flames. Flames in mixtures up to 6.5 percent were whitish or yellowish, 6.5 to 11 percent orange or reddish with deposition of carbon or tar, and beyond 11 percent blue without carbon deposition, but with formation of much aldehyde.[7]

The general form of the diagram is typical. Similar diagrams were also generated by Hsieh and Townend[7] for ether-air mixtures at reduced pressures.

Volume

Many investigators have studied the effect of reaction vessel size on autoignition temperatures. N. P. Setchkin[8] (National Bureau of Standards) early showed that as the reaction vessel became larger, the AIT was lowered.

A very interesting relationship has been discovered recently by A. Beerbower of Exxon Research and Engineering Company. He has found that a plot of AIT versus the logarithm of the reaction vessel volume seems to extrapolate (for many different materials) to a value of approximately 75°C (167°F) at a volume of 10^{15} cm^3 (3.5×10^{10} ft^3 or 2.6×10^{11} gal) (see Fig. 18.8).[9] Data obtained by the TSRL* appear to agree with this concept (see Figs. 18.9 and 18.10).

Thus, knowing the minimum AIT as determined by ASTM D-2883, one can predict the AIT for any other volume (spherical equivalent).

Table 18.4 gives some indication of the change in AIT as the vessel is increased from 200 to 500 ml.

Sufficient experimental data is not available to attempt to try to correlate cool flame temperatures with vessel size except to say that cool flames initiate at lower temperatures as the reaction volume is increased (Fig. 18.10).

It further appears that the ratio of surface area to volume is significant for cool flame initiation since surfaces act as terminating agents for the chain branching reactions dominating the cool flame phenomena.

Pressure

That increasing pressures lower the AIT has been confirmed by numerous investigators. A recent study by the TSRL obtained the following values for glacial acetic acid in 500-ml vessels (see Table 18.5):

The data given in Table 18.5 appear to fit Semenov's equation,[10]

$$\log \frac{P}{T} = \frac{A}{T} + B$$

*Technical Safety Research Laboratory, Eastman Kodak Co, Rochester, N.Y.

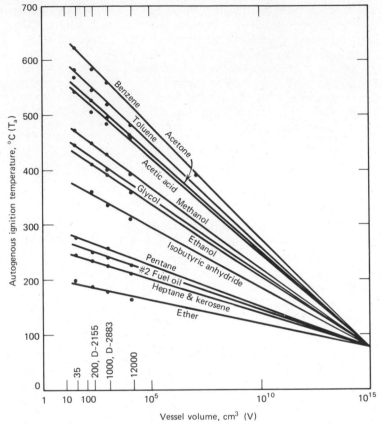

Figure 18.8. Autogeneous ignition temperature versus volume of reaction vessel: $T_D = [12/(15 - \log V)](T_a - 75) + 75$; T_D = AIT by *D*-2883.

quite well over the pressure range tested. The data for acetic acid plotted in Fig. 18.11 yield the equation

$$\log \frac{P}{T} = \frac{3390}{T} - 7.129$$

where P is pressure in atmospheres and T is the temperature in degrees K.

The effect of pressure is further illustrated by the *typical* ignition diagram shown in Fig. 18.12 where it is seen that:

1.   At any pressure greater than P_1 only autoignitions are noted.

2.   At pressures between P_1 and P_2 a cool flame is noted at temperatures below the AIT.

3.   At pressures such as P_3 or P_4 one notes first a temperature where cool flames occur, at some higher temperature ignitions occur, then at a still higher temperature cool flames are again possible, etc.

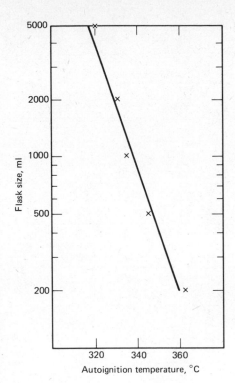

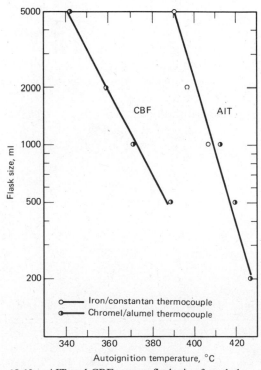

Figure 18.9. AIT versus flask size; isobutyric anhydride.

Figure 18.10. AIT and CBF versus flask size for ethylene glycol.

TABLE 18.3.   Typical Ignition Temperatures

Compound	Ignition Temperature, °C	
	AIT (NFPA 325M)	Cool Flame (EK Co.)
Methyl ethyl ketone	515	265
Methyl iso-butyl ketone	460	245
Iso-propyl alcohol	400	360
n-butyl acetate	420	225

TABLE 18.4.[a]   Effect of Volume on AIT

	AIT			
	D-2155 (200 ml)		D-2155 (500 ml)	
m-Xylene	557°C	1035°F	529°C	985°F
Propionaldehyde	207°C	405°F	199°C	390°F
Ethyl Alcohol (190 proof)	410°C	770°F	388°C	730°F
Butyl Alcohol	354°C	670°F	327°C	620°F
Butyl Ether	177°C	350°F	171°C	340°F
l-Bromopentane	238°C	460°F	229°C	445°F

[a]Data from TSRL.

For a system subjected to an ever increasing temperature, the first ignition will preclude the formation of a cool flame at a higher temperature. These diagrams are developed by rapidly raising the sample temperature to the temperature of interest. If this is done fast enough, one prevents the forma-, tion of cool flames or autoignitions at lower temperatures because of the time delays involved in their formation.

The Naval Research Laboratory[11] has reported systems wherein they have observed as many as eleven alternating zones of cool flame formation and autoignitions. The most observed by the TSRL has been four; e.g., UCON HTF X-600:

Phenomena	Temperature Range	
Cool flame	202–243°C	395–470°F
Autoignition	246–285°C	475–545°F
Cool flame	288–357°C	550–675°F
Autoignition	360°C & above	680°F & above

TABLE 18.5.   Effect of Pressure on AIT of Glacial Acetic Acid

Pressure	AIT (in air)
14.7 psia	524°C (975°F) (ASTM glass flank)
100 psia	378°C (712°F) (stainless steel vessel)
500 psia	299°C (570°F)

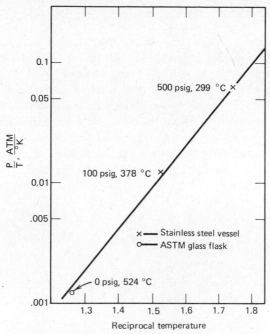

Figure 18.11. AIT of acetic acid versus pressures: equation of line = log P/T = $3390/T$ − 7.129.

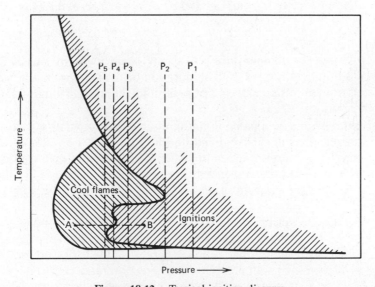

Figure 18.12. Typical ignition diagram.

18.5. TRANSITIONS

Since cool flames generate only minor temperature and pressure changes, they in themselves do not present a significant hazard. However, if there is a transition to ignition, then "all hell breaks loose."

From Fig. 18.12 one can surmise that at some pressure, P, the heat generated by a cool flame might be sufficient to raise the temperature to the ignition point. This is frequently observed, often with significant time delays.

An increase in pressure can also cause a transition from a cool flame to ignition as by a shift from point A to point B or by the temperature pulse associated with the pressure change. Such a temperature pulse may be estimated by the equation:

$$T_f = T_0 \left(\frac{P_f}{P_0}\right)^{\frac{\gamma-1}{\gamma}}$$

where γ = ratio of specific heats of the principal gases.

For an ideal gas where the process is adiabatic and reversible then

$$T_f = \gamma T_0$$

i.e., for diatomic gases (air) if $T_0 = 300$ K, $\gamma T_0 = 420$ K, and $\Delta T_m = 120$ K or 216°F.

Even though the temperature pulse may seem to be of sufficient magnitude, ignition may not result. The duration of these temperature pulses is very short (of the order of 1 sec or less) and the system time delays may preclude ignition.

Transition from a cool flame to hot ignition can also occur due to composition changes. This is clearly illustrated by Fig. 18.7. For example, a cool flame initiating at 5 atmospheres pressure, 150°C, and a hexane concentration of 28% in air can be converted to a potentially damaging hot flame by reducing the hexane concentration to 12% by volume by the admission of more air.

Thus it may be surmised that the unit operation most conducive to the generation of cool flames and potential transitions to explosive ignitions is high-temperature vacuum distillation where a sudden loss of vacuum can shift all the operating parameters: pressure, temperature, and concentration, in the direction most favorable for the initiation of such phenomena.

Any review of so-called "unexplained" industrial losses will quickly uncover numerous incidents which can now be explained as possibly being due to the initiation of a cool flame with subsequent transition to a hot ignition.

18.6. CONCLUSION

Due to the lack of available data and due to the complexity of the problem, one can only conclude that to assure safety in high temperature operations one must work at temperature sufficiently below the cool flame initiation temperature such that a sudden change in pressure cannot cause a transition to ignition.

REFERENCES

*1. J. H. Knox, *Photochemistry & Reaction Kinetics*, Cambridge University Press, 1967, pp. 250–286.

2. H. Davy, *Gmelin's Handbook of Organic Chemistry*, **8**, 179 (1812).

3. W. H. Perkin, *J. Chem. Soc.* (London) **41**, 363 (1882).

4. M. Prettre et al., *r. hebd. Séanc. Acad. Sci.*, Paris, **191**, 329, 414 (1930).

5. D. T. A. Townend, and M. R. Mandlekar, *Proc. Roy. Soc. A,* **141**, 484; **143**, 168 (1933).

*6. A. Fish, *Angew. Chem. internat. Edit.*, **7**, No. 1, 45–60 (1968).

*7. D. T. A. Townend, and M. S. Hsieh, *J. Chem. Soc.* (London), Pt. 1, 332–345 (1939).

8. N. P. Setchkin, *J. of Research of the Nat. Bu. of Stds.*, **53**, No. 1, 49–68 (1954).

9. A. Beerbower, private communication to the author (1974).

10. M. Zabetakis et al., *Autoignition of Lubricants at Elevated Pressures*, Bu. Mines RI 6112 (1962).

11. H. Carhart, private communication to the author.

12. D. H. Fine, P. Gray, and R. MacKinven, *Proc. Roy. Soc. Lond. A.*, **316**, 223–240 (1970).

*Recommended reading for further clarification of the cool flame phenomenon.

19

Safe Handling of Flammable and Combustible Materials

William S. Wood

The ubiquitous use of flammable materials in industry results in many fires and explosions because the necessary controls are not understood and heeded. Olefins, aromatics, ethers, and alcohols are examples of flammable materials that are manufactured in chemical and petrochemical processes and which may then become the raw materials for organic chemical synthesis. Such liquids may find use as the vehicles in which reactions occur or as solvents to effect separations. Solvents are widely used as thinners, cleaners, diluents, etc., in commerce and industry.

An assessment of the flammability hazard must precede utilization of these materials if injury and loss are to be minimized. It is important that the user carefully evaluate the chemical and physical properties that are pertinent to the application and to the environment in which flammables will be used.

Many of the chemicals and petrochemicals used in industry and commerce are liquids that can be easily ignited and will burn rapidly. Materials that can be ignited when they are at ambient temperatures, 100°F (38°C) or less are termed *flammable,* while those that must be heated above 100°F to be ignited by a flame or spark are considered *combustible*.

Flash point and boiling point are tests used in NFPA 30[1] to categorize flammable and combustible liquids, Table 19.1. Liquified gases such as propane and butane are not considered flammable *liquids* because of their volatility.

TABLE 19.1. Definitions from NFPA 30 (1977)

	Flammable Liquids
Class IA	Flash point below 73°F (22.8°C)
	Boiling point below 100°F (37.8°C)
Class IB	Flash point below 73°F (22.8°C)
	Boiling point at or above 100°F (37.8°C)
Class IC	Flash point at or above 73°F (22.8°C) and below 100°F (37.8°C)
	Combustible Liquids
Class II	Flash point at or above 100°F (37.8°C) and below 140°F (60°C)
Class IIIA	Flash point at or above 140°F (60°C) and below 200°F (93.4°C)
Class IIIB	Flash point at or above 200°F (93.4°C)

19.1   THE COMBUSTION OF FLAMMABLES

To grasp the meaning of flammability as a hazard one should understand the process of combustion, particularly the concept of the diffusion flame. Haessler[2] describes the mechanism whereby a hydrocarbon vaporizes, is heated to ignition temperature, and combines with oxygen in a complex chain reaction involving numerous free radicals of exceedingly short duration. These concepts are important to safe handling of flammables as well as to the extinguishment of flammable liquid fires.

The Fire Triangle

Fuel, oxygen, and heat, the basic components required for continued burning of a flame, are often represented by the fire triangle. For flammable liquids and gases the three requirements are more accurately described as:

1. Gas or vapor within a certain range of composition.
2. Air containing a minimum concentration of oxygen.
3. An ignition source of adequate temperature and energy.

A fire results and continues when all three sides of the triangle are in place, Fig. 19.1, but any two of the requisites can coexist without a fire as long as the third is not added. Since an oxygen-bearing atmosphere surrounds nearly all normal activities, the safety problem usually consists of keeping flammables separated from all ignition sources, Fig. 19.2.

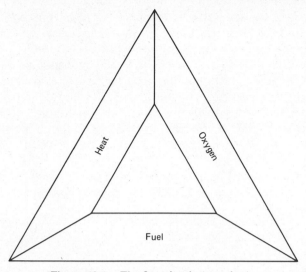

Figure 19.1. The fire triangle (complete).

The Fuel–Flammable and Combustible Materials

If even a small flame is passed over a pan of methanol at ambient temperature [68°F (20°C)], ignition occurs promptly in the vapor above the surface. The same flame over glacial acetic acid or naphthalene at ambient temperature would not ignite either material. If, however, the acetic acid is heated only slightly, enough vapor will be generated to permit ignition and,

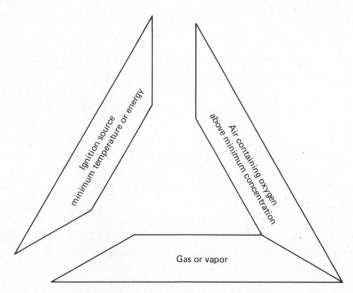

Figure 19.2. The fire triangle (incomplete).

with further heating, the naphthalene also could be burned. Liquids and solids do not burn as such. They give off vapors and gases which mix with air and can be ignited when a combustible mixture is attained.

Obviously volatility is a primary factor in determining the conditions under which an ignitable mixture can occur. Classical expressions of volatility such as vapor pressure and boiling point, while definitive, do not of themselves relate directly to flammability. The flash-point determination is an empirical but highly satisfactory method of determining the minimum temperature at which an ignitable mixture exists above the surface of a liquid.

A number of long-established procedures for flash point testing have been standardized by the American Society for Testing and Materials:

ASTM D56[3] Tag Closed Tester—for liquids having a viscosity less than 45 SUS (5.84 centistokes) @ 100°F (37.8°C) and a flash point below 200°F (93.4°C).

ASTM D93[4] Pensky Martens Closed Tester—for liquids having a viscosity of 45 SUS (5.84 centistokes) @ 100°F (37.8°C) or a flash point of 200°F (93.4°C) or higher.

ASTM D92[5] Cleveland Open Tester—for liquids having higher flash points.

ASTM D3243[6] Setaflash Closed Tester—for aviation turbine fuels.

ASTM D3828–81[7] Setaflash Closed Tester—

Prugh[8] discussed a method of estimating flash point of hydrocarbons by use of a relationship[9,10] between boiling point (BP) and flash point (FP). The following equation is applicable to hydrocarbons having initial boiling points between 200°F (93.3°C) and 700°F (371°C):

$$FP(°F) = 0.683 \times BP(°F) - 119°F$$
$$FP(°C) = 0.683 \times BP(°C) - 66°C$$

The composition of mixtures must be considered when interpreting flash point information. Chlorinated hydrocarbons blended with low flash hydrocarbons raise the flash point significantly, but upon partial evaporation the noncombustibles are lost and a low flash fraction remains. Solutions of alcohols and other polar solvents in water have finite flash points at low concentrations. For example, a 5% solution of ethanol in water has a flash point of 144°F (62°C).

Mists of high flash materials are readily ignited. Foams and froths are sometimes ignitable at temperatures lower than expected. The fact that a combustible material has been heated above its flash point and thus become readily ignitable may be overlooked by the incautious operator. Also a small amount of volatile material added to a relatively high boiling liquid will drastically lower the flash point and render the whole volume dangerous.

Dusts of combustible solids can constitute a severe explosion hazard. Finely divided polymers such as polyethylene, metals such as aluminum, elements such as sulfur, and natural products such as coal dust, flour, sugar, grain dust, and cotton fibers all have fueled serious explosions. In a typical incident, a minor explosion disperses dust that has been allowed to accumulate on flat surfaces. A second explosion, far more severe, follows immediately (see NFPA 654).

The Oxidizing Atmosphere

In nearly all potential fire situations the ambient air supplies the oxygen necessary for combustion. The normal oxygen concentration of 21% supports combustion readily but at a controlled rate. Enriched atmospheres, up to 100% oxygen, enhance ignition and increase the reaction rate dramatically.

Air deficient in oxygen can occur naturally from decay or fermentation of organic material. Inert gas such as nitrogen or carbon dioxide may be used in a closed space to inhibit combustion. For each flammable or combustible material there is a concentration of oxygen below which a flame will not be propagated at atmospheric pressure. For many substances the minimum oxygen concentration to support combustion is approximately that required to sustain human life (about 12%). For hydrogen the minimum oxygen concentration is only 5%. For hazard to life, see Chapter 25.

Other oxidizing agents such as chlorine, fluorine, ozone, or nitrogen oxides are used in chemical manufacture. Such oxidizers are vigorous and require careful control and handling. Transportation and storage of all oxidizers should adhere strictly to the vendor's recommendations (see Chap. 31.)

Flammable Limits

Lower flammable limit (LFL), also called lower explosive limit (LEL), is the leanest concentration (of vapor or gas in air) that will propagate a flame. It is determined empirically by introducing known mixtures into a cylinder and observing whether a flame front is initiated by a spark. At vapor concentrations less than the lower flammable limit the molecules of flammable vapor are so far apart that the intervening air absorbs the heat of combustion and thus prevents ignition of the unburned flammable.

Zabatakis[11] credits Jones[12] and Lloyd[13] with the following approximation for paraffinic hydrocarbons:

LFL (25°C) = 0.55 C_{st} (stoichiometric concentration of vapor in air)

Upper flammable limit (UFL) or upper explosive limit is the richest vapor or gas concentration that will propagate a flame. At vapor concentrations

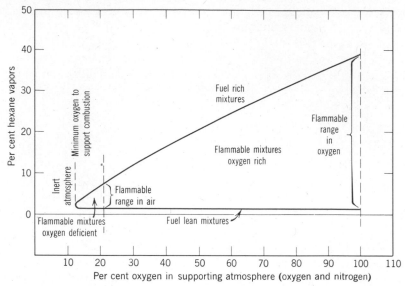

Figure 19.3. Limits of flammability of hexane in oxygen–nitrogen atmospheres.

higher than the upper flammable limit there is insufficient oxidant to burn the fuel completely, and heat release is inadequate to ignite surrounding fuel. The UFL, determined in the same apparatus as the LFL, is highly dependent upon oxygen concentration (Fig. 19.3).

Flammable range of a given vapor or gas in air comprises all concentrations between the lower and the upper flammable limits. At concentrations near the lower or upper limits, flame propagation will be slow and perhaps uncertain. At concentrations near the middle of the range, particularly at the stoichiometric concentration, propagation is rapid and the energy release may be violent.

Values for the lower and upper flammable limits for many substances are published in tables of hazard properties.[11,14]

When a flammable liquid is stored in a closed vessel, the vapor concentration in the free space will depend upon the partial pressure of the substance at the storage temperature, Fig. 19.4. It may be important to know whether the vapor–air mixture is below, above, or within the flammable range. Fig 19.5 displays the flammable ranges for some common materials and shows the temperatures at which the LFL and UFL are reached. The LFL is present above the liquid at or near the flash point. The UFL is calculated from vapor pressure data.

Detection of Flammable Vapors and Gases

Accurate, reliable detection of flammables in the air is vital to the prevention of fire or explosion. Simple, inexpensive instruments are readily available,

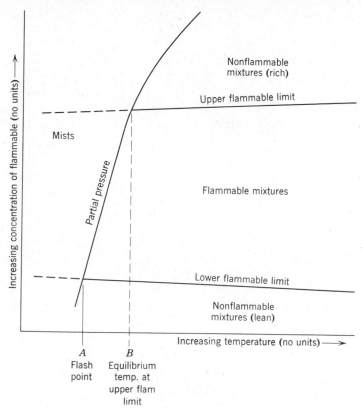

Figure 19.4. Relationship of partial pressure, flammable limits, and flash point.

but judgment and experience are needed for proper interpretation of tests[15] The "hot wire" type of meter is the oldest and most commonly used. The wiring diagram of a battery-powered portable unit is shown in Fig. 19.6. The sample passes over the heated test coil (but not the heated reference coil). Combustibles burn on the incandescent wire and raise its temperature, therefore its resistance, and the imbalance of the Wheatstone bridge is read on the meter. The usual instrument is calibrated to read the percentage of the lower flammable limit of a calibration substance such as hexane. A more sensitive portable device operating on the same principle has an additional range reading 1000 ppm full scale, Fig. 19.7.

In a vapor concentration less than LFL the needle will be on the scale when the less sensitive instrument is being used. In a vapor concentration between the LFL and the UFL the needle will remain off scale above 100%. At concentrations above the UFL the needle will swing rapidly off scale and then drop back to zero. Atmospheres deficient in oxygen result in unreliable low readings. These vagaries emphasize the need for skill in manipulation

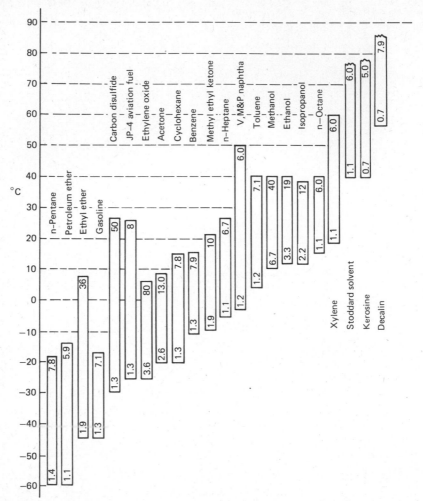

Figure 19.5. Equilibrium temperatures corresponding to flammable limits of solvents. Numbers within rectangles are lower flammable limit and upper flammable limit expressed as percent of vapor in air.

and interpretation. Such tests usually precede evaluation for "hot" work and issuance of a permit.

Similar instruments are designed for continuous sampling and may be set to sound an alarm at any desired percentage of LFL. One to sixteen or more points may be sampled by a single device.

Other principles used to detect the presence of flammables are: thermal conductivity, diffusion rate, specific gravity, and infrared spectrum. While such instruments are relatively new, they are finding acceptance for solving special problems.

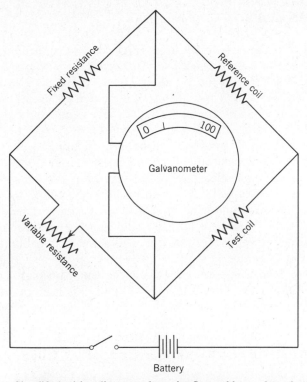

Figure 19.6. Simplified wiring diagram—hot wire flammable gas detection instrument.

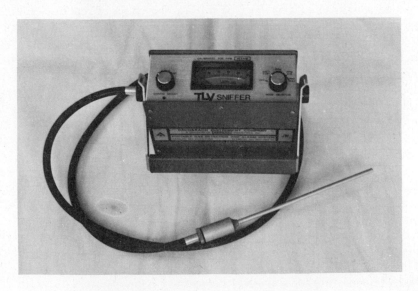

Figure 19.7. Sensitive flammable vapor detector.

19.2. PHYSICAL PROPERTIES

Specific gravity is the ratio of the unit weight of a substance compared with an equal volume of water. Thus insoluble materials with a specific gravity less than 1.0 will float on top of water, while those with specific gravity greater than 1.0 will tend to sink. The specific gravity is useful for calculating weight from volume and vice versa. Only a few flammable liquids, notably carbon disulfide, are heavier than water.

Solubility in water is important, particularly relating to the use of water for fire control. A water-miscible liquid is easily washed away. However, a water solution may still release ignitable vapors, particularly if warm. For example, 50% ethyl alcohol has a flash point below 100°F.

Vapor pressure of a fluid is another measure of the rate at which vapors can be released. It is best utilized by plotting against temperature. Zabetakis[11] used the vapor pressure curve to define flammability parameters for a number of materials.

Boiling point, the temperature at which the vapor pressure equals 1 atm, is an easily determined indication of volatility, hence, of its flammable hazard.

19.3. IGNITION OF FLAMMABLE AND COMBUSTIBLE MATERIALS

Ignition takes place when fuel in the presence of air or other oxidizing atmosphere is heated sufficiently to initiate the chain of reaction. Flames, hot surfaces, and electric sparks are the three most likely sources of ignition. The ignition source must have a sufficiently high temperature and also sufficient energy to initiate combustion of any given fuel-oxidant system.

The ignition temperature (also called autoignition or autogenous ignition temperature) is the temperature at which a small amount of a substance will ignite and burn without further heat input. A surface in contact with a flammable mixture can cause ignition if the ignition temperature is approached. See Chapter 18 on Cool Flames for a further discussion of ignition phenomena.

ASTM D2155[16] is a method of test for autoignition temperatures applicable to many flammable and combustible liquids. NFPA 325M[14] and U. S. Bureau of Mines Bulletin 627[11] lists ignition temperatures for a large number of compounds.

Kindling temperature is a term used to express the temperature to which a solid such as paper or wood must be raised to begin burning. Size, shape, and purity of the substance as well as humidity and air movement affect the kindling temperature for a given material.

Flames from a welding torch, a match, or the fire in a furnace have adequate temperature and energy to ignite gases, liquids, or solids. Flame must be prohibited where flammables may be present, and flame-producing equipment should be excluded.

Heated electrical filaments such as open heaters or broken electric lamps are effective ignitors if as little as 2 mJ of energy is available. Van Dolah et al.[17] showed that to ignite a large volume of a flammable substance, the wire temperature required is inversely proportional to wire diameter.

Husa and Runes[18] studied large metal surfaces as ignition sources for hydrocarbon vapors and showed that such surfaces must be well above the published autoignition temperatures before ignition occurs. Absence of confinement and convection currents were thought to account for the higher temperature requirement.

Friction develops heat in dry or poorly aligned bearings and seals. If flammable liquid, vapor, or gas should leak at that point, ignition is probable.

Electric sparks with energy of only 0.2 mJ can ignite a mixture of flammable vapor or gas in air.[19] Sparks occur when relays or switches are operated or when a motor with commutator is running. Accidentally broken wires or loose contacts can also produce strong sparks. A welder's arc is a potent ignition source.

Intrinsically safe instruments have been accepted for use in hazardous locations. Their circuitry is designed so that neither a break or a short will produce a spark capable of igniting a hydrocarbon–air mixture. NFPA 493[20] gives guidelines for proper use of such equipment.

Electrical equipment in an area where flammables are used or stored should be explosion-proof. This means, basically, that housings are capable of withstanding an internal explosion of a stoichiometric vapor–air mixture and, further, that exiting hot gases will be quenched below the ignition temperature. Motors, relays, switches, etc., meeting the requirements for service in hazardous atmospheres are identified with a special label designating the categories of flammables with which they may be used. For example, a motor bearing the designation "Class I, Group D" may be used in toluene service. The National Electrical Code[21] Articles 500–504 lists a number of flammable and combustible materials and the designation for each.

Static electricity is a potential ignition source, although it is sometimes improperly blamed when other causes are not apparent. Synthetic clothing worn in dry weather can generate substantial charges. Belts running over pulleys or moving surfaces rubbing a suitable nonconductor produce large potentials. Printing presses, paper calendars, rubber mills, etc., generate charges readily. NFPA 77[22] gives some of the theory of charge formation and also describes the known methods of coping with the problem.

Flow of fluids (or gases or dusts) can generate static charges, particularly if a free fall of the material is allowed to take place. Charge accumulates on metal parts of the system if they are insulated from the ground. Bringing together metal parts having different degrees of charge can cause sparks of high energy to jump the gap and ignite any flammable gas or vapor present.

Pumping of relatively clean organic fluids can cause a static charge to be accumulated at the liquid surface of the receiving vessel or tank. Klinkenberg et al.[23] demonstrated that high-velocity pumping of jet fuel resulted in explosions in the vapor space above the liquid. A dangerous static charge exists when the liquid contains just enough impurity to allow the collection of charges but not enough contamination to make the liquid so conductive that the charge can bleed off rapidly. Many common liquids are sufficiently low in conductivity that this dangerous situation prevails. Slow pumping and subsurface discharge aid in controlling the static buildup. Additives to "contaminate" the product and hasten static bleed-off have been offered and are sometimes used. Floating-roof tanks or inerted tanks are commonly used to eliminate the vapor–air space above sensitive materials. The U. S. Bureau of Mines studied this problem under a contract with the American Petroleum Institute.[24]

Lightning is a common cause of ignition for flammables stored in tanks. Arrestor systems are practical only for small tanks. If the air space above a liquid would be in the flammable range at ambient temperatures, a floating-roof tank or inert atmosphere will reduce ignition potential.

Flammables in Drums and Small Containers

Many of the uses of flammable liquids are on a limited scale where they are being handled in containers up to 55 gal capacity. The choice of containers,

Figure 19.8.   Approved safety can for flammable materials (Eagle Manufacturing Co.).

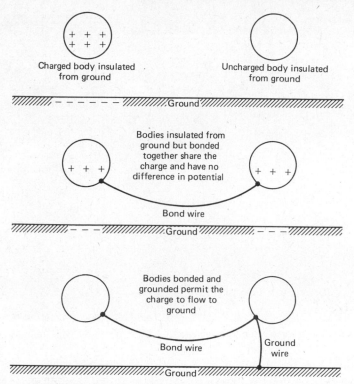

Figure 19.9. Essential features of bonding and grounding (earthing) to prevent static charge ignition.

their proper storage, and the operations involving them are vital to safety. The "Flammable Liquids Code," NFPA 30[1] gives guidelines for safe handling of containers and drums.

Glass and plastic containers must be of limited size and should be used only when metal containers cannot be tolerated. Safety cans with Underwriters Laboratory or Factory Mutual approval are the preferred dispensers in sizes up to 5 gal. The spring-loaded cap prevents loss of fluid or vapor at normal temperatures but vents if internal pressure is increased. A flame arrestor in the spout of the safety can prevents entry of a flame and thus precludes internal explosion (Fig. 19.8).

The use of plastic containers should be carefully evaluated in terms of venting and leakage from softening or melting when exposed to heat. Special storage cabinets capable of protecting the contents from an exterior fire should be used for storing containers of flammable liquids. Metal or plywood construction in accordance with NFPA 30 is acceptable.

Flammable liquid storerooms having fire-resistive walls, fire door, and a raised sill are described in NFPA 30. Explosion-proof wiring and good ventilation are specified. Automatic fire protection is recommended.

For transfers of flammables it is essential that all metal parts be electri-

Figure 19.10.   Sewer explosion from hexane spill. (Courtesy Louisville Journal Courier.)

cally bonded before the operation is begun. The static charge developed by the flow or free fall of the liquid can easily reach an igniting potential. The essentials of bonding and grounding are illustrated in Fig. 19.9.

Bulk Storage of Flammable Liquids

Materials with flash points below 100°F should not be in ordinary cone-roof tanks because the free space above the surface may be in the flammable range. Nitrogen can be used to displace the air in such tanks, but the cost may be prohibitive. A floating-roof tank is a better solution, especially for large diameters.

When cone-roof tanks are used they should be fitted with adequately sized relief vents to allow the passage of air when the temperature changes or material is pumped in or out. Conservation-type vents are designed to remain closed until a finite pressure is applied. Flame arresters on the vents are recommended in most cases, but they must be serviced regularly to keep the restricted passages open.

To prevent losses of vapor to the environment vapor recovery systems collecting and reprocessing the vapors may be used. Charcoal adsorption has been used successfully to recover vapors from both storage vessels and process units. However, adsorption can be exothermic and charcoal beds have been known to become extremely hot.

Tank overflow is a too frequent cause of spills with subsequent ignition of the vapors. When large amounts of volatile materials find their way into a plant or municipal sewer system a massive explosion is possible, Fig. 19.10.

19.4. REGULATIONS, CODES, AND SOURCES OF INFORMATION

Local, state, and Federal agencies, under a plethora of laws, have issued regulations governing the transportation, storage, and use of flammable materials.

OSHA regulation 29 CFR 1910.106 limits the size of containers, particularly glass, according to class. Storage in cabinets and solvent store rooms is specified based on NFPA 30.

The U. S. Department of Transportation regulations, 49 CFR 100–199, specify shipping containers and vehicles, require placarding, and specify minimum information on shipping papers.

New York City requires registration and special labeling of flammables sold in the domestic market. Some bridge and tunnel authorities regulate movement of hazardous materials via their facilities.

However, laws and regulations, no matter how well written or enforced, fall far short of achieving an adequate level of safety in the handling of hazardous materials. Informed and interested management utilizing a competent safety staff can institute and administer an effective program for dealing with flammable liquids.

The National Fire Protection Association, utilizing the talents of hundreds of experts on its committees, has published codes and recommended practices covering all facets of this subject.

The American Insurance Association, Industrial Risk Insurers, Underwriters' Laboratories, the American National Standards Institute and other organizations have also issued useful publications.

Much of the research on flammable properties and the development of test methods has been done by the U.S. Bureau of Mines and has been documented in its publications. Fire Research Abstracts, published by the National Academy of Sciences, is another valuable resource on new studies.

19.5. SUMMARY

Recognizing the ever present danger with flammable materials, anyone involved with their handling should:

Know the hazardous properties of the flammable material involved in the operation.

Store and transfer safely in a closed system if possible.

Avoid (or control) ignition sources in areas containing flammables.

Plan and rehearse emergency action.

In case of vapor release prevent ignition if possible. Evacuate and monitor the area.

Remember the fire triangle. Air is usually present. Keep flammables from contacting an ignition source.

REFERENCES

1. NFPA 30-1977, "Flammable Liquids Code," National Fire Protection Association, Batterymarch Park, Quincy, Mass. 02269, 1977.

2. Walter M. Haessler, "The Extinguishment of Fire," rev. ed. National Fire Protection Association, 1974.

3. ASTM D56, "Standard Method of Test for Flash Point by Tag Closed Cup Tester," American Society for Testing and Materials, Philadelphia, 1979.

4. ASTM D93, "Standard Method of Test for Flash Point by Pensky-Martens Closed Tester," American Society for Testing and Materials, 1980.

5. ASTM D92, "Standard Method of Test for Flash Points by Cleveland Open Cup," American Society for Testing and Materials, 1978.

6. ASTM D3243, "Standard Method of Tests for Flash Point of Aviation Turbine Fuels by Setaflash Closed Tester," American Society for Testing and Materials, 1977.

7. ASTM D3828–81, Standard Method of Tests for Flash Point of Liquids by Setaflash Closed Tester," American Society for Testing and Materials.

8. R. W. Prugh, "Estimation of Flash Point Temperature," *J. Chem. Ed.,* **50**(2), A85–89 (Feb. 1973).

9. Factory Mutual Corporation, *Handbook of Industrial Loss Prevention,* 2nd ed., Norwood, Mass., 1967.

10. Protectoseal Company, *Flammables Engineering,* **16** (2) (February 1969).

11. M. G. Zabetakis, "Flammability Characteristics of Combustible Gases and Vapors," U. S. Bureau of Mines Bulletin 627, 1965. (USNTIS AD701576.)

12. G. W. Jones, "Inflammation Limits and Their Practical Application in Hazardous Industrial Operations," *Chem. Rev.,* **22**, 1–26 (February 1938).

13. P. Lloyd, "The Fuel Problem in Gas Turbines," *Inst. Mech. Eng. Proc.* (War Emergency Issue No. 41), **159**, 220 1948.

14. NFPA 325M, "Fire Hazard Properties of Flammable Liquids, Gases and Volatile Solids," National Fire Protection Association, 1980.

15. R. L. Swift, "Detection of Hazardous Atmospheres," NFPA Q., **57** (2), 168–176, October 1963.

16. ASTM D2155, "Standard Method of Test for Autoignition Temperature of Liquid Petroleum Products," American Society for Testing and Materials, 1976.

17. R. W. Van Dolah, M. G. Zabatakis, D. S. Burgess, and G. S. Scott, "Review of Fire and Explosion Hazards in Flight Vehicle Combustibles," U. S. Bureau of Mines Information Circular 8137, 1963, p. 7.

18. H. W. Husa and E. Runes; "How Hazardous are Hot Metal Surfaces," *Oil Gas J.,* **61** (45), 180–182 (November 11, 1963).

19. Van Dolah, Zabatakis, Burgess, and Scott, op. cit., p. 5.

20. NFPA 493, "Standard for Intrinsically Safe Apparatus and Associated Apparatus for Use in Class I, II and III, Division 1 Hazardous Locations," National Fire Protection Association, 1980.

21. NFPA 70, "National Electrical Code," National Fire Protection Association, 1982.

22. NFPA 77, "Static Electricity," National Fire Protection Association, 1981.

23. A. Klinkenberg and J. L. van der Minne; *Electrostatics in the Petroleum Industry,* Elsevier, Amsterdam, 1958.

24. M. G. Zabetakis, G. W. Jones, G. S. Scott, and A. L. Furno, "Research on Flammability Characteristics of Aircraft Fuels," Wright Air Development Center, Technical Report 52-35, 1956.

BIBLIOGRAPHY

Ale, B. J. M., Bruning, F., and Koenders, H. A. A., "The Limits of Flammability of Mixtures of Ammonia, Hydrogen and Methane in Mixtures of Nitrogen and Oxygen at Elevated Temperatures and Pressures," *J. of Hazard. Materials*, **4**, No. 3, 283–290 (1982).

Alliance of American Insurers, *Handbook of Organic Industrial Solvents,* 5th ed., 1978. [Available from AAI, 20 N. Wacker Dr., Chicago, Ill., 60606.]

Bodurtha, F. T., *Industrial Explosion Prevention and Protection,* McGraw-Hill, New York, 1980.

Browers, S. D. "Understanding Sorbents for Cleaning Up Spills," *Plant Engineering*, **36**, No. 6, 219–221, Mar. 18, 1982.

Burgess, D., Strasser, A., and Grumer, J., "Diffusive Burning of Liquid Fuels in Open Trays," *Fire Res. Abs. Rev.,* **3**, 177–192 (1961).

Burgess, D. and Hertzberg, M., "The Flammability Limits of Lean Fuel-Air Mixtures: Thermochemical and Kinetic Criteria for Explosion Hazards," *ISA Trans.,* **14**(2), 129–136 (1975).

Calcote, H. F., Gregory, C. A. Jr., Barnett, C. M., and Gilmer, R. B., "Spark Ignition, Effect of Molecular Structure," *Ind. Eng. Chem.,* **44**(2659) (1952).

Clayton, G. D. and Clayton, F. E., eds. *Patty's Industrial Hygiene and Toxicology,* 3rd rev. ed., Vol. I, Wiley, New York, 1978, pp. 1377–1406.

Crescitelli, S., DeStefano, Russo, G., and Tufano, V., "On the Flammable Limits of Saturated Vapours of Hydrocarbons," *J. of Hazard. Materials*, **5**, No. 3, 177–188 (Feb. 1982).

Furno, A. L., Cook, E. B., Kuchta, J. M., and Burgess, D. S., "Some Observations on Near-Limit Flames," paper presented at the 13th Symposium (International) on Combustion, The Combustion Institute, Pittsburgh, 1971, pp. 593–599.

Guest, P. G., Sikora, V. W., and Lewis, B., "Static Electricity in Hospital Operating Suites: Direct and Related Hazards and Pertinent Remedies," U. S. Bureau of Mines, Report of Investigation No. 4833, 1952.

Hilado, C. J., "Flammability Tests, 1975: A Review," *Fire Technol.* **11**(4), 282–293 (Nov. 1975).

Hilado, C. J. and Clark, S. W., "Autoignition Temperatures of Organic Compounds," *Chem. Eng.,* **79**(18), 75–80 (Sept. 4, 1972).

Konst, W. M. B., "The Hazard of a Propionyl Chloride-Diisopropyl Ether Mixture," *J. of Hazard. Mater.,* **4**, No. 3, 291–298 (1982).

Kuchta, J. M., "Fire and Explosion Manual for Aircraft Accident Investigators," U. S. Bureau of Mines Report No. 4193, 1973.

Kuchta, J. M. and Cato, R. J., "Ignition and Flammability Properties of Lubricants;" *Trans. SAE,* **77**, 1008–1020 (1968).

Kuchta, J. M., Furno, A. L., Bartkowiak, A., and Martindell, G. H., "Effect of Pressure and Temperature on Flammability Limits of Chlorinated Hydrocarbons in Oxygen-Nitrogen and Nitrogen Tetroxide-Nitrogen Atmospheres," *J. Chem. Eng. Data,* **13**(3), 421–428 (1968).

Lewis, B, and von Elbe, G., *Combustion, Flames and Explosions of Gases,* Academic Press, New York, 1951.

Lovachev, L. A., "Flammability Limits: An Invited Review," *Combust. Flame,* **20**, 259–289 (1973).

Mackison, F. W., Stricoff, R. S., and Partridge, L. J. Jr., "NIOSH/OSHA Pocket Guide to Chemical Hazards," DHEW(NIOSH) Publication No. 78-210 U. S. Government Printing Office, Washington, D. C., 1979.

Mellon, I., *Industrial Solvents Handbook* Reinhold, New York, 1970.

Mullins, B. P. and Penner, S. S., *Explosions, Detonations, Flammability and Ignition,* Pergamon Press, London, 1959.

National Fire Protection Association, *Handbook of Fire Protection,* 14th ed., National Fire Protection Association, Quincy, Mass., 1976, pp. 3-18–3-38.

National Fire Protection Association; *Fire Protection Guide on Hazardous Materials,* 7th ed., National Fire Protection Association, Quincy, Mass., 1978.

NFPA 53M, "Fire Hazards in Oxygen-Enriched Atmospheres," National Fire Protection Association, 1974.

NFPA 325A, "Flash Point Index of Trade Name Liquids," National Fire Protection Association, 1972.

Phillips, H., "Differences Between Determinations of Maximum Experimental Safe Gap in Europe and U.S.A.," *J. of Hazard. Mater.,* **4,** No. 3, 245–256 (Jan. 1981).

U. S. Dept. of Transportation; "Transportation," Publication 49CFR 100-199, U. S. Government Printing Office, Washington, D.C., 1978.

Van Dolah, R. W. "Flame Propagation, Extinguishment and Environmental Effects on Combustion," *Fire Technology,* **1**(2), 138–145, (May 1965).

Welker, J. R. and C. M. Sliepcevich; "Burning Rates and Heat Transfer from Wind-blown Flames," *Fire Technol.,* **2**(3), 211–218, Aug., 1966.

Windholz, M., ed., *The Merck Index of Chemicals and Drugs,* Merck and Co., Inc., Rahway, N.J., 1976.

Britter, R. E. and Griffiths, R. F., *Dense Gas Dispersion,* Elsevier Science Publishing Co., Amsterdam and New York, 1982.

20

Fire Extinguishing Agents and Their Application

Arthur B. Guise and
Edmund D. Zeratsky

Rapid developments have occurred in the field of fire extinguishing agents and fire extinguishing equipment since the first edition of *Safety and Accident Prevention in Chemical Operations* was published in 1965. For example, two well-known types of extinguishers are no longer manufactured or listed by the Underwriters' Laboratories. The vaporizing liquid extinguishers, carbon tetrachloride and chlorobromomethane, have been discontinued because of the toxicity of their vapors and the products of decomposition. The inverting type extinguishers, soda-acid, foam, loaded stream, or antifreeze, have been discontinued because their poor design resulted in too many accidents. The vaporizing liquid types are no longer recognized in NFPA No. 10.[1] They are forbidden aboard all vessels subject to U.S. Coast Guard regulations.[2] Their use may be prohibited in almost all states and municipalities. However they are discussed in this chapter, because some units may still be in use, inadvisable as that may be.

Many changes have been noted during the last several years in hazards which require fire protection. For instance, energy requirements have resulted in an increase in the storage and transportation by ocean-going and highway tankers of liquified natural gas (LNG). The number of nuclear power plants in operation is much greater now than in 1965. These developments have led to the use of new agents or the use of a combination of agents for fire control or fire extinguishment. Without sufficient information, it is difficult for persons concerned with fire protection to make knowledgeable decisions as to what should be used to adequately protect a fire hazard. This chapter is intended to provide information that is not otherwise readily available to the director of plant protection, and to plant management.

20.1. CLASSIFICATION OF FIRES

Fires are commonly divided into four basic classifications according to the nature of the combustible material.[1] However, there are different types of fires within each classification. For example, flammable liquids may be in depth, flowing (sometimes escaping under pressure), or in relatively thin layers, as in a spill fire. Metals can be in powder form, liquids, chips or turnings, ingots or castings.

Class A Fires

Normally Class A fires are defined as "fires in ordinary combustible materials such as wood, cloth, paper, etc.," all of which produce glowing embers as the result of the formation of carbonaceous material such as charcoal. Usually overlooked is the fact that charcoal itself is a Class A material which requires special consideration, because under certain circumstances neither water nor bicarbonate-base dry chemicals are effective extinguishing agents, as will be discussed later. Also, rubber, rubberlike materials, and certain plastics are in the early stages of combustion burning more like Class B materials, but in the latter stages are definitely Class A materials.

Class B Fires

Class B fires are defined as "fires in flammable petroleum products or other flammable liquids, greases, etc." However, there are certain solids, of which naphthalene is an excellent example, which melt while burning and exhibit all of the characteristics of a flammable liquid fire and have no embers. In recent years, metal alkyls have been used more frequently in the chemical industry, and these flammable liquids present special problems because of unusually low autoignition temperatures, and in some cases, violent reactivity with water.

Technically, flammable gases do not fall in any fire classification category, but practically they should be treated as Class B materials. For many years it has been common practice to recommend against extinguishing fires in escaping gas because it was felt that if the flames were extinguished the gas would continue to flow to form an explosive mixture which might become later ignited to burn. Practical experience, however, has shown that in some cases it is necessary to extinguish the flames in order to stop the escape of gas. Gases stored in liquid form, such as LNG, propane, and vinyl chloride, present a more difficult fire when escaping in liquid state than when escaping in gaseous state.

Class C Fires

Class C fires are defined as "fires involving energized electrical equipment where the electrical nonconductivity of the extinguishing media is of importance." Where the electrical equipment is deenergized, extinguishing equipment suitable for use on Class A fires may be used unless flammable liquids such as transil oil are involved. In the latter case extinguishers suitable for Class B fires should be employed. If a combination of Class A and Class B fire materials are involved, either water spray or multipurpose dry chemical should be used as the extinguishing agent.

Class D Fires

Class D fires are defined as fires in combustible metals. Low melting point metals such as sodium and potassium present problems in extinguishment because the fire soon involves only liquid metal which, being of low specific gravity, allows most extinguishing dry powders to sink through while the liquid metal is constantly exposed to the air. These metals also offer problems in that they spontaneously react with water, sometimes violently.

High melting point metals are found in a variety of forms: powder, chips and turnings, sheet, ingots, castings, and extrusions. An extinguishing agent that might be entirely suitable for use on fires in castings might be hazardous to use on powder, or chips and turnings. A very common combustible metal, magnesium, is unusual in that it falls between the extreme low melting point metals and the high melting point metals so that it is found in all of the solid forms listed but also melts quite readily while burning to form liquid magnesium.[2]

Although the fumes from any burning metal should not be inhaled, the fumes from burning radioactive metals present extremely serious health hazards to fire fighters. Otherwise, the problems of extinguishing fires in radioactive metals are similar to those encountered in the extinguishment of fires in nonradioactive metals.

When *dry* metal hydrides are burning, they should be considered equivalent to metal fires because both hydrogen and metal are burning. It is desirable to use dry powder metal fire extinguishing agents on such fires.

20.2. EXTINGUISHING AGENTS AND METHODS OF APPLICATION

Water and Water-Based Extinguishing Agents

In this discussion, soda and acid fire extinguishers and calcium chloride solution antifreeze fire extinguishers are considered as equivalent to those containing plain water because the chemicals do not increase the fire extinguishing effectiveness. Although loaded stream fire extinguishers are also antifreeze extinguishers, this type of extinguisher is discussed separately because of its greater extinguishing effectiveness on Class A fires and because it has a small degree of effectiveness on Class B fires.

Fire extinguishers containing water, calcium chloride antifreeze solutions, or soda and acid solutions are effective on Class A fires and on fires in wood soaked with oils or greases.

These extinguishers are approved in capacities from $1\frac{1}{4}$ gal to wheeled 33 gal sizes. The water or water solutions are expelled from the extinguishers by a manually operated pump (pump tanks), by carbon dioxide from the chemical reaction of sodium bicarbonate solution and sulfuric acid (soda-acid extinguishers), by carbon dioxide from cartridges (cartridge-operated extinguishers), or by air stored under pressure in the same chamber with the water or water solution (stored-pressure extinguishers). These extinguishers are usually equipped with a hose and a nozzle designed to throw a solid stream for distances ranging from 30 to 50 ft.

SODA AND ACID EXTINGUISHERS (NO LONGER MANUFACTURED)

All soda and acid extinguishers are put into operation by first inverting to allow the acid to mix with the sodium bicarbonate solution to react to form carbon dioxide (Fig. 20.1). Because sodium bicarbonate in solution loses carbon dioxide as it slowly converts to sodium carbonate, these extinguishers should be recharged annually in order to insure that there will be a sufficient quantity of expellant gas produced when the solution is mixed with acid. Soda and acid extinguishers must be protected against freezing.

WATER (INVERTING TYPE NO LONGER MANUFACTURED)

Water cartridge-operated extinguishers are pressurized by inverting and bumping on a firm surface in order to puncture the seal of the cartridge and release the pressurizing carbon dioxide (Fig. 20.2). Pump tanks and stored-pressure water extinguishers are operated while in the upright position (Fig. 20.3). These extinguishers must be protected against freezing.

ANTIFREEZE EXTINGUISHERS (INVERTING TYPES NO LONGER MANUFACTURED)

Calcium Chloride.   Calcium chloride antifreeze solution cartridge-operated extinguishers are pressurized by inverting and bumping on a firm surface in order to puncture the seal of the cartridge and release the pressurizing carbon dioxide. Pump tanks and stored pressure extinguishers of this type are operated while in the upright position.

Because calcium chloride antifreeze solution is corrosive, approved extinguishers in which it is used are constructed of corrosion-resistant materials. Approved stainless steel extinguishers are not resistant to the corrosive effects of calcium chloride solution and it should never be used in these extinguishers. These extinguishers are suitable for use down to temperatures of $-40°F$.

Lithium Chloride.   An antifreeze extinguisher using a solution of lithium chloride, and suitable for use to temperatures as low as $-65°F$, has been developed by the U.S. D.O.D.[4]

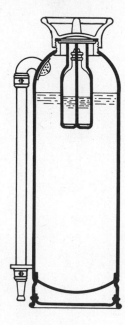

Figure 20.1. Soda and acid extinguisher—operated by inverting.

Loaded Stream.   The chemicals in water solution discussed so far have not increased the extinguishing effectiveness of the solutions. They served only to produce carbon dioxide as an expellant gas or to lower the freezing point. Other chemicals when added to water increase the extinguishing effectiveness and the best known of the extinguishers of this type is the "loaded stream extinguisher."

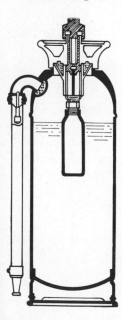

Figure 20.2. Cartridge operated water extinguisher—operated by inverting and bumping on a firm surface.

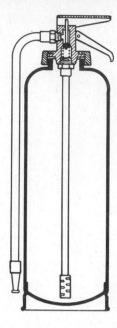

Figure 20.3.   Stored pressure water extinguisher—operated in upright position.

The extinguishing agent of a loaded stream extinguisher is a solution of potassium carbonate in water with other additives to reduce the freezing point to −40°F. These extinguishers are available in capacities from 1 to 33 gal. The older extinguishers of this type were operated by inverting which mixed acid with the potassium carbonate to form carbon dioxide expellant gas. Later models used carbon dioxide cartridges for the storage of carbon dioxide gas and were operated by inverting and bumping to puncture the cartridge seal and release the gas. More recent models have been of the stored pressure type in which air under pressure in the same chamber as the loaded stream solution expels the latter when a valve is opened with the extinguisher in the upright position.

In addition to depressing the freezing point to −40°F, the loaded stream solution improves the extinguishing effectiveness to the extent that $1\frac{3}{4}$ gal loaded stream solution is given a 2-A rating by the Underwriters' Laboratories which is obtained only by $2\frac{1}{2}$ gal plain water, soda and acid solution, or calcium chloride solution. A $2\frac{1}{2}$ gal loaded stream extinguisher is also given a 1-B rating by the Underwriters' Laboratories. On Class B fires, the loaded stream solution is most effective when applied in the form of spray. Rather specialized techniques are required to extinguish fires in volatile flammable liquids with loaded stream extinguishers.

Loaded stream solution should be used only in extinguishers in which its use has been approved. Other extinguishers may not be resistant to the corrosive effects of potassium carbonate. Corrosion might impair the operating mechanism or weaken the shell and hazardous rupture of the extinguisher could result.

Ethylene Glycol. Some owners of water extinguishers have used ethylene glycol to reduce the freezing point because their extinguishers are not designed for the use of calcium chloride or loaded stream solutions. It requires 50% by volume of ethylene glycol with water to obtain a freezing point of −40°F. It has been observed that if such a solution completely controls a fire it is equal to water in extinguishing effectiveness. However, if complete control of the fire is not obtained prior to the exhaustion of the extinguisher, the water will be easily evaporated, and the fire will rekindle to become more intense because the ethylene glycol is a flammable liquid. Antifreeze agents of the ethylene glycol type may also contribute to corrosion at the weld joints of stainless steel extinguishers. Such corrosion could result in either leaks or hazardous rupture of the extinguisher shell.

WET WATER

If freezing is not a concern, the extinguishing effectiveness may be improved by the use of a different additive, a wetting agent. This is a compound which, when added to water in proper quantities, increases its penetrating and spreading abilities.

Certain wetting agents in water solution will produce "wet water foam" when mixed with air by nozzles similar to those used in the production of mechanical foam. Wet water streams and wet water foam streams are both effective on Class A fires and wet water foam is effective on Class B fires of the hydrocarbon type, such as gasoline, but should not be used on fires in flammable liquids soluble in water, such as alcohol.

Depending upon the wetting agent, wet water streams sometimes can be used to emulsify certain flammable liquids, thereby achieving extinguishment.

Experience has indicated that Class A fires can be extinguished by wet water in a shorter time and with less water than if plain water had been used. Wet water, because of its penetrating power, is particularly effective on fires such as those in baled cotton or stacked hay.

Not all materials that can be called "wetting agents" are suitable for fire protection purposes. Although NFPA No. 18 "Standard for Wetting Agents"[5] gives detailed information on wetting agent specifications and tests, the average person responsible for fire protection will find it best to use only those wetting agents listed by the Underwriters' Laboratories. Listed wetting agents are no more corrosive than plain water to steel, brass, bronze, or copper. Wet water should be used only in extinguishing equipment or storage tanks designed for such use since listed wetting agents, although noncorrosive, exhibit a tendency to accelerate corrosion because of the cleaning and penetrating action and will penetrate and loosen unbonded coatings. For continuous storage, even galvanized iron or lead coated iron are not suitable. Approved wheeled wetting agent extinguishers are available in capacities from 10 to 50 gal.

WATER SPRAY ("FOG")

The extinguishing effectiveness of water may also be improved by the form in which it is used, such as water spray or "fog." The use of water in the form of drops rather than a solid stream was first practically recognized in the installation of the perforated pipe extinguishing systems that preceded the automatic sprinkler systems (1864).[6]

The value of spray nozzles on hose lines was recognized in the latter part of the nineteenth century, but it was only in the 1930s that hose-spray nozzles began to be used to an important extent by both public and industrial fire departments (Fig. 20.4). It was also in the 1930s that special spray, or fog, nozzles were developed for installation in systems. These spray nozzles produced directed streams of drops of water much smaller in size than those produced by the automatic sprinklers of that time. An automatic sprinkler improved in design to produce more finely divided drops of water in a better regulated pattern than the older automatic sprinklers was first used in 1952.[7]

In general, water spray may be used effectively for any one or any combination of the following purposes: extingishment of fire, control of fire, exposure protection, or prevention of fire.[8] The short range of spray nozzles is not a factor where the nozzles are installed on a well-engineered system but the short range imposes a definite limitation on the effective use of spray nozzles on hose lines. The droplet size is related to the range of the streams from spray nozzles, the heavier drops traveling farther than the very fine drops. A considerable number of carefully controlled tests with portable spray nozzles have been run to determine the effect of nozzle pressure and droplet size on the control and extinguishment of both Class A and Class B fires.[9-12] The conclusions drawn from the results of the tests may be summarized briefly as follows:

1. Increasing the nozzle pressure beyond 100 psi results in little or no increase in fire extinguishing effectiveness.
2. Very fine droplets tend to evaporate before reaching the seat of the fire and therefore control, but do not always extinguish the fire. On Class B fires, droplets having diameters of 100 to 150 μm are considerably more effective than droplets having diameters in the order of 300 μm but the very fine droplets may evaporate before reaching the surface of the burning liquid which will therefore not be cooled.
3. Although coarse drops of spray do not evaporate as readily as very fine drops, the unevaporated portion of the coarse drops penetrates the flame zone to reach the seat of the fire and exert cooling action.

Where approach to the fire permits, portable water-spray nozzles are considerably more effective on Class A fires than straight streams because the water is applied with less runoff. British tests showed that a spray gave more rapid control of a fire than a straight stream and that when straight streams were used more water was required to extinguish the test fires. Although

Figure 20.4.   Hand hose line water spray nozzle.

very small gasoline fires can be extinguished under certain conditions by the use of water-spray nozzles, the latter are usually successful only in extinguishing flammable liquids having flash points of 150°F or higher.[7]

Electrical equipment involved in a fire should be deenergized as quickly as possible, but it is not always possible to do so. Even so, either fresh or salt water may be safely applied to energized electrical equipment if proper clearances between nozzles and electrical apparatus are observed or exceeded (see Table 20.1).

Spray nozzles, either on hose lines or in systems, are sometimes used only to control fires where extinguishment is not possible or extinguishment is not desired because the fire is in flammable gases or extremely volatile flammable liquids. Sometimes such fire control is combined with the protection, by cooling, of exposed equipment, structures, or storage tanks.

Spray nozzle systems installed for extinguishment, control, or exposure protection may also sometimes be used for the prevention of fires if operated when hazardous materials have been accidentally released, but before they have become ignited.

As a guide to the quantities and rates of application of water in spray form the following may be used:

1. Extinguishment   — 0.2 to 0.75 gpm/ft²
2. Control   — 0.2 to 0.5 gpm/ft²
3. Exposure protection — 0.1 to 0.25 gpm/ft²

TABLE 20.1.   Clearance from Water Spray
Equipment to Live Uninsulated Electrical
Components[8]

Nominal Line Voltage (kV)	Minimum Clearance (in.)
To 15	6
23	8
34.5	12
46	15
69	23
115	37
138	44
161	52
196–230	63
	76
287–380	87
	98
	109
	120
500	131
	142
500–700	153
	168
	184

Water spray systems should be designed by experienced fire protection engineers to insure adequate protection for the hazard involved.

SOLID STREAMS OF WATER

When water spray streams cannot practically be used because of lack of reach because of fire intensity or fire location, or where there are dangers from falling walls or possible explosions, solid streams must be used. Solid streams are, of course, used primarily in the extinguishment of Class A fires—structures, lumber storage, and so on. However, they are also used for the extinguishment of fires in high flash point Class B materials in thin layers or fires in flammable solids such as napthalene. Solid streams are also used for the protection of exposed equipment, structures, and storage tanks when such protection is not afforded by spray-nozzle systems and are out of reach of water spray from hose lines.

Effective streams are the result of advance planning to insure adequate water supplies, of proper layout of water mains, and of training of the fire fighters to make best use of hydrants, hose, nozzles, and pumpers, if the latter are available. One of the best books for industrial and volunteer fire fighters, as well as municipal fire departments, is *Effective Streams for*

Fighting Fires by Warren Y. Kimball, for the National Fire Protection Association. Other helpful information may be found in "Mobile Fire Equipment, Organization, Management," National Fire Protection Codes, Volume 12, National Fire Protection Association. Information on water supplies and the layout of mains and hydrants may be found in "Fixed Extinguishing Equipment," National Fire Code, Volumes One and Two, and in the NFPA "Handbook of Fire Protection."

Research has been done at Syracuse University for the U. S. Navy Bureau of Yards and Docks on "Additives to Improve the Fire-Fighting Characteristics of Water." This research was aimed at evaluating the effectiveness of increasing the viscosity of water to decrease the rate of runoff and of increasing the opacity of water to transmission of infrared. It has been reported that the addition of opacifiers such as aluminum and bronze powders to water, increased in viscosity by certain chemical additives, has decreased the extinguishing time on certain standard fires to below that required for plain water.

FOAM

All foams are discussed in detail in Chapter 21. The discussion in this chapter covers only hand and wheeled portable extinguishers.

CHEMICAL FOAM (NO LONGER MANUFACTURED)

There are two types of fire-fighting foams: chemical foam in which the bubbles are filled with carbon dioxide, and mechanical foam in which the bubbles are filled with air. Chemical foam is produced by a chemical reaction which generates carbon dioxide bubbles in water solution containing a foaming ingredient. The reacting chemicals, usually sodium bicarbonate and aluminum sulfate, may be stored in separate water solutions, which are mixed when foam is wanted or they can be kept in powder form in separate containers and simultaneously mixed with water to make foam, or in powder form in a single container and mixed with water to make foam.

Portable chemical foam extinguishers all employ two solutions which are mixed when the extinguishers are inverted. Although chemical foam is used almost exclusively for the protection of Class B fire hazards, the Underwriters' Laboratories gave portable chemical foam extinguishers Class A ratings equal to those given water extinguishers of the same volumetric capacity.

Chemical foam solutions should not be exposed to temperatures lower than 40°F nor higher than 120°F. The sodium bicarbonate solution is especially sensitive to temperature conditions. The sodium bicarbonate will crystallize out of the water if the solution is frozen and will decompose to lose carbon dioxide when the solution is at temperatures exceeding 90°F. Annual recharge of portable foam extinguishers compensates for normal deterioration of the sodium bicarbonate solution.

However, chemical foam itself is also sensitive to the effect of temperature. At low temperatures the chemical reaction takes place more slowly and the foam has a lower expansion ratio as compared with the approximate ratio of eight found at normal ambient temperatures. At high temperatures the chemical reaction takes place rapidly, there is a high expansion ratio, and the foam breaks down more rapidly.

Approved hand portable chemical foam extinguishers were available in capacities of $1\frac{1}{4}$ to 5 gal, with the latter size intended primarily for use in industrial establishments where persons of ample strength could handle it. Approved wheeled extinguishers were available in capacities of 17 and 33 gal. The hose of these larger extinguishers were equipped with shut-off nozzles.

Stream ranges were in the order of 30 to 40 ft for hand portable extinguishers and up to 50 ft for wheeled extinguishers.

Mechanical (Air) Foam—AFFF

A foam agent introduced in 1964 contains a fluorochemical surfactant that creates a film over the surface of fuels similar to gasoline and thereby retards reignition even after the foam layer has broken down.[13] The foam is completely compatible with military specification potassium bicarbonate-base dry chemical. Tests have indicated that the new agent, referred to as AFFF foam, is approximately four times more effective than mechanical foams when used for the estinguishment of spill fires. An aqueous film forming foam (AFFF) $2\frac{1}{2}$-gal extinguisher is now listed by the Underwriters' Laboratories. It has a Class B rating of 20-B or 10 times greater than the discontinued chemical foam extinguisher. It has a 3-A rating, $1\frac{1}{2}$ times greater than a soda and acid or chemical foam extinguisher.

Although chemical foam is not affected, ordinary dry chemical causes mechanical foam, except for AFFF, to break down, the degree of breakdown depending to some extent on the formulation of the foam concentrate. However, there are special dry chemicals which have been examined and tested by the Underwriters' Laboratories and found suitable for use with mechanical foams which are designated by the Underwriters' Laboratories as "compatible for use with listed foam-compatible dry chemical extinguishers."[14] These compatible concentrates are generally 3% and 6% solutions of regular and low-temperature concentrate. The foam concentrates for producing mechanical foam for use on alcohol and ketones are not compatible with foam-compatible dry chemical. There are no approved portable extinguishers in the United States that produce mechanical foam except AFFF although such extinguishers are used in Europe.

Dual Agent Hoseline Systems

The complete compatibility of AFFF with dry chemicals has led to its use with them in dual system fire-suppression systems especially designed for spill fire fighting (Fig. 20.5). Like many innovations in fire fighting, this again

Figure 20.5.   Dual agent hand hose line system.

came from the Naval Research Laboratories which gave the name "light water" to AFFF because the film floats on top of the fuel. Their intent was to produce a lightweight vehicle which would have a high and quick fire suppression ability and would be the first to arrive at the scene of an aircraft crash, cut and hold a path to the aircraft, and rescue crew and passengers.

By itself the AFFF is a good extinguishant for spill fires of gasoline or heavier fuels but it is even better when used simultaneously with potassium bicarbonate dry chemical. The dry chemical serves as a rapid flame suppressant and the AFFF covers the fuel rapidly and prevents flame flashback.

The concentrate is sold in 3% or 6% strengths and may be mixed either with fresh or salt water. Usually the AFFF and dry chemical are applied by twinned hoselines and nozzles individually controlled (Fig. 20.6). The smaller units, 200-lb dry chemical and 100 gal AFFF, are frequently mounted on skid mounts for fitting on pickup truck bodies. The AFFF is premixed. Larger units are available in which the AFFF is proportioned. NFPA No. 11A of the National Fire Codes also gives details on the use of AFFF for extinguishing tank fires.[15]

Dry Chemicals

A dry chemical extinguishing agent is a finely divided powdered material that has been specially treated to be water repellent and capable of being

Figure 20.6.   Application of dry chemical/AFFF dual agents by twinned hose lines and nozzles.

fluidized and free flowing so that it may be discharged through hose lines or piping when under expellent gas pressure. The major components of five types of dry chemicals used in approved extinguishers are sodium bicarbonate, potassium bicarbonate, potassium chloride, urea-potassium bicarbonate, and ammonium phosphate (usually monoammonium phosphate).

Types of Dry Chemicals

Sodium Bicarbonate Base.   The use of sodium bicarbonate as the major component of a dry chemical extinguishing agent, expelled by gas pressure from an extinguisher, started in Germany in 1912. About 1926, the forerunner of the modern dry chemical extinguisher was first produced in the United States. The dry chemical extinguishing agent consisted of a fairly coarse sodium bicarbonate powder compounded with magnesium stearate to produce water repellancy and free-flowing properties. In 1943 it was found that the extinguishing effectiveness of dry chemical could be greatly increased by using more finely divided sodium bicarbonate. Later improvements in the method of applying the dry chemical stream to increase effectiveness even further resulted in this type of fire extinguisher becoming very important as first-aid equipment for Class B fires. In the latter half of the 1960s the size of the sodium bicarbonate particles was again reduced in some extinguishers and additional improvements were made in the method of applying the dry chemical to the fires. These changes resulted in increased ratings.

In 1930 the effectiveness of dry chemical in quickly reducing the intensity of flames, even in Class A materials, was recognized. It was also known that it would be advisable to follow up on Class A fires with the application of water to extinguish remaining smoldering embers. Improvements in agent and improvements in equipment resulted in recognition by the Factory Mutual Engineering Division in 1948, after a very thorough test program, that dry chemical extinguishers were preferable for use in cotton processing areas.[16] In 1952, extinguishers containing sodium bicarbonate-base dry chemical were submitted to the Underwriters' Laboratories for test on Class A fires. Although flames were more rapidly reduced in intensity when dry chemical was used, as compared with water, it was found that there was a greater tendency to reflash in a shorter period of time and the extinguishers were not approved.

Potassium Bicarbonate-Base.   Recognition of the great effectiveness of dry chemical extinguishers resulted in increased research and development on powdered extinguishing agents other than those of sodium bicarbonate base. The extinguishing effectiveness of potassium bicarbonate-base dry chemical on gasoline fires was first investigated by the Naval Research Laboratory and they reported in 1958 that this type of dry chemical was twice as effective as sodium bicarbonate-base dry chemical.[17] This work resulted in a military specification for the foam compatible potassium bicarbonate-base dry chemical which is now being used in Naval Air Stations in place of sodium bicarbonate-base foam compatible dry chemical.

Potassium bicarbonate-base dry chemical extinguishers were first approved in 1961 by the Underwriters' Laboratories. The results of their tests, as shown in Table 20.2, confirmed the high extinguishing effectiveness reported in 1958 by the Naval Research Laboratory.

Despite the greater effectiveness of potassium bicarbonate-base dry chemical on Class B fires, the chemical similarity between sodium bicarbonate and potassium bicarbonate indicates that the latter would be basically similar in effectiveness on Class A fires.

Potassium Chloride-Base.   This was developed in 1968 and hand portable extinguishers were first listed in the Underwriters' Laboratories "List of Approved Equipment" in 1969. One manufacturer was listed in 1969, seven in 1971, but by 1976 only two manufacturers were listed. It is said to be as effective as potassium bicarbonate-base dry chemical but Table 20.3 based on UL Ratings, does not so indicate.

Urea–Potassium Bicarbonate-Base.   Invented and developed in the late 1960s, extinguishers charged with this agent were first listed by Underwriters' Laboratories in 1972. It is a foam-compatible dry chemical. The

TABLE 20.2. Comparative Effectiveness of Dry Chemical Extinguishers on Class B Fires[a]

Nominal Charge (lb)	Sodium Bicarbonate Base	Potassium Chloride Base	Potassium Bicarbonate Base	Urea–Potassium Bicarbonate Base	Multi-purpose
1	2B				2B
2½	10B	10B	10B		10B
4½	20B		20B		
5					20B
5½			30B		
5¾					
6	30B		60B		40B
7½	40B				
8				80B	60B
9		40B			
9½			80B		
10	60B	60B			
12		60B			
17				120B	
18					80B
19		60B		120B	
19½					
20	120B		120B	160B	120B
23	120B				
30	160B				
125	160B	160B	320B		
150					240B
175			640B	480B	
300					320B
350	320B				

[a] Based on Highest Underwriters' Laboratories Listing Ratings.[14]

basic material is formed by the reaction of potassium bicarbonate and urea. Its effectiveness is due to its decrepitation when it is heated when it strikes flames. The decrepitation results in its becoming a mass of much smaller particles which are much more effective in extinguishing effectiveness than potassium bicarbonate alone. In the natural state it has a particle size distribution nearly identical to that of potassium bicarbonate-base dry chemical. Therefore it has a stream range similar to that of ordinary dry chemical.

Foam Compatible.   The United States Air Force in 1950 evaluated sodium bicarbonate-base dry chemical as an extinguishing agent for aircraft crash fires. Although the agent was found suitable from an extinguishing standpoint the metal stearate treatment of dry chemical caused rapid breakdown of mechanical foam blankets. A good solution to the problem of foam compatibility of dry chemicals was not reached until water repellency and free-flow characteristics were obtained by coating the particles of dry chemical with a silicone polymer instead of a metal stearate.

Impetus to the solution of the problem of foam compatibility was given by the promulgation of a U. S. Coast Guard regulation that marine-type dry chemical fire extinguishers over 5 lb in capacity must contain foam compatible dry chemical if manufactured after January 1, 1962.[2]

The first standard for evaluating the foam compatibility of dry chemical agents was developed by the Naval Research Laboratory and incorporated in a military specification in 1956.[18] The Underwriters' Laboratories developed a different procedure for the evaluation of foam compatible dry chemicals in 1961. The degree of foam compatibility required by the Underwriters' Laboratories is greater than that required to meet the military specification. However, dry chemical meeting the military specification had been in use for several years and under practical fire-fighting conditions had been found adequately foam compatible.

Multipurpose.   In 1956 a multipurpose dry chemical was being used in Germany but this material was of poor quality. It had limited effectiveness on both Class A and Class B fires and a pronounced tendency to cake.

In 1960 the Underwriters' Laboratories approved for the first time a multipurpose dry chemical fire extinguisher which used an extinguishing agent based upon a patented German formula. By the end of 1961 a number of other multipurpose dry chemical extinguishers had been approved, including wheeled extinguishers. The majority of approved multipurpose dry chemical fire extinguishers are very effective on both Class A and Class B fires and also are listed as suitable for use on Class C fires. Table 20.3 shows the effectiveness of the highest rated multipurpose dry chemical extinguishers as compared with water on Class A fires and as compared with sodium bicarbonate-base dry chemical on Class B fires.

It is obvious that, when considered on a pound-for-pound basis, multipurpose hand portable extinguishers are much more effective on Class A fires than water-type hand portable extinguishers. With wheeled extinguishers the difference is less; they are only twice as effective. On Class B fires they are more effective than sodium bicarbonate-base dry chemical extinguishers.

Multipurpose dry chemical extinguishers have certain obvious advantages where Class A fire hazards are to be protected:

1. For equivalent protection, they are lighter in weight which is advantageous where it is anticipated that women may use first-aid fire extinguishing equipment.

2. Where mixed Class A, B, and C fire hazards are involved, only one type of extinguisher is needed, and there is no chance of the wrong type of extinguisher being used on a fire.

3. Special models are approved for use in temperatures as low as $-65°$ F.

4. Where water damage is highly undesirable, such as in libraries, art museums, or other locations where there are articles or materials of high value that would be irreparably damaged by water, the multipurpose dry chemical extinguisher may furnish the best fire protection.

The multipurpose dry chemical extinguisher has become the most widely used extinguisher for homes, cars and boats.

PORTABLE FIRE EXTINGUISHERS

Portable fire extinguishers using all four types of dry chemicals are of two general types: one in which the fluidizing and expellant gas is stored in a separate container under high pressure and released into the dry chemical container to put the extinguisher into operation, and the other in which the expellant gas is stored in the same container with the dry chemical. Hand portable fire extinguishers of the first type have the expellant gas, usually carbon dioxide, contained in a small compressed gas cylinder commonly called a "cartridge" (Fig. 20.7). The expellant gas in the case of extinguishers approved for use at $-65°F$ is nitrogen. In almost all approved extinguishers of the cartridge type, the expellant gas is released into the dry chemical container by puncturing a metal sealing disk. Some of the cartridge operated hand portable extinguishers have the nozzle integral with the extinguisher. However, the majority are equipped with both hose and a controlling nozzle. Hand portable cartridge-type extinguishers range in capacity from 4 to 30 lb.

Wheeled extinguishers in which the gas is stored separately from the dry chemical chamber use nitrogen as the expellant gas, controlling the pressure within the dry chemical container by means of a regulator. This type of design permits the expellant gas to continue to flow during discharge of the

TABLE 20.3.   The Maximum Effectiveness of Multipurpose Dry Chemical Extinguishers as Compared with the Maximum Effectiveness of Water and Sodium Bicarbonate Base Dry Chemical Extinguishers[a]

		Class A		Class B	
				Sodium	
			Multi-	Bicarbonate	Multi-
Pounds	Gallons	Water	Purpose	Base	Purpose
2.75			2A		10B:C
6			3A		40B:C
10			10A		60B:C
10.4	1¼	1A			
11				60B:C	
12.5	1½	2A			
18			20A		80B:C
20				120B:C	
20.8	2½	2A			
41.6	5	4A			
125			40A		240B:C
150				160B:C	
275	33	20A			
300			40A		320B:C
350				320B:C	
558	67	40A			

[a]Based on highest Underwriters' Laboratories listed ratings.[14]

extinguisher thereby maintaining a fairly constant pressure. As a result there is little change in the flow characteristics throughout the entire discharge.

Wheeled extinguishers of the foregoing type range in capacity from 75 to 350 lb.

Extinguishers in which the expellant gas is stored in the same container with the dry chemical extinguishing agent are called "stored pressure" extinguishers (Fig. 20.8). Although most stored pressure extinguishers are equipped with pressure gauges there is one special type of approved stored pressure extinguisher that is not required to have a pressure gauge. Inspection of this extinguisher requires weighing to detect possible loss of gas by leakage. This type of extinguisher is sometimes called a "sealed pressure" extinguisher because the dry chemical container must be factory filled and sealed, and is nonrefillable. Therefore it is sometimes called a "disposable shell" extinguisher.

Portable stored pressure extinguishers range in capacity from 1 to 250 lb. The smaller sizes have the nozzle integral with the control valve, but the larger sizes of hand portable extinguishers are equipped with hose, in most cases controlling the flow by means of the valve on the extinguisher, but in other cases the nozzle at the end of the hose is also to control the flow. All

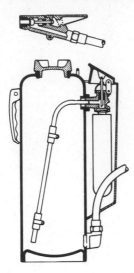

Figure 20.7.   Cartridge operated dry chemical extinguisher—operated in upright position.

wheeled stored pressure extinguishers control the flow with the nozzle at the end of the hose (Fig. 20.9).

Stream ranges of dry chemical extinguishers vary widely since even a single manufacturer may provide nozzles of two ranges for extinguishers otherwise identical in size and capacity. Hand portable extinguishers have stream ranges of 5 to 25 ft. Wheeled extinguishers have stream ranges of 10 to 45 ft.

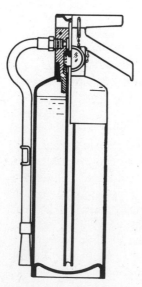

Figure 20.8.   Stored pressure dry chemical extinguisher—operated in upright position.

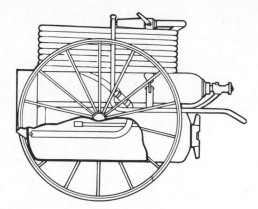

Figure 20.9. Wheeled dry chemical extinguisher with expellant gas in separate container.

Dry Chemical Systems

Hand Hose Line. Dry chemical systems of the hand hoseline type range in capacity from 150 to 3000 lb. Most use nitrogen in separate cylinders for fluidizing and expelling the dry chemical, controlling the flow with regulators. Stored pressure hand hose-line systems are approved in capacities from 150 to 250 lb. It is possible to supply as many as eight hose stations from large systems but such arrangements should be engineered by the manufacturer of the equipment.

Dry chemical hand hose-line systems are generally located where it is not necessary to have the extinguishing equipment mobile, as in the case of wheeled extinguishers. A typical fire hazard protected by such systems is a petroleum-products loading rack.

Fixed Nozzle. Dry chemical systems with piping and fixed nozzles are approved for both automatic and manual operation. Approved systems range in capacity from 10 to 500 lb. These systems may be used for the protection of specific hazards by local application of the dry chemical or installed so as to flood an entire room or building. The design and installation of a dry chemical piped system should be supervised by qualified fire protection engineers having full knowledge of the special problems of piping dry chemical to obtain the correct rate of application and the proper distribution from the nozzles. Helpful information is given by NFPA No. 17, "Standard for Dry Chemical Extinguishing Systems."[19]

Dry chemical piped systems are used for the protection of electrical equipment such as generators, circuit breakers, transformers, and for many other hazards where flammable liquids are involved. They have been notably successful in installations protecting asphalt-impregnating equipment in the manufacture of roofing and siding. Because air dilution is not such an important factor with dry chemical, piped-system protection of duct work is quite common, and there is an increasing use of dry-chemical systems for the protection of cotton gins and cotton processing such as jacquard-weave rooms, opener-picker rooms, and heavy twisting occupancies. Sodium bicarbonate-base dry chemical is the best agent and hand portable extinguishers are preferable in some cases.

Small dry chemical systems are very frequently used for the protection of kitchen ranges, deep fryers, hoods and associated duct work in restaurant kitchens, especially "fast-food" places. For these installations again sodium bicarbonate-base dry chemical is best. It is nontoxic, a requirement around food, and it saponifies the hot grease in the deep fryers, hoods and ducts, making it less flammable.

Because dry chemical is nonconductive, care should be taken in the use of dry chemical extinguishers in locations where there are many delicate low-voltage electrical contacts, such as in telephone relay rooms.

SPECIAL DRY CHEMICAL AGENTS

Because of the flexibility allowable in composition of dry chemical extinguishing agents, it is possible to compound free-flowing powdered materials for the control and extinguishment of fires in unusually hazardous chemicals. A typical example is an extinguishing agent for use on metal alkyl fires in which the agent is a mixture of sodium bicarbonate-base or potassium bicarbonate-base dry chemical and an activated adsorbent such as silica gel. The pyrophoric characteristics of short-chain metal alkyls such as triethylaluminum (which reacts violently in contact with water) allow extinguishment of the flames only during application of dry chemical. If application is discontinued, immediate reflash occurs. The special compound containing adsorbent material extinguishes the flames and at the same time the spilled metal alkyl is being adsorbed, resulting in extinguishment of the fire and allowing disposal of the now safe pyrophoric material.[20] Because of the necessity of adsorbing spilled material while controlling the fire, considerable quantities of the special agent are required. As a guide, from 8 to 10 lb of the special extinguishing agent are required for each pound of metal alkyl involved in the fire.

An alternative method of controlling metal alkyl fires is to apply an ordinary dry chemical extinguishing agent on the flames simultaneously with the application of an adsorbent from a separate extinguisher. This requires that the adsorbing agent by itself possess good flow characteristics.

Other hazardous chemicals such as fluorine and chlorine trifluoride offer unusual problems when spilled, and special dry chemical agents have been used. A good reference manual is "The Handling and Storage of Liquid Propellents" issued by the Office of the Director of Defense Research and Engineering, Washington, D.C.

Approved dry chemical extinguishing agents are not classified as toxic. Personnel of manufacturers and approval agencies have been exposed over a period of years to these dry chemicals and there have never been any ill effects reported. Persons inhaling dry chemical in high concentrations will

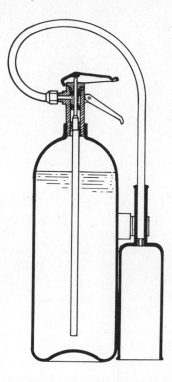

Figure 20.10.   Carbon dioxide extinguisher—operated in upright position.

experience discomfort because of the mechanical action of the dry material on moist mucous membranes.

Carbon Dioxide

Although the possibility of using carbon dioxide as an extinguishing agent was recognized as early as 1882[21] it was not until the 1920s that liquefied carbon dioxide came into general use for this purpose. The most important single factor in the successful development of portable carbon dioxide extinguishers was the use of an entrainment shield, or "horn," to limit the amount of air entrained by the carbon dioxide at the point of discharge.

PORTABLE EXTINGUISHERS

Approved hand portable carbon dioxide extinguishers are available in capacities of 2 to 25 lb (Fig. 20.10). The smaller sizes, 2 to 5 lb, usually have the discharge horn attached directly to the valve through a swivel connection and a metal tube. The larger extinguishers are equipped with both hose and horn. Most carbon dioxide wheeled extinguishers are 50, 75, or 100 lb in capacity.

Portable carbon dioxide extinguishers consist basically of DOT 3A or 3AA cylinders for the storage of the liquid carbon dioxide (under pressure of

850 psi at 70°F), a valve for controlling the flow, a horn, and either metal tubing or flexible hose connecting the horn to the valve. The extinguishers are always operated in the upright position and are provided with siphon tubes extending to close to the bottom of the cylinder so that liquid carbon dioxide is discharged from the orifice. As the liquid discharges from the orifice, within the horn, the carbon dioxide evaporates to form a mixture of gas and particles of solid carbon dioxide which forms an effective extinguishing stream as discharged from the horn.

CARBON DIOXIDE SYSTEMS

Hand Hose-Line.   There are two types of liquid carbon dioxide storage used for both hand hose-line systems and fixed-nozzle systems. In one type of storage the liquid carbon dioxide is in DOT cylinders, either single cylinders or manifolded, under a pressure of 850 psi at 70°F. These are called high-pressure systems. In the second type of storage, the liquid carbon dioxide is in insulated tanks under a pressure of 300 psi at 0°F with the reduced temperature maintained by refrigeration units. This type of system is called a low-pressure system, and it is used only where the quantity of carbon dioxide in storage is ⅜ ton or greater. Hand hose-line systems consist of a supply of carbon dioxide, either under high pressure or low pressure, connected through piping to one or more hose lines on reels or racks. Like all hand hose-line systems, carbon dioxide hand hose lines are used for the protection of specific hazards or supplementary to a fixed-nozzle system.

Fixed Nozzle.   Fixed-nozzle systems are designed for the protection of specific hazards by local application or for the protection of rooms or buildings by flooding. Although normally considered as an extinguishing agent for Class B hazards, carbon dioxide in the proper concentration will extinguish fires in Class A materials in spaces that are tight enough to prevent leakage of carbon dioxide or entrance of air after the space has been flooded. For this reason carbon dioxide flooding systems are used for the protection of record vaults, fur-storage vaults, and similar occupancies where the use of water would result in excessive damage to the contents. However, precautions must be taken to insure that the doors to such storage spaces are not opened until the contents have cooled to a point where reignition will not occur when air is admitted.

Carbon dioxide extinguishing systems should be designed by, and the installation supervised by, qualified fire protection engineers on the basis of NFPA No. 12, "Standard for Carbon Dioxide Extinguishing Systems."[22]

Spaces protected by fixed-nozzle carbon dioxide systems should be provided with an adequate audible warning system with discharge of carbon dioxide delayed to allow personnel to leave safely. There is a considerable record of fatalities where personnel have been asphyxiated in protected spaces where no audible signal was provided.

Carbon dioxide, being a gas, leaves no residue when it is used on a fire, and it is not decomposed by the flames to produce corrosive chemicals. Being a nonconductor of electricity, it is therefore an excellent extinguishing agent for use where there are delicate low-voltage relays, as in a telephone exchange, or where electronic equipment is involved in a fire.

EFFECT BY TEMPERATURE

Normally the liquid carbon dioxide is expelled from its container by its own gas pressure. Portable extinguishers charged with the specified amount of carbon dioxide are approved for use from −40° to +120°F. However, the flow rate is reduced at the lower temperatures, and the fire extinguishing effectiveness is also reduced. The Underwriters' Laboratories recommends that the extinguisher be charged with only 90% of the normal amount for use at temperatures over 120°F and up to 130°F. When the extinguishers are to be used where the temperatures will be very low, normal practice is to charge with carbon dioxide to 90% of the standard charge and supercharge with nitrogen. The same procedures may be followed with hand hose-line systems of the high-pressure type. Where low-pressure systems are used at extremely low temperatures, it may be necessary to provide heating systems as well as refrigerating systems with thermostatic control to maintain the temperature of the liquid carbon dioxide in the vicinity of 0°F.

Fixed-nozzle systems are designed to deliver carbon dioxide at specific rates of application, and where there may be considerable ranges of temperature, high-pressure storage systems should be so located that they may be heated should the temperature fall below 32°F. At temperatures of over 120°F, the carbon dioxide may be delivered from local application nozzles at too high a velocity and may splash flammable liquids as the result.

Liquified Gas Extinguishers

Halon 1211 has a boiling point of 25°F and has replaced the vaporizing liquid extinguishers, carbon tetrachloride and chlorobromomethane, because it is less toxic (see Table 20.4) and has a sufficiently high boiling point to have an effective range. Halon 1301 has a boiling point of −72°F which restricts its range and is less toxic than, and in portable extinguishers is less effective than Halon 1211. Halon 1301 is used more in total flooding systems where it is more effective than Halon 1211 (Fig. 20.11).

PORTABLE EXTINGUISHERS

The one Halon 1301 extinguisher listed by the Underwriters' Laboratories was developed for the U.S. Army and others has not been developed for civilian use primarily because of the high cost of the agent and the much greater effectiveness of dry chemical extinguishers. The $2\frac{1}{2}$ lb extinguisher has a stream range of $5\frac{1}{2}$ ft.

Figure 20.11.   Halon 1301 fixed piping and nozzle system protecting electrical equipment.

Underwriters' Laboratories listed Halon 1211 extinguishers for the first time in 1973 with Class B and C ratings slightly less than those for sodium bicarbonate-base dry chemical extinguishers. They also have Class A ratings in sizes of 9, 16, and 17 lb capacity. They have become increasingly popular because they are more effective than carbon dioxide and do not leave residues as do the dry chemicals. Stream ranges are 7 ft for the 2½ lb model to 10 ft for the 9, 16, and 17 lb models.

SYSTEMS

Halon 1301 is the nearest to a perfect total flooding agent of all of the extinguishing agents because it is nontoxic in concentrations up to 20% although it has anesthetic effects at concentrations above 7½%. However, it will extinguish fires in concentrations of less than 7½%. Its sole disadvantage is that it decomposes to toxic compounds when heated, but to a lesser degree than other halogenated compounds (see Table 28.4). The NFPA Standard No. 12A,[23] "Halogenated Fire Extinguishing Agent Systems—Halon 1301," was first adopted in 1970. Underwriters' Laboratories listing of equipment quickly followed in the same year.

The agent is suitable for the extinguishment of Class A, B, and C fires. Class B fires are readily extinguished as are surface fires in Class A materials. Deep-seated fires in fibrous or particulate materials usually require more than a 20% concentration for a sustained soaking period for extinguishment. NFPA No. 12A says, however, "it is not normally practical to maintain a sufficient concentration of Halon 1301 for a sufficient time to extinguish a deep-seated fire."

According to the Fire Protection Handbook,[24] Halon 1301 systems exist which protect as much as 300,000 ft.[3] A major use is the protection of computer rooms.

Halon 1211 is approximately twice as toxic as Halon 1301, both in the natural state and the decomposed state. NFPA Standard No. 12B[25] on Halon 1211 states that total flooding systems shall not be used in normally occupied areas. They may be used only in normally unoccupied areas that can be evacuated in less than 30 s. Although the NFPA Standard on Halon 1211 was adopted in 1973, the Underwriters' Laboratories has not listed any Halon 1211 systems as of the time of this writing.

Vaporizing Liquids (No Longer Approved)

Vaporizing liquid extinguishing agents have as their major components carbon tetrachloride or chlorobromomethane. Both have additives to depress the freezing point to $-50°F$. Carbon tetrachloride has a boiling point of $171°F$ and chlorobromomethane has a boiling point of $219°F$, and these agents are called "vaporizing liquids" because, although applied to a fire in the form of a liquid, they readily vaporize to form heavy gases that assist the extinguishing action by smothering the flames. Carbon tetrachloride-base extinguishers were first approved in 1912 and were the first extinguishers to be approved for use on electrical fires and fires in flammable liquids. Chlorobromomethane was developed as an extinguishing agent by the Germans during World War II and approved extinguishers using chlorobromomethane-base extinguishing agent first appeared in the United States in the latter part of the 1940s.

PORTABLE EXTINGUISHERS

Although no new vaporizing liquid extinguishers will be approved for sale, those now in use will remain approved with proper maintenance and inspection. Those presently approved range in size from 1 qt to $3\frac{1}{2}$ gal and the liquid is expelled either by pumping or by gas pressure from the stored pressure type using air, nitrogen, or carbon dioxide as expellant gases. Air and nitrogen are for all practical purposes insoluble in vaporizing liquids, but carbon dioxide readily dissolves in the liquids and the stream from a stored pressure extinguisher using carbon dioxide as the expellant gas is more of a spray than a straight stream because the carbon dioxide coming out of solution breaks up the stream into droplets. Underwriters' Laboratories listed vaporizing liquid extinguishers as suitable for use on incipient Class A fires and on fires in live electrical apparatus. "Specially designed" chlorobromomethane extinguishers are in addition approved for use on small flammable liquid fires.

The stream range of 1 to $2\frac{1}{2}$-qt extinguishers is 16 to 28 ft and the range of large extinguishers is 30 to 35 ft. The larger stored pressure extinguishers are equipped with nozzles permitting the use of either a straight stream or a spray stream, the latter being more effective than the straight stream for use on fires in flammable liquids.

The smaller extinguishers of this type are used primarily for protection of electrical equipment, industrial trucks, and over-the-highway trucks. The larger extinguishers are used primarily for the protection of electrical equipment. Vaporizing liquids have only a small fraction of the cooling effectiveness of water and therefore are not as effective on Class A fires as a similar quantity of water, and they are not as effective as multipurpose dry chemical extinguishers of similar weight capacity.

DANGERS OF VAPORIZING LIQUIDS

When vaporizing liquids are used on fires, corrosive fumes are liberated as the result of decomposition of the halogenated compounds. Accordingly, this type of extinguisher is not recommended where delicate electrical contacts may be involved or where fine tools or precision metal parts would be affected by corrosion.

The practical effect of the vapors released or produced when vaporizing liquids are applied to a fire has long been a matter of controversy. However, both the "Fire Protection Equipment List" of the Underwriters' Laboratories, Inc., and NFPA No. 10, "Portable Fire Extinguishers," warn against the use of extinguishers of this type in confined spaces and recommend that operators and others should take precautions to avoid effects which may be caused by breathing the vapors or gases produced. Considerable discussion and data on the toxicity of vapors of various fire extinguishing agents may be found in "The Fire Protection Handbook." Using data presented in this booklet, Table 20.4 presents the relative toxicities of agents that form gases or vapors. Carbon dioxide was assigned a value of one in this comparison because it is normally considered to be nontoxic and does not decompose when used on a fire. See chapter 25.

SPRINKLER UNITS (NO LONGER LISTED)

There used to be two vaporizing liquid automatically operated sprinkler units, one containing 2 qt carbon tetrachloride and 1 pt of ammonium hydroxide (the latter to provide the expellant pressure and to counteract corrosion) and the other containing 1 gal of chlorobromomethane pressurized with carbon dioxide. These units were intended primarily for protection where Class B fires might be expected and were limited in the volume of space that could be protected.

Dry Powders for Fires in Combustible Metals

APPROVED AGENTS

The first approved agent specially compounded for extinguishing fires in combustible metals was made available in 1940 as the result of the increasing

TABLE 20.4.   Relative Toxicity of Halogenated Agents as Compared with Carbon Dioxide[a]

Agent	Natural Vapor	Decomposed Vapor
Carbon dioxide	1.0	1.0
Halon 1301	0.8	32.9–47.0
Halon 1211	2.0	86.0
Chlorobromomethane (1011)	10.1	164.5
Carbon tetrachloride (104)	23.5	2190.0

[a]Ratio basis: carbon dioxide = 1.

use of magnesium in aircraft construction. This first approved dry powder is composed of a commercial grade of graphite to which an organic phosphate has been added. The graphite absorbs heat from the fire, and the organic material breaks down to a gas that penetrates the spaces between the graphite particles, excluding air. Because the powder cannot be discharged through pipe or hose, it is applied to a fire by means of a scoop or shovel.

A second dry powder for use on fires in combustible metals was approved in 1958. The major component of this dry powder is sodium chloride with a noncombustible thermoplastic material added to bind the particles of sodium chloride together when exposed to the heat of a fire. The dry powder contains other ingredients to produce good flow characteristics, and it is treated with metal stearates to produce moisture repellency. This agent cannot only be applied with scoop and shovel but it can also be discharged from fire-extinguishing equipment. Portable fire extinguishers ranging in size from 30 to 350 lb capacity, charged with this agent, are approved for use on fires in combustible metals.[3]

Although the two dry powders are approved for use only on magnesium, powdered aluminum, sodium, potassium, and sodium-potassium alloy (NaK), in practice they have been used successfully on fires in other combustible metals such as titanium, zirconium, and uranium. The sodium chloride-base dry powder, because of its good flow characteristics, can be applied through systems of piping and fixed nozzles and a number of such installations have been made. Such installations are used to protect liquid metal pumping, storage, and filtering systems where the liquid metal is used as a heat-transfer medium in connection with atomic power reactors. Under these conditions, fire fighting with manual equipment is impractical, and the extinguishing agent for the combustible metal must be applied through piping. A system typical of this type is designed to discharge 8000 lb of the sodium chloride base dry powder from four 2000-lb units. There has been a very limited amount of fire experience with such systems.

A third agent was listed in 1973 by the Underwriters' Laboratories for use on sodium, potassium, and sodium-potassium alloy fires where the use of a chloride is undesirable, for example in the vicinity of austenitic stainless steel where chloride stress corrosion is a potential problem.[26] The new agent is basically sodium carbonate and is equal to the sodium chloride-base agent as an extinguisher of sodium fires. It is listed for use by scooping or by application from 30 to 350-lb extinguishers, or from hose-line systems. It is now in use in several atomic plants utilizing sodium as the reactor coolant.

In these locations it is applied through piping and fixed nozzles because it has excellent flow characteristics.

Unapproved Agents

There are several nonapproved dry powders that have been specially compounded for the extinguishment of fires in combustible metals. One is primarily composed of a hard graphitic material that can be discharged from fire extinguishing equipment. Although this agent was originally compounded primarily for the extinguishment of fires in lithium, which are very difficult to extinguish because of the very low specific gravity of the liquid metal, it has been used successfully on many other combustible metals. There are two other dry powders specially compounded for use on fires in combustible metals and intended for application by scoop or shovel. The first is a shalelike material mixed with tar or pitch and ammonium chloride. The second originated with the United Kingdom Atomic Energy Authority and is a patented mixture of sodium chloride, potassium chloride, and either barium chloride or lithium chloride, in such proportions as to form an eutectic mixture having a melting point between 900° and 1100°F.[27]

Many materials have been used on metal fires with varying degrees of success. Some materials readily absorb moisture and must be kept in tightly sealed containers. Others are successful in extinguishing small fires where large masses of materials may be applied but may react with the burning metal if the fire is of considerable size. The following materials have been used on a variety of metal fires: foundry flux, lithium chloride, sodium chloride, zirconium silicate, Dolomite, soda ash, asbestos powder, talc, graphite, sand, and cast-iron borings.

Multipurpose dry chemical is not suitable for use on fires in combustible metals. Although under certain circumstances small magnesium fires may be extinguished, under other circumstances there may be a reaction between the burning magnesium and multipurpose dry chemical. The latter also reacts violently with burning lithium.

There are three liquid agents that are prepared especially for the extinguishment of fires in combustible metals. None is approved. One is a petroleum derivative applied from 2-qt and $2\frac{1}{2}$-gal stored pressure extinguishers. The liquid acts as a cooling and smothering agent, and under certain conditions it is flammable. However, the application of additional liquid, if available, is said to cool the burning material below the fire point. It is suggested for use on only small magnesium fires. A second liquid agent is composed primarily of peanut oil which is discharged from an extinguisher pressurized with carbon dioxide. The extinguisher has a capacity of $2\frac{1}{2}$-gal oil and $1\frac{1}{2}$ lb of carbon dioxide, the latter being in a separate container within the body of the extinguisher. A three-way valve allows first the discharge of the oil to act as a cooling and smothering agent with a final discharge of carbon dioxide to extinguish any fire which may result in the peanut oil. This agent also is suggested for use only on small magnesium fires.

Although no longer available, the best known liquid extinguishing agent for fires in combustible metals is trimethoxyboroxine (TMB). TMB is applied from a $2\frac{1}{2}$-gal stored pressure extinguisher which has a nozzle permitting use of either a spray or a straight stream. TMB is a flammable liquid organometallic similar in flammability characteristics to methyl alcohol. It burns with a green flame, liberating large volumes of dense white fumes. A glassy boric oxide coating remains on the metal. This coating will pick up moisture very rapidly from a humid atmosphere. The use of TMB as a fire extinguishing agent for combustible metals is patented. The Naval Research Laboratory developed the TMB extinguisher for use on magnesium castings and structural shapes under aircraft crash fire conditions. Under these conditions of application, fast extinguishment results which can be quickly followed up by foam or water. TMB extinguishers were used primarily by the United States Navy Bureau of Weapons on crash fire and rescue trucks. It is no longer used because of maintenance and storage problems with agent and hardware.

Although the flammability and heavy fuming characteristics of TMB may be undesirable under conditions of industrial fire fighting indoors, TMB is a good extinguishing agent for fires in oily or dry magnesium chips, turnings, or castings. It is also suitable for control and, in some cases, will completely extinguish fires in zirconium chips or turnings. TMB is *not* recommended for fires in sodium, potassium, sodium-potassium alloy (NaK) or lithium. TMB reacts violently with lithium and NaK.

Where combustible metals may ignite in enclosed equipment such as heat-treating furnaces, gaseous extinguishing agents may be used. Boron trifluoride and boron trichloride will control and sometimes extinguish magnesium fires in heat-treating furnaces. Boron trifluoride is considerably more effective than the boron trichloride. The gases may only control large fires which then may require the application of suitable dry powders for extinguishment. The fumes of these gaseous agents are toxic and irritating, and suitable breathing equipment should be available. See Chapter 25.

Argon, helium, and sometimes nitrogen, can also be used to extinguish fires in combustible metals in enclosed spaces. Caution should be exercised in the use of nitrogen, however, because this gas reacts readily with certain burning metals. If there is uncertainty as to whether a metal will burn in a nitrogen atmosphere, the application of a small amount of water to the residue of the metal burned in air will result in an odor of ammonia if the metal reacts with nitrogen. Argon and helium, of course, will not react.

Steam

Steam is the oldest of the smothering agents although it is seldom used in modern practice, having been replaced by carbon dioxide and other inert gases. It is most effective when used for the protection of relatively small spaces such as ovens at temperatures of 225°F and higher. The visible cloud of condensed vapor has little smothering action and will act mainly through cooling. Large volumes of steam are usually necessary because of the rapid condensation that takes place unless the walls of the enclosure are above the boiling point of water.

Steam should not be used for protection against fires in cotton, paper, wood pulp, and similar carbonaceous materials, because the charcoal produced by the fire will reignite by spontaneous heating very quickly after the steam has been shut off and air is introduced. The best information on steam smothering systems is in NFPA No. 86 "Standards on Class A Ovens and Furnaces."[28]

Steam hose lines are sometimes used for extinguishing small fires in flammable liquids such as those that may occur from a leak in a pipe line within a refinery.

20.3   MECHANISM OF EXTINGUISHMENT

Flames exist when there is a fuel, a substance which readily reacts with the fuel, and sufficient heat to speed the reaction to a point where visible and infrared radiation takes place. A familiar, and oversimplified, version of extinguishment states that it is accomplished by removal of the fuel (by blanketing with foam or interposing a layer of inert gas between the fuel and the flames), by removal of oxygen (by dilution with inert gases or vapor), or by removal of heat (by cooling with water or other extinguishing agents). All of the foregoing explanations of extinguishing action are correct but are insufficient in themselves to explain why loaded stream extinguishers are more effective than plain water extinguishers, why Halon 1301 is more effective than carbon tetrachloride, or why potassium bicarbonate-base dry chemical is more effective than sodium bicarbonate-base dry chemical.

It can be said with considerable confidence that water, whether in the form of a straight stream or a spray, with or without wetting agent, acts for all practical purposes entirely as a cooling agent for the removal of heat. Wood and similar Class A materials burn only when heated to the point where destructive distillation results in the evolution of combustible vapors. The cooling action of water stops the evolution of combustible vapors and the fire is extinguished.

Water spray and "fog" are practically effective only on flammable liquids having flash points in excess of 150°F. It is therefore obvious that even on Class B fires the cooling action of the water is of greatest importance.

The cooling action of water sometimes allows extinguishment of fires by application at very high rates of flow where low rates of flow would only increase the intensity of the fire because of chemical reaction between water and the burning material. One familiar example is the extinguishment of a fire in charcoal. If water is applied at too low a rate of flow, the well-known water-gas reaction takes place:

$$H_2O + C \rightarrow H_2 + CO$$

At high rates of application, the cooling effect of the water reduces the temperature to a point where the foregoing reaction will not proceed.

Another good example is the extinguishment of fires in magnesium castings. Applied at too low a rate, the water reacts with the burning magnesium:

$$H_2O + Mg \rightarrow H_2 + MgO$$

Applied at high rates of flow, the water cools the castings to the point where combustion no longer is sustained. Loaded stream extinguishers containing a solution of potassium carbonate are more effective than extinguishers containing plain water not only on Class A fires but on Class B fires also. There are only theories as to why the addition of potassium carbonate to water increases the extinguishing effectiveness.[29] However, in addition to the cooling action of water it is possible that the heat of the fire evaporates the solution to leave solid particles of potassium carbonate that act in the same way as dry chemical to interrupt and suppress the chain reaction of the flame.[30]

Carbon dioxide can be considered for all practical purposes to act as a diluent to both fuel vapors and oxygen. When sufficient carbon dioxide is introduced into the fire area, the reacting materials are diluted to the point where the reaction cannot proceed with sufficient rapidity for the flames to exist. Where carbonaceous material is involved in a fire, however, reignition takes place unless it is possible to maintain a sufficient concentration of carbon dioxide until the previously burning material has cooled.

Although the action of halogenated extinguishing agents, such as Halons 1301 and 1211, was once thought to be entirely because of cooling action and smothering because of the dilution of fuel vapors and air, it is now believed that the major extinguishing mechanism results from pyrolysis of the halogenated hydrocarbons and reaction of the resulting radicals with the active radicals present within the flame zone. A halogenated hydrocarbon containing bromine is a more effective extinguishing agent than a halogenated hydrocarbon containing only chlorine, but the effectiveness does not depend only upon the percent of bromine in the molecule. Published data do not satisfactorily explain why Halon 1301 (bromotrifluoromethane) has such high extinguishing effectiveness as compared with other halogenated hydrocarbons.

The extinguishing action of foam, including AFFF, is obvious in that it forms a barrier that prevents contact between fuel vapors and air.

The extinguishing action of dry chemicals has been given a great deal of study in recent years, but no one theory as to the mechanism involved has gained complete acceptance. Originally, it was thought that sodium bicarbonate decomposed upon striking flame to give off carbon dioxide and water vapor and extinguish the fire by simultaneously cooling and smothering. It is now kown that, although this may occur, it is not an important factor because equally finely divided sodium chloride treated to flow and discharge properly is an effective dry chemical extinguishing agent. One modern theory of the extinguishing action of dry chemical is that the fine particles capture sufficient of the free radicals in the flame zone to interrupt the chain reaction and suppress the flame practically instantaneously.[30] Opponents of this theory have pointed out that the mathematics involved, which relate to the diffusion of free radicals from the flame zone to the surface of the dry chemical particles, can equally well support a theory that the dry chemical extinguishing action is because of transfer of heat from the flame to the particles.

The relation between particle size and surface area of dry chemicals and the extinguishing effectiveness has been reported by several sources.[30,31] Experimental and theoretical studies have showed that the extinguishing effectiveness increases as particle size decreases and surface area increases. The most recent theory on the extinguishing mechanism of dry chemicals is that evaporation takes place and that the evaporated material acts as a flame inhibitor.[32] Using this theory also, the smaller the particle size, the more rapidly the particles will be heated to a temperature at which evaporation can proceed. The theory also takes into account "metal availability" in the salt which is inconsistent with the experimental evidence that ammonium phosphate-base dry chemical is more effective as an extinguishing agent on Class B fires than sodium bicarbonate base dry chemical.

In summary, presently there is no completely satisfactory explanation as to how powdered solids extinguish fires or completely satisfactory explanation for the greater extinguishing effectiveness of ammonium phosphate and potassium bicarbonate.

The effectiveness of monoammonium phosphate-base multipurpose dry chemicals on fires in Class A materials is probably because of a dual extinguishing action.

The flames are extinguished by an unknown mechanism as discussed in the foregoing paragraphs. The monoammonium phosphate is decomposed by the heat and leaves a residue of phosphoric acid or phosphoric anhydride on the material which had been burning. This residue prevents afterglow which would lead to renewed flaming. The use of monoammonium phosphate-base dry chemical on a wood fire results in a fire retardant coating that is very effective in preventing reignition. This was discovered in trying to dispose of the charred 2 × 4's remaining from large Class A fire tests in which 18,000 board feet of lumber was used. The 2 × 4's on which water had been used were easily disposed of by burning, but a pile of charred 2 × 4's on which multipurpose dry chemical had been used resisted reignition to such an extent that it was necessary to haul them to the public dump.

20.4. COMPARATIVE EXTINGUISHING EFFECTIVENESS OF AGENTS

Evaluation of the effectiveness of extinguishing agents can be accomplished by several methods, but the final criterion is the effectiveness in actual use. Data gathered on the laboratory bench are of value only for the preliminary screening of agents for special applications. For example, data obtained by the peak flammability explosion buret method[33] do not correlate with fire extinguishing effectiveness data obtained with fixed nozzles and a 3-ft diameter gasoline fire.[30] The data obtained in tests with fixed nozzles may be used to predict the effectiveness to be expected in practice with fixed-nozzle local application systems.

In the use of manually operated fire extinguishing equipment, the evaluation is really that of a combination of agent and equipment, and many variables can affect the results. For example, a fire fighter skilled in the use of one agent may not be adequately skilled in the use of other agents. Fire tests made with flammable liquids in depths of over 1½ in. are much more severe than fire tests that simulate spill fires. Also, an agent that may be excellent for extinguishing fires in flammable liquids in tanks or on spill fires, may be valueless when the flammable liquid is escaping under pressure or even just flowing. The extinguishment of flammable liquid fires indoors is more difficult for the less experienced operator than if the same size fire is attacked out of doors. The Underwriters' Laboratories uses wood panel fires and wood crib fires for determining the Class A rating of fire extinguishers, and observation of many tests shows that water extinguishers are more effective on the panel fires than on the crib fires whereas multipurpose dry chemical extinguishers are more effective on the crib fires.

Two fire extinguishers containing exactly the same amount of the same agent may have widely different degrees of fire extinguishing effectiveness. Two important factors that determine effectiveness are the type of stream expelled from the nozzle and the rate of application of the agent.

The rate of application of an extinguishing agent is of critical importance. If the rate is too low, the fire may not be extinguished, and if the rate is too high the supply of agent may be exhausted before the fire is extinguished.[30,34] Where unlimited water supplies are available, it is obvious that the higher the rate of application, the sooner the fire will be extinguished. However, the statements regarding rates of application are also applicable to portable fire extinguishing equipment, to hose-line systems using special extinguishing agents, and even to systems employing fixed nozzles.

TABLE 20.5. Maximum Effectiveness on Class A Fires of Underwriters' Laboratories Listed Extinguishers[14] (Weights and Volumes Denote Nominal Charges)

Underwriters' Laboratories Class A Ratings	Multi-Purpose Dry Chemical[a] [lb]	Halon 1211[a] [lb]	AFFF Foam[b] [lb (gal)]	Water, Soda-acid Foam[b] [lb (gal)]	Loaded Stream[b] [lb (gal)]	Calcium Chloride Solution [lb (gal)]	Water with Wetting Agent [lb (gal)]
1-A		9		12.5 (1½)			
2-A	2¾	13				26.3 (2½)	12.5 (1½)
3-A	6		20.8 (2½)		23 (2½)		
4-A	8						
6-A							
10-A	10						
20-A	18		275 (33)				
30-A							375 (45)
40-A	125						500 (60)

[a] Also listed as suitable for use on Class B and Class C fires.
[b] Also listed as suitable for use on Class B fires.

TABLE 20.6.   Underwriters' Laboratories Class B Test

Pan Size (ft²)	Minimum Effective Discharge Time (s)	Rating Class	Place Run
2½	8	1-B	Indoors
5	8	2-B	Indoors
12½	8	5-B	Indoors
25	8	10-B	Indoors
50	8	20-B	Indoors
75	11	30-B	Outdoors
100	13	40-B	Outdoors
150	17	60-B	Outdoors
200	20	80-B	Outdoors
300	26	120-B	Outdoors
400	31	160-B	Outdoors
600	40	240-B	Outdoors
800	48	320-B	Outdoors
1200	63	480-B	Outdoors
1600	75	640-B	Outdoors

The Underwriters' Laboratories in its test and examination of equipment submitted for listing evaluates the extinguishing effectiveness of agent and equipment as an integral unit. For example, in order to obtain a Class A rating the submitted equipment must be able to extinguish standard wood crib, wood panel, and excelsior fires. The extinguishment of these fires is a function of the rate of discharge, the amount of extinguishing agent, and the type of discharge stream. Until recently the Underwriters' Laboratories, Inc. had no standard fire tests for ratings in excess of 4-A and water extinguishers were assigned ratings based on 75% of that which would be obtained by using proportionate values given to Class A extinguishers of lower capacity. However, when wheeled multipurpose dry chemical extinguishers were submitted for examination and test, the Laboratories developed evaluation procedures using large wooden cribs made up of 2 × 4's. Tests with these large standard Class A fires verified the ratings previously given to large water extinguishers.

A fire extinguisher having a capacity of 1¼ gal water and otherwise meeting the requirements of the Underwriters' Laboratories will, when used by an expert operator, consistently extinguish a standard 1-A test fire. Table 20.5 shows the maximum effectiveness on Class A fires of Underwriters' Laboratories listed extinguishers where only the highest ratings for each size of extinguisher are used.

The Underwriters' Laboratories procedures for assigning Class B ratings are considerably more involved than the procedures used for assigning Class A ratings. The fire tests are run using square steel pans with heptane as the

TABLE 20.7. Maximum Effectiveness of Underwriters' Laboratories Listed Extinguishers[14] for Class B Fires (Weights and Volume Denote Nominal Charge)

Underwriters' Laboratories Class B Ratings	Sodium Bicarbonate Base[a] [lb]	Multi-purpose[a] [lb]	Potassium Chloride[a] [lb]	Potassium Bicarbonate[a] [lb]	Urea-Potassium Bicarbonate[a] [lb]	Carbon Dioxide[a] [lb]	Halon 1301[a] [lb]	Halon 1211[a,b] [lb]	AFFF Foam[a] [lb (gal)]	Loaded Stream[b] [lb (gal)]
1-B										28.7 (2½)
2-B	1	1					2½			
5-B	2					5		2½		
10-B	2½	2¼	2¾	2½		10		5		
20-B		5½[b]	6	4½		50			20.8 (2½)	
30-B	5¾			5¾						
40-B	6	6[b]	9½			100		13		
60-B	10	8[b]	12	6						
80-B		18[b]	22	10	9			17		
120-B	20	30[b]		20	17					
160-B			135		23				275 (33)	
240-B	150	125[b]		125						
320-B	350	300[b]								
480-B					175					
640-B				300						

[a]Also listed as suitable for Class C fires.
[b]Also listed as suitable for Class A fires.

394

**TABLE 20.8   Approximate Relative Extinguishing Effectiveness of Underwriters'
Laboratories Listed Extinguishers on Class B Fires (Basis: Comparative weights of
extinguishing agents for equal class B rating)**

Agent	Relative Effectiveness
Urea-potassium bicarbonate-base	100
Potassium bicarbonate-base	91
Multipurpose	75
Sodium bicarbonate-base	57
Potassium chloride-base	55
Halon 1211	53
Halon 1301	26
Carbon dioxide	21
AFFF	7

fuel. Ordinary motor gasoline is not used because of the variation in volatility in winter and summer. The fuel is 2 in. in depth with 6 in. of tankside above the surface of the fuel. The fire is allowed to burn for one minute before being attacked. The minimum effective discharge times vary but are no less than 8 s. Table 20.6 shows the relationship.

The Underwriters' Laboratories uses expert operators to determine the maximum area extinguished consistently by one extinguisher. To obtain the Class B rating, the maximum area extinguished as determined by test is multiplied by 0.40 to give the rating. The latter is intended to represent the flammable liquid area that might be expected to be extinguished by inexperienced operators. Fire tests with hand portable extinguishers are run indoors in a test house having a 40 by 60 ft floor area with the ceiling height in excess of 20 ft, and always with large roof vents opened. The largest indoor fire is 50 ft². Extinguishing equipment is tested on outdoor fires but only when the wind velocity is less than 10 mph. The maximum effectiveness of Underwriters' Laboratories listed extinguishers on Class B fires is shown by Table 20.7.

It will be noted that 20-B ratings are obtained with potassium bicarbonate-base dry chemical extinguishers of only 4½-lb capacity. The same rating is given to extinguishers of this type containing 18 and 27 lb of dry chemical because of design characteristics.

Using the data of Table 20.7 and taking into account the variations because of size, the extinguishers may be assigned *approximate* values to indicate relative effectiveness as shown in Table 20.8.

The Underwriters' Laboratories assigns Class C ratings to extinguishers having Class B ratings where the extinguishing agent is not a conductor of electricity.

In the past, extinguishers rated for use on both Class A and Class B fires were not rated for use on Class C fires because the extinguishing agents were

TABLE 20.9   Maximum Effectiveness of Extinguishers Listed by Underwriters' Laboratories as Suitable for Use on Both Class A and Class B fires. (Weights and Volumes Denote Nominal Charge)

Underwriters' Laboratories Class A:B Rating	Multipurpose Dry Chemical[a] [lb]	Halon 1211[a] [lb]	Loaded Stream [lb (gal)]	AFFF Foam [lb (gal)]
1A:10B	2½	9		
2A:10B	2¾	13		
2A:40B				
2A:60B		16		
3A:1B			28.7 (2½)	
3A:20B				20.8 (2½)
3A:40B	6			
3A:80B		17		
4A:60B	8			
4A:80B		20		
10A:60B	10			
20A:80B	18			
40A:240B	125			
40A:320B	300			

[a]Also listed as suitable for use on Class C Fires.

electrically conductive. Also, extinguishers rated for use on both Class B and Class C fires were not sufficiently effective on Class A fires to obtain Class A ratings. Multipurpose dry chemical and Halon 1211 being electrically nonconductive and also effective on Class A fires as well as Class B fires, have permitted the listing of estinguishers for all three classes of fires. Because there are many locations where protection must be furnished for more than one class of fire, Table 20.9 lists extinguishers approved as suitable for use on both Class A and Class B fires.

There are three agents approved for use on fires in combustible metals. Because of the complexities of determining comparative extinguishing effectiveness, no numerical ratings are assigned. Reference must be made to the respective lists of approved equipment for information as to the combustible metals for which the extinguishing agents are considered suitable and for guidance as to the quantities of agents required on various metal fires.[14,35]

The coordination of fire protection systems in a chemical operation is beyond the scope of this chapter, but the reader is referred to his insurance carrier, to the NFPA, and to reference 35a.

REFERENCES

1.   National Fire Protection Association, Standard No. 10, "National Fire Codes, 1976."
2.   *Fed. Regis.* (Sept. 6, 1958).

3. National Fire Protection Association, Standard No. 48, "National Fire Codes, 1976."

4. A. W. Bertschy, H. E. Moran, and R. L. Tuve, "Lithium Chloride Solution for Fire Extinguishers Exposed to Low Temperatures," Naval Res. Lab. Rept 4853, 1956.

5. National Fire Protection Association, Standard No. 18, "National Fire Codes, 1976."

6. Manufacturers Mutual Fire Insurance Co., "The Factory Mutuals, 1835–1935," Boston, Mass., 1935.

7. Factory Mutual Engineering Division, *Handbook of Industrial Loss Prevention,* McGraw-Hill, New York, 1959.

8. National Fire Protection Association, Standard No. 15, "National Fire Codes, 1976."

9. National Board of Fire Underwriters, "Characteristics of Water Spray Nozzles for Fire Fighting Use," 1944.

10. A. F. Matson and R. E. DuFour, "The Mechanism of Extinguishment of Fire by Finely Divided Water," National Board of Fire Underwriters Res. Rept. 10, 1955.

11. P. Nash, "Some New Techniques in Fighting Fires," *Fire,* 541 (1956).

12. D. Hird, "Extinguishing Room Fires with High and Low Pressure Water Sprays," *V.F.D.B. Zeit,* 9 (special issue), 58–61 (1960).

13. R. L. Tuve, "A New Vapor-Securing Agent for Flammable Liquid Fire Extinguishment," *Naval Res. Lab. Rpt.* 6057 (1964).

14. "Fire Protection Equipment List," Underwriters' Laboratories, Inc., Northbrook Il 1981.

15. "National Fire Codes," Standard No. 11A, National Fire Protection Association, 1977.

16. Factory Mutual Engineering Division, "Cotton Process Fires," "Loss Prevention Bulletin 12.75," Norwood, Mass., 1948.

17. R. R. Neill, "The Hydrocarbon Flame Extinguishing Efficiencies of Sodium and Potassium Bicarbonate Powders," Naval Res. Lab. Rpt. 5183 (1958).

18. H. B. Peterson, R. L. Tuve, R. R. Neill, J. C. Burnett, and E. J. Jablonski, "The Development of New Foam Compatible Dry Chemical Fire Extinguishing Powders," Naval Res. Lab. Rpt. 4986 (1957).

19. National Fire Protection Association, Standard No. 17, "National Fire Codes, 1976."

20. E. D. Zeratsky, "Special Dry Powder Extinguishing Agents," *NFPA Q.,* **54,** 154 (1960).

21. E. L. Quinn, and C. L. Jones, *Carbon Dioxide,* ACS Monograph, Reinhold, New York, 1936.

22. National Fire Protection Association, Standard No. 12, "National Fire Codes, 1976."

23. "National Fire Codes," Standard No. 12A, National Fire Protection Association.

24. *Fire Protection Handbook,* 14th ed., National Fire Protection Association, Quincy, Mass., 1976.

25. "National Fire Codes," Standard No. 12B, National Fire Protection Association.

26. "Report on Extinguishers and Agent for Use on Combustible Metal Fires, File Ex 2130, Ex 2095," Underwriters' Laboratories, Northbrook, Il., 1973.

27. L. H. Cope (to the U.K.A.E.A.) Brit. Pat. 884,946 December 20, 1961.

28. National Fire Protection Association, Standard No. 86A, "National Fire Codes, 1976."

29. C. A. Thomas and C. A. Hochwalt, "Effect of Alkali-Metal Compounds on Combustion," *Ind. Eng. Chem.,* **20,** 575 (1928).

30. A. B. Guise, "The Chemical Aspects of Fire Extinguishment," *NFPA Q.,* **53,** 330 (1960).

31. D. Hird and M. J. Gregsten, "Particle Size and Efficiency of Dry Powders," *Fire Prot. Rev.,* **19,** 473 (1956); *Fire,* **49,** 101 (1956).

32. W. A. Rosser Jr., S. H. Inami, and H. Wise, "Study of the Mechanisms of Fire Extinguishment of Propellants," *ASD Technical Rpt., 61-143,* Aeronautical Systems Division, Wright-Patterson Air Force Base Ohio, 1961.

33. H. E. Moran Jr., "Developments in Fire Extinguishers," Proceedings of the Symposium on Fire Extinguishment Research and Engineering, U.S. Navy Civil Engineering Research and Evaluation Laboratory, Port Hueneme, Calif., 1954.

34. A. B. Guise and J. A. Lindlof, "A Dry Chemical Extinguishing System," *NFPA Q.*, **49**, 52 (1955).

35a. Factory Mutual Engineering Division, "Approval Guide," Norwood, Mass., 1977, or most recent edition.

35b. *National Safety News*, entire issue on fire protection, **90**, 1 (July 1964). (revised annually.)

21
Fighting Chemical Fires with Foam

D. N. Meldrum

21.1. USES OF FIRE-FIGHTING FOAM

Effective fire protection is a basic part of safety, loss control, and accident prevention in chemical operations. Proper selection and use of fire extinguishing agents and systems can reduce significantly the losses from fire resulting from mechanical failure of equipment or from human error. Fire protection engineering technology has developed rapidly in the last decade and a half with much of this development being made in devices, agents, and systems for protection of flammable liquid processing and storage areas. Fire fighting foam has been used for more than 50 years with continuing evolution of approved foaming agents and equipment. As chemical technology has advanced with concomitant development of new compounds, mixtures, processing techniques, and methods of transport, so too has the requirement for adequate fire control and extinguishing means.

Low expansion fire-fighting foam, as defined in NFPA Standard 11, is used primarily for the control and extinguishment of fires involving flammable and combustible liquids but it also has significant value for prevention of fires involving these products by providing a vapor seal over spills. In addition, because of its water content, it can be used for extinguishment of fires involving Class A combustibles, for temporary thermal insulation of structures. High expansion foam, NFPA 11A, is primarily used for volumetric fill of enclosed spaces either for fire extinguishing purposes or for purging purposes.

Fire extinguishing foam is a quasi-stable suspension of air or gas in an aqueous suspending medium. The water used for foam production is modified by addition of surface active chemicals which confer foaming ability, heat resistance and water retention ability to the foam.

A good fire-fighting foam must be fluid enough to flow progressively across a flammable liquid surface. It must have sufficient resistance to radiant heat to prevent its destruction by flames present during the extinguishing process. It must have sufficient inherent stability and water retention ability to provide a lasting vapor seal and to absorb heat from metal surfaces with which it comes in contact. In addition, it must have sufficient chemical stability on the fuel to which it is applied so that the rate of foam coverage of the fuel exceeds the rate of destruction of the foam by the fuel. A good fire-fighting foam tends to adhere to solid surfaces and also has cohesive properties.

The extinguishing mechanism of fire-fighting foam is threefold. As it spreads over a liquid surface, it cools the liquid and adjacent solid surfaces, separates the flames and oxygen from the fuel surface and suppresses release of flammable vapors from the fuel. Because of its inherent stability as a smothering blanket, foam also prevents reflash and reignition.

Foam is considered as the primary cost-effective extinguishing agent for large flammable and combustible liquid fires and for this reason, it is particularly suited for protecting storage tanks, dip tanks, drainboards, loading racks, pump pits, and similar hazard areas where flammable or combustible liquids are stored or processed. Foam also can be used in cooling spray systems to add to the distribution of cooling agent and to enhance fire control of spilled pools of flammable liquids beneath process equipment. Foam also is used as the primary extinguishing agent for aircraft hangars and aircraft crash rescue vehicles and is used for general fire-fighting purposes in refineries, chemical plants, storage terminals and on tankers and barges.

21.2. TYPES OF FOAM CONCENTRATES

The development of fire-fighting foam paralleled the growth of the petroleum and chemical industry. The early foams were generated chemically by mixing a solution of alum with a solution of bicarbonate of soda and licorice-root extract. Since large volumes of solution and lead-lined vessels were required, storage was expensive. The development of a dry single-component system followed. This was a marked improvement, but still, 1 lb of foam powder was needed to generate 1 ft^3 of foam. Typically sized equipment required one to four 50-lb cans of powder per minute. The logistics necessary to supply the foam chemical were more difficult than actually fighting the fire. Finally, the trend to larger tanks for the storage of petroleum products made chemical foam systems impractical.

Mechanically generated foam from liquid concentrates was first developed in Germany in the 1930s. While these early concentrates produced a poor quality foam compared to today's mechanical foams, they provided a great advantage over the chemical foam powders because 1 gal produced more than 300 gal of foam. Also, the liquid concentrates could be introduced into a water system more easily than the powdered materials.

Foam systems can be characterized by the line diagrams. For most foam systems, the following pertains:

Proportioning
Water + foam concentrate → foam solution

Generation
Foam solution + air ⇆ foam

The first equation above represents what happens when the concentrate is introduced into a water supply or into a pipe line (using a proportioner). As shown, this function is irreversible. The second equation shows what happens when the solution is made into foam with air added by a foam maker or foam nozzle. The fact that this part of the system is reversible shows that most foam wants to return to the foam solution state and that proper system design requires that foam be generated at an adequate rate.

Foam systems have two other requirements. There must be a driving force or energy to move water and foam concentrate from a source or storage tank through piping to the point of discharge. A means of forming or directing the foam discharge onto a fire is also needed. This may be a nozzle (portable) or fixed-discharge device.

Modern mechanical foams are produced by mixing air with the foaming solution. Although compressed air can be used, economics limits practical fire equipment to using air inspirated by a venturi effect. The mixing needed for foam generation is done with a foam maker or foam nozzle. Expansion ranges of 4–12 (i.e., volume of foam generated divided by volume of solution used) are typical for foam designated "low-expansion." High-expansion foam generally has expansions in excess of 100 to 1.

The foam industry has developed a wide variety of foam concentrates to meet specific requirements. The following generalizations for use of low-expansion foams, however, set the application boundaries:

The hazard must be a liquid.

The hazard must be below its boiling point at the ambient condition of pressure and temperature.

Care must be taken if the bulk temperature of the liquid is higher than 100°C.

The hazard must not be unduly destructive to the foam selected.

The hazard must not be water-reactive.

The fire must be a surface fire. Three-dimensional fires cannot be extinguished by foam unless the hazard has a relatively high flashpoint and can be cooled by the water in the foam.

The quantity of foam applied and rate of application must be matched to the size, fuel and character of the fire. See Fig. 21.1.

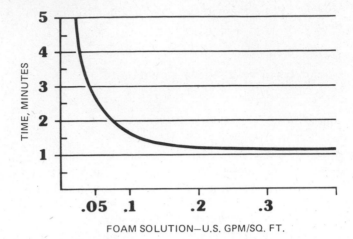

The above curve may vary, depending on fuel, type
of foam liquid and method of application. Carefully
engineered systems must be based on actual tests.

Figure 21.1.   Typical time-rate curve.

Regular protein foam is prepared from purified animal or vegetable protein, stabilizers (including iron salts), necessary solvents and an industrial germicide to protect against bacterial decomposition in storage. This material has been available since the early 1940s. It is produced in 6% and 3% concentrates (6% = 6 parts concentrate plus 94 parts water) in the United States and in a range of concentrations—primarily 4%—in foreign countries. (See Table 21.1 on characteristics of low expansion foam.)

Regular foam (3% and 6%) is suitable for use on nonpolar hydrocarbon fuels such as crude oil and petroleum products. The primary advantages of this foam are resistance to burnback, good extinguishment, and low cost. It is also biodegradable, nontoxic, noncorrosive, easy to clean up, and can be stored at 20° to 120°F. Special formulations are available with low temperature ratings of −20°, and −40°F. These foams must be used where potential storage temperatures are lower than for regular foams.

Fluoroprotein forms are a recent improvement of regular protein foam. The addition of a fluorocarbon surfactant to protein foam retains the good characteristics of the original foam and provides the additional benefits of improved dry chemical compatibility and increased fire resistance. It also makes subsurface foam application practical. Like regular foam, fluoroprotein foam is available in 3% and 6% grades. It can, therefore, be substituted in existing systems without changing the system.

The superiority of fluoroprotein foam has been demonstrated in several recent fires in which rapid extinguishment was achieved when regular protein foam or AFFF was having difficulty. The theory behind the advantages of fluoroprotein is that the highly fluorinated surfactant orients in the bubble

TABLE 21.1. Characteristics of Low Expansion Foams

Foam Concentrate	Percent Concentration	Typical Storage Temperature Limits	Application	Hazards
Regular protein	3 or 6	20°–120°F	Type I or II	Hydrocarbons
Low temperature	3 or 6	−20° or −40° to 120°F	Type I or II	Hydrocarbons
Fluoroprotein	3 or 6	20°–120°F	Type I, II, or subsurface	Hydrocarbons
Protein-based polar solvent	6	20°–120°F	Type I only	Polar solvents
Polymer-based polar solvent	20 + 3	35°–80°F	Type I or II	Polar solvents
Aqueous film-forming foam	3 or 6	35°–120°F	Type I or II	Hydrocarbons
Polymer-based AFFF	6–10	35°–120°F	Type I or II	Hydrocarbons and polar solvents

wall with the fluorine-containing "tails" outside the bubble. Since the fluorinated "tail" is also incompatible with the fuel, it tends to make the bubble
shed the fuel and, therefore, protects the integrity of the foam. This is
particularly important in subsurface foam application. As with regular protein foams, fluoroprotein foam is for use on nonpolar hazards.

Special protein-based foams (alcohol foams), have been developed for
fire control of polar solvent fires. Regular protein, fluoroprotein and surfactant foams are rapidly destroyed by solvent attack of polar solvents. A
simple way to determine if a polar solvent foam is needed is to check the
chemical composition of the fuel. If oxygen, nitrogen, or sulfur is present,
the hazard probably requires a polar-solvent type or other special foam. In
such cases, the foam manufacturer should be contacted for recommendations.

Two types of protein-based foams for polar-solvents are available. One is
an alkaline material with a pH of approximately 9. In this material, the
active system is based on zinc tetra-amine soap. The second uses a so-
lublized aluminum soap. This slightly acid material has a pH of approximately 6. These types are, of course, incompatible and should never be
mixed.

Protein-based polar-solvent foams work this way. An insoluble soap is
formed when the concentrate is diluted with water. If the soap is distributed
evenly in the bubble wall, a stabilized system is formed. As some of the foam
drains, a diluted layer is formed on the surface of the fuel. This layer helps to
protect the remaining stabilized foam.

Limitations caused by this mechanism are extremely important. If a polar
solvent fire is to be extinguished, the following must be remembered:

The foam must be applied gently. NFPA No. 11 recognizes this application as Type I. If the diluted layer is disturbed by too vigorous application,
the foam will continue to break down until the bulk of the liquid is diluted to
a less active state. When the active soap is precipitated, it will agglomerate if
too much time is allowed between mixing and foam formation. This period is
called transit time. Foams of this type have acceptable transit times of from
less than 5 s to about 2 min under average conditions. With warmer water,
the allowable transit time decreases conversely, it increases when colder
water is used.

Polymer-type, polar solvent foams were developed to extinguish fires in
aliphatic (fatty) amines. These particular compounds were extremely destructive to existing protein-based polar solvent foams. The polymer-type
foams proved tough enough to be applied by Type II or less gentle application. For the first time, foam could be practically applied in depth to polar
solvents by means of a nozzle. This was of immediate benefit to the shipping
industry, because gentle application on tankers is impossible. This meant
that large chemical tankers could be built with a practical means for deck fire
protection. Performance is extremely good on water-soluble compounds. Water

semisoluble and some water-insoluble compounds may require higher application rates.

Synthetic-hydrocarbon surfactant foams (wetting agents) have been on the market for more than 25 years, but they have not had great success. These foams are compatible with hydrocarbons and are fast-draining compared to protein foams. Hydrocarbon compatibility causes wicking of the fuel, and rapid drainage weakens the foam blanket. The result is poor sealability, poor fire-resistance characteristics, or resistance to reflash.

Some of these agents have UL listings as wetting agents or penetrants for Class A fires; a few are UL listed for Class B fires. The specifying engineer must be careful to discern which materials fall in this class and to avoid these materials except in special cases. Where wetting agent foams are required, UL-listed high expansion foams used through low-expansion foam-generating equipment generally do an excellent job.

Surfactant foams are usually corrosive to mild steel and should not be held in contact with brass for extended periods of time. Most engineering plastics are satisfactory.

Aqueous film-forming foams are products of modern chemistry. They are not found in nature and have unusual properties. By tailoring molecules to have low spreading coefficients and high resistance to hydrocarbons, aqueous film-forming foams (AFFF) were created.

AFFF has the fastest extinguishment rate of any known foam on spill fires. This quality is best applied in such areas as crash rescue for airplanes and helicopters and for petroleum processing spill fires. But because of its rapid drainage rate and relative lack of resistance to burnback, AFFF is generally less effective than protein or fluoroprotein foam for use on tank fires. As with other synthetic concentrates, particular care in choosing materials of construction for handling these concentrates is required.

21.3. SELECTION OF FOAM FOR USE

With a wide variety of different types of fire-fighting foam agents on the market, selection of which agent to use for a particular hazard becomes important. There are some foam agents which have very limited use. There are others, which are most recently developed, that can be used on most flammable and combustible liquids. Selection of the proper agent cannot only provide a higher degree of fire protection capability but can also result in significant savings in design of an effective system. In general, fire-fighting foams may be classified as follows:

1. Foams suitable primarily for use on spills of water immiscible hydrocarbons including the regular protein-base foams, fluoroprotein foams, or AFFF agents. Synthetic hydrocarbon surfactants, and certain wetting and emulsifying agents also may be used.

2.   Spills of water soluble and polar solvents require foams of the "alcohol type." Regular type or even synthetic foams can be used with limited effectiveness on spills of such materials when the application rate is high enough to dilute the fuel to 50% or more after which such foam generally becomes stable.

3.   Hydrocarbon and water immiscible product stored in depth or in areas where sufficient hot metals or other reignition sources may be present require the use of regular type or fluoroprotein foams. AFFF agents and other surfactant foams may be used but because of their lower inherent stability and less heat resistance, they may not provide sufficient vapor seal or in some cases may not even seal against hot metal surfaces.

4.   The fires involving water soluble and polar solvents in depth require the use of approved alcohol-type foams.

In recent years, foams have been developed which are highly effective on both hydrocarbon and polar solvent-type fuels. They are based on polar solvent resistant polymers and may also contain film-forming fluorinated surfactants. These foams are generally more expensive but because of their greater utility over a wider variety of fuels, they can be the most cost-effective if a variety of fuels are present in a given hazard area. The low molecular weight polar solvents such as methyl, ethyl, isopropyl and amyl alcohol, acetaldehyde, propionaldehyde, acetone, methyl ethyl ketone, methyl isobutyl ketone, methyl acetate, ethyl acetate, and diethyl ether require the use of the "universal" type foams because of the water solubility and tendency to destroy water-based foaming agents. As the molecular weight of such products increases, their flash point increases but their anti-foam tendency also increases. For this reason, even though such products may only be partially water soluble, and the cooling effect of water tends to enhance extinguishment, they also should be protected by the alcohol type or approved polymeric foams which have greater inherent stability on products of a polar nature.

Regardless of fuel, fire-fighting foam must be applied at an application rate sufficient to overcome the destructive forces working against it, and it must be applied in such a way that the foam can either be equally dispersed over the fuel or has sufficient horizontal velocity to cause it to flow the required distance across an open fuel surface. Guidelines covering the application, selection, and equipment requirements for foam systems are given in NFPA 11 for general applications. NFPA 409 gives specific design requirements for foam systems for hangar protection while NFPA 403 gives specific recommendations for use of foam in crash rescue operations. NFPA 325 provides recommendations as to general types of foam which are recommended for use on specific fuels. Underwriters Laboratories Listings also indicate the types of fuels for which foam concentrates are recommended. In any case, the selection of the most effective foam type for a given hazard should be done by persons skilled in the art of fire protection engineering. The following general guide indicates foam agents of choice for the various classes of fuels to be protected:

Fuel	Foam
A. Hydrocarbons	Fluoroprotein; regular protein; AFFF
B. Polar solvents	"Alcohol" type; approved polymeric or "universal" types
C. Mixed hazards, both hydrocarbons and polar solvents	Approved polymeric or "universal" types; some conventional "alcohol" types

21.4. FOAM PROPORTIONING EQUIPMENT

Making foam requires two steps: first, mixing foam liquid concentrate with water to form a solution and, second, mixing the solution with air to form foam. In foam terminology, the first is "proportioning" and the second is "foam generation."

Foam liquid concentrate can be proportioned into the water stream by premixing, using energy from the water stream or by using an external power source. Without proper proportioning, the foam may be too weak (if proportioning is low) or concentrate may be wasted (if proportioning is too high) (Fig. 21.2).

Venturi or line proportioners are the lowest cost proportioning devices. The operating range, however, is limited to one flow rate/pressure combination, and foam generation devices must be carefully sized to match the proportioner. In operation, water flows through the venturi creating a vacuum area. Foam liquid is introduced at this point by means of a pickup tube from the storage container. The venturi proportioner will meter accurately at a given flow rate/pressure point. Since proportioning rates vary with flows, these devices have a finite range of usage (usually expressed as a pressure limit). For practical fire-fighting purposes these devices operate satisfactorily at rated pressure ± 50%. The maximum recovered pressure, including friction loss and static head, is about 65% of the inlet pressure.

In *pressure proportioners,* a small amount of the flowing water volumetrically displaces foam liquid into the main water stream. The working pressure of the vessel must, of course, be above the maximum static water pressure encountered in the system. Water is allowed to enter the foam tank from the main stream with as little friction loss as possible, while pressure in the main stream is dropped about 10% by means of an orifice. Liquid in the tank, essentially at the initial pressure, is metered into the low-pressure area by a second orifice. Advantages of this system are low pressure drop, automatic proportioning over a range of flows, and pressures and freedom from exter-

Figure 21.2.   Standard foam pumper.

nal power. Disadvantages are long refill time (tank must be drained) and an economic limit on size. There are two types of tanks available: the first contains a flexible bladder that separates the water and foam concentrate. While this is considerably more expensive, it allows for complete recovery of the liquid if a partial tank is used. The second type depends on specific gravity and viscosity difference to keep the water and liquid separate. This is entirely satisfactory for most fire-fighting liquids, and it is widely used. It should be noted that pressure proportioning systems can become inoperable or lose efficiency if there is water leakage through the main shut-off valves. This can be easily prevented by installing a ball drip valve between the supply valve and the main tank.

Around-the-pump proportioners are used where water is to be pumped on site. Basically, the unit is a venturi proportioner that takes a portion of the main water flow, inducts a large percentage of foam liquid, and injects the flow back into the suction side of the pump. This type of proportioning has the same limitations as venturi proportioners except recovered pressure is almost 100%. (Note: This is effected at some loss in pump capacity.) Around-the-pump proportioners are generally very economical, but since they induct a fixed amount of concentrate independent of main line flow, they should not be used in a variable-flow system.

Water motor proportioners consist of a gear pump for the liquid driven by a water motor that uses pressure from the main water stream. Flows are limited to a fixed range.

Proportioning with external power is the simplest way to proportion. Compressed air or a gas cylinder is used to move liquid from a pressurized tank through an orifice into the water stream. This is done in small devices for specialty applications, but it is generally limited to less than 10 gal liquid.

Direct pumping can be used when the foam liquid flow is fixed. Normally, a gear, turbine, or piston pump is chosen in order to have positive displacement and metered discharge. Direct pumping through an orifice can be done with a centrifugal pump if the water pressure is fairly constant and if the proper pump curve is selected. Direct pumping is the lowest cost pumped system, but it is applicable only to fixed-flow systems.

Indirect orificing can be used when more than one flow is desired. In this type of system, the water supply pressure must be adjusted to a fixed value. For the lowest flow the largest bypass orifice is opened, thereby bypassing a large amount of the foam liquid concentrate and injecting the proper amount. Since more solution is required when other devices are used, the orifice size must be carefully selected to bypass the required amount of foam liquid concentrate. The largest number of different flows for bypass systems is usually three or four.

Indirect orificing can be extended to an infinite number of flows within a given range by using a variable orifice (valve) and measuring the water and foam liquid flows. This is called flow-indication proportioning. The greatest limitation here is the need for an attendant who normally observes rotometers and controls flows.

When the foam system is operated, water flow is indicated by a meter and the operator adjusts the liquid flow to obtain the desired foam liquid injection. Flow meters can be sized for a fixed injection percentage. Other percentages can be selected with the help of a simple conversion chart.

In *balanced pressure proportioning,* now most widely used, automatic proportioning is provided through a mechanical system. Two orifices discharge water and foam concentrate into a common pipe. By adjusting the area of the orifices to a particular ratio, the percent injection can be adjusted if inlet pressures are equal. Equal pressures are maintained by using a diaphragm-operated valve on the foam liquid bypass line. Water pressure is applied to the top of the diaphragm and foam liquid to the bottom. If water pressure increases above the liquid, the bypass valve will be closed, allowing less liquid to be bypassed and increasing the foam liquid pressure. The device is very simple in principle.

Under practical conditions, the automatic range in which a balanced pressure system will operate is approximately six times the lowest flow. Mechanical considerations (such as the unequal area of top and bottom diaphragms resulting from the shape of the stem diameter, for example) cause balanced-pressure systems to proportion slightly high at low flows. For example, in a nominal 3% system, the proportioning could be expected to be 2.9% at 600 gpm and 3.5% at 100 gpm.

In all cases, foam proportioning systems are designed for use with agent or concentrate of given specific gravity, viscosity, and flow characteristics. Indiscriminate substitution of one agent for another in a system usually will result in improper proportioning. In no case should the type of agent in a system be replaced by another without consultation with the system and equipment manufacturer and the authority having jurisdiction.

21.5.   FOAM GENERATING EQUIPMENT

Nozzles and fixed foam makers are devices that mix air with the proportioned foam solution. All low-expansion fire-fighting foam is generated by inspiration of air using a venturi and the foaming solution. Foam makers are divided into two classes: low back pressure, which immediately discharge foam essentially at atmospheric pressure, and high back pressure, which discharge the foam into a pipe for transfer or into the bottom of a tank for subsurface application.

Many types of low back pressure foam makers are available. Each has a specific use and can best be recommended by the manufacturer. These types include foam nozzles ranging in capacity from 30 to 4000 gpm (solution rates are always specified for capacity) (Fig. 21.3), chambers for use on storage tanks, foam makers for dike walls, special units for dip tanks, foam water sprinklers, and overhead foam sprays.

Figure 21.3.   PC-50—Roamin' Chariot foam turret nozzle.

High back pressure foam makers are similar to low back pressure in that they also induct air by venturi action of the solution. They differ in that they are designed to recover up to 40% of the applied inlet pressure. When fluoroprotein foam is used, high back pressure foam makers make practical the subsurface, or base, injection of foam into hydrocarbon tanks.

When selecting foam generating devices, be sure that they are designed and approved for the intended use.

21.6. TYPES OF FOAM SYSTEMS

Foam systems are classified as portable, fixed, or semifixed. Portable systems involve the use of hose-line equipment which can be hand operated. These may be in the form of hose stations stategically located near hazard areas, or portable equipment carried on motorized fire apparatus. For practical purposes, the maximum solution flow that can be generated per line is about 300 gpm. The flow is sufficient for fire extinguishment of fuel surface area fires of 2000–3000 ft^2 depending on the nature of the fuel and the foam used.

In fixed systems, all foam system components are permanetly stationed, requiring the provision of fixed piping for water, form solution, and foam distribution lines. Foam systems can be manually or automatically actuated. Fixed systems are most suitable for protection of unattended hazards, or

Figure 21.4.   Squrt telescopic foam tower.

where the hazard or exposures cannot afford the time lapse between ignition of fire and the response of personnel to the scene.

Semifixed systems provide the greatest flexibility and sometimes best cost-effective means for protection of hazards dispersed over a plant or tank farm area. With such systems, water hydrants and foam discharge outlets are generally permanently installed, with adequate discharge piping to the foam equipment installed so as to provide safely accessible terminal connections. Specially designed foam trucks or trailers which carry a water pump, foam concentrate, and foam proportioning equipment are brought to the scene (Fig. 21.2). Such vehicles may also carry foam turret nozzles, portable equipment, auxiliary extinguishing agent, or articulating or extendable booms for extending foam or water stream reach or accessibility into a fire area (Fig. 21.4).

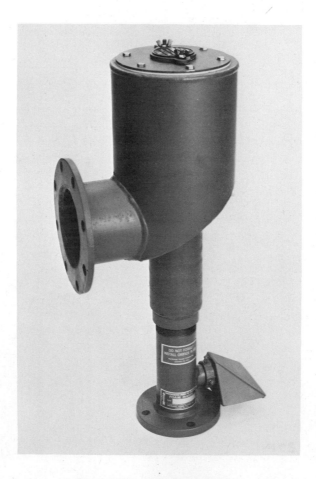

Figure 21.5. MCS foam chamber.

In all types of foam systems, the foam storage, proportioning, and discharge equipment, the water flow, the type of foam concentrate, and the method of foam application must be preengineered as a system to provide suitable protection for the specific hazards involved.

21.7.   FOAM APPLICATION REQUIREMENTS

How foam is applied to the burning surface of a hazard is very important to the success of extinguishment. The more gently the foam is applied, the better the chance of success. Three types of methods recognized by NFPA standards:

1. *Type I discharge outlet.*   An approved discharge outlet which will conduct and deliver foam gently onto the liquid surface without submergence of the foam or agitation of the surface (Figs. 21.5–21.7).

2. *Type II discharge outlet.*   An approved discharge outlet which does not deliver foam gently onto the liquid surface, but is designed to lessen submergence of the foam and agitation of the surface.

3. *Subsurface foam injection.*   Discharge of foam into a storage tank from an outlet at the tank bottom or below the liquid surface.

With protein-based, polar-solvent foam, Type I application is a necessity. Even when Type I application is not required, its use will reduce operating time.

21.8.   SYSTEM DESIGN

The initial consideration in selecting a fire protection system is the nature of the hazard. If flammable or combustible liquids are the hazard, foam is preferred. At this point, the authority having jurisdiction or the manufacturer should be contacted for recommendations on which type of foam to use. If the chemical composition and size of the hazard are known, a foam type and rate can be selected. This dictates the water demand.

Water supply is a limiting factor in many installations. Unless water can be supplied at the required rate and pressure, the foam system may be inoperable. Pumps can be used to increase pressure if the volume is sufficient. For most foam systems, application rates range between 0.1 and 0.6 gpm/ft² of hazard, depending on the hazard and application method. It is important to have good quality fresh or sea water for foam generation. Potable water is always satisfactory. Difficulties may be encountered if the water contains antifoams, oil and, in some cases, high concentrations of surfactants.

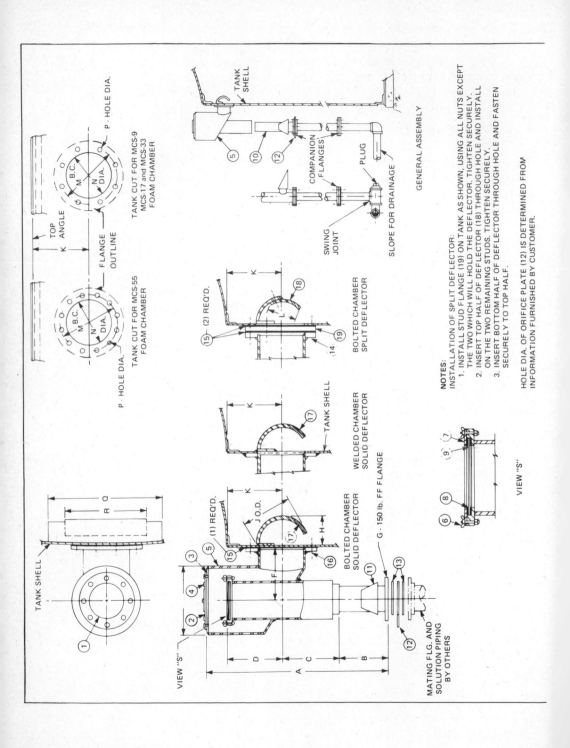

NO.	QUAN.	DESCRIPTION		NO.	QUAN.	DESCRIPTION
1	1	NAMEPLATE - ALUMINUM		11	1	AIR STRAINER - CAST IRON
2	1	INSPECTION HATCH - STEEL		12	1	ORIFICE PLATE - BRASS
3	1	INSPECTION HATCH GSKT.-1/8" ASB		13	2	RING GASKET - ASB
4	8	MCS-33, — 55, CAP SCR. - S. STL.		14	24	MCS-55 HEX NUT - STEEL
	6	MCS-9, — 17, CAP SCR. - S. STL.			16	MCS-9, 17, 33 HEX NUT - STEEL
5	1	CHAMBER BODY - STEEL		15	1	FLANGE GASKET - 1/8" ASB
6	2	WING NUT - BRASS		16	12	MCS-55 H.H. NUT, BOLTS - STL.
7	1	DIAPHRAGM RING - STEEL			8	MCS-9, 17, 33 H.H. NUT, BOLTS - STL.
8	1	CEMENT		17	1	DEFLECTOR SOLID - STEEL
9	1	DIAPHRAGM - GLASS		18	1	DEFLECTOR SPLIT - STEEL
10	1	FOAM MAKER - STEEL		19	1	STUD FLANGE W/STUDS - STL.

Table of Dimensions — All in Inches

	A	B	C	D	E	F	G	H	J	K	L	M	N	P	Q	R
MCS-9	$26\frac{1}{4}$	$7\frac{5}{16}$	$7\frac{5}{16}$	$8\frac{1}{2}$	8	7	$2\frac{1}{2}$	$3\frac{9}{16}$	5	8	$4\frac{1}{2}$	$7\frac{1}{2}$	$4\frac{3}{4}$	$\frac{3}{4}$	12	8
MCS-17	$32\frac{3}{32}$	9	$10\frac{19}{32}$	10	10	9	3	$4\frac{3}{8}$	$6\frac{3}{4}$	$9\frac{1}{2}$	$5\frac{13}{16}$	$9\frac{1}{2}$	$6\frac{3}{8}$	$\frac{7}{8}$	18	12
MCS-33	$35\frac{5}{16}$	$10\frac{1}{8}$	$9\frac{5}{16}$	$11\frac{3}{16}$	12	10	4	$5\frac{3}{8}$	9	11	$7\frac{1}{2}$	$11\frac{3}{4}$	$8\frac{3}{4}$	$\frac{7}{8}$	24	16
MCS-55	$42\frac{11}{16}$	$12\frac{13}{16}$	$12\frac{5}{8}$	$12\frac{3}{8}$	16	12	6	$6\frac{5}{8}$	11	12	$8\frac{3}{4}$	$14\frac{1}{4}$	$10\frac{3}{4}$	1	30	20

Figure 21.6. MCS foam chamber assembly.

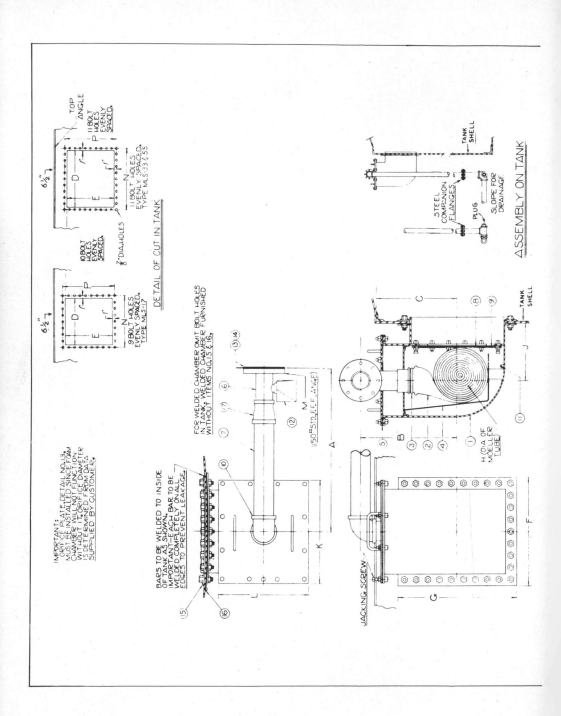

IMPORTANT:
ORIFICE PLATE DETAIL NO. 13, MUST BE INSTALLED SINCE FOAM CHAMBER WILL NOT FUNCTION WITHOUT IT. ORIFICE DIAMETER IS DETERMINED FROM DATA SUPPLIED BY CUSTOMER.

6½"

TOP ANGLE

11 BOLT HOLES EVENLY SPACED.

P

11 BOLT HOLES EVENLY SPACED. TYPE MLS-33 & 55

D

E

N

11 BOLT HOLES EVENLY SPACED.

10 BOLT HOLES EVENLY SPACED

⅞" DIA. HOLES

DETAIL OF CUT IN TANK

6½"

P

D

E

N

9 BOLT HOLES EVENLY SPACED. TYPE MLS-17

FOR WELDED CHAMBER OMIT BOLT HOLES IN TANK. WELDED CHAMBER FURNISHED WITHOUT ITEMS NOS. 15 & 16.

BARS TO BE WELDED TO INSIDE OF TANK AS SHOWN. IMPORTANT — EACH BAR TO BE WELDED COMPLETELY ON ALL EDGES TO PREVENT LEAKAGE.

TANK SHELL

STEEL COMPANION FLANGES

PLUG

SLOPE FOR DRAINAGE

ASSEMBLY ON TANK

C

8

9

TANK SHELL

13 14

6

7

2

A

M

150# STD. F.F. FLANGE

10

B

3

2

4

1

5

H (DIA. OF MOELLER TUBE)

11

K

L

15

16

JACKING SCREW

F

G

BILL OF MATERIALS

Furnished by National Foam System, Inc.

No.	Description	Material	Required	Remarks	No.	Description	Material	Required	Remarks
1	Foam Chamber Body	Steel	1		10	Elbow	Steel	1	
2	Moeller Tube	N.F. Textile	1	With Steel Ring	11	Drain Plug	Brass	1	½″ Size
3	Tube Clamp	Steel	1		12	Air Inlet Strainer	Cast Iron	1	
4	Cradle	Steel	1		13	Orifice Plate	Brass	1	
5	Cover	Steel	1	With Gasket	14	Orifice Plate Gasket	Asbestos	2	Ring Type
6	Foam Maker Body	Steel	1		15	Stud Bars	Steel	4	With Gasket
7	Inlet Tube	Steel	1		16	Gasket	Asbestos	1	
8	Transalem Closure	Asb. & Al.	1	With Gasket	17	Reducer	Steel	1	
9	Closure Frame	Steel	1	With Jacking Screws					

Table of Dimensions — All in Inches

Type	A	B	C	D	E	F	G	H	J	K	L	M	N	P
MLS-17	37	22	17¼	19	21½	23	25½	7	12¼	22	19½	3	21	23½
MLS-33	46⅜	24⅜	17½	24	22	28	26	10	12¾	27¼	20½	4	26	24
MLS-55	45⅞	25	17½	24	22	28	26	10	12¾	27¼	20½	6	26	24

Figure 21.7. MLS foam chamber.

417

TABLE 21.2. Foam Discharge Times—NFPA 11 1978, ¶ 3-3

For Tanks Containing Liquid Hydrocarbons	Type of Foam Discharge Outlet	
	Type I	Type II
Lubricating oils, dry viscous residuum (more than 50 sec. Saybolt-Furol at 122°F (50°C)); dry fuel oils, etc. with flash-point above 200°F (93.3°C)	15 min	25 min
Kerosene, light furnace oils, diesel fuels, etc. with flashpoint from 100°F to 200°F (37.8 to 93.3°C)	20 min	30 min
Gasoline, naphtha, benzol and similar liquids with flashpoint below 100°F (37.8°C)	30 min	55 min
Crude petroleum	30 min	55 min

For Tanks Containing Other Flammable and Combustible
Liquids Requiring Special Foams

Alcohol type foams require gentle application by Type I devices unless listed as suitable for application by Type II devices. The operation time shall be 30 minutes at the specified application rate, unless the manufacturer of the foam concentrate has established by fire test that a shorter time can be permitted.

Quantities of foam concentrate required by NFPA Standard 11 vary considerably depending on the type of hazard. For indoor dip tanks, for example, where the fuel level remains constant, the requirement is a two minute supply of foam at an application rate of 0.16 gpm/ft². For other indoor tanks, the supply must be 20 min. Outdoor hazards require a minimum of 0.10 gpm/ft² (see Table 21.2 on foam discharge times).

Fluid-flow calculations must be made for all foam systems. Fortunately, foam solution behaves like water when flowing through pipe and, therefore, simplified calculations and flow tables can be used. Once the flow required at each foam discharge and the water pressure/volume at the source is known, calculations can be made. The system is balanced either by use of flow losses in pipe for fixed-capacity devices (i.e., nozzles, overhead sprays, foam water sprinklers), or by changing the form-maker orifice in chambers, for example.

A wide variety of foam systems can be devised to protect flammable liquids. Typical indoor uses include dip tank systems, floor flooding, overhead sprays or foam water sprinklers, one-person nozzle stations, and oscillating monitors in hangars. Outdoor uses include storage tank systems with fixed chambers, diked area protection, and loading rack systems (Fig. 21.8). Portable equipment is designed for use either indoors or outdoors, but it requires additional time to arrive and be put in operation.

Figure 21.8. Loading rack system—or diagrammatic of semiportable system.

Ineffective foam application can result where there is:

1. Inadequate application rate in portable nozzle applications.
2. Lack of maintenance of equipment or foam concentrate.
3. Use of unapproved foam for the fuel involved.

There have been many recorded cases of successful control and extinguishment of potentially catastrophic flammable liquid fires by use of the proper foam in well-designed systems. Highly successful extinguishments have been recorded for cone roof, open floating roof, and covered floating roof tanks, dip tanks, loading racks, tankers, process areas, and many miscellaneous spill fires. As the value and variety of fuels increase, so increases the need for professional foam system design by experienced and responsible manufacturers working in concert with authorities having jurisdiction, with users and with fire protection and loss control agencies.

BIBLIOGRAPHY

Alvares, N. J. and Lipska, A. E., "The Effect of Smoke on the Production and Stability of High-Expansion Foam," *J. of Fire & Flammability*, **3**, 88–114 (April 1972).

API Pub. 2021, "Guide for Fighting Fires In and Around Petroleum Storage Tanks," American Petroleum Institute, 2101 L Street, N.W., Washington, DC 20037, 1974.

Breen, D. E., "Hanger Fire Protection with Automatic AFFF Systems," *Fire Technology*, **9**, No. 2, 119–131 (May 1973).

Burford, R. R., "Use of AFFF in Sprinkler Systems," *Fire Technology*, **12**, No. 1, 5–17 (February 1976).

DiMaio, L. R., Chiesa, P. J., and Ott, M. S., "Advances in Protection of Polar-type Flammable Liquid Hazards," *Fire Technology*, **11**, No. 3, 164–174 (August 1975).

Elliott, D. E. and Chiesa, P. J., "A New Foam Rheometer for Studying Fire Fighting Foams," *Fire Technology*, **12**, No. 1, 66–69 (February 1976).

Elliott, D. E. and Chiesa, P. J., "Automated Surface and Interfacial Measurements of AFFF's," *Fire Technology*, **9**, No. 4, 275–284 (November 1973).

Geyer, G. B., "Firefighting Effectiveness of Aqueous-Film-Forming-Foam (AFFF) Agents;" Technical Report ASD-TR-73-13, National Aviation Facilities Experimental Center, Atlantic City, NJ (April 1973).

"Guides for Fighting Fires in and around Petroleum Storage Tanks," American Petroleum Institute, API Pub. 2021, 1974.

Hiltz, R. M., "Mitigation of the Vapor Hazard from Silicon Tetrachloride Using Water-based Foams," *J. of Hazard. Mater.*, **5**(3), 169–176 (Feb. 1982).

Hird, D., Rodriques, A., and Smith, D., "Foam—Its Efficiency in Tank Fires," *Fire Technology*, **6**, No. 1, 5–12 (February 1970).

Hird, D., "The Use of Foaming Agents for Aircraft Crash Fires," *Fire*, 179–180 (September, 1974).

Jensen, R., ed., *Fire Protection for the Design Professional*, CBI Publishing Co., Inc., Boston, 1975, p. 198.

Martin, G. T. O., "Fire Fighting Foam," *The Institution of Fire Engineers Quarterly*, **32**, No. 86, 165–176 (June 1972).

McKinnon, G. P., ed., *Fire Protection Handbook*, 15th ed., National Fire Protection Association, Batterymarch Park, Quincy, MA. 02269, 1981.

Meldrum, D. N., "Aqueous Film-Forming-Foam—Facts and Falacies," *Fire Journal,* 66, No. 1, 57–64 (January 1972).

Moran, H. E., Burnett, J. C., and Leonard, J. T., "Suppression of Evaporation by Aqueous Films of Fluorochemical Surfactant Solutions," NRL Report 7247, Naval Research Laboratory, Washington, DC, April, 1971.

NFPA 11, "Standard for Foam Extinguishing Systems," National Fire Protection Association, Batterymarch Park, Quincy, MA. 02269, 1976.

NFPA 11A, "Standard for High Expansion Foam Systems," 1976.

NFPA 11B, "Standard for Synthetic Foam and Combined Agent Systems," 1977.

NFPA 16, "Standard for Foam Water Sprinkler and Spray Systems," 1974.

NFPA 325M, "Fire Hazard Properties of Flammable Liquids, Gases and Volatile Solids," 1977.

NFPA 402, "Standard for Aircraft Rescue Procedures," 1978.

NFPA 403, "Standard for Aircraft Rescue and Fire Fighting Services at Airports," 1978.

NFPA 412, "Standard for Evaluating Foam Fire Fighting Equipment," 1974.

NFPA 414, "Standard for Aircraft Rescue and Fire Fighting Vehicles," 1978.

Underwriters Laboratories, Inc., "Air Foam Equipment and Liquid Concentrates," UL-162, Northbrook, IL, 1975.

22

Safer Experimentation— Probing the Frontiers While Respecting the Unknown

Howard H. Fawcett

Recognizing the wide diversity of types of experiments and the conditions under which they are conducted, to say nothing of the variety of intellectual formal training of the experimenters, their associates, their technicians, and supporting staff, both in industry and academic situations, this chapter is divided into three sections:

Section 22.1—Safety and General Considerations.
Section 22.2—Animal Experiments and Genetic Engineering.
Section 22.3—The Literature of Health Effects on Laboratory Personnel.

The underlying theme of this chapter is that no matter how esoteric or "far-out" research may be, *the human elements should be carefully considered* and *fundamental safety and health aspects integrated into the work.*

22.1. SAFETY AND GENERAL CONSIDERATIONS

Experimentation, research and development probing of frontiers all have hazards which must be recognized if serious difficulty is to be avoided. Hazards are not confined to chemical laboratories alone; a recent survey suggested that in research and development laboratories, hazardous conditions are the rule, rather than the exception. Flammable chemicals and toxic chemicals are the regular hazards most frequently encountered. One out of eleven laboratories surveyed had no safety program at all; three out of ten no effective safety program; one out of five has had a serious work-related injury within a year.[1]

From experiments, both accidental and deliberate, new knowledge is gained, if we are sufficiently astute to combine science, common sense, and self preservation, and avoid "chemical boobytraps."[2,2a]

Knowledge and its dissemination may come unexpectedly. For example, a young chemist with one month's experience, reported to work early one August morning at the acid laboratory of an explosives plant. Surveying the work area through his new company-issued safety glasses, he noted that the night-shift chemists had not budgeted adequate time for "housekeeping and clean-up," leaving the laboratory sink half-full of dirty glassware. One 100-ml beaker was half full of a milky liquid, approximately the hue of the mixed nitric-sulfuric nitrating acids which were the main substances analyzed in this part of the laboratory. No label, marking, or other content identification appeared on the beaker.[3] Too inexperienced yet to have developed a "doubting Thomas" questioning attitude to the identity or purity of chemicals, and never warned in college about such mundane matters, the chemist did not hesitate to invert the beaker into a glass funnel, which in turn drained into a large glass waste acid jug. Turning to other matters, he walked to the other side of the laboratory just as a loud explosion occured, followed by the tinkle of breaking glass and an aerosol of brown nitrogen oxides-sulfuric mist.

Reference to the literature developed that acetone, like many ketones, is easily oxidized, passing through a peroxide, then a ketene stage, before eventually going to simple combustion products. The mixed nitric–sulfuric acid had been formulated to nitrate organics, like glycerine and toluene, as well as to oxidize other substances. When the acetone in the beaker (masked with a milky emulsion from a previous unknown use) contacted the acid, in a few seconds the heat and pressure build-up decomposed the nitric acid into the nitric oxides, and resulted in an aerosol mixed with the sulfuric, as well as acetone peroxides and ketene. Two fractured window sashes and 17 missing or fractured glass panes attested to the overpressure from the reaction.

An inquiry was shortly convened, and the whole sequence of events analyzed, since it is considered irregular to conduct unauthorized experiments in an explosive plant, especially when they result in loud, unexpected noises. From this investigation evolved recommendations to prevent future incidents:

1. Acetone was definitely identified as the liquid incompatable with the acid.
2. Several other organic solvents were identified in the laboratory that could have resulted in a similar incident.

3. No acetone or other solvents are now permitted into the acid laboratory, and all personnel duly informed.

4. Any conditions or circumstances which are questionable (such as unidentified materials) must be immediately called to the attention of the laboratory supervisor before action is instituted.

5. The night shift is not exempt from "clean-up" duties after their work period.

To the young chemist, the incident was the beginning of a series of observations which taught him never to "assume" that chemicals are what they might appear, and that small errors or incomplete analyses can have serious consequences. Had the explosion occured a few minutes later, several other chemists who worked in the same laboratory could have been seriously injured. Had the quantities of acetone and acids been greater, even more damage could have been expected.

The above "real-world" incident is introduced with the thought that even accidents can be the vehicle for learning. Eventually, the acetone-mixed acid reaction was published, adding to such items as the aluminum-perchloric-acetic anhydride acid explosion in Los Angeles in 1947 which resulted in loss of life and property.[4]

From such reports, piece by piece, has come the basic references on hazardous chemical reactions, such as the National Fire Protection Association 491 Manual,[5] the excellent work of Leslie Bretherick of BP in assembling the two editions of his compilation of over 20,000 reactions[6] the laboratory work conducted by Dr. James Flynn of Dow for the National Academy of Sciences Committee on Hazardous Materials,[7] and the most recent attention to incompatable chemicals in hazardous chemical waste disposal systems.[8] Collectively, these have evolved this relatively unattended field of chemistry into a viable data base. The increasing interest of chemists and engineers in introducing into the literature such results, even though they are often unpleasant and unprofitable, is of great significance. As a recent example, it has been recently reported that when 20 g $KMnO_4$ was added to 100 g dimethylformamide (DMF) in a bottle, shaken, and set aside, a reaction occured which resulted in a few minutes in an explosion which shattered other bottles and sprayed reaction mixture and pieces of glass throughout the laboratory.[9] While DMF has been reported and properly recorded as reacting with several other chemicals, including bromine, carbon tetrachloride (above 65°C), certain halogenated hydrocarbons (in presence of iron), chromic anhydride, 2,5-dimethylpyrrole and phosphorous oxychloride, hexachlorobenzene, magnesium nitrate, methylene diisocyanate, organic nitrates, phosphorus trioxide, and triethyl aluminum, the potassium permanganate reaction had not been previously reported.[5]

Experiments can never be completely "safe," since by the very nature of the operation, unknown factors can be expected to arise. Many of the misunderstandings which have developed when a health and safety professional points out obvious deficiences to a dedicated researcher have arisen from the lack of appreciation on both sides—the researcher not wanting to admit the risk, and the safety professional not wanting to be close-eyed to very obvious errors. A few years ago, an unescorted visitor entered the chemical research laboratory at a highly prestigious university. No one questioned his mission. No one was in sight, although the lights were all on, the experiments all running, and it was still daylight. Standing at the doorway of one laboratory, he stopped short and observed four ringstands-tripods on the floor in front of the hood, with glass experimental equipment positioned on the holders, with plastic tubing meandering into the vicinity of the hood itself. Had anyone, including emergency personnel, entered that laboratory, especially if the lights were out and the air contaminated by fumes, there is no doubt that they would have accidentally upset the floor-oriented equipment, spilling the several liters of chemicals, and creating even worse conditions. By thinking through such situations, which are not abstract, but real, it is possible for laboratory management or instructors to take constructive actions to minimize the risk, both to the individual chemist and to the facility and equipment, as well as to enhance the potential benefits which may be derived. For a one-time experiment on the subgram or gram scale, it is unlikely most researchers will ever pause to analyze all the potential hazards which *might* occur, and with justification, but certainly even on a small scale any methodical and thorough scientist should temper his academic enthusiasm with pause to consider fundamentals before the reaction proceeds for scale-up and goes to the pilot plant stage.[10]

If one or two subgram or gram-scale reactions have shown promise, and extensive investigations are planned involving a system of chemicals, certain fundamental health and safety information, if available, should be developed. It is fully recognized that only limited data exists for many chemicals, but where it is available, the prudent investigator or laboratory supervisor would be wise to begin the assembly of the data, which will be needed eventually for production and marketing.

For example:

1. If the material is a gas or liquid at STP, what is its flammability with air, and with other gases, including oxygen, over what range? For liquids, flash point data (noting the method used), boiling point, and freezing point should be assembled. Examples of limits of flammability for several common gases and liquids are cited. (Figs. 22.1 and 22.2).

2. If flammable, what is the auto- or self-ignition temperature? (See Chaps. 18 and 19).

3. If not flammable, does the material undergo decomposition into flammable or toxic products?

4. What is the vapor density compared to air (useful in determining ventilation requirements)?

5. What are the hazards and/or effects of inhaling the gas or vapor; what results from skin contact? (See Chap. 14.)

6. What bioelimination mechanism is known? (See Chap. 15.)

7. What respiratory protection, if any, is required? (See Chap. 25.)

8. Should the experiments be conducted only under a hood with a known monitored air flow, or can it be set up on an open bench?

9. What first-aid treatment is known, both for on-the-scene and for use by paramedics, nurse, and physicians who will respond to any serious emergency? (As an example, penicillin injections are not the proper medical treatment for hydrofluoric acid burns.) It must never be assumed that even medical professionals have extensive knowledge of the treatment of exposures to chemicals in general, and to research chemicals in particular. (See Chaps. 9, 10, and 11.)

10. Can the gas or liquid vapor be detected by odor, and if so, at what concentration?[11] (See Chap. 25.)

11. What analytical instrumentation is available for determining injurious concentrations of the gas or liquid vapors in air—assuming it is determined in advance what is a harmful concentration for short term exposures?[12,12a,12b]

12. What is known of the proper fire extinguishing agent to use, and of an acceptable method of clean-up for spills? (See Chap. 20.)

13. Will the waste from the reaction mixture, and the purification steps, be considered hazardous, under the Resource Conservation and Recovery Act of 1976? If so, how will it be disposed? (See Chaps. 28 and 29.)

Although all of the above items have importance, at any one time it is difficult to rate them. It is unlikely that complete information will be available. It is only in the recent past that sufficient interest in such matters (considered by many as nonacademic) justified the expense and time to develop such data. An active and sincere attempt should be made to develop as much of this data as practical and possible.

A selection can now be made, based on even the data accumulated to date, as to some basic aspects of the experiment.

If the liquid involved is relatively free of serious fire and health hazards, such as the higher alcohols, higher ethers, and the higher ketones, the normal care which an experienced, prudent, and cautious chemist normally exercises should suffice. (Chemists differ in their understanding and application of these terms.)

However, if serious health or fire hazards are present, as with the lower ethers, lower alcohols, lower aldehydes, nitro and amino compounds, and

LIMITS OF FLAMMABILITY OF GASES AND VAPORS

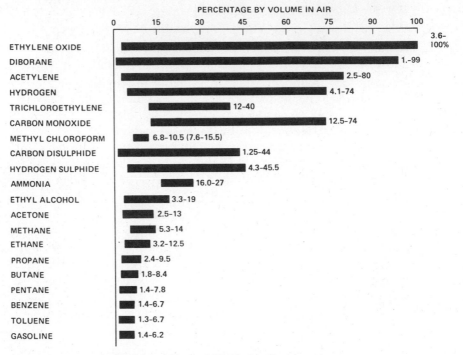

Figure 22.1. Flammability limits of gases and vapors.

many aromatics (including benzene), and common monomers, handling and reaction procedures should be carefully reviewed and developed to include protection of personnel from excessive exposures to either vapor, gas or liquid, or to eliminate exposure completely by proper facilities and procedures. Note Figs. 22.1 and 22.2 which illustrate the wide range of flammability for common substances. Closed systems should be utilized as much as possible, and personal protective equipment, including face shield, gloves, apron, and, if needed, respiratory equipment, should be available on standby. As an example, the reaction of acetylene and OF_2 was to be investigated for the first time in a laboratory. The researcher obtained from the safety office a heavy-duty reinforced "nitrometer" mask as a precaution. On the first attempt at the reaction, a violent explosion occured. The researcher was seriously cut on the arms and hands, but escaped face injuries completely, due to foresight in anticipating that the reaction might not be safe. As experience is gained, the procedures should be changed to further refine or change measures related to safety as the necessity indicates. For exam-

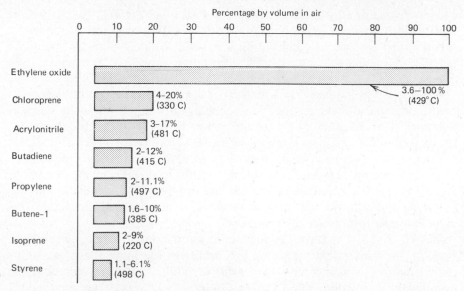

Figure 22.2. Flammability limits and autoignition temperatures of monomers. Adapted from NFPA data.

ple, adequate protection may have been taken against the gross concentrations associated with fires and explosions, but experience may show that sufficient vapor or gas is being released to the laboratory air to cause headaches, occasional dizziness, and upset stomach. Obviously, additional attention must be given to reducing releases or to improve the ventilation system, including the efficiency and proper use of hoods. *Reliance on a hood to remove or to disperse flammable vapors for disposal purposes may also be a serious error.* An undergraduate student was instructed by a graduate instructor to boil off excess toluene using an open beaker and a Bunsen burner in the fume hood. After the student attempted this for a time, he was instructed to pour the contents of the beaker into a smaller beaker so that it would boil faster. During the transfer of the hot solvent from one beaker to another, with the flame nearby, there was an explosion which burned the undergraduate. A distillation apparatus or a rotary evaporator is obviously the preferred method of solvent removal. Similar incidents have been recorded involving other flammable solvents, including ethanol, methanol, and ethyl acetate. (The Alpha Chi Sigma professional chemical fraternity safety program was established in 1942 as a direct result of the death of a fully qualified chemist who was transferring ethanol during a distillation, in the presence of an open flame. Several of the safety posters in the original series, which were made in 1948, stress the danger of fire, which remains a major hazard in many laboratories even today.)

The vapors from a flammable solvent can travel surprisingly long distances, and flash back from an ignition source.[3] *The evaporating of spent or waste solvents as a disposal method should not be attempted.* Not only does the ignition potential exist, but cross-contamination may travel into other laboratories, as well as contributing to overall air pollution. Collection of spent or waste solvents is a desirable alternative, in spite of its expense. (See Chaps. 28 and 29.)

Once it is determined that a material has properties which classify it as having unusual effects on life, on the environment, or produces unusual reactions, it is essential that these properties be given full consideration by the experimenter and pilot plant management before large-scale manufacture is attempted. As a case in point, a new insecticide was developed about 1950. That the process itself was not without hazard was documented in the 1952 U.S. Patent, which mentioned the use of liquid SO_3, sodium hydroxide solution, and sulfuric acid neutralization as reactants with hexachlorocyclopentadiene to produce the product.[14a]

In 1961, a study was reported in a major journal on the acute and subacute toxicity of 22 insecticides proposed as additives to chicken feed as potentially effective against house fly larvae in poultry manure, including the insecticide in question. The LD_{50} for female chicks was determined as 480 mg/kg for the substance in question. 440 ppm in the feed produced mortality of 11/11. In discussing the results, the report noted "affected chicks exhibited a typical syndrome characterized by a violent shaking of the entire body. The onset of the syndrome was rapid, the intensity varying with dosage level. At the highest levels, the chicks were unable to walk; they just sat and trembled. At the lower dosage levels, the animals walked in an ataxic gait, yet appeared to feed and drink at the normal rate. The intensity of the syndrome gradually diminished with time." In the general discussion, the report went on to state "This was the only insecticide (of the 22 tested) that caused a distinctive syndrome in all chicks, regardless of the concentration level. The continuous ingestion of feed containing smaller concentrations resulted in delayed symptoms which increased in intensity with time. This demonstrates the cumulative effect. The gradual diminution of the syndrome after withdrawal indicates that although the effects of this material were reversible, the chick metabolized and eliminated it slowly."[14b]

For several years, this product was manufactured. By 1975, questions were raised when the city into whose effluent the plant waste drained found its sewage treatment plant inoperative, and the city was eventually fined by the state. In 1976, a report by the bioassay group of the National Cancer Institute produced even more discussion.[14c] National attention was directed to the substance and its manufacture when the lower James River was closed to fishing for many months, and a Senate panel investigated.[14d,14e]

The ban on the fishing was lifted in mid-1980, but several former employees are claiming disability from exposure to the material at this writing.[14f]

We cite the above to note that had the 1961 report, cited above, been fully appreciated, and the material recognized as requiring unusual care in handling and disposal, this unfortunate sequence of events could have been prevented. The experimental chemist should have looked over the shoulder and followed the production more closely, in our opinion.

The product and eventual fate of the reaction from the experimental program may also be questioned in terms of:

1. Impact sensitivity (see Chap. 16).
2. Stability on storage and transport, especially under excessive moisture or heat, as from a fire or fire-fighting situation.
3. Bioaccumulation or biodegradability to suggest disposal methods which do not create long-term problems.
4. Melting and freezing point (including flash point as specified in RCRA).
5. Acute and chronic dosages as required by the Toxic Substances Control Act.
6. Fire hazard, including the ease of ignition, and the nature of the combustion or pyrolysis products.
7. Is there any data or reason to suspect the material is carcinogenic, mutagenic, teratogenic, or interferes with reproduction?
8. In handling this substance, has anyone experienced any abnormal effects to their health or skin?
9. Have first-aid and clinical treatment for possible exposures been developed, in case of misuse or spills which may affect personnel? (See Chaps. 9 and 11.)
10. What is the fire extinguishing agent of choice, and what other fire-control aspects should be considered (for example, the effect of water or moisture on the substance?) (See Chap. 20.)

Usually a review of these items will be made during the transition of the process from the laboratory to the pilot plant, and as much of this, and the data suggested on the reactants and the reaction, passed on with the process *before* the pilot plant or operating personnel attempt to conduct the process. From this data should evolve a warning label, if needed, which should clearly define in simple language the outstanding hazards of the compound and which should define the outstanding hazards of the compound and which should be displayed prominently on every container, regardless of size. The failure to transmit all the known information about a process or product can result in serious problems. For example, during the transfer of a pilot plant for preparation of diborane under Army contract from one company to another, inadequate attention was given to warning the new company that a serious potential hazard was known to exist in the use of carbon tetrachloride (at that time a commonly accepted fire extinguishing agent) and the diborane, under fire conditions. A fire occurred, the carbon tetrachloride

TABLE 22.1. **Range of Acute Toxicity Ratings for Several Common Chemicals**[a]

Chemical	LD50	Toxicity Rating
Parathion	2	1
Arsenicals	10	2
Kerosene	20	2
Nicotine	55	2
Caffeine	200	2
2, 4, 5-T	300	3
2, 4-D	500	4
Aspirin	750	4
Ammate	1600	4
Table salt	3320	4
Hyvar-X	5200	5
Tordon	8200	5
Amitrol	25,000	5

[a] Rating Scale 1–6: 1 = extremely toxic; 2 = very toxic; 3 = moderately toxic; 4 × slightly toxic; 5 = almost non-toxic; 6 = non-toxic.

was applied by the unsuspecting chemical engineers, with the result that the explosion was fatal to two. While carbon tetrachloride extinguishers are no longer the vogue, and are not now approved by any agency, (See Chap. 20) the possibility of a reaction between water, carbon dioxide, or dry chemical powder (several chemical variations exist) should be investigated, for protection not only of the laboratory and plant personnel, but for outside emergency response sources who may become involved. The classic example of the fire department not taking the recommendations of reputable sources during an emergency occurred at Somerville, Massachusetts on April 3, 1980 when during a control of a rupture of a phosphorous trichloride tank car, water was clearly not indicated but was used, which forced evacuation of 23,000 persons from the acid fumes-fog. This shows that even in modern times with well-known chemicals, the inability to communicate adequately in an emergency can have very serious consequences. (See Chap. 34.) See also *Hazardous Materials Emergency Response Guidebook,* DOT P5800.2, 1980 or most recent edition, U.S. Dept. of Transportation, Washington, D.C. 20590.

It is difficult to generalize as to when exhaustive and expensive tests should be undertaken as the experiment is scaled-up. In the toxicological area, under the premarketing provisions of the Toxic Substances Control Act, data must be developed and submitted to establish that the new material does not pose an unreasonable risk. (See Chap. 10.) Under Sec. 8(a), General Assessment Information Reporting, of TSCA, as this is written, it is believed that a rule will be made to require chemical manufacturers and processors to supply information on their products, including exposures, by-products and toxicity, to further classify chemicals for possible regulation (Table 22.1). The effect upon innovation and advancement of chemical sciences and the confidentiality of information is raised by such proposed action. As noted previously, the chemist, even the researcher, can no longer dissociate himself from the legal aspects of chemistry and its products (see Chaps. 1, 2, and 9.) Whether or not the proposed rules are actually put into effect, it becomes more critical every day that as complete information as possible be developed as early as possible in a proposed process or new experimental material.

When the new product is first isolated, the usual elementary and structure analyses, GC, AA, and other analytical techniques will at least suggest possible structures. Particular attention should be directed to those groups which create explosive tendencies, known as "plosophores." These include:

$-ONO_2$ nitrate $R-NO_2$ aliphatic nitro

$-NH-NO_2$ primary nitramine $Ar-NO_2$ aromatic nitro

$-N-NO_2$ secondary nitramine

as well as organic salts of the following:

chlorates
perchlorates
picrates
nitrates
iodates

Less powerful, but quite sensitive compounds contain:

$-N_3$ $-O-O-$ peroxide

$-NO$ nitroso $=N-X$ halamines

$-N=N-$ diazo $-C\equiv C-$ acetylides

$-N=N-S-N=N-$ diazosulfide

In considering the explosive potential, in general the more plosophore groups in the molecule, the more powerful the explosive.

Explosives are not always recognized as such if they are formed in solutions intended for another purpose. A case in point is the widespread use of electrolyte solutions containing perchloric acid in electropolishing metallic samples for metallographic examination. Perchloric acid with organic liquids deserve special consideration. The hazards of perchloric acid of 70% or higher concentrations, especially when heated in contact with organic material, have been extensively investigated and reported. The explosion in Los Angeles in 1947, mentioned previously, in which perchloric acid–acetic anhydride mixture decomposed in the presence of aluminum and a plastic used as a tank lining, clearly showed the power of the perchlorate decomposition.[15] Previously, the hazards of mixing perchlorates and chlorates with reducing materials had been published.[16] However no reference to relatively dilute perchloric acid in ethyl alcohol had been made other than a statement that electrolytes of the DeSy and Haemer type, used in a certain commercial electropolisher, were *"completely harmless."*[17] The writer investigated the mixture of

62 ml perchloric acid, 70% sp. gr. 1.67
137 ml distilled water
700 ml ethanol
100 butylcellosolve

prepared according to the instructions: "The perchloric acid should be added to the previously prepared mixture of ethanol and water. This mixture must be delivered in a separate 1 liter flask and the butylcellosolve in another flask, and not be added to the rest of the mixture until immediately before use."

The flash point as determined with the Fisher-Tag Cleveland Open Cup tester (ASTM D-92) was 81°F (27.2°C). The fire point coincided with the flash point, producing a faint blue flame typical of ethyl alcohol fires. In view of this, the mixture should be treated as a flammable liquid. However, the corrosive nature of the mixture prevents the use of metal safety cans and makes the use of glass or plastic bottles and beakers mandatory. Adequate protection to protect the glass or plastic from being broken, fractured or otherwise damaged to release the mixture is obviously necessary.

Depending on the temperature and rate of evaporation, 10-ml portions of the solution in an open evaporating dish were observed to either spontaneously ignite with a bright flash and continue burning, or else to actually decompose with explosive force. Evaporations up to 7 min on a moderate hotplate usually ignited spontaneously and continued to burn. If the evaporations were allowed to proceed on a low heat over 8 min, explosions observed. These explosions ranged from "mild" to "violent," but in no cases were the evaporating dishes broken. The reason for the above action is the formation of small amounts of ethyl perchlorate, which is recognized as a violent explosive, coupled with the very rapid action of the strong oxidating effect of the acid at acid concentrations over 70% as evaporation of the water proceeded.

Both as a method of disposal and as a means of studying burning characteristics, several outdoor fires were observed, each involving 1 liter of used mixture (following the practice that mixtures should be discarded one week after the butylcellosolve was added). The liquid was poured in a metal waste can lid, and ignited by throwing an ignited paper towel into the liquid. The blue alcohol flame was almost invisible in the early stages of burning, but as the burning progressed, the areas where organic matter from the paper had entered became small eruptions resembling pyrotechnics in color, noise, and smoke. The fire could be extinguished using a carbon dioxide extinguisher, but tended to reignite. Smothering with another lid was likewise only partly effective since reignition occurred. Dry chemical from an Ansul bicarbonate base extinguisher gave the best extinguishing with no reignitions.

Paper towels of the type widely used in laboratories burned much more rapidly when contaminated with the mixture, burned approximately twice as fast as uncontaminated towels, indicating the importance of keeping the mixture from paper or clothing.

Tests using rabbits were made to appraise and evaluate the degree of hazard which the mixture produced to the eye. One drop of the mixture was instilled into the eye without any washing out. Twelve hours later ulcerative keratitis of the entire cornea, with necrosis of the conjunctiva, was observed. Four days later the entire area had healed with no scar. In another eye experiment, one drop in the eye was followed after 5 min with washing with normal saline. Twelve hours later ulcerative keratitis, with complete necrosis of the conjunctiva was observed. Four days later scar tissue was present in preplay area, and the eye was not completely healed. From the above, the strict use of approved eye protection, such as side-shield safety glasses or a face shield, was recommended. No obvious effects were observed when the mixture was applied to rabbit skin, other than reddening and irritation. No obvious effects from skin absorption was observed in the animals.[18] The conclusions drawn clearly suggested that the mixture is not *"completely harmless."*

The one most frequently identified molecule which over the years has caused serious injury to chemists is the peroxide. Organic peroxides, which vary widely in their sensitivity and fire potential, have been well recognized, but are often mishandled. Less obvious even today is the peroxide formation in various chemicals, including the ethers. Although diisopropyl ether is perhaps the most hazardous from a standpoint of peroxide formation, several other compounds have been shown to have peroxide formation, especially when the can or bottle is opened and time passes. For that reason, it is important that all substances known or suspected of peroxide formation be recorded on arrival, be carefully labeled and dated, and within a year disposed of safely regardless of how much or how little of the material remains. Numerous references are available on peroxide formation, and should be familiar to chemists and others who use ethers and other compounds.[19]

Although the various aspects of explosion protection are beyond the scope of this chapter, certain fundamentals as recommended by the National Fire Protection Association may be of interest. If a laboratory or laboratory unit contains an explosion hazard (such as high-pressure equipment or experiments with rocket motors or the equivalent) protection should be provided to protect occupants of the laboratory and the occupants of the surroundings. Explosion hazard protection designs should be based upon the hazard defined in terms of: (1) blast effects (including considerations of impulse, rate of pressure rise, peak pressure developed, and duration of pressure, velocity of the pressure wave propogation, and residual overpressures), and (b) missiles, classed by mass, shape, and velocity (see Chap. 23).

Protection should be provided by one or more of the following means: (1) special preventive or protective measures for the reactions, equipment, or materials themselves, such as explosion suppression, high speed fire detection with deluge sprinklers, or explosion-resistant enclosures; (2) providing blast- and missile-resisting shields or blast hoods between small scale explosion hazards and laboratory personnel; (3) using remote control to minimize personnel exposure; (4) carrying out experiments in a detached or isolated building, or outdoors; (5) providing explosion-resistant walls or barricades for the hazardous laboratory; (6) limiting amounts of flammable or reactive chemicals exposed by experiments; (7) providing explosion venting in outside walls sufficient to maintain the integrity of the walls separating the hazardous laboratory unit from adjoining areas; and/or (8) disallowing the use of hazardous laboratories for other operations.

Laboratory personnel may be protected by explosion-resisting shields or hoods from hazards with an equivalence of a fraction to a few grams of TNT. Specially designed and explosion vented hoods may provide protection for slightly larger explosion hazards. Explosion resistant construction, isolated location and other protective means should be considered for larger scale explosion hazards.

When explosion-resistant construction is used, adequately designed explosion resistance shall be achieved by one of the following methods:

1. Reinforced concrete walls.
2. Rodded and filled concrete block walls.
3. Steel walls.
4. Steel plate walls with energy-absorbing lining.
5. Barricades, such as those used for explosives operations constructed of reinforced concrete, sand-filled wood sandwich, wood-lined steel plate, earth or rock berm.

Explosion-resistant construction should be based upon the anticipated explosion impulse defined in terms of the peak pressure impulse and duration and the worst case shrapnel mass, shape and velocity. It is possible to achieve velocities of 10,000 ft/s with missiles of a mass equal to the TNT equivalence of a detonated explosive. Although these velocities may not be achieved during accidental explosions, velocities of 1000 to 4000 ft/s would be expected (see Chap. 23).

The NFPA recommends that a laboratory be classified as to explosion hazard on the basis of the substances present which have been rated in the NFPA 704-M system. Using this criteria, a laboratory shall be considered to contain an explosion hazard if the violent accidental release of energy in the following items could cause serious or fatal injuries to personnel within the laboratory in an explosion:

1. The storage of materials with a reactivity of 4 on the NFPA 704 scale (Table 22.2).
2. The use or formation of materials with a reactivity of 3 or 4.
3. Highly exothermic reactions, for example, polymerizations, oxidations, nitrations, peroxidations, hydrogenations or organo-metallic reactions.
4. The use or formation of materials whose structure indicate a potential hazard but whose properties have not been established.
5. High pressure reactions, usually considered over 200 lbs/in.
6. High pressure experimental and test vessels.
7. High rotational velocity equipment, such as a centrifuge.

Three types of chemical reaction explosions are recognized:

thermal explosions
deflagrations
detonations

Thermal explosions are exothermic self-heating accelerating reactions (often decompositions) which occur throughout the substance (no separate distinct reaction zone is obvious in the material). Substances which can cause a thermal explosion are organic peroxides, nitro compounds, and nitrates, as well as mixtures of oxidizing and reducing agents.

When the exothermic rate exceeds the cooling rate, the temperature of the material rises exponentially with time and a thermal explosion results with a sudden evolution of very hot vapors and decomposition gases (without combustion).

A test which measures the minimum unsafe storage temperature at which a self-accelerating temperature will occur in the material is called the SADT (self-accelerating decomposition temperature) test.

A test which determines the rate of evolution of decomposition gases and vapors is called the pressure-vessel test. The rate is given by the maximum venting-orifice diameter which is just insufficient to prevent rupture of a 100 psi burst disk when the sample is heated.

The homogeneous explosion test measures the peak pressure and rate of rise of pressure when a sample is heated and expands into a large expansion (1000 : 1) vessel. These values are a measure of the energy and power of the thermal explosion and are scaled to MJ/kg and MW/kg, respectively.

TABLE 22.2.

Identification of Health Hazard Color Code: BLUE		Identification of Flammability Color Code: RED		Identification of Reactivity (Stability) Color Code: YELLOW	
Type of Possible Injury	Signal	Susceptibility of Materials to Burning	Signal	Susceptibility to Release of Energy	Signal
Materials which on very short exposure could cause death or major residual injury even though prompt medical treatment were given.	4	Materials which will rapidly or completely vaporize at atmospheric pressure and normal ambient temperature, or which are readily dispersed in air and which will burn readily.	4	Materials which in themselves are readily capable of detonation or of explosive decomposition or reaction at normal temperatures and pressures.	4
Materials which on short exposure could cause serious temporary or residual injury even though prompt medical treatment were given.	3	Liquids and solids that can be ignited under almost all ambient temperature conditions.	3	Materials which in themselves are capable of detonation or explosive reaction but require a strong initiating source or which must be heated under confinement before initiation or which react explosively with water.	3

2	Materials which on intense or continued exposure could cause temporary incapacitation or possible residual injury unless prompt medical treatment is given.
1	Materials which on exposure would cause irritation but only minor residual injury even if no treatment is given.
0	Materials which on exposure under fire conditions would offer no hazard beyond that of ordinary combustible material.

2	Materials that must be moderately heated or exposed to relatively high ambient temperatures before ignition can occur.
1	Materials that must be preheated before ignition can occur.
0	Materials that will not burn.

2	Materials which in themselves are normally unstable and readily undergo violent chemical change but do not detonate. Also materials which may react violently with water or which may form potentially explosive mixtures with water.
1	Materials which in themselves are normally stable, but which can become unstable at elevated temperatures and pressures or which may react with water with some release of energy but not violently.
0	Materials which in themselves are normally stable, even under fire exposure conditions, and which are not reactive with water.

Source. National Fire Protection Association.

The peak pressure and rate of rise of pressure for thermal explosions are *directly proportional to the amount of material per unit volume of container.* This is quite unlike gas or vapor explosions where the loading density is normally fixed by the combustible mixture at one atmosphere. Combustion of 5% propane–air mixture, for example, has a peak pressure of 96 psi with a rate of rise of 1700 psi/sec. These values may be compared to the totally confined thermal explosion of tert-butyl peroxy isopropyl carbonate:

Loading	kg/m^3	7.5	0.75	75
Peak Pressure	psi	125	12.5	1251
Rate of rise of pressure	psi/s	1631	163.1	16,313

Adequate venting, however, can reduce the values of both these explosions.

Additional tests useful in characterizing thermal explosions are DTA–TGA studies, and the determination of kinetic parameters of the reactions.

The Frank-Kamenetskii theory is useful for evaluation of the critical mass in the thermal explosion of solids.

Deflagrations are characterized by a progressing reaction zone (flame front) with a velocity of 10^{-5} to 100 m/s (subsonic velocity). Energy is transferred ahead of the flame front by conduction, convection and radiation.

The rate of burning, R, is significantly increased by an increase in pressure as

$$R = a + b\,p^{\gamma}$$

This fact makes possible the very high burning rate of propellant in the cartridge of firearms without producing a detonation. The transition of a deflagration into a detonation is influenced by compression of combustion products under confinement by the advancing flame front and the accelerating influence of pressure on the combustion rate. Experimental measurements of deflagrations are of two types, the Crawford bomb tests and deflagration tube tests.

Detonations are characterized by a constant supersonic rate of propagation of the reaction zone through the substance ($1\frac{1}{2}$–9 km/s). A shock wave leads the reaction front which produces gaseous reaction products of very high temperature and pressure (3000–5500°C, 1–4 psi $\times$ 10^6) which sets up a shock wave in the surrounding media.

The rate of pressure rise of the shock wave in air is of the order of microseconds in close to the charge with a duration of a few milliseconds. The air-shock velocity is also supersonic. This is to be compared with deflagration and thermal explosion subsonic rates of propagation with rate of pressure rise of ten to hundreds of milliseconds.

Safety tests basic to evaluation of a detonation are the measurement of the supersonic shock velocity propagating in the reacting substance such as the steel-tube test. Since a detonable substance has a critical diameter below which a detonation cannot propagate, there is a possibility that the tube diameter chosen is not large enough for a suitable test. Detonation sensitivity tests are:

Impact (drop weight)

Friction

Shock (card-gap)

Explosion venting may be utilized to lower the peak pressure of a thermal explosion or deflagration. This does not apply to detonations because of their much more rapid rise of pressure rise and supersonic shock wave.

High-pressure experimental reactions should be conducted behind fixed barriers able to withstand the calculated lateral forces. The barricades should be firmly supported at the top and bottom to take these loads. One wall, at least, should be explosion vented. (See NFPA 68-1974, "Guide for Explosion Venting," National Fire Protection Association, Quincy, Mass. 02269.)

Experimental reactions that are known to involve materials inherently unstable, such as reactions with acetylenic compounds and certain oxidations such as halogenations and nitrations, should be barricaded.

Routine reactions where pressures and temperatures are expected between certain predetermined limits based on long experience or routine work need not be conducted behind barricades if the vessels comply with the following: (1) vessels should be built of suitable materials of construction and have an adequate safety factor; and (2) vessels are provided with pressure relief in the form of a dependable tested safety relief valve or a rupture disk.

The explosive energy in a vessel containing fluid (gas or liquid) under pressure is the sum of the energy of pressurization in the fluid and the strain energy in the cylinder, due to pressure-induced expansion.

In pressurized gas systems, the energy in the compressed gas represents a large proportion of the total energy released in the event of a pressure vessel rupture. In liquid pressure systems, the strain energy in the containment vessel becomes a significant portion of the total explosion energy available.

Liquid systems with pressures over 5000 psi, large volumes at lower pressures, or vessels made of materials exhibiting high elasticity should be evaluated. This should not imply that nonelastic vessels are preferred. Liquid systems in which air or any other gas is entrapped will store more energy and are more hazardous than that presented by the liquid alone.

The blast criteria applicable to detonations of small charges (0.1–100 g weights) are shown in Table 22.3.

Classic courses in chemistry at some schools often use as teaching aids

TABLE 22.3. Blast Criteria Applicable to Detonations of Small Charges

Blast Effect	Range Indicated Explosive Yield (ft)				Criteria[a]
	0.1 g	1.0 g	10 g	100 g	
1% eardrum rupture	1.1	2.4	5.2	11	$P_i = 3.4$ psi
50% eardrum rupture	0.47	1.0	2.2	4.7	$P_i = 16$ psi
No Blow-down	0.31	1.3	6.9	~30	$I_i + I_q = 1.25$ psi · ms $V_{max} = 0.3$ ft/s
50% Blow-down	<0.1	0.29	1.1	4.1	$I_i + I_q = 8.3$ psi · ms $V_{max} = 2.0$ ft/s
1% serious displacement injury	<0.1	<0.2	<0.5	~1.1	$I_i + I_q = 54$ psi · ms $V_{max} = 13$ ft/s
Threshold lung hemorrhage	<0.1	<0.2	<0.5	~1.8	$I_i + I_q = 26$ psi · ms
Severe lung hemorrhage	<0.1	<0.2	~0.5	1.1	$I_i + I_q = 52$ psi · ms
1% mortality	<0.1	<0.2	<0.5	~1	$I_i + I_q = 85$ psi · ms
50% mortality	<0.1	<0.2	<0.5	<1	$I_i + I_q = 130$ psi · ms
50% big (16–25 ft²) windows broken	0.26	1.1	5.7	~30	$I_r = 3$ psi · ms
50% small (1.3–6 ft²) windows broken	0.17	0.49	1.9	9.9	$I_r = 8$ psi · ms

[a] P_i—peak incident overpressure (psi); V_{max}—maximum translational velocity for an initially standing man (ft/s); I_i—impulse in the incident wave (psi·s); I_r—impulse in the incident wave (psi·m); I_q—dynamic pressure impulse in the incident wave (psi·ms); I_r = the impulse in the incident wave upon reflection against a surface perpendicular to its path of travel (psi·ms).

experiments which are at least questionable from a safety viewpoint, and which often give students an inadequate understanding of the importance of safety and health, as part of their training. The classification of experiments (Table 22.4) which was published originally in *Education in Science,* 16–18 (April 1980) and reprinted in *Chem 13* (October 1980) is a critical evaluation of experiments and contains an interesting graduation of hazard potential, by which either academic or industrial chemists can evaluate their experiments. As noted, the categories of restrictions, while directed to the academic situation, can easily be adapted to other environments. The categories are:

N   Unsuitable. The experiment is considered unsafe for use in schools.

T   For teacher demonstration only. Teachers should be thoroughly familiar with the technique to be used. It is assumed that these experiments will have been rehearsed before being done in front of a class for the first time.

S   Considered suitable for supervised senior pupils. Some of these could perhaps be entrusted to responsible pupils in the final year of a 0-level, 0-grade, CSE, or similar course.

O   Considered safe as a class experiment in the last two/three years of 0-level, 0-grade, or CSE and similar courses. It is essential here for the teacher to exercise his or her discretion as to the responsibility of a particular class. Any experiment listed here may present dangers to irresponsible pupils. The use of a fume cupboard is recommended. Teachers may have to use their discretion and allow experiments classified in this way to be carried out in a very well-ventilated room with small quantities of materials.

Hazards of Laboratory Chemicals

Although intended primarily for guidance of academe, the study by the *Education in Science,* 19–26 (April 1980) and reprinted in *Chem 13* (November 1980) gives a gradation of hazards, in terms of schools (Table 22.5). As listed, the categories of restriction are:

X   Chemicals to be excluded from schools.

N   Chemicals not recommended to be normally held or stored.

E   Chemicals restricted to small quantity for observation or exhibition only.

R   Chemicals restricted to small quantities—in storage.

T   Teachers use only.

S   Senior pupils, i.e., post '0' Grade and post '0' level use.

O   '0' grade, '0' level, or CSE and above (meaning the last two/three years of these and similar courses).

TABLE 22.4.

Experiment	Restriction
Ammonia, oxidation using oxygen in an enclosed apparatus	N Use air (T). Oxygen may be used in an open vessel (T).
Ammonium dichromate(VI), heat ("Volcano experiment")	T,F A fume cupboard is needed to avoid possible inhalation of chromate (VI) dust
Ammonium dichromate(VI), heat with aluminum or magnesium powder	N
Ammonium nitrate, heat	T Heating a mixture of ammonium chloride and sodium nitrate is considered safer. Use safety screens
Ammonium nitrite, prepare and heat	T In solution only, concentration less than molar
Aryl and acyl halides, reactions	S,F
Cadmium iodide, electrolysis of molten	N Lead bromide preferable
Carbon monoxide, reductions with	T,F Use safety screens
Carbonyl chloride, preparation	N
Chlorine, reaction with ammonia	N
Chlorine, reaction of a mixture with hydrogen	N This refers to the gas syringe and similar experiments. It is possible to demonstrate the reaction in, for instance, a plastic bag. Burning hydrogen at a jet in chlorine is safe for teacher demonstration.
Chlorine, reaction with ethyne	N The reaction where the gases are generated simultaneously by adding dilute hydrochloric acid to a mixture of bleaching powder and calcium dicarbide is acceptable as a teacher demonstration (T, F Use safety screens).
Chlorine, preparation	S,F See Potassium manganate(VII), reaction with concentrated hydrochloric acid.
Chlorine, reaction with metals	S,F
Chlorine oxides, preparation	N
Crude oil, distillation	O,F
Cyanogen, preparation	N
Ethene or ethyne, explosion of a mixture with oxygen	N
Ethene or ethyne, igniting in a gas jar or test-tube	T

TABLE 22.4. *(Continued)*

Experiment	Restriction
Explosives (*e.g.*, mixtures of chlorates, manganates(VII) or nitrates with combustible substances)	N
Hydrogen, large scale generation and collection	T
Hydrogen, generation and testing for on a test-tube scale	O
Hydrogen, burning in air	T ⎤
Hydrogen, burning in chlorine	T ⎥ Use safety screens.
Hydrogen, explosion with air	T ⎥
Hydrogen, explosion with oxygen	T ⎦
Hydrogen, reductions using	T Use safety screens for the normal scale experiment. Reduction of metal oxides may be performed on a test-tube scale using, for example, a mixture of zinc powder and calcium hydroxide in the same tube as the oxide to generate the gas. Such experiments may be classified 0.
Hydrogen cyanide, preparation	N
Hydrogen sulphide, preparation	T,F
Hydrogen sulphide, use of gas	S,F
Hydrogen sulphide, use of aqueous solution	O
Iodine, heating in air	T or S,F
Lead bromide, electrolysis	T or O, F in which case the fume cupboard is essential.
Lead(II) carbonate, heating	O
Lead(II) nitrate, heating	O,F
Lead oxides, heating	O
Lithium, heating	T Use safety screens
Mercury, heating	T,F The fume cupboard is essential and must be left on for the duration of the experiment (*i.e.* while the mercury is above room temperature).
Mercury(II) oxide, heating	T,F The fume cupboard is essential.
Natural gas, enrichment for reductions	T If ethanal tetramer (metaldehyde or "meta fuel") is used as the enriching agent, in a test-tube scale experiment, it may be classified 0.

TABLE 22.4. *(Continued)*

Experiment	Restriction
	The large-scale experiment has proved dangerous, probably because of the extra dead volume introduced into the apparatus.
Nitrations, organic	S,F In some cases when only a mild nitrating agent, such as dilute nitric acid, is necessary, a fume cupboard is not needed (*e.g.*, nitration of phenols)
N-nitrosamines, preparation from amines	N See previous article on carcinogens (*Education in Science*, September 1979)
Oxygen mixture, use of	T See 'Potassium chlorate(V) and manganese(IV) oxide, heating mixture'
Phosphine, preparation	T,F
Phosphorus halides, reaction with water	S,F
Phosphorus, red, burning	O,F
Phosphorus, white, burning	T,F
Plastics: heating polyurethanes and polystyrene	O,F
Plastics: heating PVC	T,F
Plastics: polymerization and depolymerization of acrylics	S,F ⎫ The fume cupboard is essential
Plastics: polymerization of phenylethene	S,F
Plastics: preparation of nylon "rope"	O Note that if a solution of dioyl chloride in tetrachloromethane is used this must be classified T,F. 1,1,1-trichloroethane may be used as solvent if the solution is freshly prepared.
Potassium, reaction with water	T Use safety screens
Potassium chlorate(V) and manganese (IV) oxide, heating mixture	T Many safer alternatives for oxygen preparation. Use demonstration as illustration of catalysis only. Use safety screens.
Potassium manganate(VII), heating	O Eye protection essential. Heat in small test-tubes fitted with a loose ceramic wool plug to prevent splitting
Potassium manganate(VII), reaction with concentrated hydrochloric acid	S,F Cover the manganate(VII) with water first. This experiment is

TABLE 22.4. (*Continued*)

Experiment	Restriction
	highly dangerous if sulphuric acid is used by mistake instead of hydrochloric. It is safer to use bleaching powder or sodium chlorate(I) and dilute hydrochloric or sulphuric acid.
Rocket fuels, preparation	N
Silicon(IV) oxide, reduction with magnesium or aluminum	S The reactants must be dry. Use safety screens
Sodium, reaction with water	S Use safety screens
Sodium hydroxide (molten), electrolysis	T,F
Sodium peroxide, preparation of oxygen from	T Use safety screens
Sulphur and zinc, reaction	T Do not confine the mixture in any way, *i.e.* heat the mixture on a ceramic centered gauze or mineral fibre paper. Use safety screens.
Sulphuric acid, concentrated, reactions	O With close supervision, otherwise S. Use a fume cupboard if corrosive or toxic gases are likely to be evolved
Thermite reaction	T Use safety screens (or perform outdoors). Fe_2O_3, Mn_3O_4, Cr_2O_3 are safe oxides to use. Do not use CuO, MnO_2 or CrO_3.
Zinc, burning	O,F

t   Years 1 and 2 of secondary school (11/12 years upward) with close teacher supervision.

F   To be used in fume cupboard (hood).

(F)   To be used in fume cupboard (hood) if in open vessels on a scale other than small.

L   Short safe shelf-life.

A listing of 32 chemicals which the Consumer Products Safety Commission considers too hazardous for use in schools was recently published (see *Chem. & Engr. News*, **60**, No. 5, 5–6 (Feb. 1, 1982). Another list of hazardous substances is being considered for regulation by the state of California (see *Notice of Public Hearing, Nov. 20, 1981*, Department of Industrial Relations, P.O. Box 603, San Francisco, CA 94101).

TABLE 22.5.

Name	Restriction
Acetonitrile (methyl cyanide)	R, S, F
Aerosol sprays	R, Ot, F
Alcohols other than ethanol	R, O, (F)
Aldehydes other than methanal	R, O, (F)
Alkyl halides	R, Ot, (F)
All unlabeled bottles which contain substances of unknown composition	X
Aluminum powder	Rt
Aluminum bromide, anhydrous	R, S, L
Aluminum carbide	R, T, L
Aluminum chloride, anhydrous	R, S, L
4-aminobiphenyl	X
Ammonia '880'	Ot, (F)
Ammonical silver nitrate (Tollens reagent)	X Prepare as required and immediately dispose of excess
Ammonium chlorate(VII) (perchlorate)	X
Ammonium dichromate(VI)	S
Ammonium nitrate	R, St
Ammonium peroxodisulphate(VI) (persulphate)	St avoid raising dust
Ammonium sulphide	Ot, (F)
Anhydrone	N, L alternatives are calcium sulphate, calcium chloride (both anhydrous) and molecular sieves
Anthracene	E
Antimony	E
Antimony compounds	S
Aromatic amines (except aminobiphenyls or naphthalenamines)	R, St, F
Arsenic	E
Arsenic compounds	N if Marsh's Test needs to be shown use impure zinc
Asbestos, soft forms paper fibre mats platinised centred gauzes	X ceramic wool forms of paper, wool and platinized wool available. Calcium silicate matrix heat resistant mats and ceramic centred or stainless steel gauzes available.
gloves	use heat resistant leather

446

TABLE 22.5. (*Continued*)

Name	Restriction
Azo dyes	R, Ot See article on Carcinogens in *Education in Science,* September 1979, p. 17
Barium metal	T
Barium compounds, solid	S
Barium compounds, dilute solution	t
Barium chromate(VI)	N should not be isolated from any preparation
Barium peroxide	R, T, L
Barium sulphate	O
Benedict's solution	O
Benzamide	S
Benzene, use as a solvent	X use methylbenzene as alternative solvent. Use cyclohexane for change of state experiment.
Benzene, use as reagent	R, St, F (small scale preparative work) (in hood)
Benzenecarbaldehyde (benzaldehyde)	R, S, (F)
Benzenecarbonyl chloride (benzoyl chloride)	R, S, F
Benzene-1,3-diamine (m-phenylenediamine)	R, S, F
Benzene-1,4-diamine (p-phenylenediamine)	R, S, F
Benzene-1,2-diol (catechol)	S
Benzene-1,3-diol (resorcinol)	Ot, F
Benzene-1,4-diol (quinol or hydroquinone)	S
Benzenesulphonic acid	St
Benzene-1,2,3-triol (pyrogallol)	S
Benzonitrile	R, S, F
Beryllium	E
Beryllium compounds	X
Biphenyl-4,4'-diamine (benzidine)	X
Bis (4-isocyanatophenyl) methane (Caradate 30)	R, St, F
Bismuth	S
Bismuth compounds	O
Bleaching powder	R, O, L
Bromates	R, S
Bromine, element	R, St, F
dilute solutions	O

TABLE 22.5. *(Continued)*

Name	Restriction
Bromobenzene (phenyl bromide)	S
1-bromobutane (n-butyl bromide)	R, S, (F)
2-bromobutane (sec-butyl bromide)	S, (F)
Bromoethane (ethyl bromide)	R, S, (F)
Bromomethane (methyl bromide)	R, T, F
2-bromo-2-methylpropane (t-butyl bromide)	R, S, (F)
Bromopropane (propyl bromides)	R, S, (F)
3-bromoprop-1-ene (allyl bromide)	R, S, F
Butanal	R, O, (F)
Butane cylinder	N not recommended in lab.
Butanoic acid	R, S
Butan-1-ol (n-butanol)	O, (F)
Butan-2-ol (sec-butanol)	O, (F)
Butanone (methyl ethyl ketone)	O, (F)
Cadmium	R, St
Cadmium compounds	R, St
Calcium, metal turnings	R, Ot
Calcium dicarbide	R, St, L
Calcium hydride	R, St, L
Calcium oxide	R, Ot, L
Calcium phosphide	N
Calcium sulphide	R, S, F
Carbon dioxide, solid	Ot
Carbon monoxide	X small scale preparation St, F
Carbon disulphide	N use dimethyl-benzene as solvent for preparing rhombic sulphur and ethyl cinnamate for prism experiment.
Camphor	S
Chlorates(I) (hypochlorites)	R, Ot, L purchase sodium salt as solution; ensure cap is vented
Chlorates(III) (chlorites)	N
Chlorates(V) (chlorates)	R, St
Chlorates(VII) (perchlorates)	N small quantities may be isolated as in fractional crystallisation of disproportionation products of potassium chlorate (V) - not to be stored
Chloric(VII) acid (perchloric acid)	X
Chlorine cylinder	N

TABLE 22.5. (*Continued*)

Name	Restriction
Chlorobenzene	S
Chlorobutanes (butyl chlorides)	R, S, (F)
Chloroethane (ethyl chloride)	R, S, (F)
Chloroethane (vinyl chloride monomer)	X
(Chloromethyl) benzene (Benzyl chloride)	R, S, F
Chloropropanes (propyl chlorides)	R, S, (F)
Chlorosulphonic acid	R, St, F
Chromates(VI) and dichromates(VI)	
solutions	O
solids	S
Chromium(VI) oxide (chromium trioxide)	R, S
Chromium(III) compounds (chromic compounds)	O
Cleaning mixture (dichromate(VI)/conc. sulphuric acid)	T Do not store. Decon 90 good alternative
Crude Oil	Ot, F
Cyanates	R, T
Cyanides	N
Cyclohexane	O, (F)
Cyclohexanol	S, (F)
Cyclohexanone	S, (F)
Cyclohexene	O, (F)
DDT	N
Decanedioyl dichloride (sebacoyl chloride)	Ot, F
Devarda's alloy	O
N,N′-dialkylphenylamines (N, N′-dialkylanilines)	R, S, F
Di(benzenecarbonyl)peroxide (benzoyl peroxide)	N di(dodecanoyl)peroxide is alternative polymerisation catalyst
1, 2-dibromoethane (ethylene dibromide)	N may be formed as in test for unsaturation, but should not be isolated
1, 2-dibromopropane (propylene dibromide)	R, S, F
Dichlorobenzenes	R, S, F 1,2-isomer most toxic of the three isomers
Dichlorobiphenyl-4, 4′-diamines (chlorobenzidines)	X

TABLE 22.5. *(Continued)*

Name	Restriction
Dichlorodimethylsilane	R, T, F
1,2-dichloroethane (ethylene dichloride)	R, T, F
Dichloroethanoic acid (dichloroacetic acid)	R, S
2,4-dichlorophenols	R, S
Di(dodecanoyl)peroxide (lauroyl peroxide)	R, Ot
Diethylamine	R, Ot, F
Diethylamine, dilute solution	O
Diethyl sulphate	X
Diiodine hexachloride (iodine trichloride)	R, St, F
3,3'-dimethoxybiphenyl-4, 4'-diamines	X
Dimethylamine	R, Ot, F
Dimethylamine, dilute solutions	O
3,3'-dimethylbiphenyl-4,4'-diamine (o-tolidine)	X
Dimethyl formamide	R, S, F
Dimethyl sulphate	X
Dinitrobenzenes	R, St
3,5-dinitrobenzoic acid	R, S
4,4'-dinitrobiphenyl	X
2,4'-dinitrobromobenzene	R, T
2,4-dinitrochlorobenzene	N
2,4-dinitrofluorobenzene	N
Dinitrophenols	R, St
2, 4-dinitrophenylhydrazine	R, S
Dioxan	R, St, L
Dipentene	R, O
Diphenylamine solution for redox indicator	R, St otherwise N
Esters (general)	R, O, (F)
Ethanal (acetaldehyde)	R, O, (F)
Ethanal tetramer (metaldehyde)	Ot
Ethanal trimer (paraldehyde)	R, S, F
Ethane cylinder	N
Ethane-1,2-diamine (ethylene diamine)	R, S, F

TABLE 22.5. *(Continued)*

Name	Restriction
Ethanedioic acid and salts (oxalic acid and salts)	
solid	R, S
dilute solution	R, Ot
Ethanoic acid, glacial (acetic acid, glacial)	Ot, (F)
Ethanoic anhydride (acetic anhydride)	R, S, (F)
Ethanol	O, (F)
Ethanoyl chloride (acetyl chloride)	S, F
Ethene cylinder (ethylene)	N
Ethers (general)	R, S, F
Ethoxyethane (diethyl ether)	R, S, F, L
Ethoxyethanol (cellosolve)	S, F
Ethylamine	R, Ot, F
Ethylamine, dilute solution	O
Ethylbenzene	O, (F)
Ethyl benzoate	O, (F)
Ethyl carbamate	N
Ethyl ethanoate (ethyl acetate)	O, (F)
Ethyl methanoate (ethyl formate)	O, (F)
Ethyne cylinder (acetylene)	X
Fehlings solution No. 2	Ot Use water bath. Alternative is Benedict's or Barfoed's solution
Fluorine	N
Fluorene	N
Fluorenone	N
Fluorides, solid	R, T
Fluorides, solutions	Ot
Germanium tetrachloride	R, S, F
Heptane	R, O, (F)
Hexacyanoferrates(II) (ferrocyanides)	S
Hexacyanoferrates(III) (ferricyanides)	S
Solutions of hexacyanoferrates	Ot no heating, no addition of strong acids
Hexamethylcosane (squalene)	O
Hexamine	R, S, F
Hexane	R, O, (F)
Hexane-1, 6-diamine (hexamethylenediamine)	Ot

TABLE 22.5. *(Continued)*

Name	Restriction
Hexanedioic acid (adipic acid)	O
Hexanedioyl dichloride (adipoyl chloride)	Ot, F
Hexenes	O, (F)
Hydrazine	X
Hydrazine chloride	R, St
Hydrazine hydrate	R, T
Hydrazine sulphate	R, St
Hydrides, metal	R, St
Hydriodic acid	R, S
Hydrobromic acid	R, S
Hydrocarbons, aliphatic	O small scale, (F)
Hydrocarbons, aryl	S, (F)
Hydrochloric acid, conc	Ot
Hydrofluoric acid	X
Hydrogenation catalysts	R, St
Hydrogen cyanide	X
Hydrogen cylinder	T
Hydrogen peroxide 20 volume	t, L
Hydrogen peroxide 100 volume	R, T, L
Hydrogen sulphide gas	S, F aqueous solutions may be used outside fume cupboard
2-hydroxybenzoic acid (salicylic acid)	O
Indicator powders	T
Indium compounds	S
Iodates(V)	S
Iodic(V) acid	R, T
Iodine, solid	O, lower levels t, if heated t, (F)
Iodine(V) oxide (iodine pentoxide)	N
Iodoethane (ethyl iodide)	R, T, F
Iron(III) chloride, solid	O
Iron(II) sulphide	O
Isocyanates	R, T, F
Lead alkyls	N
Lead(II) bromide	Ot, if fused for electrolysis F
Lead(II) chloride	Ot

TABLE 22.5. *(Continued)*

Name	Restriction
Lead(II) chromate(VI)	N
Lead(II) ethanoate (lead acetate)	Ot
Lead(II) methanoate (lead formate)	Ot
Lead oxides	all levels t
Lithium	R, Ot if heated T
Lithium compounds	R, S lithium borohydride or sodium borohydride are more
Lithium aluminium hydride	R, St, F, L stable and are suitable for some applications
Lithium hydride	R, T, F, L
Lithium hydroxide	R, S
Magnesium powder	Ot
Magnesium ribbon	t
Mercury	R, St in well ventilated room on spillage tray
Mercury alkyls	X
Mercury salts, solid	St
Mercury salts, solution	Ot
Methanal (formaldehyde, formalin)	R, Ot in F unless very dilute. Do not use in presence of hydrogen chloride
Methanoic acid (formic acid)	Ot, (F)
Methanol	Ot, (F)
2-methoxyphenylamine (o-anisidine)	R, S, F
4-methoxyphenylamine (p-anisidine)	R, S, F
Methylamine	S, F
Methylamine, dilute solution	O
Methylbenzene (toluene)	Ot, (F)
3-methylbutanol (iso-amyl alcohol)	O
3-methylbutyl ethanoate (isoamyl acetate)	S, (F)
Methyl ethanoate (methyl acetate)	O, (F)
Methyl ethyl ketone peroxide	Ot
Methyl methanoate (methyl formate)	O, (F)
Methyl 2-methylpropenoate (methyl methacrylate)	R, S, F, L
Methylphenols (cresols)	R, S
N-methylphenylamine (N-methylaniline)	R, St, F
Millon's reagent	R, S alternatives are Albustix, Cole's Modification of Millon's reagent or Sakaguchi Test

TABLE 22.5. *(Continued)*

Name	Restriction
Molybdenum	R, S
Naphtha	R, T, F liquid paraffin is preferred for storing alkali metals
Naphthalen-1-amine (1-naphthylamine)	X
Naphthalen-2-amine (2-naphthylamine)	X
Naphthalene	O For cooling curves use hexadecan-1-ol, octadecan-1-ol, hexadecanoic or octadecanoic acid
N-naphthylethane-1, 2-diamine as solution (N.E.D. or N-naphthylethylenediamine)	R, St
Naphthylthiourea (ANTU)	R, T
Nessler's reagent	R, S
Nickel, dust	R, St
Nickel salts, solid	S avoid raising dust
Nickel salts, solution	O
Ninhydrin, solid	R, T
Ninhydrin, aerosol spray	R, Ot, F
Nitric acid, conc	Ot
Nitric acid, fuming	R, St, F
Nitrobenzene	R, S, F
4-nitrobiphenyl	X
Nitrocellulose	X
Nitrogen dioxide	O, F
Nitrogen triiodide	X
Nitromethylbenzenes (nitrotoluenes)	N
Nitronaphthalenes	X
Nitrophenols	R, St
4 [(4-nitrophenyl) azo] benzene-1,3-diol solution (Magneson 1)	O
4 [(4-nitrophenyl) azo] napthalen-1-ol solution (Magneson II)	O
Nitrosamines	X
Nitrosophenols, 2- and 3-isomers	X
4-nitrosophenol	R, St
Octane	R, O, (F)
Oct-1-ene	R, S, (F)
Oleum	N

TABLE 22.5. *(Continued)*

Name	Restriction
Orthophosphoric acid	S
Orthophosphoric acid, dilute	O
Osmic acid	N alternative stains for microscope work are the Sudan dyes
Oxygen mixture (potassium chlorate(V)/manganese(IV) oxide)	N alternative preparation is decomposition of 20 vol hydrogen peroxide catalyzed by manganese(IV) oxide
Paraffin oil	O
Paraquat	R,T
Pentane	O, (F)
Pentan-1-ol and 2-ol (n- and sec-amyl alcohol)	O, (F)
Pentan-3-one (diethyl ketone)	O, (F)
Pentyl ethanoate (amyl acetate)	R, O, (F)
Peroxides, inorganic (excluding H_2O_2)	S
Peroxodisulphates(VI) (persulphates)	St
Petroleum ether, BP below 80° C	S, (F)
Petroleum ether, BP above 80°C	Ot, (F)
Phenols	St
Phenols, dilute solutions (*e.g.,* indicators)	O
Phenylamine (aniline)	R, St, F
Phenylammonium salts	R, S (F)
Phenylethene (styrene)	R, S, F
Phenylhydrazine and salts	R, S
Phenylthiourea (P.T.U.) also phenyl-thiocarbamide (P.T.C.)	Ot see *Education in Science,* September 1979, page 17.
Phosphides, metal	N
Phosphorus, red	R, Ot, F
Phosphorus, white	R, T, F, L
Phosphorus(V) oxide	R, St, F, L
Phosphorus pentabromide	R, St, F, L
Phosphorus pentachloride	R, St, F, L
Phosphorus tribromide	R, St, F, L
Phosphorus trichloride	R, St, F, L
Phosphorus trichloride oxide	R, St, F, L
Photographic developer	t
Potassium	R, T, L
Potassium amide (potassamide)	N, L

TABLE 22.5. *(Continued)*

Name	Restriction
Potassium hydrogen-sulphate (potassium bisulphate)	O
Potassium hydroxide, solution <2 molar	Ot
Potassium hydroxide, solid, melt or concentrated solution	St
Potassium manganate(VII) (potassium permanganate)	t
Potassium nitrate	t
Potassium nitrite	R, S
Potassium sulphide	R, Ot
Propanal (propionaldehyde)	O, (F)
Propanoic acid (propionic acid)	O, (F)
propan-1-ol and -2-ol (n-propyl and isopropyl alcohols)	O, (F)
Propanone (acetone)	O, below O if t, (F)
Propylamine	S, F
Propylamine, dilute solution	O
Propyl ethanoate (propyl acetate)	O, (F)
Pyridine	R, St, F
Quinine	R, T cold tea is alternative for taste buds experiment
Selenium and compounds	R, S, F
Silicon tetrachloride	R, St, F, L
Silver nitrate	R, S
Silver nitrate solution	Ot
Sodamide (sodium amide)	N, L
Sodium	R, St
Sodium amalgam	R, S
Sodium azide	R, S
Sodium chlorate(I) solution (hypochlorite)	Ot Ensure container is vented
Sodium hydroxide, solution <2 molar	Ot
Sodium hydroxide, solid, melt or concentrated solution	St
Sodium hydrogen sulphate (bisulphate)	O
Sodium nitrate	t
Sodium nitrite	R, S
Sodium pentacyanonitrosylferrate(II) (sodium nitroprusside)	S

TABLE 22.5. *(Continued)*

Name	Restriction
Sodium peroxide	R, St, L
Sodium sulphide	R, Ot
Strontium	R, T
Sulphides, heavy metal	Ot
Sulphur chlorides	R, St, F
Sulphur dioxide canister	St, F
Sulphuric acid, conc.	Ot
Sulphuric acid, dilute	t
Tellurium metal and compounds	E
Tellurium compounds	X
Tetrachloromethane (carbon tetrachloride)	R, St, F 1,1,1-trichloroethane is less harmful substitute for solvent applications
Thallium and compounds	X
Thermite mixture	T Do not store
Thiocyanates, solid	S Do not heat to decomposition or add strong acids
Thiourea (also thiocarbamide)	R, S *See Education in Science,* September 1976, p. 17
Tin(II) chloride (stannous chloride)	O
Tin(IV) chloride (stannic chloride)	R, S, F, L
Titanium(IV) chloride (titanium tetrachloride)	R, T, F
1, 1, 1-trichloroethane as solvent	t, (F)
Trichloroethanoic acid (trichloroacetic acid)	S
Trichloroethane (trichloroethylene)	N 1,1,1-trichloroethane is a less harmful substitute for solvent applications
2,2,2-trichloroethanediol (chloral hydrate)	R, S, F
Trichloromethane (chloroform)	R, St, F 1,1,1-trichloroethane is a less harmful substitute for solvent applications
3,4,5-trinitrobenzoic acid (gallic acid)	R, S
2,4,6-trinitrophenol (picric acid)	R, S
Turpentine	R, t, F
Uranium compounds, solid	R, T
Uranium compounds; solution	S
Vanadium(V) oxide (vanadium pentoxide)	T

TABLE 22.5. (*Continued*)

Name	Restriction
Vanadates(V)	S
Xylene cyanol solid	R, S
Zinc powder	R, t
Zinc chloride	S
Zinc chromate(VI)	X

In view of our present knowledge of the potential hazards of chemicals, it is prudent to reevaluate the purchase, handling, storage, reactions, and disposal of *all* chemicals—not just those which have been singled out for special attention. The average laboratory, both in industry as well as academe, has often prized itself on the large number of bottles, jars, cans, and other containers of materials. While long time periods may ensue before an abnormal situation develops, such as electrical failure, an uncontrolled reaction, a fire, or other emergency, eventually the possibility of chemicals being in circumstances where they are not adequately controlled can occur. In one university laboratory, for example, perchloric acid was being used and evaporated in a standard hood on the third floor of a building. On the wooden floors of the building were more than a dozen standard-size acid bottles containing nitric, sulfuric, and hydrochloric acids and one large glass bottle containing metallic sodium. In viewing the scene after the explosion in the hood, the independent observer noted that the main hood sash, which had been blown completely away from its normal position, had missed the group of acid bottles on the floor by a matter of 3 in., as it landed from its flight. Had the acids been released by rupture of the bottles, on the wood (all this on the third floor), it is likely the chemistry building would have been destroyed. An analogous result in another chemistry building is shown in Fig. 22.3. Such situations are not academic; they are real, and suggest that more attention must be given to the proper handling and storage, as well as use of even common chemicals. Among the several ways which are useful in assuring that chemicals are known is the posting of a list (Fig. 22.4), revised frequently, on each door or doorway, which indicates what chemicals are in the laboratory. While it might be argued that such listing requires time and effort, the value of such a list to emergency personnel, as well as to maintenance personnel, would seem to offset its cost. In addition, a computerized record of all chemicals may be kept, with amounts on hand and where they are located, to assist in daily requirements for information as to who has what chemical in what quantity. The economics of proper inventory control may offset the cost of the up-to-date information.

Figure 22.3. Chemistry building on fire from chemical misadventure. Loss was $150,000. (Photo courtesy N.F.P.A.)

22.2   ANIMAL EXPERIMENTS AND GENETIC ENGINEERING

The June 16, 1980 decision of the Supreme Court in the General Electric Case, which held that a live, human-made microorganism is patentable on the heels of the requirements for more detailed animal experiments to establish the relative toxicity of substances, has given a great emphasis to biologically oriented chemistry. The Court gave approval, on narrow legal grounds, for the patenting of life-forms, but it did not endorse research in genetic engineering, per se. It did not deal directly with the problems posed by recombinant DNA technology, which is the subject of an application filed in 1974 by Stanley Cohen and Herbert Boyer.

The current bioassay program of the National Cancer Institute for evaluation of chemicals, has developed carcinogenic effects data on 300 chemicals. The NIOSH Registry of Toxic Effects of Chemical Substances, in the 1980 edition, lists at least some data on 45,000 substances, of which 2500 are suspected carcinogens, 2700 suspected mutagens, and 370 suspected teratogens. This is still significantly less than the 55,103 chemical substances (including 686 generic names) listed in the cumulative supplement to the EPA

CHEMICAL INVENTORY REPORT

INVENTORY NUMBER	BLDG	ROOM	CAS-NUMBER	ARRG	CHEMICAL NAME	H	F	R	C	QUANTITY ON HAND	WASTE CODE
4469	1	041	007446700		ALUMINUM CHLORIDE	2	0	2	3	.2500 LBS	
1324	1	041	NO CAS		AMMONIUM IODIDE	1			3	.2500 LBS	
4470	1	041	NO CAS		AMMONIUM META-VANADATE				3	.2500 LBS	
1537	1	041	NO CAS		BISMUTH NITRATE	2			3	.5000 LBS	
4471	1	041	010108642		CADIUM CHLORIDE	3	3	1	3	1.0000 LBS	
3445	1	041	010043524		CALCIUM CHLORIDE	1	0	0	3	1.0000 LBS	
4472	1	041	007789186		CESIUM NITRATE	2	3	2	3	.2500 LBS	
3585	1	041	007647010		HYDROCHLORIC ACID	3	0	0	3	3.0000 LITERS	
2118	1	041	007553562		IODINE	3			3	.2500 LBS	
1163	1	041	007697373		NITRIC ACID	3	0	1	3	3.0000 LITERS	
876	1	041	007789233		POTASSIUM FLUORIDE	3	0	0	3	.0030 KG	
4473	1	041	001310583		POTASSIUM HYDROXIDE	3	0	1	3	1.0000 LBS	
880	1	041	007681110		POTASSIUM IODIDE	2			3	4.0000 LBS	
881	1	041	007757791		POTASSIUM NITRATE	1	0	1	3	1.0000 LBS	
882	1	041	NO CAS		POTASSIUM PHOSPHATE MONOBASIC				3	.1000 KG	
878	1	041	007646937		POTASSIUM SULFATE	1			3	1.0000 LBS	
883	1	041	NO CAS		PRASEODYMIUM	1	2	0	3	.0200 KG	
884	1	041	NO CAS		QUINALIZARIN				3	.0050 KG	
888	1	041	007791119		RUBIDIUM CHLORIDE	1	3	0	3	.2500 KG	
887	1	041	NO CAS		RUTHENIUM CHLORIDE	2	2		3	.0100 KG	
886	1	041	NO CAS		RUTHENIUM TRICHLORIDE	2	2		3	.0030 KG	
889	1	041	NO CAS		SAFANIN				3	.0100 KG	
890	1	041	NO CAS		SAMARIUM OXIDE		2		3	.0100 KG	
891	1	041	NO CAS		SCANDIUM NITRATE				3	.0020 KG	
892	1	041	NO CAS		SEA SAND				3	.0001 LBS	
893	1	041	007783008		SELENIOUS ACID	3			3	.1250 LBS	
894	1	041	NO CAS		SILICON FLUID				3	.1250 LBS	
4474	1	041	020667123		SILVER OXIDE	1			3	.2500 LBS	
897	1	041	000506616		SILVER POTASSIUM CYANIDE	3			3	.1000 KG	
898	1	041	007631892		SODIUM ARSENATE	4			1	.0001 LBS	
900	1	041	000497198		SODIUM CARBONATE	1	0	0	3	.0001 LBS	
901	1	041	007775113		SODIUM CHROMATE	3	0	0	2	.5000 LBS	
899	1	041	000144332		SODIUM CITRATE	1			3	.0002 LBS	
903	1	041	001310732		SODIUM HYDROXIDE	3	0	1	3	.0001 LBS	
906	1	041	007681552		SODIUM IODATE	2			3	.1250 LBS	
904	1	041	007681825		SODIUM IODIDE	2	0	0	3	.0001 LBS	
905	1	041	007757746		SODIUM METABISULFITE	3			3	.5000 LBS	
416	1	041	007631994		SODIUM NITRATE	2	0	2	3	.2500 LBS	
907	1	041	007790285		SODIUM PERIODATE	2			3	.1250 LBS	
4475	1	041	010102188		SODIUM SELENITE	4			1	.1000 KG	
909	1	041	007757826		SODIUM SULFATE ANHYDROUS	0			3	.0002 LBS	
910	1	041	NO CAS		SODIUM TELLURATE	2			3	.1020 KG	
908	1	041	007772987		SODIUM THIOSULFATE	1	0	0	3	.0001 LBS	
885	1	041	NO CAS		SODIUM TRICHLORIDE HYDRATE				3	.0010 KG	
912	1	041	013472452		SODIUM TUNGSTATE	1			3	.5000 LBS	
914	1	041	010476854		STRONTIUM CHLORIDE	1			3	.0001 LBS	
913	1	041	010042769		STRONTIUM NITRATE	1	0	1	3	.5000 LBS	
915	1	041	007704349		SUBLIMED SULFUR	2	1	0	3	.0010 KG	
916	1	041	001314610		TANTALUM PENTOXIDE	0	2		3	.0100 KG	
918	1	041	NO CAS	D	TARTARIC ACID	0	1		3	.0001 LBS	
920	1	041	007446073		TELLURIUM OXIDE	2	2	0	3	.0250 KG	
919	1	041	NO CAS		TELLURIUM TETRACHLORIDE	2			3	.0250 KG	
922	1	041	NO CAS		TERBIUM OXIDE	1	2		3	.0005 KG	
923	1	041	NO CAS		THORIUM NITRATE	1	0	1	3	.5000 LBS	
924	1	041	NO CAS		THULIUM OXIDE				3	.0001 KG	

Figure 22.4. Chemical inventory report as posted outside a laboratory to alert emergency personnel to unusual hazards. It is revised monthly.

Chemical Substances Inventory under TSCA. In 1978 alone, carcinogenic test data has been assembled on 664 compounds, of which 21 are inorganic, 590 organic compounds, and 53 unclassified compounds.[22]

The bioassay of chemicals for toxicity, carcinogenic, mutagenic and teratogenic effects, involving large numbers of carefully bred animals and reptiles, including mice, rats, hamsters, monkeys, dogs, chickens, larger mammals up to apes, gorillas, and chimpanzees, introduces many potential hazards to the investigators. Animal laboratories are characterized by a complex interaction of biology with human endeavor. Ventilation of cage rooms must be carefully balanced; temperature, light, and water must be carefully controlled, litter for animal cage bedding must be changed and disposed of safely (with the potential of disease or other infections from the urine and fecal matter); insects and other vermin, which are often present, must be controlled (preferably without the use of pesticides which might poison the experimental animals), and general cleanliness maintained without sweeping. Mixing of feed with test material is a potentially dangerous routine usually requiring respiratory protection. Pathology laboratories, which use

Figure 22.5. Kilo quantities of suspect carcinogens in bioassay laboratory. (Photo by H. Fawcett.)

formaldehyde and other volatile chemicals, require special attention to insure adequate ventilation.[21,23]

In the older facilities, the ventilation system may be suspect, insofar as ventilation, air movement, and balance is concerned, and air intakes may be closer to exhausts than the various wind conditions would safely permit.[22] Chemicals under test, often in multikilo quantities, are identified by test numbers which would be meaningless to emergency personnel in case of fire, flood, or other situations requiring entry during off hours. All substances in the containers in Fig. 22.5, 22.6, and 22.7 were suspected carcinogens.

In newer laboratories, fortunately, much better control has been engineered into the control systems, including properly designed and operated incinerators for on-site disposal of bedding, carcasses, and other wastes and residues, instead of off-plant disposal in dumps or so-called landfills.

In contrast to bioassay laboratories, some of which have been observed to display limited or virtually no sincere concern with the more enlightened aspects of personnel health and safety, recombinant DNA and other biological engineering have been the subject of much interest by the scientific community, and by the public at large. The National Institutes of Health have established strict guidelines for DNA research after much dispute over safety even within the scientific community itself. For details of these regu-

Figure 22.6. Careless storage of suspect carcinogens. (Photo by H. Fawcett.)

Figure 22.7. Inadequate control of suspect carcinogens. (Photo by H. Fawcett.)

lations, see reference 24. Whether these will insure safety to the researcher and the public, if they are actually followed, remains for future evaluation.

22.3. THE LITERATURE OF HEALTH EFFECTS ON LABORATORY PERSONNEL

The subject of health hazards to laboratory workers has been approached in the literature from both a hazard-specific and a laboratory-specific point of view. Much has been written on each of many specific hazards that may be encountered by personnel employed in any of a variety of laboratories, such as exposure to commonly used toxic chemicals, dangers involved in the handling of radioactive materials, risk of infection and bites from laboratory animals, exposure to pathologic viruses and bacteria, injuries resulting from accidents with glassware, etc. Likewise, surveys, epidemiologic studies, and other types of investigations have been carried out for the purpose of assessing accident rates and morbidity or mortality of patterns among employees or for obtaining a qualitative description of hazards specific to a particular type of laboratory. Results of this type of study have been reported for chemical laboratories, medical or clinical laboratories, microbiological laboratories, and radiological laboratories.

In this section, an attempt was made to provide the most thorough review of health hazards to laboratory personnel whose work involves the use of chemical compounds. This necessitated a synthesis both of articles concerning chemical hazards to workers in all types of laboratories and of articles reviewing a variety of hazards related to work in a chemical laboratory, whether or not such hazard arose from the presence of chemical compounds (for example, risk of bites and infection in a laboratory engaged in animal toxicologic testing of chemical substances). Secondary emphasis was placed on review of microbiological and radiological laboratories and their associated hazards. Two final sections discuss some hazards involved in the handling of experimental animals and characteristics of laboratory accidents.

Chemical Hazards

Among the literature dealing with chemical hazards to laboratory workers, several types of articles have appeared: (1) general surveys/reviews of the types of hazards associated with work in a chemical laboratory; (2) case reports of exposure to a specific chemical or group of chemicals commonly utilized by the laboratory workers; and (3) epidemiologic studies of mortality patterns within specific groups of professionals who are exposed to chemicals in the laboratory. General reviews and surveys of health risks to personnel working in the chemical laboratory are also available in handbooks on laboratory safety.[25,26] Although information of the type available in general reviews and case reports does not allow quantitative assessments of health risks to workers, it does provide a description of the types of hazards associated with exposure to chemicals in the laboratory. Epidemiologic investigations are generally relied upon to provide a more quantitative determination of health risks to laboratory workers.

The Teddington, England, National Chemical Laboratory's "Safety Measures in Chemical Laboratories"[25] outlines several of the hazards to which laboratory workers are exposed. These include the risk of fire and explosion from flammable substances and hazardous reactions, dangers connected with the use of compressed gas cylinders, radiation hazards from the use and handling of radioactive materials, and the risk of burns, dermatitis, irritation, and systemic effects from exposure to specific chemical substances.

The Chemical Manufacturers Association[26] presents a more detailed discussion of the hazards involved in handling toxic chemicals. Localized irritation and burning from skin or eye contact, systemic effects resulting from subsequent absorption; inhalation of toxic gases and fine dusts with the possibility of their absorption into the general body circulation; swallowing of contaminated air, food, or beverage; and injection occurring through mechanical injury with contaminated glassware, careless handling of hypodermic needles, and small leaks in high-pressure systems are mentioned as possible means of exposures to toxic substances in the chemical laboratory.

A generalized discussion of the irritant effects of corrosive materials is also presented. Liquid irritants, which act directly on the skin either by chemically reacting to it, by dissolving some of its essential components, by denaturing its proteins, or by disturbing the membrane equilibrium of the skin cells, are stated to be the physical form of an irritant most likely to cause immediate injury. This class of substances includes mineral and organic acids, organic solvents, and solutions of solid irritants such as caustics, oxidizing and reducing salts, salts of organic acids, anhydrides, halogenated organics, etc.

The corrosive action of solid irritants is said to depend largely upon the specific irritant's solubility in the skin's moisture. In addition to a direct corrosive action, serious damage to the skin may also result from the substance's thermal heat of solution. Caustic alkalies are said to pose the greatest potential for harm, primarily because of their wide applicability. The elements sodium, potassium, and phosphorus, and the salts of antimony, arsenic, calcium, chromium, mercury, silver, and zinc are listed as agents that produce irritation by thermal burns.[26]

Irritants in the gaseous phase pose the "most serious hazard associated with irritants in general." A wide variety of localized and systemic effects (both chronic and acute) may result depending upon the specific substance's mode of action, solubility, and site or structure affected. In general, highly soluble substances affect the upper respiratory tract; intermediately soluble substances produce irritation in respiratory tract structures as deep as the bronchi; inhalation of minimally soluble substances may cause little primary irritation but frequently results in delayed pneumonitis.[26]

In another review of laboratory hazards, Fawcett[27] mentions the risk from use of cryogenic (low temperature) materials and chemical carcinogens, as well as the dangers associated with fire, explosion, toxic substances, some radioactive materials, and high pressure gases. Equally important are the observations concerning the attitudes of many laboratory personnel. Fawcett notes that inadequate attention is often given to the toxic potential of substances being handled, to the use of fume hoods, proper ventilation, or respiratory protective equipment. In addition, he reports a "great disbelief" among laboratory personnel that small amounts of substances, particularly the carcinogens and mutagens, may be harmful.

Cancer hazards to laboratory personnel have been reviewed by Heuper[28] who stresses a "proper awareness of the exposed individuals, of the existence, types and sources of occupational cancer hazards" as the most essential requirement of successful prevention. A few of the carcinogenic materials commonly encountered by laboratory personnel are briefly discussed, including benzene, aromatic amines such as beta-naphthylamine, benzidine, distillate and combustion products of coal and petroleum, arsenic, nickel, chromium, asbestos, and ultraviolet and ionizing radiations.

Additional chemical carcinogens utilized by laboratory workers, especially those employed in a microbiological laboratory, are reviewed by Wood and Spencer[29] and Collier.[30] Wood and Spencer list naphthylamines, benzidine, beta-propriolactone, isoniazid and aflatoxins as carcinogenic substances, and selenium as a possible teratogen. Collier also stresses the carcinogenic hazard of benzidine and raises the question of whether orthotolidine, dianisidine and other aromatic amines may be carcinogenic as well.

Although it would appear from the preceding accounts that the potential for exposure of laboratory personnel to toxic chemicals is great, the available literature contains few reports of injury or toxic effects specifically among laboratory workers from exposure to chemical compounds. Among the recent literature, five case reports of injury to laboratory personnel involving the use of chemical substances have been published.[30,31,32,33] Burke et al.[31] reported a case of systemic fluoride poisoning in a laboratory technician who had accidentally incurred a fluoride skin burn while performing an experiment with 100% anhydrous HF. The rupture of a connecting tube generated a stream of HF which struck the technician on the right side of the face, neck, and right arm. The author noted that "Despite specific safety regulations, the sliding glass door of the exhaust hood had been half open and a personal face shield had not been in use at the time of the accident."

Two technicians employed at a quality control laboratory in which the carbon dioxide content of beer was measured with a Martin's apparatus were discovered to have "dangerously high" levels of urinary mercury.[32] The author cited improper techniques among the laboratory workers (i.e., vigorous raising and lowering of the mercury reservoir, which produced a stream of mercury that spilled over the top of the apparatus and rapid expulsion of the test liquid through the side arm, which carried an amount of mercury with it) and a lack of awareness of the dangers associated with mercury exposure as the major cause of this hazardous situation. Analysis of mercury levels in the air of the laboratory revealed concentrations from 12 to 15 times the current TLV. The laboratory was subsequently overhauled to remedy the unsafe conditions, and over the next three months the technicians' urinary mercury levels were reported to decrease significantly.

Oliver[33] reported the case histories of three young laboratory workers who suffered minimal, transient exposure to 2,3,7,8-tetrachlorodibenzo 1,4-dioxin (dioxin). Two of the scientists were reported to develop typical chloracne; all three were found to have raised serum cholesterol; and two suffered delayed symptoms (occurring up to 2 years later) such as personality changes, other neurological disorders, and hirsutism. The author noted that even though each scientist had employed rigorous precautions to avoid skin contact or inhalation of dioxin, some exposure had nevertheless occurred, an indication that normal laboratory precautions were obviously inadequate.

Occupationally induced allergy to *Bandeiraea simplicifolia* seeds in a research biochemist was reported by Kanellakes and Mathews.[34] The allergy apparently first developed in this biochemist after almost 4 years of laboratory work involving the grinding, extracting and analyzing of *B. simplicifolia* seeds. The authors also reviewed four cases of occupational-induced allergy to ragweed among a botanist, a meteorologist, and two pharmacists, and one case of an allergic response to elm pollen by an allergist.

Pye and Burton[35] reported the development of severe eczema among two laboratory technicians following the preparation of 2,4-dinitrochlorobenzene (DNCB). One technician felt that no DNCB had come in contact with his skin, but that some of the prepared solution had spilled on his examination paper, which he subsequently folded and put in his back pocket. It was reported that some of the DNCB solution may have come in contact with the hands of the second laboratory technician. The authors observed that although both technicians had possibly worked with DNCB on prior occasions, neither had been warned of its sensitizing properties. The authors additionally discuss the possibility that occasional exposure to DNCB may also sensitize a person to the drug chloramphenicol with which it cross-reacts.

Legg,[36] Mirvish et al.,[37] and Braymen and Songer[38] have investigated potential sources of chemical exposure for laboratory workers. Legg[36] measured the fumes produced during aspiration of trichloroacetic acid into a flame emission spectrometer. The continuous and intermittent aspiration of these fumes, when performed in the absence of a fume hood possessing efficient extractor fans, had been reported to cause symptoms of headache and lachrymation among laboratory staff. The author's analysis of the effluent vapor revealed chloroform and phosgene in concentrations of 2–3 μg/liter and approximately 2 μg/liter, respectively.

Mirvish et al.[37] discuss the carcinogenic activities of some *n*-nitroso compounds and subsequent implications for the safety of laboratory personnel who are handling these compounds or working with animals on which these compounds are being tested. The authors note that some *n*-nitroso compounds, when injected or painted onto the skins of animals, can be expired unchanged by the animal, and that untreated mice have developed nasal tumors when kept in the same room as mice whose skins had been painted with the *n*-nitroso compound methyl-*n*-butylnitrosamine.

Disinfectant solutions that are allowed to stand in open pans present another potential hazard to laboratory personnel.[38] Measurement of phenol concentrations at various distances from standing containers of 5% solutions revealed that the occupational TLV was exceeded only at close proximity to the disinfectant pans and when the solutions were hot. The authors cautioned, however, that allergic reactions among the laboratory staff may develop at the concentrations measured, and recommended a reduction of exposure through use of ventilation, exhaust hoods, or respiratory protection.

Few epidemiologic investigations of health risk specifically among chemical laboratory workers have been carried out. Hence, the studies reviewed here cover a variety of groups, such as pathologists and clinical and hospital laboratory workers, who have a potential for exposure to chemicals in the laboratory. Ott et al.[39] developed background mortality data on 8171 male employees, 18 to 64 years of age, within the research department, corporate headquarters, and manufacturing division of a Dow Chemical Company facility. Personnel with a history of exposure to asbestos or arsenicals were excluded; socioeconomic level and job category were determined; vital status was traced through the Social Security Administration; and expected death rates, both age-specific and cause-specific, were calculated on the basis of the U.S. white male population. The authors reported that mortality, by cause and age, was generally lower than expected. Personnel within the job category of nonprofessional laboratory employees were not observed to exhibit an increased mortality risk. This group's standardized mortality ratios (SMRs) for death from all causes, from cardiovascular disease, and from all other causes were reported as 74, 101, and 48, respectively. No SMR for malignant neoplasms was calculated since fewer than five deaths occurred in this category (three deaths from malignant neoplasms were observed, compared with an expected 5.3 deaths). Hoar has conducted an epidemiologic study of 3686 chemists working for a major chemical company.[40]

Taylor and Marks[41] surveyed urinary mercury excretion among five small study groups with varying potential for exposure to mercury. Two of these subgroups were characterized by no known exposure to mercury—hospital outpatients and some hospital laboratory staff. The remaining three subgroups had experienced some exposure to mercury—hospital and university laboratory staff, assemblers of mercury hollow cathode lamps, and dentists and dental surgery assistants. Of the five groups surveyed, only the dental workers were observed to show significantly increased levels of urinary mercury. The authors concluded that monitoring of mercury excretion was an important step in identifying personnel who may ultimately develop symptoms of mercury toxicity.

Early manifestations of exposure to mercury vapors among laboratory workers have also been investigated by Lauwerys and Buchet.[42] Occupational exposure levels and blood and urine samples were collected from a group of 40 technicians employed in chemical and biological laboratories where there was a history of exposure to metallic mercury vapors. Similar measurements were performed on a control group of 23 technicians employed at biological laboratories where metallic mercury was not handled. Analysis of airborne mercury concentrations revealed that 80% of the 40 technicians were exposed at levels below 0.04 mg/m^3. The average exposure level was 0.028 mg/m^3 with a range from 0.002 to 0.124 mg/m^3. The authors reported significantly higher mercury concentrations in the blood and urine of exposed technicians compared to the control group. The exposed group was additionally observed to exhibit significantly higher plasma galactosidase and catalase activities and reduced RBC cholinesterase activity. A weak significant correlation was reported between time-weighted mercury exposures and urinary mercury concentrations among the exposed technicians. Mean urinary mercury content of those workers exposed to more than 0.04 mg/m^3 mercury was reportedly three times greater than that of workers exposed at lower concentrations; hence, the authors concluded that on a group scale, urinary mercury determinations were a valuable index of mercury vapor exposure.

Other epidemiologic investigations have described mortality patterns among pathologists, medical laboratory technicians[40] and chemists.[41-43] Harrington and Shannon[43] identified 2079 pathologists and 12,944 medical laboratory technicians from membership lists of professional bodies. Cause of death was determined for 151 of the 156 deaths among pathologists and for all 154 of the deaths among technicians. The authors reported a lower mortality rate from all causes among their study population compared with the general population (SMR for pathologists: 60; SMR for technicians: 67). However, among male pathologists, a higher than expected death rate from neoplasms of the lymphatic and hemopoietic tissue was observed (8 observed deaths versus 3.3 expected, $p < 0.01$). The authors could not explain this excess, and noted that it was not due to Hodgkin's disease or leukemia. Among both pathologists and technicians, the authors observed a higher-than-expected rate of suicide (SMR for pathologists: 265; SMR for technicians: 191). Among female technicians, suicide was reportedly the commonest cause of death. The authors postulated a ready access to lethal chemicals at work as a possible factor in explaining the high proportion of suicide by poisoning. In addition, they noted that suicide rates among pathologists and medical laboratory technicians were higher than among all medical practitioners (SMR of 176) and all laboratory technicians (SMR of 69) respectively.

Increased rates of suicide and cancer among chemists were reported by Li and associates[44,45] in a study of ACS members whose deaths were announced in *Chemical and Engineering News* between April 1948 and July 1967. Deaths among ACS members reported during this period numbered 4644, and death certificates were obtained for 3637 (78%) of the decedents. Occupation, derived from death certificates, was listed as chemist, engineer, teacher, or scientist for 71% of the study population. The remainder were usually categorized as manager or administrator in the chemical industry. Causes of death were examined among 2152 male chemists dying between the ages of 20 and 64, among 1370 male chemists dying after age 64, and among 115 female chemists. Relative frequencies of specific causes of death were calculated and compared with expected rates. The expected rates for male chemists 20–64 years of age were based on cause-specific proportionate mortality of 9957 U.S. professional men of similar age in 1950; those for chemists dying after 64 years of age were based on mortality rates of U.S. white males of similar age in 1959; while mortality rates for female chemists were compared with those of the U.S. female population in 1959.

Among male chemists, significantly increased mortality from cancer, particularly malignant lymphoma, was reported. Among male chemists 20–64 years of age, cancer of the pancreas was also significantly increased while anemia and cancer of the pancreas were observed to be moderately increased among male chemists over 64 years of age. Female chemists exhibited a rate of suicide five times that of females in the general U.S. population. In addition, a higher-than-expected number of deaths from breast cancer were observed but were explained by the authors as possibly due to the higher socioeconomic status and larger percentage of unmarried women in the study group.

In discussing imperfections in their study design that may have limited the interpretation of results, Li et al.[45] observed that: (1) 22% of the death certificates were not located; (2) deaths from cancer may have been preferentially reported in *Chemical and Engineering News;* (3) the relative-frequency method may be more misleading than a population-based study; and (4) differences in mortality between the study and control groups may have been partly due to the fact that the groups were not strictly comparable with respect to calendar time. In summary, the authors state that while not conclusive, a possibility has been raised of increased risk of lymphoma and pancreatic cancer among chemists.

The increased suicide rate among female chemists is discussed in greater detail in a second paper by Li.[44] In addition to an age-adjusted frequency five times that of the general U.S. female population in 1959, the author noted that the suicide rate (11%) among female chemists was nearly three times that of male chemists. The author further noted that 49% of ACS members committing suicide used solid or liquid poisons compared with only 6% of suicides within the white male population of 1959.

A significant excess of cancer deaths among chemists was also reported by Dr. R. Olin (reviewed in Rawls[46]) who followed through 1974 a group of

538 male chemistry students graduating from the Swedish Royal Institute in Stockholm between 1930 and 1950. Overall mortality was not significantly different from expected (58 observed deaths versus 67 expected); however, 22 deaths attributable to cancer were observed compared with only 13 expected. In addition significantly increased mortality rates for malignant lymphoma (1.7 deaths expected versus 6 observed), Hodgkin's disease (0.3 deaths expected versus 3 observed), and cancer of the urinary organs (1 death expected versus 3 observed) were reported. A comparison of findings between the Olin study of cancer mortality among chemists and the study by Li et al.[45] is made and, as reported by Rawls,[46] a similarity in results is noted by Olin which is significant considering the differences in methods of study.

From the preceding discussion of chemical hazards involved in laboratory work and health patterns among laboratory workers who sustain some degree of occupational exposure to chemical substances, two notes of importance should be made. First, the attitudes and personal habits of laboratory personnel may be a significant factor in the etiology of laboratory-related accidents and diseases. This point has been touched upon in some way by Burke et al.,[31] Fawcett,[27] Harrington,[32] Heuper,[28] and Legg,[36] and has been given greater recognition by Phillips,[47,48] (see following section on Accident Characteristics). Secondly, epidemiologic study of mortality patterns among chemistry graduates in Sweden, pathologists and medical laboratory technicians, and American Chemical Society members (a diverse occupational group, but one which, nonetheless, would be expected to include many laboratory employees whose occupation entails the use and handling of a variety of chemical substances) has resulted in observation of higher-than-expected rates of suicide, and, among males, increased incidence of deaths from malignant lymphoma, from neoplasms of the hemopoietic tissue, and from cancer of the pancreas.[45] Although there has been speculation on specific causal factors, no substantive conclusions have been offered as to the observed association between exposure to laboratory chemicals and mortality patterns. It is generally agreed that further investigation is needed.

Microbiological Laboratory Hazards

The primary risk involved in microbiological research laboratories is the transmission of infectious diseases. In a survey by Sulkin and Pike[49] data was collected from 5000 laboratories by means of a questionnaire to disclose information regarding laboratory acquired infections during the period 1930–1950. Response to the questionnaire was received from better than one-half of the solicited laboratories, and provided information on 875 cases not reported in the literature. Sulkin and Pike identified 1342 infections, 39 (3%) of which resulted in death. The highest case fatality rate was reported for viral infections, with the most frequently identified viral agents belonging to the *Brucella* group. Of the infections, 75% occurred among "trained scientific personnel." Of the type of laboratories surveyed, research and diagnostic work laboratories accounted for nearly 90% of the reported cases of infectious disease. Among the accidents of known cause, infection from the needle and syringe accounted for over a quarter of the cases reported.

One of the most lengthy works, the Fort Detrick epidemiological study[47] provides data compiled from the literature, three laboratory institutions, and those of the Fort Detrick laboratory. A total of 133 disease agents were identified, 79 of which are controlled by a Federal or other authorized agency. Of interest is Phillips[48] comment that "effective vaccines, toxoids or drug therapy are available for only 37 of the 133 possible diseases (27%)." It has been determined that 58% of the disease agents are viruses and of these only eight have a known effective vaccine. Fifty-eight of the 78 virus agents have been cited in the literature as being responsible for laboratory–acquired infections; and, based on literature reports, 81% of the total 133 disease agents have been observed to cause disease among laboratory workers.

Known accidents were identified for only 20% of the infections reported. For the remaining 80% of reported infections, no incidents or accidents were recorded prior to the reporting of the infections, hence they are classified as being of "unknown" or "unrecognized" cause. This percentage of unknowns has remained fairly consistent in the various infection surveys. Of the known causes of infection, five are frequently mentioned: accidental oral aspiration by pipet, accidental needle and syringe innoculation, animal bites, syringe sprays, and centrifuge accidents.[50-55]

In a microbiological safety report by Reitman and Wedum[51] some 28 laboratory procedures were analyzed by the degree of risk of infection. It was found that aerosols containing infectious agents were more often created during hand grinding (mortar and pestle) of tissues, decanting after centrifugation, resuspending packed cells, withdrawing a culture from bottles, and shaking and blending cultures in high-speed mixers. Accidents that occurred while centrifuging or handling dried cultures were more apt to cause widespread infection in the laboratory. Wedum has suggested that the fewer number of infections found among housekeeping personnel (i.e., personnel who do not directly handle microbiologic cultures) indicates that there is a higher risk of contracting infection for those workers who directly handle the pathologic cultures.

In a continuation of the work begun in 1950, Pike[53] analyzed and reviewed 3921 cases of overt laboratory–acquired infections. Worldwide literature reports were obtained for review, and a record was maintained by the author of all laboratory associated disease cases from 1950 through 1974. A more detailed tabulation of the sources of infection was possible from this data and is presented in Table 22.6. Pike also observed that fewer cases of infection were recorded between 1955 and 1964 than in the previous 10 years. Even fewer cases were recorded between 1965 and 1974, and Pike suggested several factors responsible for the decline: an increasing awareness of the

TABLE 22.6.   Distribution of Cases According to Proved or Probable Source of Infection

Sources	Number	Percent
Accident	703	18
Animal or ectoparasite	659	17
Clinical specimen	287	7
Discarded glassware	46	1
Human autopsy	75	2
Intentional infection	19	0.5
Aerosol	522	13
Worked with the agent	827	21
Other	16	0.5
Unknown or not indicated	767	20
Total	3921	100.0

Source.   R. M. Pike, Laboratory-Associated Infections: Summary and Analysis of 3921 cases, *Health Lab. Sci.* **13**(2):105–114, 1976.

hazards of working with these agents; increased literature on safety procedures, laboratory devices, and laboratory design with subsequent implementation into the research laboratory; and finally, a decline in the experimental use of the Brucella viruses.

Although there has been an increase in oncogenic (tumor producing) virus research, there is no evidence in the published literature of a single case of laboratory-related human malignancy.[53] The hazard, unfortunately, is a distinct possibility. A report of Baldwin[54] demonstrated that the present centrifuge models were not intended for research involving oncogenic viruses, and proper safety controls were not engineered into the design by the manufacturer. However, laboratory practices can allow for safe operation. Measurement of exposures in dose per milliliter of preparation placed centrifuging hazards below that of the risks of infection posed by general laboratory accidents (i.e., cuts, spills, and splattering, etc.).[55]

Two reports of an epidemiology study performed in Denmark appear in the literature.[56–57] A questionnaire was sent to clinical chemistry departments requesting information on dangerous episodes and injuries. Information on 2600 persons was obtained, accounting for more than 90% of those employed in clinical chemistry laboratories in Denmark. The data revealed that hepatitis, burns, and cuts were most often the cause of injury; and the centrifuge was noted to be the cause of most dangerous episodes. The infection rate for hepatitis in the laboratory population sampled was seven times that of the incidence rate for the general population.

Radiological Hazards

Some radiological hazards have been identified with medical laboratory personnel due to the use of various radiopharmaceuticals. LeBlanc et al.[58] found that employee exposures in a nuclear medical laboratory had risen from an annual average in 1965 of 43 mrem/year/person to an annual average in 1969 of 423 mrem/year/person. This increase in annual exposure was attributed to an increased number of hand contamination cases, mostly from technetium (^{99m}Tc) which was introduced into the laboratory in 1966. Possible routes of exposures included contaminated glassware and cotton swabs, splattering from IV infection containers, and improper use of disposable gloves.

Another use of medical radionuclides, radioimmunoassay measurements, has also been associated with exposure of laboratory personnel. Elevated levels of ^{125}I were found in laboratory workers' urine and thyroids.[59] Most cases of contamination occurred during the ionization of serum globulins even though the process was conducted in radiochemical hoods with proper protocol. The authors suggested that the chemical hoods are not adequate to confine gaseous forms of iodine during nonspecific labeling of large molecules. Xenon (^{135}Xe), used to assess lung ventilation, was frequently found to contaminate examination rooms by leaking from patients' mouthpieces during ventilatory testing.[60]

Two cases of accidental inhalation were reported from the Savannah River Laboratory.[61] One incident involved exposure to ^{244}Cm of a laboratory employee engaged in removing contaminated waste from a decontamination chamber. The second exposure occurred while shields were being removed from a holding cell designed for handling CmO_2 and AmO_2. The biological elimination and retention of americium and curium was monitored in both cases.

There was one case report of an exposure which occurred during preparation of a target for mass spectroscopic analysis.[62] An error made by a technician while using a special glove box resulted in contamination of gloves, clothing, and face of the technician and the subsequent inhalation of ^{242}Cm. No cases of exposure while using mass spectroscopy equipment were identified in the literature when proper protocol was observed.

Increasing attention and interest is being paid to hazards associated with nonionizing electromagnetic radiation. This area was not reviewed for purposes of this chapter.

Hazards Involved in the Handling of Laboratory Animals

In a review article by Griesemar and Manning[63] accidents and infectious diseases appeared to be the major risks to laboratory personnel working with experimental animals. Self-innoculation from the needle and syringe were determined to be the most hazardous of the reported accidents. Other accidents mentioned include cuts from improperly designed or repaired cages; thermal and chemical burns from steam and concentrated disinfectants used to clean the facilities; electric shocks from deteriorated electrical insulation; and falls on slippery floors. The authors also identify the noise level as a possible hazard to the quality of life of both the animals and the workers. Although animal bites and scratches occur daily, they do not contribute to any significant clinical injuries or illnesses.

TABLE 22.7.   Reported Causes of 1218 Laboratory Accidents

Stated Cause	Number of Accidents	Percent
Employee at fault	564(14)[a]	46.3
Equipment at fault	239 (8)	19.6
Combined human and equipment failure	112 (2)	9.2
Supervisor at fault	62	5.1
Another work group at fault	50	4.1
Unknown	191(23)	15.7
Total	1218(47)	100.0

Source. G. B. Phillips, Causal Factors in Microbiological Laboratory Accidents and Infections, Report No. MP-2, p. 181, U.S. Army Biological Laboratories, Frederick, Maryland, 1965.

[a]Parentheses denote lost-time accidents.

There exists a possibility of exposure to infected animals, and a note is made that 30%–40% of the infections acquired in a laboratory involve animals or animal tissue.[63] Transmission of infection is generally from animal or ectoparasite bites or during autopsy, at which time infectious aerosols may be emitted or cuts incurred. Two papers have identified airborne transmission as a source of infection. One paper described a case of airborne rabies virus infection which resulted in the death of the technician involved.[64] Biggar et al.,[65] in an epidemiology study, identified 57 cases of lymphocytic choriomeningitis transmitted from animal tissue.

The final hazard referred to in the literature is hypersensitivity to the epidermal scales or dander of animal hair. Lutsky and Neuman[66] collected questionnaires from 1293 workers at 39 animal facilities to establish that acquired laboratory animal dander allergy was indeed an occupational disease. Their data showed that 70% of the affected individuals had no previous history of allergies, and that it was necessary for these individuals to change jobs because of the affliction.

Accident Characteristics

An accident can be the result of either a dangerous action by the technician or a dangerous condition in the laboratory. Stark[67] has concluded in a review of laboratory hazards that the most important factor in maintaining a safe laboratory is the individual worker. However, Uldall[56] suggests that an effective solution to the accident problem would first involve change in laboratory conditions, and next, a motivation to safe behavior in the technician.

The most extensive work identified in the literature that demonstrates accident causes is the four year retrospective epidemiology study performed at the U.S. Army Biological Laboratories, Fort Detrick conducted by Phillips.[47] Although the majority of the information contained in the report is most relevant to microbiological laboratories, the conclusions reported from

TABLE 22.8.   Annual Average Employment and Injury and Illness Incidence Rates for Miscellaneous Service Employees (SIC 89 and 892)

Year	SIC Code	Annual Average Employment	Incidence Rates per 100 Full-Time Workers— Illness and Injuries			
			Total Cases	Lost Workday Cases	Nonfatal Cases without Lost Workdays	Lost Work-days
1974[a]	89	852,100	2.3	0.7	1.6	6.5
1975[b]	89	877,800	2.2	0.7	1.5	7.8
1975[b]	892	139,800	3.5	1.2	2.3	14.5

[a]U.S. Bureau of Labor Statistics, Bulletin No. 1932, 1976 p. 29, 40.

[b]Personal communication, Jerry Faulkner, U.S. Bureau of Labor Statistics, San Francisco, June 1977.

the section of the study that interviewed 33 accident-involved and 33 accident-free persons are relevant to all laboratory workers. The data provided on accidents classified according to laboratory conditions and accidents classified as human error vs mechanical error also can be generalized.

Elicited by questionnaire and interview, accident-involved individuals were more likely to smoke and/or drink, to have been divorced and to have less family involvement or few close family ties. Data more directly related to accident behavior evinces the following characteristics of the accident-involved group: (1) a lack of ability to perceive accident conditions or to realize the meaningfulness of safety regulations; (2) accident-involved workers tend to be less critical of other's behavior; (3) they may harbor hostile feelings toward the general safety program; (4) they are prone to taking excessive risks, e.g. intentionally violating safety protocol; (5) they tend to work at an excessive rate of speed to complete assignments; and (6) are generally unwilling to change work habits that have been identified as unsafe.

A summary of the findings on the causes of accidents, contrasting personal versus mechanical failures, is presented in Table 22.7. Phillips noted that 82% of the accidents occurring to accident-involved workers were due to personal failures, while 65% of accidents were due to such failures among both accident-involved and accident-free groups considered as a whole.

It has also been reported that younger workers tended to have a higher accident rate than older experienced workers, and that the technical scientific staff exhibited a higher accident rate compared to housekeeping or caretaking personnel.[47,68]

In a current survey of National Health Service Public Health Laboratories, Harrington[68] identifies the following in regard to accident conditions and actions: two-thirds of the laboratories surveyed allow mouth pipeting, only 54% formally instructed new employees in safety, and both centrifuges and safety cabinets were infrequently or never serviced.

There is one report in the literature concerning the effectiveness of a

TABLE 22.9.   Incidence Rates for Injuries and Incidence Rates for Illnesses for Miscellaneous Service Employees (SIC 89 and 892)

Year	SIC Code	Incidence Rates per 100 Full-Time Workers					
		Injuries			Illnesses		
		Total Cases	Lost Work-day Cases	Nonfatal Cases without Lost Workdays	Total Cases	Lost Work-day Cases	Nonfatal Cases Without Lost Workdays
1974[a]	89	2.1	0.7	1.4	0.2	[c]	0.1
1975[b]	892	3.3	1.1	2.2	0.2	0.1	0.1

[a]U.S. Bureau of Labor Statistics, Bulletin No. 1932, 1976 p. 29, 40.

[b]Personal communication, Jerry Faulkner, U.S. Bureau of Labor Statistics, San Francisco, June 1977.

[c]Indicate incidence rates less than 0.05 per 100 full-time workers.

research laboratory safety program.[69] During 1965–1975 reported injuries were lowered from 96/month to 30/month. Total serious injuries also decreased from 64/month in 1965 to 19/month in 1974.

No precise data are available on accident rates of all types specifically among U.S. laboratory personnel. However, some indirect measure can be obtained through rates, provided by the U.S. Bureau of Labor Statistics, for injury and illness incidence within the Standard Industrial Classifications

TABLE 22.10.   Injury and Illness Incidence Rates, by Industry Division, United States, 1972–1974[a]

Industry	Incidence rates per 100 full-time workers		
	1972	1973	1974
Private sector	10.9	11.0	10.4
Agriculture, forestry, and fisheries	—	11.6	9.9
Mining	—	12.5	10.2
Contract construction	19.0	19.8	18.3
Manufacturing	15.6	15.3	14.6
Transportation and public utilities	10.8	10.3	10.5
Wholesale and retail trade	8.4	8.6	8.4
Finance, insurance, and real estate	2.5	2.4	2.4
Services	6.1	6.2	5.8

[a]Estimates for 1973 and 1974 include data for agricultural production (SIC 01), all of mining (SIC 10-14), and railroads (SIC 401). With the exception of oil and gas extraction (SIC 13), data for these activities were not included in the 1972 estimates. In addition, data for agricultural services, forestry, and fisheries (SIC 07-09) were included in the services division for 1972.

(SIC) 89—Miscellaneous Services, and 892—Noncommercial, Educational, Scientific, and Research Organizations. While it is likely that many laboratory workers would be included in this SIC group 892 (establishments primarily engaged in noncommercial research into information for public health, education or general welfare), a number of laboratory personnel are nevertheless excluded, including those employed at commercial and hospital laboratories.

Published data are available for 1974 for the SIC group 89; and a personal communication with Jerry Faulkner, U. S. Bureau of Labor Statistics, San Francisco, identified the rates for 1975, as yet unpublished, for both the SIC code 89 and the more specific SIC designation 892. These figures are presented in Tables 22.8–22.10.

Every new avenue of science and technology appears to open unique hazards, some of which are overlooked until serious problems develop. The frontier is not without hazards.

ACKNOWLEDGEMENT:

Gordon Siegel, M.D. contributed major input to the medical aspects of this chapter, which is gratefully acknowledged.

REFERENCES

1. R. R. Jones, "Readers Reveal the Dangerous Lives of R & D Scientists," *Ind. Res. Devel.*, 127–130 (July 1980).
2. H. H. Fawcett, "Chemical Boobytraps." *Ind. Eng. Chem.* **51**(4), 89A–90A (April 1959).
2a. H. H. Fawcett, "Chemical Boobytraps," 16-mm sound movie 10 min., General Electric Audio-Visual Center, Scotia, New York.
3. "Labeling Manual," developed by LAPPI Committee, Manufacturing Chemists Association (Now Chemical Manufacturers Association, 2501 M St. N. W., Washington, D. C. 20037) now ANSI Z 129.1-1976, "Standard for Precautionary Labeling of Hazardous Industrial Chemicals," American National Standard Inst., New York, 1976.
4. F. A. Herr, "Los Angeles Plating Plant Explosion," *Metal Finish.*, **45**(3), 72–73, 107 (1947).
5. "Hazardous Chemical Reactions," NFPA No. 491M. National Fire Protection Association, Quincy, Mass. (1976 or later edition).
6. L. Bretherick, *Handbook of Reactive Chemical Hazards*, 2nd ed., Butterworths, London, 1979.
7. M. D. Morrisette, "Hazard Evaluation of Chemicals for Bulk Marine Shipment," *J. Hazard. Mater.* **3**(1), 33–48 (1979).
8. H. K. Hatayama et al., "A Method for Determining Hazardous Wastes Compatibility," Grant No. R 804692, Municipal Environmental Research Laboratory, Office of R & D, U.S. EPA, Cincinnati, 1980.
9. J. B. Finley, "$KMnO_4$ + DMF Explosion," Chem. Eng. News, Feb. 23, p. 9, April 6, p. 3, Apr. 27, p. 47, Aug. 3, p. 3, 1981.
10. "Pilot Plant Safety Manual," American Institute of Chemical Engineers, New York, 1978.
11. W. H. Stahl, "Compilation of Odor and Taste Threshold Values Data," DS 48, Committee E-18, American Society for Testing and Materials, Philadelphia, 1973.

12. "Instrumentation for Environmental Monitoring," AIR, Lawrence Berkeley Laboratory, Berkeley, California, 1975.

12a. *ACGIH, Air Sampling Instruments for Evaluation of Atmospheric Contaminants,* 4th ed., American Conference of Governmental Industrial Hygienists, Cincinnati, 1972.

12b. M. Katz, *Methods of Air Sampling and Analysis,* 2nd ed., APHA Intersociety Committee, American Public Health Association, Washington, D. C., 1977.

13. G. M. Kintz and H. Browne, "Flame Propagation," 16-mm sound color movie, 30 minutes, Kennedy Productions, Chicago, Ill. [Available from American Gas Association, Rosslyn, Virginia.] (Note: the rate of flame propagation is defined as the speed at which a flame progresses through a combustible gas-air mixture under the temperature, pressure and mixture conditions existing in the combustion space, burner or piping under consideration; the flame spreading from layer to layer independently of the source of ignition.)

14. a. U.S. Patents 2,616,825 and 2,616,928. Gilbert, E. E. and Giolito, S. L., November 1952. 1 mole hexachlorocyclopentadiene in 1.5 moles liquid SO_3 forms dark red viscous intermediate which on hydrolysis in an excess of 6% NaOH followed by neutralization with H_2SO_4 gives the product.

   b. M. Sherman and E. Ross, "Acute and Subacute Toxicity of Insecticides to Chicks," *Toxicol Appl. Pharm.* **3**, 521–533 (1961).

   c. Report on Bioassay of Technical Grade Chlordecone, National Cancer Institute, 1976.

   d. "Senate Panel Probes *Kepone* Disaster," *Chem. Eng. News,* **54**(5), 17–18 (February 2, 1976).

   e. S. B. Cannon, "Epidemic *Kepone* Poisoning in Chemical Workers," *Am. J. Epidemiol.* **14**(7) 534 et seq. (1978).

   f. R. J. Huggett and M. E. Bender, *Kepone* in the James River, *Environ. Sci. Technol.* **14**(8), 918–921 (1980).

15. O'Connor Electro-Plating Corp., Los Angeles, California, February 20, 1947. Seventeen fatalities and serious building damage resulted. (See also reference 4.)

16. I. Kabik, "Hazards from Chlorates and Perchlorates in Mixtures with Reducing Materials," Report I.C. 7340, U. S. Bureau of Mines, Washington, D. C., 1945.

17. E. Knuth-Winterfeld, "Perchloric Acid Hazard," *Metals Prog.* **61**(3), 80 (1952).

18. H. H. Fawcett, "Perchloric Acid—Ethyl Alcohol Electropolishing Mixtures," Research Laboratory Report RL-1185, Class 1, General Electric Co., Schenectady, New York, 1954.

19. a. D. Swern, *Organic Peroxides,* 3 vols, Wiley—Interscience, New York, 1970–1980.

   b. Noyes Data Corp., *Organic Peroxide Technology,* Park Ridge, N.J., 1973.

   c. "Fire and Explosion Hazards of Peroxy Compounds," Special Technical Publication 394, American Society for Testing and Materials, Philadelphia, 1965.

   d. Th. M. Groothuizen and J. Romijn, "Heat Radiation from Fires of Organic Peroxides as Compared with Propellant Fires." *J. Hazard. Mater.,* **1**(3), 191–198, (1976).

   e. D. N. Treweek, et al. "Use of Simple Thermodynamic and Structural Parameters To Predict Self-Reactivity Hazard Ratings of Chemicals," *J. Hazard. Mater.* **1**(3), 173–189 (1976).

   f. R. M. Johnson and I. W. Siddiqi, *Determination of Organic Peroxides,* Pergamon Press, New York, 1970.

   g. "F.M. Data Sheet on Organic Peroxide Storage," Loss Prevention Sheet 7-80, Factory Mutuals Research Corp., Norwood, Mass., 1972.

   h. N. Steere, ed., *Handbook of Laboratory Safety,* 2nd ed., Chemical Rubber Publishers, Cleveland, 1979.

   i. "Scientist Injured in Ether Peroxide Explosion," NBS Safety Bulletin 130, National Bureau of Standards, Washington, D.C., September 1977.

j. V. K. Mohan, K. R. Becker, J. E. Hay, "Hazard Evaluation of Organic Peroxides," *J. Hazard. Mat.,* **5,** 3, 197–220 (Feb. 1982).

20. a. Supreme Court Decision on Patenting of Microorganisms, Supreme Court of the U. S., No. 79-136, June 16, 1980, Sidney A. Diamond, Commissioner of Patents and Trademarks, Petitioner, v. Ananda M. Chakrabarty et al. (assigned to the General Electric Co.)

    The invention included "a bacterium from the genus *Pseudomonas* containing therein at least two stable energy-generating plasmids, each of said plasmids providing a separate hydrocarbon degradative pathway." Chakrabarty combined the genetic materials from five different oil-eating bacterial species into one "superbug" that would gobble up slicks by itself. The bug, *Pseudomonas aereoginosa* reliably gobbled up oil slicks.

    b. See also "Court Decision Spurs Genetic Research," *Ind. Res. Devel.,* 45–46 (August 1980).

21. M. L. Simmons, "Danger! Experimental Animals," *Occu. Health and Safety,* 30–33, March 1982. See also R. T. Hughes and A. A. Amendola, Recirculating Exhaust Air, Guidelines, Design, Parameters and Mathematical Modeling, *Plant Engineering,* **36,** No. 6, 235–239, March 18, 1982 and *Animal Resources: A Research Resources Directory,* Fourth Revised Edition, 1981, Research Resources Information Center, National Institutes of Health, Bethesda, MD. 20205, NIH Pub. No. 82-1431.

22. Franklin Research Center, Survey of Compounds which have been tested for carcinogenic activity, 1978 volume, NIH Publication No. 80-453, U. S. Department of Health, Education, and Welfare, National Institutes of Health, U.S. Government Printing Office, Washington, D.C.

23. E. B. Sansone, A. M. Losikoff, and R. A. Pendleton, "Potential Hazards from Feeding Test Chemicals in Carcinogen Bioassay Research," *Toxicol. Appl. Pharmacol.,* **39,** 435–450 (1977).

24. a. "Recombinant DNA Research; Proposed Actions Under Guidelines," *Fed. Regis.,* **45,** 79386–79387 (November 28, 1980); "Actions Under Guidelines," *Fed. Regis.,* **45,** 77372–77408 (November 21, 1980); *Fed. Regis.* **46,** 59368-425 (December 4, 1981); *Fed. Regis.* **46,** 59737 (December 7, 1981).

    b. *Recombinant DNA Technical Bulletin,* (issued occasionally), **3**(2) September, 1980); **3**(1) (August, 1980); **2**(4), (April 1980), National Institutes of Health, Bethesda, Maryland.

    c. *Recombinant DNA Research,* Vol. 1, Guidelines Relating to "NIH Guidelines for Research Involving Recombinant DNA molecules," February 1975; June 1976.

    d. A supplement to the NIH Guidelines for Recombinant DNA Research, Laboratory Safety Monograph, July 1978.

    e. Administrative Practices, Supplement to the NIH Guidelines for Research, Laboratory Safety Monograph, July 1978.

    f. Administrative Practices Supplement to the NIH Guidelines for Research Involving Recombinant DNA Molecules Nov. 1980; April 1980; June 1979; National Institutes of Health, Bethesda, Maryland.

    g. Ruth B. Kundsin, "Airborne Contagion," *Ann. N.Y. Acad. Sci.,* **353** (1980).

    h. R. Burt, "The U.S. Fight Against Chemical War," *The Wall Street Journal,* January 4, 1982, p. 31.

25. Teddington, England National Chemical Laboratory, "Safety Measures in Chemical Laboratories," HMSO, London, England, 1964.

26. Manufacturing Chemists Association, *Guide for Safety in the Chemical Laboratory,* 2nd. ed., Van Nostrand Reinhold Company, New York, 1972.

27. H. H. Fawcett, Exposures of Personnel to Laboratory Hazards, *Am. Ind. Hyg. Assoc. J.* **38**(8), 559–567 (1972).

28. W. C. Heuper, "Occupational Cancers with Special Reference to Occupational Cancer Hazards to Laboratory Personnel," *Am. J. Med. Technol.*, **27**, 157–164 (1961).

29. J. M. Wood and R. Spencer, "Carcinogenic Hazards in the Microbiology Laboratory," *Society for Applied Bacteriology Technical Series, No. 6. Safety in Microbiology Symposium*, D. A. Shapton and R. G. Board, eds., London, England, October 28, 1970, Academic Press, New York, New York, 1972, pp. 185–189.

30. H. B. Collier, "Are Orthotolidine and Dianisidine Health Hazards to Laboratory Workers?," *Clin. Biochem.*, **7**(1), 3–4 (1971).

31. W. J. Burke, U. R. Hoegg, and R. E. Phillips, "Systemic Fluoride Poisoning Resulting from a Fluoride Skin Burn," *J. Occu. Med.*, **15**(1), 39–41 (1973).

32. J. M. Harrington, "Risk of Mercurial Poisoning in Laboratories Using Volumetric Gas Analysis," *Lancet* **1**, 86 (1974).

33. R. M. Oliver, "Toxic Effects of 2, 3, 7, 8 Tetrachlorodibenzo 1,4 Dioxin in Laboratory Workers," *Br. J. Ind. Med*, **32**(1), 49–53 (1975).

34. T. U. Kanellakes and K. P. Mathews, "Immediate-Type Hypersensitivity to a Bean Lectin Source with Commentary on Occupational Allergy in Allergy-Immunology Laboratory Research Workers," *J. Allergy Clin. Immunol.*, **56**(5), 407–410 (1975).

35. R. J. Pye and J. L. Burton, "DNCB, Chemical Laboratory Workers, and Chloramphenicol," *Br. Med. J.*, **2**(6044), 1130–1131 (1976).

36. E. F. Legg, "Phosgene Production during Aspiration of Trichloroacetic Acid into a Flame Emission Spectrophotometer," *Clin. Chim. Acta*, **50**(1), 157–159 (1974).

37. S. S. Mirvish, P. Issenberg, and H. C. Sornson, "Air Water and Ether Water Distribution of N Nitroso Compounds, Implications for Laboratory Safety, Analytic Methodology and Carcinogenicity for the Rat Esophagus, Nose and Liver," *J. Nat. Cancer Inst.*, **56**(6), 1125–1129 (1976). See also R. A. Scanlan and S. R. Tannenbaum, eds. *N-Nitroso Compounds*, ACS Symposium Series 174, American Chemical Society, Washington, D.C., 1981, and J.-P. Anselme, ed., *N-Nitrosamines*, ACS Symposium Series 101, American Chemical Society, Washington, D.C., 1979.

38. D. T. Braymen and J. R. Songer, "Phenol Concentrations in the air from Disinfection Solutions," *Appl. Microbiol.*, **22**, 1166–1167 (1971).

39. M. G. Ott, B. B. Holder, and R. R. Langner, "Determinants of Mortality in an Industrial Population," *J. Occup. Med.*, **18**(3), 171–177 (1976).

40. S. K. Hoar, Epidemiologic Studies of Chemists, presented to American Chemical Society 182nd National Meeting. New York, N.Y., August 27, 1981.

41. A. Taylor and V. Marks, "Measurement of Urinary Mercury Excretion by Atomic Absorption in Health and Disease, *Br. J. Ind. Med.*, **30**(3), 293–296 (1973).

42. R. R. Lauwerys and J. P. Buchet, "Occupational Exposure to Mercury Vapors and Biological Action," *Arch. Environ. Health*, **27**, 65–68 (1973).

43. J. M. Harrington and H. S. Shannon, "Mortality Study of Pathologists and Medical Laboratory Technicians," *Br. Med. J.* **4**(5992), 329–332 (1975).

44. F. P. Li, "Suicide of Chemists," *Arch. Environ. Health*, **19**, 518–520 (1969).

45. F. P. Li, J. F. Fraumeni, N. Mantel, and R. W. Miller, "Cancer Mortality among Chemists," *J. Nat. Cancer Inst.*, **43**, 1159–1164 (1969).

46. R. L. Rawls, "Cancer Death Rate for Chemists (Swedish Lab)," *Chem. and Eng. News*, **54**(27), 17 (1976). See also Olin under Bibliography below.

47. G. B. Phillips, "Causal Factors in Microbiological Laboratory Accidents and Infections," Report No. MP-2, Army Biological Labs, Frederick, Maryland, 1965.

48. G. B. Phillips, "Control of Microbiological Hazards in the Laboratory," *Am. Ind. Hyg. Assoc. J.*, **30**, 170–176 (1969).

49. S. E. Sulkin and R. M. Pike, "Survey of Laboratory-Acquired Infections," *Am. J. Public Health,* **41**(1), 769–781 (1951).

50. W. V. Powell, "Biological Hazards in the Research Environment," *J. Am. Coll. Health Assoc.,* **19**(4), 205–210 (1971).

51. M. Reitman and A. G. Wedum, Microbiological Safety, *Public Health Rep.* **71,** 659–665 (1956).

52. A. G. Wedum, "Laboratory Safety in Research with Infectious Aerosols," *Public Health Rep.,* **79,** 619–633 (1964).

53. R. M. Pike, "Laboratory-Associated Infections: Summary and Analysis of 3921 Cases," *Health Lab. Sci.,* **13**(2), 105–114 (1976).

54. C. L. Baldwin, "Biological Safety Recommendations Associated with Oncogenic Virus Perification in Large High-Speed Centrifuges," *Centrifuge Biohazards Symposium,* Frederick Cancer Research Center, Frederick, Md., pp. 61–70, 1973.

55. A. G. Wedum, "Microbiological Centrifuging Hazards," *Centrifuge Biohazards Symposium,* Frederick Cancer Research Center, Frederick, Maryland, pp. 5–16, 1973.

56. A. Uldall, "Occupational Risks in Danish Clinical Chemical Laboratories." *Scand. J. Clin. Lab. Invest.,* **33**(1), 21–25 (1974).

57. P. Skinhoj, "Occupational Risks in Danish Clinical Chemical Laboratories, Part 2, Infections," *Scand. J. Clin. Lab. Invest.,* **33**(1), 27–29 (1974).

58. A. Leblanc, P. C. Johnson, and M. F. Jahns, "Relative Utility of Various Methods of Personnel Monitoring in a Nuclear Medicine Laboratory," *Health Phys.,* **21**(2), 332 (1971).

59. C. J. Paperiello and J. J. Gabay, "Personnel Contamination from Iodine-125 used in Biological and Medical Laboratories," *Trans. Am. Nucl. Soc.,* **18,** 100 (1974).

60. A. D. Leblanc and P. C. Johnson, "Survey of Xenon Contamination in a Clinical Laboratory," *Health Phys.,* **28**(1), 81–83 (1975).

61. S. M. Sanders Jr., "Excretion of 241-Am and 244 Cm Following two Cases of Accidental Inhalation," *Health Phys.,* **27**(4), 359–365 (1974).

62. J. R. Vaane and E. M. DeRas, "Analysis of a Case of Internal Contamination with Curium-242," *Health Phys.,* **21**(6), 821–826 (1971).

63. R. A. Griesemer and J. S. Manning, "Animal Facilities," *Conference on Biohazards in Cancer Research,* held at Asilomar Conference Center, Pacific Grove, California, June 22–24, 1973 (Biohazards in Biological Research), A. Hellman, M. N. Oxman, and R. Pollack, eds., Cold Spring Harbor Laboratory, Cold Spring Harbor, New York, pp. 316–326.

64. W. G. Winkler, T. R. Fashinell, L. Leffingwell, P. Howard, and J. P. Conomy, "Airborne Rabies Transmission in a Laboratory Worker," *J. Am. Med. Assoc.,* **226**(10), 1219–1221 (1973).

65. R. J. Biggar, R. Deibel, and J. P. Woodall, Implications Monitoring and Control of Accidental Transmission of Lymphocytic Choriomeningitis Virus within Hamster Tumor Cell Lines, *Cancer Res.* **36**(2 pt 1), 537–553 (1976).

66. I. Lutsky and I. Neuman, "Laboratory Animal Dander Allergy, Part I. An Occupational Disease," *Ann. Allergy,* **35**(4), 201–205 (1975).

67. A. Stark, "Policy and Procedural Guidelines for Health and Safety of Workers in Virus Laboratories," *Am. Ind. Hyg. Assoc. J.,* **36**(3), 234–240 (1975).

68. J. M. Harrington, "Occupational Health and Safety in Great Britain—1973," *Br. J. Ind. Med.* **32**(3), 247–250 (1975).

69. R. H. Winget, "Laboratory Services Series: A Safety Program for Service Groups in a National Research and Development Laboratory (1965–1974)," Oak Ridge National Laboratory, Tennessee, 1975.

BIBLIOGRAPHY

Accident Prevention Manual for Industrial Operations, 7th ed., National Safety Council, 444 No. Michigan Avenue, Chicago, 1978. (Revised biennially.)

Albrecht, J. "Infections and Female Personnel in the Laboratory," Army Biological Labs, Frederick, Maryland, Report No. Trans-961, 1963.

Alzvert, L. G., "Determination of Atrolactic Acid in Urine as a Test for Exposure to Alpha-Methylstyrene," *Gig. Tr. Prof. Zabol.,* **3,** 38–41 (1975).

American Chemical Society, "Safety in Academic Chemical Laboratories," 3rd. ed., American Chemical Society, Washington, D.C., 1979.

American Chemical Society Division of Chemical Health and Safety, 183rd National Meeting, Las Vegas, Nev., March 28—April 2, 1982.

Jameson, C. W. and Walters, D. B., co-chair, Symposium on Chemistry and Safety for Toxicity Testing of Environmental Chemicals (36 papers). See Meeting Abstracts for details.

Young, J. A. and Fawcett, H. H., Education for a Safe Professional Life, Symposium on Education for a Professional Life, 183rd ACS National Meeting, Las Vegas, Nev., April 2, 1982.

J. A. Young, chair, Symposium on Fire Toxicity (5 papers). See Meeting Abstracts for details.

W. P. Taggert, chair, Symposium on Laboratory Waste Disposal (12 papers). See Meeting Abstracts for details.

Anthony, H. M., and Thomas, G. M., "Tumors of the Urinary Bladder. An Analysis of the Occupations of 1030 patients in Leeds, England," *J. Nat. Cancer Inst.,* **45,** 879–895 (1970).

Baldwin, C. L., "Biological Safety Recommendations Associated with Oncogenic Virus Purification in Large High-Speed Centrifuges," *Centrifuge Biohazards Symposium,* Frederick Cancer Research Center, Frederick, Maryland, 1973, pp. 61–70.

Bandal, S. K., Goldberg, L., Marco, G., and Leng, M., *The Pesticide Chemist and Modern Toxicology,* ACS Symposium Series 160, American Chemical Society, Washington, D.C., 1981.

Bebie, J., *Manual of Explosives, Military Pyrotechnics and Chemical Warfare Agents—Composition, Properties, Uses,* Macmillan, New York, 1943.

Beller, K., "Infection in the Laboratory with the Lansing Virus," Army Biological Labs, Frederick, Maryland, Report No. Trans-891, 1963.

Belousuv, A. Z., "State of Health and Immunobiologic Resistance in Undergraduates Mastering the Trade of Laboratory Assistant in Chemical and Oil Industry," *Gig. Sanit,* **37**(1), 45–50 (1972).

Benner, Ludwig, Jr., "Accident Investigations—A Case for New Perceptions and Methodologies," Preprint 800387, SAE/SP-80/461/$02.50, Society of Automotive Engineers, Inc., 1980.

Berger, J. D., and Cloutier, R. J., "Health Physics Practices in Laboratories using Tritium and Carbon-14 Labeled Tracers," *Lipids,* **7**(9), 604–610 (1972).

Blumberg, J. M., "Laboratory safety," Quality Control in Microbiology Symposium, J. E. Prier, J. Bartola, and H. Friedman, eds., University Park Press, Baltimore, 1973.

Brooks, G. F., "Health Surveillance of Lunar Receiving Laboratory Personnel during Apollo-II Quarantine Period," *Am. J. Public Health* **60**(10), 1956 (1970).

Brown, S. S., "Laboratory Design and Chemical Hazards," *Ann. Clin. Biochem.,* **8**(4), 125–129 (1971).

Burt, R. "The U.S. Fight Against Chemical War," *The Wall Street Journal,* January 4, 1982, p. 31.

Butman, A. B., and Kovach, R. I., "Methodological Principles of a Hygienic Evaluation of the Degree of Effect of Diffusely Reflected Laser Radiation on Eyes," *Gig. Tr. Prof. Zabol.*, 18(2), 45–46 (1974).

"Centrifuge Biohazards Symposium," *Cancer Research Safety Monograph Series,* Vol. 1, Frederick Cancer Research Center, Frederick, Maryland, 1973.

Chemical Carcinogens, C. E. Searle, ed., *American Chemical Society Monograph 173*, Washington, 1976.

Cherpak, V. V., and Fedorchuk, A. A., "Hygienic Evaluation of Conditions When Working with Treated Seeds in Seed Control Laboratory," *Gig. Sanit.*, 38(4) (1973).

Choudhary, G., *Chemical Hazards in the Workplace*, ACS Symposium Series 149, American Chemical Society, Washington, D.C.

Cole, P., "Epidemiologic Studies and Surveillance of Human Cancers among Personnel of Virus Laboratories," *Conference on Biohazards in Cancer Research*, A. Hellman, M. N. Oxman, and R. Pollack, eds., held at the Asilomar Conference Center, Pacific Grove, California, June 22–24, 1973. (Biohazards in Biological Research). Cold Spring Harbor Laboratory, Cold Spring Harbor, New York, pp. 309–315, 1973.

Collins, C. H., Hartley, E. G., and Pilsworth, R., "The Prevention of Laboratory Acquired Infection," *Public Health Laboratory Service Monograph No. 6*, HMSO, London, 1974.

"Control of Hazardous Materials Spills," *Proceedings of the National Conference on Control of Hazardous Materials Spills,* May 13–15, 1980, Louisville, Ky., sponsored by U.S.EPA, U. S. Coast Guard, and Vanderbilt University. Annual conference.

Cotruvo, J. A., Trihalomethanes in Drinking Water, *Env. Sci. Technol.* 15 (3), 268–270 (March 1981).

"Cryogenic Fluids in the Laboratory—Data Sheet I-688-80," National Safety Council, Chicago, 1980.

Fringer, J. M. et al., "A Safe Practices Manual for the Manufacturing, Transportation, Storage and Use of Pyrotechnics," 210-77-0145, NIOSH, Morgantown, W.V., 1979.

Chemical Hazards to Human Reproduction, Council on Environmental Quality, 1981, Washington, D.C. Government Printing Office.

DiGrazia, H. X., "Development of a Hazards Control Department," Report No. CONF-730552-1, California University, Lawrence Livermore Laboratory, Livermore, California, 1973.

"DOD Ammunition and Explosives Safety Standards," DOD 5154.45, Department of Defense, Office of the Assistant Secretary of Defense (Installations and Logistics), Washington, D.C., March 1976.

Dollberg, D. D. and Verstuyft, A. W., *Analytical Techniques in Occupational Health Chemistry, ACS Symposium Series 120,* American Chemical Society, Washington, D.C., 1980.

Drouhet, E., Searetain, G., and Mariat, F., "Coccidioido Mycosis Infection of a Laboratory Worker," *Bull. Soc. Fr. Mycol. Med.* 3(2), 163–165 (1974).

Dunlap, J. H., "Radiation Protection Considerations in the Cardiac Catheterization Laboratory," *Health Phys.*, 29, 415–416 (1975).

Ellis R. E., "Radiation Hazards in Laboratories," *Ann. Clin. Biochem.*, 8(4), 147–149 (1971).

Evans, C. GT., Harris-Smith, R., and Stratton, J., eds., "The Use of Safety Cabinets for the Prevention of Laboratory Acquired Infection." Society for Applied Bacteriology Technical Series, No. 6, Safety in Microbiology Symposium, D. A. Shapton and R. G. Board, eds., London, England, October 28, 1970, Academic Press, New York, 1972, pp. 21–36.

Farina, G. F., Alessio, L., and Forni, A., "Hematological Changes in 2 workers Occupationally Exposed to Chloramphenicol," *Med. Lav.* 63(1–2), 52–56 (1972).

Fawcett, H. H., "Past General Chairman offers options for Handling Wastes," *Chemical Newsletter,* National Safety Council, Chicago, Ill., May 1980, pp. 1–2.

Fawcett, H. H., "Supplementing the Chemical Curriculum with Safety Education," *J. Chem. Ed.,* **26**(2), p. 108 (February 1949).

Fedoroff, B. T. et al., *Encyclopedia of Explosives and Related Items,* 3 volumes, Picatinny Arsenal, Dover, N.J., 1966.

"Feltman Research Laboratories Safety Manual," Feltman Research Laboratories, Picatinny Arsenal, Dover, New Jersey, March 1976.

Fire Safety Aspects of Polymeric Materials, 10 volumes, Technomic Publishing, Westport, Conn., (1978–1980).

Flury, P. A., "How Safe is Your Laboratory? Take a Close Look Before the Occupational Safety and Health Act Does." *Am. J. Med. Technol.,* **37**(3), 155–157 (1975).

Garmet, S., "Visiting the Front in the Battle Against Regulation," *The Wall Street Journal,* December 4, 1981, p. 34.

Gennaro, T., Peirone, F. G., and Fenoglio, S., "Occupational Risk of Certain Deep Dermatophytoses," *Minerva Med.,* **62**(42), 2148–2151 (1971).

Gerber, P., "Occurrence of Epstein Barr Virus in Human Leukocyte Culture," *In Vitro* **10**(5-6), 247–252 (1974).

Gresikova, M. and Sekeyova, M., "Noninfectious Tick-Borne Encephalitis Antigen," *Acta Virol.,* **20**(3), 260–262 (1976).

Griggs, J. H., Weller, E. M., Palmisano, P. A., and Niedermeier, W., "The Effect of Noxious Vapors on Embryonic Chick Development," *Ala. J. Med. Sci.,* **8**(3), 342–345 (1971).

Grist, N. R., *Proc. R. Soc. Med.,* **66**, 795 (1973).

"Guidelines for the Laboratory Use of Chemical Substances Posing a Potential Occupational Carcinogenic Risk," Laboratory Chemical Carcinogen Safety Standards Subcommittee of the DHEW (now DHHR) Committee to Coordinate Toxicology and Related Programs, Revised Draft, March 1980, Office of Research Safety, National Institutes of Health, Bethesda, Maryland.

Hanel, E. Jr. and Kruse, R. H., "Laboratory-Acquired Mycoses," Fort Detrick, Frederick, Maryland, Report No. Miscellaneous Publication-28, 1967.

Harmon, M. and J. King, "A Review of Violent Monomer Polymerization," Operations Research, Inc., Silver Spring, Md., prepared under Contract No. CCT-45-74-74, National Research Council, Washington, D.C., October 31, 1974. [Available as AD A-017-443 from National Technical Information Service, Springfield, Va.]

Harrington, J. M., "Health of Laboratory Workers," *Proc. R. Soc. Med.,* **68**(2), 94 (1975).

Harrington, J. M. and Shannon, H. S., "Incidence of Tuberculosis, Hepatitis, Brucellosis, and Shigellosis in British Medical Laboratory Workers," *Br. Med. J.,* **1**(6012), 759–762 (1976).

Hartley, C. L., Petroche, V., and Richmond, M. H., "Antibiotic Resistance in Laboratory Workers," *Nature,* **260**(5551): 558 (1976).

Hartung, M. and Salfelder, K., "Histoplasmosis with Fatal Results as an Occupational Disease of a Mycologist," Report No. Trans-897, Army Biological Labs, Frederick, Maryland, 1963.

"Hazardous Materials in the Laboratory," Occupational Safety & Health Program, U.S. EPA, Washington, D.C., 1980.

"Health Factors in the Handling of Chemicals," Safety Guide, SG-1, rev., Chemical Manufacturing Association, Washington, D.C., 1979.

Hoffman, R., *Radiation and Human Health,* Univ. of Calif. Press, Berkeley, 1982.

Hughes, D., "Radiation Accidents in University Laboratories," *Accid. Anal. Prev.,* **2**(1), 29–34 (1970).

Husak, V. and Kleinbauer, K., "Radiation Hazard Associated with the Use of Technetium-99m in Nuclear Medicine Laboratories," *Acta. Univ. Palacki. Olomuc. Fac. Med.*, **74**, 207–219 (1975).

Iammarino, R. and Saslow, A. R., "Viral Hepatitis in Clinical Chemistry Laboratory Workers," *Clin. Chem.*, **20**(4), 514–515 (1974).

"An Investigation of Fifteen Flammable Gases or Vapors with Respect to Explosion-Proof Electrical Equipment," Bulletin of Research No. 58, Underwriters' Laboratory, Northbrook, Ill., 1970.

Illinger, K. H., ed., Biological Effects of Nonionizing Radiation, ACS Symposium Series 157, American Chemical Society, Washington, 1981.

Jemski, J. V. and Phillips, G. B., "Microbiological Safety Equipment," *Lab. Anim. Care*, **13**(1), 2–12 (1963).

Kaskevich, L. M. and Bezuglyi, V. P., "State of Health of Persons Working with Tetramethyl Thiuram at Seed Control Laboratories and Seed Farms," *Gig. Tr. Prof. Zabol.*, **17**(3), 49–51 (1973).

Kawamata, J. and Yamanduchi, T., "Protection Against Bio Hazards in Animal Laboratories." *Proceedings of the VITH International Symposium*, Yohei Ito and Ray M. Dutcher, eds., Bibliotheca Haematologica, No. 40. Comparative Leukemia Research 1973, Leukemogenesis, Nagoya and Ise-Shima, Japan, 1973.

Kreutzer, H. H. and Rechsteiner, J., "Hepatitis B Antigen in Clinical Chemistry Control Sera," *Clin. Chim. Acta,* **60**(1), 117–120 (1975).

Kubias, F. O., Chairman, Chemical Information: A Directory of Chemicals Used in Industry—properties, labels, hazards, precautions, Chemical Section, National Safety Council, Chicago, Illinois 60611, 1979.

"Laboratory Animals and Public Health," *WHO Chron.*, **27**(9), 352–355 (1973). World Health Organization, Geneve, Switzerland.

"Lab Safety at the Center for Disease Control," Manual, Office of Biosafety, Center for Disease Control, Atlanta, Ga.

Leblanc, A. and Johnson, P. C., "Medical Radiation Exposure Survey in a Hospital with a Nuclear Medicine Laboratory," *Health Phys.*, **19**, 433 (1970).

Lehmann, P., "Laboratory Safety Today," *Job Saf. Health,* **3**(5), 4–9 (May 1975).

Levy, D. A., "Allergenic Activity of Proteins from Mice," *Int. Arch. Allergy Appl. Immunol.*, **49**(1/2), 219–221 (1975).

Lewis, H. F., ed. *Laboratory Planning for Chemistry and Chemical Engineering,* Reinhold, New York, 1962; see Chapter VII, Snow, D. L. and Fawcett, H. H. "Occupational Health and Safety."

Lobstein, T., "Laboratory Studies of the Office Environment," *Ergonomics,* **18**(4), 473 (1975).

Long, E. R., "The Hazard of Acquiring Tuberculosis in the Laboratory," *Am. J. Public Health,* **41**, 782–787 (1951).

Magrath, I. M., "Occupational Infections in Laboratory Workers," (Letter), *Br. Med. J.,* **1**(6016), 1020 (1976).

Martin, G. M., Schwartz, B. R., and Derr, M. A., "Human Lymphoid Cell Lines—Potential Hazards to Laboratory Workers," *Lancet* **2**(7676), 772 (1970).

Martin, G. M., Schwartz, B. R., and Derr, M. A., "Laboratory Safety and Human Cell Lines," *Lancet,* **1**(7700), 650 (1971).

Martini, G. A., *Dtsch. Med. Wochenschr.,* **93**, 559 (1968).

Mayer, C. E., "Report Paints a Grim Picture of Asbestos," *Washington Post,* December 30, 1981, pp. D8–10.

Mazzella, D. I. and Bosco, M., "Some Cases of Occupational Intoxication Due to Dinitro

Phenols Binapacryl 4,6 Dinitro-o Cresol Karathane in an Agricultural Laboratory," *J. Eur. Toxicol.,* **3**(5), 325–331 (1970).

MCA Guide for Safety in the Chemical Laboratory, Van Nostrand Rheinhold Co., New York, 1972.

Miller, J. A., "Are Rats Relevant?" *Sci News,* **112**, 12–13 (July 2, 1977).

Morton, W. I., "Safety Techniques for Workers Handling Hazardous Materials," *Chem. Eng.,* **83**(22), 127–132 (1976).

Bretherick, L., ed., *Hazards in the Chemical Laboratory,* 3rd ed., The Chemical Society, London, 1981.

Fox, J. L., Biotechnology: a high-stakes industry in flux, *Chem. & Engr. News,* **60**, n. 13, 10–15 (Mar. 29, 1982).

Mukhordov, F. G., Krasnov, A. V., and Dokukina, A. G., "Clinical and Laboratory Characteristics of the Restorative Period of Viral Hepatitis in Workers of Chemical Enterprises," *Ter. Arkh.* **47**(2), 92–95 (1975).

National Research Council, *Prudent Practices for Handling Chemicals in Laboratories,* National Academy Press, Washington, D.C., 1981.

National Cancer Institute, "Biohazard Control and Containment in Oncogenic Virus Research," U.S. Government Printing Office, Washington, D.C., 1969.

NIH Guidelines for the Laboratory Use of Chemical Carcinogens, NIH Ref. No. 81-2385, May 1981, avail. from Frederick Cancer Research Laboratory, Frederick, MD. 21701.

Nauck, E. G. and Weyer, F., "Laboratory Infections with Q-fever," Report No. Trans 957, Army Biological Labs, Frederick, Maryland, 1963.

Neame, K. D. and Homewood, C. A., *Liq. Scintill. Counting,* (1974).

Neuman, I. and Lutsky, I., Laboratory Animal Dander Allergy, Part 2: Clinical Studies and the Potential Protective Effect of Disodium Cromoglycate," *Ann. Allergy,* **36**(1), 23–29 (1976).

"Occupational Diseases Acquired from Animals," University of Michigan, School of Public Health Continued Education Series No. 124, Ann Arbor, Michigan, 1964.

"Occupational Exposure Limits for Airborne Toxic Substances," No. 37, International Labour Office, Geneva, Switzerland, 1977.

Office of Management and Budget, Executive Office of the President, "Standard Industrial Classification Manual," U.S. Government Printing Office, Washington, D.C., 1972.

Olin, R., "Leukaemia and Hodgkin's Disease among Swedish Chemistry graduates," *Lancet,* **2**(7991), 916 (1976).

Olin, G. R. and A. Ahlbom, The Cancer Mortality among Swedish Chemists Graduated during Three Decades, *Environ. Research* **22**, 154–161 (1980).

Olin, G. R., Health Hazards in the Chemistry Profession: A Question of Awareness and Facilities, presented to 182nd American Chemical Society National Meeting, New York, August 28, 1981.

Ortel, S., "Listeriosis During Pregnancy and Excretion of Listeria by Laboratory Workers," *Zentralbl. Bakteriol. Parasitenktt. Infektionskr. Hyg. Abt. Orig. Reihe. A,* **231**(4), 491–502 (1975).

Palmer, K. N., *Dust Explosions and Fires,* Chapman and Hall, London, 1973.

Parker, H. D., Thaxter, M. D., and Briggs, M. W., "Current Status of Curium Inhalation Exposure to Humans," UCRL—9361.

Perkins, F. T. and Hartley, E. G., "Handling of Primates. Infections and Immuno-Suppression in Subhuman Primates Symposium," H. Balner, and W. I. B. Beveridge, eds., Williams & Wilkins, Baltimore, 1978, pp. 167–172.

"Peroxidizable Compounds," Chemical Safety Data Sheet 655, National Safety Council, Chicago, 1976. See also "Ethyl Ether," Data Sheet 396, and "Chemical Safety References," Data Sheet 486A, 1977.

Phillips, G. B., "Microbiological Hazards in the Laboratory," *J. Chem. Educ.*, **42**, A43–A48 (1965).

Phillips, G. B., "Microbiological Safety in U.S. and Foreign Laboratories," U.S. Army Corps, Biological Laboratories, Technical Study No. 35, Fort Detrick, Frederick, Maryland, 1961.

Phillips, G. B., and Jemski, J. V., "Biological Safety in the Animal Laboratory," *Lab. Anim. Care,* **13**(1), 13–20 (1963).

Proper, G. H., Jr., *New York State Chemical Hazards Data,* Division of Fire Safety, Albany, 1970.

Purchase, I. F. H. and Van der Watt, J. J., "Carcinogenicity of Sterigmatocystin to Rat Skin," *Toxicol. Appl. Pharmacol.,* **26**(2), 274–281 (1973).

Radiological Health & Safety Program, Occupational Safety & Health Programs, U.S. EPA, Washington, D.C., 1979.

Radford, E. P., "Cancer Risks from Ionizing Radiation," *Technology Review (M.I.T.)* V. 84, no. 2, 66–78 (1981).

Rappaport, S. M. and Campbell, E. E., "The Interpretation and Application of OSHA Carcinogen Standards for Laboratory Operations," Am. Ind. Hyg. Assoc. J., **37**(12), 690–696 (1976).

Registrar General, "Registrar General's Decennial Supplement England and Wales, Occupational Mortality Tables (1959–1963)," HMSO, London, England, 1971.

Reid, D. D., *Br. Med. J.* **2**, 1 (1957).

"Reports of the Department of Defense Explosives Safety Board," Washington, D.C., annual.

"Research Laboratory Safety"—15 min. 80-slide presentation, produced by NIH, 1980. [Available from National Safety Council, Chicago, Illinois, 60611.]

Review and Outlook, "Highest Priority," *The Wall Street Journal,* January 7, 1982, p. 20.

Reynolds, John M., "Controlling the Use of Hazardous Materials in Research and Development Laboratories," J. Haz. Mat., **2**(4), 299–308 (1978).

Roberts, J. L., and Gallese, L. R., "Home Values Hurt by Foam Insulation Ban," *Wall Street Journal*, 33, February 26, 1982.

Robinson, T. R., "Medical Monitoring in Industry," *J. Occup. Med.*, **15**(2), 127–128 (1973).

Royal Cancer Hospital, London, Chester Beatty Research Institute, "Precautions for Laboratory Workers who Handle Carcinogenic Aromatic Amines," 1966 (reprinted with additional notes 1971).

Safety Manual for Research and Development, ed., by W. S. Wood, Sun Oil Co., Marcus Hook, PA, 1970.

"Safety Standards for Research Involving Carcinogens," Occupational Safety & Health Program, U.S. EPA, Washington, D.C., 1979.

Sansone, E. B. and Slein, M. W., "Application of the Microbiological Safety Experience to work with Chemical Carcinogens," *Am. Ind. Hyg. Assoc. J.,* **37**(12), 711–720 (December 1976).

Scott, Ralph, *Toxic Chemicals and Explosives Facilities,* American Chemical Society Symposium No. 96, American Chemical Society, Washington, D.C., 1979.

"Shell Chemical Safety Guide," Shell Oil Co., Houston, TX, 1979.

Sichak, S., *The Laboratory Safety Deskbook: A Guide to OSHA Standards,* Chicago Section, American Chemical Society, Chicago, 1979.

Snee, R. D., and Smith, P. E., Jr., "Statistical Analysis of Interlaboratory Studies," *Am. Ind. Hyg. Assoc. J.*, **33**(12), 784–790 (1972).

Spencer, R. C., Brucellosis and Laboratory Workers, *Lancet,* **1**(7905), 528 (1975).

Sprout, W. L., Neeld, W. E., Jr., and Woessner, W. W., "Management of Chemical Cyanosis by Oxygen Saturation Readings," *Arch. Environ. Health,* **30**(6), 302–306 (1975).

Steerman, J. J. and Sanders, L. J., "Simplified Breath Sampling for Tritium Bioassay," *Health Phys.*, **27**(6), 1974.

Steere, N. V., *Handbook of Laboratory Safety*, Chemical Rubber Co., Cleveland, 1978.

Stocks, C. J., "Handling and Disposal of Mercury," *J. Am. Med. Technol.*, **35**(3), 164–167 (1973).

Strehlow, R. A. and Baker, W. E., "The Characterization and Evaluation of Accidental Explosions," prepared for Aerospace Safety Research and Data Institute, Lewis Research Center, NASA, Cleveland, Ohio, NASA CR 134779, AAE 75-3, UILU-ENG 75 0503, June 1975.

Tabor, M., "Uses and Limitations of Male/Female Data," *Occu. Health and Safety*, **50**(11), 38–40 (Nov. 1981).

New Day Films, *In Our Water* (Pollution of drinking water from toxic waste), 16-mm sound film, 58 min., P.O. Box 315, Franklin Lakes, N.J. 07417.

Tarasenko, V. D. and Kozin, Y.A. B., "Rate Fixing of Medical Personnel Work in Radiodiagnostic Laboratories," *Med. Radiol.*, **20**(2), 43–47 (1975).

Teyssie, A., Gutman, L., Hostein, B. A., and Barrerao, J. G., Junin Virus Antibodies in Laboratory Workers. Neutralization Tests in Tissue Culture (SP), *Medicine*, **3C**(1), 34 (1970).

Thomas, R. H., "Implementing the Requirement to Reduce Radiation Exposure to as Low as Practicable at the Lawrence Berkeley Laboratory," *Health Phys.*, **30**(3), 271–279 (1976).

Tinsley, T. and Harrup, K., "Serological Studies and Surveys of Laboratory Workers. Baculoviruses for Insect Pest Control: Safety Considerations," M. Summers, et al., eds., American Society for Microbiology, Washington, D.C., 1975.

Tomlinson, A. J. H., "Infected Airborne Particles Liberated on Opening Screw-capped Bottles," *Brit. Med. J.* **2**, 15 (1957).

U.S. Bureau of Labor Statistics, "Employee Compensation and Payroll Hours; Commercial Research and Development Laboratories, 1967," Report No. 363, U.S. Department of Labor, Washington, D.C., 1969.

U.S. Bureau of Labor Statistics, "Occupational Injuries and Illnesses in the United States, by Industry 1974," Bulletin No. 1932, U.S. Department of Labor, Washington, D.C., 1976.

U.S. Bureau of Labor Statistics, "Occupational Safety and Health Statistics," Report No. 459, U.S. Department of Labor, Washington, D.C., 1976.

Waldholz, M., "Toxic Tragedy: Lead Poisoning Takes A Big, Continuing Toll As Cures Prove Elusive," *The Wall Street Journal*, May 27, 1982, pp. 1, 20.

Walters, D. B., *Safe Handling of Chemical Carcinogens, Mutagens, Teratogens, and Highly Toxic Substances*, 2 volumes, Ann Arbor Science, Ann Arbor, Mich. 1980. (Stresses, Laboratory Control, and Safety Precautions.)

Wedum, A. G., "Bio-Hazard Control," *Handbook of Laboratory Animal Science*, Vol. I, E. C. Melby and N. H. Ahman, eds., CRC Press, Cleveland, 1974.

Wedum, A. G., "Disease Hazards in the Medical Research Laboratory," *Am. Assoc. of Ind. Nurses J.*, 1964.

Whalen, R. P. and Davies, S., "Americium Contamination Incident in a New York State Health Department Laboratory," *Radiat. Data Rep.*, **13**(5), 249–253 (1972).

Willis, J. B. et al., "A Safe Practices Manual for the Manufacturing, Transportation, Storage and Use of Explosives," 210-77-0145, NIOSH, Morgantown, W. Va., 1979.

Yurchenk, N. D., "Studies of Central Measurement Laboratory Worker," *Meas. Tech. R.*, **1970**(10), 1592–1593 (1970).

Zabetakis, M. G., *Safety with Cryogenic Fluids*, Plenum Press, New York, 1967.

Zabetakis, M. G. and Carbill, R. S., "Safety Manual #4—Laboratory Safety," U.S. Department of the Interior, Mining Enforcement and Safety Administration, Washington, D.C., 1978.

23

Design of Blast-Resistant Buildings in Petroleum and Chemical Plants

D. J. Forbes

Nomenclature

f'_c 28-day cylinder crushing strength, psi

f_y static yield of reinforcement, psi

F_a AISC allowable compressive stress, psi

F_y static yield strength of structural steel, psi

I blast impulse per unit area, psi-msec

M_e equivalent mass of structural element

$P, P(t)$ blast load pulse, psi

P_a average uniform roof loading, psi

P_o peak overpressure, psi

P_r peak reflected pressure, psi

q_u static ultimate soil bearing pressure, ksf

r/c reinforced concrete

$R, R(x)$ dynamic resistance, equivalent static load, psi

t time, msec

t_a duration of average roof loading, msec

t_o effective duration of blast load, msec

T period of structural element in fundamental mode of vibration, msec

x structural displacement, in.

x_m maximum dynamic displacement, in.

x_u ultimate displacement at incipient collapse, in.

x_y effective yield displacement, in.

α energy absorption factor

δ displacement ratio, x_m/x_y

η dynamic resistance factor, P/R

τ duration factor, t_o/T

Recent catastrophic explosions in the petroleum and chemical industries have refocused attention on the issue of blast protection for vital facilities in process plants. For example, as a result of the Flixborough incident[1] the Chemical Industry Association recently developed a standard on blast-resistant design for control rooms for consideration by the government of the United Kingdom as a Code of Practice. A similar standard has been prepared in the Netherlands, and in the United States the Chemical Manufacturers' Association is establishing some guidelines.

In the mid-1960s, because of incidences of plant explosions and the trend towards centralizing computerized process control in a single building, blast protection became an important consideration in plant layout and design.[2] The central control building, together with the expensive equipment it houses, represents a significant plant investment. Moreover, in the event of an explosion, loss or severe damage to a central control building could result in long downtime in otherwise unaffected units.

These factors led to the development of simplified static design criteria for blast-resistant control houses.[3] The intent then was to provide a reasonable level of blast resistance in structures by employing criteria familiar to most designers at a cost only 10–20% more than nonblast-protected construction. However, these criteria did not reflect the dynamic response of structures to blast loading. Experience indicates that the technically more appropriate dynamic design criteria can be adapted, without much difficulty, to the design of blast-resistant buildings in refineries and chemical plants. This paper provides the background and some details on one approach to blast-resistant design.

23.1. BASIC DESIGN CONSIDERATIONS

To establish rational structural design criteria it is necessary to have a blast protection philosophy. Requirements for blast-resistant buildings in a refinery or chemical plant are based on the following considerations.

  *1.   A blast-resistant building should be capable of withstanding an external plant explosion of realistic magnitude in order to protect personnel, instruments, and equipment it houses from the damaging effects of the blast.

2. Structural damage is tolerable if it is not detrimental to the safe operation of the facility during and after the accident.

3. In the event of an explosion of intensity (damage potential) in excess of the design values, the structure should "fail" by excessive deformation without any significant loss of its load-carrying capacity, thereby providing an adequate margin of safety against catastrophic collapse.

4. The building is expected to resist a major explosion only *once* in its life. It should, however, have the capacity to safely support the post-explosion conventional design loads with some minor repairs.

To design a blast-resistant structure rationally and economically requires some knowledge of the magnitude and characteristics of the blast, the loading it causes, and of the response of the structure to this loading.

23.2. CHARACTER OF EXPLOSIONS—LEVEL OF PROTECTION

An explosion results from a sudden release of a large amount of energy such as the ignition of an accumulation of flammable hydrocarbon vapor, the rupture of a high-pressure reactor or the detonation of high-yield explosives. An explosion causes a wall of compressed air to move outward from the source at supersonic speed. The excess pressure in this blast wave over the ambient atmospheric pressure is called the overpressure. Typical blast wave profiles are shown in Fig. 23.1. The overpressure phase of the blast wave is followed

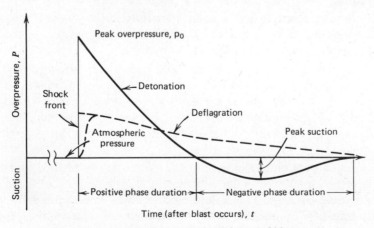

Figure 23.1. Typical pressure-time history of blast wave.

by a suction phase. In addition, the motion of air particles (blast wind) exerts a drag pressure on any object in the path of the blast wave. These parameters decrease with distance from the explosion and time after the arrival of the blast wave.

Structural damage is caused primarily by the blast wave, although other hazards such as fire and projectiles are also important factors in blast protection. The intensity of an explosion, or its damage potential is measured in terms of the peak overpressure or unit impulse (area under the overpressure—time curve) of the generated blast wave. These parameters depend on the energy released at the explosion source and on the distance between the explosion and the structure.

In a process plant the most damaging explosion probably results from the detonation of an unconfined vapor cloud from an accidental release of a large volume of flammable hydrocarbons. However, very little experimental or theoretical data are available which can be used to quantity the loading which could result from such explosions.[4] On the other hand, there is a vast amount of published data available on the character and effects of large and small scale detonation of high yield explosive such as TNT.[5] Consequently, it is customary to characterize hydrocarbon explosions in terms of TNT equivalence based on their damage potential. The values of the blast wave parameters (peak pressures, impulse, durations, etc.) for a specified TNT equivalent explosion can be readily determined from the published plot of these parameters versus distance.[5]

Based on the limited data available,[6-8] some differences in TNT and vapor-cloud blast waves which could have some impact on design are as follows:

1. Vapor-cloud explosion can result from a relatively slow deflagration (burning), detonation (rapid chemical decomposition), or more likely from a combination of both processes. These processes result in different blast waves for the same released energy.

2. Point-source explosion scaling laws may not be valid for gas-cloud deflagration but are appropriate outside the cloud for detonation in the low overpressure range (2–50 psi).

3. Deflagration results in a blast wave with lower pressure and longer duration compared with a detonation which produces a sharp pressure–time profile. The overpressure does not decay with distance and time as rapidly as in TNT explosions. Moreover, there is a significant rise time to the deflagration pressure pulse in the near range.

4. For a detonation explosion the blast wave in the far field (outside the explosive material) is dependent only on the amount of blast energy released. A vapor-cloud detonation can be represented by a point source (TNT equivalence) for obtaining the blast wave properties outside the cloud. Inside the cloud the overpressure is assumed con-

stant and its impulse increases linearly from zero at the center of the cloud to the boundary value.

5.   For a given impulse, the blast wave with the higher pressure (detonation) causes more structural damage.

Until more quantitative data are available on vapor-cloud explosions, the TNT equivalence is a reasonable approach to the design and evaluation of plant facilities.

An explosion equivalent in damage potential to 1 ton of TNT near the ground surface at a distance of 100 ft is used as a realistic basis for blast-resistant design. Such an explosion could result, for example, from the detonation of an unconfined cloud of ethane (6%) in air, 200 ft in diameter by 13 ft high.[2] The details of the structural design criteria also reflect aspects of vapor-cloud blast wave characteristics noted above. For example, a blast resistant control house is located no closer than 100 ft from a potential explosion source to place it outside the cloud. In addition, it is assumed that in a typical plant an explosion could occur in any direction from the protective building. Therefore, all walls are considered equally vulnerable to the blast effects.

23.3.   TOLERABLE STRUCTURAL DAMAGE—MARGIN OF SAFETY AGAINST COLLAPSE

An explosion is an extremely severe event and occurs quite infrequently. It imposes loads on a building far in excess of those for which it would normally be designed. It is reasonable, therefore, to design blast-resistant structures expecting some structural damage in a major explosion, provided this does not impede the safe operation of the housed facilities during and after the explosion. Damage such as cracking of concrete walls and slabs, and some permanent distortion in reinforced concrete or steel frames is considered tolerable under blast conditions. The incremental cost for blast protection would be prohibitively high if such damage was unacceptable and conventional factors of safety on stresses and deformations were required.

The specified explosion magnitude (1 ton of TNT at 100 ft) is a *design* value for which moderate structural damage is tolerable. It is recognized that process plant explosions can and have exceeded 1 ton TNT equivalence as evidenced by reports on recent explosions (Table 23.1). However, the design criteria provide a significant margin of safety to allow for larger explosions. Table 23.2 shows the range of blast protection the suggested criteria provide as the distance from the source and the levels of damage are varied. The protection provided by a conventionally designed building is shown as reference. Thus, a structure designed to withstand 1 ton of TNT at 100 ft with moderate damage could survive an explosion of the order of 50 tons TNT equivalence at 400 ft.

TABLE 23.1.   TNT Equivalence of Some Vapor-Cloud Explosions[6,8]

Incident	Material	Estimated TNT Equivalence (tons)
Beek, Netherlands, 1975	—	8–10
Antwerp, Belgium, 1975	—	7
Flixborough, U.K., 1974	Cyclohexane	18–45
Decatur, Ill., 1974	Propane	5–10
East St. Louis, Ill., 1972	Propylene	1–2.5
Pernis, Netherlands, 1968	Hydrocarbons	18–25

23.4.   BLAST LOADS

The overpressure is of primary consideration in structural design of enclosed buildings for blast resistance and the suction and drag pressure are of secondary effect. The blast loads depend on the geometry of the building as well as on the blast wave characteristics. First, the pressure pulse strikes the wall nearest the explosion source and is reflected. Then the roof and side walls are loaded as the blast wave travels across the building. Finally, the face of the building remote to the explosion is loaded some time after the front was initially loaded.

The design blast loading for a rectangular box-shaped building is listed in Table 23.3 and illustrated in Fig. 23.2. The blast loading is approximated by a triangular pressure pulse having a shock front (instantaneous pressure rise) with peak pressure and duration dependent on the structural element being considered as indicated in Fig. 23.2. The wall and roof slabs are designed for the reflected and incident pressure pulse, respectively. The structural framing is designed for the lateral loading on any one side plus the average uniform roof loading. In view of the serious consequences of the failure of columns and foundations under the vertical blast loading, they should be designed for the total vertical load capacity of the structural elements they support.

In addition to the overpressure loading, a structural member is stressed in the opposite direction as a result of the suction of the blast wave and the structural rebound. Usually the suction loading is much smaller than the required rebound resistance; and therefore, it can be safely ignored.

A blast resistant building must, of course, satisfy the requirements for conventional loadings such as provided by ACI or AISC in addition to the special requirements for blast resistance. For the blast loading condition the blast effects are combined with the normal sustained (dead and live) load without the use of any load factors. However, the design wind or earthquake loads are not to be combined with blast loads since the chance of their simultaneous occurrence is extremely remote.

TABLE 23.2. Level of Blast Protection in Process Plant Buildings

Design Basis	Structural Damage Level[a]	Explosion Magnitude (tons of TNT) Versus Distance from Source				
		50 ft	100 ft	200 ft	400 ft	600 ft
Basic blast resistance	Negligible	0.1	0.3	1.5	5	10–20
	Slight–moderate	0.4	1.0	3–5	15–25	40–60
	Severe	1.5	4	15	50–60	>100
	Incipient collapse	3	6–8	20–25	80–100	>100
Conventional nonblast	Glass breakage	~1 lb	~10 lb	~100 lb	~500 lb	~1.0
	Severe	0.1	~0.3	~1	3–5	5–10
	Incipient collapse	0.2	~0.5	2–3	5–10	10–20

[a]Damage description

Negligible: no apparent structural damage; elastic design; "cosmetic" repairs only.

Moderate: cracking of walls and roof slab; some spalling of concrete at supports and midspan; some permanent structural distortion.

Severe: significant cracking and spalling; pronounced frame distortion; no danger of immediate collapse, but major repair needed; economic decision—abandon versus repair.

Incipient collapse: distortion capacity of some elements reached; potential danger of flying debris; detailed evaluation for repair program—abandonment likely.

Glass breakage: overpressure ≤ 0.2 psi.

TABLE 23.3.   Design Blast Loading for Explosion Equivalent to 1 Ton of TNT at 100 ft

Structural Component	Peak Pressure (psi)	Duration (ms)
Wall	25	20
Roof		
Slab	10	20
Beams	10	20
Structural Frame		
Lateral load	25	20
Vertical load depends on span—		
≤10 ft span	10	20
20 ft	5.5	35
40 ft	4.0	50
≥60 ft	3.0	70
Foundation		
Lateral load ~ Equivalent static load on walls and structural frame		
Vertical load ~ Ultimate capacity of vertical-load-carrying members		

23.5.   STRUCTURAL DESIGN

A structure under blast loading is a complex dynamic system with infinite degrees of freedom of motion. The dynamic design problem is to determine the structural resistance required to keep the deflection of each structural member within specified limits when it is subjected to the transient blast loading pulse.

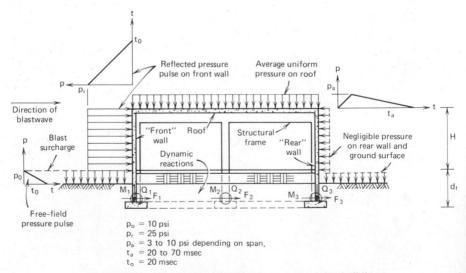

Figure 23.2.   Design blast loading for a rectangular building.

The ultimate strength dynamic design (USDD) method is appropriate for designing or appraising blast resistant structures. This method takes into account the dynamic response of the building due to the high-intensity short-duration blast loading and the capacity of the structural members to absorb blast energy due to their ductility. Also, the full structural strength of each element is mobilized to withstand the "once in a lifetime" blast load.

For the purpose of design each structural element is represented by a single-degree-of-freedom system as illustrated in Fig. 23.3. The simplified lumped-mass model approximates the response of the structural element under the blast loading in the principal response mode which is usually the fundamental mode of vibration. Structural damping is conservatively ignored. The system is excited by the blast loading as the forcing function, $P(t)$.

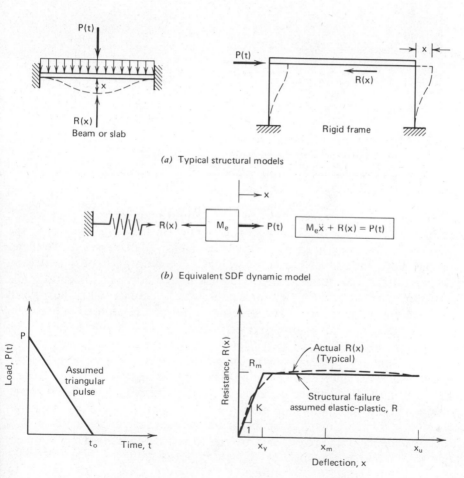

(a) Typical structural models

(b) Equivalent SDF dynamic model

(c) Load and resistance functions

Figure 23.3. SDF dynamic model for structural elements.

TABLE 23.4. Allowable Stresses and Deflections for Blast-Resistant Design

Material/ Loading Mode	Allowable Stress for Dynamic Loading	Displacement Factor $(\delta = x_m/x_y)$
Reinforced Concrete (ACI-318)		
Cylinder strength (dynamic)	$1.25\,f'_c$	
Axial compression		1.0
Tied column	$0.8\,f'_c$	
Spiral column	$0.9\,f'_c$	
Flexure	f'_c	3.0
Shear		1.5
Direct	$0.2\,f'_c$	
Diagonal tension	$2\sqrt{f'_c}$	
Bond	$0.15\,f'_c$	
Bearing	$0.85\,f'_c$	
Reinforcement yield	$1.2\,f_y$	3.0
Structural Steel (AISC)		5.0
Dynamic yield strength	$1.2\,F_y$	
Tension and bending	$1.2\,F_y$	
Compression	$2\,F_a \leq F_y$	
Shear	$0.6\,F_y$	
Girder (R/C or Steel)	$1.2\,F_y$	1.0
(part of main frame supporting both vertical and lateral loads)		

The dynamic resistance $R(X)$ of a structural member such as a beam, column, frame, slab, or shear wall is computed in the same manner as the static ultimate strength, except that the dynamic yield stresses of the material are used instead of the static values. The resistance of a structural element to dynamic loads is higher than it is to static loads because most structural materials have yield strengths which increase with the rate of loading. Values appropriate for structural steel and reinforced concrete in various modes of loading are provided in Table 23.4. It should be noted that no dynamic increases are allowed for brittle loading modes.

The ultimate strength dynamic design method introduces several additional design parameters over those encountered in static design. The required structural resistance depends not only on peak loads but also on the duration of the blast pulse, the natural period of vibration and the maximum acceptable dynamic displacements for the various structural elements. It is this dependence of the required structural resistance on dynamic parameters that makes the specification of a single uniform static load unrealistic for blast resistant design of structures.

The structural design parameters can be grouped conveniently in dimensionless form as follows:

Dynamic Resistance Factor (η)—the ratio of the peak applied load to the required dynamic resistance (P/R).

Duration Factor (τ)—the ratio of the duration of the applied equivalent triangular load pulse to fundamental period of vibration of the structural element (t_o/T).

Displacement Factor (δ)—the ratio of the maximum permissible dynamic displacement to the effective yield displacement of the element (x_m/x_y).

Ductility Factor (μ)—the ratio of the ultimate displacement (maximum attainable value, without collapse) to the yield value (x_u/x_y).

The time to maximum displacement t_m is a parameter which is often discussed in the literature; however, it does not enter directly in the design process. The ratio of the energy absorbed at the maximum displacement to that at the yield displacement is referred to as the energy absorption factor, $\alpha = 2\delta - 1$.

The dynamic design method establishes the relationships among the parameters discussed above. The resistance factor can be expressed approximately in terms of τ and δ as follows[9]:

$$\eta = \frac{\sqrt{\alpha}}{\pi\tau} + \frac{\alpha\tau}{2\,\delta\,(\tau + 0.7)}$$

This is an empirical relationship derived from the numerical solution of the undamped single-degree-of-freedom (SDF) model used to represent the structural member as shown in Fig. 23.3. It expresses the structural resistance (R) required for the specified blast loading in terms of the other design parameters. This resistance can be considered as an equivalent static load in designing the structural member. However, this load depends on the dynamic characteristics (period, ductility) of the member. Figure 23.4 shows this dependence in the form of plots of η versus τ for various values of δ. These curves are divided into three ranges according to what feature of the blast wave the structure responds—impulse, pressure, or both. The structural response is also presented in Fig. 23.5 in the form of normalized pressure–impulse curves (P–I diagrams). These curves show the various pressure–time loadings which cause a particular deformation or level of damage in a structural member.

When a structure is loaded dynamically, and reaches a maximum deflection, the stored energy tends to cause deflection in the opposite direction. This tendency exists even in the case where there is some positive load still acting on the structure at the time it reaches maximum deflection. In general, this rebound is elastic and is reduced by the effect of structural damping but is augmented by the suction in the late stage of the blast wave. The required rebound resistance (R_r) can be conservatively taken as $0.75R$.

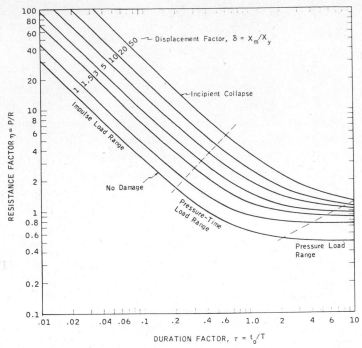

Figure 23.4.   Dynamic response of SDF model for elastic–plastic structural elements.

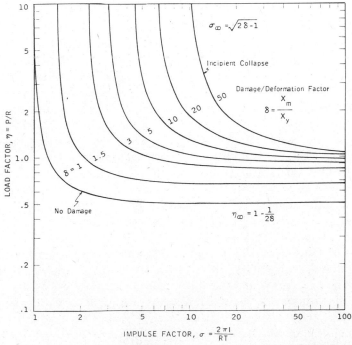

Figure 23.5.   Normalized pressure-impulse diagram for elastic–plastic SDF structural model.

23.6. PERMISSIBLE DEFLECTIONS

The dynamic design method requires that maximum deflection limits be established for the structural elements. The deflection permitted in an element depends on its ductility and importance in the overall structural performance of the building under blast loading and also on operational considerations. The displacement factor (δ) can be considered a measure of tolerable blast damage, with $\delta \leq 1$ corresponding to no significant damage and $\delta = \mu$ representing incipient collapse of the structural member.

The design is strongly affected by the amount of deformation the structure can take without critical damage. To effect economy, deflection limits are set as high as possible, consistent with ductility, post-explosion safety and operational considerations, to provide for maximum energy absorption in an explosion. Tests of reinforced concrete and steel structures by the Atomic Energy Commission, the Bureau of Mines, and others have shown that under blast loading they can withstand relatively large permanent deformations without seriously affecting their ability to carry normal loads.[10,11] The values of δ appropriate for the tolerable damage at the design loads are listed in Table 23.4. These limits are set so as to assure a safety factor of at least 2.5 on the energy absorption capacity of the structural elements.

23.7. FOUNDATION DESIGN

In an explosion, the foundation is subjected to transient loads of very short durations, but of magnitudes considerably higher than conventional static loads. These loads are transmitted directly from the blast pressure wave and through the superstructure. The transmitted vertical and lateral loads depend on the dynamic characteristics of the structural elements and are usually severely distorted and attenuated. These dynamically modified loads, together with the conventional static dead loads and live loads, must be considered in blast-resistant design. The blast surcharge around the structure has a beneficial effect and can be conservatively ignored.

Excessive movement of the foundation elements under the combined effects of the blast and long-term static loading constitute "failure." Under the vertical blast loads, the foundation elements tend to punch into the underlying soils, whereas the lateral loads cause sliding and tilting of the structure. The amount of movement depends on the soil characteristics, the magnitude and duration of the transmitted blast loads, and on the mass, size, and shape of the foundation elements.

Model and full-scale tests of footings on both cohesive and cohesionless soils conducted by the U. S. Army indicate that strengths considerably greater than the static bearing capacities can be realized under impulsive loads.[12] The foundation movements and stress patterns under blast loading differ radically from those under static loads. This is attributed to the effects of rate and duration of loading on the underlying soils and on differences in the modes of failure (punching versus general shear failure). Based on these experimental and analytical data, simple empirical formulas have been developed to predict the maximum dynamic foundation displacements due to the blast loadings.[13]

A conservative design for the foundation is warranted because its blast damage is not always visible and its repair is difficult and costly compared with that for the superstructure. The foundation must be designed to mobilize the full capacity of the structure to resist the explosion effect. This means that excessive foundation movement which could cause severe structural distortions must be prevented.

Generally, if the foundation is sized for the total sustained plus blast loads using a bearing capacity approaching the ultimate static bearing pressure (q_u), the displacement will not be excessive. It is suggested that under blast conditions the allowable bearing pressure be limited to q_u for cohesive soils and $0.8\,q_u$ for cohesionless soils. If passive resistance of the foundation is required in addition to friction to resist sliding, it should be at least 1.5 times the unbalanced lateral load. This load is the horizontal reaction to the blast load less the frictional resistance.

23.8. GENERAL CONSTRUCTION GUIDELINES

The design of blast-resistant buildings requires the use of good design and construction practices as well as a knowledge of the characteristics of the blast loads and the behavior of structures under these loadings. The blast-resistant quality of a building is influenced by its shape, layout, and orientation relative to potential explosion sources. In addition, certain structural schemes and types of construction materials are preferable for blast resistance while others should not be used.

Building Layout and Appropriate Structural Schemes

Nonstructural considerations such as safety, operation, architecture, cost, and owner-preference may dictate the shape, orientation, and layout of a plant building. In establishing these, however, the designer should also consider the requirements for blast resistance.

Generally, for a given building volume, the cost of blast resistance increases with the building height. A low-profile building experiences lower blast loads and overturning effect compared with a tall structure. Buildings over two stories in height are, therefore, not recommended as protective structures. Moreover, other factors being equal, a one-story building is preferable to a two-story one, because foundation design for the taller structure is generally more troublesome due to the more severe overturning effects of the lateral blast loading.

The plan outline and elevation profiles of a blast resistant building should be as "clean" and simple as possible. Reentrant corners and offsets, in particular, should be avoided. Such features, beside being difficult to detail, create local high concentrations of blast loading. The orientation of the building should be such that the blast-induced loads are reduced as much as possible. This requires that as small an area of the building as possible should face the most probable source of explosion.

Controlroom layout considerations often require large floor area unobstructed by interior structural walls or columns. There is, however, a cost premium for large span structural framing. Interior columns should, therefore, be used where feasible so that the maximum span does not exceed 40 ft in either frame or shear wall type buildings. Arches and igloo type structures are very effective for larger spans. These structures, when used for large spans, usually require much less structural material compared with conventional frame or shear wall types. However, the cost saving in material must be weighed against the high erection costs of special shell construction.

Satisfactory Types of Construction Materials

The most important feature of blast-resistant construction is the ability of the structural elements to absorb large amounts of blast energy without catastrophic failure in the structure as a whole. Construction materials in blast-protective structures must therefore have ductility as well as strength. For a structure to exhibit any measure of blast resistance, its frame and foundation must be capable of sustaining the large lateral loading. This requirement is similar to that for earthquake-resistant design. In general, structures which are earthquake resistant are also, to some degree, blast resistant.

Reinforced concrete is an excellent material for blast resistant construction. Its mass provides inertia to help resist the transient blast loads and a reinforced concrete building possesses continuity and lateral strength. Rigid-frame, shear wall, and shell construction are all appropriate for blast-resistant design. In addition to its inherent structural strength, reinforced concrete is effective in providing protection against fire and flying debris which usually accompany an explosion. The amount, placement, and quality of the steel used in reinforced concrete construction should be chosen to assure ductile behavior.

Structural steel is a ductile, high-strength material which is especially suitable for the structural framing in blast-resistant buildings. Structural and intermediate grades, with assured ductility, are preferable. High-strength steel with marginal ductility should not be used. The structural steels appropriate for blast-resistant design are as designated in Part 2 of the "AISC Specifications for Plastic Design."

In both reinforced concrete and structural steel construction all joints and connections should be capable of developing the full (ultimate or plastic) strengths of the structural members. Joints are usually the weak points in conventional construction.

Unacceptable Construction

Certain types of construction commonly used in ordinary buildings are not recommended for blast-resistant structures. The principal basis for evaluating such construction is its mode of failure if severe overloading occurs. *Brittle construction* is not suitable for blast resistant structures. Besides being vulnerable to catastrophically sudden failure under blast overload, it provides a source of debris which can cause major damage when hurled by the blast wind. Unreinforced concrete, brick, timber, and masonry are examples of this type of construction. These normally should not be used in the exterior shell of blast-resistant structures. If in an otherwise ductile structure, brittle behavior of some elements cannot be avoided, as is the case for axially loaded reinforced concrete columns or for shear walls, the margin of safety for these elements should be increased; that is, their capacity should be downgraded. Details which may result in unnecessary hazards in the event of an explosion should be avoided. For example, seemingly harmless architectural details such as parapets, copings, signs and falsework become dangerous in an explosion.

A typical design for a blast-resistant building is provided as an appendix and summarized in Fig. 23A.1. It consists of 8-in. reinforced concrete walls and 5-in. roof slab supported on structural steel beams and frames.

23.9.   CONCLUSIONS

Based on safety and economic considerations, vital buildings in a process plant should be protected from the hazard of accidental explosions. The blast wave from the ignition and detonation of a flammable hydrocarbon vapor cloud is most damaging to such buildings. Although there are differences between vapor cloud and TNT explosions, for the purpose of structural design, TNT blast wave data can be used to determine blast loading based on the concept of a TNT equivalent explosion. A blast-resistant building is designed to withstand an explosion equivalent to 1 ton of TNT at 100 ft with minor structural damage.

A dynamic technique is suggested for the structural design or appraisal of blast-resistant buildings in refineries and chemical plants. The dynamic approach ensures the appropriate distribution of structural strength and ductility for a building to withstand the blast wave effects. It generally will result in a more efficient and safer design compared with the purely static overpressure approach which does not take into account the dynamic response of the structural elements of a building to the transient blast loading. In the dynamic criteria an equivalent static load is calculated for each element

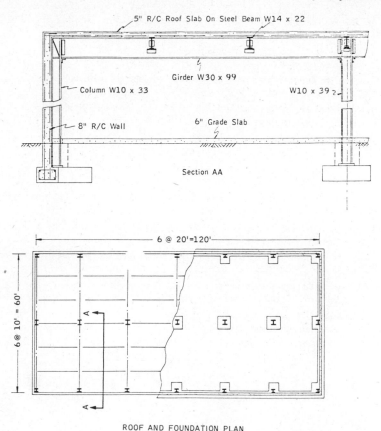

Figure 23.A.1. Example of blast-resistant building.

based on its dynamic characteristics and its ability to absorb the explosion energy.

Generally, the cost of blast resistant construction by the recommended design criteria has ranged from 15% to 30% higher than conventional building cost. The increased cost is due primarily to the choice of structural system (reinforced concrete and structural steel versus masonry), special doors and hardware, etc. At building costs of 100 dollars per square foot this represents an investment of up to $200,000 for protecting a typical plant control building from severe damage by accidental explosions.

REFERENCES

1. T. A. Kletz, "The Flixborough Cyclohexane Disaster," *Loss Prev.,* **8,** 106–118 (1975).
2. W. J. Bradford, and T. L. Culbertson, "Design of Control Houses to Withstand Explosive Forces," *Loss Prev.,* **1,** 28–30 (1967).

3. B. W. Burns, "Design Control Centers to Resist Explosions," *Hydrocarbon Process.*, **46** (11) (November 1967).

4. R. A. Strehlow, "Unconfined Vapor-Cloud Explosions—An Overview," *14th International Symposium on Combustion,* Combustion Institute, 1973.

5. Department of Defense Manual, "Structures to Resist the Effects of Accidental Explosions," TM 5-1300/NAVFAC P-397/AFM 88-22, June 1969.

6. R. A. Strehlow and W. E. Baker, "The Characterization and Evaluation of Accidental Explosions," NASA Report CR-134779, NTIS, June 1975.

7. D. S. Allan and P. Athens, "Influence of Explosions on Design," *Loss Prev.,* **2,** 103–109 (1968).

8. V. C. Marshall, "The Siting and Construction of Control Buildings—A Strategic Approach," Symposium on Process Industry Hazards, Livingston, U.K., September 1976.

9. ASCE, "Design of Structures to Resist Nuclear Weapons Effects," Manual of Engineering Practice No. 42, 1964.

10. S. Glasstone, ed., *The Effects of Nuclear Weapons,* U.S. Atomic Energy Commission, U.S. Government Printing Office, Washington, D.C., 1964.

11. C. H. Norris et al., *Structural Design for Dynamic Loads*, McGraw-Hill, New York 1959, p. 302.

12. H. M. Taylor, "Dynamic Response of Rectangular Footings on Clay and Sand," TR S-72-6, U.S. Army Engineer Waterways Experiment Station, Vicksburg, May 1972.

13. P. F. Hadala, "Dynamic Bearing Capacity in Soils—Investigation of a Dimensionless Load-Displacement Relation for Footings on Clay," TR 3-599-4, U.S. Army Engineer Waterways Experiment Station, Vicksburg, June 1965.

Appendix:

Design Example

Design for a one-story blast-resistant building based on the following data (see Fig. 23A.1):

Dimensions: 120 × 60 × 16 ft

Structural Scheme

Walls—r/c slab spanning between foundation and roof

Roof—r/c slab supported on steel beams 10 ft o/c

Structural framing—two-bay column-girder bents with roof slab acting as diaphragm to transmit lateral loads to side shear walls

Foundation Soil—cohesive

Cohesion, $c = 1.0$ ksf

Adhesion, $c_a = 0.5$ ksf

Loading

Live load—Roof—20 psf; grade floor—150 psf

Wind load—ANSI, $V_{30} = 90$ mph, Exposure C

Blast Load (from Table 23.3)	P (psi)	t_o(msec)	Displacement Factor (δ)
Wall	25	20	3
Roof Slab	10	20	3
Beams	10	20	5
Girder (30-ft span)	4.8	42	1

Design Summary

MATERIAL

All concrete—$f'_c = 3000$ psi
Reinforcement—structural grade deformed bars ($f_y = 42$ ksi)
Structural steel—A36
Bolts—A325

WALL SLAB

Thickness—8 in.
Vertical bars—#7 @ 6½ in. inside face
            #7 @ 8 in. outside face
Horizontal bars—#4 @ 12 in. Placed alternately on inside and outside faces

ROOF SLAB

Thickness—5 in.
Main bars—#5 @ 12 in. top and bottom
Secondary bars—#4 @ 18 in. placed alternately top and bottom perpendicular to main bars
Rebound anchors—#3 @ 24 in. welded to purlins

STEEL FRAMING

Purlins—W14 × 22 @ 10 ft spacing
Girders—30 × 99 @ 20 ft spacing
Columns—W10 × 33 outside
        W10 × 39 center

FOUNDATION

Footing thickness—18 in. throughout
Wall footings 2-ft wide, 4—#6 bars longitudinally
            #5 @ 10 in. ties
Outer column footings 4 × 6 ft—#6 @ 8 in. bottom, both directions
Center column footings 5 × 5 ft—#6 @ 8 in. bottom, both directions

24

An Insurer's Perspective of the Chemical Industry

Gail P. Norstrom II

The history of the underwriting of property damage and business interruption insurance in the chemical industry is characterized by the complex interaction between the business cycles of the chemical and allied industries and those of the insurance industry. As the earnings potential of the chemical industry is tied to consumer demand, raw material cost, labor costs, other costs, interest rates, taxation policy, and other regulations and other overhead, so it goes with the insurance industry as well. When these cycles reinforce each other, the chemical industry is expanding its investment base in plant and equipment and earning potentials are at their peak, while the insurance market is able to provide the necessary coverages at proper pricing levels.

However, when these cycles are not in phase, then shortages of insuring dollar capacity for a reasonable, or for that matter any, cost versus the needs of the chemical industry, develop financial exposures which are abhorrent to responsible managers and shareholders. This situation and other considerations have led to the development of many forms of risk self-assumption mechanisms such as the use of higher deductibles, self-insurance, captive or subsidiary insurance companies, risk pooling, etc., to serve as an alternative or complement to commercial insurance.

This chapter will not delve into the specifics of insuring risks of the chemical process industries, but rather, will relate the property underwriters' point of view as respects; the character of the risk, the nature of exposures, how exposures are analyzed in the light of loss experience in the class, and some elaboration upon the theory of insuring such risks. Also, observations are advanced concerning past difficulties in providing coverage for chemical risks and insights presented as to future consideration.

24.1. CHARACTER OF THE RISK

The insurers' overview of the chemical industry reveals a spectrum of capital intensive facilities with many very large insurance exposures.[1] Of necessity, there are large insurance premiums involved with the industry to support the loss experience. The character of the management and ownership is increasingly multinational and financially oriented as opposed to the technicians/entrepreneurs of the past. By nature of its importance, size, visibility, and the variety of its products, the industry is subject to the caprices of an evermore complex pantheon of government regulation and potential litigation which can dramatically impact loss experience and the underwriters' assessment of exposures.

24.2. NATURE OF THE EXPOSURES

The nature of the exposures involved in the chemical and allied industries are characterized by fast-changing technologies, the development of larger and larger so-called "world scale" plants, a huge capital investment base and large earnings potentials associated with single line or very compact facilities. This plant and equipment can be exposed by large quantities of materials of process involving moderate to severe flammability, reactivity/explosibility, incompatibility, toxicity, radioactivity, and pyrophoricity hazards.

24.3. EXPOSURE EVALUATION

As complete as possible understanding of the nature of these exposures is essential to the insurance underwriter so that an appraisal of a particular risk may be made in a consistent, logical fashion which will permit risk-to-risk comparison, and absolute evaluation as well. From the property damage and business interruption point of view, this will generally involve an evaluation of some or all of the following considerations[2]:

1. Plant Site
    a. Topography
    b. Utilities
    c. Off-site exposures
    d. Waste disposal/treatment
    e. Climate

2. Plant Layout
   a. Congestion (spacing considerations)
   b. Drainage
   c. In-plant access roads
   d. Ignition sources and prevailing winds
3. Buildings, Structures, Equipment
   a. Construction
   b. Suitability for occupancy
   c. Windstorm design
   d. Earthquake design
   e. Lightning exposure
   f. Hail exposure
   g. Equipment evaluation
4. Materials
   a. Hazards evaluation program
   b. Materials monitoring program
   c. Special precautions
   d. High unit value
   e. Damageability
5. Process Hazard Evaluation
   a. Description of raw materials, processes, products
   b. Safeguards and protection
   c. Loss potential
   d. Business interruption analysis
6. Operator Training
   a. Manuals program
   b. Training—original and review
   c. Emergency procedures
7. Equipment Design, Testing, and Maintenance
   a. Storage tank venting devices
   b. Unfired pressure vessels
   c. Metals inspection
   d. Fired equipment
   e. Electrical equipment
   f. Pipelines
   g. Nonmetallic equipment
   h. Maintenance philosophy
8. Loss Prevention and Protection
   a. Loss prevention program
   b. Private property protection
   c. Public property protection and other outside assistance
   d. Plant security
   e. Disaster program
   f. Loss record

24.4. LOSS EXPERIENCE AND ANALYSIS

Given the nature of the exposures at a particular risk, the underwriter then proceeds to assess the insurability of the risk as related to a particular underwriting philosophy. As a general rule, this philosophy involves a consideration of the loss experience of the particular Insurer involved. The 1978–1980 inflationary economy produced a very "soft" insurance market, i.e., the market is able to provide coverage for most any risk at a reasonable cost. However, there are many Underwriters who will not insure chemical risks due to a combination of adverse past, or perceived future loss experience and the large exposures involved. There are probably only four or five viable commercial property insurance markets which will make the necessary insurance dollar capacity available to the industry on a consistent basis.

A review of recent property loss experience will yield yet more insight into this reluctance on the part of the Insurance Industry to insure chemical risks.

The data which will be considered is the Industrial Risk Insurers' recent (1978–1980) experience in the chemical and allied classes. During this period there were 1028 occurrences which resulted in a total dollar loss of approximately $152,000,000 nominal dollars net of insured's deductible or other self-retention. It is important to note that during this period there was only one catastrophic loss, that is, loss greater than $20,000,000. Hence, the data is instructive, in that it conveys information concerning the "normal losses" in the class to be expected by the underwriter (including the infrequent catastrophe).

Peril Analysis

The first exhibit (Table 24.1) to be examined is the peril analysis. The data suggests that while fire is by far the most common occurrence, it is the explosion peril[3] which causes the greatest amount of dollar loss. This is more explicitly shown on the bar graph (Fig. 24.1) where the relationship between an element of frequency to an element of dollar loss is clearly much more severe in the case of the explosion peril vis-à-vis that of the fire peril.

TABLE 24.1. Peril Analysis of Losses in the Chemical Industry; IRI Chemical, and Allied Classes

Peril	Frequency	$ Loss (%)
Explosion	23.4	66.2
Fire	41.6	20.6
Windstorm	14.1	8.1
All other	20.7	5.1
	100	100

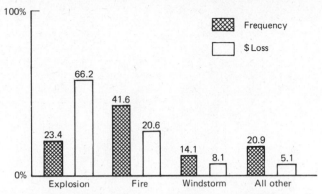

Figure 24.1. Losses in the chemical industry, IRI chemical and allied classes, 1978–1980 peril analysis: Frequency loss, cross-hatched bar; dollar loss, open bar.

Occupancy Analysis

The next portion of the data which will be considered is the occupancy analysis (Table 24.2). Industrial Risk Insurers (IRI) classes this subset of its business into seven occupancy groups. To generalize, the extra heavy hazard chemical class is that involving organic peroxide and explosive manufacture, nitrations, Claisen & Grignard reactions, and other very high hazard processes. The petrochemical occupancies are processes utilizing petroleum hydrocarbon or similar feed stocks (olefin facilities predominate this occupancy class). The heavy hazard class includes most polymerization processes, as well as solvent extraction, sulfonation, low pressure hydrogenation and other processes involving flammable or combustible materials. The light hazard chemical occupancies involve primarily inorganic operations. The remaining classes are self-explanatory.

The bar graph in (Fig. 24.2) indicates that the vast bulk of the dollar loss is being produced in the petrochemical and heavy hazard classes. That is, for an increment of insured value in the chemical and allied classes, there is

TABLE 24.2. Occupancy Analysis of Losses in the Chemical Industry; IRI Chemical, and Allied Classes

Occupancy Class	Frequency (%)	$ Loss (%)
63—Extra heavy hazard	4.9	4.2
64—Petrochemicals	8.9	19.9
65—Heavy hazard	32.1	59.6
66—Light hazard	22.9	9.6
67—Paint, dyestuffs, inks	11.4	3.3
68—Soaps and vegetable oil	6.2	.9
69—Pharmaceutical & fine chemicals	13.6	2.5
	100	100

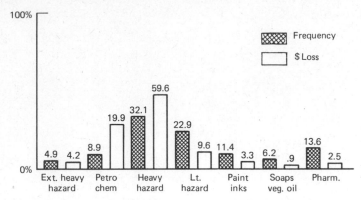

Figure 24.2. Losses in the chemical industry, IRI chemical and allied classes, 1978–1980 occupancy analysis. Frequency loss, cross-hatched bar; dollar loss, open bar.

approximately a 41% chance that the loss will occur in either a petrochemical or heavy hazard chemical unit. However, incidents will account for approximately 80% of all dollar loss.

One item of particular note is the excellent relationship of an element of frequency to an element of dollar loss enjoyed by the pharmaceutical, soap, and vegetable oil classes during this "normal" loss period. While this can be explained to some degree as being due to current self-retention practices of firms with these occupancies, there is no question that this subsegment of the chemical industry presents a much less severe exposure to property insurance underwriters than the remainder of the class. Of course, this would apply only to property damage and business interruption insurance and may not be applicable to other types of insurance.

Size of Loss Analysis

The next area which will be examined is the size of the loss analysis. It is noted in Table 24.3 that 52.6% of all occurrences reported resulted in no dollar loss, while a mere 1.2% of all occurrences resulted in a loss greater than $2.5M. That 12.7% of all occurrences producing 94.7% of total dollar loss will be reviewed in more detail later in the chapter. *Clearly, major losses in the Chemical Industry can be characterized as having relatively low frequency, but enormous dollar loss potential associated with them.*

This is again illustrated on the bar graph of Fig. 24.3. This is more evidence in support of Murphy's Law ("That which can happen . . . will"). That is, examination of the trend curves indicates that ever-decreasing frequency produces an ever-greater loss.

This exhibit also leads to the conclusion that class or industry underwriting is the only statistically valid approach to insuring this class. No single firm can accurately predict its catastrophic exposures from its own loss experience. However, for that "normal or expected" loss (losses less than

TABLE 24.3. Size of Loss Analysis—in the Chemical Industry; IRI Chemical, and Allied Classes

Size	Occurrences (%)	$ Loss (%)
No claim	52.6	0
$ 0–$ 9,999	14.2	.6
$ 10,000–$ 24,999	9.9	1.1
$ 25,000–$ 49,999	5.7	1.3
$ 50,000–$ 99,999	4.9	2.3
$ 100,000–$ 249,999	4.4	4.6
$ 250,000–$ 499,999	3.7	8.8
$ 500,000–$ 999,999	2.1 → 12.7%	10.2 → 94.7%
$1,000,000–$2,499,999	1.3	15.3
$2,500,000 & over	1.2	55.8
	100	100

$100,000 in 1981 dollars—87.3% frequency and 5.3% dollar loss) a given firm's experience can be very instrumental in the decision as to the proper level of self-retention. This will be discussed in more detail.

24.5. LARGE LOSSES

Next to be examined will be a typical six-quarter segment of 1978 and 1979 experience. In an in-depth analysis of larger losses, that is, occurrences which have produced dollar loss greater than $100,000 combined property

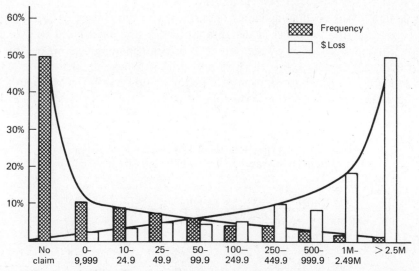

Figure 24.3. Losses in the chemical industry, IRI chemical and allied classes, 1978–1980 size of loss analysis. Frequency loss, cross-hatched bar; dollar loss, open bar.

TABLE 24.4   Typical Six-Quarter Period Losses
Greater Than $100,000 in the Chemical Industry;
IRI Chemical, and Allied Classes

Type of Loss	Frequency (%)	$ Loss (%)
Explosion	51.4	73.2
Fire	33.3	20.1
Windstorm	1.5	0.2
All other	13.8	6.5
	100	100

damage and time element, the critical factors which are of importance to underwriters as well as safety and property conservation engineers are isolated. Explosion and fire losses will be specifically examined from the point of view of the type, cause, location, and key contributing factors which are associated with these more severe occurrences. It should be noted that proximate cause (fire, explosion or other) determines the statistical classification of the frequency and severity.

During this period, contrary to the overall data discussed earlier, in the case of larger losses the greatest frequency of loss is due to explosion. This is not surprising, as the explosion peril produces the bulk of the large dollar loss. It is important to note in Table 24.4 that for these larger losses, fire and explosion perils produced some 84.7% of frequency and 93.3% of dollar loss.

24.6.   LARGE LOSS ANALYSIS—EXPLOSION

Explosion losses producing damage greater than $100,000 are examined first. The violence and widespread instantaneous damage associated with explosions, further aggravated by the potential of large ensuing fire, makes the job of preventing or controlling the loss potentials of these incidents one of the main tasks of the loss prevention professional and a matter of great concern to the insurance underwriter as well.

TABLE 24.5.   Type Analysis of Explosion Losses Greater
Than $100,000 in the Chemical Industry; IRI Chemical, and
Allied Classes

Type of Loss	Frequency (%)	$ Loss (%)
Chemical reaction	44.2	85.9
Boiler explosion	17.7	5.3
Furnace explosion	5.8	0.8
Explosion—other	32.3	8.0
	100	100

Type Analysis

It is noted below in Table 24.5, that the explosion losses involving a chemical reaction developed 44.2% of frequency and 85.9% of total dollar loss. Fired equipment explosions produce another 23.5% of frequency and all other explosions, examples of which would be free-air vapor-cloud explosions, steam explosions, etc., comprise the other 32.3% of frequency. However, the dollar loss associated with these occurrences, not involving chemical reaction, produce only 14.1% of dollar loss. Now the principal reason for the analysis of this period is apparent—there were no large vapor-cloud explosion losses. Such an occurrence weights the data in such a fashion that insights into the more predictable loss are lost. There have been other treatments[4-9] of this subject in the literature and reader is urged to consult the cited references.

Cause Analysis

Examining the chemical reaction loss cause analysis, Table 24.6, it should be noted that accidental and uncontrolled chemical reaction are the causes in 73.3% of all cases. Recalling that chemical reactions are responsible for 44.2% of all major explosions, it is apparent that about one-third of all major explosions in the chemical industry will be due to either accidental or uncontrolled chemical reaction.

TABLE 24.6. Cause Analysis of Major Chemical Reaction Losses Greater Than $100,000 in the Chemical Industry; IRI Chemical, and Allied Classes

Cause	Frequency (%)
Chemical reaction (accidental)[a]	33.3 ⎤
Chemical reaction (uncontrolled)[a]	40.0 ⎦ → 73.3
Decomposition of unstable materials	13.3
Other	13.4
	100

[a] Definitions:

Chemical reaction (accidental). Term for incidents caused by accidental contact of materials. Can occur within a vessel and could include the introduction of a wrong material, excessive quantity of a material or improper sequence of introduction of materials. May also occur outside of a vessel, such as accidental spilling of an oxidizing material such as sodium hypochlorite, contact with an oily floor and resultant ignition of the oil.

Chemical reaction (uncontrolled). Term for a reaction which is an intentional combination of materials which becomes uncontrolled as a result of agitation, loss of coolant, a contaminant, instrumentation failure, etc.

TABLE 24.7. Location Analysis of Explosion Losses Greater Than $100,000 in the Chemical Industry; IRI Chemical, and Allied Classes

Location	Frequency (%)
Enclosed process building	66.7
Open process structure	20.1
Other	13.2
	100

Location Analysis

Next to be examined is where in the plant a major explosion would be more likely to occur. We see (Table 24.7) that the odds are approximately 2 to 1 or that 66.7% of all losses will occur in an enclosed process building as opposed to an open process structure or other areas, such as a tank farm, laboratory building, etc. In other words, the greatest normal explosion loss potentials exist in enclosed process areas and chemical risks should be evaluated from that point of view.

Occupancy Analysis

The occupancy analysis, Table 24.8, indicates that in this enclosed building, the type of process which will most probably be involved will be that of a batch chemical reaction as contrasted to a continuous chemical reaction system or other type of unit.

Contributing Factors

The last area of analysis as respects major explosion losses, will be that of contributing factors (Table 24.9). Needless to say, in any given loss one or more contributing factors may be involved in determining the magnitude of loss. Accordingly, for the purpose of this discussion, we will only identify the four major contributing factors which have been found to be involved in

TABLE 24.8. Occupancy Analysis of Explosion Losses Greater Than $100,000 in the Chemical Industry; IRI Chemical, and Allied Classes

Occupancy	Frequency (%)
Chemical reaction process—batch	60.0
Chemical reaction process—continuous	13.6
Recovery unit	6.6
Evaporation unit	6.6
Other	13.2
	100

TABLE 24.9. Major Contributing Factor Analysis to Explosion Losses Greater Than $100,000 in the Chemical Industry; IRI Chemical, and Allied Classes

Contributing Factor	Frequency (%)
Rupture of vessel or equipment or mechanical failure	40.0
Human element—carelessness	20.0
Inadequate venting	20.0
Improper location of equipment or congestion	20.0
	100

large explosion losses during this period. The major factor is the rupture of a vessel or equipment, or the mechanical failure of piping, valves, agitators, etc. Three other elements which find their way into 20% of all losses during this period are the human element and carelessness, inadequate venting (which could also have contributed to the rupture of vessel or equipment), and finally the improper location of equipment, or congestion.

24.7. LARGE LOSS ANALYSIS—FIRE

Occurrences producing fire losses greater than $100,000 will be analyzed next. The large potential for the destruction associated with major fires, causes the underwriter and engineer to consider the means by which such occurrences can be maintained at the lowest possible frequency, and when unavoidable fires do occur, the means to minimize their damage to human life and property.

Cause Analysis

Table 24.10 indicates that the principle cause of these major fire losses is that of a flammable liquid or gas release, or overflow. The next most prevalent cause is the overheating of hydrocarbon materials, followed by the failure of

TABLE 24.10. Cause Analysis of Fire Losses Greater Than $100,000 in the Chemical Industry; IRI Chemical, and Allied Classes

Cause	Frequency (%)
Flammable liquids or gas release, overflow	28.5
Overheating	24.3
Failure of pipe or fitting	14.3
Static electricity or electric spark	9.5
Electrical or mechanical breakdown	9.4
All other	14.0
	100

TABLE 24.11. Location Analysis of Fire Losses Greater Than $100,000 in the Chemical Industry; IRI Chemical, and Allied Classes

Location	Frequency (%)
Enclosed process building	38.5
Outdoor process structure	33.3
Other	28.2
	100

a pipe or fitting, static electricity or electric sparks, electrical or mechanical breakdown and "all other."

Location Analysis

As in the case of the explosion peril, the data in Table 24.11 attempts to place the location in the plant where such incidents will occur. It should be noted, however, that in contrast to explosion, the frequency of loss in an enclosed process building versus an outdoor process structure is approximately equal in the case of major fire losses.

Contributing Factors

Examining the major contributing factors Table 24.12 we see that *by far* the lack of automatic sprinkler protection or water spray is the key factor which has been involved in almost 40% of these fire losses.

24.8. CONCLUSIONS DRAWN FROM LOSS EXPERIENCE

An examination of these data and similar studies yields the following conclusions:

1. Taken together, explosion and fire losses provide the greatest frequency and severity of property incident in the chemical industry.

TABLE 24.12. Major Contributing Factor Analysis to Fire Losses Greater Than $100,000 in the Chemical Industry; IRI Chemical, and Allied Classes

Contributing Factor	Frequency (%)
Sprinklers or water spray needed	38.1
Human element	19.0
Vapor-laden atmosphere	14.3
Excessive residue or deposits	9.5

2. The explosion loss is of much more severe character than that of the fire loss.

3. Major explosion losses will most likely be caused by either accidental or uncontrolled chemical reaction.

4. Major explosion losses will occur in enclosed process buildings involving batch chemical reactions.

5. The magnitude of loss will be contributed to by the rupture of vessels or equipment.

6. In the case of major fire losses, the cause will be due to the release of either a flammable liquid or gas and will occur with equal frequency in enclosed buildings or outdoor process structures.

7. The principal contributing factor to major fire loss will be the lack of automatic sprinklers or water spray.

24.9. UNDERWRITING CONSIDERATIONS

An examination of this and similar data led the Industrial Risk Insurers to the development and publication of its "Guiding Principles for Protection of High Hazard Chemical and Petrochemical Plants" in 1977.[10] These principles outline protection recommendations in 12 major categories:

Administration	Separation
Construction	Duplication of facilities
Water supplies	Inspection and maintenance of equipment
Automatic fire protection	Process design
Manual fire protection	Materials hazard evaluation
Drainage	Operator training

Other underwriters have developed similar criteria.

To the extent that the criteria suggested by these guidelines or other accepted standards are met in a given facility, the Insurer will make the appropriate underwriting decisions. These will involve consideration of the following:

1. Capacity
2. Deductibles
3. Rate
4. Restrictions

Capacity

Capacity in insurance parlance refers to the amount of insuring dollars which may be made available to provide financial protection for a given facility.

This amount of insurance may be used to cover only direct damage exposure, or may be used to cover time element coverages involved with the loss of earnings as well. Each underwriter has finite resources at his or her disposal, based upon the insurance company's assets and underwriting practices, which can be devoted to risks in the chemical industry. In general, the greater degree of compliance with protection standards enables the underwriter to provide the greatest amount of insuring dollars available to the industry.

There are usually cost trade-offs involved between the provision of protection and the decision to purchase insurance. However, insurance alone can never be the sole and sufficient financial justification for the provision or nonprovision of protection. There should be an overriding management thrust as to what constitutes a minimum or basic level of protection for any given operation. Considerations above and beyond these minimum requirements are then more the realm of the dollar trade-offs involved in the insuring decisions.

Deductibles

As indicated earlier in this chapter, one goal of risk management is to eliminate the diseconomies of the "dollar trading" associated with the purchase of insurance with low deductibles. While this might be an unavoidable necessity to the firm which cannot stand the strain of a large self-assumed retention, there are premium savings available to the firm which can assume a portion of its own risk.

The first concept in a discussion of deductibles for chemical risks is that of the minimum deductible. Certain underwriters, based on their own experience, will not insure chemical risks unless the deductible is at least a certain amount. This may be a dollar deductible, waiting period for time element coverage, or both.

As the firm can assume risk above this minimum deductible, there are rate credits (reductions) allowed to the base rate (to be discussed later) which can lower insurance costs. As the high frequency of loss is in the low severity portion of the loss curve, there are higher credits available per $1000 of

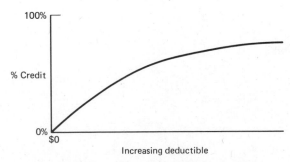

Figure 24.4. Typical deductible credit curve.

retention initially which become progressively smaller with increasing self-retention (Fig. 24.4).

There has been a general increased use of higher deductibles in the chemical industry in the past decade. This, plus competitive pressures, has led to declining average rate levels.

24.10. RATE

The rate (cents per $100 insured value) that is charged to insure a chemical risk is comprised of the following components:

1. Loss cost
2. Underwriting expenses
3. Taxes and other Assessments
4. Commissions

It is apparent from the discussion of deductibles that it is only the loss cost component which is affected by deductibles for a given risk. To the extent that the "normal" losses are removed by a deductible or other self-retention, the loss cost component is reduced to the catastrophic burning rate for the class as a whole.

It is also apparent that an improvement in the risk which reduces this loss expectancy should cause a reduction in rate. Conversely, any characteristic of the risk which leads the underwriter to anticipate a higher loss expectancy should result in an increased rate.

In practice, the underwriter defines a standard risk to which is associated a base rate for the occupancy involved. In the case of the IRI, this means a risk in substantial compliance with the guidelines provided by the "Guiding Principles."

The risk at hand is then compared to this standard risk and adjustments are made in the so-called "base rate" reflecting the positive or negative aspects of a given risk.

The base rate is based upon the loss expectancy of the standard risk in the occupancy of the risk being rated. If the protection criteria for the standard risk is high (e.g., Guiding Principles) the base rate is low and downward adjustments are small for additional protection. However, large upward adjustments are possible to reflect the lack of loss prevention features.

24.11. RESTRICTIONS

There may be the occasion where the underwriter would be able to provide coverage for a facility, except for one particular facet of the risk. In this case, rather than reject the entire risk, the underwriter will attempt to tailor the coverages so that the insured enjoys the maximum possible protection, while at the same time the underwriter retains that measure of safety which will permit the insuring of the bulk of this exposure.

Examples of this type of approach would be:

1. Sublimit of Liability on a poorly protected portion of the risk.
2. Exclusion of coverage on a particularly dangerous operation.
3. Specific application of high deductibles.
4. Requirements for the provision of a protective feature prior to the granting of insurance coverage.

There are many other approaches. The underwriter seeks to use restrictions only where necessary in order to provide a product that is competitive in the insurance marketplace, and serves the needs of the chemical industry.

24.12. HISTORY

The cycles of the insurance industry are, in a very simple sense, caused by the interaction of industry loss experience and the economy as a whole. In periods devoid of catastrophic loss and relatively high interest rates, the insurance industry gains surplus. This surplus gain yields additional insuring capacity, either in the form of existing companies increasing their commitment to underwrite risks or the creation of new companies.

From 1977 to 1981 the industry was in this growth mode. Of course, as in any other industry, expanding capacity yields competitive forces between underwriters which tends to depress rate levels without regard to the intrinsic character of the risk.

This was not always the case. The recent past has witnessed several periods of the inability of the insurance industry to fulfill the needs of insuring chemical risks. In the aftermath of the disaster at the Cities Service Refinery in Lake Charles, Louisiana in 1967 and other severe losses, most particularly in the polyvinyl chloride subsegment of the industry, many insurers and reinsurers (the insurers of insurance companies whose function is to spread the risk) flatly refused to provide capacity for chemical risks. This led to much competition for available risk transfer markets with the consequence of increased rates and more selectivity on the part of underwriters. This also led to the creation of other underwriting and risk transfer entities.

More recently, the disastrous loss experience of the 1974–1976 period resulted in some shrinkage of capacity and heightened underwriting standards. However, this was not nearly as restrictive a period as early ones due to a general expansion of the insurance industry in the mid-1970s.

24.13. FUTURE CONSIDERATIONS

As discussed above, these cycles in the insurance industry tend to be driven by the three-pronged pressure of rates reduced by competitive pressures, interest rate fluctuation, cycles in catastrophic loss experience. Inasmuch as these variables will fluctuate, the financial managers of chemical risks should prepare themselves for periods of relatively less available insuring capacity by building into their risks the highest level of loss prevention possible.[11]

As the Chemical Industry moves to the next generation of yet larger world scale plants utilizing the most modern technology, the demand to provide financial protection for the investors in the form of insurance will be great. Underwriters will be able to satisfy the desire for risk transfer so long as the chemical industry takes those steps to maintain their exposures at a manageable level for Insurers.[12,13]

24.14. SUMMARY AND CONCLUSION

The perspective of the chemical industry to insurers is one of capital intensive, high-technology risks presenting high loss exposure with attendant high premium opportunities. An examination of one underwriter's experience reveals the basis for risk evaluation. To the extent that risks meet the standards set by the analysis of loss experience, the underwriting decisions as to capacity, deductible credit and rate are made.

Risks in the chemical industry present unusual, but not insurmountable, challenges to the underwriter. The cooperation between the chemical and insurance industries has been great, and is increasing in the face of heightened awareness of risk and exposure by both industries. This trend will undoubtedly continue and will serve as the focal point of increased knowledge in the field of chemical risk loss prevention.

GLOSSARY

BUSINESS INTERRUPTION: Coverage which is designed to reimburse a business property owner for specified fixed costs and profits resulting from damage caused by an insured peril.

CAPTIVE INSURANCE COMPANY: A captive insurance company is an insurance company formed for the purpose of insuring the loss exposures of the firm which establishes it. A captive insurance company may also be established by a group of companies, usually in the same line of business, to underwrite the exposures of all participants.

DEDUCTIBLES: A policy provision that the insurer will pay only that amount of any loss which is in excess of a specified amount. May be either (1) a specified dollar amount, or (2) a specified percentage of the loss or of the value of the property, or (3) a specified waiting period, usually in hours, for time element losses.

Example: With a $10,000 deductible, gross loss of $100,000 results in a net insurance recovery of $90,000.

EXPOSURES: In property insurance, the neighboring premises which may cause loss to the property because of perils or losses in them. In all lines of insurance, the amount of loss the company could conceivably pay under the terms of the policy.

PERIL: Cause of loss.

Example: Fire, lightning, explosion.

PROPERTY DAMAGE: A term ordinarily including fire insurance and inland marine insurance; covers loss to the property of the insured as contrasted with liability coverage.

RISK POOLING: An organization made up of a group of companies that join voluntarily or involuntarily in order to supply insurance to one or more persons or businesses.

SELF-RETENTION: A method of handling risk by the firm itself. This retention may be passive or active. The retention is passive when the firm is not aware that the risk exists and consequently does not attempt to handle it. A related form of passive retention occurs when the firm has properly recognized the risk but has underestimated the magnitude of the potential losses. A firm actively retains the risk when it considers other methods of handling it and consciously decides to pay the potential losses out of its own resources.

SELF-INSURANCE: A form of self-retention in which the firm chooses to account for contingencies with funded reserves or other financial mechanisms.

Example: A firm may choose to fund fire losses for a large number of small warehouses (where values do not exceed $100,000 per unit) based upon a predictable loss frequency.

SUBSIDIARY INSURANCE COMPANY: A captive insurance company which is a subsidiary of the firm which establishes it.

GLOSSARY REFERENCES

R. B. Holtom, *Underwriting Principles and Practices,* The National Underwriter Company, Cincinnati, Ohio, Glossary, pp. 690–691.

Wm. H. Rodda, J. S. Trieschmann, B. A. Hedges, *Commercial Property Risk Management and Insurance*—Volume I, Captive Insurance Companies, pp. 98–99.

REFERENCES

1. H. S. Robinson, "Loss Risks in Large Integrated Chemical Plants," *Loss Prev.*, **1**, 1967, pp. 18–22.

2. "Hazard Survey of the Chemical and Allied Industries," Technical Survey No. 3; American Insurance Association, New York, 1979.

3. W. H. Doyle, "Industrial Explosions and Insurance," *Loss Prev.*, **3**, 1969, pp. 11–17.

4. J. A. Davenport, "A Survey of Vapor Cloud Incidents," *Chem. Eng. Prog.* (September 1977).

5. R. W. Nelson, "Know Your Insurer's Expectations," *Hydrocarbon Process.* (August 1977).

6. *A Review of Catastrophic Property Damage Losses—30 Years—Worldwide—Oil, Gas, Chemical and Petrochemical Industries,* M&M Protection Consultants, Chicago, Ill., 1981.

7. S. R. Brinkley, "Determination of Explosion Yields," *Loss Prev.*, **3**, 1969, pp. 79–82.

8. W. C. Brasie and D. W. Simpson, "Guidelines for Estimating Damage Explosion," *Loss Prev.*, **2**, 1969, pp. 91–102.

9. T. A. Kletz, "Unconfined Vapor Cloud Explosions," *Loss Prev.*, **11**, 1977, pp. 50–58.

10. *Loss Prevention and Protection for Chemical and Petrochemical Plants,* 2nd ed., Industrial Risk Insurers, Hartford, Conn., 1980.

11. T. A. Kletz, "Plant Layout and Location: Methods for Taking Hazardous Occurrences into Account," *Loss Prev.*, **13**, 1980, pp. 147–153.

12. W. Wohleben and F. Vahrenholt, "Precautions Against Accidents in Chemical Facilities," *J. Hazardous Mater.* 5 (1–2), 41–48 (1981).

13. R. A. Cox, "Improving Risk Assessment Methods for Process Plant," *J. Hazardous Mater.* **6**(3), 249–260 (1982).

25

Respiratory Hazards
and Protection

Howard H. Fawcett

Humans, plants, and animals are addicted to "air," a unique mixture of gases that is the least appreciated but most essential and valuable basic resource of planet Earth. It has been recognized as elemental by the Ancient Greeks, who held air, along with water, earth, and fire, as basic to life. Without air, life is measured in minutes. As the term is commonly used, "air" is approximately 78% by volume nitrogen, 21% by volume oxygen, less than 1% by volume of other gases including helium, argon, neon, krypton, xenon, and usually small amounts of carbon monoxide, carbon dioxide, sulfur dioxide, oxides of nitrogen, ozone, and moisture as well as particulates (aerosols).

We note these ingredients which normally comprise "air" to stress that even "air" fully suitable for human consumption usually contains varying concentrations of many substances in addition to the essential life-sustaining oxygen.

In the real world, "air" is constantly being modified or changed by numerous agents, processes and activities, including natural forces such as precipitation which contributes moisture; the winds, which are carriers of dusts, pollen, seeds, spores, salt and other particulates; changes or stagnation in atmospheric pressures which contribute to air movement or lack of movement (stagnation or lapse conditions); lightning which contributes ozone and oxides of nitrogen; volcanic activity (such as the 1980 eruption of Mt. St. Helens in Washington State) which contributes sulfur dioxide and ash; decay of radioactive nuclides of long half-life which contribute radon gas and other "daughter" products[1]; fire (from both natural and person-related causes) which contributes carbon dioxide, carbon monoxide, and particulates in smoke; biological processes, such as fermentation and decay, which contribute carbon dioxide, and occasionally, carbon monoxide, hydrogen sulfide, and methane; by action of bacteria, enzymes, and other agents; and the sun, contributing the ultraviolet, visible and infrared radiation. Solar radiation is an important ingredient of this complex mixture by providing light and heat, the energy for plant photosynthesis (important in the oxygen–carbon dioxide balance), as well as in catalyzing the action of oxidizers, such as ozone and oxides of nitrogen on hydrocarbon gases and vapors with particulate to synthesize "smog" (smoke and organic matter). Gases dissolved in the water of natural springs and spas evolve carbon dioxide (carbonated water) and hydrogen sulfide (mislabeled "mineral waters or radium waters" in some locations). Temperature extremes, in which air is too cold or too hot to breathe without tempering, may also contribute to respiratory stress.[1-3] Table 25.1 reflects the composition of unpolluted air.

To this limited inventory of natural sources which contribute to atmospheric contamination must be added products from the many activities of humans. For our purposes, it may suffice to state that most humans can tolerate a significant deviation from "pure breathing air," regardless of the source of the contamination, since most humans have amazing powers and ability to adjust or to tolerate "insults." It is only when the summation of the various "pollutants" in the breathing zone reaches an "intolerable" level that human life is threatened by respiratory hazards, whether of artificial or natural origin.[4] Serious effects may be observed in seconds from a few lungsful of some gases, such as arsine or phosgene, or decades after chronic exposures to an aerosol containing Bis-dichlorodiethyl ether or to asbestos.[5]

Aerosols, consisting of solids or liquids suspended in air, are characterized in Table 25.1. It is recognized that the human system in generally good health has remarkable tolerance or ability to cope with assaults from

TABLE 25.1.   The Gaseous Composition of Unpolluted Air (Dry Basis)

	Percent (Vol)	Parts per Million (Vol)
Nitrogen	78.09	780,900
Oxygen	20.94	209,400
Water	—	—
Argon	0.93	9,3001
Carbon dioxide	0.03	325
Neon	Trace	18
Helium	Trace	5.2
Methane	Trace	1.0–1.2
Krypton	Trace	1.0
Nitrous oxide	Trace	0.5
Hydrogen	Trace	0.5
Xenon	Trace	0.08
Organic vapors	Trace	~0.02

Source.   A. C. Stern, H. C. Wohlers, R. W. Boubel, and W. P. Lowry, *Fundamentals of Air Pollution,* Academic, New York, 1973, p. 21.

inhalation insults, but this tolerance has an upper limit beyond which injury will eventually result. Hence the term "threshold," applied to limits above which airborne contaminants may be expected to produce injury or ill effects of some type. However, this "threshold" is not a go no-go concept. For a small, but very real percentage of persons, who constitute a high-risk population, including the very young, the very old, and persons with severe respiratory diseases, such as bronchial asthma, chronic bronchitis, pulmonary emphysema, or pneumonia, exposures which may be "permissible" or "acceptable" to most healthy persons may have serious effects.

Respiratory hazards have been studied over a century, especially in Germany and England, but even today much is not completely known.[6] We are limited in our understanding about the synergistic or combined effects of air contaminants. We do know that several substances have enhanced toxicity when inhaled with a "carrier." However, studies of "real-world" air, including the concentrations of the various gases, vapors and particulates encountered in daily exposures, especially when aggavated by smoking, alcohol, and physiologically active aerosol drugs are seldom directly applicable to an individual, because our data base is usually too incomplete.[7-9]

Due to the importance of aerosols to respiratory studies, working definitions of certain terms may be useful:

PARTICLES: any minute piece of solid or liquid. Many particles that are important in studies of air pollution are unstable—they can change or even disappear on contact with a surface. Examples are a raindrop striking a surface and coalescing, a loose aggregate of carbon black disintegrating on contact with a surface, and an ion losing its charge after contact with a surface or an oppositely charged particle.

AEROSOLS: a mixture of gases and particles that exhibit some stability in a gravitational field. In atmospheric aerosols, this gravitational stability excludes particles[10,11] with a diameter greater than several hundred micrometers. See Table 25.2 for names and characteristics of aerosol particles.

Under the Federal Clean Air Act, standards have been set for six air pollutants: carbon monoxide, sulfur dioxide, nitrogen dioxide, particulate matter, gaseous hydrocarbons, and photochemical oxidants as ozone. In California, standards are also set for lead, hydrogen sulfide, and "visibility reducing particles."

The main class of chemically identified carcinogens in the air is the polycyclic aromatic hydrocarbons (PAHs). Benzo(2)pyrene (BP) is the PAH which has been most widely studied, but it is not recognized as either a complete or conclusive indicator of the carcinogenic potential of pollution, since BP is not a carcinogen as such. It must be metabolically activated to a chemically reactive electrophilic species. An enzyme system believed involved in these changes is the aryl hydrocarbon hydroxylase (AHH) system.[10]

TABLE 25.2. Names and Characteristics of Aerosol Particles

Name	Unique Physical Characteristics	Effects	Origin	Predominant Size Range (μm)
Coarse particle			Mechanical process	>2
Fine particle			Condensation	<2
Dust	Solid	Nuisance, ice nuclei	Mechanical dispersion	>1
Smoke	Solid or liquid	Health and visibility	Condensation	<1
Fume	Solid	Health and visibility effects	Condensation	<1
Fog	Water Droplets	Visibility reduction	Condensation	2–30
Mist	Water Droplets	Visibility reduction; cleanse air	Condensation or atomization	5–1000
Haze	Exists at lower RH than fog—hygroscopic	Visibility reduction		<1
Aitken or condensation nuclei (CN)		Nuclei for condensation at supersaturation <300%	Combustion, atmospheric chemistry	<0.1
Ice nuclei (IN)	Have very special crystal structure	Cause freezing of super-cooling water droplets	Natural dusts	>1
Small ions	Stable particle with an electric charge	Carry atmospheric electricity	All sources	>0.0015
Large particles	Special name			0.1–1
Giant particles	Special name			>1

Eye irritation is one of the more conspicuous and obvious effects of the aerosols which constitute "smog." Olefins are especially reactive when oxides of nitrogen react photochemically with the hydrocarbons to produce "smog." Solvents play a major role in such formation, and have been rated for their tendency to produce a smog of eye irritation capability at 20 ppm and at 5 ppm concentrations. At the 20 ppm level, the following ranking of substances has been reported (in decreasing order of irritation):

methyl isobutyl ketone

trichloroethylene

xylene

methyl ethyl ketone

hydrocarbon fractions (six representative samples with boiling ranges from 110°–264°F to 355°–488°F constituted the hydrocarbon fractions in this research)

methanol

toluene

hexane

ethanol

isopropanol

It is recognized that the composition of gases and aerosols in the "air" may be measured and monitored by well-established analytical methods, and hence a cause–effect relationship of relative safety or potential hazards, such as oxygen deficiency or carbon monoxide concentrations, can be performed relatively easily. On the other hand, most particulate or aerosol matter is measured by its total suspended particulate content, which includes all filterable particles. Molecular and ionic species are often lumped together, precluding cause–effect studies where chemical composition is important. For this reason alone, one must examine carefully any analytical data on air for its implications as "breathing air."

Often overlooked is that aerosols may travel thousands of miles and affect local conditions. For example, during September and October, 1960, radioactive debris or "fallout" particles from open-air nuclear weapon testing in Nevada caused a highly significant increase in the radiation levels of the Northeast, including the tri-city area of Albany, Schenectady, and Troy, New York. Recently, the World Climate Conference in Geneva heard reports that suggested smog and dust from industrial Europe and China may account for the haze that persists over Alaska, Greenland, and the Arctic Ocean every spring. "Acid rain" is another example.

To appreciate the complexity of evaluating the actions and effects[12,13] of respiratory hazards, and protection from them, whether from "air pollu-

Particle Size	Nasal Penetration
microns	%
1.0	100
1.2	100
1.5	78
2.0	57
2.5	41
3.0	31
3.5	26
4.0	22
5.0	17

Figure 25.1.   Particle size vs. nasal penetration.

tion'' or from "industrial airborne contamination," a short description of the human respiratory system is in order. Breathing "air" normally enters the body through the nostrils. Larger particles or aerosols may be retained on the unciliated anterior portion of the nose, while other particles or aerosols, as well as the gases, pass through a web of nasal hairs, then they flow through the narrow passages around the turbinates. The inspired "air" is warmed, moistened, and partially depleted of particles with aerodynamic diameters greater than 1 μm by sedimentation and impaction on nasal hairs and passages (Fig. 25.1). The beating action of the cilia propels the inspired "air" towards the pharynx. Deposited insoluble particles are transported by the mucus; soluble particles may dissolve in it.

Particles inhaled through the nose and deposited in the nasopharynx or particles inhaled through the mouth and deposited in the mouth and oropharynx are swallowed in minutes, into the gastrointestinal tract.

The inspired "air" now proceeds down the trachea, or windpipe, which divides into two branches, or primary bronchi leading in turn to upper and to low lobar bronchi for the right and left (Fig. 25.2). The airway diameter decreases, but the number of tubes increase, the total cross section for flow increases and the air velocity decreases.

At the low velocities in the smaller airways, particles deposit by sedimentation and diffusion. Inert nonsoluble particles deposited on normal ciliated airways are cleaned within one day by transport on the moving mucus to the larynx. Soluble particles are cleaned much faster, presumably by bronchial blood flow.[14]

Gas exchange occurs in the acini of the lung parenchyma, or the portions of the lungs from the first order of respiratory bronchioles down to the alveoli. These respiratory bronchioles originate from the terminal bronchioles which are the smallest airways not concerned with gas exchange. The system of airways leading to the acini does not participate in the gas exchange and is called the "dead space." Inhaled particles may be deposited either in the lung parenchyma (the bronchioles, atrial sacs, and alveoli) or in the dead space. Smaller inhaled particles may be breathed in and out of the

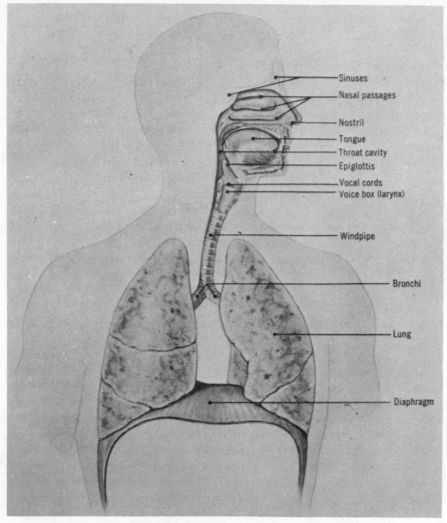

Figure 25.2. The respiratory tree.

respiratory tract without deposition.[15] Examples of respiratory hazards and the protective devices, by classes, which protect against them, are shown in Fig. 25.3.

25.1. DETECTION OF ATMOSPHERIC CONTAMINANTS

Odor is the simplest sense for detecting certain contaminants in "air," since "pure air" should smell "clean" or "odorless." Odor is not a dependable criteria of air quality, however, since it is often misleading or unreliable.

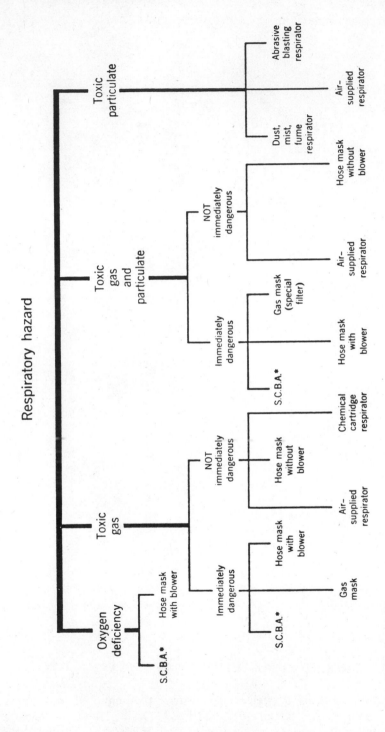

Figure 25.3. Outline for selection of respiratory protective services.

*S.C.B.A. = Self-contained breathing apparatus

Certain substances such as acrolein, ammonia, bromine, chlorine, formaldehyde, acetic acid, acetic anhydride, sulfur dioxide and trioxide, and hydrogen chloride have distinctive odors at relatively low concentrations. Some gases and vapors, such as fluorine and hydrogen fluoride, are so immediately corrosive to the upper respiratory tract that odor is not a safe warning. Certain gases, such as arsine, phosphine, stibine, hydrogen cyanide, nitric oxide, and hydrogen sulfide may be above permissible "safe" inhalation limits before a "fresh" nose can respond. The ability of several of these gases including hydrogen sulfide to cause olefactory fatigue is well recognized. The olfactory nerves become rapidly overcome, and can no longer sense changes in concentrations. A concentration of 1000 ppm hydrogen sulfide causes loss of consciousness in seconds from a few breaths (Table 25.3d). This ability of hydrogen sulfide to rapidly overcome detection is doubtlessly a factor in the continuing fatal exposures which have occurred over the years from exposure to this gas.[16,17]

At the other end of the scale of odors, carbon monoxide is, for all practical purposes, odorless and colorless. Inhalation can often result in significant, and even fatal, consequences (Table 25.3a). The action of CO is due to its ability to compete with oxygen for hemoglobin binding sites in the blood, and subsequent transfer to the body tissues. The degree of competition of oxygen versus CO is reflected in that CO has an affinity for the hemoglobin complex that is over 200 times greater than the oxygen–hemoglobin affinity. Once carboxyhemoglobin is formed, a considerable amount of time is required for the complex to dissociate. Until this happens, oxygen binding or transport is impossible. Thus, the main effect of increased blood CO levels is tissue hypoxia, or oxygen deficiency.

As the tissue demand for oxygen increases, increased stress on the heart is experienced because of an increased cardiac output. Therefore, the heart is the first target organ during CO inhalation; this is especially significant for those individuals who may already be experiencing cardiovascular difficulty. The other organ system most rapidly affected by hypoxia is the central nervous system (CNS).

As the carboxyhemoglobin level increases, a progression of symptoms is noted. Nausea and headache symptoms begin at blood carboxyhemoglobin levels of 15%. At levels of 25% changes in the electrocardiogram occur, and at 40% unconsciousness will usually ensue. Blood carboxyhemoglobin levels of 66% or higher are usually considered fatal, although death has resulted from extended exposures at percentage levels in the thirties, such as during attempted escape from fire gases which may contain high percentages of CO.

After intense, high-level exposure to CO has ceased, several pathologic events can occur. The most serious is the development of cerebral edema, which can be life-threatening if left untreated. This excessive accumulation of fluid in the brain is a result of increased permeability of the capillaries due to the change in oxygen tension. Within several days, confusion and other signs of mental deterioration can ensue. Serious mental deficits that are often

TABLE 25.3. Symptoms of Exposures to Selected Gases

Effect	Concentration (ppm)
a. Carbon Monoxide [a]	
Slight headache in some cases	0–200
After 5–6 hr, mild headache, nausea, vertigo, and mental symptoms	200–400
After 4–5 hr, severe headache, muscular incoordination weakness, vomiting, and collapse	400–700
After 3–5 hr, severe headache, weakness, vomiting, collapse	700–1000
After 1.5–3 hr, coma, breathing still fairly good unless poisoning has been prolonged	1100–1600
After 1–1.5 hr, possibly death	1600–2000
After 2–15 min, death	5000–10,000
b. Ammonia [a]	
Least detectable odor	53
Least amount causing immediate irritation to the eyes	698
Least amount causing immediate irritation to the throat	408
Least amount causing coughing	1720
Maximum concentration allowable for prolonged exposure	100
Maximum concentration allowable for short exposure (½-1 hr)	300–500
Dangerous for even short exposure (½ hr)	2500–4500
c. Sulfur Dioxide [a]	
Least amount causing detectable odor	3–5
Least amount causing immediate eye irritation	20
Least amount causing immediate throat irritation	8–12
Least amount causing coughing	20
Max. concn. allowable for prolonged exposure	10
Max. concn. allowable for short (30 min) exposure	50–100
Amount dangerous for even short exposure	400–500
d. Hydrogen Sulfide [b]	
Eye and respiratory tract irritation after exposure of 1 h	50–100
Marked eye and respiratory tract irritation after exposure of 1 h	200–300
Dizziness, headache, nausea, etc., within 15 min; loss of consciousness and possible death after 30–60 min exposure	500–700
Rapidly produces unconsciousness and death occurs a few minutes later	700–900
Death is apparently instantaneous	1000–2000

Apparent death from H_2S is not irreversible as prompt and efficient artificial respiration may restore life

[a]*Sources.* Kirk-Othmer, *Encyclopedia of Chemical Technology,* 2nd ed., Vol. 2, p. 291; Vol. 4, p. 442; Vol. 19, p. 417. Wiley, New York, 1980.

[b]*Source. International Oil Tanker and Terminal Safety Guide,* The Institute of Petroleum, London, England, 1980.

irreversible may result from prolonged intense exposure. In addition, transient cardiac arrhythmais and enzyme elevations may also develop. The danger of asphyxiation in the unconscious person is increased by such serious complications as aspiration pneumonia and laryngeal edema.[18,a,b]

While hydrogen sulfide and carbon monoxide are extremely serious problems, other gases are of concern. Sulfur dioxide (Table 25.3c), oxides of nitrogen, and ammonia (Table 25.3b) are also serious potential gases insofar as respiration is concerned, the first two acidic in nature, and the third, ammonia, basic. If an individual is occupationally exposed to these materials, the entire respiratory tree is affected, with reactions ranging from sneezing and coughing to severe bronchoconstriction and cessation of respiration. Because of this induced change in normal respiratory physiology, cardiovascular responses which may follow include increased blood pressure and an increase in pulse rate.

Acute, intense exposure may also produce pulmonary edema, an increase in intercellular and interstitial fluid. This edematous conditon may result in a variety of physiologic malfunctions, including decreased lung compliance, hypoxemia, ventilation/perfusion mismatch, and respiratory alkalosis followed by metabolic acidosis.

As continuous exposure to these irritant gases increase, definite clinical disease patterns may be seen, such as bronchiolitis, bronchitis, and pneumonia as well as adverse changes in the teeth, eyes, and skin. An individual who is experiencing respiratory disease and who continues to be exposed to these irritant gases may ultimately experience irreversible respiratory pathology.

Table 25.3, parts b, c, and d, present a summary of some of the inhalation data regarding these gases.

The effect of exposure to gases is enhanced by factors such as heavy labor, high environmental temperature, and increased altitude (over 2000 ft). Susceptibility is greatest in the aged, the very young, those with cardiac or chronic respiratory disease, and pregnancy.

The odor intensity and physiological response produced by paraffin hydrocarbons have been studied in terms of potential warning of possible concentration buildup leading to an explosive mixture with air. The odors of heptane and hexane are easily noticed in concentrations below their lower flammable limits; heptane and hexane vapors produce distinct symptons. The odor of pentane is indistinct, and that of ethane and propane practically absent in lower flammable limit mixtures.

Historically, animals such as canaries and Japanese waltzing mice have been used as warning for "bad air" in mines and other confined operations. The Davy mine safety lamp, an indicator of oxygen deficiency or methane atmospheres, has been largely displaced by portable as well as fixed-station instrumentation to detect and record low or changing concentrations of gases, vapors, dusts, and aerosols. The availability of gas detector tubes for use with personnel samplers worn by workers has been a major advance. In addition, personnel gaseous "film badges" have been developed and are in use for monitoring exposures to many gases and vapors.

The "Odor Threshold Manual" of the American Standards for Testing and Materials is an excellent compilation of published data on odors.[19] It must always be recognized that some persons have virtually no ability to detect an odor, while others have a high sensitivity, which may produce serious physiological as well as emotional effects.[20]

25.2. PROTECTION

To prevent or reduce exposures to airborne substances, the substances should be confined to closed systems, or removed from the breathing zone by properly designed vents or fume hoods, and moved to outside the building away from air intakes and other inhabited areas, after proper dilution or filtration or scrubbing to reduce the level of discharged effluent at the stack to an acceptable level. The use of filters and other pollution control devices should be carefully engineered.[21] To illustrate, discharge to the roof level may be satisfactory most of the time, but if the wind-directional rose indicates that, even for a small percentage of time, that the effluent air may be recycled or reach a populated area, serious concern should be given the air movements. In one specific operational problem, ethylene dibromide was being exhausted from an animal inhalation laboratory at a concentration of 20 ppm. Since it was near a populated area, it was desired to reduce the stack concentration to 1 ppm. Activated charcoal bed filters were installed into the vent system at a cost of $10,000, part of which was for changes made to the duct system to accommodate the filters. The filters performed successfully, and frequent monitoring was instituted to insure that saturation or breakthrough did not occur between the scheduled monthly filter replacements.

If confinement or removal by adequate properly engineered and operated ventilation systems is not possible, personal protective equipment on a temporary basis should be considered. We stress the word "temporary" since respiratory protection can seldom be relied on for long periods of time in hazardous exposures, unless highly unusual control procedures are established and rigorously enforced. Where the risks are known to be high, as in the possibility of breathing plutonium or dioxin, or with aerosols containing carcinogens, programs of respiratory protection have been successfully instituted, but in areas or situations where less obvious hazards exist, respiratory protection is successful only if well-planned and implemented programs of education, maintenance, and enforcement are in place.[22,23]

25.3. AIR SUPPLY

Where air-supplied respiratory protection is used, the very important problem of the purity of the breathing air must be considered. In the past, fatal accidents have occurred when an air-supplied respirator was plugged into a plant air supply, which had, without warning, been changed to a nitrogen supply while work was being done on the compressor. In other cases, cylinders of "breathing air" were, in fact, filled with nitrogen, carbon dioxide, and butane. In some instances, "air" has been produced by mixing compressed oxygen with compressed nitrogen, but all such mixtures should be suspect until it can be shown that they do, in fact, contain a mixture which is within the recognized limits for breathing purposes. While oxygen deficiency is the more obvious potential error, the problems associated with oxygen above 25%, from a fire hazard viewpoint, should not be overlooked. Air-supplied hoods are frequently used for short duration tasks; a typical example is shown in Fig. 25.4.

Legal standards exist for air supply from a compressor: OSHA specification 1910.94 (6) notes that the air for abrasive-blasting respirators shall be free of harmful quantities of dusts, mists, or noxious gases, and shall meet the requirements for air purity set forth in ANSI Z9.2-1960. The air from the

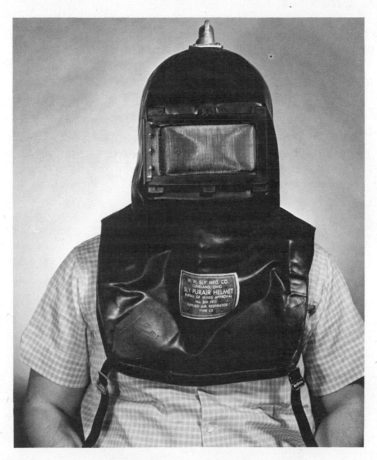

Figure 25.4. Air supplied hood.

regular compressed air line of the plant may be used for the abrasive-blasting respirator if (1) a trap and carbon filter are installed, and regularly maintained, to remove oil, water, scale, and odor, (2) a pressure-reducing diaphragm or valve is installed to reduce the pressure down to requirements of the particular type of abrasive-blasting respirator, and (3) an automatic control is provided to either sound an alarm or shut down the compressor in case of overheating. In OSHA 1910. 134, the quality of "air" is specified. Oxygen shall meet the requirements of the United States Pharmacopoeia for medical or breathing oxygen.[24] Breathing air shall meet at least the requirements of the specification for Grade D breathing air as described in Compressed Gas Association Commodity Specification G-7.1-1966. Compressed oxygen shall not be used in supplied-air respirators or in open circuit self-contained breathing apparatus (SCBA) that have previously used compressed air. Oxygen must never be used with air-line hoods or respirators.[25] If a compressor is used, it shall be of breathing air type, constructed and situated so as to avoid entry of contaminated air into the system and suitable in-line air purifying sorbent beds and filters installed to further assure breathing air quality. If an oil-lubricated compressor is used, it shall have a high-temperature or carbon monoxide alarm, or both. If only a high temperature alarm is used, the air from the compressor shall be frequently tested for carbon monoxide to insure that it meets the specifications.[24] Where air is used for diving, the specifications of the U.S. Navy Diving Manual, *NAV-SHIPS 250-538*, should be followed. The CGA and Navy Standards are summarized below:

	Maximum Allowable Level	
Contaminant	Compressed Gas Association-Grade D	U.S. Navy
Carbon monoxide (ppm)	20	20
Carbon dioxide (ppm)	1000	500
Oil vapor (g/m³)	5	5

The stages of California, New Hampshire, New Jersey, New York, and Washington have even lower limits, and specify that SCBA air be free from odors and "other contaminants."

Classification and Description of Respiratory Protective Devices

Based on the mode of operation, several types of respiratory protective devices are available. Table 25.4 describes the general classification and the mode of operation. Basically, these are:

TABLE 25.4. Classification and Description of Respirators by Mode of Operation

Atmosphere-Supplying Respirators

A respirable atmosphere independent of the ambient air is supplied to the wearer.

Self-Contained Breathing Apparatus (SCBA)

Supply of air, oxygen, or oxygen-generating material carried by wearer. Normally equipped with full facepiece, but may be equipped with a quarter-mask facepiece, half-mask facepiece, helmet, hood, or mouthpiece and nose clamp.

1. Closed-Circuit SCBA
   (oxygen only)
   a. Compressed or liquid oxygen type equipped with a facepiece or mouthpiece and nose clamp. High-pressure oxygen from a gas cylinder passes through a high-pressure reducing valve and, in some designs, through a low-pressure admission valve to a breathing bag or container. Liquid oxygen is converted to a low-pressure gaseous oxygen and delivered to the breathing bag. The wearer inhales from the bag, through a corrugated tube connected to a mouthpiece or facepiece and a one-way check valve. Exhaled air passes through another check valve and tube into a container of carbon dioxide removing chemical and reenters the breathing bag. Make up oxygen enters the bag continuously or as the bag deflates sufficiently to actuate an admission valve. A pressure relief system is provided and a manual bypass system and saliva trap may be provided depending upon the design.
   b. Oxygen-generating type. Equipped with a facepiece or mouthpiece and nose clamp. Water vapor in the exhaled breath reacts with chemical in the canister to release oxygen to the breathing bag. The wearer inhales from the bag through a corrugated tube and one-way check valve at the facepiece. Exhaled air passes through a second check valve breathing tube assembly into the canister. The oxygen release rate is governed by the volume of exhaled air. Carbon dioxide in the exhaled breath is removed by the canister fill.

2. Open-Circuit SCBA (compressed air, compressed oxygen, liquid air, liquid oxygen). A bypass system is provided in case of regulator failure except on escape-type units.
   a. Demand type.[a] Equipped with a facepiece or mouthpiece and nose clamp. The demand valve permits oxygen or air flow only during inhalation. Exhaled breath passes to ambient atmosphere through a valve(s) in the facepiece.
   b. Pressure-demand type.[b] Equipped with facepiece only. Positive pressure is maintained in the facepiece. The wearer may have the option of selecting the demand or pressure-demand mode of operation.

Supplied-Air Respirators

1. Hose mask, equipped with a respiratory-inlet covering (facepiece, helmet, hood, or suit), nonkinking breathing tube, rugged safety harness and a large diameter heavy-duty nonkinking air supply hose. The breathing tube and hose are securely attached to the harness. A facepiece is equipped with an exhalation valve. The harness has provision for attaching a safety line.

TABLE 25.4. *(Continued)*

a. Hose mask with blower. Air is supplied by a motor-driven or hand-operated blower. The wearer can continue to inhale through the hose if the blower fails. Up to 300 ft (91 m) of hose length is permissible.

b. Hose mask without blower. The wearer provides motivating force to pull air through the hose. The hose inlet is anchored and fitted with a funnel or like object covered with a fine mesh screen to prevent entrance of coarse particulate matter. Up to 75 ft (23 m) of hose length is permissible.

2. Air-Line Respirator. Respirable air is supplied through a small-diameter hose from a compressor or compressed air cylinder(s). The hose is attached to the wearer by a belt and can be detached rapidly in an emergency. A flow control valve or orifice is provided to govern the rate of airflow to the wearer. Exhaled air passes to the ambient atmosphere through a valve(s) or opening(s) in the enclosure (facepiece, helmet, hood or suit). Up to 300 ft (91 m) of hose length is permissible.

a. Continuous-flow class. Equipped with a facepiece, hood, helmet or suit. At least 115 liters (4 ft^3) of air per minute to tight-fitting face-pieces and 170 liters (6 ft^3) of air per minute to loose-fitting helmets, hoods and suits are required. Air is supplied to a suit through a system of internal tubes to the head, trunk and extremities through valves located in appropriate parts of the suit.

b. Demand type.[a] Equipped with a facepiece only. The demand valve permits flow of air only during inhalation.

c. Pressure-demand type.[b] Equipped with a facepiece only. A positive pressure is maintained in the facepiece.

Combination Air-Line Respiratory with Auxillary Self-Contained Air Supply

Includes an air-line respirator with an auxiliary self-contained air supply. To escape from a hazardous atmosphere in the event the primary air supply fails to operate, the wearer switches to the auxiliary self-contained air supply. Devices approved for both entry into and escape from dangerous atmospheres have a low pressure warning alarm and contain a self-contained air supply.

Air-Purifying Respirators

Ambient air, prior to being inhaled is passed through a filter, cartridge, or canister which removes particles, vapors, gases, or combination of these contaminants. Breathing action of wearer operates non-powered type. Powered type contains a blower, stationary, or carried by wearer, which passes ambient air through an air-purifying component and then supplies purified air to the respiratory-inlet covering. Nonpowered type is equipped with a facepiece or mouthpiece and nose clamp. Powered type is equipped with a facepiece, helmet, hood, or suit.

Vapor- and Gas-Removing Respirator

Equipped with cartridge(s) or canister(s) to remove a single vapor or gas (for example: chlorine gas), a single class of vapors or gases (for example: organic vapors), or a combination of two or more classes of vapors or gases (for example: organic vapors and acid gases) from air.

TABLE 25.4. *(Continued)*

Particulate-Removing Respirator

Equipped with filter(s) to remove a single type of particulate matter (for example: dust) or a combination of two or more types of particulate matter (for example: dust and fume) from air. Filter may be a replaceable part or a permanent part of the respirator. Filter may be single-use type or reuseable type.

Combination Particulate-, Vapor, and Gas-Removing Respirator

Equipped with cartridge(s) or canister(s) to remove particulate matter, vapors, and gases from air. The filter may be a permanent part or a replaceable part of a cartridge or canister.

Combination Atmosphere-Supplying and Air-Purifying Respirators

An atmosphere-supplying respirator with an auxiliary air-purifying attachment is to provide protection in the event the air supply fails. An air-purifying respirator with an auxiliary self-contained air supply which is used in case the atmosphere unexpectedly exceeds safe conditions for use of an air-purifying respirator.

[a] Equipped with a demand valve that is activated on initiation of inhalation and permits the flow of breathing atmosphere to the facepiece. On exhalation, pressure in the facepiece becomes positive and the demand valve is deactivated.

[b] A positive pressure is maintained in the facepiece by a spring-loaded or balanced regulator and exhalation valve.

The capabilities and limitations of various types of respirators are listed in Table 25.5 and discussed in detail below.

1. Atmosphere-supplying (or air-supplying) devices.
   a. Self-contained (SCBA).
   b. Supplied air.
   c. Combination self-contained and supplied air.
2. Air purifying respirators.
   a. Gas and vapor.
   b. Particulate (aerosols including dust, fog, fume, mist, smoke, and sprays).
   c. Combination gas, vapor, and particulate.
3. Combination atmosphere-supplying and air-purifying respirators.

The self-contained breathing apparatus is generally understood to furnish the maximum protection against respiratory hazards. It is designed to supply complete respiratory protection (except for gases or vapors which may penetrate the skin). When used in atmospheres where gases or vapors with significant action of toxicity through the skin is involved, such as hydrogen cyanide, the respiratory protective device must be supplemented by complete skin protection of an impervious type, in addition to respiratory protection. Until a full recognition and agreement on a standard for testing impervious clothing, suspicion of the effectiveness of protection should al-

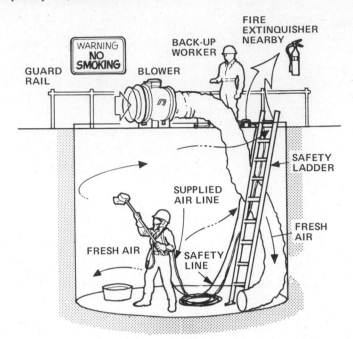

Figure 25.5. Entry into confined spaces should include SCBA. Photograph courtesy of *Plant Engineering*.

ways be raised, especially when details of what is involved and the concentration may be lacking. (See Chapter 27.)

Since the SCBA has no external connection to an air supply, it is obviously the device of choice for emergency situations, and entry into confined spaces (Fig. 25.5).

The first self-contained breathing apparatus to be approved by the U.S. Bureau of Mines in 1916 was the 2-hr rebreathing type. It was intended for mine rescue work, and is still used today for that purpose (Fig. 25.6). Originally a mouth breathing facepiece with a nose clip to seal the nose was used, but in more recent times a full facepiece for these devices has been approved. When the NIOSH was formed in 1971, the approval system which had been with the U.S. Bureau of Mines (and the Department of Agriculture) was gradually moved to NIOSH, and the certification of devices became an activity at NIOSH's laboratory in Morgantown, W. Va. The early mine-rescue devices carried a carbon dioxide absorbing chemical to remove the carbon dioxide from the exhaled air, and a cylinder of oxygen to enrich the CO_2-free exhalation gas for rebreathing (note Fig. 25.6).

The most widely used SCBA today is a demand open circuit type, consisting of an air cylinder, an air-line and regulator, and facepiece (Fig. 25.7). Depending on the size and pressure in the cylinder, the supply may last from 5 to 30 min. In recent times, it has been recognized that the demand principle, i.e., air was available only when the wearer inhaled, and was cut off

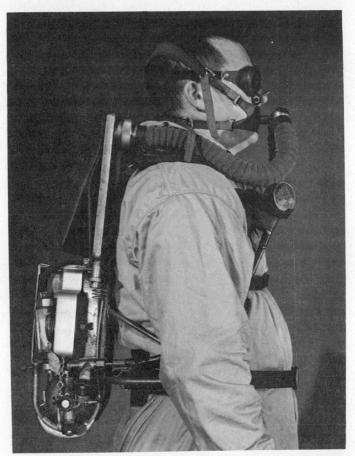

Figure 25.6. Mine-rescue rebreathing apparatus, the Mc Caa, approved by Bureau of Mines in 1916. Note absorbing CO_2 canister and oxygen cylinder carried on back. Photograph courtesy of General Electric Research Laboratory.

when the wearer exhaled (often creating a negative pressure in the facepiece which could encourage leakage from the hazardous external atmosphere) is less desirable than a positive pressure device, which has a continuous flow of air into the breathing zone to insure against negative pressure at any time. The positive pressure device has now been recommended for the fire services and any other exposure where extreme care must be taken to prevent toxic gases from entering the facepiece. SCBA devices are usually of 30-min duration, but smaller devices with air supplies of 5, 10, 15, and 20 min have been available at various times, and may have specific application for quick entry and inspection purposes.[25] The impact of the NASA space program development on the SCBA should be noted, since the aluminum cylinder, wound with fiberglass reinforcement, is significantly lighter than the traditional steel cylinder, and is coming into wide use.

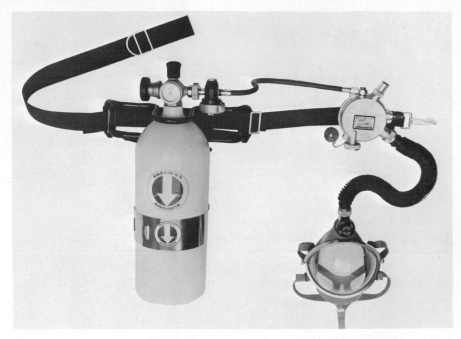

Figure 25.7. Components of a self-contained breathing apparatus (SCBA): full facepiece, regulator, air cylinder. Photograph courtesy of U.S. Divers.

Closed circuit SCBA have evolved and developed from the mine-rescue models, and are now available with up to a 4-hr service life. Closed circuit equipment obtains this long life from the oxygen supply of a liquid, gaseous, or chemical solid. During World War II, an oxygen breathing apparatus (OBA) was developed, using potassium superoxide as the oxygen source, which also adsorbed carbon dioxide.[26] These devices were further developed, and were rated at times of 30, 45, and 60 min (Fig. 25.8). A chlorate starting "candle" was added to aid in starting, especially in very cold weather.[27] In this writer's opinion, the devices using the superoxide demand much more careful training, especially in cold climates. The very real explosion potential of a carelessly discarded canister in contact with oil or other combustibles must not be overlooked. Once a canister has been used, even for a short period, it should be promptly and carefully destroyed, and not used for "practice" or "training." The disposal must follow exactly the instructions, if a fire or explosion is to be avoided.

Supplied-Air Respirators

If a limited area is to be entered, and a hose or other connection is not a problem, air-supplied respirators or hoods have real advantages. The hose

Figure 25.8. A self-contained breathing apparatus using chemical solid oxygen source, potassium superoxide. This is a "re-breather" type. Some models can be started by a "candle." Photograph courtesy of General Electric Research Laboratory.

mask, which has been used for years for certain operations where long-time is important consists of a large diameter hose, anchored in uncontaminated air and connected to the facepiece. If the diameter is sufficiently large, and the work requirements not excessive, the wearer experiences little resistance to breathing (Fig. 25.8). For some applications, a positive blower may be included, either manually operated by another person (who also serves as a back-up or buddy) or by an electric powered fan or blower, using power from a completely dependable source. The maximum flow of approved air masks is 150 liters/min. This flow will not maintain a positive pressure within the facepiece during periods of exertion by the wearer. Powered hose masks should be considered negative pressure devices.

Air-line respirators deliver breathing air to the facepiece from a source

Figure 25.9. Air line respirators may be supplied from cylinder "breathing air." Note wearer on left has "escape" cylinder for entry or escape. Photograph courtesy of U.S. Divers.

which supplies pressure. Normally, a compressor is used to supply the air, but systems may be engineered by which "breathing air" in cylinders may be used (Fig. 25.9). They may be demand, pressure demand, or continuous flow. The demand device has an airflow regulator between the facepiece and the supply, so inhalation negative pressure activates the diaphragm, which opens the air supply. The negative pressure required to activate the air supply may be as much as 50 mm of water, and unless a constant flow type is used, the possibility of facepiece leakage may be real. The facepieces may be quarter, half, or full facepieces. NIOSH certification requires that a minimum of 115 liters/minute be delivered to the facepiece on demand. One objection to air-line respirators, in addition to the limiting area, is the noise of the air passing into the facepiece.

Combination Self-Contained and Supplied Air

For situations where prolonged times are necessary, and back-up escape essential, the SCBA may be supplemented with an air-line respirator, which can be plugged in or unplugged rapidly, and which serves as the air supply when engaged, thus conserving the breathing air in the cylinder.

Air-Purifying Respirators

If the air to be encountered is known to be within the limits of the protection factor (see Table 25.6) for air-purifying devices and has an adequate oxygen content (over 19%), an air-purifying respirator may be used. It must be noted that the unit will protect only against the specific substance or combinations in certain concentrations for which it was designed (for example, aerosol filter-type respirators will afford no protection against gases and vapors), and that the canister or filter mask must be maintained in proper condition attached to a properly maintained facepiece which fits snugly all over the area of contact to preclude leakage of contaminated air. (Fig. 25.10) In addition, the very serious question can arise of one person using another person's respirator, of inadequate instruction, of vague records of canister use, and of the unknown endpoint as to when a material has begun to penetrate the canister. In one specific case which occurred during an inspection of a carcinogen bioassay laboratory in which respiratory protection was required for entrance to the area, a visitor was furnished a respirator of the chemical cartridge type which had been used by other employees. During the inspection, which required 3 h, the visitor was exposed to unknown (but supposedly small) concentrations of numerous suspect carcinogens, including ethylene chlorohydrin. No data were available as to how rapidly ethylene chlorohydrin penetrates a chemical cartridge respirator. It may or may not have been coincidence that the visitor suffered acute pulmonary edema that same evening, resulting in congestive heart failure, and seven months of disability. Unless very careful control is maintained of a respirator program, such incidents can occur.

Aerosol filter-type respirators require frequent changing in filters when breathing resistance becomes uncomfortably high when worn in areas containing excessive dust, mists, or fumes. Wearer acceptance will be a problem unless great care is taken to instruct, properly fit, and maintain devices. Recently OSHA has ruled that respirators need not be worn continuously during exposures to cotton dusts, but only to the extent necessary to reduce the total daily exposure level to the limits accepted. This rule was a direct result of the protest of workers who felt the respirators were too uncomfortable to wear for the whole work period.

Chemical cartridge respirators will safely protect only against the specific gases and vapors for which they were certified in nonemergency situations. Exposures to extremely toxic materials, such as acrolein, acrylonitrile,

TABLE 25.5. Capabilities and Limitations of Respirators

Atmosphere-Supplying Respirators

Atmosphere-supplying respirators provide protection against oxygen deficiency and toxic atmospheres. The breathing atmosphere is independent of ambient atmospheric conditions.

General Limitations: Except for some air-line suits, no protection is provided against skin irritation by materials such as ammonia and hydrogen chloride, or against sorption of materials such as hydrogen cyanide, tritium, or organic phosphate pesticides through the skin. Facepieces present special problems to individuals required to wear prescription lenses. Use of atmosphere-supplying respirators in atmospheres immediately dangerous to life or health is limited to specific devices under specified conditions.

Self-Contained Breathing Apparatus (SCBA)

The wearer carries his own breathing atmosphere.

Limitations: The period over which the device will provide protection is limited by the amount of air or oxygen in the apparatus, the ambient atmospheric pressure (service life is cut in half by a doubling of the atmospheric pressure), and work. Some SCBA devices have a short service life (less than 15 min) and are suitable only for escape (self-rescue) from an irrespirable atmosphere. Chief limitations of SCBA devices are their weight or bulk or both, limited service life, and the training required for their maintenance and safe use.

1. Closed-Circuit SCBA. The closed-circuit operation conserves oxygen and permits longer service life at reduced weight. A negative pressure in the respiratory-inlet covering is created during inhalation in most closed-circuit devices and may permit inward leakage of contaminants.
2. Open-Circuit SCBA. Demand and pressure-demand. The demand type produces a negative pressure in the respiratory-inlet covering during inhalation whereas the pressure-demand type maintains a positive pressure in the respiratory-inlet covering during inhalation and is less apt to permit inward leakage of contaminants.

Supplied-Air Respirator

The respirable air supply is not limited to the quantity the individual can carry and the devices are light weight and simple.

Limitations: Limited to use in atmospheres from which the wearer can escape unharmed without aid of the respirator.

The wearer is restricted in movement by the hose and must return to a respirable atmosphere by retracing route of entry. The hose is subject to being severed or pinched off.

1. Hose Mask. The hose inlet or blower must be located and secured in a respirable atmosphere.
   a. Hose mask with blower. If the blower fails, the unit still provides protection, although a negative pressure exists in the facepiece during inhalation.
   b. Hose mask without blower. Maximum hose length may restrict application of device.

TABLE 25.5. *(Continued)*

2. Air-Line Respirators (Continuous Flow, Demand, and Pressure-Demand Types)

The demand type produces a negative pressure in the facepiece on inhalation whereas continuous flow and pressure-demand types maintain a positive pressure in the respiratory inlet covering and are less apt to permit inward leakage of contaminants.

Air-line suits may protect against atmospheres that affect the skin or mucous membranes or that may be absorbed through the unbroken skin.

Limitations: Air-line respirators provide no protection if the air supply fails. Some contaminants, such as tritium, may penetrate the material of an air-line suit and limit its effectiveness.

Other contaminants, such as fluorine, may react chemically with the material of an air-line suit and damage it.

Combination Airline Respirator With Auxiliary SC Air Supply

The auxiliary self-contained air supply on this type device allows the wearer to escape from a dangerous atmosphere. This device with auxiliary self-contained air supply is approved for escape and may be used for entry when it contains at least a 15-m auxiliary self-contained air supply.

Air-Purifying Respirators

General Limitations: Air-Purifying respirators do not protect against oxygen-deficient atmospheres nor against skin irritations by, or sorption through the skin of airborne contaminants.

The maximum contaminant concentration against which an air-purifying respirator will protect is determined by the design efficiency and capacity of the cartridge, canister, or filter and the facepiece to face seal on the user. For gases and vapors, the maximum concentration for which the air purifying element is designed is specified by the manufacturer, or is listed on labels of cartridges and canisters.

Nonpowered air-purifying respirators will not provide the maximum design protection specified unless the facepiece or mouthpiece/nose clamp, is carefully fitted to the wearer's face to prevent inward leakage. The time period over which protection is provided is dependent on canister, cartridge, or filter type, concentration of contaminant, humidity levels in the ambient atmosphere, and the wearer's respiratory rate.

The proper type of canister, cartridge, or filter must be selected for the particular atmosphere and conditions. Nonpowered, air-purifying respirators may cause discomfort due to a noticeable resistance to inhalation. This problem is minimized in powered respirators. Respirator facepieces present special problems to individuals required to wear prescription lenses. These devices do have the advantage of being small, light, and simple in operation.

Use of air-purifying respirators in atmospheres immediately dangerous to life or health is limited to specific devices under specified conditions.

TABLE 25.5. *(Continued)*

Vapor- and Gas-Removing Respirators

Limitations: No protection is provided against particulate contaminants. A rise in canister or cartridge temperature indicates that a gas or vapor is being removed from the inspired air.

An uncomfortably high temperature indicates a high concentration of gas or vapor and requires an immediate return to fresh air.

Should avoid use in atmospheres where the contaminant(s) lacks sufficient warning properties (that is: odor, taste, or irritation).

1. Full Facepiece Respirator. Provides protection against eye irritation in addition to respiratory protection.
2. Half and Quarter Mask Respirator. Not for use in atmospheres immediately dangerous to life or health unless a powered type respirator with escape provisions. A fabric covering (facelet) available from some manufacturers shall not be used.
3. Mouthpiece Respirator. Shall be used only for escape applications. Mouth breathing prevents detection of contaminant by odor. Nose clamp must be securely in place to prevent nasal breathing.

A small, lightweight device that can be donned quickly.

Particulate-Removing Respirators

Limitations: Protection against nonvolatile particles only. No protection against gases and vapors.

1. Full Facepiece Respirator. Provides protection against eye irritation in addition to respiratory protection.
2. Half and Quarter Mask Respirator. Not for use in atmospheres immediately dangerous to life or health unless a powered type respirator with escape provisions.
3. Mouthpiece Respirator. Shall be used only for escape applications. Mouth breathing prevents detection of contaminant by odor. Nose clamp must be securely in place to prevent nasal breathing.

A small, lightweight device that can be donned quickly.

Combination Particulate- and Vapor- and Gas-Removing Respirators

The advantages and disadvantages of the component sections of the combination respirator as described above apply.

Combination Atmosphere-Supplying and Air-Purifying Respirators

The advantages and disadvantages, expressed above, of the mode of operation being used will govern. The mode with the greater limitations (air purifying mode) will mainly determine the overall capabilities and limitations of the respirator since the wearer may for some reason fail to change the mode of operation even though conditions would require such a change.

Source. ANSI Z88.2-1980, Standards for Respiratory Protection.

554

Figure 25.10. Canister air purifying mask with canister worn on back. Especially useful for exposures where wearer must be in close proximity to gases or vapors, as in tank rodding. Photograph courtesy of General Electric Research Laboratory.

aniline, dimethylaniline, arsine, diborane, hydrogen cyanide, methylfluorosulfate, carbon disulfide, phosphine, and other gases and vapors with high toxicity and low warning odor levels or unknown characteristics should not be assumed safe when a chemical cartridge respirator is worn. Chemical cartridge respirators should not be used against gases and vapors which are odorless, or whose odor threshold is high, since odor is the only in place warning of failure of the canister, facepiece leakage or of concentrations which are above the protection factor limits of the respirator. Methyl chloride is an example of a gas whose warning properties are inadequate for practical warning purposes. Substances which are highly irritating to the eyes, such as sulfur dioxide and ammonia, as noted in Tables 25.3b and c, require eye protection such as a full facepiece, gas-tight goggles or an air-supplied hood. Several lacrimatory (tear-producing) substances, such as benzyl chloride, are in the same classification, and a respirator alone is clearly inadequate protection. Carbon monoxide cannot be stopped by a chemical cartridge, except by the type known as Type N universal mask (a seriously misleading term), and the miner's self-rescuer unit, both of which contain Hopcalite, a mixture of oxides of manganese, copper, cobalt, and silver. This mixture catalyzes the oxidation of carbon monoxide to carbon dioxide, provided it is properly activated, is dry, and the carbon monoxide concentration is not excessive. These qualifications have lead to serious problems in the past, and for fire fighting, and other services where *unknown* concentrations of carbon monoxide and other substances may be present, the Type N mask is no longer recommended.

Combination cartridges in masks and in respirators have the advantage that they provide some protection against more than one class of hazard, but the serious disadvantage that the shorter service life for any particular hazard makes them less desirable.

Industrial canister-type gas maks, with canisters designed for specific substances or specific combinations, have the same limitations in general as chemical cartridge respirators, except that they are effective in concentrations of any specific gas or vapor, or members of the group of gases or vapors for which they were designed, of not more than 2% concentration in air or a 2% total concentration for a mixture of gases and vapors for which the canister is designed. The industrial-type ammonia canister mask is certified for a higher concentration.

Depending on the size of the canister, and the service for which it was designed, the service time varies. Protection against a combination of various gases and vapors, such as acid gases, organic vapors and ammonia can all be obtained in one canister, but the service life of such a combination is shorter then when used with one substance, that is, the life of an equivalent canister designed for the specific substance alone.

The only warning that a canister or cartridge is "exhausted" or spent usually is sensory detection of vapor or gas passing through the canister or cartridge to the wearer's nose. For carbon monoxide, which is odorless, canisters may include a timer or color indicator window. The self-rescuer, using the Hopcalite, does not have this warning feature. The only tangible indication that high concentrations are being encountered is the temperature increase, since the action of the CO on the catalyst is highly exothermic. Unless we have a complete knowledge of the actual concentrations of what material are encountered, there may be serious false sense of security in using canister-type or cartridge-type masks and respirators.

Persons who need corrective lenses in order to work properly when wearing respiratory protection, especially full facepieces, should consider incorporation of corrective lenses inside a facepiece in such a manner that the facepiece seal is not compromised. One method consists of wire frames which fit around the circumference of the facepiece sight area and hold 50-mm round lenses. A second approach suspends wire-frame goggles with

Figure 25.11.   Prescription glasses attached to post inside facepiece where lenses are required. Photograph courtesy of General Electric Research Laboratory.

short bows between holders molded into the facepiece. A third approach is a center post built inside the facepiece, on which can be attached 40 mm rimless glasses. Such arrangements seldom encourage perfect alinement of the prescription lenses to the viewer's eyes, but, with adjustment, provide a sufficiently accurate fitting for most persons (Fig. 25.11). Contact lenses offer a possible solution, but the hazards of contact lenses in increasing or aggravating a chemical burn in the eye, especially if the lenses are not immediately removed before irrigation with water, must be recognized. It is the writer's belief that persons with serious limitation to their vision without lenses should not rely on respiratory devices which compromise their sight.

Protection Factor

The protection factor (PF) is a relatively new concept which has introduced new understanding to the respiratory field. The work largely was done with the development group at Los Alamos which became concerned about this problem as related to protection from highly toxic aerosols, such as plutonium dust. Edwin C. Hyatt has published the basis for the PF.[28] The formula for calculation of the PF is

$$\text{Protection factor (PF)} = \frac{\text{ambient air concentration}}{\text{concentration inside facepiece}} \text{ or enclosure}$$

The PF concept is based largely on the premise that one must know how efficient the device is in actual use, and how much protection it really delivers. Over the years it has been observed that many wearers of respirators deliberately slip the straps so the face seal is loose, and permits leakage, or may even entirely negate the device as suggested by Fig. 25.12. For these reasons, we emphasize the importance of the fit test to insure that the wearer has an adequate seal all around the face, and that the device, be it air-purifying or air-supplied, is functioning as it was designed. The original American designs, facial measurements, and relationships of facial features were based on American male anthropometric panels in World War I, when military gas masks were being developed after the first use of chlorine and phosgene at Ypres, Belgium in 1916. The proper fitting of the current wearers of respiratory protective devices has been made more acute by the increasingly high percentage (41%) of women workers, as well as the influx of workers from Asia whose measurements and proportions differ significantly from the American male norms. The fact is that many facepieces have not fitted properly, and any effective respiratory program must assign fitting a high priority. Hyatt found a wide variation between different facepieces, respiratory efficiency, and protection factors. Protection factors were noted as low as 5 (for single use and quarter masks for dusts) to as high as 10,000 for open circuit pressure demand full facepiece self contained breathing apparatus. The protection factors from the ANSI Z88.2-1980 standard are shown in Table 25.6.

Fit Tests

Two types of fit tests have been recognized. In the qualitative test, a person wearing a respirator is exposed to an irritant smoke (such as titanium tetrachloride smoke), to an odorous vapor (such as isoamyl acetate vapor), or other test agent. An air-purifying respirator must be equipped with one or more air-purifying elements which effectively remove the test agent from inside the facepiece. If the wearer is unable to detect penetration of the agent into the facepiece, the wearer is judged as having a satisfactory fit. It should be noted that persons vary widely in their ability to sense or smell odors, and this factor should be considered before giving a complete endorsement of the qualitative test.

In the quantitative fit test, which is the test recommended by NIOSH, the

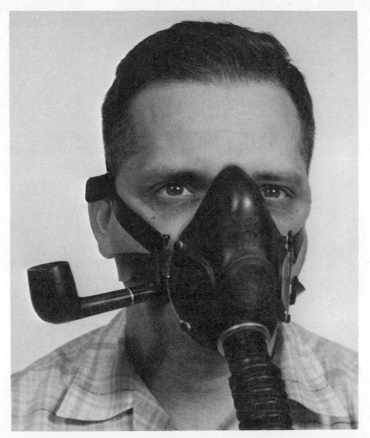

Figure 25.12. Breathing protection may be negated by improper introduction of foreign objects. Photograph courtesy of General Electric Research Laboratory.

wearer enters an atmosphere containing a test aerosol or gas, such as polydisperse sodium chloride aerosol, polydisperse DOP (dioctylphthlate) or a gas such as dichlorodifluoromethane (F-12). Questions have been raised about the safety of the use of DOP aerosol, and it is likely this test agent will not be used widely in the future.

Test chambers and detailed procedures for both tests are detailed in ANSI Z88.2-1980, Appendix A6. The exercises used insure that the respiratory device is, in fact, providing proper protection under all conditions likely to be encountered at work.

Education and Training

It is recognized that respirators are not a comfortable addition to the face.[29] The importance of the respirator as temporary protection is significant. If the physical condition of the wearer is good, no respiratory or cardiovascular

TABLE 25.6. Respirator Protection Factors[a]

Type of Respirator	Permitted for Use in Oxygen Deficient Atmosphere	Permitted for Use in Immediately Dangerous to Life or Health Atmosphere[f]	Respirator Protection Factor	
			Qualitative Test	Quantitative Test
Particulate-filter, quarter-mask or half-mask facepiece [b,c]	No	No	10	As measured on each person with maximum of 100 if dust, fume, or mist filter is used, or maximum of 3000 if high efficiency filter is used, whichever is less
Vapor- or gas-removing, quarter-mask or half-mask facepiece[c]	No	No	10, or maximum use limit of cartridge or canister for vapor or gas, whichever is less	As measured on each person with maximum of 3000, or maximum use limit cartridge or canister for vapor or gas, whichever is less
Combination particulate-filter and vapor- or gas-removing, quarter-mask or half-mask facepiece [b,c]	No	No	10, or maximum use limit of cartridge or canister for vapor or gas, whichever is less	As measured on each person with maximum of 100 if dust, fume, or mist filter is used, maximum of 3000 if high efficiency filter is used, or maximum use limit of cartridge or canister for vapor or gas, whichever is less

Particulate-filter, full facepiece[b]	No	No	100	As measured on each person with Maximum of 100 if dust, fume, or mist filter is used, or maximum of 3,000 if high efficiency filter is used, whichever is less
Vapor- or gas-removing, full facepiece	No	Yes, if concentration of hazardous substance does not exceed product of respirator PF and permissible concentration, and/or maximum use limit of cartridge or canister for vapor or gas	100, or maximum use limit of cartridge or canister for vapor or gas, whichever is less	As measured on each person with a maximum of 3,000, or maximum use limit of cartridge or canister for vapor or gas, whichever is less.
Combination particulate-filter and vapor- or gas-removing, full facepiece[b]	No	No	100, or maximum use limit of cartridge or canister for vapor or gas, whichever is less	As measured on each person with maximum of 100 if dust, mist or fume filter is used, maximum of 3,000 if high efficiency filter is used, or maximum use limit of cartridge or canister for vapor or gas, whichever is less.
Powered particulate-filter, any respiratory-inlet covering[b,c,d]	No	No (yes, if escape provisions are provided[a])	N/A	N/A

No tests are required due to positive pressure operation of respirator; the maximum protection factor is 100, if dust, fume or mist filter is used, or 3,000 if high efficiency filter is used, whichever is less

TABLE 25.6. (Continued)

Type of Respirator	Permitted for Use in Oxygen Deficient Atmosphere	Permitted for Use in Immediately Dangerous to Life or Health Atmosphere[f]	Respirator Protection Factor	
			Qualitative Test	Quantitative Test
Powered vapor- or gas-removing, any respiratory-inlet covering[b,c,d]	No	No (yes, if escape provisions are provided[a])	N/A	No tests are required due to positive pressure operation of respirator; the maximum protection factor is 3000, or maximum use limit of cartridge or canister for vapor or gas, whichever is less.
Powered combination particulate-filter and vapor- or gas-removing, any respiratory-inlet covering[b,c,d]	No	No (yes, if escape provisions are provided[a])	N/A	No tests are required due to positive pressure operation of respirator, the maximum protection factor is 100 if dust, fume or mist filter is used, 3000 if high efficiency filter is used, or maximum use limit of cartridge or canister for vapor or gas, whichever is less
Air-line, demand, quarter-mask or half-mask facepiece, with or without escape provisions[c,e]	No	No	10	As measured on each person with maximum of 3000, whichever is less
Air-line, demand, full-facepiece, with or without escape provisions[c,e]	No	No	100	As measured on each person with maximum of 3000, whichever is less
Air-line, continuous flow or pressure-demand, any facepiece, without escape provisions[e]	No	No	N/A	No tests are required due to positive pressure operation of respirator; the maximum protection factor is 3000.

Air-line, continuous flow or pressure-demand, any facepiece, with escape provisions[c,e]	Yes[f]	Yes	N/A	No tests are required due to positive pressure operation of respirator; the maximum protection factor is 10,000 plus[a]
Air-line, continuous flow, helmet, hood, or suit, without escape provisions	No	No	N/A	No tests are required due to positive pressure operation of respirator; the maximum protection factor is 3000
Air-line, continuous flow, helmet, hood, or suit, with escape provisions[e]	Yes[f]	Yes	N/A	No tests are required due to positive pressure operation of respirator; the maximum protection factor is 10,000 plus[a]
Hose mask, with or without blower, quarter-mask or half-mask facepiece[c]	No	No	10	As measured on each person with maximum of 3000, whichever is less
Hose mask, with or without blower, full facepiece	No	No	100	As measured on each person with maximum of 3000, whichever is less
Self-contained breathing apparatus, open-circuit demand, or closed-circuit, quarter-mask, or half-mask facepiece[c]	Yes[f]	No	10	As measured on each person with maximum of 3000, whichever is less
Self-contained breathing apparatus, open-circuit demand or closed-circuit, full facepiece or mouthpiece/nose clamp[c]	Yes[f]	No (yes, if respirator is used for mine rescue)	100	As measured on each person with maximum of 3000, whichever is less

TABLE 25.6. (*Continued*)

Type of Respirator	Permitted for Use in Oxygen Deficient Atmosphere	Permitted for Use in Immediately Dangerous to Life or Health Atmosphere[f]	Respirator Protection Factor	
			Qualitative Test	Quantitative Test
Self-contained breathing apparatus, open-circuit pressure-demand, quarter-mask or half-mask facepiece, full facepiece, or mouthpiece/nose clamp[c]	Yes[f]	Yes	N/A	N/A
			No tests are required due to positive pressure operation of respirator. The maximum protection factor is 10,000 plus[g]	
Combination respirators not listed.	The type and mode of operation having the lowest respirator protection factor shall be applied to the combination respirator.			

Source. ANSI Z88.2-1980, Standards for Respiratory Protection.

[a] A respirator protection factor is measure of the degree of protection provided by a respirator to a respirator wearer. Multiplying the permissible time weighted average concentration or the permissible ceiling concentration, whichever is applicable, for a toxic substance, or the maximum permissible airborne concentration for a radionuclide by a protection factor assigned to a respirator gives the maximum concentration of the hazardous substance for which the respirator can be used. Limitations of filters, cartridges, and canisters used in air-purifying respirators shall be considered in determining protection factors.

[b] When the respirator is used for protection against airborne particulate matter having a permissible time weighted average concentration less than 0.05 milligram particulate matter per cubic meter of air or less than 2 million particles per cubic foot of air, or for protection against airborne radionuclide particulate matter, the respirator shall be equipped with a high efficiency filter(s).

[c] If the air contaminant causes eye irritation, the wearer of a respirator equipped with a quarter-mask or half-mask facepiece or mouthpiece and nose clamp shall be permitted to use a protective goggle or to use a respirator equipped with a full facepiece.

[d] If the powered air-purifying respirator is equipped with a facepiece, the escape provisions means that the wearer is able to breath through the filter, cartridge, or canister and through the pump. If the powered air-purifying respirator is equipped with a helmet, hood or suit, the escape provisions shall be an auxiliary, self-contained supply of respirable air.

[e] The escape provisions shall be an auxiliary, self-contained supply of respirable air.

[f] See definition for oxygen deficiency—immediately dangerous to life or health.

[g] The protection factor measurement exceeds the limit of sensitivity of the test apparatus. Therefore, the respirator has been classified for use in atmospheres having unknown concentrations of contaminants.

Note. Respirator protection factors for air-purifying type respirators equipped with a mouthpiece/nose clamp form of respiratory-inlet covering are not deficiencies exist, and the facepiece is properly fitted by tests, day-by-day implementation requires full understanding of the advantages of using the device. If the facial hair, such as the male beard, is allowed to grow more than two days (the average facial hair grows about 1/72 in./day), a proper seal may be difficult, if not impossible. Other excessive facial or head hair, including hair-dos which interfere with the proper use of respiratory protective devices must be avoided.

25.3. OTHER FACTORS AND CONSIDERATIONS

In implementing and maintaining an effective respiratory protection program, in addition to selecting, fitting, and maintaining the proper equipment for a particular work environment, other factors may be important. For example:

The cost differences between various types of devices.

The length of time a device must be worn and its relationship to the wearer's comfort and acceptance.

Availability of breathing air of known certified purity for air-line use, and for the refilling of cylinders for self-contained breathing apparatus.

The coordination of the respiratory protection with other personal protective equipment needed for the job, such as face shield, welding goggles, ear muffs and hard hats (Fig. 25.13).

Temperature extremes which may make it desirable to supply warmed or cooled air to the respiratory inlet covering.

Use of prescription lenses by workers who need proper corrective lenses to perform their assigned tasks.

Stability of the respiratory inlet covering on the face while performing certain movements, for example, a half-mask respirator with two-point strap suspension is more unstable on the face than a mask with four-point suspension; a full facepiece is more stable than a quarter or a half-mask.

Cartridge or canister capability and length of time it may be used before contaminant breakthrough can occur. (If test data are not available, the manufacturer of the device should be required to furnish it or an air supplied device used instead).

Communication between workers. The use of in-mask microphones or throat masks external to the mask should be considered, and either amplifiers or two-way radio equipment used where necessary (see H. H. Fawcett, "Speech Transmission Through Respiratory Protective Devices," *Am. Ind. Hyg. Assoc. J.*, **22**(3) 170–174 (June 1961).

Physical limitations and capabilities of each individual user, including work requirements versus physical characteristics.

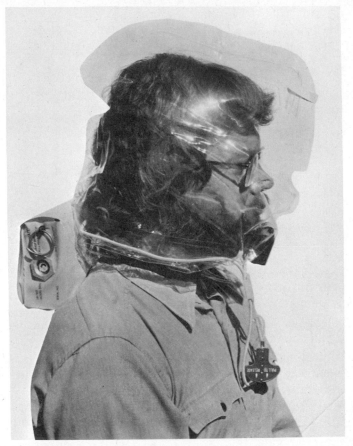

Figure 25.13.   An escape mask/hood which can be donned in seconds, supplying "breathing air" from high pressure coils in packet in rear. Usual life for escape is 5 minutes. Hood is tough plastic. Photograph courtesy of Robertshaw.

Vision restriction, mobility, and other safety factors. Provision for escape, as from the engine room of a ship, in case of serious fire emergency, should be considered. The mask/hood shown (Fig. 25.13) is an example of an escape device intended to be donned in seconds, and to give 5 min air supply for escape purposes.

REFERENCES

1a.   "The Radon Question: Government, Independent Studies Show Home Insulation May be Hazardous," *Consumers' Research Magazine,* **63**(11), 12–14 (November 1980).

1b.   "The Air You Breathe," EPA J. **4**(9) (October 1978). U.S. Environmental Protection Agency, Washington, D.C. 20460.

2. J. Dorigan, B. Fuller, and R. Duffy, "Scoring of Organic Air Pollutants: Chemistry, Production and Toxicity of Selected Synthetic Organic Chemicals," MITRE Technical Report MTR-7248, for U.S. Environmental Protection Agency, September 1976.

3. B. G. Ferris, Jr., "Literature Review of Health Effects of Air Pollution," American Thoracic Society, American Lung Association, New York, 1973.

4. McFadden, J. E., Beard, J. H. III, and Moschendreas, D., Survey of Indoor Air Quality, Health Criteria and Standards, Final Report, GEOMET EF-595, Aug. 1977, for U.S. Environmental Protection Agency and U.S. Housing and Urban Development, Washington, D.C.

5. J. H. Comroe, *Physiology and Respiration,* Yearbook Medical Publishers, Chicago, Ill., 1965.

6. "An Investigation of Self-Contained Breathing Apparatus for Use in Mines, Report of a Committee of the South Midland Coal Owners upon an Investigation Conducted in the Mining Department in the University of Birmingham into Self-Contained Breathing Apparatus for Rescue and Recovery Work in Mines after Underground Fires and Explosions," Cornish Brothers Ltd., Birmingham, England, 1910–1911.

7. Marshall Sitting, ed., "Priority Toxic Pollutants—Health Effects and Allowable Limits," Noyes Data Corp., Park Ridge, N.J., 1980.

8. Marshall Sittig, ed., "Hazardous and Toxic Effects of Industrial Chemicals," Noyes Data Corp., Park Ridge, N.J., 1979.

9. "Principles of Toxicological Interactions Associated with Multiple Chemical Exposures," Panel on Evaluation of Hazards Associated with Maritime Personnel Exposed to Multiple Cargo Vapors, Board on Toxicology and Environmental Health Hazards and Committee on Maritime Hazardous Materials, National Research Council, Washington, D.C., 1980.

10. "Particulate Polycyclic Organic Matter, Division of Medical Sciences," Committee on Biologic Effects of Atmospheric Pollutants, National Research Council, Washington, D.C. for U. S. Environmental Protection Agency, NTIS No. PB 212 940, 1972.

11. John M. Dement., Estimates of Pulmonary and Gastrointestinal Deposition for Occupational Fiber Exposures, DHEW (NIOSH) Pub. No. 79-135, April 1979.

12. Ember, L. R., "Acid Rain Focus of International Cooperation, *Chem. and Eng. News,* 15–17 (December 3, 1979).

13. J. E. McFadden and M. D. Koontz, "Sulfur Dioxide and Sulfates Materials Damage Study," GEOMET Report No. ES-812, for U. S. Environmental Protection Agency, Research Triangle Park, N.C., February 1980.

14. Ruth Porter, ed., Breathing: Hering-Breuer Centenary Symposium," A Ciba Foundation Symposium, J. & A. Churchill, London, 1970.

15. Atherley, G. R. C., *Occupational Health and Safety Concepts,"* Applied Science Publishers, London, 1978, pp. 25–55.

16. Fawcett, H. H., "Hydrogen Sulfide—Killer that may not stink," *J. Chem. Ed.* **9**(25) 511 (1948).

17. Smith, R. P. and Gosselin, R. E., "Hydrogen Sulfide Poisoning," *J. Occup. Med.* **21**(2), 93–97 (1979).

18a. National Academy of Sciences/National Research Council, "Carbon Monoxide: Effects of Chronic Exposure to Low Levels of CO on Human Health, Behavior and Performance," 1969, Washington, D.C. 20418.

18b. Lindgren, S. A., "A Study of the Effect of Protracted Occupational Exposure to Carbon Monoxide," *Acta Med. Scand. Stockholm,* Supplement 356(1961).

19. W. H. Stahl, ed. "Compilation of Odor and Taste Threshold Values Data," DS 48, publication Code Number 05-048000-36, American Society for Testing and Materials, Philadelphia, Pa., 1977.

20. Temple, T., "On the Cutting Edge," *EPA J.* **6**(9), 12–15 (October 1980).
21. Little, A. D., "Recommended Industrial Ventilation Guidelines," National Institute for Occupational Safety and Health, (NIOSH) 76-162, Cincinnati, January 1976.
22. W. E. Ruch and B. J. Held, *Respiratory Protection: OSHA and the Small Businessman,* Ann Arbor Science Publishers, Ann Arbor, 1975.
23. Douglas, D. D., "Respiratory Protective Devices," *Patty's Industrial Hygiene and Toxicology,* 3rd rev. ed. Vol. 1, G. D. Clayton and F. E. Clayton, eds., Wiley-Interscience, 1978, pp. 993–1057.
24. Federal Specification BB-A-1034a, "Air, Compressed for Breathing Purposes," June 21, 1968.
25. Interim Federal Specification GG-B-675d, "Breathing Apparatus, Self-contained," September 23, 1976.
26. Military Specification MIL-0-1563c, "Oxides, Oxygen Producing," September 25, 1964.
27. Military Specification MIL-E-83252, "Emergency Oxygen Supply, Chlorate Candle," Aircraft CRU-74/P, February 20, 1970.
28. Hyatt, E. C., "Respirator Protection Factors," Los Alamos Scientific Laboratory Report No. LA-6084-MS, January 1976.
29. "Bibliography of Journal Articles, Reports and Other Publications on Physiology and Psychology of Respirator Use, and Physiology and Psychology of Respirator Use," International Respirator Research Workshop, National Institute for Occupational Health and Safety, Morgantown, W. Va., September 8–11, 1980.

BIBLIOGRAPHY

"American National Standard Practices for Respiratory Protection," ANSI Z88.2-1980, and "Respiratory Protection for the Fire Services," ANSI Z88.5-1980, American National Standards Institute, New York, 1980.

Australian Standards CZ 11 and Z 18-1968, "Respiratory Protective Devices," Standards Association of Australia, North Sydney, N.S.W., 1968.

Bidstrup, P. L., *Toxicity of Mercury and Its Compounds,* Elsevier, Amsterdam, 1964.

Birkner, L. R., *Respiratory Protection: A Manual and Guideline,* Celanese Corporation, American Industrial Hygiene Association, Akron, 1980.

Braker, W. and Mossman, A. L., "Effects of Exposure to Toxic Gases—First Aid and Medical Treatment," Matheson Gas Products, E. Rutherford, 1970.

Buchanan, W. D., *Toxicity of Arsenic Compounds,* Elsevier Publishing Co., Amsterdam, 1962.

Burrows, B. *Respiratory Insufficiency,* Yearbook Medical, Chicago 1975

Comroe, J. H., Jr., *Lung: Clinical Physiology and Pulmonary Function Tests,* 2nd ed. Yearbook Medical, Chicago 1962.

Cotes, J. E., *Lung Function,* 3rd ed. Lippincott, Philadelphia, 1975.

Crofton, J. and Douglas, A., *Respiratory Diseases,* 2nd ed., Lippincott, Philadelphia, 1975.

Ebling, P. R., "Phosphorus, A Respiratory Poison," Dissertation, University of Cincinnati, Cincinnati, 1960.

Encyclopedia of Occupational Health & Safety, "Respiratory Protective Equipment," 1214–1220, "Respiratory System" (1220–1223), McGraw-Hill, New York 1972.

Ferris, B. G., Jr., "Literature Review of Health Effects of Air Pollution," American Thoracic Society, American Lung Association, New York, 1973.

Florent nee Jarlet Patricia, "Intoxications par les vapeurs de fuel lourd, Thèse pour le doctcrat en Medecine, University of Paris," 1976.

Flury, F. and Zernick, G., *Schadliche Gase* (Noxious Gases, Vapors, Mist, and Smoke-and-Dust Particles), Verlag von Julius Springer, Berlin, 1931.

Gerarde, H. W., *Toxicology and Biochemistry of Aromatic Hydrocarbons,* Elsevier, Amsterdam, 1960.

Geschickter, C. F., *Lung in Health and Disease,* Lippincott, Philadelphia, 1973.

Glass, N. R., Effects of Acid Precipitation, *ES&T,* **16,** no. 3, 162A–169A (March 1982). See also *An Updated Perspective on Acid Rain,* Edison Electric Institute, Washington, D.C., 1981.

Goldwater, L. J., *Mercury, A History of Quicksilver,* York Press, Baltimore, 1972.

Hack, A., Hyatt, E. C., Held, B. J., Moore, T. O., Richards, C. P., and McConville, J. T., "Selection of Respirator Test Panels Representative of U.S. Adult Facial Sizes," Los Alamos Scientific Laboratory LA-5488, U.C.41, reporting date December 1973, issued March 1974.

"Handbook of Hazardous Materials," Technical Guide No. 7, American Mutual Insurance Alliance, Chicago, 1974, or most recent supplement.

Heitzman, E. R., *Lung Radiologic—Pathologic Correlations,* Mosby, St. Louis, Mo., 1973.

Henderson, Y. and Haggard, H. W., *Noxious Gases,* American Chemical Society Monograph Series, American Chemical Society, Washington, D.C., 1943.

Hepple, P., *Lead in the Environment,* Institute of Petroleum, London, 1972.

Henry, Norman W. III, "Respiratory Cartridge and Canister Efficiency Studies with Formaldehyde, *Amer. Indust. Hyg. Asso. Journal,* **42**(12), 853–857 (Dec. 1981).

Hunsinger, J., *Respiratory Technology,* 2nd ed., Reston, Reston, Va., 1976.

Hyatt, E. C. and White, J. M., "Respirators and Protective Clothing, Safety Series No. 22, International Atomic Energy Agency, 1967.

Hygienic Guides, American Industrial Hygiene Association, Akron, Ohio 1964–1980.

Industrial Safety Data Sheets on Specific Chemicals, National Safety Council, Chicago, 1982.

"International Symposium, Environmental Health Aspects of Lead," Proceedings, Amsterdam, October 2–6, 1972, Commission of the European Communities, CID, Luxembourg, May 1973.

"Instrumentation for Environmental Monitoring, AIR, Part 2," 1st ed. Environmental Instrumentation Group, LBL-1 Vol. 1 Part 2, Lawrence Berkeley Laboratory, University of California, Berkeley, Calif. 1975.

James, R. H., "Breathing Resistance and Dead Space in Respiratory Protective Devices," DHEW (NIOSH) Pub. No. 77-161, October 1976.

Junod, A. F. and DeHaller, R., "Lung Metabolism: Proteolysis and Antiproteolysis, Biochemical Pharmacology, Handling of Bioactive Substances," *Proceedings of International Symposium,* Davos, Switzerland, October 1974, Academic Press, New York, 1976.

Kamon, E., "Cooling Efficiency of Different Air Velocities in Hot Environments," DHEW (NIOSH) Pub. No. 79-129, March 1979.

"Lead, Environmental Health Criteria 3," World Health Organization, Geneva, 1977.

Labows, J. N., Jr., "What the Nose Knows: Investigating the Significance of Human Odors, *The Sciences* 10–13, (November 1980).

Lee, K. P. and Trochimowicz, H. J., Induction of Nasal Tumors in Rats Exposed to Hexamethylphosphoramide (HMPA) by Inhalation, *J. Nat. Cancer Inst.,* **68**(1), 157–171 (Jan. 1982)

Leithead, C. S. and Lind, A. R., *Heat Stress and Heat Disorders,* F. A. Davis Co., Philadelphia, 1974.

Lindsay, D. B. and Stricoff, R. S., *A Feasibility Study of the State-of-the-Art of Personnel Monitors,* Arthur D. Little, Inc., Cambridge, Mass. 02140, DOT-CG-73211-A, 1978, NTIS Number AD A 072992, National Technical Information Service, Springfield, Va., 1978.

"Mercury, Environmental Health Criteria 1," World Health Organization, Geneva, 1976.

NAS-NRC, "Physiological and Toxicological Aspects of Combustion Products," International Symposium, March 18–20, 1974, Committee on Fire Research, National Research Council, Washington, D.C., 1976.

Publications of the National Academy of Sciences–National Research Council, Washington, D.C.

"Chlorine and Hydrogen Chloride," 1976.

"Fluorides," 1971.

"Selenium," 1976.

"Lead—Airborne Lead in Perspective," 1972.

"Nickel," 1975.

"Manganese," 1973.

"Chromium," 1974.

"Carbon Monoxide: Effects of Chronic Exposure to Low Levels of CO on Human Health, Behavior, and Performance" 1969.

No$_x$(nitrogen oxides) Emission Controls for Heavy-Duty Vehicles: Toward Meeting a 1986 Standard, 1981.

Montague, K. and Montague P., *Mercury,* Sierra Club, San Francisco, 1971.

NFPA *Hazardous Chemical Data,* "49," *Manual of Hazardous Chemical Reactions,* 491 M, National Fire Protection Association, Quincy, Mass., 1975.

Publications of the National Institute for Occupational Safety and Health, Cincinnati, Ohio: Pritchard, J. A., "A Guide to Industrial Respiratory Protection," DHEW (NIOSH) Publication No. 7, 6-189, June 1976.

"Abrasive Blasting Respiratory Protective Practices," NIOSH 74-104

"An Air-Supplied Respirator for Underground Coal Miners," CONTR HSM-99-71-43

"Anthropometry for Respirator Sizing," CONTR HSM-99-71-11

"Breathing Resistance and Dead Space in Respiratory Protective Devices," NIOSH 77-161

"Design Specifications for Respiratory Breathing Devices for Firefighters," NIOSH 76-121

"Determination of Respirator Filtering Requirements in a Coke Oven Atmosphere," CONTR CDC-99-74-88

"Development of a Prototype Service Life for Organic Vapor Respirators," NIOSH 78-170

"Development of Improved Respirator Cartridge and Canister Test Methods," NIOSH 77-209

"Engineering Control of Welding Fumes," NIOSH 75-115

"An Evaluation of Organic Vapor Respirator Cartridges and Canisters Against Vinyl Chloride," NIOSH 75-111

"Evaluation of the NIOSH Certification Program Division of Safety Research Testing and Certification Branch," NIOSH 80-113

"Evaluation of Two-Way Valves for Respiratory Testing," NIOSH 75-123

"Evaluation of Two-Way Valves for Resting Level Respiratory Testing," NIOSH 77-212

"Exhalation Valve leakage Test"

"Human Variability and Respirator Sizing," NIOSH 76-146

"Industrial Health and Safety Criteria for Abrasive Blast Cleaning Operations," NIOSH 75-122

"NIOSH Certified Equipment List as of July 1," 1980 NIOSH 80-144 (revised annully)

"Performance Evaluation of Respiratory Protective Equipment Used in Paint Spraying," NIOSH 78-177

"Respiratory Protection—A Guide for the Employee," NIOSH 78-193B

"Respiratory Protection—An Employer's Manual," NIOSH 78-193A

"Suggested Research and Development Programs for Gas and Vapor Respirators," CONTR 210-73-0080

"Survey of Personal Protective Equipment Used in Foundries," NIOSH 80-100

NIOSH CERTIFIED Equipment List, revised annually, National Institute for Occupational Safety and Health, (NIOSH) No. 80-144 (October 1980) or most recent revision.

Editorial, Clean Air Realism, Wall Street Journal, page 6, Dec. 31, 1981

Moeschlin, S., "Outstanding Symptoms of Poisoning," pp. 644–678, in *Diagnosis and Treatment*, Gruen & Stratton, New York, 1965.

Nitrogen, Data Sheet, Chemical Section, National Safety Council, Chicago, Ill., 1982.

Editorial, Air Pollutants Nox and Ozone, *Environ. Sci. Tech.* **15**(3), 253–254 (March 1981).

"Oxides of Nitrogen," Environmental Health Criteria 4, World Health Organization, Geneva, 1977.

Padour, J. and Shaw, A., "Respiratory Testing," pp 16–19; Horvath, E., "Respiratory Testing," pp. 20–31; Held, B. J. and Richards, C. B., "Respiratory Protection," pp 32–35; Murphy, A. J., "Respiratory Testing Role of OHN," pp 36–40, Occup. Health Safety, **46**(5) (September/October 1977).

Pesticides Enforcement Division, "Pesticides Inspection Manual," U.S. Environmental Protection Agency, Washington, D.C., 1980.

Proctor, N. H. and Hughes, J. P., *Chemical Hazards of the Workplace*, Lippincott, Philadelphia, 1978.

"Respirator Studies for the National Institute for Occupational Safety and Health, January 1–December 31, 1977," Los Alamos Scientific Laboratory, UC-41, LA-7317-PR, HEW Publication No. (NIOSH) 78-161, June 1978.

"Respiratory Protection," Occupational Safety and Health Program, U. S. Environmental Protection Agency, Washington, D.C.

Review & Outlook, *Time, Air, and Money, Wall Street Journal*, 26 (March 3, 1982).

Practices for Respiratory Protection During Fumigation, Z88.3, Draft August 18, 1981, American National Standards Institute, New York, 1981.

Respiratory Protection against Radon Daughters, Z88.1-1975, American National Standards Institute, New York, 1975.

"Respiratory Protective Equipment," Information Sheet No. 9, International Labour Office, Geneva, 1964.

"Safety Precautions for Oxygen, Nitrogen, Argon, Helium, Carbon Dioxide, Hydrogen, Acetylene, Ethylene Oxide, and Sterilant Mixtures," Publication F-3499C, Linde Division, Union Carbide Corp., New York, March 1976.

Staub, N. C., *Lung Water and Solute Exchange (Lung Biology in Health and Disease Series)*, Dekker, New York. 1978.

Stern, A. C. *Air Pollution*, 2nd ed., 3 volumes, Academic Press, N.Y., 1968.

Strauss, M. J., *Lung Cancer: Clinical Diagnosis and Treatment, Clinical Oncology Monograph*, Grune & Stratton, New York, 1977.

Sykes, M. K. *Respiratory Failure*, 2nd ed., Lippincott, Philadelphia, 1976.

Tuck, A., Johnson, J. W. and Moulton, D. C., *Human Responses to Environmental Odors*, Academic Press, New York, 1974.

WHO Scientific Group, "Respiratory Viruses," Technical Report Series No. 408, World Health Organization, Geneva, 1969.

International Asso. of Fire Fighters, *Life Support, A Fire Guide To Self-contained Breathing Apparatus*, Washington, D.C. 20006, 1981.

International Society for Respiratory Protection, P.O. Box 7567, St. Paul, Minnesota 55119.

Anon, Chemical-Biological Warfare in Afghanistan, *Wall Street Journal*, 20 (June 7, 1982).

26

Eye Safety in Chemical Operations

Joseph Nichols

Control of injuries to the eyes of workers in the chemical industries through prevention of accidents is not necessarily a specialized branch of professional accident-prevention work. In approaching the problem, one should apply the same general principles as would be applicable in any problem of industrial accident prevention. A logical order of procedure is suggested in this chapter.

26.1. IDENTIFYING THE HAZARD

A hazard is a condition or changing set of circumstances which present a potential injury. Examples: the unshored trench, the toxic chemical, the jaws of a power press. In the chemical industries, the examples are legion and include such typical hazards as exposure to mists and fumes; splashes of liquids when opening pipe flanges; contacts with hot process equipment; mistakes in identification of toxic or corrosive substances. Many of these same hazards, of course, are found in industries not usually classified as chemical industries, such as metal plating, acid cleaning of masonry walls, operating degreasing tanks, and pest-control work.

26.2. EVALUATING THE RISK

A risk is the probability of injury resulting from a hazard. There may be negligible risk from a hazard because adequate control measures have been taken to reduce the degree of risk to zero. There may also be no satisfactory way of reducing the risk of injury and the use of alternate or supplementary measures may have to be employed. Example: a solid fuel rocket propellant can be made of atomized aluminum which is manufactured by machines equipped with remote controls because of the recognized extreme hazard of explosion in the process. The worker is therefore removed from the area to a point where the risk of injury from an explosion is eliminated. The hazard is recognized and identified and the risk is evaluated and minimized. Other steps, of course, are taken to minimize the hazard through the use of process controls.

Sometimes the hazard may be a well-recognized one, but the degree of risk is poorly defined. Example: ultraviolet radiation is known to be hazardous in excessive amounts as well as a potential cause of injury to the eyes. The intensity of radiation however is not easily evaluated and because of this we tend to "overprescribe" in terms of the recommended eye protection. Example: Common sunburn. Everyone knows that excessive exposure to sunlight is a hazard and that the risk of injury is severe with some people while others are apparently able to avoid injury through their own bodily protecting processes.

26.3. MINIMIZING THE DEGREE OF RISK

When a hazard cannot be removed or eliminated, steps must be taken to protect those exposed to the hazard. Very often this is done by the use of personal protective equipment. It is usually preferable to reduce the risk of injury without the use of personal protective equipment if this can be done. Modern eye-protective equipment is good and the range of choice in the marketplace is wide, indeed. Designs now available include eye protective equipment for almost any kind of eye hazard to which a worker may be exposed.

To minimize the degree of risk it is necessary to interpose a barrier between the hazard and the eyes or face of the worker. The hazard may be flying objects, dust particles, liquids (both cold and hot), radiant energy, fumes, smoke, or the possibility of striking against something. The simplest protection (although not always suitable for industrial use) may consist of plastic "visitor's spectacles" such as are commonly provided for visitors or others who need only temporary protection, to face shields covering the entire facial area and attached to a headband or to a helmet and worn with plano or prescription ground safety spectacles over the eyes. Such equipment can minimize the degree of risk and injury. Hazards encountered with fumes or contaminated atmosphere may require even more protective equipment such as a hood covering the entire head and with a "window" in the front to permit viewing. Severe atmospheric contamination in the workplace will require a mask or hood which provides for a noncontaminated air supply and eye protection at the same time. It is clear that the use of personal protective equipment for the eyes is secondary to the removal of the hazard and reduction of the risk wherever that is possible.

26.4. SELECTING THE PROTECTIVE EQUIPMENT

Every device designed to be worn for the prevention of eye injuries to the industrial worker creates a degree of handicap to the freedom of motion or to optimum visibility, or both. It is therefore most important that the hazard be accurately identified and the protective equipment be suited to the wearer, fitted for maximum comfort, have the greatest effectiveness for the job at hand, and designed to offer the minimum obstruction to vision and mobility. Common safety spectacles meeting the requirements of the ANSI Standard Z87.1 (1979), may be suitable for a wide range of exposures, but even this protection can be enhanced with the use of side shields attached to the spectacles. Where necessary these spectacles can be optically fitted to the individual corrective prescription of the wearer.

Where the hazard is not severe and the risk is minimal, the use of a simple plastic face shield may be suitable. Such shields offer a minimum of encumbrance, are lightweight, easily adjustable, inexpensive, and relatively comfortable. They have the disadvantage of poor resistance to scratching. In chemical laboratories such shields can be placed near hazardous jobs for use by anyone who must perform that job. For a worker who must wear corrective lenses, it is desirable that they be safety spectacles and oftentimes a plastic safety shield should be worn in combination with them. Even in the case of a window in a laboratory hood, it should not be assumed that the window alone offers enough protection because such windows are notorious for being inoperative, etched, clouded, off-the-track, broken, just plain dirty, or simply unused thus leaving the worker without needed protection.

The equipment available for eye safety on the market today can be had in a very wide range of styles, colors, sizes, materials and prices. In selecting the proper equipment for each particular job, "let the buyer beware" is as true as with anything else that is purchased. Equipment which purports to meet the requirements of ANSI Standard Z87.1 (1979) is usually acceptable. It is well to remember that Standard Z87.1 (1979) is intended to set forth the requirements for *minimum* standards of performance and quality. Performance standards greater than the minimum may often be required.

Beware of the claim or warranty that safety spectacles or eye shields are "shatter-proof," or that they offer "complete protection." Even the best of such equipment made of either glass or plastic can be shattered or broken if the impacting force is great enough.

With today's scenario of the increasing importance of consumer safety, the buyer of eye protective equipment would do well to be extremely careful in the selection of equipment for employee or personal use. The number of lawsuits based on unfulfilled warranties is increasing and a significant number of such cases are resulting in large awards for the plaintiff.

Almost every type of "chemical industry" has some hazards of eye injury in the use of chemicals. There is eye safety equipment available for every recognized hazard that will at least reduce the degree of risk if not eliminate or control it completely.

It is frequently necessary to use more than one such device for maximum safety. Maintenance work, especially on pipelines, presents severe hazards with high risks of flowing materials under pressure such as steam, liquids, corrosives, or gases escaping from broken or leaking pipe joints or fittings. This requires eye protection against the hazards of splashing fluids (cold or hot), perhaps escaping steam or other gas and/or corrosive substances. Frequently the use of a face shield over the goggles is necessary. The goggles may also have to be fitted over the prescription glasses of the wearer. Obviously, for such exposures, other personal protective equipment will probably have to be provided such as a helmet and protective clothing.

It should be kept in mind that the use of eye safety equipment should be considered as a secondary means of preventing injury. Removal or control of the hazard by other means should always be effected when possible. While almost any recognized hazard can be removed or controlled, the cost of such effort must be balanced against the use of personal protective equipment.

26.5.   APPLYING THE PROTECTIVE EQUIPMENT

The story is told of the industrial plant manager who reluctantly conceded that he would furnish safety boots for the use of his workers in certain jobs in his plant. He ordered a supply—all of the same size (large)—and later was unable to understand the difficulty in getting his employees to wear them.

No two human bodies are exactly the same. When it is necessary, as is usual with advancing age, for a person to wear corrective lenses to compensate for the changes in visual ability, such prescription lenses must be made especially for that individual's needs. Such lenses can of course be made for safety purposes as well as for corrective needs.

To provide standards for manufacturers, safety spectacles are made in accordance with the requirements of ANSI Standard Z87.1 (1979). Spectacles so made are very often high in their cosmetic quality, too, so that they can be worn for street wear as well as at work. When safety spectacles, either prescription ground or not, must be worn with side shields, they may frequently need to be worn with supplemental protection such as the use of a face shield.

In selecting the equipment for individual workers, optimum results will be obtained when the equipment is fitted most carefully. Facial contours vary through such a wide range of measurements that, like safety boots or shoes, different sizes and shapes of spectacles must be available. Because of the necessary thickness of the lens material, safety spectacles are usually heavier than ordinary street wear glasses. This can be partially compensated for by having the safety spectacle lens made of plastic instead of glass, but the plastic material can become scratched where the glass is much more scratch resistant.

Safety spectacles can be made in a wide range of tints and colors. Such tinted glasses sometimes are worn as sunglasses. Use of tinted glasses indoors, however, should be restricted to those who are actually exposed to excessive glare. Tinting of glass for eyewear cuts down the visibility in direct proportion to the amount of the tinting, thus increasing the degree of risk of injury. Welders, of course, are among those exposed to the most severe hazard of excessive glare and need carefully prescribed protection as dictated by the type and duration of the welding work they are to do.

Eye protective equipment should be provided on an individual basis. Even where nonprescription eyewear is provided it is not good practice to permit the equipment to be shared indiscriminately among workers. The standards of proper hygiene should be observed.

Wearing of safety eyewear is a highly personal matter. In introducing the use of such equipment, it is well to plan for an educational program for the users. The monetary value of good eyesight cannot be measured and the loss of sight in any degree is a major lifetime handicap. Promotion of the proper use of safety equipment should be based on the recognized hazards, the evaluated degree of risk, and the institution of control measures. For protection against any remaining risk the complete and continuing cooperation of the worker is necessary.

Safety spectacles and goggles available today are high in quality, designed from the results of many years of use and research, and nearly always well suited to the special needs of the chemical industry. ANSI Standard Z87.1 (1979) is an important document and is the concensus of trained and experienced people motivated to produce a standard that is the best available for the guidance of manufacturers and users.

With the hard usage encountered in industrial exposures, lenses get scratched, chipped, spattered with welding spatter, and attacked by corrosive chemicals. Spectacle bridges get bent out of shape. Temple pieces get bent, broken, and cracked. Hinges become loosened, and of course, eye glasses get dirty with ordinary usage. Proper maintenance of equipment is important, not only to insure maximum protection and visibility, but also to help sustain the morale of the wearer in the belief that safety equipment is important to his or her welfare. A significant scratch on a heat-treated lens will impair its ability to resist an impact, therefore reducing its effectiveness in preventing injury. Scratched lenses should be replaced. Welding spatter for some reason will stick to glass more readily than it will to plastic, and it causes permanent damage to safety glass lenses.

Eye protective equipment serves its purpose only when it covers the eyes. All too frequent are the examples of supervisors and workers, too, who habitually wear their safety spectacles or goggles over their forehead, in the belief that they will pull them down over their eyes when needed. Such instances reflect a fault in the program planning for the use of eye safety equipment.

Professional safety practitioners have much to say about the advantages and disadvantages of a so-called 100% eye-safety program. Such a title is intended to convey the idea that a program is in force which assures the use of proper eye safety equipment by every employee in a plant. Such a state of affairs is, in practice, seldom achieved. There are always too many workers who feel that they should be exempted from such a requirement. Office workers, for example, are first to make the claim that such a program is not needed for them. Perhaps they are right—if they are never exposed to the conditions out in the plant. Supervisors are often next in line to claim that most of their time is spent at a desk and that they should be exempt from the rule. Whatever the merits of these claims, a 100% eye safety program can not be accomplished over night and once established as a matter of company safety policy it will need careful and continuous enforcement.

As pointed out earlier, there is a logical method for approaching the problem: (1) identifying the hazard(s) and locating those personnel who are exposed to the hazard(s). (2) evaluating the risk—a certain chemical transfer process may present a known hazard of severe proportions, but if the process takes place only once each month the risk of injury is obviously not as large as if the same process were to take place twice a day. The needs for eye protective equipment can be evaluated accordingly. (3) Minimizing the degree of risk. Again, as in the above example, if the process takes place only once each month, perhaps it can be done at night or between shifts, whereas this same remedy might be highly impractical for the same process done twice a day.

To achieve the 100% goal of eye protection, it might be feasible to provide light plastic visor-type shields for those who are only occasionally exposed such as office workers who are in the production areas of the plant infrequently. Supervisors, however, should always set examples for their workers by proper use of eye protective equipment (as well as other personal protective safety equipment).

The National Society for the Prevention of Blindness of New York sponsors the "Wise Owl Club" which is an organization devoted to the prevention of blindness by encouraging the use of eye protective equipment. In their files are authenticated records of over 50,000 actual incidents in industry in which the vision of workers was saved in one or both eyes by the proper use of eye protective equipment. They also have records of many actual instances of eyesight being saved for the *second time* by the same worker who had previously saved his eyesight. Companies in the chemical industries number large among those who have a chartered chapter in the Wise Owl Club.

26.6. TRAINING AND INSTRUCTING EMPLOYEES

As good as the present day eye safety equipment is, it cannot accomplish its purpose if it is not worn. A thorough plan for training and instructing workers must be instituted and carried out to make the program effective. Workers who wear their eye protective equipment only when the boss is around are not only endangering themselves, they are also presenting a handicap to the success of the program for the other workers.

There are reasons for nonacceptance of such a program. One important reason is that the prescribed equipment does not fit or is uncomfortable. The fitting of safety glasses or goggles requires skill and patience to achieve optimum acceptance. For those requiring prescription lenses, the fitting is usually done by the examining practitioner and there should be no excuse for misfitting. For those who require the added protection of glasses with side shields or the close fitting goggles, even more care is required to prevent discomfort. For those who must wear equipment such as plastic face shields with safety spectacles underneath them, the problem can become difficult. The use of any prosthetic device such as false teeth, a wooden leg, an artificial hand, or even a glass eye can never be as pleasant and acceptable as the facility which it replaced. Workers must be convinced that the use of eye safety equipment even if it is not as pleasant as they might wish is clearly preferable to the loss of vision in any degree.

26.7. SOURCES OF SUPPLY

The marketplace is crowded with vendors of eye safety equipment. Many of the leading and well-known vendors of a general line of safety equipment offer a complete line of eye safety equipment in their catalogs. Some of these vendors have their own force of sales people who demonstrate their wares to their own customers. Such sales people are usually well versed in the qualities and capabilities of their own lines of merchandise. If the prospective buyer can accurately describe the requirements for a job and its hazards and the degree of risk, the vendor can usually provide suitable equipment. He does not do fitting of the glasses to the wearer. Most commercial vendors of eye wear could probably supply safety glasses on order but would be less likely to have a proper understanding of the requirements of the job.

It is advisable to de-emphasize the purchase and use of safety eye wear on the basis of its cosmetic attractiveness. Protection, proper fit, and maximum comfort for the wearer should be the governing considerations. Cost of the best equipment is often very little different from that of inferior grades whose sales appeal is based on features of dubious value or price alone. All such equipment should of course, meet the minimum requirements of the ANSI Standard Z87.1 (1979).

26.8.   TREATING THE INJURY

While the primary goal of an eye safety program is to prevent eye injuries, it might well also include a program for improved eye health and hygiene. Total prevention of eye injuries is a worthwhile goal but may be difficult to achieve. When injuries do occur there must be adequate facilities available for treatment of the injury.

The most basic of these facilities is the eye-wash fountain. For most industrial eye injuries the most useful and effective remedy is the use of plain water for flushing the injured eye by the prompt use of an eye-wash fountain. A variation of this method is the use of specially prepared solutions for eye flushing. Such special solutions should be approved for use by your company physician or ophthalmologist.

There are also emergency deluge showers which will deluge the entire body of the user, and there are small eye wash fountains either portable or fixed. In certain high-risk areas it is common to find eye wash "stations" equipped with bottles of specially prepared eye wash solutions ready to be used with sprays, cups, or other means. Such "stations" must be regularly serviced and the solution renewed or changed routinely. For any but the most superficial eye injuries, it is advisable to obtain the service of trained personnel such as a medical doctor for treatment. Certainly all eye injuries no matter how slight should be reported to the plant medical department as soon as possible.

The use of these special solutions is best confined to plants or to areas where the chemical or substance likely to be splashed into the eyes is always the same all over the plant. Pure water is still considered to be the best all-round material for irrigating the eyes on the job when this must be done without the assistance of trained medical personnel.

Contact lenses should never be considered to be a substitute for protective safety eye wear. Foreign bodies such as dirt and dust can get behind the contact lens and be the cause of damage to the cornea of the eye before it is discovered and removed. Even worse, splashes of any liquids can easily get behind the contact lens and very quickly do serious damage if the substance is acid or corrosive. Anyone wearing contact lenses can be fitted with eye protective equipment.

On June 12, 1981, ANSI Z-358.1-1981 was approved for emergency eye-washes and showers, to which the reader is referred.

26.9.   OCCUPATIONAL SAFETY AND HEALTH ACT OF 1970 (OSHA)

The requirements of Federal and state laws and regulations must be complied with. The Occupational Safety and Health Act of 1970 became effective on April 27, 1971, and on January 1, 1972, a regulation by the Federal Food and Drug Act became effective requiring all eyeglasses (whether designed for safety purposes or for street wear) to be fitted with impact-resistant lenses. With the various school eye safety laws and the still newer Federal Consumer Product Safety Act signed December 26, 1972, these make a powerful combination of forces for change and have created a major impact on the entire field of occupational safety.

OSHA regulations make the use of personal protective equipment mandatory when, in the opinion of the employer, its use is necessary for the prevention of injury to employees at work. Its provisions are pervasive and the burden of enforcement is placed upon the employer. For eye wear, the regulations (*Federal Register*, Oct. 18, 1972, Part II), read as follows:

> Part 1910—Occupational Safety and Health Standards, Section 1910.133, Eye and Face Protection
> (a) General
> 1. Protective eye and face equipment shall be required where there is a reasonable probability of injury that can be prevented by such equipment. In such cases, employers shall make conveniently available a type of protector suitable for the work to be performed, and *employees shall use such protectors.* No unprotected person shall knowingly be subjected to a hazardous environmental condition. Suitable eye protectors shall be provided where machines or operations present the hazard of flying objects, glare, liquids, injurious radiation, or a combination of these hazards.
> 2. Protectors shall meet the following requirements:
> (i) They shall provide adequate protection against the particular hazards for which they are designed.
> (ii) They shall be reasonably comfortable when worn under the designated conditions.
> (iii) They shall fit snugly and shall not unduly interfere with the movements of the wearer.
> (iv) They shall be durable.
> (v) They shall be capable of being disinfected.
> (vi) They shall be easily cleanable.
> (vii) The protectors should be kept clean and in good repair. (See also sections 3, 4, 5 and 6 under this part).

The law now backs the employer.

ANSI Standard Z87.1 (1979) is a voluntary concensus standard for industry and others interested in standards for safety eye wear. As written, all such standards were intended for voluntary acceptance by the user, but by the mass adoption of scores of such codes and standards, OSHA gave them the status of becoming enforceable as law-something for which they were neither planned nor intended when written.

A copy of this standard should be obtained for reference and for guidance in organizing a program for eyesight conservation. The *National Safety News* of November, 1973, pages 53–66 contains some valuable and entirely practicable ideas, suggestions, and information on the subject of industrial eye safety and the management of a sight conservation program, all of which would be useful in the chemical industries.

27

Other Personal Protective Equipment

Howard H. Fawcett

In addition to the previously discussed personal protective equipment (respiratory protection, Chapter 25 and eye protection, Chapter 26), several other types of personal protective equipment play an important role as "secondary" or back-up equipment where chemicals and other hazardous materials or substances are involved. It is generally agreed that proper engineering, adequate attention to safety and health practices, and education of both personnel and management so most control is reduced to a standard operating procedure, carefully adhered to, cannot completely insure employee or plant safety alone, since unusual incidents, spills, and failures, both human and mechanical, do in fact occur in the real world.

Personal protective equipment is then called upon to provide that extra measure of operational safety and health.

The injurious effects of body, skin, respiratory, and eye contact with materials have been well documented elsewhere, but should be noted as ranging from acute irritation or injury, such as skin irritation, inflammation or lung involvement which might even result in pulmonary edema, or cardiac impairment, to more chronic injuries which may manifest themselves years later, including carcinogenic, teratogenic, mutagenic, or reproduction-related effects. It is not clear even with the present knowledge whether one exposure to certain materials might induce highly undesirable effects years later.

Exposure or contacts may range from one-time splashes or inhalation, to continuous wettings of liquids, or chronic inhalation of background "odors," as well as contact with solids. Clothing, as well as breathing protection, should be effective for more than a few minutes, to permit the wearer to retreat from the area, remove protective gear and clothing, and then thoroughly wash or flush with water and detergent. Under no circumstances should work clothing be worn off the site; some provision must be made to insure that a high level of personal hygiene, including, if possible, shower facilities and clothing change, is included. A portable facility, such as the shower and toilet facilities in a recreational vehicle or camper is a possibility for on-site control of contamination, if operations are in an area remote from normal building facilities.

Protective clothing, including coveralls or protective suits, gloves, boots or rubber shoes, and face and head protection deserve critical attention. Unfortunately, no comprehensive performance criteria have been established or recognized for protective gear worn by persons in chemical waste situations, and the present tables of protection provided by the makers and vendors of various equipment are less than adequate. Most are based on tests which may or may not apply to the exposure at hand. This lack of standards is especially disturbing, since, in most cases the waste site may be an unknown terrain. Until specifics are established, the unknown nature of the potential exposures must be recognized and respected.

A variety of materials from which protective clothing (suits and gloves) can be fabricated is available, from paper suits to plastic and other carefully designed suits and gloves, but each material fabrication and combination has limitations of use which must be considered. Normally, for routine occupational exposures, it would be recommended that any materials used in gloves, suits, or footwear to protect the wearer from contaminants should be tested for permeation before use, but when the nature of the contaminants and the degree or amount of exposures are unknown, the prudent wearer must rely on whatever test data is available for general classes of materials as analogous as possible to the match which is known. For example, if it is suspected that that benzene may be encountered, good data on the permeability of benzene is available in the literature. For many other materials, and for combinations of chemicals which may be encountered in the field, corresponding data should be sought but often will be found not available. In that case, the opinions of an industrial hygienist competent in this area should be sought and followed. The use of protective suits in sampling sewer systems is illustrated in Fig. 27.1.

Protective equipment, especially impervious clothing and respiratory protective devices, are not always given the proper attention since, in addition to cost, they may be uncomfortable when worn for long periods of work. This discomfort will be especially severe when the devices and clothing are worn in hot or humid atmospheres. Various alternatives to minimize such discomfort are available, and may be considered, such as the use of air-supplied circulation to the suits to permit more nearly normal body heat and perspiration to be dissipated. In one model, a vortex air injector provides both internal air as well as cooling.

If an air-supply is considered for breathing or cooling of an impervious suit, the problems associated with air purity should be carefully analyzed, Diaphram compressors or other devices which cannot generate carbon

Figure 27.1. Application of full protection in sampling sewer system for toxic substances. Suits were selected after tests to determine suitability for application. Photo courtesy Geomet Technology, Inc.

monoxide or excessive oil (from oil-lubrication) should be noted. Cross connections to plant air supply (including the ever-present danger of some gas other than air being introduced into the system) must be considered. The quality of the air to the compressor or vortex must meet or exceed breathing air standards. (See Chap. 25.)

27.1. HEAD PROTECTION

For chemical operations, especially in facilities with extensive overhead piping, tanks, columns, and related equipment which may occasionally leak or rupture, some form of head protection is clearly indicated. In addition, head protection may offer protection against impact, flying particles, electric shock. If properly designed and engineered, they can protect the scalp, face, and neck as well as the head.

As noted by the recent study by NIOSH, safety helmets are rigid headgear made of varying materials, such as polycarbonate plastic and even aluminum.[1-3] Two basic types are noted: full-brimmed, and brimless, with peak. Both types are further classified into four classes:

Class A: Limited voltage resistance for general service.
Class B: High voltage resistance.
Class C: No electrical voltage protection (i.e., metallic).
Class D: Limited protection for fire-fighting use.

Class A helmets are made without holes in the shell, except for those used for mounting suspensions or accessories, and have no metallic parts in contact with the head. They must pass voltage tests of 2200 V AC (rms) at 60 Hz for 1 min, with no more than 9 mA leakage. In addition, they should not burn at a rate greater than 3 in./min. After 24-h immersion in water, the shell should absorb no more than 5% by weight. They are designed to transmit a maximum average force of not more than 850 lb. Class A helmets do not weigh more than 15 oz, including suspension, but excluding winter liner and chin strap.

Class B helmets are designed specifically for use around electrical hazards, with the same impact resistance as Class A helmets. They pass voltage tests of 20,000 V AC (rms) at 60 Hz for 3 min, with no more than 9 mA leakage. In addition, no holes are allowed in the shell for any reason, and no metallic parts can be used in the helmet. The burning rate of the thinnest part is not greater than 3 in./min. After 24 h in water, absorption of the shell will be no more than 0.5% by weight. The maximum weight of Class B helmets is 15.5 oz.[1,3,3a]

Class C metallic helmets do not afford the same degree of protection, but are preferred by many because of their lighter weight. They should not be used in the vicinity of electrical equipment, or near acids, alkali, and other substances that are corrosive to aluminum and alloys.

Brims are an optional feature of helmets; a brim completely around the helmet affords the most complete head, face, and back of neck protection. The brimless with peak type of helmet may be used where a brim may be objectionable and the brimless may be equipped with lugs to support a face mask or face shield to provide face and eye protection.

The key to the protection from impact is the proper helmet suspension, since the suspension distributes and absorbs the impact force. Examples of suspensions include a compressible liner or a cradle formed by the crown straps and headband or a combination of the two. To function properly, the adjustment of the straps must keep the helmet a minimum distance of 1.25 in, above the head.

In order to keep the helmet from dislodging during normal use, various leather, fabric, and elastic chin straps are used. Helmet liners are optional equipment, often used in colder weather.

In recent times, the "bump" cap has been introduced to be worn by persons who are working in cramped quarters. In general, it is thin-shelled, and made of plastic. No specifications exist for these caps, and they are not a substitute for the helmets discussed above.

Another aspect of head protection which must be considered is the length of head hair in both males and females. This is especially critical where moving parts and machinery are involved. Hair nets or snoods are often specified, but worn only with reluctance by many workers. When combined with a cap, they have more appeal and hence increased utility. In areas where sparks or hot metals, such as in welding or cutting operations, are encountered, the nets or caps should be flame-resistant, and have a visor sufficiently long to afford protection from the exposures coming from above on an angle.

A wide variety of plastics, fiberglass, and other materials are available for helmet construction, and many are available in colors. The use of colors often contribute to immediate identification of various crafts, such as millwrights, electricians, pipefitters, and also is a type of security clearance to the worker, especially in a large facility. The wearing of the employee's name on the hat is also a useful incentive for each person to keep his or her hat in proper condition of cleanliness. Ear muffs or ear protectors are often integrated into hats, and, when properly adjusted are sufficiently comfortable for long periods of wearing.

27.2. HEARING PROTECTION

The question of what exposure to what level of noise of what frequency causes hearing loss is not completely resolved. However, we live in a world which seems to become noisier, as heavy traffic, subways, radios, televisions, stereos, public address systems, outboard motors, power lawn mowers, jet engines, rocket engines, helicopters, and supersonic booms compete for our attention. Within the chemical industry, it is possible to note agitators, grinders, mixers, pulverizers, compressors, fans, blowers, and other equipment which are noisier than the generally accepted standard of 90 dBA. At what point for which periods of time the noise becomes a problem can only be determined by measurements, analyzing both the frequency and the length of exposure. Although industrial hygiene engineers usually have the qualifications and equipment for making surveys, frequently an acoustical engineer is required to make the final determinations and recommendations to the plant physician or occupational health nurse. The medical department may elect to institute a regular program of audiometric testing, during which the employee is tested at regular intervals, such as yearly or semiannually, for his or her response to various frequency tones which correspond to normal speech. The records of such test then constitute an important part of the employee's overall medical records, and can be referred to in the future to determine if a tendency toward hearing loss is occurring, and whether it is from occupational or nonoccupational causes. Since disco music is often in the range of 100 to 120 dBA, and exposure may

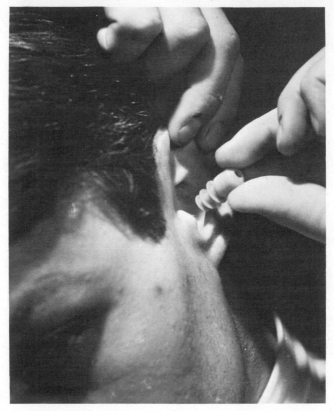

Figure 27.2. Ear plugs or valves must be carefully fitted to be effective and safe. Photo courtesy Norton Co.

be for significant times after work, it may be difficult to pinpoint the exact cause of hearing impairment.

Once a program is agreed upon, the medical department may recommend the proper fitting of protective devices, such as ear plugs, ear muffs, helmets, or combinations of these, so the individual is matched for the specific job requirements. It is vital that the individual understand the importance of this program to him or her as an individual, and that he or she actually cooperate in the fullest. Ear plugs, which are the simplest and least expensive control, have only limited ability to attenuate sound and must be carefully fitted, as illustrated in Fig. 27.2, and for the higher sound levels, they must be supplemented by muffs, or helmets with built-in muffs. (See Chap. 11.) To facilitate voice communications, built-in earphones and attached microphones are available to permit the use of telephones or two-way radio equipment.

27.3. SAFETY FOOTWEAR

Safety shoes and safety boots have become widely accepted in the chemical industry, since the intrinsic value of the steel safety cap, which weighs little more than an ounce, has been demonstrated many times. In addition, safety shoes and boots, in styles both for women and men, are attractive, and represent an excellent value. Many companies encourage the use of safety shoes by giving partial or total credit for shoe purchase, and the shoe vendors often have mobile shoe vans or trucks which make on-the-job fitting most convenient.

According to NIOSH, protective footwear may be classified into five categories:

1. Safety-toe shoes (the most commonly accepted).
2. Conductive shoes (to leak off static electricity).
3. Foundry or molders' shoes.
4. Explosive-operations shoes (nonsparking).
5. Electrical hazard shoes.

Although not a class as such, acid shoes, that is, safety shoes or boots made of materials such as neoprene which resist attack by acids, alkali, and other corrosives, are important to acid-handling operations, including tank car loading and discharging.

OSHA requires the use of safety-toe shoes for persons who work with heavy material, since the shoes afford protection against impact and rolling objects, and against the hazard of inadvertently striking sharp sheet metal. Safety-toe shoes have been divided into three groups, namely 75, 50, and 30, which represent the minimum requirements for both compression and impact. The usual material of construction for the safety toe is iron. For additional protection against extremely heavy impacts, metatarsal (or over-feet) guards may be used in conjunction with safety-toe shoes. These guards are made of heavy gauge metal, flanged, and corrugated to protect the feet to the ankles.

In locations where there is a potential for a fire or explosion, such as the liquefaction of hydrogen, methane, or in the handling of highly volatile substances, such as ether, conductive shoes should be worn. These shoes are engineered to dissipate static charges, reducing the possibility of static spark, when used in connection with a conductive flooring, and the shoes and floors are properly maintained to insure real conductance. There has been much controversy as to whether or not a static discharge from a human would institute or cause ignition; the evidence is sufficiently strong that no prudent person can argue against the use of every precaution.[5] Periodic tests should be performed, using proper equipment, to ensure that the maximum allowable resistance of 450,000 ohms is not exceeded.

Foundry shoes protect against molten metal splashes, and are made so they may be removed easily and rapidly if a spill occurs. The tops of the shoes should be closed by the pants leg, spats, or leggings to preclude splashes entering from the top.

In hazardous locations, such as those areas where the floors are not conductive and grounded, and in which highly flammable and volatile gases and liquids are handled, such as gasoline tanks, nonsparking shoes should be worn. These shoes do not have conductive soles, and have nonferrous eyelets and nails. The metal toe-boxes are coated with nonferrous metal, such as zinc or aluminum.

Electrical hazards, such as those presented by the potential for contact with an electrical current running from the point of contact to ground, may be minimized by the wearing of electrical shoes. These shoes contain no metal, except for the toe-box, which is insulated from the remainder of the shoe. Dampness or significant wear will definitely decrease the protection provided by these shoes.

27.4. HAND PROTECTION (GLOVES)

Although gloves were discussed briefly in a previous section of this chapter, the extensive use of various types of gloves, for protection against a variety of insults including chemical, should be mentioned. In addition to toxics, gloves are available for protection against abrasions, cuts, heat, and cold.

Two basic methods are used in manufacturing liquid-proof gloves, the latex-dipped process and the cement-dipped process. In the former, particles of rubber, suspended in water, are coagulated onto a glove form. Since numerous tiny air pockets may develop that have resulted from the incomplete contact of rubber particles with the form, the latex-dipped gloves have a higher penetration or permeation rate than do the cement-dipped gloves. In the cement-dipped process, the rubber is dissolved in an organic solvent before it is deposited on the form. This permits a more complete joining of rubber particles resulting in fewer air pockets and greater resistance to permeation.[6]

NIOSH has noted that other factors that affect the rates of permeation are glove thickness and solvent concentration. (To this should be added the type solvent, since some materials may be highly resistant to a solvent, while others are not.) Permeation rates are indirectly proportional to glove thickness and directly proportional to solvent concentration.[6] Naturally, time of contact and area involved should also be considered.

As noted in the Protective Clothing section, the first consideration in selection of gloves is to ascertain, insofar as possible, the exact nature of the substances to be encountered, and to note any possible reactions which might occur between the solvents and the glove materials. The manufacturers of gloves have available to prospective buyers chemical resistance charts. These charts should be examined critically, and the test data supporting the classification should be requested from the laboratory which developed the chart. (As noted previously, no generally agreed test method exists for gloves or other protective clothing.) Since most gloves are supplied in several different thicknesses, the choice must be made as to most complete protection versus flexibility. In general, the thickest glove practical should be specified, if it is not too stiff and will not introduce an additional hazard into the work. The length of the glove is important, since it should be sufficiently long to protect the forearms as well as hands. If liquids are handled, long gloves permit the wearer to turn down the top of the gloves to form cuffs, thereby preventing the liquid from passage down the arms when the hands are raised.[6]

27.5. SAFETY BELTS, LANYARDS, LIFELINES, AND SAFETY NETS

Where work is to be performed at significant heights, safety belts and lifelines should be used. During normal operations, such operations as hoisting or lowering the wearer, or providing him or her with steady support while at work, the belt or lifeline may be considered part of the routine equipment and procedure for the job. (This includes the lowering of personnel into utility holes or tanks where no ladder is available or cannot be used for some reason.) Under these normal operations, comparatively mild stresses are applied to the belt, and are usually less than the total static weight of the wearer. In emergency use, however, such as preventing the wearer from falling further than the slack in the line, the belt or line is subjected to impact loading, which can amount to many times the weight of the user. The amount of impact force developed in stopping a fall depends on three factors:

1. The weight of the user.
2. The distance of the fall.
3. The speed of deceleration or stopping.

To limit the distance of the fall, the wearer should never tie off below waist level. Some shock absorber or decelerating device (such as a spring or elastic rope) should be incorporated into the system, to bring the fall to a gradual stop. The impact load on the equipment and the wearer will thus be reduced.

Four classes of safety belts have been recognized:

Class I — Body belts for limited movement and positioning to restrict the worker to a safe area in order to prevent a fall.

Class II — Chest harnesses used where freedom of movement is paramount, and only limited fall is possible. These are not recommended for vertical free fall hazard situations.

Class III — Body harnesses are used when the worker must move about at dangerous heights. In event of a fall, the harness distributes the impact force over a wide body area, reducing the injury potential.

Class IV — Suspension belts, used when working from a fixed surface. The user is completely supported by the suspension belt. These are used in shipboard and bridge painting and maintenance, stack maintenance, and tree-trimming.

Of the four classes, the body-harnesses are preferred for many operations, since they more completely absorb impact and will keep the wearer upright during a fall. When considering belts and harnesses, all aspects of the hazard should be recognized: the possible distance to fall, the limitations against restricted mobility, and the health and stability (both physically and emotionally) of the wearer are considerations in the selection of personnel and belting protection.

To judge the safety of a belt, NIOSH recommends several criteria. First, the belt should be of sufficient strength to withstand the maximum possible free-fall of the wearer. The lanyard of the safety belt should be a minimum of 0.5-in. nylon (or equivalent) with a maximum length that limits a fall to 6 ft or less. The rope should have a nominal breaking strength of 5400 lb. The belt should be equipped with some form of shock absorber to limit the impact loading. The stopping distance of the belt should be such that it will prevent the user from striking some dangerous obstruction before the fall is arrested. Finally, there must be a sufficient safety margin on all of the above to cover all unknowns, including the weight of the wearer, the distance of the fall, physical condition of the wearer, distance to dangerous obstructions, variations in the strength or elasticity of materials, and deterioration of the materials due to wear or other causes.[1,2]

Lifelines, which attach the user of a safety belt to an anchorage, such as window washing and similar maintenance, should be secured above the point of operation. This positioning will limit the fall of the wearer, thus reducing the impact loading. All lifelines should be capable of supporting a minimum deadweight of 5400 lb. The most important criteria for lifelines is their shock loading strength and energy absorption ability. Shock absorbers should be used to assist the lifeline in these critical areas. Lifelines should be secured so the wearer is subjected to as little shock as possible; this will limit the free fall.

Numerous materials are available for lifelines including nylon, dacron, and manila. Each has specific characteristics, and lifelines should be selected based on the specific situation in which they will be used.

Safety nets represent another approach to minimization of fall injuries. Safety nets should be used wherever the use of safety belts and lifelines is impractical or infeasible where the protection from falls is important.

Safety nets are usually designed for specific purposes and are therefore custom-made from various natural and synthetic materials. These materials are used in the form of rope or webbing and in numerous combinations of rope or webbing in various mesh sizes. Each net and section of net should have boarder ropes of the same material, but of larger diameter. The Associated General Contractors and the U.S. Corps of Engineers have established minimum sizes for safety net materials, as well as specific performance standards for nets of various sizes and dimensions.[7]

Where the use of safety nets is considered, extra care should be exercised to arrange the nets so sufficient clearance exists to prevent the nets from contact with surfaces or structures below or to each side when the anticipated impact load is applied. When using safety nets near electrical power lines, this factor becomes especially critical.[7]

When more than one net is employed and are joined to form a larger net, they should be laced or otherwise secured so they perform properly. For all nets, perimeter suspension systems should be designed and installed in such a manner that the suspension points are either level or slope toward the building so a rebounding load will be directed into a protected area. Perimeter nets should always follow the work upward as it progresses, and should never be more than 25 ft below the working level, and if possible, closer.

In order to ensure that nets perform as anticipated, daily inspections are essential. Inspections should be made prior to and after installation, after any alterations and after impact loading occurs. Recommended test procedures are available.[8] All defects should be repaired properly and promptly.

27.6. SUMMARY

Selection, regular use, care, cleaning, and maintenance of personal protective equipment must be based on complete understanding of the hazards to which the worker is exposed. For example, a worker exposed to electrical hazards as well as foot and head injuries should be equipped with a Class B, electrical-resistant safety helmet, and with special electrical hazard shoes that have no soles of flexible metal. In addition, the worker must be adequately and periodically instructed in proper use of the equipment, the importance of the equipment to his or her health and safety on the job, and the proper methods of adequate maintenance. He or she must understand that a protective device for one job is not always adequate for another, and hence he or she should not loan equipment to another worker. A paint-spray respirator, for example, is not effective against chemical acid gases; gloves which may be adequate protection for paint dipping may not be proper protection for wearing around a trichloroethylene degreaser. (See Chapter 25.)

No protective equipment is without some disadvantage, but should be as comfortable as possible, and properly fitted and maintained to encourage use. The equipment should not increase or introduce additional hazards to the job; visibility, weight, and appearance are often factors in the real-world use. A full-brimmed hard hat used in close quarters where mobility is limited, for example, may obstruct the wearer's vision, and make him or her prone to other hazards.

Unless the protective devices are accepted for their intrinsic value, *and actually used*, a false sense of security, which may result in serious problems, will result from an inadequate program.

REFERENCES

1. J. Chapman and G. W. Pearson, *Safety Information Profile, Personal Protective Equipment Usage in Industry*, National Institute for Occupational Safety and Health, Morgantown, W. Va., 1980.
2. *Accident Prevention Manual for Industrial Operations*, 8th ed., National Safety Council, Chicago, Ill., 1977.
3. "Safety Requirements for Industrial Head Protection," ANSI Z89.1-1969, American National Standards Institute, New York, 1969.
    a. W. I. Cook and D. W. Groce, "Report on Tests of Class B Industrial Helmets," National Institute for Occupational Safety and Health, Cincinnati, Ohio, 1975.
4. "Safety Requirements for Industrial Protective Helmets for Electrical Workers, Class B," ANSI Z89.2-1971, American National Standards Institute, New York, 1971.
5. R. W. Johnson, "Ignition of Flammable Vapors by Human Electrostatic Discharges," Alleghany Ballistics Laboratory, Hercules, Inc., Cumberland, Md., 1980.
6. W. H. Figard, "Intensifying the Efforts of Proper Glove Selection," *Occup. Health Safety*, **49**(7), 30, 42, 43 (1980).
7. Construction Hazards Committee, "Construction Management Bulletin CM-4.0, Safety Nets," American Insurance Association, New York, 1972.
8. "Safety Nets, Data Sheet 608," National Safety Council, Chicago, Ill., 1967.

BIBLIOGRAPHY

Anon, New Data Announced on Safety Glove Permeation, *Update,* Premier Issue, pp. 1–8, May 1982, Lab Safety Supply, PO Box 1368, Janesville, WI 53547.

Blackman, W. C., Jr., "Enforcement and Safety Procedures for Evaluation of Hazardous Waste Disposal Sites," *Management of Uncontrolled Hazardous Waste Sites,* U.S. EPA National Conference, October 15–17, 1980, Washington, D.C., pp. 91–106.

Fourt, L., "Heat and Moisture Transfer through Fabrics: Biophysics of Clothing," AD 684-949, November 1959.

Federick, E. B. and Henry, M. C., "A Study of Seam Leakage in Coated Fabrics: Summary Report," AD708 874, August 27, 1979. National Technical Information Center, Springfield, VA.

Hart, John A. H., "Cellular Resin Foams Resistant to the Passage of Noxious Chemicals in Liquid and Vapor Forms," Canadian Patent No. 878560, May 4, 1979.

Mychko, A. A., Efremov, V. A. and Andkhin, V. V., "Methods for Determining the Acid Permeability of Special Materials," *Iv. Vyssh. Uchen*, Zaved. Tekhnol. Legk. Promsti, **78**, 21 (6), 8–11 (1980).

Schnabel, G. A., Christofano, E. E., and Harrington, W. H., "Safety and Industrial Hygiene During Investigations of Uncontrolled Waste Disposal Sites," *Management of Uncontrolled Hazardous Waste Site*, U.S. EPA National Conference, October 15–17, 1980, Washington, D.C., pp. 107–110.

Votta, F., Jr., "Flow of Heat and Vapor through Composite Perm-Selective Membranes," AD 671-681, January 1968. National Technical Information Center, Springfield, VA.

Noise and Hearing Protection

Burns, W. and Robinson, D. W., "Hearing and Noise in Industry," Department of Health and Social Security, Her Majesty's Stationery Office, London, 1970.

"Compendium of Materials for Noise Control," Contract No. HSM 99-72-99, HEW Publication No. (NIOSH) 75-165, June 1975.

"Industrial Noise Manual," American Industrial Hygiene Association, Akron, Ohio, 1979.

Lund, A. O., "Noise Control Enclosures for Industrial Equipment," *Am. Ind. Hyg. Assoc. J.,* **40**(11), 961–969 (1979).

Michael, P. L., "Industrial Noise and Conservation of Hearing," *Patty's Industrial Hygiene and Toxicology*, Vol. 1, 3rd rev. ed., Clayton and Clayton, eds. Wiley-Interscience, New York, 1978.

Moselhi, M., "A Six-Year Follow Up Study for Evaluation of the 85 dBA Safe Criterion for Noise Exposure," *Am. Ind. Hyg. Assoc. J.,* **40**(5), 424–426 (1979).

"Noise Control: A Guide for Workers and Employers, Occupational Safety and Health Administration," U.S. Department of Labor, Washington, D.C., 1980.

Reischl, U., "Fire Fighter Noise Exposure," *Am. Ind. Hyg. Assoc. J.,* **40**(6), 482–489 (1979).

Royster, L. H. and Thomas, W. G., "Age Effect Hearing Levels for a White Nonindustrial Noise Exposed Population (NINEP) and Their Use in Evaluating Industrial Hearing Conservation Programs," *Am. Ind. Hyg. J.,* **40**(6), 504–511 (1979).

"Safety Guidelines and OSHA Summaries, Noise and Vibration Analysis, Noise and Vibration Reduction, Hearing Protection," *Best's Safety Directory, Industrial Safety, Hygiene, Security,* Vol. 1, A. M. Best Co., Oldwick, N. J., 1979, Chap. 5, pp 401–459.

Schmidek, M. E., "Survey of Hearing Conservation Programs in Industry," HEW Publication No. (NIOSH) 75-178, June 1975.

Sataloff, J., "Noise Regulation: Resounding Need for Joint Cooperation," *Occup. Health & Safety,* 10–11 (Jan. 1982).

Heffler, A. J., "Audiometry Does Not Equal Hearing Conservation," *Occup. Health & Safety,* 38–40, 42, 45 (March 1982).

Hynes, G., "To Buy or Not to Buy: The Dosimeter vs. Sound Level Meter Question," *Occup. Health & Safety,* 18–25, 57–60 (March 1982).

Thunder, T. D. and Lankford, J. E., "Relative Ear Protector Performance in High vs Low Sound Levels," *Am. Ind. Hyg. Assoc. J.,* **40**(12), 1023–1029 (1979).

"Were noise controls 'techniologically feasible' "? *Occup. Hazards,* 37–38 (January 1981). (See also p. 160, same issue.)

Clothing

ASTMF 739-81, Test Method for Resistance of Protective Clothing Materials to Permeation by Hazardous Liquid Chemicals (Part 46, 1982), ASTM, Philadelphia, PA 19103.

Ahmed, I., "Criteria Used by Companies in Choice of Chemical Protective Clothing (Results of a Survey)," National Safety Council, Chicago, Ill., 1980.

Arons, G. N. and MacNair, R. N., "Laminated, Highly Sorbent, Activated Carbon Fabric," U.S. Patent No. 46733, June 7, 1979.

"ASTM Standard Test Method for Resistance of Protective Clothing Materials to Permeation by Hazardous Liquid Chemicals, Draft of Test Method," ASTM D WXYZ-80, American Society for Testing and Materials, Philadelphia, Pa., 1980.

Berger, M. R., "Safety Clothing—A Matter of Personal Protection," *Natl. Saf. News,* 63–67 (September 1976).

Bienvenue, G. R. and Michael, P. L., "Permanent Effects of Noise Exposure on Results of a Battery of Hearing Tests," *Am. Ind. Hyg. Assoc. J.,* **41**(8), 535–541 (August 1980).

Bosserman, M. W., "How to Test Chemical-Resistance of Protective Clothing," *Natl. Saf. News,* 51–53 (September 1979).

Coplan, M. J. and Lopatin, G., "Preparation of Activated Carbon-Filled Microporous Hollow Multifilament: A Summary Report," AD-A067663, 1979.

"Dressing Right for Safety," *Natl. Saf. News*, 104–107 (March 1980).

Houston, Davis E., "The State of the Art—Safety Wearing Apparel," *Natl. Saf. News*, 115–120 (March 1978).

"If They Have to Wear It, Tell Them to Wear It Right," *Natl. Saf. News*, 136–142 (March 1970).

Lynch, P., "Matching Protective Clothing to Job Hazards," *Occup. Health & Safety*, 30–34 (Jan. 1982).

Michael, P. L. and Bienvenue, G. R., "Hearing Protector Performance—An Update," *Am. Ind. Hyg. Assoc. J.*, **41**(8), 542–546 (August 1980).

Mihal, C. P., Jr., "Effect of Heat Stress on Physiological Factors for Industrial Workers Performing Routine Work and Wearing Impermeable Vapor-Barrier Clothing," *Am. Ind. Hyg. Assoc. J.*, **42**(2), 97–103 (February 1981).

Morrow, R. W. and Hamilton, J. H., "MOCA Permeation of Protective Clothing," Oak Ridge Y-12 Plant, Department of Energy, Y-DK-109, November 7, 1979.

Moshkovich, L. G. and Aleksandrova, T. M., "New Semiwool Fabric with Acid-Resistant Finishing for Protective Clothing," *Tekst. Promst.* (*Moscow*), **78**(2), 52–53 (1979).

NIOSH Technical Report, "Development of Performance Criteria for Protective Clothing Use Against Carcinogenic Liquids," Gerard C. Coletta, DHHS (NIOSH), October, 1978.

Oliver, T. N., "How Protective Clothing Industry Is Coping," *Natl. Saf. News*, 105–110 (July 1974).

"Programming Personal Protection—Everybody Needs Protection," *Natl. Saf. News*, 104–106 (November 1975).

"Protective Clothing and Skin Contact—A Bibliography," First Draft 6-23-80, by I. Ahmed, National Safety Council, Chicago, Ill.

Royster, L. H., "An Evaluation of the Effectiveness of Two Different Insert Types of Ear Protection in Preventing TTS in an Industrial Environment," *Am. Ind. Hyg. Assoc. J.* **41**(3), 161–169 (1980).

Sansone, E. B. and Tewari, Y. B., "The Permeability of Protective Clothing Materials to Benzene Vapor," *Am. Ind. Hyg. Assoc. J.*, **41**(3), 170–174 (1980).

Sansone, E. B. and Tewari, Y. B., "Differences in the Extent of Solvent Penetration through Natural and Nitrile Gloves from Various Manufacturers," *Am. Ind. Hyg. Assoc. J.*, **41**(7), 527 (July 1980).

"Standards Are in Store for Protective Clothing," *Chem. Eng.*, **63**, 76–79 (April 21, 1980).

Tabor, M., "Women in Coal Mines: PPE, Where Are You?" *Occup. Health & Safety*, 22–26, Jan. 1982. (A Safety and Health Personal Protective Equipment Checklist for Women is on p. 25.)

Vo-Dihn, T. and Gammage, R. B., "The Lightpipe Luminoscope for Monitoring Occupational Skin Contamination," *Am. Ind. Hyg. Assoc. J.*, **42**(2), 112–120 (February 1981).

Weeks, R. W., Jr. and McLeod, M. J. "Permeation of Protective Garmet Material by Liquid Benzene," LA - 8164 - MS, Los Alamos Scientific Laboratory, December 1979.

Williams, J. R., "Permeation of Glove Materials by Physiologically Harmful Chemicals," *Am. Ind. Hyg. Assoc. J.*, **40**(10), 877–882 (1979).

Williams, John R., "Chemical Permeation of Protective Clothing," *Am. Ind. Hyg. Assoc. J.*, **41**(12), 884–887 (December 1980).

Head Protection

"Controversy Swirls Around Head Protection," *Occup. Hazards*, **18**, 22–26, 163 (January 1981).

28

Chemical Wastes: New Frontiers for the Chemist and Engineers— RCRA and Superfund

Howard H. Fawcett

Waste is a by-product of most chemical research and production, but until quite recently has received relatively little attention. Some wastes are hazardous, others are not, and the disposal has not always been conducted in a manner which considered the effects, or potential effects, on humans, the environment, the biota, and the long-term economy of the area.

Resource Conservation and Recovery Act

Several incidents came to public attention during the 1970s which created a demand that a more organized approach to waste be structured. A glimpse of poor practices is shown in Fig. 28.1. Congress responded to this demand for adequate control and utilization of discarded materials and wastes by passing the Resource Conservation and Recovery Act (RCRA), as Public Law 94-580, on October 21, 1976.[1] This law, which is cited as 42 USC 6901, the Solid Waste Disposal Act of 1976, gave the incentive to those generating, transporting, treating, storing, and disposing of solid waste to consider carefully their management practices. (As specified in the law, the wastes may be gaseous, liquid, or solid, or any variation of the physical state.) RCRA directed the EPA to promulgate regulations to protect human health and the environment from improper management of hazardous (Subtitle C) and nonhazardous wastes (Subtitle D). It further provides technical and

Figure 28.1. "Real World" practices which are no longer acceptable. Photo courtesy U.S.E.P.A.

financial assistance for the development of management plans and facilities for the recovery of energy and other resources.

Under Subtitle C of RCRA, EPA is given the authority to impose a "cradle to grave" control system for hazardous waste. EPA has defined hazardous waste using the dual approaches of identifying general characteristics and listing specific hazardous waste, sources, and process waste streams. The law establishes the minimum amount of waste generated on a monthly basis that will require adherence to RCRA regulations. Any facility that stores (for more than 90 days), processes, or disposes of hazardous waste must meet the design and operating standards set forth in the regulations. Minimum performance standards are also proposed for incineration processes. An important feature of the act is that it requires the complete tracking of hazardous waste from its point of generation, through each step of processing to the actual disposal of the waste. The tracking will be accomplished by a manifest system that will document each movement of the waste until it is ultimately disposed. Table 28.1 notes several pertinent sections of the RCRA regulations and the date on which they were promulgated by publication in the *Federal Register*.

Subtitle D of the law requires an inventory of all nonhazardous waste disposal sites. Under the regulations promulgated for Subtitle D all open dumps will be closed or upgraded to meet the requirements of a sanitary landfill as defined by the regulations.

Subtitle F of RCRA requires that all Federal agencies comply with all Federal, state, interstate and local regulations stemming from RCRA unless exempted by the President. RCRA must be seen in the context of other laws. For example, it has an interface with the Safe Drinking Water Act (SDWA), PL 95-523, with the Clean Water Act (CWA), PL 95-217, with the Clean Air Act (CAA), PL 95-95, and with the Hazardous Materials Transportation Act (HMTA), PL 95-403. The application of these laws to any facilities will depend on the actual waste treatment and disposal practices. The following examples show how the laws may overlap. If a pathological incinerator is being used, its stack emissions must meet the standards promulgated under CAA. Any transporting of hazardous waste will be regulated by HMTA, as well as by the manifest system established by RCRA. The sludges from water treatment facilities must be treated as solid waste. Some disposal methods may be prohibited or restricted; for example, landfilling or deepwell injection could contaminate ground water and surface water supplies. Any treatment or disposal facility will have to be permitted by EPA, thus ensuring that it meets the requirements of the environmental laws. In addition to the legislation and regulations mentioned above, every Federal agency is required by the Executive Order No. 12196 to meet their responsibilities under the Occupational Safety and Health Act of 1970, 29 CFR Part 160.

For any facility that generates solid waste there are common problems which must be faced by the party or parties responsible. Safety and health considerations should include the facility personnel, contractor personnel, and the community at large. Suitability of the treatment or disposal method must be evaluated. Energy requirements must be a major factor in the management of a waste treatment or disposal facility. The methods of disposal must meet all Federal, state, and local regulations, and must be permitted by the U.S. EPA, state EPA, or other appropriate state or local agency. Other considerations include those associated with the actual managing of a facility, such as identification of waste and quantity, manpower requirements, equipment limitations and operating costs.

EPA estimates that 10% to 15% of the annual production of about 344 million metric tons (wet) of industrial waste is hazardous. Quantities of hazardous waste are expected to increase by 3% annually. EPA estimates that 90% of hazardous waste is managed by practices that will not meet new Federal standards.

TABLE 28.1. Relevant Portions of the Resource Conservation and Recovery Act and Their Status

RCRA Section	Federal Register Date	Status
Section 3001 Identification and listing of hazardous waste	5/19/80	Promulgated
Section 3002 Standards applicable to generators of hazardous waste	2/26/80	Promulgated
Section 3003 Standards applicable to transporters of hazardous waste	2/26/80	Promulgated
Section 3004 Standards applicable to owners and operators of hazardous waste treatment, storage, and disposal facilities	5/19/80	Promulgated
Section 3004 Interior status standards applicable to owners and operators of hazardous waste treatment, storage, and disposal facilities	5/19/80	Promulgated
Section 3005 Permits for treatment, storage and disposal of hazardous waste	5/19/80	Promulgated
Section 3006 Guidelines for authorized state hazardous waste programs	5/19/80	Promulgated
Section 3010 Preliminary notification of hazardous waste activity	2/26/80	Promulgated

Additional Final Rules and Regulation Interpretation
(*with date of publication in* Federal Register)

40 CFR Part 122 November 10, 1980
Consolidated Permit Regulations and Hazardous Waste Management System
40 CFR 261 November 12, 1980
Hazardous Waste Management System: Identification and listing of hazardous waste—Finalizing the lists of hazardous wastes (261.31 and 261.32) and proposal to amend 261.32
40 CFR 122, 260, 264 and 265 November 17, 1980
Hazardous Waste Management System: Suspension of Rules and Proposal of Special Standards for Wastewater Treatment Tanks and Neutralization Tanks
40 CFR 261 November 19, 1980
Hazardous Waste Management System: Mining and Cement Kiln Wastes Exemptions; Small Quantity Generator Standards; Generator Waste Accumulation Amendment; Hazardous Waste Spill Response Exception, and clarification of interim status requirements

TABLE 28.1. *(Continued)*

40 CFR 261, 262, and 265 November 25, 1980
  Hazardous Waste Management System: Clarification of Regulations on Hazardous
  Waste in Containers; Exception of Certain Treated-Wood Wastes; Final List of
  Commercial Products which are Hazardous Wastes if Discarded (S261.33); Exclu-
  sions in Response to Delisting Petitions
Hazardous Waste Management System: Addition of General Requirements for
  Treatment, Storage and Disposal Facilities (40 CFR Part 264); Amendment of
  Interim Status Standards Respecting Closure and Post-Closure Care and Financial
  Responsibility (40 CFR Part 265) and Conforming Amendments to the Permitting
  Requirements (40 CFR Part 122), *Federal Register*, 46(7), 2802–2892 (January 12,
  1981).
Closure, Tank, and Waste Pile Standards for Owners and Operators of Hazardous
  Waste Facilities (40 CFR Parts 264-265), *Federal Register,* 46(7), 2893–2897 (1981).
Hazardous waste: Definitions of "existing hazardous waste management facility,"
  Federal, state or local approvals or permits, and permit prior to construction
  requirement; interim rule and request for comments *Federal Register,* 2344–2348
  (January 9 1981) (40 CFR Parts 122 and 260)

Major hazardous waste generators, among 17 industries EPA has studied in
detail, are (1977 estimates):

Organic chemicals	11.7 million metric tons (wet)
Primary metals	9.0
Electroplating	4.1
Inorganic chemicals	4.0
Textiles	1.9
Petroleum refining	1.8
Rubber and plastics	1.0
Misc. (seven sectors)	1.0
Total	34.5

Seventy to 80% of these industries' hazardous waste is disposed of on the
generator's property: 80% is disposed of in nonsecure ponds, lagoons, or
landfills; 10% is incinerated without proper controls; and 10% is managed
acceptably as compared to proposed Federal standards, i.e., by controlled
incineration, secure landfills, and recovery.

About 50% of hazardous waste is in the form of liquid or sludge.

Ten states generate 65% of all hazardous waste: Texas, Ohio, Pennsyl-

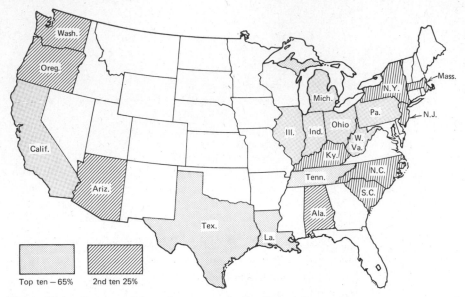

Figure 28.2. Sources of hazardous waste in the United States. Photo courtesy *Science*, A.A.A.S.

vania, Louisiana, Michigan, Indiana, Illinois, Tennessee, West Virginia, and California (Fig. 28.2).[2]

In another survey, conducted by the Subcommittee on Oversight and Investigations of the House Committee on Interstate and Foreign Commerce[3] 53 of the largest domestic chemical companies were asked to give information on their waste generation and disposal (both nonhazardous and hazardous). This survey disclosed that approximately 66 million tons of chemical process wastes were generated in 1978 by the 1605 chemical plants of the 53 companies. No conclusion could be drawn as to what percent of all chemical wastes this amount represents, nor is it known in the absence of final Federal definitions [not available at the time of the survey (1978–1979)] what percent of the 66 million tons of chemical process waste would be classified as hazardous. It was noted that EPA estimated that about 379 million tons of industrial wastes were generated in 1977 by all industry, of which EPA then estimated approximately 39 million tons were hazardous. Thus, the 66 million tons of chemical process wastes generated in 1978 by the 1605 facilities participating in the survey relates to about 17% of the 379 million tons of industrial wastes which EPA estimated were generated in 1977. (An adjustment was made between the two studies to reflect tons, from the metric tons used originally by the EPA study. One metric ton is 2240 lb.)

The House Committee survey revealed that approximately 762 million tons of chemical wastes generated by the 53 participating companies since 1950, or earlier, have been disposed in 3383 locations known to the companies. These sites do not necessarily pose threats to the public health or the environment. Of these 762 million tons, 94% were disposed of on the immediate property of the chemical plants; 6% were sent off-site for disposal (Fig. 28.2).

The major hazards which have been observed from improper disposal have been noted by EPA. Major routes for damage are:

1. Direct contact with toxic wastes.
2. Fire and explosions.
3. Ground water contamination via leachate.
4. Surface water contamination via runoff or overflow.
5. Air pollution via open burning, evaporation, and wind erosion.
6. Poison via the food chain (bioaccumulation).

EPA has documented over 500 cases of damage to health or the environment due to improper hazardous waste management.[4-7]

The Department of Transportation has proposed regulations pursuant to the Hazardous Materials Transportation Act pertaining to transportation of hazardous waste, which were published in the *Federal Register* of May 25, 1978.

To the chemist or engineer, as well as to management and legal staff, the question of what is or is not a hazardous waste is paramount. Two routes or approaches determine the answer. Certain wastes are clearly defined by official notes in the *Federal Register*, as previously noted. Other wastes are either hazardous or nonhazardous depending on whether they do or do not meet the testing criteria[8,8a] for certain characteristics:

1. Ignitability
   a. For liquids, has a flash point less than 60°C (140°F) by the specified methods.
   b. Nonliquids are listed on the basis of being capable of ignition under normal conditions of spontaneous and sustained combustion.
   c. An ignitable compressed gas per Department of Transportation regulations.
   d. An oxidizer per Department of Transportation regulations.
   The EPA hazard code for ignitability is "I," with EPA Hazardous Waste number D001.

2. Corrosivity
   pH is less than 2 or greater than 12.5.
   corrodes steel at a rate greater than one-fourth in./year.
   EPA Hazard Code is "C"; EPA Hazardous Waste Number D002.

3. Reactivity

   Normally unstable—reacts violently.

   Reacts violently with water.

   Forms explosive mixture with water.

   When mixed with water, generates toxic gases, vapors, or fumes.

   Contains cyanide or sulfide and generates toxic gases, vapors, or fumes at pH between 2 and 12.5.

   Capable of detonation if heated under confinement or subjected to strong initiating source.

   Capable of detonation at standard temperature and pressure.

   Listed by Department of Transportation as Class A or Class B explosive.

   EPA Hazardous Waste Number D003.

4. EP Toxicity

   If extract of waste contains concentrations (in mg/liter) greater than:

Arsenic	5.0
Barium	100.0
Cadmium	1.0
Chromium	5.0
Lead	5.0
Mercury	0.2
Selenium	1.0
Silver	5.0
Endrin	0.02
Lindane	0.1
Methoxychlor	10.0
Toxaphene	0.5
2,4–D	10.0
2,4,5–TP	1.0

   The concentrations listed above are 100 times concentrations in the Pure Drinking Water Standards.

   No extract is necessary if waste is liquid containing less than 0.5% filtrate solids.

   EPA Hazardous Waste Number of each of the above constituents D004-D017.

   EPA Hazard Code "E".

   The criteria for listing of hazardous waste as acutely toxic or hazardous:

small doses can cause human mortality

Oral LD_{50}—less than 50 mg/kg

Inhalation LC_{50}—less than 2 mg/liter

Dermal LD_{50}—less than 200 mg/kg

Otherwise, cause or contribute to increased mortality, cause serious irreversible illness, or serious incapacitating reversible illness.

Radioactive and infectious wastes are treated under separate sections.

In addition to the legal aspects of hazardous wastes, the potential economic aspects of improper disposal should be considered. Table 28.2 from the report prepared by the Library of Congress for the Committee on Environment and Public Works of the U.S. Senate, gives graphic illustration of the potential financial problems which have already surfaced.

Although land disposal in "approved" secure sites, where the leachate cannot leave the site due to proper barriers, is a relatively simple and popular approach, there are several other alternatives which might be considered. These include:

Storage under the proper conditions above ground, so the condition of the containers can be checked frequently, and any potential problems removed before serious loss of material to the environment.

Landfilling, which is a form of land disposal, can be useful if the geology and other aspects are correct. However, it should be recalled that Love Canal began as a simple landfilling operation (Fig. 28.3 and Table 28.3).[9] Laboratory analysis, so the exact composition of the wastes is known, is essential before any safe or legal disposal is attempted.

Solar evaporation has limited application in most parts of the United States, but may be a useful approach where the combination of limited rainfall and long cloudless days make the evaporation possible. One achieves some evaporation even of aqueous wastes when the lagoon is in the sun and wind, and the humidity is low.

Acid neutralization is a time-tested procedure used by many industries for acid substances. Normally this involves reacting the acid with a base such as lime or limestone, forming a neutral solid. Every hazardous waste control facility which handles acid wastes should consider this. In addition, of course, highly alkaline materials can be neutralized by acid to bring them into a nonhazardous category.

Chemical fixation is a solidification process, by which liquid wastes are mixed with a solid material which then is disposed as a solid.

Treatment and detoxification involves chemical reactions to change the chemical composition to that of a more innocuous material. Oxidation is a

TABLE 28.2. Selected Cases Illustrating Financial Aspects of Improper Disposal

State	Location	Type Pollution (Date Discovered)	Source	Type of Injury	Compensation Sought	Compensation Obtained
Alabama	Pickwick Reservoir and impounded tributaries near Mobile	Mercury in water (1970)	Dumping	Latent: possible effects from eating fish contaminated with Hg	$157 million	Settled for undisclosed amount
Alabama/Georgia	Lake Weiss, Coosa River	PCB's in water (1976)	Plant in Rome, Ga.	Latent: possible CA, MU, TER effects from eating fish contaminated with PCB	$1.6 billion	Private plaintiffs settled for undisclosed amount; Alabama settled its $100 million suit for $67,900
California	Lathrop	Pesticides in ground; DBCP, alpha-BHC and lindane, and radioactivity in ground water (1979)	Company plant	Acute: none reported unless worker sterility also due to ground-water pollution Latent: possible CA, MU, TER from pesticides and radioactivity in ground water Nat. res: contamination of groundwater Related: 3 domestic wells closed, other wells under study. Possible property value decrease	$30 million civil penalties $15 million clean-up	None to date (June 1980)

Florida	Youngstown	chlorine in air (1978) derailed rail tank car	Acute: 8 dead, 138 injured Latent: possible chronic injury to respiratory tract $1,089,000 property damage; pollution of water in wells; injured persons unable to work	About $750 million	None to date
	(for details, see RAR-78-7, National Transportation Safety Board, Washington, D.C. 20594)				
Michigan	Montague	Toxic chlorinated hydrocarbons in groundwater (1976) Plant waste disposal	Latent: Possible CA, MU, TER Nat. res: 2×10^9 gal of groundwater contam. About 12 wells contaminated Possible property value decline	Fine and $15 million clean-up	Settled for $15 million clean-up and $1 million fine. Performance bond posted
Michigan	Lake Erie	Mercury in lake (1969) Discharges from several companies	Latent: possible TER effects from eating fish contaminated with Hg	About $60 million	Claimed $59,350,000, won $120,000

Source: Adapted from "Six Case Studies of Compensation for Toxic Substances Pollution: Alabama, California, Michigan, Missouri, New Jersey, and Texas," prepared under the supervision of the Congressional Research Service of the Library of Congress, 96th Congress, 2d Session, Serial No. 96-13, Government Printing Office, Washington, D.C., June 1980.

Note: CA = carcinogens; MU = mutagens; TER = teratogens.

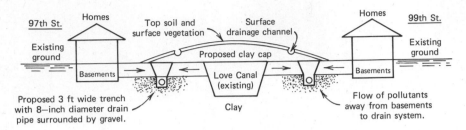

Figure 28.3. Love Canal Remedial Construction Plan: Note leachate drains.

frequently used method of change. Among materials which lend themselves to such treatment are the mercaptans, sulfides, formaldehyde, and cyanide. Oxidation is accomplished by such agents as chlorine, ozone, hydrogen peroxide, and permanganates.

Hydrolysis is the process by which reaction with moisture causes the change in composition by evolution of a toxic gas, such as HCl or HBr. Organophosphate esters lend themselves to hydrolysis by strong alkali.

Deep-well injection is a disposal method which has been widely used in those areas where the material can be filtered, and then injected under

TABLE 28.3. **Concentrations of Some Toxic Organic Chemicals Found in Houses Near Love Canal Dump Site**

Chemical	Highest concentration observed[a] ($\mu g/m^3$)
Chloroform	24
Benzene	270
Trichloroethylene	73
Toluene	570
Perchloroethylene	1,140
Chlorobenzene	240
Chlorotoluene	6,700
Xylene (meta + para)	140
Xylene (ortho)	73
1,3,5-Trichlorobenzene	74
Total organics (one sample, dining room)	12,919

Source: "In the Matter of the Love Canal Chemical Waste Landfill Site," Report of the Commissioner of Health of the State of New York, August 1978. *Chemical and Engineering News,* 6, August 7, 1978.

[a]Except as noted for total organics, concentrations were measured in basements of homes.

pressure into a wellhead into a permeable zone deep beneath the aquifer, into an impermeable zone. A permeable zone may be sandstone or sand; an impenetrable zone is rock or other strata between the aquifer, sealing it off. The main objection to this process is that the material may eventually leak, causing contamination in water supplies or stream.

Farming is a method of disposal in which certain materials can be spread on land, or crops, permitting soil microorganisms to attack it, hoping that what is left is a fertilizer. Much attention must be given to heavy metals and to chlorinated hydrocarbon concentrations.[9a] The method is losing favor at present.

Dewatering is an important step, especially if the material is to be evaporated or transported. Filtration by diatomous earth or activated carbon is often used.

Incineration under proper conditions is the most attractive method of disposal if the material has no further value. Even highly halogenated materials can be disposed of in this manner, if the proper temperatures, residence time, turbulence, and afterburners, with appropriate scrubbers, are installed and maintained properly.[9b,9c] The residue or "ashes" from incinerators must, of course, be disposed of, usually by approved landfilling (Fig. 28.3). If landfilling or burial is undertaken, the proper procedures, including a leachate collection system, should be followed.

Recycling, which is the process of reusing wastes, has become of much interest, since with treatment, materials may often be used more than once in the cycle, or by another organization with completely different needs. As an example, #5 fuel oil may be reclaimed from hazardous waste and burned as fuel. Automobile crank-case drainings, with proper treatment, are also highly useful after reclaimed. Solvents from processes as varied as painting and printing, are being successfully and economically recycled. One authority has stated that between 50 to 80 percent of hazardous chemical wastes can be recycled, with economic gain as well as environmental plus. Figures 28.4 and 28.5, 6 suggest how information about waste change can be of benefit. There has been serious discussions of the possibility of a Recycling Institute, which would make technical information available to encourage recycling of many wastes which currently must be disposed. The Midwest Industrial Waste Exchange, 10 Broadway, St. Louis, Mo. 63102, is an organization which lists availability of materials, and directs the potential user to a source. An annual catalog of wastes is issued to members. Another group is the Pennsylvania Waste Exchange, which is a service of the Pennsylvania Chamber of Commerce, Harrisburg, Pa. (Table 28.4).

Recognizing that the RCRA regulations will present many problems in compliance, the U.S. EPA has established a toll-free service to provide assis-

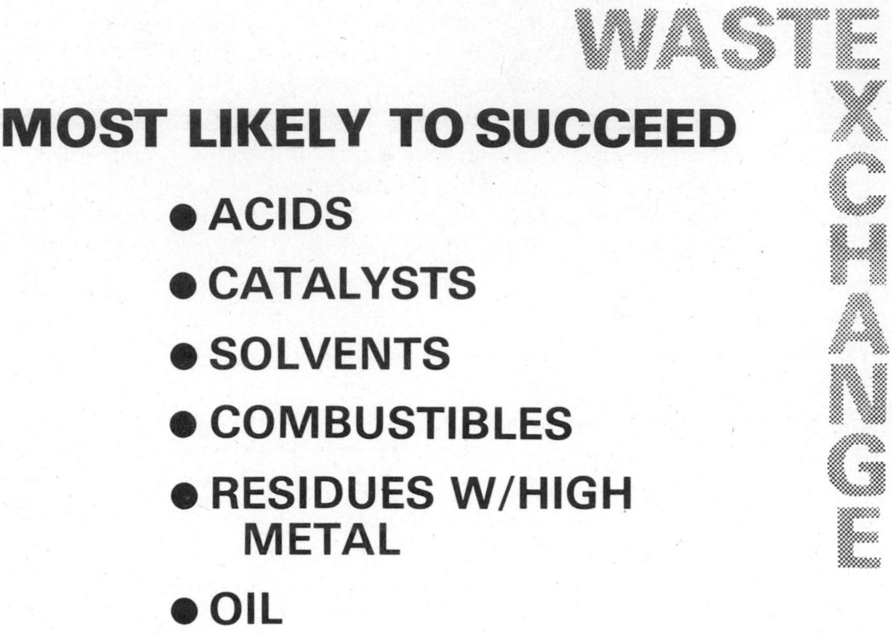

WASTE EXCHANGE
MOST LIKELY TO SUCCEED
• ACIDS
• CATALYSTS
• SOLVENTS
• COMBUSTIBLES
• RESIDUES W/HIGH METAL
• OIL

Figure 28.4. Waste Exchange: Substances most likely to be exchanged. Source: U.S. Environmental Protection Agency, SW-887.1, *Solid Waste.*

STEPS TO SUCCESS
WASTE EXCHANGE
OPEN
CONTACT
INQUIRE
PUBLISH
CODE
LIST
IDENTIFY

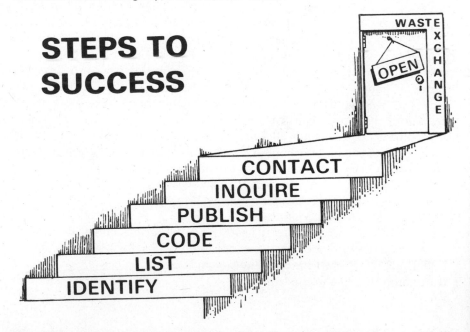

Figure 28.5. Steps to organization and operation of a waste exchange. Source: U.S. Environmental Protection Agency, SW-887.1, *Solid Waste.*

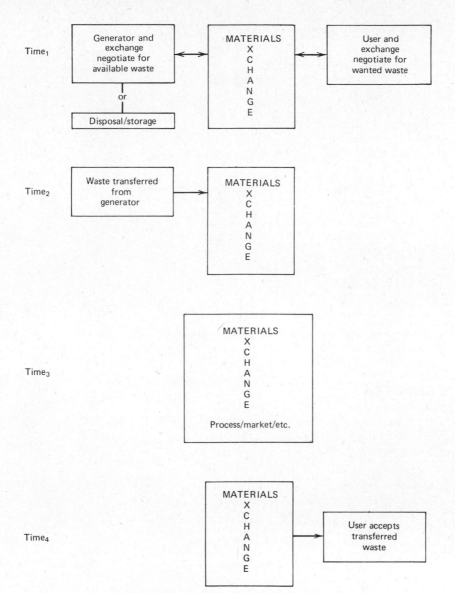

Figure 28.6. Transfer via materials exchange.

TABLE 28.4. United States Waste Exchanges, 1980

California

Department of Health Services
  Hazardous Materials Management Section
  2151 Berkeley Way
  Berkeley, Calif. 94704
Zero Waste Systems, Inc.
  2928 Poplar Street
  Oakland, Calif. 94608

Connecticut

World Association for Safe Transfer and Exchange (WASTE)
  130 Freight Street
  Waterbury, Conn. 06702

Georgia

Georgia Waste Exchange
  Georgia Business & Industry Association
  181 Washington St. SW
  Atlanta, Ga. 30303

Illinois

Environmental Clearinghouse Organization—ECHO
  3426 Maple Lane
  Hazel Crest, Ill. 60429
American Chemical Exchange (ACE)
  4849 Golf Rd.
  Skokie, Ill.
Waste Materials Clearinghouse, Environmental Quality Control, Inc.
  1220 Waterway Boulevard
  Indianapolis, Ind. 46202

Tennessee

Tennessee Waste Swap
  Tennessee Manufacturers Association
  708 Fidelity Federal Building
  Nashville, Tenn. 37219

Texas

Chemical Recycle Information Program
  Houston Chamber of Commerce
  1100 Milam Building, 25th Floor
  Houston, Tex. 77002

Washington

Information Center for Waste Exchange
  2112 Third Avenue, Suite 303
  Seattle, Wash. 98121

TABLE 28.4. *(Continued)*

West Virginia

Union Carbide Corporation
  Invest Recovery Department
  P.O. Box 8361
  Building 3005
  South Charleston, W. Va. 25303

Iowa

Iowa Industrial Waste Information Exchange
  Center for Industrial Research & Service
  201 Building E
  Iowa State University
  Ames, Io. 50011

Massachusetts

The Exchange
  63 Rutland Street
  Boston, Ma. 02118

Michigan

American Materials Exchange Network
  19489 Lahser Road
  Detroit, Mich. 48219

Minnesota

Minnesota Association of Commerce & Industry (MACI)
  200 Hanover Building
  480 Cedar Street
  St. Paul, Minn. 55101

Missouri

Midwest Industrial Waste Exchange
  920 Main Street
  Kansas City, Mo. 64105

New Jersey

Industrial Waste Information Exchange
  New Jersey State Chamber of Commerce
  5 Commerce Street
  Newark, N.J. 07102

New York

Enkarn Research Corporation
  P.O. Box 590
  Albany, N.Y. 12201
The American Alliance of Resources Recovery Interests, Inc. (AARRII)
  111 Washington Avenue
  Albany, N.Y. 12210

TABLE 28.4.   *(Continued)*

North Carolina

Mecklenburg County Waste Exchange
  Mecklenburg County Engineering Department
  1501, I-85 North
  Charlotte, N.C.

Ohio

ORE Corporation—"The Ohio Resource Exchange"
  Columbus Industrial Association
  1646 West Lane Avenue
  Columbus, Ohio 43221
Industrial Waste Information Exchange
  Columbus Industrial Association
  1646 West Lane Avenue
  Columbus, Ohio 43221

Oregon

Oregon Industrial Waste Information Exchange
  Western Environmental Trade Association
  333 SW 5th Suite 618 .
  Portland, Ore. 97204
or
  Resource Conservation Consultants
  1615 N.W. 23 Suite One
  Portland, Ore. 97204

Pennsylvania

Pennsylvania Waste Information Exchange
  Pennsylvania Waste Information Exchange
  (A service of the Pennsylvania Chamber of Commerce)
  222 North Third Street
  Harrisburg, Pa. 17101

Source: "Waste Exchanges, Background Information," SW-887-1 Office of Water and Waste Management, U.S. Environmental Protection Agency, Washington, D.C. 20460, December 1980.

tance to industry and interested citizens. The number is 800-424-9346. Those in the Washington, D.C. area should use 202-554-1404. Calls will be answered by trained professionals Monday through Friday from 9 a.m. to 4:30 p.m. EST. By mail, the assistance may be obtained by addressing Industry Assistance Office, EPA Office of Solid Waste (WH-565), Washington, D.C., 20460.

It should be noted this is the second toll-free "hot-line" which has been established by EPA to extend assistance to industry. The first, for inquirers about the Toxic Substances Control Act, is 800-424-9065. (In the Washington, D.C. area, call 202-554-1404.) Mail inquiries regarding TSCA should be directed to Industry Assistance Office, Office of Toxic Substances (TS-799), U.S. Environmental Protection Agency, Washington, D.C. 20460. In addition to answering specific inquiries, this office maintains an extensive mailing list, so documents and other regulatory reviews may be mailed on a routine basis.

"SUPERFUND" and Its Implications for the Chemist and Engineer

With the passage of PL 96-510, the Comprehensive Environmental Response, Compensation & Liability Act of 1980, signed by President Carter on December 11, 1980, a five-year, $1.6 billion trust will be established to clean up hazardous chemical spills and abandoned waste dumps, and to respond, with remedial action, to the threat of hazardous material release. According to the bill, industry fees contribute 87.5% or $1.38 billion; appropriations of general revenues supply $220 million. Fees are collected on 42 specified substances; 65% of industry's share comes from taxes on petrochemicals, 20% on inorganic chemicals, and 15% on crude oil. Liability is equivalent to that in section 311 of the Clean Water Act, which courts have interpreted as strict liability; common law has established joint and multiple liability for hazardous waste activities, liable party responsible for costs of removal or remedial action incurred by government, and for restoring lost natural resources up to $50 million. Operators of hazardous waste disposal sites will be taxed to set up $200 million post-closure fund to cover costs of monitoring and maintenance of disposal sites after closure. An agency for toxic substances and disease registry will be set up within the Public Health Service. Failure to notify proper agency of existence of disposal facility can elicit fines of up to $10,000 and/or up to one year in jail; failure to notify agency of a release can incur fines of up to three times the cost of clean-up.

With this amount of money, and the public demand for prompt action, it is inevitable that the chemist and engineer will be called upon to analyze hazardous waste itself (relatively easy if a properly equipped laboratory and trained personnel are available to carry out the test procedures outlined by the implementation of the RCRA act), and to pass judgment on the questions which doubtlessly will arise from past practices of waste disposal. While final regulations and criteria have not been issued for disposal site evaluations, and doubtlessly will vary from state to state insofar as implementation is concerned, fundamental knowledge can be applied in a systematic method to more clearly elucidate the problems and their potential solutions. In the final report "Methodology for Rating the Hazard Potential of Waste Disposal Sites," May 5, 1980, prepared for the EPA Office of Research and Development by JRB Associates, several aids are offered to the rating and evaluation of sites insofar as waste disposal is concerned. Included are a worksheet for waste disposal sites and a rating form. This material is quoted in Tables 28.5–28.8. Included is material on the relative persistence or biodegradability of certain organic compounds, quoted from E. F. Abrams report. The reader may also find of value the book by A. DeBruin, *Biochemical Toxicology of Environmental Agents,* Elsevier, Amsterdam, 1976.

One of the more interesting aspects of waste disposal, which brings into focus a largely overlooked field, is that of waste compatibility, or incompatibility. The subject has developed in importance over the years, as more and more chemical combinations are possible, and chemists are apparently not aware of the consequences. Fawcett published a preliminary list of then recognized incompatible substances in 1952.[12] The National Fire Protection Association recognized the problem in 1953, and the author organized and chaired the committee which reviewed and expanded on the work of the late George Jones. From this resulted NFPA 491.M "Manual of Hazardous Chemical Reactions," which has undergone five revisions and lists over two thousand reactions known to be hazardous.[13] The excellent exploratory work which the Dow Chemical Company did under contract with the National Research Council, in behalf of the Committee on Hazardous Materials, provided guidance to the U.S. Coast Guard in regulating the storage and handling of hazardous cargoes on barges and ships.[14] The publications of two editions of the book by Leslie Bretherick, of BP Research Laboratory, *Handbook of Reactive Chemical Hazards,* was a major improvement in available data base.[15] All the above assumed that relatively pure Chemical A would react or not react with Chemical B under certain conditions, often with spectacular results, as pictured in the motion picture "Chemical Boobytraps."[16] However, extensive as our knowledge base is, it is relatively poor insofar as information about the reactions of chemicals in the complex matrices of wastes. Many factors greatly influence waste component reactions, among these are temperature, catalytic effects of dissolved or particulate metals, soil reactions, and reactions with surfaces of transport vehicles or containers. For these reasons, a study of the problem by the California Department of Health Services for the Solid and Hazardous Waste Research Division of the Municipal Environmental Research Laboratory should be of great interest, and the essence of that study is included from their final report[17] (Table 28.5). If the above assignments seem unproductive, they should be viewed in terms of risk/benefit, of human health, and environmental impact which may be significant over many years, to say nothing of the potential legal and financial liability and public relation aspects to the generator, transporter, and disposal-site owner. The alternative is to reengineer the process or operation so hazardous waste is not a by-product, or is part of a closed system. It is the responsibility of chemists and engineers to insure that such illicit dumping of hazardous wastes, as shown in Fig. 28.7 will not reoccur.

TABLE 28.5. Rating Factors and Scales for Each of the Four Generic Areas (Receptors, Pathways, Waste Characteristics, and Waste Management Practices)

Rating Factors	Rating Scale Levels			
	0	1	2	3
Receptors				
Population within 1000 ft	0	1–25	26–100	Greater than 100
Distance to nearest drinking-water well	Greater than 3 miles	1–3 miles	3,001 ft–1 mile	0–3000 ft
Distance to nearest off-site building	Greater than 2 miles	1–2 miles	1,001 ft–1 mile	0–1000 ft
Land use/zoning	Completely remote (zoning not applicable)	Agricultural	Commercial or industrial	Residential
Critical environments	Not a critical environment	Pristine natural areas	Wetlands, floodplains, and preserved areas	Major habitat of an endangered or threatened species
Pathways				
Evidence of contamination	No contamination	Indirect evidence	Positive proof from direct observation	Positive proof from laboratory analyses
Level of contamination	No contamination	Low levels, trace levels, or unknown levels	Moderate levels or levels that cannot be sensed during a site visit but which can be confirmed by a laboratory analysis	High levels or levels that can be sensed easily by investigators during a site visit

TABLE 28.5. (Continued)

Rating Factors	Rating Scale Levels			
	0	1	2	3
Type of contamination	No contamination	Soil contamination only	Biota contamination	Air, water, or foodstuff contamination
Distance to nearest surface water	Greater than 5 miles	1–5 miles	1,001 ft–1 mile	0–1000 ft
Depth to ground water	Greater than 100 ft	51–100 ft	21–50 ft	0–20 ft
Net precipitation	Less than −10 in.	−10–+5 in.	+5–+20 in.	Greater than +20 in.
Soil permeability	Greater than 50% clay	30–50% clay	15–30% clay	0–15% clay
Bedrock permeability	Impermeable	Relatively impermeable	Relatively permeable	Very permeable
Depth to bedrock	Greater than 60 ft	31–60 ft	11–30 ft	0–10 ft
Waste Characteristics				
Toxicity	Sax's level 0 or NFPA's level 0	Sax's level 1 or NFPA's level 1	Sax's level 2 or NFPA's level 2	Sax's level 3 or NFPA's levels 3 or 4
Radioactivity	At or below background levels	1–3 times background levels	3–5 times background levels	Over 5 times background levels
Persistence	Easily biodegradable compounds	Straight chain hydrocarbons	Substituted and other ring compounds	Metals, polycyclic compounds, and halogenated hydrocarbons
Ignitability	Flash point greater than 200° or NFPA's level 0	Flash point of 140°F, to 200°F, or NFPA's level 1	Flash point of 80°–140°F, or NFPA's level 2	Flash point less than 80°F, or NFPA's levels 3 or 4

	NFPA's level 0	NFPA's level 1	NFPA's level 2	NFPA's levels 3 or 4
Reactivity	NFPA's level 0	NFPA's level 1	NFPA's level 2	NFPA's levels 3 or 4
Corrosiveness	pH of 6–9	pH of 5–6 or 9–10	pH of 3–5 or 10–12	pH of 1–3 or 12–14
Solubility	Insoluble	Slightly soluble	Soluble	Very soluble
Volatility	Vapor pressure less than 0.1 mm Hg	Vapor pressure of 0.1–25 mm Hg	Vapor pressure of 25–78 mm Hg	Vapor pressure greater than 78 mm Hg
Physical state	Solid	Sludge	Liquid	Gas
Waste Management Practices				
Site security	Secure fence with lock	Security guard but no fence	Remote location or breachable fence	No barriers
Hazardous waste quantity	0–250 tons	251–1000 tons	1001–2000 tons	Greater than 2000 tons
Total waste quantity	0–10 acre ft	11–100 acre ft	101–250 acre ft	Greater than 250 acre ft
Waste incompatibility	No incompatible wastes are present	Present, but does not pose a hazard	Present and may pose a future hazard	Present and posing an immediate hazard
Use of liners	Clay or other liner resistant to organic compounds	Synthetic or concrete liner	Asphalt-base liner	No liner used
Use of leachate collection systems	Adequate collection and treatment	Inadequate collection or treatment	Inadequate collection and treatment	No collection or treatment
Use of gas collection systems	Adequate collection and treatment	Collection and controlled flaring	Venting or inadequate treatment	No collection or treatment
Use and condition of containers	Containers are used and appear to be in good condition	Containers are used but a few are leaking	Containers are used but many are leaking	No containers are used

Figure 28.7. ''Valley of the Drums'' illustrates the economic, as well as human and environmental waste. Photo courtesy Louisville-Courier-Times paper.

Information Requirements

Data availability is the key to the usefulness of a rating system. Experience has shown that systems that rely on generally available information are the most widely used. In addition, systems that can tolerate data gaps are used more commonly than those systems that cannot. As a result, this rating methodology was designed to be used primarily with readily available information. Sources of this information are identified in Section 28.1. The methodology has also been made flexible enough to allow sites to be rated even if some data is not available. Section 28.2 describes two methods for resolving missing-data problems when rating a site.

28.1. DATA SOURCES

The first consideration in designing the rating methodology was who would be using the system, and where would the necessary information be found. As a result of this consideration, we have identified and compiled data sources for each rating factor. Table 28.6 presents a list of the 31 rating factors together with possible sources of information for each. The various sources of data can be grouped into four categories:

Standard references and indices.

Other published sources.

Contacts with knowledgeable parties.

Site visits.

"Standard reference and indices" includes Sax's *Hazardous Properties of Industrial Materials*, the National Fire Protection Association's *Guide to the Hazardous Properties of Chemicals*, CRC's *Handbook of Chemistry and Physics*, Lang's *Handbook of Chemistry*, the *Handbook of Environmental Data on Organic Chemicals*, and similar publications. "Other published sources" includes the *Federal Register*, technical journals, maps, and reports from the United States Geological Survey, the National Oceanographic and Atmospheric Administration, and the Soil Conservation Service, environmental impact statements for projects in the area of the site, and other publications from environmental groups such as the National Wildlife Federation, and state and local governments. "Contacts with knowledgeable parties" (via telephone, mail or in person) includes such individuals as those in industry (waste generators, transporters, or managers), in Federal, state, and local governments, or in environmental action groups as well as some nearby residents. "Site visits" refers to cases where the rater or some other trained investigator actually visited the site in question.

The availability and quality of these data sources will vary, but should generally be adequate.

28.2. DATA GAPS

Most of the information required for implementing this rating methodology will be available from the readily available data sources listed in Table 28.6. The bulk of this information should be obtained as part of an office-based data gathering effort performed prior to a site visit. The methodology has been designed so that all factors should be rated by the time the site visit is completed. However, to develop a preliminary rating (in order to determine the preferential order of site visits) it is sometimes necessary to handle rating

TABLE 28.6. Sources of Information for Each of the Rating Factors

Rating Factor	Sources of Information
Receptors	
Population within 1000 ft	Local housing officials or census officers
	Current topographic maps or aerial photos
Distance to nearest drinking water well	Information obtained from knowledgeable sources such as Public Health Departments, water supply companies, well drillers, residents
Distance to nearest offsite building	Local housing officials or census officers
	Current topographic maps or aerial photos
Land use/zoning	Land use or zoning maps
	Aerial photos
Critical environments	National Wildlife Federation and other National environmental groups
	State and local environmental groups
	U.S. Fish and Wildlife Service
	State departments of Fish and Game
Pathways	
Evidence of contamination	Information obtained from knowledgeable parties
Level of contamination	Information obtained from knowledgeable parties
Type of contamination	Information obtained from knowledgeable parties
Distance to nearest surface water	USGS topographic maps or reports
	Maps and reports from state or local Highway Departments or from universities or state geological surveys
Depth to ground water	USGS water supply papers, ground water bulletins and geologic reports
	Local well drillers, water suppliers, and universities (geology departments)
Net precipitation	See Fig. 28-A1
	NOAA annual weather summaries
	General precipitation and evapotranspiration maps
Soil permeability	See Fig. 28A-2.
	USDA Soil Conservation Service county maps and reports
	USGS soil maps and reports
Bedrock permeability	USGS water supply papers, groundwater bulletins, and geologic reports
	Local well drillers, water suppliers, and universities (geology departments)
Depth to bedrock	USDA Soil Conservation Service county maps and reports
	USGS soil and geologic maps and reports

TABLE 28.6. *(Continued)*

Rating Factors	Sources of Information
Waste Characteristics	
Toxicity	*Hazardous Properties of Industrial Materials* by N.I. Sax
	National Fire Protection Association's Guide on Hazardous Materials
	Registry of Toxic Effects of Chemical Substances
Radioactivity	Information obtained from knowledgeable parties
Persistence	Appendix A
	Partition Coefficients (see "Partition Coefficients and Bioaccumulation of Selected Organic Chemicals," *Environmental Science and Technology,* **II** (5), 475 (May 1977)
Ignitability	*NFPA Guide on Hazardous Materials*
	Lang's *Handbook of Chemistry*
Reactivity	*NFPA Guide on Hazardous Materials*
	Proposed RCRA Regulations, *Federal Register* (December 18, 1978)
Corrosiveness	Information obtained from knowledgeable parties
Solubility	*CRC Handbook of Chemistry and Physics*
	Lang's *Handbook of Chemistry*
	Merck Index
	Handbook of Environmental Data on Organic Chemicals
Volatility	*CRC Handbook of Chemistry and Physics*
	Lang's *Handbook of Chemistry*
	Handbook of Environmental Data on Organic Chemicals
Physical state	Information obtained from knowledgeable parties
Waste Management Practices	
Site security	Information obtained from knowledgeable parties
Hazardous waste quantity	Information obtained from knowledgeable parties
Total waste quantity	Information obtained from knowledgeable parties
Waste incompatibility	Appendix A
Use of liners	Information obtained from knowledgeable parties
Use of leachate collection systems	Information obtained from knowledgeable parties
Use of gas collection systems	Information obtained from knowledgeable parties
Use and condition of containers	Information obtained from knowledgeable parties

TABLE 28.7. Worksheet for Rating Disposal Sites

WORK SHEET FOR RATING DISPOSAL SITES

NAME OF SITE _____ ACTIVE: INACTIVE: INACTIVE AND ABANDONED (CIRCLE ONE)

LOCATION_____

OWNER/OPERATOR _____

COMMENTS: _____

_____

_____

_____

PREPARED BY: _____ ON _____ 19 _____

FACTOR	OBSERVATION
RECEPTORS	
POPULATION WITHIN 1,000 FEET	
DISTANCE TO NEAREST DRINKING-WATER WELL	
DISTANCE TO NEAREST OFF-SITE BUILDING	
LAND USE/ZONING	
CRITICAL ENVIRONMENT	
USE OF SITE BY RESIDENTS	
USE OF NEAREST BUILDINGS	
PRESENCE OF PUBLIC WATER SUPPLIES	
PRESENCE OF AQUIFER RECHARGE AREA	
PRESENCE OF TRANS-PORTATION ROUTES	
PRESENCE OF IMPORTANT NATURAL RESOURCES	
OTHER:	
PATHWAYS	
EVIDENCE OF CONTAMINATION	
TYPE OF CONTAMINATION	
LEVEL OF CONTAMINATION	
DISTANCE TO NEAREST SURFACE WATER	
DEPTH TO GROUND WATER	
NET PRECIPITATION	
SOIL PERMEABILITY	
BEDROCK PERMEABILITY	
DEPTH TO BEDROCK	
EROSION AND RUNOFF PROBLEMS	
SUSCEPTIBILITY TO FLOODING	
SLOPE INSTABILITY	
SEISMIC ACTIVITY	
OTHER:	

TABLE 28.7. *(Continued)*

WASTE CHARACTERISTICS	
TOXICITY	
PERSISTENCE	
RADIOACTIVITY	
IGNITABILITY	
REACTIVITY	
CORROSIVENESS	
SOLUBILITY	
VOLATILITY	
PHYSICAL STATE	
INFECTIOUSNESS	
BIOACCUMULATION POTENTIAL	
CARCINOGENICITY, TERATO-GENICITY, AND MUTAGENICITY	
OTHER:	
WASTE MANAGEMENT PRACTICES	
SITE SECURITY	
HAZARDOUS WASTE QUANTITY	
TOTAL WASTE QUANTITY	
WASTE INCOMPATIBILITY	
USE OF LINERS	
USE OF LEACHATE COLLECTION SYSTEMS	
USE OF GAS COLLECTION SYSTEMS	
USE AND CONDITION OF CONTAINERS	
LACK OF SAFETY MEASURES	
EVIDENCE OF OPEN BURNING	
DANGEROUS HEAT SOURCES	
INADEQUATE WASTE RECORDS	
INADEQUATE COVER	
OTHER:	

TABLE 28.7. *(Continued)*

RATING FORM FOR WASTE DISPOSAL SITES

NAME OF SITE _____ ACTIVE: INACTIVE: INACTIVE AND ABANDONED (CIRCLE ONE)

LOCATION _____

OWNER/OPERATOR _____

COMMENTS: _____

_____

_____

_____

PREPARED BY: _____ ON _____ 19 _____

RATING FACTOR	SOURCE AND BASIS OF INFORMATION	SITE RATING (CIRCLE ONE)				MULTI-PLIER	SITE SCORE	MAXIMUM POSSIBLE SCORE
RECEPTORS								
POPULATION WITHIN 1,000 FEET		0	1	2	3	12		36
DISTANCE TO NEAREST DRINKING-WATER WELL		0	1	2	3	8		24
DISTANCE TO NEAREST OFF-SITE BUILDING		0	1	2	3	8		24
LAND USE/ZONING		0	1	2	3	6		18
CRITICAL ENVIRONMENTS		0	1	2	3	6		18
ADDITIONAL POINTS FOR OTHER RECEPTORS								50

NUMBER OF MISSING AND ASSUMED VALUES = _____ OUT OF 5.

PERCENTAGE OF MISSING AND ASSUMED VALUES = _____ %.

SUBTOTALS

SUBSCORE (SITE SCORE DIVIDED BY MAXIMUM SCORE AND MULTIPLIED BY 100.)

RATING FACTOR	SOURCE AND BASIS OF INFORMATION	SITE RATING (CIRCLE ONE)				MULTI-PLIER	SITE SCORE	MAXIMUM POSSIBLE SCORE
PATHWAYS								
EVIDENCE OF CONTAMINATION		0	1	2	3	2		6
LEVEL OF CONTAMINATION		0	1	2	3	7		21
TYPE OF CONTAMINATION		0	1	2	3	5		15
DISTANCE TO NEAREST SURFACE WATER		0	1	2	3	8		24
DEPTH TO GROUNDWATER		0	1	2	3	7		21
NET PRECIPITATION		0	1	2	3	6		18
SOIL PERMEABILITY		0	1	2	3	6		18
BEDROCK PERMEABILITY		0	1	2	3	4		12
DEPTH TO BEDROCK		0	1	2	3	4		12
ADDITIONAL POINTS FOR OTHER PATHWAYS								25

NUMBER OF MISSING AND ASSUMED VALUES = _____ OUT OF 9.

PERCENTAGE OF MISSING AND ASSUMED VALUES = _____ %.

SUBTOTALS

SUBSCORE (SITE SCORE DIVIDED BY MAXIMUM SCORE AND MULTIPLIED BY 100)

TABLE 28.7. *(Continued)*

NAME OF SITE _____

WASTE CHARACTERISTICS								
TOXICITY		0	1	2	3	7		21
RADIOACTIVITY		0	1	2	3	7		21
PERSISTENCE		0	1	2	3	5		15
IGNITABILITY		0	1	2	3	3		9
REACTIVITY		0	1	2	3	3		9
CORROSIVENESS		0	1	2	3	3		9
SOLUBILITY		0	1	2	3	4		12
VOLATILITY		0	1	2	3	4		12
PHYSICAL STATE		0	1	2	3	4		12
ADDITIONAL POINTS FOR OTHER WASTE CHARACTERISTICS								20

NUMBER OF MISSING AND ASSUMED VALUES = _____ OUT OF 9.   SUBTOTALS

PERCENTAGE OF MISSING AND ASSUMED VALUES = _____ %.   SUBSCORE
(SITE SCORE DIVIDED BY
MAXIMUM SCORE AND MULTIPLIED BY 100.)

WASTE MANAGEMENT PRACTICES								
SITE SECURITY		0	1	2	3	7		21
HAZARDOUS WASTE QUANTITY		0	1	2	3	7		21
TOTAL WASTE QUANTITY		0	1	2	3	5		15
WASTE INCOMPATIBILITY		0	1	2	3	5		15
USE OF LINERS		0	1	2	3	3		9
USE OF LEACHATE COLLECTION SYSTEMS		0	1	2	3	3		9
USE OF GAS COLLECTION SYSTEMS		0	1	2	3	2		6
USE AND CONDITION OF CONTAINERS		0	1	2	3	2		6
ADDITIONAL POINTS FOR OTHER WASTE MANAGEMENT PRACTICES								30

NUMBER OF MISSING AND ASSUMED VALUES = _____ OUT OF 8.   SUBTOTALS

PERCENTAGE OF MISSING AND ASSUMED VALUES = _____ %   SUBSCORE
(SITE SCORE DIVIDED BY
MAXIMUM SCORE AND MULTIPLIED BY 100)

NUMBER OF MISSING AND ASSUMED
VALUES = _____ OUT OF 31.   TOTAL SITE SCORE _____

TOTAL MAXIMUM POSSIBLE SITE SCORE _____

PERCENTAGE OF MISSING AND
ASSUMED VALUES = _____ %   OVERALL SCORE _____
(TOTAL SCORE DIVIDED BY MAXIMUM SCORE
AND MULTIPLIED BY 100)

factors for which no data is available. There are two ways to approach this problem. First, raters can use their best technical judgment. For example, if there was no information available on a waste's physical state, a rater could either make a guess based on the constituents in the waste or assume that the waste was a liquid (i.e., make a conservative, yet reasonable, assumption). Second, raters can delete factors for which there is no data available on which to base a technical assumption. The resulting score can then be mathematically normalized so that it is comparable to scores for other sites.

In most cases, both of these methods are needed to fill data gaps. It is important that raters note which factors have been evaluated on the worksheet for rating disposal sites (Table 28.7).

CHAPTER 28

Appendix *A*

Guidance for Assessing the Rating Factors

RECEPTORS

Population within 1000 feet is a rough indicator of the potential hazard exposure of the residential population near the site. It is measured from the site boundaries on all sides. Where only houses can be counted (e.g., from an aerial photograph), it can be computed using 3.8 individuals per dwelling unit.

Distance to nearest drinking water well is the distance between a site and the nearest downgradient building that is unlikely to be served by a public water supply (e.g., farmhouses).

Distance to the nearest offsite building is a direct indicator of potential property damage and an indirect measure of potential human exposure. The building's use is considered under the additional point system, and not in assessing this factor.

Land use/zoning is intended to indicate the nature and level of human activity in the vicinity of the site. Impending change in the zoning or land use should be noted in rating the site.

Critical environment is any environment which contains important biological resources, or which is a fragile natural setting that will suffer an especially severe impact from pollution.

PATHWAYS

Evidence of contamination indicates how confident the rater is of his or her contamination information.

Level of contamination signifies how readily apparent and severe a sites hazards are. In addition, it provides a rough indication of the concentration of contaminants at the site.

Type of contamination indicates what media have been contaminated by a site. Since this methodology is for land-based disposal sites, it is assumed that when air or water is contaminated, then so is the biota. And when the biota is contaminated, then so is the soil.

Distance to the nearest surface water is the shortest distance from the perimeter of the site to the nearest body of surface water (e.g., lake or stream) which periodically contains water. This measure provides an indication of the ease with which pollutants may flow overland to surface water bodies.

Depth to ground water is measured vertically from the lowest point of the filled wastes to the highest point of the seasonal water table. This factor provides an indication of the ease with which a pollutant can contaminate ground water.

Net precipitation provides an indication of the potential for leachate generation at a site. It is equal to the mean annual precipitation minus the annual evapotranspiration.

Soil permeability indicates the speed at which a contaminant could migrate from a site. In addition, when considered with soil thickness and permeability, can give an indication of how likely the contaminant is to be attenuated through soil filtration. The following chart gives approximate permeabilities of various soil textural classes and the level assigned to each:

JRB Level	Textural Classes	Approximate Permeability (cm/sec)
3	Sand, loamy sand, sandy loam	$>10^{-2}$
2	Loam, Silty loam, silt	$10^{-2} - 10^{-4}$
1	Sandy clay, clay loam, silty clay loam, sandy clay loam	$10^{-4} - 10^{-5}$
0	Clay, silty clay	$<10^{-6}$

A field guide for determining soil texture is included as Fig. 28A.1.

Bedrock permeability is another indicator of the ease and rapidity with which pollutants can migrate from a site. The following will aid the rater in evaluating this factor:

JRB Level	Rock Types	Approximate Permeability cm/sec
3	Any well-fractured rock, carbonates and evaporites (in humid climates), volcanic igneous rocks (except basalt)	$>10^{-2}$
2	Any slightly fractured rock, uncemented and weakly cemented sandstones, and conglomerates	$10^{-2}-10^{-4}$

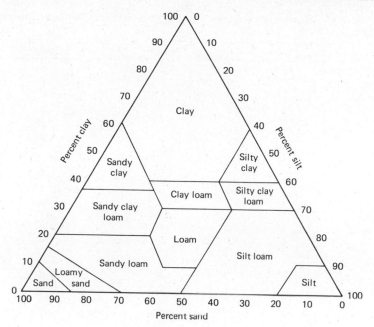

| 1 | Siltstones, moderately and well cemented sandstones, and conglomerates | 10^{-4}–10^{-5} |
| 0 | Unfractured shales, metamorphic and igneous rocks (except volcanics), carbonates and evaporites in an arid climate | $<10^{-6}$ |

Depth to bedrock is the thickness of the overburden (soil and weathered rock) at the site. This provides a measure of the potential for pollutant attenuation through filtration and soil reactions when coupled with soil permeability.

WASTE CHARACTERISTICS

In order to simplify assessing a site's waste characteristics, Table 28.A.1 has been prepared listing ratings for several common chemical compounds. *Toxicity* levels are excerpted from: N.I. Sax, *Hazardous Properties of Industrial Materials* and the *National Fire Protection Association's Guide on Hazardous Materials*.

TABLE 28.A.1.   Waste Characteristics Ratings for Several Common Chemical Compounds

Chemical/Compound	Toxicity	Persistence	Ignitability	Reactivity	Solubility	Volatility
Acetaldehyde	2	0	3	2	3	3
Acetic acid	2	0	2	1	3	1
Acetone	1	0	3	0	0	3
Aldrin	3	3	1	0	0	0
Ammonia	3	0	1	0	1	3
Aniline	3	1	2	0	2	1
Benzene	2	1	3	0	1	3
Carbon tetrachloride	2	3	0	0	0	3
Chlordane	3	3	0	0	0	0
Chlorobenzene	2	2	3	0	0	1
Chloroform	2	3	0	0	3	3
Cresol-0	3	1	2	0	2	1
Cresol-M&P	3	1	1	0	2	1
Cyclohexane	1	2	3	0	0	3
DDT	3	3	0	0	0	0
Dioxin	3	3	0	0	0	0
Endrin	3	3	1	0	0	0
Ethyl Benzene	2	1	3	0	0	1
Formaldehyde	2	0	2	0	2	3
Formic acid	3	0	2	0	3	2
Hexachlorobenzene	1	3	0	0	0	1
Hydrochloric acid	3	0	0	0	2	3
Isopropyl ether	2	1	3	1	3	3
Lindane	2	3	1	0	0	0
Methane	1	1	3	1	2	2
Methyl ethyl ketone	2	0	2	3	3	2
Methyl parathion	3	3	3	2	1	2
Naphtalene	2	1	2	0	0	1
Nitric acid	2	0	0	1	0	2
Parathion	3	3	1	0	0	0
PCB	3	3	0	0	0	0
Petroleum	1	1	3	0	0	1
Phenol	3	1	2	0	2	1
Sulfuric acid	3	0	0	2	1	0
Toluene	2	1	3	0	0	2
αTrichloroethane	2	2	3	2	1	3
Trichlorobenzene	2	3	0	1	0	1
Xylene	2	1	3	0	0	1

Sax's Toxicity Ratings

"In Section 9 in [Sax] (Sax, N.I., Dangerous Properties of Industrial Materials, 5th ed., Section 9, pp. 271–272, Van Nostrand Reinhold, New York, 1979), the following system of toxicity ratings is used to indicate the relative hazard:

U = UNKNOWN

This rating has been assigned to chemicals for which insufficient toxicity data were available to enable a valid assessment of hazard to be made. These compounds usually are in one of the following categories:

(a)   No toxicity information could be found in the literature and none was known to the authors.

(b)   Limited information based on animal experiments was available but in the opinion of the authors this information could not be applied to human exposures. In some cases this information is mentioned so that the reader may know that some experimental work has been done.

(c)   Published toxicity data were felt by the authors to be of questionable validity.

0 = No Toxicity

This designation is given to materials which fall into one of the following categories:

(a)   Materials which cause no harm under any conditions of normal use.

(b)   Materials which produce toxic effects on humans only under the most unusual conditions or by overwhelming dosage.

1 = Slight Toxicity

(a)   Acute local. Materials which on single exposures lasting seconds, minutes, or hours cause only slight effects on the skin or mucous membranes regardless of the extent of the exposure.

(b)   Acute systemic. Materials which can be absorbed into the body by inhalation, ingestion, or through the skin and which produce only slight effects following single exposures lasting seconds, minutes, or hours, or following ingestion of a single dose, regardless of the quantity absorbed or the extent of exposure.

(c)   Chronic local. Materials which on continuous or repeated exposures extending over periods of days, months, or years cause only slight and usually reversible harm to the skin or mucous membranes. The extent of exposure may be great or small.

(d)   Chronic systemic. Materials which can be absorbed into the body by inhalation, ingestion, or through the skin and which produce only slightly usually reversible exposures extending over days, months, or years. The extent of the exposure may be great or small.

In general, those substances classified as having "slight toxicity" produce changes in the human body which are readily reversible and which will disappear following termination of exposure, either with or without medical treatment.

2 = Moderate Toxicity

(a)   Acute local. Materials which on single exposure lasting seconds, minutes, or hours cause moderate effects on the skin or mucous membranes.

These effects may be the result of intense exposure for a matter of seconds or moderate exposure for a matter of hours.

(b)   Acute systemic. Materials which can be absorbed into the body by inhalation, ingestion, or through the skin and which produce moderate effects following single exposures lasting seconds, minutes, or hours, or following ingestion of a single dose.

(c)   Chronic local. Materials which on continuous or repeated exposures extending over periods of days, months, or years cause moderate harm to the skin or mucous membranes.

(d)   Chronic systemic. Materials which can be absorbed into the body by inhalation, ingestion, or through the skin and which produce moderate effects following continuous or repeated exposures extending over periods of days, months, or years.

Those substances classified as having "moderate toxicity" may produce irreversible as well as reversible changes in the human body. These changes are not of such severity as to threaten life to produce serious physical impairment.

3 = Severe Toxicity

(a)   Acute local. Materials which on single exposure lasting seconds or minutes cause injury to skin or mucous membranes of sufficient severity to threaten life or to cause permanent physical impairment or disfigurement.

(b)   Acute systemic. Materials which can be absorbed into the body by inhalation, ingestion, or through the skin and which can cause injury of sufficient severity to threaten life following a single exposure lasting seconds, minutes, or hours, or following ingestion of a single dose.

(c)   Chronic local. Materials which on continuous or repeated exposures extending over periods of days, months, or years can cause injury to skin or mucous membranes of sufficient severity to threaten life or cause permanent impairment, disfigurement, or irreversible change.

(d)   Chronic systemic. Materials which can be absorbed into the body by inhalation, ingestion or through the skin and which can cause death or serious physical impairment following continuous or repeated exposures to small amounts extending over periods of days, months, or years.

NFPAs Toxicity Ratings

"0   Materials which on exposure under fire conditions would offer no health hazard beyond that of ordinary combustible material.

1   Materials only slightly hazardous to health. It may be desirable to wear self-contained breathing apparatus.

2   Materials hazardous to health, but areas may be entered freely with self-contained breathing apparatus.

3   Materials extremely hazardous to health, but areas may be entered with extreme care. Full protective clothing, including self-contained breathing apparatus, rubber gloves, boots and bands around legs, arms and waist should be provided. No skin surface should be exposed.

4   A few whiffs of the gas or vapor could cause death; or the gas, vapor, or liquid could be fatal on penetrating the fire fighters' normal full protective clothing which is designed for resistance to heat. For most chemicals having a Health 4 rating, the normal full protective clothing available to the average fire department will not provide adequate protection against skin contact with these materials. Only special protective clothing designed to protect against the specific hazard should be worn.''

Radioactivity is evaluated in terms of background radioactivity. Background radioactivity is the level of radiation due to natural sources such as cosmic rays, building materials, and naturally radioactive materials in soil and rock. To determine background radiation levels, the investigator should use radiation detection devices such as scintillation detectors and measure the radiation levels at some location close to the site, but not affected by wastes deposited in the site. The elevation of the location where background is measured should be similar to that of the site. This background level is then compared with that found on the site and rated accordingly (see Chapter 37).

Persistence is evaluated on the biodegradability of the wastes. In rating this factor, the most persistent compound should be chosen. A guide for evaluating this factor for organics can be found in Table 28.A.2.

Ignitability provides an indication of the threat posed by fire at a site, and is based on the flammability classification of the NFPA.

JRB Level		NFPA Level
3	4	Very flammable gases, very volatile flammable liquids, and materials that in the form of dusts or mists readily form explosive mixtures when dispersed in air. Shut off flow of gas or liquid and keep cooling water streams on exposed tanks or containers. Use water spray carefully in the vicinity of dusts so as not to create dust clouds.
	3	Liquids which can be ignited under almost all normal temperature conditions. Water may be ineffective on these liquids because of their low flash points. Solids which form coarse dusts, solids in shredded or fibrous form that create flash fires, solids that burn rapidly, usually because they contain their own oxygen, and any material that ignites spontaneously at normal temperatures in air.

TABLE 28.A.2. Persistence (Biodegradability) of Some Organic Compounds

Level 3: Highly Persistent Compounds

Aldrin
Benzopyrene
Benzothiazole
Benzothiophene
Benzyl butyl phthalate
Bromochlorobenzene
Bromoform butanal
Bromophenyl phenyl ether
Chlordane
Chlorohydroxy benzophenone
Bis-chloroisopropyl ether
m-chloronitrobenzene
DDE
DDT
Dibromobenzene
Dibutyl phthalate
1,4-dichlorobenzene
Dichlorodifluoroethane
Dieldrin
Diethyl phthalate
Di(2-ethylhexyl)phthalate
Dihexyl phthalate
Di-isobutyl phthalate
Dimethyl phthalate
4,6-dinitro-2-aminophenol
Dipropyl phthalate
Endrin

Heptachlor
Heptachlor epoxide
1,2,3,4,5,7,7-heptachloronorbornene
Hexachlorobenzene
Hexachloro-1, 3-butadiene
Hexachlorocyclohexane
Hexachloroethane
Methyl benzothiazole
Pentachlorobiphenyl
Pentachlorophenol
1,1,3,3-tetrachloroacetone
Tetrachlorobiphenyl
Thiomethylbenzothiazole
Trichlorobenzene
Trichlorobiphenyl
Trichlorofluoromethane
2,4,6-trichlorophenol
Triphenyl phosphate
Bromodichloromethane
Bromoform
Carbon tetrachloride
Chloroform
Chloromochloromethane
Dibromodichloroethane
Tetrachloroethane
1,1,2-trichloroethane

Level 2: Persistent Compounds

Acenaphthylene
Atrazine
(Diethyl) atrazine
Barbital
Borneol
Bromobenzene
Camphor
Chlorobenzene
1,2-bis-chloroethoxy ethane
b-chloroethyl methyl ether
Chloromethyl ether
Chloromethyl ethyl ether
3-chloropyridine
Di-t-butyl-p-benzoquinone
Dichloroethyl ether
Dihydrocarvone
Dimethyl sulfoxide
2,6-dinitrotoluene

Cis-2-ethyl-4-methyl-1,3-dioxolane
Trans-2-ethyl-4-methyl-1,3-dioxolane
Guaiacol
2-hydroxyadiponitrile
Isophorone
Indene
Isoborneol
Isopropenyl-4-isopropyl benzene
2-methoxy biphenyl
Methyl biphenyl
Methyl chloride
Methylindene
Methylene chloride
Nitroanisole
Nitrobenzene
Tetrachloroethylene
1,1,2-trichloroethylene
Trimethyl-trioxo-hexahydro-triazine isomer

TABLE 28.A.2. *(Continued)*

Level 1: Somewhat Persistent Compounds

Acetylene dichloride	Limonene
Behenic acid, methyl ester	Methyl ester of lignoceric acid
Benzene	Methane
Benzene sulfonic acid	2-methyl-5-ethyl-pyridine
Butyl benzene	Methyl naphthalene
Butyl bromide	Methyl palmitate
e-caprolactam	Methyl phenyl carbinol
Carbon disulfide	Methyl stearate
o-cresol	Naphthalene
Decane	Nonane
1,2-dichloroethane	Octane
1,2-dimethoxy benzene	Octyl chloride
1,3-dimethyl naphthalene	Pentane
1,4-dimethyl phenol	Phenyl benzoate
Dioctyl adipate	Phthalic anhydride
n-dodecane	Propylbenzene
Ethyl benzene	1-terpineol
2-ethyl-n-hexane	Toluene
o-ethyltoluene	Vinyl benzene
Isodecane	Xylene
Isopropyl benzene	

Level 0: Nonpersistent Compounds

Acetaldehyde	Methyl benzoate
Acetic acid	3-methyl butanol
Acetone	Methyl ethyl ketone
Acetophenone	2-methylpropanol
Benzoic acid	Octadecane
Di-isobutyl carbinol	Pentadecane
Docosane	Pentanol
Eicosane	Propanol
Ethanol	Propylamine
Ethylamine	Tetradecane
Hexadecane	n-tridecane
Methanol	n-undecane

Source. Abrams, E. F., "Identification of Organic Compounds in Effluents from Industrial Sources," EPA-560/3-75-002, April 1975.

2	2	Liquids which must be moderately heated before ignition will occur and solids that readily give off flammable vapors. Water spray may be used to extinguish the fire because the material can be cooled to below its flash point.
1	1	Materials that must be preheated before ignition can occur. Water may cause frothing of liquids with this flammability rating number if it gets below the surface of the liquid and turns to steam. However, water spray gently applied to the surface will cause a frothing which will extinguish the fire. Most combustible solids have a flammability rating of 1.
0	0	Materials that will not burn.

Reactivity is a measure of the explosion threat of a site, and is also based on the classification of the NFPA. See also report by H. K. Hatayama, "A Guide for Determining Hazardous Wastes Compatibility," Grant No. R804692, U.S. Environmental Protection Agency, Municipal Environmental Research Laboratory, Cincinnati, 1980.

JRB Level		NFPA Level
3	4	Materials which in themselves are readily capable of detonation or of explosive decomposition or explosive reaction at normal temperatures and pressures. Includes materials which are sensitive to mechanical or localized thermal shock. If a chemical with this hazard rating is in an advanced or massive fire, the area should be evacuated.
	3	Materials which in themselves are capable of detonation or of explosive decomposition or of explosive reaction but which requires a strong initiating source or which must be heated under confinement before initiation. Includes materials which are sensitive to thermal or mechanical shock at elevated temperatures and pressures or which react explosively with water without requiring heat or confinement. Fire fighting should be done from an explosion-resistant location.
2	2	Materials which in themselves are normally unstable and readily undergo violent chemical change but do not detonate. Includes materials which can undergo chemical change with rapid release of energy at normal temperatures and pressures or which can undergo violent chemical change at elevated temperatures and pressures. Also includes those materials which may react violently with water or which may form potentially explosive mixtures with water. In advanced or massive fires, fire fighting should be done from a protected location.

| 1 | 1 | Materials which in themselves are normally stable but which may become unstable at elevated temperatures and pressures or which may react with water with some release of energy but not violently. Caution must be used in approaching the fire and applying water. |
| 0 | 0 | Materials which are normally stable even under fire exposure conditions and which are not reactive with water. Normal fire-fighting procedures may be used. |

Corrosiveness is based on the pH of the wastes and indicates the potential for damage to equipment, fixtures, and any organisms that may contact these wastes.

Solubility signifies how easily contaminants can mix with water, and thus, how readily they can migrate from the site by way of surface and ground waters. It is based on the *CRC Handbook of Chemistry and Physics* as follows:

JRB Level	CRC Level
3	V—Very soluble, infinitely soluble, or miscible
2	S—Soluble
1	δ/S^h—Slightly soluble
0	I—Insoluble

Volatility is a measure of a disposed material's tendency to change from a liquid, solid, or semisolid state, directly to a gaseous state under normal ambient conditions of temperature and pressure. Volatility thus provides a means of rating the potential for air pollution problems due to the disposed material. The material's vapor pressure is the most readily available relative measure of this tendency. Unknown chemical odors at a site should be rated at least a "1" in the absence of more definite data.

Physical state refers to the state of the wastes at the time of disposal. Gases generated by the wastes in a disposal area should not be considered in rating this factor.

Waste Management Practices

Site security is an indication of what positive actions have been taken to limit the exposure of people and animals to waste-related hazards.

Hazardous waste quantity indicates a greater potential hazard for sites with large amounts of hazardous wastes. As an aid in estimating this value, amounts are expressed in three units below. On occasion, raters may have to convert several pieces of quantity data to a common unit in order to add them together.

TABLE 28.A.3. Hazardous Waste Compatibility Chart

Reactivity Group No.	Reactivity Group Name	1	2	3	4	5	6	7	8	9	10	11	12	13
1	Acids, Mineral, Non-oxidizing	1												
2	Acids, Mineral, Oxidizing		2											
3	Acids, Organic	G/H		3										
4	Alcohols and Glycols	H	H/F	H/P	4									
5	Aldehydes	H/P	H/F	H/P		5								
6	Amides	H	H/GT				6							
7	Amines, Aliphatic and Aromatic	H	H/GT	H	H			7						
8	Azo Compounds, Diazo Compounds, and Hydrazines	H/G	H/GT	H/G	H/G	H			8					
9	Carbamates	H/G	H/GT						G/H	9				
10	Caustics	H	H	H	H					H/G	10			
11	Cyanides	GT/GF	GT/GF	GT/GF					G			11		
12	Dithiocarbamates	H/GF/F	H/GF/F	H/GF/GT		GF/GT	U	H/G					12	
13	Esters	H	H/F						H/G		H			13
14	Ethers	H	H/F											
15	Fluorides, Inorganic	GT	GT	GT										
16	Hydrocarbons, Aromatic	H/F												
17	Halogenated Organics	H/GT	H/F/GT					H/GT	H/G		H/GF	H		
18	Isocyanates	H/G	H/F/GT	H/G	H/P			H/P	H/G		H/P	G	U	
19	Ketones	H	H/F						H/G		H	H		
20	Mercaptans and Other Organic Sulfides	GT/GF	H/F/GT						H/G					
21	Metals, Alkali and Alkaline Earth, Elemental	GF/H/F	GF/H/F	GF/H/F	GF/H/F	GF/H/F	GF/H	GF/H	GF/H	GF/H	GF/H	GF/H	GF/GT/H	GF/H
22	Metals, Other Elemental & Alloys as Powders, Vapors, or Sponges	GF/H/F	GF/H/F	GF					H/F/GT	U	GF/H			
23	Metals, Other Elemental & Alloys as Sheets, Rods, Drops, Moldings, etc.	GF/H/F	GF/H/F						H/F/G					
24	Metals and Metal Compounds, Toxic	S	S	S			S	S			S			
25	Nitrides	GF/H	H/F/E	H/GF	GF/H/E	GF			U	H/G	U	GF/H	GF/H	GF/H
26	Nitrites	H/GT/GF	H/F/GT	H					U					
27	Nitro Compounds, Organic	H/F/GT			H				H/E					
28	Hydrocarbons, Aliphatic, Unsaturated	H	H/F		H									
29	Hydrocarbons, Aliphatic, Saturated	H/F												
30	Peroxides and Hydroperoxides, Organic	H/G	H/E	H/F	H/G		H/GT	H/F/E	H/F/GT		H/E/GT	H/F/GT		
31	Phenols and Cresols	H	H/F						H/G					
32	Organophosphates, Phosphothioates, Phosphodithioates	H/GT	H/GT					U	H/E					
33	Sulfides, Inorganic	GT/GF	HF/GT	GT	H		E							
34	Epoxides	H/P	H/P	H/P	H/P	U		H/P	H/P		H/P	H/P	U	
101	Combustible and Flammable Materials, Miscellaneous	H/G	H/F/GT											
102	Explosives	H/E	H/E	H/E				H/E	H/E					H/E
103	Polymerizable Compounds	P/H	P/H	P/H				P/H	P/H	H	U			
104	Oxidizing Agents, Strong	H/GT		H/GT	H/F	H/F	H/F/GT	H/E	H/F		H/E	H/F/GT	H/F/GT	H/F
105	Reducing Agents, Strong	H/GF	H/F/GT	H/GF	GF/F	GF/H	H/F	GF	H/G			H/GT	H/F	
106	Water and Mixtures Containing Water	H	H						G					
107	Water Reactive Substances	EXTREMELY REACTIVE! DO NOT MIX WITH ANY CHEMICAL OR WASTE MATERIAL! EXTREMELY REACTIVE!												

Source. *A Method for Determining Hazardous Wastes Compatibility,* by H. K. Hatayama et al., Grant No. R804692, for Municipal Environmental Research Laboratory, Office of Research and Development, U.S.E.P.A., Cincinnati, OH 45268, 1980.

Reactivity Code	Consequences
H	Heat generation
F	Fire
G	Innocuous and non-flammable gas generation
GT	Toxic gas generation
GF	Flammable gas generation
E	Explosion
P	Violent polymerization
S	Solubilization of toxic substances
U	May be hazardous but unknown

Example:

H
F
GT   Heat generation, fire, and toxic gas generation

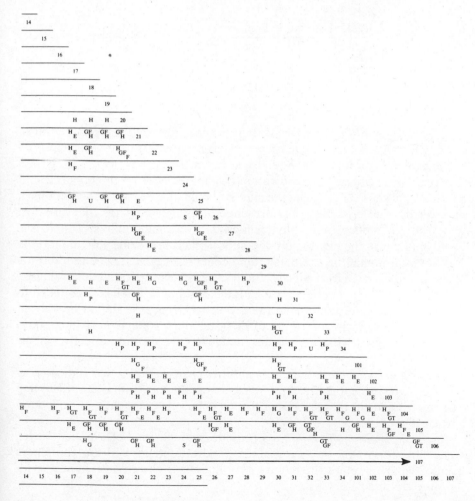

Level	Volume (yd³)	Weight (short tons)	Drums (55 gal)
3	>2370	>2000	>650
2	1191–2370	1001–2000	326–650
1	300–1190	250–1000	80–325
0	<300	<250	<80

Total waste quantity is used as an indication of a site's size, and in conjunction with *hazardous waste quantity* as an estimate of what proportion of the total wastes are hazardous. Additional volume units are given below. Raters can estimate a site's total waste quantity by multiplying the area of the site (in acres) by the average thickness of the wastes (in feet) to obtain acre-feet.

JRB Level	Acre-Feet	Cubic Yards
3	>250	>400,000
2	100–250	161,000–400,000
1	10–100	16,000–161,000
0	<10	<16,000

Waste incompatibility can lead to explosion, fire, and the production of toxic fumes. A table of incompatible wastes is included in Table 28.A.3 as an aid in evaluating this factor. See also Chapter 22.

Use of liners provides an indication of improvement on the impermeability of natural soils occurring at the site. Although there are wide variations in liner quality and durability, the presence of a liner indicates that an effort was made to reduce or eliminate migration of contaminants, which is a positive factor in the evaluation of disposal sites. The scale shown in Table 28.A-1 assumes that the site being rated contains organics, which represents a "worst case" scenario. Sites without organics can be rated accordingly.

Use of leachate collection systems indicates that action has been taken to prevent groundwater contamination problems. Hence, it should be viewed as a positive factor in the rating of disposal sites. Raters must judge on a case-by-case basis what constitutes "adequate" collection and treatment. See Fig. 28.3.

Use of a gas collection system is another positive factor that indicates an effort is being made to control the release of pollutants from a site. Sampling and rater judgment may be necessary to determine adequacy of collection and treatment.

Use and condition of containers provides an indication of whether wastes were dumped in bulk or in containers, and the integrity of the containers used. Containers retard the release of pollutants to the environment providing they are intact.

REFERENCES

1. Resource Conservation and Recovery Act, Public Law 94-580, October 21, 1976.
2. Waste Alert: (17 articles), *EPA J,* 5(2) (February 1979).
3. "Waste Disposal Site Survey," Report Together with Additional and Separate Views by the Subcommittee on Oversight and Investigations of the Committee on Interstate and Foreign Commerce, House of Representatives, 96th Congress, First Session, Committee Print 96-IFC 33, October 1979.
4. "Six Case Studies of Compensation for Toxic Substances Pollution: Alabama, California, Michigan, Missouri, New Jersey and Texas," a report prepared under the supervision of the Congressional Research Service of the Library of Congress for the Committee on Environment and Public Works, U.S. Senate, 96th Congress, 2d Session, Serial No. 96-13, June 1980.
5. "Health Effects of Toxic Pollution: A Report from the Surgeon General and A Brief Review of Selected Environmental Contamination Incidents with a Potential for Health Effects," reports prepared by the Surgeon General, Department of Health and Human Services and the Congressional Research Service, Library of Congress, for the Committee on Environment and Public Works, U.S. Senate, 96th Congress 2d Session, Serial No. 96-15, August 1980.
6. "Interim Report on Ground Water Contamination: Environmental Protection Agency Oversight," Twenty-Fifth Report by the Committee on Government Operations, Union Calendar No. 874, House Report No. 96-1440, September 30, 1980.
7. "Management of Hazardous Chemical Wastes," Smithsonian Science Information Exchange, Washington, D.C., August 1979.
8. "Test Methods for Evaluating Solid Waste (SW-846), Waste Analysis Program," Office of Solid Waste, U.S. Environmental Protection Agency (WH-565), Washington, D.C., 1980.
   a. "Technical Update, Physical/Chemical Methods, Test Methods for Evaluating Solid Waste, Revision A," SW-846, Office of Water and Waste Management, U.S. Environmental Protection Agency, Washington, D.C., August 8, 1980.
9. R. Fung, "Protective Barriers for Containment of Toxic Materials," Pollution Technology Review No. 66, Noyes Data Corp., Park Ridge, N.J., 1980.
   a. D. J. DeRenzo, "Biodegradation Techniques for Industrial Organic Wastes, Pollution Technology Review No. 65, Noyes Data Corp., Park Ridge, N.J., 1980.
   b. H. H. Fawcett, "Experimental Facility for the Disposal of Waste Solvents," Report No. 61-GP-216, General Electric Research Laboratory, Schenectady, New York, 1962.
   c. R. D. Ross, "Incineration—A Positive Solution to Hazardous Waste Disposal, presented to American Chemical Society Division of Chemical Health and Safety, Washington, D.C., September 10, 1979.
10. "Methodology for Rating the Hazard Potential of Waste Disposal Sites," prepared by JRB Associates for Office of Research and Development, U.S. Environmental Protection Agency, Washington, D.C., May 5, 1980.
11. A. DeBruin, *Biochemical Toxicology of Environmental Agents,* Elsevier, Amsterdam, 1976.
12. H. H. Fawcett, "Safety and Industrial Hygiene in the Laboratory," *Chem. Eng. News,* 30 (251), 2588–2591 (June 23, 1952).
13. NFPA, *Manual of Hazardous Chemical Reactions,* 5th ed., 491M, National Fire Protection Association, Boston, Mass., 1975.
14. M. D. Morrissette, "Hazard Evaluation of Chemicals for Bulk Marine Shipment," 3(1), 33–48 (1979) and Navigation and Vessel Inspection Circular No. 4-75, U.S. Coast Guard, December 1975.

15. Leslie Bretherick, *Handbook of Reactive Chemical Hazards,* 2nd ed. Butterworth, London, 1979.
16. "Chemical Boobytraps," 10 min, 16 mm sound color movie, available from General Electric Co., Audiovisuals Section, Scotia, New York, 1959.
17. H. K. Hatayama, "A Method for Determining Hazardous Wastes Compatibility," Grant No. R804692, Municipal Environmental Research Laboratory, Office of Research and Development, U.S. Environmental Protection Agency, Cincinnati, Ohio.

BIBLIOGRAPHY

American Chemical Society Division of Chemical Health and Safety, 183rd National Meeting, Las Vegas, Nev., March 28–April 2, 1982.

    (a)   Jameson, C. W. and Walters, D. B., co-chair, Symposium on Chemistry and Safety for Toxicity Testing of Environmental Chemicals (36 papers). See Meeting Abstracts for details.

    (b)   Young, J. A., chair, Symposium on Fire Toxicity (5 papers). See Meeting Abstracts for details.

    (c)   Taggart, W. P., chair, Symposium on Laboratory Waste Disposal (12 papers). See Meeting Abstracts for details.

Acute and Chronic Toxicity of HCN to Fish and Invertebrates, NTIS, PB-293 047/7BE, 1979.

Adams, J. and Detjen, J., Bibliography of Chemical Waste Disposal 1969–1979, Developed during research for "Warning: Toxic Waste," November 25, 1979, *Courier-Journal,* Louisville, Ky.

Aerial Reconnaissance of Hazardous Substances Spills and Spill-Threat Conditions, National Technical Information System, PB 294 980/8BE, 1979. Springfield, Va.

Alter, H. and Dunn, J. J., *Solid Waste Conversion to Energy: Current European and U.S. Practice,* Marcel Dekker, New York, 1980.

Aquatic Toxicology (Third Symposium), American Society for Testing and Materials, Philadelphia, Pa., STP 707, 1980.

"Aquatic Toxicology," STP 667, American Society for Testing and Materials, Philadelphia, Pa., 1980.

"Aquatic Toxicology and Hazard Evaluation," STP 634, American Society for Testing and Materials, Philadelphia, Pa., 1980.

Azad, H. S., *Industrial Wastewater Management Handbook,* McGraw-Hill, New York, 1976.

"Bacterial Indicators/Health Hazards Associated with Water," STP 635, American Society for Testing and Materials, Philadelphia, Pa., 1980.

Berry, E. E. and MacDonald, L. P., "Experimental Burning of Used Automotive Crankcase Oil in a Dry-Process Cement Kiln," *J. Haz. Mater.* 1 (2), 137–156 (1976).

"Biological Data in Water Pollution Assessment: Quantitative and Statistical Analyses," STP 652, American Society for Testing and Materials, Philadelphia, Pa., 1980.

"Biological Monitoring of Water and Effluent Quality," STP 607, American Society for Testing and Materials, Philadelphia, Pa., 1980.

Brown, M., *Laying Waste,* Pantheon Press, 1980.

Buikema, A. L., Jr. and Cairns, John, Jr., "Aquatic Invertebrate Bioassays," ASTM/STP 715, American Society for Testing and Materials, Philadelphia, Pa., 1980.

Cally, A. G. *Treatment of Industrial Effluents,* Hodder and Stoughton, 1978.

Chappell, C. L. and Wellets, S. L., "Some Independent Assessments of the SEALOSAFE/STABLEX Method for Toxic Waste Treatment, *J. Haz. Mater.* 3(4), 285–292 (1980).

"Characteristics of Hazardous Waste Transportation and Economic Impact Assessment of Hazardous Waste Transportation Regulations," National Technical Information Service, PB-296 497/1BE, Springfield, Va., 1979.

Chen, E., *PBB: An American Tragedy,* Prentice-Hall, Englewood Cliffs, N.J., 1979.

Cheremisinoff, P. N. and Morresi, A. C., *Energy from Solid Wastes,* Marcel Dekker, New York, 1976.

"Control of Hazardous Material Spills," *Proceedings of the 1980 National Conference on Control of Hazardous Material Spills,* May 13–15, 1980, Louisville, Kentucky, sponsored by the U.S. Environmental Protection Agency, U.S. Coast Guard and Vanderbilt University.

Cook, R., *Fever*, Putnam Press, N.Y., 1982.

Davies, D. R., and Mackay, G. A., "Recent Developments in the Transport of Liquid Wastes," *J. Haz. Mater.* **1**(3), 199–214 (1976).

Dawson, G. W., "The Acute Toxicity of 47 Industrial Chemicals to Fresh and Saltwater Fishes," *J. Haz. Mater.* **1**(4), 303–318 (1977).

Dawson, G. W., "Treatment of Hazardous Materials Spills in Flowing Streams with Floating Mass Transfer Agents, *J. Haz. Mater.* **1**(1), 65–82 (1975).

"Destroying Chemical Wastes in Commercial-Scale Incinerators," SW 122, 6 volumes, U.S. Environmental Protection Agency, Washington, D.C., 1977.

"Estimating the Hazard of Chemical Substances to Aquatic Life," STP 657, American Society for Testing and Materials, Philadelphia, Pa., 1980.

Environmental Law Handbook, 6th ed. Government Institutes, Washington, D.C., 1979.

"Environmental Statutes, 1981," Government Institutes, Inc., Washington, D.C., 1981.

Publications available from U.S. Environmental Protection Agency Office of Water and Waste Management, Washington, D.C. 20640.

> SW-826 Everybody's Problem: Hazardous Waste," 1980
> "SW-800 Waste Alert! A Citizen's Introduction to Public Participation in Waste Management," 1979.
> SW-814 "What Is Waste Alert?," 1979.
> SW-830 "Operating A Recycling Program: A Citizen's Guide," 1980.
> SW-832 "Source Separation/Waste Reduction," 1980.
> SW-831 "Cleaning Up America's Dumps: State Solid Waste Management Plans Under RCRA," 1980.
> SW-833 "Public Participation Requirements for Federal, State, and Local Agencies," 1980.
> SW-834 "Solid Waste Management Programs under RCRA," 1980.
> SW-737 "Hazardous Waste Facts," 1979.

Finley, S., *Hazardous Waste Options,* 16 mm sound movie, 28 min., 1981, Stuart Finley, Inc., 3428 Mansfield Rd., Fallschurch, Va. 22041.

Gilmore, W. R., "Radioactive Waste Disposal—Low and High Level," Noyes Data Corp., Park Ridge, N.J., 1977.

Hanson, D. J., Progress under Superfund criticized, defended, *Chem. and Engr. News,* **60**(23), 10–15 (June 7, 1982).

"Hazardous Materials Spill Monitoring Safety Handbook and Chemical Hazard Guide: Part A," PB 295 853/6BE; "Part B," PB 295 854/4BE, 1979.

"Hazardous or Difficult to Handle Waste Survey Report," National Technical Information Service, PB-292-968/5BE, 1979.

"Hazardous Substances Summary and Full Development Plan," National Technical Information Service PB-289 923/5BE, 1979.

"Hazardous Waste Site Investigation Training Manual," prepared by the FIT National Project

Management Office, Ecology and Environment, Inc., Arlington, Va. under EPA contract No. 68-01-6056 (TDD: HQ-8008-04), 1980.

Jones, C. J., "The Ranking of Hazardous Materials by Means of Hazard Indices," *J. Haz. Mater.* **2**(4), 363–389 (1978).

Jones, C. J., "Absorption of Some Toxic Substances by Waste Components," *J. Haz. Mater.* **2** (3), 219–226 (1978).

Jones, C. J., "The Leaching of Some Halogenated Organic Compounds from Domestic Waste, *J. Haz. Mater.* **2**(3), 1978.

Jones, C. J., "An Investigation of the Evaporation of Some Volatile Solvents from Domestic Waste Landfill Conditions, *J. Haz. Mater.* **2**(3), 259–290 (1978).

Jones, C. J., "Evaporation of Mercury from Domestic Waste Leachate," *J. Haz. Mater.* **2**(3), 253–258 (1978).

Jones, C. J., "An Investigation of the Degradation of Some Dry Cell Batteries under Domestic Waste Landfill Conditions, *J. Haz. Mater.* **2**(3), 259–29 (1978).

Jones, C. J., "The Combustion and Pyrolysis of Some Halogenated Organic Compounds in a Laboratory Tube Furnace, *J. Haz. Mater.* **2**(3), 291–296 (1978).

Lowrance, W. W., ed. *Assessment of Health Effects at Chemical Disposal Sites,* William Kaufmann, Inc., Los Altos, Calif., 1981.

Lazar, E. C., "Damage Incidents from Improper Land Disposal, *J. Haz. Mater.* **1**(2), 157–164 (1976).

"Liners for Sanitary Landfills and Chemical and Hazardous Waste Disposal Sites," National Technical Information Service, PB-293 335/6BE, 1979.

"Measurement of Organic Pollutants in Water and Wastewater," STP 686, American Society for Testing and Materials, Philadelphia, Pa., 1980.

"Metal Bioaccumulation in Fishes and Aquatic Invertebrates, A Literature Review," National Technical Information Service, PB-290 659/2BE.

"Methodology for Biomass Determinations and Microbial Activities in Sediments," STP 673, American Society for Testing and Materials, Philadelphia, Pa., 1980.

"Methods and Measurements of Periphyton Communities: A Review," STP 690, American Society for Testing and Materials, Philadelphia, Pa., 1980.

Meyer, E., *Chemistry of Hazardous Materials*, Prentice-Hall, Inc., Englewood Cliffs, NJ, 1976.

National Wildlife Federation, "The Toxic Substances Dilemma, A Plan for Citizen Action," National Wildlife Federation, Washington, D.C., 1980.

"Native Aquatic Bacteria: Enumeration, Activity and Ecology," STP 695, American Society for Testing and Materials, Philadelphia, Pa., 1980.

Neissen, W., *Combustion and Incineration Processes*, Marcel Dekker, New York, 1978.

"Oil and Hazardous Materials: Emergency Procedures in the Water Environment, CWR 10-1," U.S. Federal Water Pollution Control Administration, Northeast Region, North Atlantic Water Quality Management Center, Edison, N.J., October, 1968.

Parves, D., *Trace-Element Contamination of the Environment,* Elsevier, Amsterdam and New York, 1977.

Patrick, P. K., "Treatment and Disposal of Hazardous Wastes in Western Europe, *J. Haz. Mater.* **1**(1), 45–58 (1975).

Powers, P. W., "How to Dispose of Toxic Substances and Industrial Wastes," Noyes Data Corp., Park Ridge, N.J., 1976.

Quinlivan, S. C. et al., "Sources, Characteristics and Treatment and Disposal of Industrial Wastes Containing Hexachlorobenzene," *J. Haz. Mater.* **1**(4), 343–360 (1977).

Renolds, J. M., "Controlling the Use of Hazardous Materials in Research and Development Laboratories, *J. Haz. Mater.* **2**(4), 299–308 (1978).

"Report of the Interagency Ad Hoc Work Group for the Chemical Waste Incinerator Ship Program," U.S. Environmental Protection Agency, U.S. Dept. of Commerce Maritime Administration, U.S. Coast Guard and National Bureau of Standards, September 1980.

Resource Recovery and Conservation (An International Journal) Quarterly, Elsevier, Amsterdam and New York.

Rogers, H. W. et al., "Problems of Waste Chemical Handling at a Large Biomedical Research Facility, *J. Haz. Mater.* **4**(2), 191–205 (1980).

Rohlich, G. A., "Summary Report, Drinking Water and Health," Committee on Safe Drinking Water, National Research Council, Washington, D.C., 1977.

Roulier, M. H. et al., "Current Research on the Disposal of Hazardous Wastes," *J. Haz. Mater.* **1**(1), 59–64 (1975).

Schnabel, G. A., Christofano, E. E., and Harrington, W. H., "Safety and Industrial Hygiene During Investigations of Uncontrolled Waste Disposal Sites," pages 107–110, and Blackman, W. C., Jr. "Environmental and Safety Procedures for Evaluation of Hazardous Waste Disposal Sites," pages 91–106; in *Management of Uncontrolled Hazardous Waste Sites,* U.S. EPA National Conference, October 15–17, 1980, Washington, D.C.

Silvestri, A. et al., "Development of a Kit for Detecting Hazardous Material Spills in Waterways," EPA-600/2-78-055, Environmental Protection Technology Series, U.S. Environmental Protection Agency, Industrial Environmental Research Laboratory, Cincinnati, Ohio, 1978.

Sittig, J., "Toxic Metals, Pollution Control and Waste Protection," Noyes Data Corp., Park Ridge, New Jersey, 1976.

Small, W. E., *Third Pollution: The National Problem of Solid Waste Disposal,* Praeger Publishers, New York, 1970.

Smith, A. J., *Managing Hazardous Substances Accidents,* McGraw-Hill, New York, 1981.

Smith, W. E. and Smith, A. M., *Minamata, The Story of the Poisoning of a City* (with Mercury) and of the People who Choose to Carry the Burden of Courage, Alskog-Sensorium Book, Holt, Rinehart & Winston, New York, 1975.

Spiro, T. G. and Stigliani, W. M., "The Chemical Bases of Environmental Issues," ACS Audio Course, cassettes with manual, American Chemical Society, Washington, D.C., 1980.

"Sulfur Bacteria," by Fjerdingstod, E., STP 650, American Society for Testing and Materials, Philadelphia, Pa., 1980.

Taylor, J. M. and Parr, J. F., "Considerations in the Land Treatment of Hazardous Wastes: Principles and Practices," Biological Waste Management and Organic Resources Laboratory, Agricultural Environmental Quality Institute, Science and Education Administration, U.S. Dept. of Agriculture, Beltsville, Maryland.

Thorne, P. F., "The Dilution of Flammable Polar Solvents by Water for Safe Disposal, *J. Haz. Mater.* **2**(4), 321–331 (1978).

Throop, W. M., "Alternate Methods of Phenol Wastewater Control, *J. Haz. Mater.* **1**(4), 319–330 (1977).

Valovic, T. S., "Hazardous Waste Management: Overview of Proposed EPA Regulations and Their Impact Upon Industry," CU Risk Management, Inc., Boston, Mass., April 1980.

Voss, G., "Energy and Resource Recovery from Solid Waste," AAAS Intergovernmental Research and Development Project, September 24–26, 1979, Lanham, Md., American Association for the Advancement of Science, Washington, D.C. 20036.

Wakeman, R. J., "Filtration Post—Treatment Processes," *Chemical Engineering Monographs,* Vol. 2, Elsevier, Amsterdam and New York, 1975.

Willmann, J. C., "Case History: PCB Transformer Spill," *J. Haz. Mater.* **1**(4), 361–372 (1977).

Wulfinghoff, Max, "Disposal of Process Wastes, Liquids, Solids, Gases," a symposium presented at the ACHEMA Meeting, 1964, Frankfurt/Main, Germany, Chemical Publishing Co., New York 1968 (translation).

RCRA and SUPERFUND

Alsop, R., "Chemical, Gas Firms Seek Legal Immunity For Helping Clean Up Hazardous Spills," *Wall Street Journal*, 29 (March 18, 1982).

Anon, "EPA Hazardous Waste Site Ranking Blasted," *Chem. & Engr. News*, 6 (Nov. 6, 1981).

Baasel, W. D. et al., "Multimedia," *Environ. Goals*, 37–51 (October 1980).

Baker, R. A., "Taste and Odor in Water: A Critical Review," Final Report F-A2333, Manufacturing Chemists Association, 1978.

Carlson, G. A. and Collin, R. L., "Toxic Materials in the Environment," Technical Paper No. 52, Department of Environmental Conservation, New York State, Albany, N.Y., June 1978.

Gibbs, Lois M., *Love Canal: My Story*, State Univ. of New York Press, Albany, 1982.

Giffin, C. E. and Jahnsen, V. J., "Chemical Analysis of Waste Sites Using GC/MS Instrumentation," presented at the American Chemical Society Meeting, Las Vegas, Nevada, August 29, 1980.

Interim Status Standards for Owners and Operators of Hazardous Waste, Treatment, Storage, and Disposal Facilities, *Fed. Regis.*, **46**(221), 56592–56596 (November 17, 1981). U.S. EPA rule on "Lab-Pacs."

Metry, A. A., *The Handbook of Hazardous Waste Management*, Technomic Publishing, Westport, Conn., 1980.

Mintz, M., "Jail Terms Sought for Business Health, Environment Violators," *Wash. Post*, A-1, A-13 (25 Nov. 1979).

Proceedings of the U.S. EPA National Conference on Management of Uncontrolled Hazardous Waste Sites, October 15–17, 1980, Washington, D.C.

Roland, R. A., "Chemist or Culprit: A Look Ahead," *The Chemist of AIC*, 4, 18 (August/September 1980).

Senkan, S. M. and Stauffer, N. W., "What To Do With Hazardous Waste," *Technology Review (M.I.T.)* **84**, No. 2 (Nov./Dec. 1981).

Silvestri, A., "Development of a Kit for Detecting Hazardous Material Spills in Waterways," EPA–IAG-0546, EPA-600/2-78-055, Edison, N.J., March 1978.

Thibodeaux, L. J., "Estimating the Air Emissions of Chemicals from Hazardous Waste Landfills," *J. Hazard. Materials*, **4**, No. 3, 235–244 (Jan. 1981).

"Those Toxic Chemical Wastes," *Time*, 58–69 (September 22, 1980).

Touhill, C. J., Shuckrow, A. J., and Pajak, A. P., "Hazardous Waste Management at Abandoned Dump Sites—Evolving Perspectives," *J. Hazard. Materials*, **6**(3), 261–265 (1982).

Cumberland, R. F., "The Control of Hazardous Chemical Spills in the United Kingdom," *J. Hazard. Materials*, **6**(3), 277–287 (1982).

29

Disposal of
Hazardous Materials

Richard D. Ross

The safe disposal of hazardous material is important not so much because of a special group or category of wastes which could be legally classified hazardous, but the very fact that waste disposal involves the handling and final disposal of a wide variety of materials about which very little may be known. The fact that very little is known about the waste materials that are handled on a daily basis by industrial plants and contract organizations means that they are in fact hazardous. Any time it is necessary to handle materials about which there is little published information, we have a hazardous situation; any time someone with limited knowledge of a particular material must process it, we have a hazardous situation. A bomb demolition expert must pass through many rigorous days of training before being allowed to defuse a bomb. The lion tamer at the circus has spent years learning the profession, and those who jump from planes to fight forest fires have had months of training for their assignment. All of these occupations or vocations are hazardous, yet people with only a sixth-grade education are permitted to haul away chemical wastes and dispose of them in a presumably safe manner so that the wastes cannot affect our environment, and we don't even bother to tell the haulers whether these wastes are toxic, flammable, explosive, corrosive, or the many other aspects which should be known in order to do the job properly. There is also a good chance that even if they did know the hazards of the waste material being handled they would not know how to protect themselves, other people, or the environment both during and after its disposal.

This, therefore, is what really makes wastes hazardous. We could single out a few substances and say that this waste is toxic and this waste is a suspected carcinogen, or that another waste will burn the eyes and the skin, and yet another waste will contaminate groundwater. But sometimes the hazards are much more subtle than this. Sometimes the hazards do not exist until two or more chemicals are mixed together, possibly at the disposal site, creating a third chemical which is toxic or even lethal.

29.1. DEFINITION

To examine what is meant by a hazardous waste a definition or a series of conditions are necessary to identify the hazardous waste as opposed to the nonhazardous waste. To do this we must first ask the questions—who, what, where, why, and when?

The first question should deal with who or what is affected, or might be affected by the hazard. Injury or harm can be sustained by humans, animal life, vegetation, property, or the general environment. We should not only be concerned with human health and property, but for all other living organisms and the environment. The environment is somewhat of a broad-based term, which if taken literally includes human, animal, and vegetable life. Even if we look at it selfishly, any change in the environment may, in the long run, affect our own way of living.

The second question that must be asked is, why is the material hazardous? It must be hazardous for some reason. For example, is the waste toxic or does it contain toxic substances which might be ingested or absorbed by humans, animals, or plants leading to their ultimate destruction. The material might be flammable, it might be explosive, or highly reactive. It might cause irritation or sensitization. It might be corrosive or radioactive. It might involve bioconcentration, or even have genetic change potential. It is considered hazardous for a specific reason.

Finally, we must determine when the material is hazardous. Some materials are not hazardous until their threshold of exposure has been exceeded. For example, minute quantities of ozone are created during a thunder and lightning storm, yet it has a threshold value of 0.1 ppm, indicating that it is an extremely toxic substance. Everyone knows that sulfuric acid is very corrosive, but in very weak concentrations it will not burn the skin. However, if we were to put a drop of concentrated sulfuric acid on our hand, it would be very painful and cause a serious injury. Therefore, the hazard is not only function of the chemical formula, but frequently of the concentrations as well. It is important, therefore, to know when things are hazardous and when they are not. Some materials are quite safe to handle and store, but they may have a significant hazard associated with them when mixed with other materials. For example, sodium sulfide is not a particularly hazardous material in itself. It may be handled as a generally nonhazardous substance, but if a spent sodium sulfide liquor should be mixed with a strong acid at a disposal site, the result would be hydrogen sulfide gas, which is extremely toxic, and in relatively low concentrations is lethal. Other materials may be hazardous and become more hazardous with time. Ethyl ether, which has been used for years as an anesthetic and a solvent, is hazardous because of its flammability. However, when stored for a long time, it becomes more hazardous because of its tendency to form peroxides, which are extremely explosive. The waste ether might autoignite or detonate without warning.

It is, therefore, not enough to just say waste is hazardous, we must define why it is hazardous, situations where it may be hazardous, and conditions when it is hazardous, such as, concentration, form, and degree. With this information in mind, we can tackle some of the areas or aspects of hazardous waste which make the waste hazardous in the first place, and this should lead us to possible solutions.

Wastes may produce effects not immediately obvious. For example, the higher incidence of respiratory problems in areas such as London, which at one time burned high sulfur coal, is probably also a direct etiological effect of the sulfur dioxide in the atmosphere. The genetic effect of the use of thalidomide was malformed children. We rely on the genetic effect of certain insecticides to change certain types of troublesome insects in such a way that the species will disappear. There is the possibility of hazardous wastes causing similar etiological and genetic effects, if they reach the atmosphere or drinking water, or the food chain.

29.2.   THE HANDLING OF HAZARDOUS WASTE MATERIAL

Once we have defined the why's of the hazard, we then must determine a system which will permit safe handling of the hazardous waste between its point of creation and its ultimate disposal. Actually, the problem of handling a hazardous waste is perhaps more difficult than disposing of the same material. This is perhaps because we have many more options in the handling of such materials than we do in their ultimate disposal. Waste materials can be packaged in drums, bottles, cartons, cans, trucks, tanks, cylinders, etc. The variety of possible containers is almost unlimited. Obviously, the transportation of hazardous waste virtually eliminates gaseous wastes. Most gaseous wastes which are hazardous in nature must be treated on the site, and therefore, transportation is not involved, with the exception that it may have to be piped from one section of the plant to another for its disposal. Liquid wastes, on the other hand, are very often transported to disposal sites. The bulk of hazardous materials handled in the United States today are those in liquid form. There are some hazardous solids, but the preponderance of solid waste materials, industrial and otherwise, is not hazardous.

The handling of hazardous wastes will, of course, depend upon the type and degree of hazard. All waste material should be carried in containers which are immune to the corrosive aspects of the waste. Those people handling these waste containers should have proper personal protection. This might involve special suits of clothing, or only the use of special eyeglasses and gloves. In any event, if the material is severely corrosive, the individuals handling the material should be protected from any physical harm, which might be caused in the event of a spill or leak in one of the containers. (See Chaps. 25, 26, and 27.)

Highly flammable or explosive materials should be handled in special areas that are equipped with explosive-proof electrical fixtures, and spark-proof machinery with special provisions for the elimination of all static discharge. (See Chap. 19.)

Toxic wastes or bioconcentrative materials, or materials which are suspected of being carcinogens, should be transferred in sealed containers from the point of inception to the disposal site; and where it is necessary on occasion for workers to be exposed to these materials or their vapors, they should have proper respirators as well as complete body cover so that the materials cannot come in contact with the skin or the respiratory tract. This is also true where irritating or sensitizing chemicals may be used. Even though the degree of protection may not be quite as important, respiratory protective devices definitely should be utilized. (See Chap. 25.)

Hazardous waste materials should always be transported in the largest feasible container unless there is a serious chance of a violent reaction or explosion, then perhaps the smallest feasible container makes more sense. The most difficult disposal problem today is dealing with the hundreds of relatively small containers that we often use to transport our waste. Laboratory chemicals which are no longer useful are often discarded in their original containers, which are usually small glass bottles. These laboratory containers could be virtually anything, from acids and bases, either strong or weak, solvents, very reactive materials, organic dyes, even explosives, such as iodides and peroxides. Unlabeled, they constitute one of the most serious hazardous waste disposal problems in the United States today. Even when they are properly labeled, they constitute a serious problem because many uniformed or ignorant people feel that they can be dumped into a landfill or an incinerator with complete impunity, because the quantities are so small nothing serious could result. Anyone who has ever seen metallic sodium thrown into water quickly realizes that a very small amount of sodium makes a big flame. Imagine a bottle of metallic sodium or potassium, which is equally reactive, thrown into a chemical landfill where flammable solvents might be present. The resulting fire could shut down the landfill for weeks or months, and someone could be seriously injured in the bargain.

To properly dispose of such wastes, each bottle or container must have some process established for its ultimate demise. As you can see, this could become very expensive. Such materials as acids, bases, and certain organic materials could be reasonably combined at the time they are discarded in a way that their disposal process would be the same. Then the number of individual steps would be greatly reduced.

If we go to larger containers, such as drums, which are perhaps the most common receptacle for industrial waste liquid material, we still have a fairly serious disposal problem. One drum absolutely full holds only 55 gal of material, and each drum that is filled with a hazardous waste becomes a potential hazard itself. If the drum is made of suitable material which will not corrode from the waste which it contains, it may yet have to contend with atmospheric corrosion before the ultimate disposal of its contents if it is stored outdoors for any period of time. In some circumstances, a drum will begin to leak its contents onto the ground terminating any thought of proper disposal. Each drum in a truckload of drums full of waste hazardous material is a potential hazard. Each drum has the capability of leaking by itself, and creating a separate hazardous situation both to the people handling the material as well as to those transporting and receiving it, not to mention the general public.

The best approach to any type of liquid waste disposal, hazardous or otherwise, if it is feasible, is to use a tank truck or railroad tank car. These vehicles were designed for such purposes, and come equipped with a variety of corrosion-proof linings and can be ordered for handling almost any type of corrosive hazardous waste. Rubber-lined, glass-lined, and plastic-lined tank trucks and tank cars are available in most areas. If a tank truck leaks, of course, we have a rather serious situation, or if a railroad car or a tank truck is in an accident where the tank is broken or leaks in the accident, then a very serious situation results. The incidence of this type of problem is much less than it is in the use of drums for carrying waste materials (see Chapter 31).

Solid waste materials can be carried in a variety of containers, usually by truck or rail car. If they are hazardous, then the options are somewhat limited. Waste cyanides from heat treating operations could be considered to be a hazardous waste. They are toxic to a limited degree, but of greater concern, they can react with acids to produce hydrogen cyanide gas, which is deadly. Such materials should be shipped in the largest feasible closed containers, which are impervious to moisture or any material which may be outside the container. Often this is best effected by enclosing the solid waste in some type of impervious plastic envelope, and then placing this envelope in the drum. "Fiber pack" drums have often been used for waste filter cake which is hazardous in nature, especially if the waste material is organic and lends itself to incineration, the advantage being that the entire package can be burned as a unit. This is often ideal from a final disposal standpoint, but not too acceptable from a transportation safety standpoint. These fiber packs are somewhat vulnerable to mechanical damage by forklift trucks, or other conveying devices. They cannot be left outdoors for any extended period of time because the fiber eventually deteriorates with water contact, exposing or leaching out the contents. Often such fiber drums shipped in steel drums make a satisfactory double container arrangement. The steel drums, of course, must be the type with removable tops, so that the fiber packs can be removed prior to incineration. Large returnable bins, or other types of sealed containers, might be preferable for certain types of wastes. Each waste will have its own character and will demand a separate type of container for transport.

Many times solid waste materials have only one acceptable mode of disposal—landfill. If they cannot be landfilled in bulk because of their hazardous nature, which most likely will be toxicity, then they can be buried in drums or fiber packs if they are immediately covered with a sufficient quantity of earth in a proper "dedicated" approved landfill where no leaching of the material by groundwater can take place. (See Chap. 28.)

Identification

One of the biggest problems in handling a hazardous waste material, especially when it is being shipped off the plant site where it was created, is in its proper identification. In other words, what is it? What are its hazardous characteristics? What should be done with it at the disposal site? And what should be done with it in the event of an accident along the road or railroad during the course of its transportation? Since Nov. 1980, when the RCRA law went into effect, no generator of hazardous waste may offer to ship; no person may transport and no disposal site receive hazardous waste unless it is accompanied by a manifest which properly describes the waste, its hazards and identifies the disposal site. Furthermore all containers must be properly labeled and all vehicles properly placarded in accordance with D.O.T. Regulations. This manifest labeling and placarding greatly reduces the danger of accidental spills and has improved the safe handling of hazardous waste. (See Chapter 28.)

Storage and Disposal

Disposal of a hazardous waste is not unlike the disposal of a nonhazardous waste of similar physical and chemical properties. Actually, the act of disposal is quite simple and straightforward. The major hazard involves the handling of the material from the time it is brought to the disposal site until the time that it is ultimately disposed of. First, we should look at the problem of on-site storage of hazardous wastes. If the wastes are highly flammable, they should be stored in accordance with rules and regulations for flammable materials. This means that the storage tanks should be remotely located from buildings and other facilities; that they should have breather type vents (the discharge of the breather lines preferably run through activated carbon beds); that any pumps or agitators or other electric equipments in the immediate area be of Class I, Group D, Division 2, explosive proof construction, and that the tanks be diked in accordance with petroleum refinery practice. Large quantities of extremely flammable material should not be stored at the disposal site any longer than absolutely necessary. (See Chap. 19.) Usually flammable material is incinerated, and it should be incinerated with the greatest dispatch.

Incineration may also be applicable to explosive materials or highly reactive organic materials. These materials should not be stored, unless absolutely necessary, at the disposal site. They should be incinerated properly upon receipt. While this may not be possible in every case, any storage at the disposal site provided for explosive materials should be remote from the rest of the facility, and properly bunkered and diked to prevent the spread of a fire in the event of an explosion.

Materials which are reactive and form lethal gases upon reaction should be carefully separated from each other at the disposal site so that no inadvertent mixing of the two can occur. Acids and inorganic cyanides and sulfides should be located on opposite sides of the disposal site in a way that an inadvertent mixing cannot happen.

Toxic or irritating wastes should be stored in tanks or containers which suit the corrosive properties of these wastes and breather vents should be trapped with activated carbon. Such wastes should never be stored in open lagoons or ponds.

While drum storage at disposal sites is inevitable, it would certainly be desirable to have drums stored on a concrete pad which has provision for adequate drainage into a sump, which can be analyzed and treated before the runoff can get to the groundwater. It would be ideal to have all drums stored under roof so that they would not be open to the weather. Fiber pack drums must be stored in this manner or their contents may ultimately contaminate the ground and the water table.

The final disposal of hazardous wastes is essentially the same as the disposal of any nonhazardous wastes. The options which are open are limited. The first, and probably the most desirable, yet not the least costly, is incineration. Incineration provides for complete destruction of hazardous wastes if the system is properly designed and the ultimate temperature is high enough. A consideration of the combustion characteristics of the waste must be given, and if the waste does not have a high enough calorific value of its own, then auxiliary fuel must be used.

Halogenated wastes will produce halogen acids which must be scrubbed before discharging the final effluent into the atmosphere. Wastes containing sulfur must also be scrubbed before release. But in all cases, incineration at final temperatures above 1800°F is recommended for complete destruction of the waste. Residence times in the incinerator will vary depending upon the specific material. All organic materials are applicable for incineration. Incineration in special ships at sea has been demonstrated, even for highly halogenated agents, in tests monitored by the U.S.E.P.A. Inorganic–organic mixtures will present difficulties because of the final effluent which must be removed by high-energy scrubbing. In the case of an inorganic–organic mixture it is usually desirable to incinerate the organic material if it is hazardous, rather than using some other approach.

The second possibility is the chemical landfill. Chemical landfills should have a design such that any leachate from the landfill can be trapped in some type of a sump or lagoon and further treated before release to the groundwater. The design of the chemical landfill will greatly depend upon the geology of the disposal site. An impervious clay soil which is suitable as a barrier between the chemical landfill and the groundwater is available in certain parts of the United States. In most areas some type of artificial barrier or membrane will have to be constructed between the soil and the contents of the landfill to be totally satisfactory. Such a landfill should be monitored at the perimeters with wells which are used for testing the possible contamination of the groundwater. This procedure is generally covered in Chapter 28 but is extremely important for hazardous chemical waste.

Only two types of hazardous wastes should be deposited in a chemical landfill. One is a waste which will normally biodegrade, thus removing the present and potential hazard. The other type of waste is one which is impossible to degrade in any manner. This type of waste placed in a chemical landfill will never disappear. Oxides or sulfides of heavy metals which are highly insoluble in water and which do not present a severe leachate problem may be deposited in a well-designed chemical landfill, or they may find their final resting place in a dedicated approved landfill which presumably can never be opened once it is closed. A dedicated landfill could consist of a large concrete vault into which drums or other containers of this type of material are placed and then covered with concrete. The question concerning the use of a dedicated landfill is whether this is not a very poor use of the land and whether in generations to follow it may not come back to haunt us. There are, nevertheless, certain types of inorganic waste, which, if they are not recycled, cannot be destroyed either biologically or by incineration, and therefore, they must find a resting place for perpetuity. This has been true, of course, for radioactive wastes, especially those with high levels of radiation and long decay periods. Such materials, until now, have been placed in underground tanks or vaults to remain there for hundreds, or even thousands, of years. Earthquakes or geological changes in the earth's crust at that particular location might cause serious problems. In those cases, sites have been selected to obviate such circumstances.

A third possibility for the disposal of hazardous wastes is the use of the deep well. Deep wells have been used for this purpose in Louisiana, Texas, Ohio and Michigan. The injection zone for wastes is usually sand or porous sandstone located many feet below any existing aquifers and capped by several layers of granite or shale. A well-designed deepwell is a safe storage area for aqueous acids and bases. It is not a good choice for the disposal of organic liquids.

Biological systems may be used for the disposal of certain types of hazardous wastes which are in dilute form, if they are organic and if they lend themselves to biological degradation. Care should be given to see that refractory types of biological materials are not present in quantity because they may destroy the biosystem.

Another possibility for the destruction or disposal of hazardous wastes is offshore disposal. This particular form of disposal method has not been proven positive or negative with respect to hazardous wastes. Certain radioactive materials have been encapsulated in concrete and dropped to deep sections of the ocean, presumably to remain there forever. If the casing for one reason or another should break open, the dilution factor of the ocean would be such that it would render the waste ultimately harmless. However, there might be a serious effect in the immediate vicinity for some time after the rupture occurred. It is not a final solution to the problem, unless we are certain the hazardous waste will decompose in the sea water within a reasonable period of time. Studies now under way to determine the effect of dumping both municipal and industrial wastes at sea should give us guidelines concerning the types of wastes and quantities of wastes we can expect to dispose of in this manner. As noted in Chapter 28, waste exchange is another approach worthy of further attention.

To reiterate the emphasis of this chapter, the disposal of hazardous wastes is not so much a difficult *disposal* problem as it is a difficult *handling* problem. In most cases, hazardous wastes can be disposed of safely, if all people involved in the handling and disposal of wastes are acutely aware of what it is, and how it should be handled, and the disposal or treatment regimen which must be used for safe final disposal. It is always desirable to dispose of hazardous wastes at the point nearest to where they are created, rather than to give them to a third party for disposal, but it is recognized that this is not always possible. When transportation to an ultimate point of disposal is necessary, then complete information should follow the containers describing the hazards in the simplest of terms so that those who handle the material are aware of the hazards of the waste.

BIBLIOGRAPHY

Alter, H. and Dunn, J. J., Jr., *Solid Waste Conversion to Energy, Current European and U.S. Practice*, Marcel Dekker, Inc., New York, 1980.

Anon, "Insurance Is Required by EPA at Facilities for Hazardous Wastes," *Wall Street Journal*, p. 38 (April 12, 1982).

Bonner, T., "Hazardous Waste Incineration Engineering," *Poll. Tech. Rev.* **88**, (1981) Noyes Data Corp., Park Ridge, N.J. 07656.

Cheremisinoff, P. N., "Disposal of Hazardous Wastes: Treat or Truck," *Pollut. Eng.*, 7(5), 30–39 (May 1975).

Cheremisinoff, P. N. and Holcomb, W. F., "Management of Hazardous and Toxic Wastes," *Plant Eng.*, **30**(9), 259–263 (April 1976).

Cheremisinoff, P. N. and Morresi, A. C., *Energy from Solid Wastes*, Marcel Dekker, Inc., New York, 1976.

DeRenzo, D. J., ed., "Biodegradation Techniques for Industrial Organic Wastes," *Poll. Tech. Rev.* **65** (1980), Noyes Data Corp., Park Ridge, N.J. 07656.

Dillon, A. P., "Pesticide Disposal and Detoxification: Processes and Techniques." *Poll. Tech. Rev.* **81** (1981), Noyes Data Corp., Park Ridge, N.J. 07656.

Fung, R., ed., "Protective Barriers for Containment of Toxic Materials," *Poll. Tech. Rev.* **66** (1980), Noyes Data Corp., Park Ridge, N.J. 07656.

Hanson, D. J., Progress under Superfund Criticized, Defended, *Chem. and Engr. News*, 60(23), 10–15 (June 7, 1982).

"Hazardous Waste Management Facilities in the United States," EPA/530/SW-429 (1977).

Helsing, L. B., "Science and Law: New Lab Partners at ETC," *Waste Age*, pp. 42–46 (October 1981).

Hess, L. Y., "Reprocessing and Disposal of Waste Petroleum Oils," *Poll. Tech. Rev.* **64** (1979), Noyes Data Corp., Park Ridge, N.J. 07656.

Hickman, H. L., "Building a Communications Network for Waste Managers," *Waste Age*, pp. 17–18 (March 1982).

Hooper, G. V., "Offshore Ship and Platform Incineration of Hazardous Wastes," *Poll. Tech. Rev.* **79,** Noyes Data Corp., Park Ridge, N.J. 07656.

Johnson, C. A., *Resource Recovery Decision-Makers Guide*, National Solid Waste Management Asso., Washington, D.C., 1980.

Koppenaal, D. W. and Manahan, S. E. "Hazardous Chemicals From Coal Conversion Processes," *Environ. Sci. Technol.*, **10**(12), (November 1976).

Lindsey, A. W. "Ultimate Disposal of Spilled Hazardous Material," *Chem. Eng.*, (October 17, 1975).

McConnaughey, W. E., Welsh, M. E., Lakey, R. J., and Goldman, R. M., "Hazardous Materials Transportation," *CEP*, **66**(2), (February 1970).

Morton, W. I., "Safety Techniques for Workers Handling Hazardous Materials," *Chem. Eng.*, Deskbook Issue, (October 18, 1976).

National Solid Waste Management Asso., publications on the following: (1) *Mismanagement is the real hazard in hazardous wastes* (1980); (2) *Incineration* (1980); (3) *Secure Landfill, Siting and Site Design*, 1980; *Secure Landfill Operation*, 1981; *Secure Landfill Closure and Post-Closure Care*, 1981; (6) *Deep injection wells for disposal of hazardous waste . . . but where does the waste go?* 1981; (7) *Transportation of Hazardous Waste*, 1981; (8) *Fixation Stabilization and Solidification of Hazardous Waste*, 1981.

"Pesticides and Pesticide Containers," *Fed. Reg.* **39**(85), Part IV (1974).

Sanitary Landfills: A Bibliography EPA/530/SW-46, 1974.

Sittig, M., "Incineration of Industrial Hazardous Wastes and Sludges," *Poll. Tech. Rev.* (1979), Noyes Data Corp., Park Ridge, N.J. 07656.

Sittig, M., "Landfill Disposal of Hazardous Wastes and Sludges," *Poll. Tech. Rev.* **62** (1979), Noyes Data Corp., Park Ridge, N.J. 07656.

Sittig, M., "Metal and Inorganic Waste Reclaiming Encyclopedia," *Poll. Tech. Rev.* **70** (1980), Noyes Data Corp., Park Ridge, N.J. 07656.

"State Program Implementation Guide: Hazardous Waste Transportation Control," EPA/530/SW-512, 1976.

Quincy, M. E., 1-hour television program on hazardous waste disposal, aired by NBC-TV 10 p.m. (EST) December 23, 1981. Script available from NBC, New York, N.Y. 10019.

30

Fire and Explosion Investigations on Chemical Plants and Oil Refineries

Arthur D. Craven

Any accident is more tragic if human experience is none the richer for it. This is certainly true of chemical and petroleum process accidents where fires and explosions may result in enormous financial losses and, worse still, inflict injury, perhaps fatal, on personnel (see Fig. 30.1). A child learns to walk by falling and it is similarly true to say that the process industries today are safer by virtue of their experience in past disasters. It is also a fact that the process industries would have been even safer sooner if some of the past disasters had been investigated properly and the findings published more widely. The purposes of investigations should be extended beyond apportioning blame, although this may be important. In addition to assisting with the avoidance of a recurrence, invaluable data relating to the effects of fires and explosions may emerge. In addition, those areas where future research would be profitable are often highlighted.

In the field of fire and explosion technology, our knowledge had persistently lagged behind experience, usually in the form of disasters. The effects of the detonation wave, for example, were first observed as a result of the accidental ignition of a gas leak in a sewer running under the length of Tottenham Court Road in the city of London in 1880.[1] Prior to that time it had been widely accepted that flames could not travel much faster than those observed on the inner cone of the Bunsen burner. Soon after this accident, however, the nature of the detonation wave was elucidated by a theory which was established and well tested by the turn of the century.[2] In the next 70 years, during the development of the reciprocating internal combustion

Figure 30.1. Example of devastation which may face the investigator after a process disaster. Flixborough, United Kingdom, January 1974.

engine, the jet engine and rockets, with the added technological impetus of two world wars, our understanding of combustion processes still left us completely unprepared for the effects of the unconfined vapour cloud explosions such as those which occurred at Lake Charles in Louisiana (1967), at Pernice, near Rotterdam in the Netherlands (1968), and at Flixborough in the United Kingdom (1974).

Tests designed to reproduce the effects of fires and explosion, particularly on the large scale, are very expensive. It is essential, therefore, that the maximum information be obtained from accident investigations. Valuable information may be obtained from investigations into minor accidents and *near misses*. This is becoming widely recognized in many chemical and petroleum companies and more detailed investigation and wider publicity is now given to the results of investigations into accidents which by good fortune have proved less serious. However, people are more inclined to pay attention to the lessons which may be apparent from major disasters than to those which may be deduced from extrapolations of data obtained from small-scale incidents. With purely theoretical predictions based on data from laboratory experiments, the common reaction is "It may never happen."

The explosive properties of ammonium nitrate were well known at the time of World War I but many serious questions regarding its storage and use were never asked until 9,000,000 lb exploded in Germany in 1922 and completely destroyed the town of Oppau, killing 1100 people. After this incident it emerged that huge outdoor stockpiles of ammonium nitrate based fertilizer which had tended to form a solid crust on exposure to weather, had been quarried using conventional explosives, prior to transportation.[3]

30.1. SELECTION OF AN INVESTIGATOR

After any accident the selection of an investigator should be expedited as quickly as possible. A formal selection process should be part of the contingency plans of every organization. Such a process must be geared to dealing with different types of accidents and the severity will be only one factor influencing the selection. A major disaster resulting in damage extending beyond the company's boundaries will clearly merit a different approach to a local process fault involving a few hours outage. Generally speaking, the more extensive the damage the greater will be the problem of investigator selection.

In the event of a major disaster it may be considered necessary to instruct several investigators of differing disciplines to form a team. This should not prevent the selection of one individual whose task will be to collect and collate all the evidence with a view to preparing a final report. If the system produces only a team with no leader, the team should sit down first and elect one. This process may be obviated by selecting one investigator who may then call upon others to assist with both the investigations and the interpretations.

Considerations regarding selection may be grouped under the headings *expertise* and *affiliation*. Some account may also be taken of personality and past experience. Professional expertise and qualifications are not of paramount importance. In the case of fires and explosions, however, some knowledge of the fundamentals of combustion and flame processes and their effects is a definite advantage. It is fair to say that in no other field of technology is credence so readily given to uninformed opinion. Even where the primary cause of a fire or explosion is known to be metallurgical failure and an investigator is selected accordingly, it is advisable to enlist the assistance of somebody with experience in the field of fire and explosions.

The biggest problems regarding investigator selection are often associated with affiliation. Any organization failing to face up to this fact is taking an incalculable risk. At the heart of the problem lies the question of the in-house investigator versus the outsider. The problems facing the outsider may be familiarization, availability, and confidentiality. It will clearly take the outsider longer to arrive on site and start the investigation than the in-house investigator. This problem may be overcome by appointing a temporary site investigator who will hand over to the outsider upon arrival. A company may, however, be reluctant to involve any outsider in an investigation which may lead to their normal procedures being questioned. This is understandable. An outsider, on the other hand, may inject new ideas into any organization and this may be of great value in the aftermath of disaster.

The advantages of the in-house candidate are obvious. He or she will clearly get things going quickly and for a small incident involving little loss of property and only slight personal injury such a selection might be adequate.

The biggest problem facing the in-house investigator in the event of a major incident will be psychological. The emotional pressures may be very high. The investigator may have been responsible for certain aspects of the running of the process involved, or may have friends who were either injured, killed, or who had responsibilities in some technical or managerial capacity. Recognition of such influences need not arise for considerations of personal integrity. With the best will in the world, the in-house investigator will find psychological pressures that may cloud his or her judgment. In this age when companies spend large sums of money on training courses in objectivity, such considerations may be considered unnecessary. It should be recognized, however, that these adopted disciplines will be subjected to the most severe test by the emotive pressures generated in the aftermath of disaster. This is clearly a matter for debate. Such an important debate should not, however, be conducted in haste among the smoldering debris. An agreed procedure for the appointment of investigators should be as much a part of any company's contingency plans as the insurance policy and the first-aid kit.

30.2. FORMAL APPROACH TO INVESTIGATION

Having decided who is to carry out the investigation it is essential that the investigator should maximize his or her activities by adopting a formal approach. The method described below has stood the test of time and is highly recommended. It involves collecting as much evidence as is reasonably possible before any theory regarding causation is considered. The alternative approach involving inspired guesses into causation based on past experience is too frequently adopted. Hastily based theories are then put to the test by the selection of evidence and other evidence is ignored.

The following seven steps to diagnosis are recommended:

1. Remit
2. Brief survey
3. Determine preincident facts
4. Detailed examination of damage
5. Interviews with witnesses
6. Research and analysis
7. Preparation of report.

Steps 1, 2, and 3 should be carried out in strict sequence. Steps 4, 5, and 6 need not be taken in any particular order and investigations may run concurrently under these three headings. As the investigation progresses draft forms of step 7 may be in preparation when further information from any of the preceding steps may be obtained by going back over the ground. This is more clearly evident when these are considered in more detail.

Remit

This stage in the investigation concerns both the investigator and those instructing him or her. After the investigator has been appointed and the remit defined, he or she should be left to get on with the job. Until his remit is defined, however, the investigator should do nothing. Unfortunately, the task often seems so self-evident that reference is sometimes never made to remit. The investigator may later be surprised to learn that he or she has taken some aspects of the investigation beyond the limits envisaged by those who instructed him or her and on the way some evidence, the value of which was not appreciated, was not properly recorded.

Where possible, the investigator should not be asked initially to consider what steps should be taken to prevent a recurrence of the incident. The remit should be restricted to matters relating purely to the incident. Decisions regarding future production can only be made in the light of both technical and financial considerations. When the investigation is complete, the investigator may then be required to make recommendations regarding future practice but this should not be written into the remit of an accident investigation and he or she should not consider this problem until the investigation into the cause is completed.

The purpose of defining the remit and the scope of the investigation is for the benefit of the investigator. It should not be designed to impose rigid limits beyond which he or she must not transgress. The remit should not act like the blinkers on a horse. On the other hand, the investigator should take a broad view of all the evidence and ensure that as much as possible of this is recorded. A good investigator should make a note of any manifestation which cannot be explained immediately even when this appears to have no diagnostic value. At the end of the exercise he or she must be satisfied that as far as possible all the observed effects can be accounted for on the basis of the conclusions. How much of the detail must be recorded will be a matter of judgment. The importance of the photographic record cannot be over-emphasized.

It may transpire that evidence emerges during the course of the investigation which requires reconsideration of the remit. This is often the case and the matter should be discussed again with those who instructed him or her. By defining these responsibilities higher management will then be in a better position to deal with the many other problems which require their attention. This is particularly of value in the event of a major disaster. When considering remit, the formal strategy of fire and explosion loss prevention science should be recognized. Four quite independent lines of defense can usually be identified. The first line of defense should be associated with either preventing or limiting the generation of flammable media. The second involves the elimination of potential ignition sources. On the third line of defense, matters relating to reducing the possible effects of fire and explosion are considered, such as the strength of buildings, explosion relief, fire and explosion suppression, spacing, etc. These are the built-in features as compared with the fourth category under which such matters as rescue, fire fighting, and other emergency procedures are considered. After any accident, the remit of the investigator may become clearer if the questions How? Why? What? are posed under the above four headings.

Brief Survey

There is a temptation with the inexperienced investigator to dive right into the problem and to start digging in the debris. This is wrong, even if pressed by anxious members of management requesting instant diagnosis. If comments are made at an early stage they may be bitterly regretted later. The main purpose of the brief survey of the damage is to enable him or her to understand more easily the value of the information which will be gathered together in Step 3. In many cases, the brief survey may take only a few minutes. It may comprise nothing more than a conducted survey from an office window. In fact, an observation of the area involved from an elevated, stand-off position may be of more value than a scramble through the debris at this stage.

Preincident Facts

Where the investigator is not familiar with normal conditions prevailing at the scene prior to the incident, time spent in reviewing the preincident facts is well worthwhile. This will save time at later stages, particularly when the damage is being examined and witnesses are being interviewed. The establishment of the normal condition is best considered with steps 4 and 5 in mind. This can be conveniently divided into two sections under the headings "hardware" and "process."

HARDWARE

There are several sources of information which may be available to the investigator. Engineering drawings and perhaps old photographs taken in connection with other activities prior to the incident may be available. The investigator may be able to see a similar process on a nearby plant of almost identical design. It may also be possible to discuss this matter with witnesses who were not involved at the time of the accident but who may be familiar with the hardware. On process plants such as oil refineries and chemical plants which operate 24 hours a day, the members of another shift may be of some assistance in respect of both hardware and process.

PROCESS

The investigator must be completely familiar with the process in question and properties of the materials involved. From a basic knowledge of the equipment an attempt should be made to build up a mental picture of the way in which it should have been used. Later, witnesses who are injured and distressed may have to be interviewed. These people may be prepared to give their accounts of what happened but they should not be expected to explain process details which should be available from other sources. The investigator should also be ready to adopt the local lingua franca, and must be on the alert to process colloquialisms which tend to develop and become accepted, particularly on old plants. During interviews the precise meaning of certain words may be very important.

Examination of Damage

The main purpose of the detailed examination of the damage initially is to locate first the area of origin and perhaps the location of the fault. In the case of major disasters, however, serious consideration should be given to the secondary effects since such information may only be available after the catastrophe. In the case of explosions the location of fragments and the effects of blast remote from the point of origin may yield valuable evidence regarding the cause. The value of such evidence may not be immediately obvious and this should be recorded photographically before anything is disturbed (see Fig. 30.2). Once this has been done, suspected items may be labeled as to location, salvaged, and set aside for subsequent examination. The extent to which the damage and the debris should be studied is a matter of judgment but the temptation to ignore apparently inexplicable phenomena should be resisted. In this respect evidence recorded photographically is invaluable. Where possible the investigator should take the photographs and for this purpose much can be achieved with a good quality small reflex camera with interchangeable lenses. A useful combination comprises a wide angle lens (28–35 mm) for general views and a normal lens (55–80 mm) for detailed photographs, with attachments for close-up studies capable of producing a negative image of near object size. If greater magnification is required, specimens should be removed for laboratory photomicroscopy. The diagnostic features of fire and explosion damage are dealt with in more detail in Aids to Diagnosis.

Interviews with Witnesses

Eyewitness evidence is very valuable. Where possible, evidence should be considered from members of the public in addition to those on site. All witnesses should be encouraged to write down their observations as soon as possible after the incident. This should be done before they have had an opportunity to discuss the incident with other witnesses. Witnesses who

Figure 30.2. Example of evidence not necessarily related directly to cause but recorded photographically in the interests of posterity. Road tanker trailer crushed by hydrostatic pressure generated during an unconfined vapor cloud explosion.

have not prepared statements beforehand should be asked to give their own account of the incident without any prompting from the interviewer. When questioned later any changes of opinion should be recorded but the original statements and views expressed should never be destroyed. Whenever possible, witnesses should be taken out of the interview room and encouraged to describe their experience either on site or in reconstructed situations. In many cases eyewitness evidence will be contradictory. This should always be recorded, however, and considered later in the light of the technical evidence. No weight factor should ever be placed on eyewitness evidence in relation to apparent levels of intelligence or position in a company hierarchy.

Research and Analysis

As the facts emerge it may be considered necessary to carry out further research. If the chemical analysis of samples is envisaged these should be taken as soon as possible. Ad hoc tests, on the other hand, should not be conducted until all the evidence is available to ensure that they are as realistic as possible. In cases where long-term experimentation is considered, however, the investigator may be required to reach working conclusions before this is completed.

The Report

The form of the report will vary according to the intended circulation. It should, however, take the reader stepwise from the circumstances prior to the incident, through the evidence, to the logical conclusions. Voluminous details which may be distracting to the reader may be relegated to appendices. A formal method of presentation should always be adopted. Although views differ as to the best form of report writing, the following list of headings is highly recommended.

INTRODUCTION

This section should contain a brief reference to the nature, time, and location of the incident. It should also contain an account of the remit of the investigator and an outline of the course investigation with dates.

PROCESS

This section should contain a description of the process in sufficient detail to allow the reader to understand what follows in the report. Simplified sketches are very valuable. These may be based on more detailed drawings to which reference may be made.

INCIDENT

The events leading up to incidents should be followed by a brief description of the event and a catalog of the consequences. This section should only contain facts which are initially beyond doubt and should not anticipate the outcome of the investigation. Any unusual features of phenomenology may be recorded here and illustrated by photographs.

INVESTIGATION

All the material evidence relating to the investigation should be presented in this section. The manner in which this evidence was obtained should be outlined. Short captions with arrows may be superimposed on photographs to illustrate salient points. If detailed consideration is to be given to conflicting evidence, this should also be recorded without qualification in this part of the report. This will be the most voluminous chapter in any report and consideration should be given to subdivision, particularly where technical evidence based on differing disciplines is presented.

EVIDENCE OF WITNESSES.

A brief review of the eyewitness evidence which is considered germane to the investigation should be presented in this section. Reference may be made to formal statements which may be appended.

DISCUSSION

The discussion should take the reader through all the material and eyewitness evidence to the author's conclusions. This logical analysis should form the main burden of this section and alternative theories and conclusions should not be interposed during this analysis. The author should, however, put forward any alternative theories which may have emerged during the investigation. The evidence which supports these theories should be considered and the author's reasons for either rejecting them or considering them less likely must be given. If such a detailed consideration tends to interfere with the smooth presentation of the report, a major proportion of the conflicting evidence may be relegated to an appendix and dealt with separately.

Conclusions

This section should only contain a short list of the main points arising from the previous section. A synopsis of the whole report is sometimes required and this is very often presented on a page at the front of the report. The purpose of this is to inform the busy reader whether or not the report is likely to contain anything of value to him or her. This may contain the conclusions in a slightly differing form.

30.3. AIDS TO DIAGNOSIS

The disciplines involved in the understanding of the causes and effects of fires and explosions are many and varied. Engineering (civil, electrical, and mechanical), chemistry (thermodynamics and kinetics), physics (properties of matter, heat, and dynamics), and pathology are but examples. Clearly it would be beyond the scope of this book, let alone this chapter, to deal with any of these matters in depth. In this respect this section must only be regarded as an introduction into the techniques available while at the same time pointing out some of the pitfalls which may be awaiting the investigator.

Fires

The distribution of damage may contain valuable evidence regarding the area of origin of a fire. The charring in woodwork will normally be deepest where a fire has been burning longest. If the time of extinguishment can be established by interviews with the fire fighters, the area of origin may be established from depth of char measurements. Experience would indicate that most types of wood char to a depth of an inch in 40 min regardless of the nature of the fire, provided that the flames are of the classical diffusion type normally associated with fires. In the case of a torching flame, supported by a rapid discharge of gas or vapor impinging on a piece of wood,[4] a more rapid rate of charring could be expected.

The post-fire condition of containers may also give valuable evidence regarding the time scale of fire in any area. The liquid burden inside any container will normally prevent damage to the paintwork whereas the paintwork on the walls above this level will be blistered. If the original liquid level is known the approximate time of exposure to fire can be estimated by equating the quantity of liquid evaporated to a heat input rate across the wetted area of the container of 11 W/cm². This corresponds to an average rate of heat input which could be expected to occur with a small container standing upright on a surface surrounded by flames. Since the maximum rate of heat input equal to 17 W/cm² may occur in very severe circumstances the evidence relating to the type of fire must be studied carefully. If there is doubt the use of 17 W/cm² would lead to a minimum *possible* exposure time in the absence of torching flames.[5]

Most fires on process plants will not generate temperatures much in excess of 1000°C. Fire temperatures are similar whether flammable liquids or solid combustibles such as wood, plastics, paper, card, or textiles are involved. Significantly higher temperatures are not normally encountered other than by the combustion of escaping jets of gases or vapors in highly turbulent flame brushes. Since the melting point of pure copper is 1080°C, therefore, it usually survives the effects of fire although copper wires tend to waste away by surface oxidation. Alloys of copper, such as brass and bronze (MP 800–1000°C), will usually melt and the appearance of beads of such alloys adhering to metal surfaces after the fire is not significant. Where beads of pure copper are to be found on conductors, however, it can usually be taken as strong evidence to indicate that the heat of the fire had been augmented by the passage of electric current during the fire. Since copper conductors tend to waste away as the fire temperature rises to a maximum, beads are not normally produced by the effects of fire alone even where local temperatures in excess of the melting point of copper are produced. In addition to copper beads, the pitting of the end surfaces of adjacent conductors due to vaporization during arcing may survive the subsequent fire. Such evidence may not necessarily be associated with the primary ignition source of a fire or an explosion but where the run of cables is such that fuses and the like would be expected to cut off the supply at an early stage, arcing and pitting on copper conductors may give valuable evidence regarding the area of origin of a fire. In a major fire over a wide area where the electrical supply is distributed to local fuse boards and circuit breakers, a detailed examination of the system in relation to the presence of cables showing signs of arcing is usually worthwhile.

The effects of fire on structural steelwork is not a very valuable tool in relation to fire investigation. It may even be very misleading. Iron and steel (MP 1300–1500°C) will not normally melt in fires. However, structural steel suffers a significant deterioration in strength at temperatures between 550 and 600°C. Since such temperatures are usually exceeded in the first few minutes of any fire it is only a matter of time before local stresses may produce quite spectacular contortions. These are often eye-catching and may receive undue attention, forming the basis for unrewarding discussion (see Fig. 30.3).

Figure 30.3. Example of heat generated locally by burning cans causing eye-catching contortion to steel column. Evidence not related to cause.

Aluminum and its alloys (MP 650–660°C) seldom survive the effects of fire other than in the form of very large castings. Plate aluminum may even burn in a fire although this is not a property peculiar to the low melting point metals such as aluminum and magnesium. Even thin steel in the form of metal cans may burn in a fire involving initially only cardboard boxes.

Explosions

A classification of the effects of various types of explosions is useful, particularly in relation to the manner in which the containing vessel may burst. A comparison of the blast effects in the stand-off position is of less value for the purposes of failure analysis since the expanding blast wave rapidly assumes a standard form regardless of the nature of the explosion. An analysis of the magnitude of blast effects is useful, however, as a check on the energy potential of any postulated event. This will be dealt with in Blast Effects.

CLASSIFICATION OF EXPLOSIONS

A number of factors influence the manner in which a container may fail due to internal pressurization. In addition to the strength in relation to static loading, much will depend upon the natural frequency of the vessel walls compared with rate of pressure rise. At one extreme, the detonation of a condensed phase may produce brittle failure at pressures in excess of those associated with the static loading. With a slow gaseous deflagration, on the other hand, the mode of failure may be purely ductile and similar to a physical overpressurization resulting in static loading. A deflagration will, however, tend to produce more fragments than static loading since the pressure continues to rise after the instant of primary failure.

Great care must be taken when looking at fragments. A slow pressure generation process inside a vessel which produces a classical ductile failure initially may give rise to a tear which accelerates to brittle fracturing. It is important, therefore, to locate the point of primary failure. Detonation in the gaseous phase, on the other hand, may not necessarily produce brittle failure since the propagating wave, although supersonic in relation to the gaseous phase, may be travelling slower than the velocity of sound in steel. Cause and effect may, however, be related as follows.

Physical Overpressurization.   The bursting of vessels due to physical overpressurization such as boilers and sealed vessels surrounded by fire is quite common. The pressurization may be either pneumatic or hydraulic. Hydraulic overpressurization is less likely to produce missiles than pneumatic overpressurization. With physical overpressurization, the mode of failure will always reflect the latent weakness of the vessel. In cases of vessels surrounded by fire this weakness may be induced by the heat of the fire.[5]

The so-called "BLEVE" resulting from the effects of fire on pressurized storages is a familiar manifestation of this process.

Single Volume Gaseous Phase Explosions.   If a flammable gaseous mixture is ignited inside either a process vessel, a tank, or a building, the flame process will result in a rising pressure which may ultimately cause the enclosure to burst. Such a failure will always be attended by blast effects. If only one volume is involved, the flame processes will usually be subsonic (referred to as deflagration) and the stresses will always be applied uniformly to give a mode of failure similar to that resulting from physical overpressurization.

The vessel will fail at its weakest point. This may be due to combustion in a pocket or layer anywhere inside the vessel. However, no deductions regarding the position of either the pocket or the ignition source is possible from the pattern of bursting. This axiom applies equally well to explosions inside single compartment buildings.

Multivolume Gaseous Phase Explosions.   When gaseous phase explosions propagate between interconnecting volumes the maximum damage is almost invariably observed, remote from the ignition source. As the flame propagates from one vessel to the next it tends to accelerate thereby increasing the rate of pressure rise and hence the damaging power. Similarly, with dust explosions in multivolume complexes, the maximum damage is usually observed remote from the primary ignition source. In gaseous explosions pressure piling may be responsible for the most spectacular effects in interconnecting volume systems and may even lead to gaseous detonation. By these mechanisms vessels may be suddenly subjected to pressures equal to over 100 times the normal process pressure. Single volume explosions on the other hand seldom give pressure rises in excess of 10 times the initial pressure.

When unusually violent explosions are observed involving apparently single compartment buildings these may be due to the multicompartment effect. Thus, if ignition occurs inside a cabinet and this bursts, the rapid spread of flame into the building may accentuate the build-up of pressure to a value higher than one might normally expect.

Gaseous Detonation.   Gaseous detonations which usually occur in pipelines and vessels of high aspect ratio will also inflict their maximum damage remote from the ignition source. In a particularly vicious type of pseudodetonation which is known to occur in compressed air systems[6] the area of origin, in the vicinity of the compressor final stage outlet valves, is often completely unaffected whereas piping, filters, and air receivers downstream may be torn asunder. The most characteristic feature of gaseous detonation in process plant manifests itself as sudden changes in direction. Evidence of bursting and general overpressurization may also be found at many isolated points along a process line through which a detonation wave has traveled. Detonation may produce longitudinal splitting by hoop stress failure which may propagate for considerable distances thereby converting a section of pipe run into a ribbon (see Fig. 30.4). When this mode of failure occurs due to either simple overpressurization or deflagration the split is seldom longer than a few pipe diameters.

Explosive Condensed Phase Reactions.   The local effects of gaseous detonation may be difficult to differentiate from those associated with the detonation of condensed phases. In general, however, the latter are capable of producing even more localized effects and the orientation of damage may reflect the tendency of condensed explosives to accumulate at low levels. The possible existence of condensed phase explosives may be eliminated by process considerations. It should be remembered, however, that a wide variety of oxidizing agents such as nitric acid, nitrates, chlorates, and liquid oxygen and nitrogen oxides, are all capable of producing sensitive and powerful condensed phase explosives when mixed with organic materials.

Unconfined Vapor Cloud Explosions.   In recent years the ignitions of large unconfined vapor–air clouds have resulted in the generation of pressure effects. An analysis of the major damage resulting from these explosions is of no direct value in establishing either the source of the vapor release or

Figure 30.4. (*a*) Classical example of "ribbon" resulting from a longitudinal tear along a piece of piping due to the passage of a detonation wave. (*b*) Note tear occurred a few inches away from the original weld (x).

the position of the ignition source. At the present time no unifying theory exists to relate the magnitude of these explosions to the type or quantity of fuel involved. Since experimental work on this scale is very costly, it is essential that as much factual information as possible is obtained from such incidents and that this is compared with existing knowledge. *Unconfined Vapor Cloud Explosions,* by K. Gugan, 1979, Published by Gulf Publishing Co., Houston, Texas is an excellent review.

FAILURE ANALYSIS

If the mode of failure of any vessel is to be used to determine the nature of the pressure generation process, it is important to identify the point of initiation. Where a vessel produces fragments due to brittle failure, chevron marks may be produced which point back to the point of initiation.[7] In most ductile failures the maximum thinning is usually to be found near to the point of initiation. The fragments are first collected and assembled as shown in Fig. 30.5. The graphical reassembly illustrated in Fig. 30.6 can then be made indicating a longitudinal split up the cylindrical section along XY. The area of primary failure is located by using a long reach micrometer screw gauge to

Figure 30.5. Fragments of vessel assembled for examination.

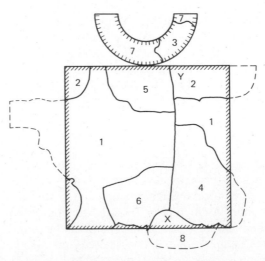

Figure 30.6. Reconstruction of fragments of vessel in relation to original plate. Numbers correspond to fragments illustrated in Fig. 30.5. Thickness measurements carried out on the tear XY.

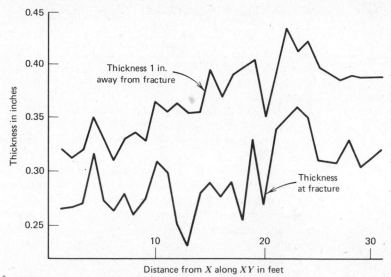

Figure 30.7.   Thickness of metal along main fracture (*X* to *Y* in Fig. 30.6).

take thickness readings adjacent to the fragment edges. The results are given in Fig. 30.7.

BLAST EFFECTS

An analysis of the blast effects in the stand-off position from any explosive event may yield valuable evidence regarding the cause. From the extent of the blast damage a TNT equivalent can be estimated and it should be possible to relate this approximately, in terms of energy, to the event from which, it is postulated, the blast wave originated. The blast effects from the detonation of an explosive charge are inversely proportional to what is known as the scaled distance *Z*, which is the actual distance *D* between the charge and the target divided by the cube root of the weight of explosive.

$$Z = (D/W)^{1/3} \text{ ft/lb}^{1/3} \tag{30.1}$$

By considering the blast effects from an accidental explosion, the scaled distance *Z* can be estimated from Table 30.1. A number of apparent TNT equivalents can then be calculated from the relationship:

$$W = \left(\frac{D}{Z}\right)^3 \text{ lb TNT} \tag{30.2}$$

from which an average value can be obtained. Due to shock reflections and inconsistencies in the response of individual structures, some values may be

TABLE 30.1.   Comparison of Blast Effects at Various Scaled Distances

Scaled Distance (Z ft/lb$^{1/3}$)	Observed Effect
300	Annoying noise (137 dB).
250	Occasional breakage of large glass windows already under strain
200	Loud noise (143 dB); sonic boom; some glass failure
150	Approximately 10% window damage
100	Limit of serious damage; limit of normal missiles; damage to house ceilings
80	Limited minor structural damage to normal buildings
40	Windows shattered; occasional damage to window frames
30	Partial damage to houses: rendered uninhabitable
25	Corrugated asbestos shattered; fastenings to steel, aluminium and wood cladding fail and panels buckle; steel frames distort slightly
20	Partial collapse of walls and roofs of houses
17	50% of brickwork of houses demolished
14	Steel framed buildings distorted and pulled away from foundations
12	Oil tanks ruptured
10	Wooden utility poles snapped
7	Loaded train box cars demolished; buildings most probably totally destroyed
1.4	Crater lip

disregarded in the estimation of a realistic average value. Damage to houses at a scaled distance $Z = 40$ is shown in Fig. 30.8.

A comparison of the observed TNT equivalent can now be made with the value which can be expected from any accidental explosion. The method of comparison will depend upon the nature of the postulated event. In the case of bursting process vessels an estimate of the TNT equivalent can be obtained if the static failure pressure is known. It is assumed in the first instance that the failure pressure will be the same whether the mechanism of pressure generation is physical overpressurization or combustion. Experience would indicate, however, that the calculated pressure could be doubled if the vessel failed during gaseous detonation. The mode of failure which determines the bursting pressure P_f from cylindrical containers is usually taken as that associated with hoop stress and is calculated from:

$$P_f = \frac{YD}{2t} \text{ psi} \qquad (30.3)$$

where Y is the yield stress of the material of construction, D is the vessel diameter, and t is the wall thickness. For failure due to detonative loading

Figure 30.8. Typical damage to houses at scaled distance $Z = 40$. Windows shattered, some window frames broken. These were 1700 ft away from epicenter of NYPRO explosion. Substitution in equation (30.2) gives $W = (1700/40)^3 = 34$ tons TNT.

$$P_f = \frac{YD}{t} \text{ psi} \qquad (30.4)$$

Various formulae have been proposed for estimating the TNT equivalent from the bursting pressure P_f psi and the volume V ft^3 but[5] for the purpose of diagnostics, isentropic decompression energy is assumed to be imparted to the shock wave. Thus

$$W = \frac{1.2 \times 10^{-4}\, P_f V}{\gamma - 1} \left[1 - \left(\frac{P_a}{P_f} \right)^{(\gamma-1)/\gamma} \right] \text{ lb TNT} \qquad (30.5)$$

where P_a is the atmospheric pressure and γ is the ratio of the specific heats of the exploding gas. Although γ is not a constant, a value of 1.4 can be assumed which, used in conjunction with Eq. (30.5), gives reasonable agreement with observation.

If a reactive exothermic condensed phase is thought to be responsible for an explosion, the resulting TNT equivalent may reflect the total chemical energy involved. In other words the material involved may have behaved like a classical explosive. Under these circumstances the TNT equivalent should be estimated from the total thermal decomposition energy available on the basis that 1 lb of TNT is equivalent to 550 kcal. In the case of unconfined flammable exploding vapor clouds, experience would indicate that only a fraction of the combustion energy is converted to blast. If the total calorific value of the vapor release is calculated then the TNT equivalent should not exceed 10% and for releases less than 10 ton it may be significantly less. In the case of the BLEVE the contribution of the fireball to the blast effect is usually negligible.

Evidence Regarding the Cause of Death

It is unfortunate that the precise manner in which loss of life occurs in major disasters is often not considered by accident investigators. The reluctance to deal with this matter in depth arises from a natural revulsion to the subject and out of respect for the feelings of relatives and friends. This is unfortunate since valuable evidence regarding both the cause of the accident and the lessons to be learned regarding the future safety of personnel may be lost. The detailed examination of the bodies of the deceased must be left to the qualified pathologist although some liaison with the accident investigator may be valuable at an early stage.

Death occurs during fire or explosion for a variety of reasons such as (1) suffocation in the products of combustion, (2) exposure to heat, (3) pressurization, (4) injury by missiles, or (5) being thrown against solid objects by blast. In fires, death in the first few minutes of the incident is usually due to suffocation and never by heat on the body surface. It is important to distinguish between death by carbon monoxide poisoning as deduced from the carboxyhemoglobin content of blood samples from deep-seated tissues and that caused by the inhalation of flame. Care should be exercised with regard to deductions made from the positions of the bodies of victims killed in fires. The first effects of carbon monoxide poisoning manifest themselves on the centers of higher thought. Victims may wander about completely confused before they collapse. (See Chap. 25.)

30.4. INTENTIONAL FIRES AND EXPLOSIONS

It is only in recent years that the incidence of intentional fires and explosions in the process industries has reached significant proportions. Although intentional fires are uncommon on process plants, stores, utilities, and warehouses are common targets. Unfortunately, site management are often very reluctant to accept evidence which indicates that they may have an incendiary employee on their payroll. Conclusive proof is difficult. In the case of fires it may be difficult to differentiate between intent and carelessness where the source of ignition cannot be identified. The investigator should nevertheless be fully aware of this possibility.

Motivation varies. Fires are often started out of malice or mischief. The target may be an individual as well as a company. Fires are sometimes lit in the interests of self-aggrandisement where the individual responsible may try to draw attention by then fighting the fire. Small fires started for these reasons sometimes get out of control and those responsible may involve themselves at great personal risk. To accuse such a person at a later stage on the basis of technical and circumstantial evidence is difficult. Fires started for the purposes of sabotage and fraud are not common in process plants. With the increase in urban guerrilla activity in recent years, however, the question "Was it a bomb?" is being asked more frequently. Motivation in this field is most commonly political and some organization will usually claim responsibility. Regardless of these claims, however, the investigator must analyse all the evidence available to ensure that no one is making political capital from a genuine accident.

30.5. NOTES ON IGNITION SOURCES

If the area of origin of a fire or explosion is positively identified, the material first ignited may be obvious. This may give some indication of the nature of the ignition source. If woodwork, textiles, lagging, or some other solid material is first ignited, the source of ignition would clearly involve something much more energetic than the small spark which could ignite a gas or vapor–air mixture. In this respect the primary source of ignition responsible for starting a classical fire may be more easy to identify than that associated with gaseous phase ignition. In general, fires started in solid materials require an ignition source which persists for some time compared with, for example, a static electrical discharge which may ignite a gaseous mixture in a fraction of a second. Common causes of fires on process plants are electrical faults on motors and cables, contamination of lagging by flammable liquids[10] and causes associated with human activity. Although it has been shown that a discarded lighted cigarette will not normally ignite petrol vapor,[11] unless oxygen concentrations are higher than in air, fires can be started when a burning cigarette end lands in a substrate which is capable of smoldering. As the smoldering process accelerates, the mass may burst into flame at a later stage. Welding and cutting sparks are equally not as incendive as one might imagine. If these are suspected to be the cause of a fire or explosion, simulated tests should be considered and the possible role of higher oxygen should not be overlooked.

The identification of the primary ignition source is often considered to be the ultimate objective in any investigation. While this might be possible in relation to fires, a high proportion of the sources of ignition in accidental gaseous phase explosions are never established with any degree of certainty. This was well illustrated in the investigations which followed a series of explosions in the empty cargo spaces of supertankers.[12] These explosions occurred on the ballast voyage during tank washing. Despite exhaustive enquiries and a world-wide research program at an estimated cost of over $1 million, the tanker industry was left with a series of least improbables which is still continuing to change with time in the light of subsequent experience. The sudden admission of air into distillation equipment often results in explosions which would indicate that either residues or metal surfaces which have been exposed to hydrocarbons for lengthy periods may assume a reduced state which forms a local hot spot as soon as it is exposed to the oxygen of the atmosphere. This phenomenon is not always explicable in terms of iron sulfide formation. The ease with which certain discharges to atmosphere apparently ignite may be a further manifestation of this and related processes. It is against this background that the reappraisal of remit may be necessary.

The failure to elucidate the precise nature of the ignition source should not deter the investigator. In many cases where this is so, sufficient evidence has been obtained from a well-conducted investigation to prevent a recurrence. In terms of the future safety of a process, nothing more is required.

REFERENCES

1. *J. Gas Light.*, **36**, 63 (1880).

2. D. L. Chapman, *Philos. Mag.*, **47**, 90 (1899).

3. M. A. Cook, *Science of High Explosives*, Reinhold, New York, 1958.

4. A. D. Craven, Institute of Chemical Engineers Symposium Series No. 33, Institute of Chemical Engineers, London, 1972.

5. A. D. Craven, "Fire and Explosion Hazards Associated with Small Scale Spillages," *Institute of Chemical Engineers Symposium Series No. 47,* 1972 Institute of Chemical Engineers, London, 1972.

6. J. H. Burgoyne and A. D. Craven, "Fire and Explosion Hazards in Compressed Air Systems," *Loss Prev.*, **7**, 79–87 (1973).

7. R. B. Jacobs, W. L. Bulkley, J. C. Rhodes, and T. L. Speer, "Destruction of a Large Refining Unit by Gaseous Detonation," *Chem. Eng. Prog.*, **53**, 12 (1957).

8. C. S. Robinson, *Explosions: Their Anatomy and Destructiveness*, McGraw-Hill, New York, 1944.

9. S. Glasstone, *The Effects of Nuclear Weapons*, U.S. Atomic Energy Commission, 1962.

10. K. Gugan, "Lagging Fire: The Present Position," Institute of Chemical Engineers *Series No. 39*, London, 1974.

11. J. R. Yockers and L. Segal, "Cigarette Fire Mechanisms," *NFPA Q*. **49**, 3 (1956).

12. Report of Court (The Merchant Shipping Act 1894) re: S.S. MACTRA O.N. 337004, before The Hon. Mr. Justice Brandon, M.C., October 1971 to March 1972, pp. 158–217.

31
Transportation of Hazardous Materials*

William S. Wood

Transportation of hazardous cargo is one of the most complex and important operations in the chemical process industries. Gases, solids, and liquids are transported across the country in railroad cars and in trucks, via waterways in barges, overseas in ships, and underground in pipelines.

Many of the materials are flammable; some are toxic; others are corrosive; and still others explosive.[1] An added complication is that some cargo must be shipped at elevated temperatures or under cryogenic conditions.

Because of the increasing volume of material being transported, carriers are using jumbo railway cars, larger trailers, large-diameter pipelines and huge tankships. Some transportation vessels are specifically designed for a special cargo.[2-5]

A serious transportation accident[6] is a problem to a broad spectrum of people. Most affected, perhaps, is the exposed public.[7] Police and fire personnel, unfamiliar with the transported material and inadequately informed as to the scope of the problem, may find it difficult to cope with a bad incident (see Chapter 34). However, through specific training programs, they are learning effective and safer procedures.

31.1. EMERGENCY SERVICES

The local fire department usually is called upon to deal with a major transportation disaster. While train crewmen have a large measure of responsibil-

---

Chemicals in transit require specific safeguards not always directly applicable to the movement of most other commodities. Due to industry awareness and major advances in cargo protection and identification, the past decade has recorded a significant improvement in accident prevention and incident control.

ity in rail accidents, many of them have little opportunity to gain experience in the handling of emergencies. Many fire departments—particularly in rural areas—have inadequate training in dealing with the products that are nowadays shipped in such profusion in large volumes.

Rural fire fighters are mostly volunteers, hence they give of their own time to attend classes or obtain training of a specialized nature. Cities have paid fire departments, and many of their fire and police personnel receive special training programs.[8]

Police, civil defense, and other such organizations, while equipped with radio and some respiratory protection, have relatively little capability other than that of controlling traffic, evacuating residents, and assisting in communications.

Through the Fire Marshals Association of North America and the International Association of Fire Chiefs, fire departments in industrial areas have strongly supported improved placarding of transportation vehicles (see present truck placards and railway placards in Fig. 31.1). Anxious to learn about the proper handling of emergencies, they attend meetings, seminars and training schools. *Disaster Planning for Fire Chiefs*, published by *Federal Emergency Management Administration*, Washington, D.C., 1981, is a valuable training aid.

Increasingly, fire departments (both paid and volunteer) are being staffed by better-trained individuals, some of whom have degrees in chemistry and engineering. Technically trained officers are invaluable in teaching their force how to handle complex problems so that the fire fighters themselves can take proper measures. Fire services are also becoming increasingly aware of the aid they can obtain from CHEMTREC, and are calling that office for information during emergencies.

What CHEMTREC Does

A number of train wrecks, tank truck incidents, and explosive disasters prompted the Department of Transportation and other agencies to request the Manufacturing Chemists Association, now the Chemical Manufacturers Association, to set up an emergency service, where immediate information could be obtained about materials involved in emergency situations.

Out of this request, the Chemical Transportation Emergency Center (CHEMTREC) emerged. The center was set up in Washington, D.C., with a nationwide toll-free telephone number.; the service began operation on September 5, 1971 on a 24-hour basis.

John Zercher, formerly a transportation manager for Celanese Corporation, was chosen as director. Under his guidance, an extensive information file has been built up on forms sent in by member companies on all regulated materials they ship. Products are cross-referenced by all synonyms and also by various subsidiary companies.

The "communicators" who handle the calls are trained in answering

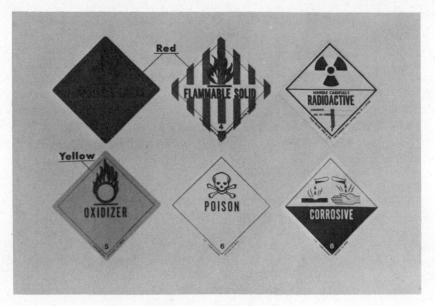

Figure 31.1.   DOT Placards.

questions from the file, and in securing additional help or information under the stress of the moment.

When fire chiefs or other users call CHEMTREC, they are asked to report the location, type of accident, material involved, etc. The duty officer then refers to the file and reads back (often within as little as 10 sec) the material's chief hazard properties and the action to be taken in case of fire, spillage, exposure, etc. Next the duty officer offers to contact a shipper's representative, and if this is desired, he places a call from a prearranged list.

The representative is then asked to call back to the scene of the accident with supplemental information, and offer help. Shippers often send personnel to the scene to advise on critical aspects. By 1982 CHEMTREC had received over 155,000 calls and had given assistance in 21,700 transportation emergencies.

It is realized that emergency personnel unfamiliar with the product involved and under pressure may have difficulty remembering and digesting the information obtained from CHEMTREC. "Hard copy" transmission is being installed between CHEMTREC and major shippers. Relay of readable information to a printout machine near the accident scene will be possible when the necessary equipment becomes available.

CHEMTREC is unanimously regarded by involved government, carrier, and fire-service people as a highly valuable service. In March, 1980, the Department of Transportation officially recognized CHEMTREC as the central emergency response service for dealing with incidents involving the transportation of hazardous materials. The toll-free CHEMTREC number in the

48 continental states is 800-424-9300. For Alaska and Hawaii, the number is 202-483-7616.

Major chemical companies respond to emergencies when they are notified by CHEMTREC that one of their products is involved. In addition, the manufacturers of chlorine, vinyl chloride monomer, and certain pesticides will cooperate with information and, if necessary, with on-site consultation, when called by CHEMTREC.

The Federal National Response Center should also be called on 800-424-8802 when any spills or releases of hazardous materials occur, whether or not related to transportation.

31.2   RAILROAD TRANSPORTATION

Among the carriers, railroads handle a major proportion of the hazardous materials conveyed over land. While most box cars are owned by the railroads, it is generally true that shippers either own the tank cars they use or lease them from rental companies.

Railroads have not always been blameless in mishaps; cars have been bumped, and equipment not altogether fit for the road has escaped inspection and has been accepted for shipment. Nevertheless, there are factors over which railroads have minimum control.

For instance, they are committed to put into the train any tank car that is submitted with the proper papers; yet, any one car can jeopardize the whole train. In other instances, such as the location of hazardous-materials cars on the train, the railroads have some control, but may often be limited in their ability to follow recommended practices. The review that follows on how some railroads operate gives some insight on the problems involved.

Railroad Procedures

The manager of hazardous materials for a major rail system is employed full time to deal with hazardous materials. The manager and part-time co-workers are responsible for training and also for seeing that proper reports on incidents involving hazardous materials[9] are submitted.

One important subject taught is the positioning of cars containing hazardous substances. Code of Federal Regulations Title 49, Paragraphs 174.88–174.93, specifies the positions of explosive, poisonous, flammable, and other hazardous cargo with respect to the locomotive and the occupied caboose. If possible, flammable or explosive materials must have five cars of unregulated material between them and the locomotive or caboose.

Training is also given to those responsible for car inspections. Placarded cars must be inspected frequently and may not be moved from the siding without prior inspection. Cars for explosives are even more tightly regulated. The railroad must certify on the shipping papers that the cargo has been loaded properly, and once again the railroad has the car inspected to certify that the loaded car is ready.

Such precautions do much to prevent a faulty car from creating a hazardous situation. Unfortunately, a car loaded with nonregulated material may be the one that fails, causing a pile-up involving hazardous materials.

Most employees work a lifetime with the railroad without being involved in a hazardous-material incident. Therefore, because it is impractical—if not impossible—to keep all employees well versed in hazardous-material handling, it is necessary to simplify instructions so that employees can take proper actions without having to become experts.

Employees are told how they can obtain the necessary information and instructions in case of an accident. Basically, the instructions consist in first calling CHEMTREC, whose telephone number is carried by every train worker who could possibly be involved in a hazardous-material incident. A second call must be made as early as possible to the Bureau of Explosives, which is an office of the Association of American Railroads in Washington.

The majority of hazardous-material incidents are not due to road accidents but to defective cars. Some cars leak because a valve is not closed tightly or a gasket has aged and has not been replaced.

Sometimes, overloaded cars cause failure of the relief device (e.g., rupture disk). If caps and plugs are not replaced when required, leakage is harder to control. However, when a serious problem occurs, shippers are extremely cooperative and are most willing to respond and assist the railroad.

The Southern Railway System

Operating approximately 6% of the rail ton-miles in the U.S., Southern Railway System has taken progressive action in the handling of emergencies.

Because hazardous materials represent 20% of Southern's business, the railway company became particularly concerned when, within a period of two weeks, it had three wrecks involving hazardous substances. One of these wrecks was the disastrous Laurel, Mississippi, incident, following which the National Transportation Safety Board investigated and made recommendations.

Two of these recommendations were of particular importance. One of them limits the size of new tank cars to 34,000-gal capacity, or 263,000 lb. gross on the rail; the other observed that most carriers have no emergency action plans, and recommended that they be made.

Southern Railway developed its plan, and now has a "go team" equipped with tools and protective gear fitted into an automobile-drawn trailer. The team drives if the accident is within 100 miles; otherwise the appropriate equipment is taken with the team by air.

The team then dons the necessary protective clothing and other devices, and does whatever is necessary to control the situation. They also are able to give competent guidance to those responsible for handling the clean-up. Pollution prevention is incorporated as a necessary part of the emergency action plan. To get in contact with knowledgeable people—particularly with the shipper of the problem cargo—CHEMTREC is used extensively.

The traditional responsibility of the carrier in all cases of railway accidents has been successfully challenged in the courts. On the basis of some important decisions, the shipper shares the responsibility for damage and pollution control. (See *Wall Street Journal*, p. 29, March 18, 1982.)

By regulation (Title 49, Paragraphs 171.15–171.16), the Department of Transportation requires a carrier to report every hazardous leak, intentional or not.

Southern looks with some concern at regulations requiring preferred routing for some of the more hazardous types of cargo. Many such routes could be much longer, to avoid centers of population. However, based upon statistics, the poorer roadbeds on these less-used tracks increase the probability of derailment.

Union Pacific, Southern Pacific, and Family Lines also have instituted excellent response systems in recent years for hazardous cargo control.

Improving Railroad Safety

The director of the AAR (Association of American Railroads) research center in Chicago, described some of the work the Research Progress Institute of the AAR has undertaken during the past decade to improve the operational integrity of railway components.[10] Some of the more important work programs are as follows:

1.   Engineers were assigned to develop safer railway couplers. They have come up with a "shelf coupler" that will have less tendency to disengage—due to vertical motion—at the time of a derailment (Fig. 31.2). If the couplers remain together, they are not likely to puncture a car.

2.   Another study dealt with truck-component safety, where field-failure data are analyzed, and field tests conducted to alter the specifications so as to improve the components just mentioned.

3.   A 10-yr study of train/track dynamics involves people from various industries who have been made available to the AAR for this particular study.

4.   A survey made on train make-up—which included the studying of orders to railroad employees with respect to arrangement of cars and other operating rules—furnished important data concerning safety in the operation of railway equipment.

5.   Computer models were devised to simulate train response and operation under both normal and emergency conditions.

6.   Stopping distances were studied, using both ordinary-service and emergency brakes.

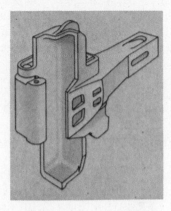

Figure 31.2.   Shelf Coupler.

7.   The sensitivity of train crews to train behavior was analyzed. Crews were asked about such things as interpretation of vibrational phenomena and indications of drag or resistance.

8.   Teams have studied the psychological response to various stimuli. Human engineering is thus becoming involved to determine the factors affecting crew sensitivity and behavior under different conditions.

9.   Study of some special problems, such as the "sudden wide gauge"—the apparent separation of rails under great stress—yielded valuable results. Another problem is "harmonic rock" (the sway or roll of high-center-of-gravity cars).

10.   Wheel design has been studied with a view to improving dependability and useful life.

11.   In conjunction with the Iron and Steel Institute, steels have been studied to combine the desired toughness that will withstand various stresses with the wear-resistance of harder and more brittle steels.

12.   Track-maintenance criteria were studied to balance effort with service derived, to obtain the greatest roadbed improvement with minimum expenditure.

13.   Another subject has to do with tankcar design to reduce the penetration in case of derailment.[11] In several wrecks, couplers and other similarly shaped objects have penetrated the head of the tank, with release of product. Shields have been devised that can reduce the probability of penetration; a number of different configurations have been evaluated.[11] Head shields have been installed on many LPG-type cars. Another aspect of this project concerns the application of thermal insulation[12] to tankcars to reduce the heat-transfer coefficient of cars that are subjected to flames.

In 1978 an Inter-Industry Task Force on Rail Transportation of Hazardous Materials was established by the chemical, railroad, and associated industries. Work of the Task Force includes: expediting installation of protective equipment on special tank cars, improved emergency communications training programs and, broader participation by industry in mutual aid.

Substantial progress has been made in installation of shelf couplers and insulation as well as head shields.

31.3 LABEL AND PLACARD REQUIREMENTS

Code of Federal Regulations 49, Transportation, Parts 100–199, regulates shipment of hazardous cargo by rail, highway, waterway, and air. The October 1, 1978, revision of CFR 49, Section 172.101 lists regulated substances with quantity limitations for various methods of shipment.

Label requirements are also listed in 172.101. Sections 172.400–172.450 give label specifications and details of their application to packages for shipment. The labels are based upon the United Nations labeling system and are authorized for domestic and foreign shipments. U.N. Hazard Class Numbers which are required for international transport may be shown in the lower corner of the diamond-shaped label (Table 31.1).

Sections 172.500–172.558 contain regulations for use of placards on rail cars and trucks carrying hazardous cargo. Placards are similar to the above labels, conforming generally to the U.N. graphic symbols. Placards are readable from a distance of about 75–100 ft and the symbols are recognizable at much greater distance. Placards are required on both sides and both ends of each car or truck. When two or more materials having different hazards

TABLE 31.1.   United Nations Hazardous-Class Number

Hazard-Class Number	Type of Material
1	Explosives—Class A, B, and C explosives
2	Nonflammable and flammable gases
3	Flammable liquids
4	Spontaneously combustible substances
5	Oxidizing materials and/or organic peroxides
6	Poisonous materials—Class A, B, and C poisonous or toxic substances
7	Radioactive materials—White I, Yellow II, or Yellow III
8	Corrosive materials (acids, corrosive liquids or solids, and alkaline caustic liquids)
9	Miscellaneous hazardous materials (substances that, during transport, present a danger not covered by other classes—no specific label authorized)

are carried, a "Dangerous" placard is permitted. This can lead to confusion in emergency conditions.

General Safety Recommendations

As of now, DOT favors instructing fire services to await the necessary information before combating an emergency. Personnel should be protected as extensively as possible,[7] and should avoid exposure.

Experience gained in railroad tank-car failures due to pressure build-up in fire exposures indicates that relief devices, as required now, are not sufficient in capacity. Generally, the Railroad Administration favors relief valves rather than rupture disks, because the valve closes once the pressure is reduced. (See Chap. 7.).

Figure 31.3 shows present standard fitting components for acid-carrying tank cars. These cars are generally lined with rubber or polyvinyl chloride to protect the tank and fittings from corrosion.

Desirable controls that should be exercised in shipping hazardous materials include prevention of overloading. All too often, a tank car is loaded at low temperature up into the dome, leaving very little space for thermal expansion. Pressure generated when the contents warm up can cause failure of the rupture disk.

Physical inspection of tanks for dents, gouges, or other damage is not

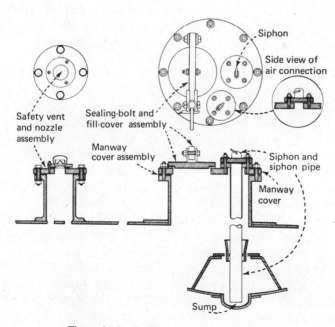

Figure 31.3. Railway Acid-Car Fittings.

performed often enough. The running gear should also be checked frequently to make sure there is no incipient cause of failure.

The practice of bottom loading and unloading is not particularly desirable from the standpoint of safety, but it seems inevitable in light of ecological pressures. Figure 31.3 shows how a standard acid car has a sump but no bottom valve.

31.4. TANK-TRUCK OPERATIONS

Highway transportation of bulk cargo is extremely big business. For example, one of the major haulers has 3,000 trailers for 20 different types of shipments. This company makes about 2000 shipments per day, or one-half million per year (Fig. 31.4).

How Tank-Truck Carriers Operate

Tank-truck carriers are perhaps unique in the transportation of bulk cargo because their drivers customarily are expected to load and unload. Drivers are therefore trained in loading procedures, and equipped with fittings to make up almost any connection for cargo transfer. Three companies use training trailers for new drivers at bulk plants or terminals. These trailers are equipped with training aids, projectors, and with displays of valves, fittings, hoses, plus all of the equipment the drivers need.

A driver should have at least two years of accident-free driving on semi-trailer units before being hired to operate tanker rigs. Whenever a driver is asked to carry a new or different hazardous cargo, he or she is trained for that particular operation.

Size and weight restrictions on the roads are the factors limiting the size of the rigs. These vary from state to state; it is the more conservative ones that tend to set the pattern for interstate commerce.

Flammable and toxic compounds that would contribute to air pollution are being loaded through the bottom in many cases. Benzene is one of these.[13] There is concern about bottom transfer because this necessarily leaves the loading lines of the tank full. Malfunctions or accidents could result in appreciable spills. (See Chaps. 13, 14, and 19.)

In addition, when the connection is broken, there may be liquid spill and exposure of the driver. The method has the advantage, though, that it gets the driver off the top of the rig during loading. However, it is still necessary for the driver to go to the top for gauging and sampling.[14]

There are two trends in the use of trailers. One is the universal-use trailer, in which a tank is made of a material that, by resisting corrosion by any of the materials hauled, will minimize contamination of cargo. Although there is a different cargo for nearly every haul, a cleaning operation prevents any mixture or undesirable reaction between consecutive cargos.

The other trend consists of dedicating a trailer for one use during its entire

Figure 31.4.   Tank Truck in Chemical Service.

life. This is done when a single cargo is hauled on an extended contract, or when the hauling of the material is very frequent.

There is a strong tendency to go to 316 stainless steel instead of 304 for better corrosion resistance, particularly to acetic acid. Aluminum trailers are being extensively used for petrochemical and dry-bulk cargo because the ligher weight of the vehicle means that more actual freight can be transported. For acid cargos, it is not uncommon to use Corro-Seal or polyvinyl chloride linings to maintain desired product purity. Hose and pump for loading and unloading, which must be adapted to the cargo handled, are usually supplied by the carrier.

A large operator used at least eight different routine cleaning methods, depending upon the cargo involved, and has developed still other cleaning methods to meet special conditions. A caustic wash seems to be the most widely used. Wash liquors constitute a disposal problem, since they must be treated to reduce their pollution potential. (See Chap. 28 and 29.)

Shipping papers frequently do not contain sufficient information to identify the cargo. The tariff name is not adequate, and frequently neither is the trade name. *It is highly desirable that the generic name of the material be given so that its hazardous properties can be readily ascertained.* See *Emergency Response Guidebook for Hazardous Materials*, DOT P-5800.2, Materials Transportation Bureau, U.S. Department of Transportation, Washington, D.C. 20590. This guide should be carefully monitored for technical accuracy.

Tank-Truck Safety Features

National Tank Truck Carriers, Inc., (NTTC), estimates that there are 60,000 semitrailer tank vehicles for hire, and about 30,000 more used exclusively by their owners. The carriers for hire make about 110,000 drops of materials each day, and over half of these are considered hazardous substance that are DOT regulated.

According to NTTC, one of the most difficult types of accidents to control is the single-vehicle accident, where the truck rolls over, generally due to too high speed on tight curves. Some of the cloverleaf ramps on limited-access highways are too tight for trucks to maneuver. Drivers are being trained to use better speed judgment, and to take the curves at less than the posted speed.

DOT regulations limit the number of hours a driver may work to 60 hr/week. In peak periods, drivers often work 15 hr/day, which means that a driver can only work four days/week, and must then be off for approximately three days before returning to work. Thus, it is often necessary to assign relatively inexperienced drivers. Efforts are being made to obtain some relief so that experienced drivers may be used more effectively.

Most of the carriers question the value of some of the available safety accessory times, such as anti-jackknife devices. These are now fairly well developed but are expensive and impose limitation on the maneuverability of the rig. Pressure to improve the overall safety of tank trailers will lead to the gradual acceptance of such devices. Manufacturers of tank trucks are continually improving them through better design and through correction of faults, as reported by the carriers.

Since 1974 the chemical industry has been working with NTTC to improve safety in truck movement of hazardous cargo. Jointly developed safety seminars were conducted. This liaison group has task forces addressing equipment design, emergency response, improved accident reporting, and interpretation of safety data. Other ideas being developed include stronger enclosures for valves atop tank trucks and locking chocks to ensure inspection before release. Significant improvement has been made.

Function of the Motor Vehicle Administration

The Bureau of Motor Carriers of DOT has 187 field inspectors to inspect vehicles operated by 168,000 private and common motor carriers. These carriers can be single-vehicle operators or operators of large fleets. Obviously, this small number of inspectors is spread rather thin over the vast number of trucks to be monitored.

Regulations for highway transportation of hazardous materials are covered in Title 49, Code of Federal Regulations. The list of hazardous commodities in Paragraph 172.101 applies to motor vehicle as well as rail transportation.

A hazardous-materials incident report is required by the Hazardous Materials Regulations Board, DOT, in case of any accident occurring in the transportation of regulated materials. In addition, a motor-vehicle accident report is required by the Bureau of Motor Carriers of DOT, if the vehicle itself was involved in an accident. Both reports would have to be made if a motor-vehicle accident involved hazardous materials.

The Bureau collaborated with National Tank Truck Carriers, Inc., to study stress-corrosion cracking of tank trucks that carry anhydrous ammonia. There were indications that when such tanks are quenched and tempered, they become predisposed to stress-corrosion cracking, which could contribute to the failure of a tank in transit. Stress-corrosion cracking can best be detected by the wet, fluorescent-particle test, as described in a standard of the Compressed Gas Association.

31.5. WATER AND PIPELINE TRANSPORT

CFR, Title 49, Part 176, contains rules promulgated by the U.S. Coast Guard for safe transportation of hazardous materials and other cargo by water. In contrast with the Railroad and Motor Vehicle offices of DOT, the Coast Guard has no current plan to require placards on water-transport vessels.

Safety Emphasis in Water Transport

The Coast Guard is well aware of the rapid growth of traffic in chemicals and other reactive or hazardous materials,[15-18] and has initiated a number of studies aimed at improving marine transportation safety through better regulations,[19,20] more information, etc.[21] Increasing costs of insuring marine vessels and cargo is proof of the need of improved maritime safety.[22-24]

Early in 1964, the Committee on Hazardous Materials was established in the Division of Chemistry and Chemical Technology of the National Academy of Sciences-National Research Council (NAS-NRC), and was charged to advise the Coast Guard on scientific and technical questions relating to safe maritime transportation of hazardous substances.

Rapid growth of the chemical industry, coupled with major advances in shipping technology, convinced the Coast Guard of the need for such a body.[16] The function of the Committee was to visualize problems created in marine transportation with regard to safety and public health, and to formulate engineering solutions.

Made up of able scientists and advisers (scholastic, industrial and military), the Committee addressed itself to various special problems such as pressure relieving systems[25,26] risk analysis and hazard evaluation; reactivity of mixed chemical cargoes[1]; electrical equipment classification; and shipping of liquefied gas.[20,27-31]

Some of the deliberations relate to the high pollution potential of some cargo.[32] Such concern with ecology may require quite drastic changes in the construction of ships. For example, for some materials a double-hull vessel could be required so that a collision or grounding would affect only the outer shell and not damage the actual container of the chemical cargo.[17,33,34]

One of the Committee's most important publications was *Evaluation of the Hazard of Bulk Water Transportation of Industrial Chemicals (Publication No. 1465)*. The last edition provides hazard properties of 335 commodities that may be transported by water.[35] The committee was abolished in 1975.

There is improved rapport with other countries in the shipping of hazardous materials, and we are approaching a coordination of regulations and transportation equipment, particularly with respect to the prevention of water pollution.

J. S. Gardenier, II, Operations Research Analyst in the R & D Office of the U.S. Coast Guard, pointed out in publications the advantages of the systems approach to risk analysis in marine operations.[36,37] With increasing utilization, the systems technique should generate new concepts of risk management for marine and other modes of transportation.[33,38–40]

The Coast Guard *CHRIS Manual*, published in 1979 is in four volumes: (1) *Condensed Guide to Chemical Hazards*; (2) *Hazardous Chemical Data*; (3) *Hazardous Assessment Handbook*; and (4) *Response Methods Handbook*. The manual is available from Supt. of Documents, Government Printing Office, Washington, D.C. 20402.

Transporting Through Pipelines

Pipelines carry more than 18% of the total ton-miles of freight in the United States and a much larger proportion of the flammable material. Most lines convey natural gas under high pressure, liquefied petroleum gas, or petroleum crude or products (mainly gasoline and fuel oil).

Gas lines as large as 48–50 in. are in operation; crude lines are as large as 36 in. One line constructed for gasoline transportation from Oklahoma to the Chicago area has a 40-in. diameter. These pipelines operate at pressures up to 1200 psi, and are located 30–36 in. underground.

Booster stations required to maintain pressure and volume of transmission may be 100 miles apart on level ground, but in mountainous territory they must be close together to overcome hydrostatic pressure. Most booster stations are unstaffed, except for occasional visits.

To offset the absence of operators, a multiplicity of safety devices has been installed. Fire detection equipment with automatic alarm and shutdown; automatic extinguishing systems; and closed-circuit television are used to enable the distant dispatcher to monitor the operations adequately. Because instrument readouts are available to the dispatcher, he or she receives indications of any leakage that may take place in the line. For example, each meter measuring the flow should read the same all along the line, and pressure should remain constant. Any variation indicates leakage, which is handled by shutting down the line and locating the leak.[41,42] Computers are extensively used for dispatching as well as monitoring pipeline shipments.

Most of the problems of leakage or other accidents to pipelines come about from outside influences, the most common of which is excavation or ditching equipment.[43,44] Most contractors are aware of the damage potential and, when they are in the vicinity of the pipeline, will call the operating company to let it know what construction they are planning to do. Pipeline operators will send representatives to the site to help equipment operators avoid the pipeline.

Unfortunately, not all contractors make the necessary call, and many farmers digging a ditch on their property strike the line and breach it. There may then be a fire or accumulation of oil; a telephone call alerts the dispatcher to shut down the line.[45,46]

Line walkers have been largely replaced by pilots who fly over the line, making weekly inspections. They can tell by the appearance of the ground or changes in vegetation where leakage is occurring, and can also report any apparent activity, such as construction or digging, that might affect the pipeline.

DOT's Office of Pipeline Safety has regulations (49 CFR, Subchapter D) that detail the reporting requirements, as well as construction and operating features regarding pipelines. Covered by the regulations are: materials to be used; pipe design; design of pipeline components; welding and other joining methods; construction requirements; consumer meters; regulators and service lines; erosion control; test requirements; and maintenance.[47-51]

The National Transportation Safety Board (NTSB), after investigating a number of accidents involving pipeline failures of various sorts, concluded that a systems approach to safety in pipeline operation could predict failure, and enable action to be taken to control hazards before an accident occurred.[52] System safety analysis is recommended as being applicable to pipeline operation, and is seen as the best available method of improving operation and eliminating future accidents.

31.6. LOADING AND UNLOADING OPERATIONS

Technical bulletins of the Chemical Manufacturers Association and the American Petroleum Institute provide excellent guidance for the loading and unloading of tank cars and tank trucks. CMA also has bulletins recommending methods for the loading and unloading of solids.

One general safety requirement is safe access to the top of the vehicle. This is particularly important for top loading or unloading, but may be necessary for operations such as gauging or sampling. Thus, loading racks with suitable ladders, platforms, gangways, or railings permanently affixed to the vehicles are required. (See Chapter 16, pages 263–272, Handling and Transport of Hazardous Materials, in *Toxic Chemical and Explosives Facilities*, ed., by R. A. Scott, Jr., ACS Symposium Series 96, American Chemical Society, Washington, D.C. 20036, 1979.)

Safety in Liquid Loading or Unloading

Keeping liquid-loading facilities usable in adverse weather conditions, such as icing, may be difficult, but every effort must be made to keep them safe.[53] Good general illumination, especially at night, is by far preferable to providing the operator with an extension light, which might be dropped and broken.

An emergency shower and an eyewash device should be available at each loading location. Preferably, these devices should be tied to an alarm system that would bring help to anyone making use of them.

Fire extinguishers of adequate size and type for the material being handled must be distributed throughout the area. Drainage from the loading-rack area should collect liquids at a point away from the rack to permit recovery, or proper disposal.

Where air pressure is used for discharging tankers, air connections should be close enough to allow air hoses to be attached without extensive runs. And when other gases are used, such as nitrogen, their identity must be indicated. It is desirable to use different connecting devices to avoid confusing the various gases. Air lines used for discharging tankers should have reducing valves set at no more than 20 psi, and should have relief valves set at slightly higher pressure, perhaps 25 psi.

When cargos require steaming to warm the contents for unloading, the steam connections and steam lines should be so located as to coincide with the connections on the car; and again, unnecessary runs of hose must be avoided. Steam pressure should not be more than 50 psi; relief valves for protection of the steam coils may be set even lower in some cars.

Cargo should not be heated more than necessary to assure proper flow. Pipe, hose, valves, and fittings should be of sufficient size to permit adequate flow without excess pressure. Water lines and hose are necessary for washing down spills or other materials from the area.

Personnel responsible for unloading must stay in the immediate vicinity of the operation at all times, and should ascertain that all conditions are normal.

Adequate personal protective equipment should be furnished for those involved in the loading and unloading of tank cars. For materials that are corrosive to the skin or that may be absorbed through it, full protective clothing with face masks, rubber gloves, rubber shoes, etc., is required. When materials that have toxic vapor or gas are unloaded, personnel should be equipped with airline respirators or self-contained breathing equipment.[54] Protective equipment not available at the site should be obtained, even though this may mean some delay in making the transfer. (See Chaps. 25, 26, and 27.)

In all cases, personnel should be thoroughly trained and should be given written instructions suitable for each material they will handle.

Common mistakes are transferring into the wrong tank or transferring into a tank that does not have sufficient remaining capacity to contain the contents of the transportation vehicle. In the first case a violent reaction may occur; in the second an overflow is the result.

First aid and medical procedures should be worked out in advance and posted in the unloading area. Where unusually toxic substances are handled, medical personnel should have information on the characteristics of the material and on the medical management needed for any material they may encounter. (See Chaps. 9, 10, and 11.)

Some tank cars and tank trucks have interior linings of rubber or plastic of various kinds. When such cars are being loaded or unloaded, special care must be taken to prevent damage to the lining.

Damage to the exterior paint, metal, or stenciled markings is evidence that the unloading operation was permitted under conditions that could have been hazardous to employees and property. This can usually be attributed to careless operation, defective fittings, defective or improper connections, or disconnecting lines before they are completely drained.

Any spillage on the tank car exterior should be washed off promptly with water to avoid damage and possible injury to persons coming in contact with the tank car upon its return. Consignees are responsible for such damage.

Tank-truck operators also run into serious problems in loading and unloading. In one case, nitric acid was pumped into a trailer containing a substantial amount of glycerin; the nitroglycerin formed by the reaction had to be removed with extreme care.

NTSB has reported another case,[55] in which two drivers delivering a cargo of sulfide solution followed the explicit directions of the plant supervisor, but nevertheless discharged the cargo into the wrong tank. Reaction with acid in the tank generated substantial amounts of hydrogen sulfide that resulted in fatalities of plant employees. There was evidence the drivers had been cautious, but there was no external sign or other identification on the three possible discharge lines.

Equipment for Liquid Loading/Unloading

Because connections for tank truck transfer can be of almost any pipe size and terminate in almost any fitting, it is customary for drivers to carry a substantial number of adapters and fittings.

Operating out of a large petroleum products terminal, a fleet of compartmented tank trucks of 8050-gal capacity is used primarily for deliveries of gasoline of three grades. Each of the compartments has its own line to the loading manifold; the truck can be loaded from one location by simply switching loading lines from one connection to the other (Fig. 31.5).

When a driver breaks the connections after loading, there are no more than a few drops of gasoline lost from the line. Loading is done through preset meters, which are connected with computers in the office of the terminal. The driver presets the meter, opens the valve and fills the com-

Figure 31.5.   Tank Truck Loading Manifold.

partment. The meter then shuts off the flow, but because the meter could malfunction, there is—on top of each compartment—a sensing device that trips the foot valve when the tank capacity is reached.

A key device on the inlet to the truck is the API-designed fitting[56] with check valve (flush with the point of breakage), which retains all gasoline in the fill line of the truck (Fig. 31.6). A similar flush fitting is operated by the tightening mechanism of the mating connection of the loading truck. At the moment of disconnecting the fill line, there are two flat metal surfaces almost touching, which squeeze out almost all liquid from between the two mating pieces.

The compartments of the tank truck are all connected through a passage at the top of the truck to a small vapor tank at the truck's rear. This tank is not obvious because it has the same diameter and is really another compartment, otherwise unused, of the tank truck itself. Before filling, the vapor tank is connected to the terminal's storage system. Then, as the tank truck

Figure 31.6.   API Flush Disconnect Valve.

fills, the vapors pass to the vapor tank and then out—through the vapor connection—to the recovery system.

Bottom-Loading Features

No gasoline odor could be detected at the loading terminal. The new bottom-loading system has proved to be a far cleaner operation than top loading.

The entire loading-rack operation—with five drive-through positions—is equipped with a water spray system that is actuated manually by a button at each of the loading positions.

One unexpected benefit from this loading system is that the driver is free from distraction. With top loading, the drivers are on elevated platforms, changing the dip pipes from one compartment to another. This modern operation not only is more efficient but also reduces air pollution and fire hazards.

The following case history attests to the integrity of the filling connections, even under accident conditions:

A driver felt his truck behaving strangely, as it were not at full power. On glancing in the rear-view mirror, he saw that a smoking automobile was sticking out from under the truck and was being dragged along at a fair rate of speed. He stopped the truck, took a fire extinguisher and went under the truck to extinguish the fire in the automobile and extricate the driver of the car.

The car driver got out without help and ran away from the scene because he had been driving a stolen automobile. Because there were no keys to turn off the ignition switch, it was necessary to raise the hood and pull off the ignition wire to kill the engine and stop the rear wheels from turning.

Despite all these happenings, no major accident occurred. The reason for the truck's safe behavior was that, while the filling lines were actually damaged, the foot valve in each compartment secured the gasoline in the tank. The only fluid that escaped was that spilled from the few feet of fill line.

Solids Loading and Unloading

Increased tonnages of chemicals and other hazardous materials are being shipped as dry cargo in hopper cars or hopper highway vehicles. Among these solids are caustic soda, tetrachlorobenzene, asbestos powder, magnesium oxide, sodium nitrate, sodium cyanide, lead oxide, and sodium metasilicate.

Unloading is easily achieved by a pneumatic system that transfers the solids—dispersed in air or inert gas—through a closed pipe.[57] Selection of the conveying medium (air or other gas) depends upon whether the product is affected by humidity or contact with air; whether quality control requires inert gas; but, most of all, whether air would form an explosive mixture with the conveyed substance.

Storage of these dry solids is usually in silos of sufficient size to hold at least 1.25 times the volume of the largest anticipated car or truck. An important feature of a silo is a high-level alarm to alert the unloading operator that the silo is nearly full, and that action must be taken to prevent overfilling.

The principal components of an unloading system include a pickup device or truck; a high-capacity blower to move the conveying gas; a cyclone to separate conveyed material from the gas; and a filter to prevent escape of conveyed material into the atmosphere.

Safety features include high-pressure alarms and interlock controls to shut down the system (should personnel fail to do so manually), and pressure-relief devices, in case the interlock shutdown fails. Bonding and grounding provisions are needed if static electricity could cause personnel hazards due to shock, explosion hazards, or poor product-flow characteristics.

Sometimes a vibrator is used on a car to encourage the flow of material into the discharge opening. A chain or cable should be attached to restrain the vibrator's movement, should it become disengaged.

Periodic inspection, testing and maintenance of safety devices and accessories is essential. A written and posted procedure for preunloading and post-unloading inspection of the car and equipment is strongly recommended. If inert gas is used for conveying, the lines and vessels should have warning signs so that maintenance personnel will not be jeopardized by oxygen deficiency.

31.7.   REGULATING ORGANIZATIONS

Of the government agencies and private organizations that issue regulations or otherwise influence the methods of transporting, loading and unloading hazardous materials, the ones that follow are the most important.

Department of Transportation

Rulemaking at DOT* is time consuming. On the basis of recommendations from knowledgable people, one of the administrations devises a regulation to meet a particular need. The regulation may be published as an order, or as a notice to be followed by an order, or can be a preliminary notice requesting comments to develop a more equitable regulation.[2,4]

DOT is involved in a number of research projects and studies, many of them in cooperation with other agencies. For example, DOT sponsored a research project at Cornell University on relief devices. The results call into question some of the factors that have been accepted for the calculation of relief devices, particularly the terms $A^{0.82}$ and 34,500 Btu/ft²/hr in the formula for heat input under fire conditions:

$$Q = 34,500A^{0.82}$$

where Q is the heat input, Btu/hr, and A is the exposed wetted surface in square feet.[58] (The Flammable Liquids Code, NFPA 30-1977 uses four formulae for heat input.) (See Chap. 7.)

Other research has been conducted jointly with the Railway Progress Institute of the Association of American Railroads. This involves shield protection for tank cars to prevent couplers from penetrating tank heads.

NBS conducted a research project on steels for tank-car construction to find one that will hold up under the severe conditions of collision and resultant fire exposure. The protection that fire-retardant coatings might offer in case of fire exposure was evaluated. Both intumescent (foam formed upon heating) and other more conventional insulations showed promise.

National Transportation Safety Board

Although NTSB† appears on the Department of Transportation organizational chart, it is responsible directly to Congress rather than to the Secre-

---

*Department of Transportation, 400 Seventh St., SW, Washington, D.C. 20590.

†National Transportation Safety Board, 800 Independence Ave., SW, Washington, D.C. 20592.

tary of Transportation. Its function is primarily to investigate transportation accidents and make recommendations for steps to minimize the chances of recurrence.[59]

The conclusions reached by NTSB in a special study[52] on pipeline safety reflect its advocacy of system safety analysis for all modes of transportation. It is not enough to investigate and make recommendations after an accident has occurred, but NTSB strives to develop a methodology for preventing accidents.

The big problems are determining what risks are involved, and then measuring and evaluating them so that decisions can be made as to their acceptability. Studies of comparative risks must include possible effects upon third parties and harm to the ecology, as well as destruction of property, damage to public facilities, and the involvement of crew members, drivers, etc. A recognition that hazards exist in all the things we do is essential to the philosophy of NTSB.

A significant advance in the study and reporting of spill incidents is the inclusion of maps using the USGS 15' base, in the report showing the magnitude and direction of gas and vapor clouds released for different time periods after the spill. For example, when a tank vehicle ruptured, releasing 18,300 gallons of anhydrous ammonia, the vapor traveled over a mile from the accident scene in 340 minutes (over 5 hours).[60]

The Bureau of Explosives

Tariff BOE6000B, the accepted code for the shipment of hazardous materials by surface transportation, is identical to Code of Federal Regulations, Title 49. This document lists a large number of commodities and specifies the placarding or labeling required. It also designates specifically the types of transportation equipment that are suitable. Tariff is published by the Bureau of Explosives, which is part of the Association of American Railroads, 1920 L St., N.W., Washington, D.C. 20036.

Although the Bureau of Explosives Laboratory at Edison, New Jersey, has the primary function of examining new explosives and classifying them for transportation, it will also test flammable materials for classification (a fee is charged to shippers for this service).

The Bureau has field inspectors whose duty it is to inspect loading and unloading facilities, to check on the proper placarding of cargo, and on the billing and other documents for complete descriptions. They check on the suitability of cars for the transportation job, paying particular attention to the gasketing of closure devices because this is a frequent source of leaks.

They also check to see if cars are tested at the proper intervals, if correct closure devices are used, and if the carriers use derails, stop flags, and so forth, when loading or unloading. The correct placing of hazardous materials in the trains is also checked.

The Bureau of Explosives has expressed concern over the very high tonnage that is carried on 70-lb rail, since rails and ties are not replaced frequently enough to avoid derailments due to their failure.

Jumbo cars were developed to meet the need of shippers for a more economical way of transporting low-cost commodities. A few very high-capacity cars (approximately 50,000 gal) were built, but are no longer considered optimal.[3]

At present, the maximum size for many commodities is a 34,500-gal. car. Accidents with these cars have been few, considering the great tonnage they have transported.

National Safety Council

Three of the 29 industrial sections of the National Safety Council* are devoted to safety in cargo transportation. These are the marine, railroad, and commercial vehicle sections. Working with the Council staff, the executive committees of these sections draft safety data sheets and other training materials[60] aimed primarily at improving employee and public safety. All sections present programs at the National Safety Congress each year, and some schedule training seminars at other times. The Chemical Section frequently collaborates in such programs.

Movies and slide presentations on transportation safety that have been produced by government, industry, and carriers are distributed or listed through the Council. Driver training is offered in courses at various universities, in collaboration with the American Trucking Association. In addition, a one week seminar at NSC headquarters is held at frequent intervals.

Some publications are written for fleet operators, others are pointed directly to the driver. A successful development is a driver-to-driver letter, where a working driver writes down his or her method of handling a given situation and sends it to the Council. After just enough staff editing to assure clarity, the driver's letter is duplicated and sent to the carriers for distribution to their drivers.

National Fire Protection Association

The Railroad Section of NFPA† reviews fires and other incidents in railway transportation, and provides a forum for exchange of information among fire-safety professionals.

NFPA has a number of codes (Table 31.2) that provide guidance in the transport of hazardous materials.[61] Of particular value is NFPA 385 on tank vehicles for flammable and combustible liquids. NFPA 58 covers the transportation, as well as the storage and handling, of liquefied petroleum gas. NFPA 87 has valuable information on piers and wharves, particularly relat-

---

*National Safety Council, 444 N. Michigan Ave., Chicago, Ill. 60611.
†National Fire Protection Association, Batterymarch Park, Quincy, Mass. 02269.

ing to solid, combustible cargo. NFPA 305, the control of gas hazards on vessels, refers to the functions of cleaning and repairing. A list of publications can be obtained upon request.

With DOT funding, the NFPA developed a course to train fire fighters in handling emergencies involving hazardous cargo. Hundreds of police officers and fire fighters have attended this course.

American Petroleum Institute

API,* a trade association comprising essentially all of the petroleum-producing and refining companies in the United States, is vitally interested in the transportation of hazardous materials because petroleum products constitute the greatest tonnage of regulated materials transported.

The Institute became very much involved when it was demonstrated that static electricity could be built up by pumping petroleum products through a pipeline into a tank,[62,63] and this static charge could then be so concentrated as to produce a delayed explosion within the tank ships,[64] in each case with a material of medium volatility, such as jet fuel. This static problem has been substantially overcome by closer control of pumping rates and sometimes by an additive in the fuel to increase conductivity.

Acting as a standard-making body, API called together tank-truck manufacturers and hardware people to work out a single adapter for tank trucks so that one fitting could be used on any tank truck at any loading rack. A characteristic of this device was the almost complete elimination of any motor fuel loss when the connection is broken. When agreement was

TABLE 31.2.   NFPA Codes on Bulk-Transportation Fire Safety

Code No.	Title of Publication
385	Tank Vehicles for Flammable and Combustible Liquids (1979)
386	Portable Shipping Tanks (1979)
704M	Fire Hazards of Materials (1975)
58	Liquified Petroleum Gas Storage and Handling (1979)
49	Hazardous Chemicals Data (1975)
495	Explosive Materials (1973)
498	Explosives Motor Vehicle Terminals (1976)
70	National Electrical Code (1978)
77	Static Electricity (1977)
87	Piers and Wharves (1975)
513	Motor Freight Terminals (1978)
306	Control of Gas Hazards on Vessels (1975)
307	Operation of Marine Terminals (1967)
512	Truck Fire Protection (1978)
59	Liquified Natural Gas Storage and Handling (1979)

*American Petroleum Institute; 2101 L St., NW, Washington, D.C. 20037.

reached, "API Recommended Practice 1004" was published,[56] giving a procedure for bottom unloading and also detailing the adapter that had been approved by the various parties.

Chemical Manufacturers Association

The Chemical Manufacturers Association,* founded in 1872 as the Manufacturing Chemists Association, is one of the oldest trade associations in the western hemisphere. Its 200 member companies account for 90% of basic chemical manufacturer in the United States and Canada. Through its Occupational Safety and Health Committee CMA presents symposia on safety relating to chemicals. As a support function for CHEMTREC, described elsewhere in this chapter, CMA conducts regional workshops for leaders of emergency response teams. Participants learn and practice procedures for dealing with chemical spills.

REFERENCES

1. Fawcett, H. H., Non-Compatible Chemicals as Cargoes in Industry on Road, Rail and Water, *Transp. Plann. Technol.*, **1**, 86–88 (1972).

2. "Transportation of Hazardous Materials . . . (Ethylene Oxide)," *Fed. Regis.*, **37**, (66), 6871–6872 (April 5, 1972).

3. "New Design of Rail Tank Cars," *Pet. Times,* **76**(1932), 8, (February 25, 1972).

4. "Hazardous Materials Regulations Board, Dept. of Transportation . . . Specifications for Tank Cars," *Fed. Regis.*, **37**(29), 3058 (February 11, 1972).

5. "Transportation of Hazardous Materials . . . (Propylene)," *Fed. Regis.* **37**(3), 149 (January 6, 1972).

6. "Derailment of Missouri Pacific Train at Houston, Texas, October 19, 1971," National Transportation Safety Board, Report No. NTSB-RAR-72-6, December 13, 1972.

7. Benner, Ludwig, Jr., "Safety, Risk and Regulation, Basic Concepts for 'Safety' Risk Analysis," Transportation Research Forum, 13th Annual Meeting Proceedings, Vol. XIII, No. 1, 1972.

8. Wall, W. L., "LNG . . . A Fire Service Appraisal," *Fire J.,* **66**, 15–20, (January 1972).

9. "Guide for Preparing Hazardous Materials Incident Reports," Office of Hazardous Materials, DOT, Washington, D.C. (revised August 1972).

10. "How Much Damage Should a Tank Car Be Able to Stand Without Failing?" *Butane-Propane News,* **3**(7), 21–24 (July 1971).

11. "Hazardous Material Tank Cars, Tanker Head Protective 'Shield' or 'Bumper' Design," Federal Railroad Administration, DOT, August 1971.

12. "Fire Protection of Railroad Tank Cars Carrying Hazardous Materials," Naval Ordnance Laboratory, Silver Springs, Md., July 21, 1972.

13. *Handling Hazardous Materials—Guidebook*, American Trucking Asso., Washington, D.C. 20036, 1982.

14. "Why Cities Service Likes Bottom Loading," *Nat. Pet. News*, **63**(7):82–83 (July 1971).

15. Rogers, J., "Tank Coatings for Chemical Cargoes," *Trans. Inst. Mar. Eng.* **83**, 139 (1971).

*Chemical Manufacturers Association, 2501 M St., NW, Washington, D.C. 20037.

16. de Talhouet, L., "Sea Transportation of LPG and Ammonia," *Tanker Bulk Carrier,* **18**(12), 10, 12, 14 (April 1972).

17. The DuoKleen Tanker . . . Safety and Economic Factors, *Tanker Bulk Carrier,* **18**(12), 21–23 (April 1972).

18. "A Tanker That Had Recently Hauled Benzene Has Disappeared and Probably Exploded," *Chem. Eng.,* **79** 25 (Feb. 21, 1972).

19. "Transportation or Storage of Explosives or Other Dangerous Articles or Substances, and Combustible Liquids on Board Vessels," *Fed. Regis.,* **37**(154), 15994–15995 (August 9, 1972).

20. Jansky, C., "LNG and the U.S. Coast Guard," Soc. Naval Architects and Marine Engineers Metropolitan Section, New York, Oct. 29, 1970, Paper No. 20p. Abstract: *Br. Ship Res. Assoc. J.,* **26**(10), 735–736 (October 1971).

21. "Large (Tankers) Safe," *Oil Gas J.,* **69**(20), 236 (May 17, 1971).

22. Ballaloud, H., "Insurance: The increasing, Difficulty of Covering Large Risks (in the Petroleum Industry)," *Petroleum Inf.,* No. 1222: 6–10 (March 24, 1972).

23. Clemmetsen, O. M., Lloyd's Register Rules for Chemical Carriers, *Shipp. World Shipbuild.* **164**, 1197 (October 1971).

24. "Lloyd's Register of Shipping (Has Issued) Provisional Tanker Classification (Rules)," *Eur. Chem. News,* **20**(499), 10 (September 24, 1971).

25. "Pressure Relieving Systems for Marine Bulk Liquid Cargo Containers," Committee on Hazardous Materials, National Academy of Sciences, Washington, D.C., 1971.

26. A High Velocity Venting Valve, *Tank Bulk Carrier,* **18**(2), 38 (June 1971).

27. Vrancken, P. L. L., "Current LNG Tanker Designs," *Gas J.,* No. 349, 173–174, 177 (March 15, 1972).

28. "LNG Importation and Terminal Safety, Conference Proceedings," Committee on Hazardous Materials, National Academy of Sciences, Washington, D.C., 1972.

29. "Study Finds No Danger of LNG Exploding if Spilled on Water," *Oil Gas J.,* **70**(9), 24 (February 28, 1972); *Chem. Eng. News,* **50**(9), 57 (February 1972); *Chem. Eng.,* 57 (March 6, 1972).

30. Park, R. R. (Texas Eastern Transmission Corp.), "Ocean Transport of LNG," *Gas,* **47**(8), 48–51 (August 1971).

31. Dynamic Loading and Fatigue Testing on LNG Carriers, *Shipbuild. Shipp. Rec.,* **117**, 40 (January 29 to February 5, 1971). Abstract: *British Ship Research Assn. J.,* **26**(5), 310 (May 1971).

32. Wilder, I., and Lafornara, J. (Environmental Protection Agency), "Control of Hazardous Material Spills in the Water Environment: An Overview," 162nd ACS National Meeting, Washington, D.C., September 12–17, 1971, ACS Division of Water, Air, Waste Chemicals, Preprint 11, No. 2, 47–53, 1971.

33. Evans, K. G., "Designing Ships of the Future," Paper No. 1367, Institute of Engineering and Shipbuilding, Scotland Meeting, Feb. 15, 1972. Abstract: *Br. Ship Res. Assoc. J.,* **27**(6), 396 (June 1972).

34. Guilhem, J., "Construction of Large Methane Tankers . . . Why A Secondary Barrier," 88th Assoc. Technical Industrial Gaz of France Congress (Evian, June 22–25, 1971), Paper No. 4p; Abstract: *Gas Abstr.,* **27**(10), 237–238 (October 1971).

35. "Evaluation of the Hazard of Bulk Water Transportation of Industrial Chemicals," National Academy of Sciences, Publication 1465, Washington, D.C. (1970 with additions to 1972).

36. Gardenier, John S., II, "Concepts for Analysis of Massive Spill Accident Risk in Maritime Bulk Liquid Transport," U.S. Coast Guard, Office of Research and Development, June 1972.

37. Gardenier, John S., II, "A Search for a Practical Spill Risk Model," Transportation Research Forum, 13th Annual Meeting Proceedings, Vol. XIII, No. 1, 1972.

38. Dickson, A. F. (Shell International Mar. Ltd.), "Navigation Problems (Tankers)," 2nd International Chamber of Shipping Tanker Safety Conference, October 1971. Abstract: *Br. Ship Res. Assoc. J.,* **27**(5) (May 1972).

39. Pointon, E. F., "(Safe) Mooring," paper No. 9, ibid.

40. "Doppler Docking Sonar Gives Precise Speed Measurement," *Marine Engineering Log.,* No. 75, 60 (November 1970). Abstract: *Br. Ship Res. Assoc. J.,* **26**(2), 165 (February 1971).

41. "Pipe Line News Annual Automation Symposium," *Pipe Line News,* **43**(11), 4, 11–21, 24–25 (October 1971).

42. "A Pipeline Failure Warning System," *Chem. Process. Eng. (London),* **52**(9), 72 (September 1971).

43. "Increased Fatality Rate From Gas-Pipeline Leaks and Failures," *Oil Gas J.,* **70**(11), 48 (March 13, 1972).

44. "A Pipeline Safety Panel," *Mech. Eng.,* **93**(11), 81–82 (November 1971).

45. Traverse, D. K., (Southern California Gas Co.), "Planning for Emergencies," American Gas Association Distribution Conference, Chicago, 1971. Adaptation: *Pipeline Gas J.,* **198**(10), 37–38 (August 1971).

46. Gober, W. H. (Northern Natural Gas Co.), American Gas Association Transmission Conference, Houston, 1971, *Pipe Line News,* **43**(7), 14–15 (July 1971).

47. "Safety Regulations—1. Minimum Federal Safety Standards for Liquid Pipelines," *Pipeline Gas J.,* **199**(4) (handbook issue), 5–9, 12, 14, 16, 18 (March 15, 1972).

48. "Safety Regulations—2. Minimum Federal Safety Standards for Gas Lines," *Pipeline Gas J.,* **199**(4) (handbook issue), 19–20, 23, 24, 27, 30, 32, 34–36, 38–40, 42, 44, 53, 55–60, 62, 64, 72, 76, 82 (March 15, 1972).

49. "Safety Regulations—3. Leak Reporting Requirements for Gas Lines," *Pipeline Gas J.,* **199**(4), (handbook issue), 79, 81, 83, 85–86, 88, 90, 95 (March 15, 1972).

50. Murphy, W. E. (Mueller Co.), "A Manufacturer Looks at Pipeline Safety Regulations," American Gas Association Distribution Conference, Chicago, 1971. Adaptation: *Pipeline Gas J.,* **198**(10), 40–42 (August 1971).

51. "Corrosion Prevention/Control Requirements Set for Safety Regulations. . . . August 1, 1971, Effective Date," *Pipeline Gas J.,* **198**(10), 70, 72–73, 76, 78, 80, 82–84 (August 1971).

52. "Special Study. A Systematic Approach to Pipeline Safety. Adopted: May 25, 1972," U.S. National Transportation Safety Board, Report NTSB-PSS-72-1, 1972.

53. "An Ideal Gangway," *Tanker Bulk Carrier,* **18**(4), 27 (August 1971).

54. Sherwood, R. J., "Evaluation of Exposure to Benzene Vapour During the Loading of Petrol, Esso Eur. Inc., *Br. J. Med.,* **29**(1) 65–69 (January 1972).

55. "Accidental Mixing of Incompatible Chemicals During a Bulk Delivery, Berwick, Me., Apr. 2, 1971," National Transportation Safety Board, Report NTSB-HAR-71-7, August 26, 1971.

56. "Tank Vehicle Bottom Loading and Unloading," 2nd ed., API RP 1004, American Petroleum Institute, Washington, D.C., June 1972.

57. "Powder-Conveying System Slashes Air Requirements," *Chem. Eng.,* 56 (April 3, 1972).

58. "Pressure Relieving Systems for Marine Bulk Liquid Cargo Containers," National Academy of Sciences, Washington, D.C., 1971.

59. "Risk Concept in Dangerous Goods Transportation Emergencies," National Transportation Safety Board, Report No. NTSB-STS-71-1, January 27, 1971.

60. "Hazardous Material Releases from Railway Tank Cars near Ridgefield, WA. Jan. 14, 1980," NTSB-HZM-Map-81-3, 1981, National Transportation Safety Board, Washington, D.C. 20594.

61. Stannard, J. H., Jr., "New NFPA 59A-1971 LNG Code," *Pipeline Gas J.,* **198**(10), 39 (August 1971).

62. Mahley, H. S. (Mobil Research and Development Corp.), "Static Electricity in the Handling of Petroleum Products," 64th AIChE Annual Meeting, Paper No. 52d, San Francisco, Calif., Nov. 28, 1971.

63. "Electrostatic Discharge May Cause Tanker Blasts," *Hydrocarbon Process.,* **50**(5), 17,19 (May 1971).

64. "API To Probe Tanker Explosions," *Petro/Chem. Eng.,* **43**(5), 5 (June 1971).

BIBLIOGRAPHY

Hermann, S. L., "Hazardous Materials Transportation Accidents, Illustrated," *Explosives Research Institute, Inc.*, P.O. Box 2103, Scottsdale, AZ 85252 (1982).

"Safety Report—Status of DOT Hazardous Materials Regulations Program NTSB-SR-81-2, "National Transportation Safety Board," Washington, D.C., 1981.

"Transporting Chemicals Safely: What We're Doing About It," Chemical Manufacturers Assoc., Washington, D.C. 20037 (1981).

"Transport of Dangerous Goods Recommendations, UN Committee of Experts on the Transport of Dangerous Goods," Internat. Regulations Publishing and Distrib. Org., Chicago, IL 60626 (1981).

Miller, J. M., "Commercial Vehicle Occupational Health Hazards," (University of Michigan), SAI Automotive Engineering Congress, Paper No. 720264, Detroit, Michigan, January 10–14, 1972.

Parkinson, G. S. (Shell-Mex. and British Pet. Ltd., London), "Benzene in Motor Gasoline. Possible Health Hazards in and Around Filling Stations and in Normal Transport Operations," *Ann. Occup. Hyg.,* **14**(2), 145–153, 155–157 (1971).

Canadian Standards Association, Proposed Preliminary Standard B338, Highway Tanks and Portable Tanks for the Transportation of Dangerous Goods, (April 1981); Canadian Standards Association, Rexdale, Ontario, Canada M9W 1R3.

"Pipeline Accident Report . . . Mobil Oil. Corp. High-Pressure Natural Gas Pipeline, Near Houston, Texas, September 9, 1969," U. S. National Transportation Safety Board, Report NTSB-PAR-71-1, 1971.

"Pipeline Accident Report. Phillips Pipe Line Co. Propane Gas Explosion, Franklin County, Missouri, December 9, 1970," U.S. National Transportation Safety Board, Report NTSB-PAR-72-1, May 24, 1972.

Siegfried, C. G. (Ebasco Services Inc.), "(Hazards Due to) Multiple Use of Rights-of-Way for Pipelines (and Power Transmission Lines), American Gas Association Transportation Conference, Houston, Texas, May 1971, *Pipe Line News,* **43**(8), 8–9, 12–15 (August 1971).

Aertssen, G., "Extreme Ship Motions of (Mammoth Tankers) in Coastal Waters," *Transport,* **8**(1), 9, 1970. Abstract: *Br. Ship Res. Assoc. J.,* **25**(11), 783–784 (November 1970).

Lucktretz, R. T. and Schneider, A. L., "Decision Making in Hazardous Materials Transportation," *J. Haz. Matls.,* **4**(2), 129–144 (1980).

Morrisette, M., "Hazard Evaluation of Chemicals for Bulk Marine Shipment," *J. Haz. Matls.* **3**(3), 33–48 (1978).

"Chemical Data Guide for Bulk Shipments by Water, CIM-16616.6," U.S. Coast Guard, Washington, D.C. (1982) (Replaces CG-388).

"'Cavendish' . . . Automated LPG Carrier for Ocean Gas Transport (Ltd.), *Shipp. World Shipbuild.,* **164**, 1305 (November 1971).

Betz, G. M., Recovering Spill Cleanup Costs, *Plant Engineering*, **36**(10), 157–160 (1982).

Harris, W. J., "A Shorter History of Tankers—2. Enter the Leviathan," *Petroleum News,* 25(297), 310–314 (September 1971).

Hughes, J. F. (Southampton University), "Electrostatic Hazards in Supertanker Cleaning Operations," *Nature,* 235(5338), 381–383, (February 18, 1972).

"Liquid Nitrogen Plants for Tankers," *Tanker Bulk Carrier,* 18(2), 24–25 (June 1971).

Maybourn, R. and Mason, P. F., "(Safety) Training," paper No. 3, 2nd International Chamber of Shipping Tanker Safety Conference, October 1971.

Morita, Y., "Accidents in Tankers Due to Static Electricity," *Navigation,* 32, 42 (September 1970). Abstract: *Br. Ship Res. Assoc. J.,* 26(1), 109 (January 1971).

Pierce, E. T. (Stanford Research Institute), Society of Automotive Engineers—U.S. Air Force, Lightning and Static Electricity Conference, SAE Paper No. 700922, San Diego, California, December 9–11, 1970.

Robinson, G. M. O. and Neal, P., "Damage Control (in Tankers)," paper No. 2, 2nd International Chamber of Shipping Tanker Safety Conference, October 1971.

"Tankers: A Special Survey/Computer-Controlled Tanker May Lead to Unmanned Operation," *Oil Gas International,* 11(2), 46–48 (February 1971).

"Tankers in Distress," *Tanker Bulk Carrier,* 19(1), 20–21, 27 (May 1972).

"Ultrasonic Emulsification of Oil-Tanker Cargo," Sonica International, Inc., April 1970. Abstract: *Br. Ship Res. Assoc. J.,* 26(8), 555 (August 1971).

Liquefied Natural Gas

Kampschaefer, G. E., Havens, F. E., and Sarno, D. A. (Armco Steel Corp.), Society of Naval Architects and Marine Engineers Gulf Section Meeting, February 5, 1971, *Mar. Technol.,* 9(3), 297–301 (July 1972).

Kotcharian, M. and Pauthier, J., "Loading, Unloading, Gassing, Degassing, and Evacuation of Vaporized LNG From Very Large Methane Tankers," 2nd International Liquefied Natural Gas Congress and Exhibition, Paris, Oct. 19–23, 1970, Proc. 1, Session 3, paper No. 6. Abstract: *Gas Abstr* 27(6), 150 (June 1971).

LNG Plant Explosion," *Fire J.,* 66, 38–39 (July 1972).

Wilcox, D. C., "Empirical Vapor Dispersion Law for an LNG Spill," American Gas Association, LNG Safety Program, Reprint No. 5, 1971.

Witte, L. C., and Cox, J. E., "Questions About LPG Explosions," *Hydrocarbon Process.,* 51(3), 67, 69 (March 1972).

Coast Guard, "Publications on LNG.," U.S.C.G., Dept. of Transportation, Washington, D.C., 1981.

Miscellaneous

Student, P. J., Emergency Handling of Hazardous Materials in Surface Transportation, Bureau of Explosives, Association of American Railroads, Washington, D.C., 20036, 1981.

"Hazard Information System, Hazardous Materials Regulations Board, Docket HM-103," *Fed. Regist.* 37(124) (June 27, 1972).

"Highway Transportation and Industrial Traffic, Report on Annual Conference," Division of Transportation, American Petroleum Institute, Washington, D.C., November 1972.

"Labeling of Hazardous Materials, Hazardous Materials Regulations Board, Docket HM-8," *Fed. Regist.,* 38(38) (February 27, 1973), and 38(44) (March 7, 1973).

"Risk Concept in Dangerous Goods Transportation Emergencies," National Transportation Safety Board, Report No. NTSB-STS-71-1, January 27, 1971.

"A Study of Transportation of Hazardous Materials," National Academy of Sciences, Washington, D.C., 1969.

Audio-visual (slides and cassette) for training drivers and dispatchers, in 8-parts, "Transport of Flammable and Combustible Liquids," Video Systems, Inc., 12530 Beatrice St., Los Angeles, CA 90066, 1982.

32

Loss Prevention Evaluations for Chemical Operations

George L. Gorbell

32.1. INTRODUCTION

The tendency of the chemical industry in the past few years has been to build larger production units involving processes using flammable and highly reactive materials at high pressures and temperatures, thus creating large loss exposures. This, of course, coupled with an increase in occurrence of high loss of life and property incidents, has created a serious problem in the procurement of insurance. Insurance companies are reluctant to provide coverage and generally do so only at substantially higher insurance rates. (See Chaps. 24 and 30.)

The purpose of this chapter is to discuss methods for evaluating potential sources of losses from injury to personnel, damage to structures, equipment, raw materials, finished products, public property, the environment, and injury to the public, in order to determine means to eliminate loss sources or to control them sufficiently to minimize their effect, not only on insurance costs but on community relations.

The degree of emphasis placed on loss prevention depends largely on the nature of the risk potentials of proposed or existent operations; therefore it is essential that risk evaluations be conducted to develop estimated dollar loss figures for capital investment considerations of loss control measures. Insurance underwriters require risk analyses and evaluations for rating purposes, whereas management needs such information for loss prevention evaluation. Since this chapter is primarily directed to loss prevention evaluations, no attempt will be made to examine risk evaluation methods in depth other than to suggest reference to available literature. An excellent paper by Arthur Spiegelman of the American Insurance Association, entitled "Risk Evaluation of Chemical Plants"[1] contains invaluable information for anyone planning to establish a risk evaluation procedure.

Concerning the completion of loss prevention evaluations, the method presented in this chapter is a Monsanto Company procedure. Other companies have utilized approaches designed to develop certain criteria to accomplish desired objectives. Many smaller companies with limited numbers of technically qualified people resort to consultants or, if their insurance carriers have competent staffs, they use their services. Whatever method is used the evaluation itself depends on the judgment of people qualified through experience and technical capabilities.

32.2.   RISK EVALUATION

Purpose

The purpose of risk evaluation for both insurance underwriters and management is to estimate the probable property damage and business interruption costs if a loss occurs with all loss control facilities (i.e., sprinklers, pressure release vents, etc.) functioning as intended to limit the damage and loss. Reference is suggested to an MCA pamphlet[2] which presents guidelines for risk evaluations including a list of the major factors which influence risk evaluation.

Risk Evaluation or Loss Exposure Estimates

Methods of conducting risk evaluations or loss exposure estimates vary in companies and the results are greatly dependent on the judgment of those making the estimates. A suggested procedure is outlined below:

1.   Make an estimate of the replacement value of the property in each fire zone or unit of a total facility which should be considered a single separate loss exposure.

2.   Tabulate the annual business interruption loss that would result from total loss of this unit or segment of the facility. If the total property breaks down into physical units which constitute separate property exposure risks but which operate in a series, the business interruption for each will probably be the same as the business interruption exposure for the total facility. If the separate property units operate in parallel, and are separated so as not to be involved in a common incident but can operate independently, the business interruption exposure for each will probably be an appropriate portion of the total. Business interruption value is defined as sales value of production less raw materials, operating wages, purchased utilities, repair wages and materials, operating supplies, and royalties.

3. Having defined the exposed values, proceed with a brief description of an imaginary incident involving the maximum foreseeable property damage to the facility or portion of the facility in question. This might be a fire, an explosion, a rupture of a particular operating vessel, etc., when protection facilities fail to function.

4. Make an estimate of the probable maximum property damage in the concerned areas which could result from this incident. This should include determination of the maximum value of the largest fire area and the largest unit of storage in any fire area and is made from experience, knowledge of the operation and of loss control features such as sprinkler systems, etc.

For petrochemical operations the MCA pamphlet "Guidelines for Risk Evaluation and Loss Prevention in Chemical Plants"[2] includes a risk evaluation worksheet for estimating maximum probable losses which is a Western Actuarial Bureau Rating Schedule form. The worksheet provides a method of establishing explosion—fire (E/F) factors which are used to modify probable maximum property damage estimates.

5. Based on the assumed property damage in the item above, make an estimate of the probable downtime for the facility. This may be controlled by the extent of the damage, or by probable delivery of critical pieces of equipment assumed to have been damaged in the incident.

6. Calculate the business interruption value by multiplying the daily business interruption exposure by the estimated downtime.

7. The sum of the probable maximum property damage and the equivalent business interruption loss gives the probable maximum loss for the facility in question under the conditions assumed in the incident. If the maximum probable loss is low and the maximum foreseeable loss is high, then a review and a reengineering approach may be necessary to consider means of reducing the maximum foreseeable loss.

Fire and explosion hazards and electrical or mechanical breakdown hazards are often considered separately for a given facility with the largest probable maximum loss controlling as a measure of the risk exposure in the facility.

32.3. LOSS PREVENTION EVALUATION

Loss prevention evaluations as covered in this chapter concern the determination of potential sources of injury to personnel and damage to property in prospective and existent operations. The loss prevention review team and checklist approach was developed by the Monsanto Company. For information on another company's procedure it is suggested that the Dow Chemical Company's "Safety and Loss Prevention Guide, Hazard Classification and Protection" be consulted.[3]

Loss Prevention Reviews

The purpose of a loss prevention review is to foresee hazards and to attack them before accidents occur. The savings resulting from accident prevention justifies the cost of holding loss prevention reviews.

Several kinds of loss prevention review should be conducted by a plant, such as: process development loss prevention review: loss prevention review for plant modifications to be handled within the plant; pre-start-up loss prevention review; and periodic loss prevention reviews for existing plants.

PROCESS DEVELOPMENT LOSS PREVENTION REVIEW

The research group responsible for development of a process is responsible for the safety of the research and for considering the impact of process hazards on the ultimate design of the plant. The choice of solvents, reactants, and process conditions will also have a major impact on the design and cost of the ultimate operation. Consultation with engineers who will later be engaged in reducing the process technology to commercial practice may be helpful. The review should be made preparatory to initiation of the bench scale or pilot plant phase of development. The purpose of this review is to:

1. Explore potential plant hazards if the process should develop to the commercial state and decide which of these hazards should be given particular attention during the development work. (See Chap. 22.)
2. Determine whether all hazards are adequately provided for in the proposed design and operation of the bench scale or pilot plant unit under normal and abnormal conditions.

Substantial savings in the plant costs can be realized if consideration is given to the safety of the ultimate operation during the early stages of process development.

The loss prevention review at this stage of process development should include all functions concerned with the project. The write-up of the results of the review and the ultimate resolution of the problems uncovered should be incorporated in the safety section of a tentative process report.

PRELIMINARY LOSS PREVENTION REVIEW

As the process development progresses and the final project is initiated, engineering personnel become responsible for loss prevention consideration. The project manager should consider each recommendation and comment on the action to be taken. Although at this stage of a project the design is usually not far enough advanced to get into detail, the recommendations are important and alternative solutions and judgment differences should be resolved.

PROJECT DESIGN LOSS PREVENTION REVIEW

Since it is generally necessary to submit a cost proposal to the company's executive group for approval to proceed with the project, it is essential that the cost figure include funds for appropriate loss prevention provisions. These are estimated from the deliberations of a preliminary loss prevention review made by representatives of concerned groups, including division (department), plant, safety, a chemical engineer from an outside department, and corporate engineering personnel. A report of this meeting's transactions will usually include loss prevention recommendations and a risk evaluation indicating the probable maximum property damage and business interruption loss which could be incurred by a foreseeable incident or accident. Approval of capital for the project should be contingent on maximum loss probabilities.

The project manager is responsible for obtaining appropriate consideration of safety and loss prevention during the design stage of a project, and should arrange with the safety department to have all phases of loss prevention reviewed during design, that is, electrical, fire protection, instrumentation, mechanical, process and personnel safety. The project manager should be certain that the needed loss prevention has been included when design is complete and before so many commitments have been made that corrections become difficult or expensive.

Use of a checklist by project engineers will enable them to ensure that all aspects of loss prevention receive attention. An example of such a checklist is found in Table 32.1.

PRE-START-UP LOSS PREVENTION REVIEW (MANUFACTURING)

At the time of completion of the plant operating manual and before plant startup, a loss prevention review should be held. This is done to:

1. Make sure that operating personnel are thoroughly familiar with the hazards involved and the controls that have been engineered into the process (see Chapter 33).
2. Determine if any additional process hazards are involved, including those which could develop under abnormal conditions (see Chapters 4, 5, 7, 16, and 17).
3. Check the safety adequacy of operating procedures, emergency procedures, and startup plans (see Chapters 12, 13, 34 and 35).

The manufacturing superintendent involved is responsible for carrying out this review. Tables 32.1 and 32.2, though developed primarily for periodic reviews, give suggestions which can be profitably used for pre-start-up reviews.

Most of the items in the checklist will have been given consideration, and hazards will have been eliminated or adequately controlled at this stage but

TABLE 32.1. Loss Prevention Review Checklist

Section I—Boiler & Machinery Review Checklist

A. *Boilers*
  1. ASME Code constructed and stamped (Section I for 15 psi and above, Section IV for less than 15 psi).
  2. Safety Valves
      a. Long and large vent lines supported.
      b. Vent line drain connection provided.
      c. First drum valve set to relieve at or below boiler design pressure.
      d. The last drum valve set to relieve at or below 103% of boiler design pressure.
      e. Superheater outlet safety valve set to relieve first in order to insure flow of steam (for cooling) through superheater.
  3. Blow-off Piping
      Use steel piping of next higher gauge than required for boiler pressure, avoid sharp radius ells, slope all lines and drain all low points in the lines.
  4. Feedwater Piping
      The bypass around feedwater regulator should be accessible from the operating level and located where the drum level gauge glass can be seen.
  5. Steam Outlet Piping
      a. Both nonreturn and header stop valves where two or more boilers discharge into the same piping system.
      b. A visible free blow and drain in piping between nonreturn and header stop valves.
      c. Condensate drain provisions for all section of piping.
      d. Adequate piping expansion flexibility and piping supports.
  6. Drum Water Level
      a. Provide both high- and low-water alarms.
      b. Provide a single low water cutoff of gas or oil burners on attended boilers (if drop or loss of plant steam pressure does not jeopardize process safety.)
      c. Remote drum level gauge is independent of drum level controls.
      d. Provide all fire-tube boilers and all unattended water tube boilers with two independent low-water level switches interlocked with gas or oil burner safety shutoff valves.
  7. Gas Burner Control and Piping—General
      a. Plug cocks provided for manual shutoff service.
      b. In-line strainer in gas line ahead of all regulating and safety shutoff valves.
      c. Provide for stable gas pressure regulation at all loads. (This may require a small regulator in parallel with the full-sized regulator for startup or low fire service.)
      d. Provide double safety shutoff and vent valve arrangement.
      e. Provide automatic fuel–air ratio control.
      f. Separate pressure regulation of pilot gas.
      g. Safety control circuit DC, or 120V AC with the safety controls in the ungrounded circuit.
      h. Insure positive, tamper-proof time period to provide minimum of six air changes in combustion chamber before light-off. Air flow rate during purge should be at least 25% of maximum fan capacity.

TABLE 32.1. (*Continued*)

      i.   Controls provided to prevent burner firing rate from being reduced below minimum stable flame.

      j.   Controls or interlocks installed to prevent burner light-off when insufficient combustion air flow is present.

      k.   Interlock to assure low-fire burner light-off when light-off cycle is automatically programmed.

      l.   Monitoring of main burner flame for alarm and/or automatic burner shutoff in event of flame failure.

      m.   Shutdown interlocks on low burner gas pressure and low combustion air flow.

8. *Additional* gas burner controls and interlocks for unattended operation.
   a. Automatic flame failure monitoring and shutdown interlocks.
   b. Other interlocks for safety shutdown of burners.
      i. High gas supply pressure.
      ii. Low boiler water (double switches).
   c. Tamper-proof programmed light-off sequence to purge, light and prove pilot, light and prove main flame.
   d. Positioning type fuel–air ratio controls.
   e. Consider a self-checking feature for flame scanner and flame scanner relay circuitry.
9. Oil Burner Controls and Piping—General
   a. Oil line strainer.
   b. Oil pressure control.
   c. Heater for heavy oil.
   d. Single safety shutoff valve.
   e. Startup recirculating line (for heavy oil).
   f. Positive fuel–air ratio control.
   g. Low oil pressure and low combustion air flow shutdown interlocks.
   h. Low oil temperature alarm or interlock (for heavy oil).
   i. Low atomizing steam pressure alarm or interlock.
   j. Positive purge cycle and low fire start controls.
   k. Interrupted pilot.
10. *Additional* oil burner controls and interlocks for unattended operation.
    a. Automatic flame failure monitoring and shutdown interlocks.
    b. Other interlocks for safety shutdown of burners are:
       i. Low oil temperature—for heavy oils.
       ii. Low atomizing steam pressure.
       iii. Low boiler water—double switches
    c. Provide a tamper-proof programmed light-off sequence.
    d. Use positioning fuel–air ratio controls.

B. *Pressure Vessels* (see Chapter 8).
1. Unfired pressure vessels, ASME Code, Section VIII, Divisions 1 or 2, constructed and stamped.
2. In the state of Texas, unfired pressure vessels generating steam shall be ASME Code, Section I constructed and stamped.
3. Organic fluid vaporizers, ASME Code, Section I constructed and stamped.

TABLE 32.1. *(Continued)*

    4. Other fired pressure vessels not generating steam ASME Code Constructed and stamped. Review with S&PP Branch for Section I versus Section VIII jurisdiction.

C. *Piping and Valves*

    1. Piping systems analyzed for stresses and movement due to thermal expansion.

    2. Piping systems supported and guided.

    3. Piping systems provided for antifreezing protection, particularly cold water lines, instrument connections and lines in dead-end service such as piping at standby pumps.

    4. Do not install cast iron valves in piping subjected to strains or shock service.

    5. Avoid nonrising stem valves wherever possible.

    6. Use double block and bleed valves on emergency interconnections where possible cross-contamination is undesirable.

    7. Provide for draining and trapping steam piping.

D. *Machinery*

    1. Provide adequate piping supports and flexibility to keep forces on machinery due to thermal expansion of piping within acceptable limits.

    2. Is there adequate separation of critical and operating speeds?

    3. Check valves should be positive and fast acting to prevent reverse flow and reverse rotation of pumps, compressors, and drivers.

    4. Provide adequate service factors on speed changing gears in shock service.

    5. Are there full-flow filters in lube-oil systems serving aluminum bearings?

    6. Provide for draining and trapping steam turbine inlet and exhaust lines.

    7. Provide separate visible-flow drain lines from all steam turbine drain points.

    8. Driven machines must be capable of withstanding tripping speed of turbine drivers.

    9. Use nonlubricated construction or nonflammable synthetic lubricants for air compressors with discharge pressures of greater than 75 psig to guard against explosion.

    10. Make provisions for emergency lubrication of critical machinery during operation and during emergency shutdowns.

    11. Make provisions for spare machines or critical spare parts for critical machines.

    12. Provide vibration switches on interlock shutdown service for cooling tower fans.

    13. Provide vibration monitoring and rotor axial movement monitoring of critical rotating machinery.

    14. Is this a first-of-a-kind machine, with respect to speed and horsepower rating and other major design features?
Obtain data on experiences of other users of similar machines. Avoid, if possible, the installation of an unproved machine for "testing and development" in a Monsanto plant.

    15. Have the specifications for critical machinery and their lube oil systems been reviewed with mechanical specialists in the Engineering Technology and Services and Security and Plant Protection branches of Central

TABLE 32.1. *(Continued)*

      Engineering Department or in plants where this type of machinery is in service?

  16. Provide dependable liquid knock-out facilities on the suction side of gas compressors and on waste process gas piping to boilers and fired heaters.

Section II—Electrical Safety Review Checklist

A.   What degree of power supply reliability is required for:
    1. Instrumentation power?
        Is there any real need for the instrument power to be available on loss of plant power? What would be the difference in damage or downtime if the instrument power was lost at the same time as plant power for:
        a. 1 min
        b. 10 min
        c. 2 hr
    2. Total unit power?
        What would be the damage and downtime for power outages of:
        a. 1 min
        b. 10 min
        c. 2 hr
    3. Are there any individual power users that, if served continuous power, would reduce this damage and/or downtime?

B.   Are there any special limits of power supply quality?
    1. What would be the effect of a voltage variation 5% above or 5% below normal at motor control center bus?
    2. What would be the effect of a 5% frequency variation high or low?
    3. Are there any pieces of equipment proposed that may be damaged by over voltage peaks of 1½ times normal for 5 μsec
    4. Are there any pieces of equipment proposed that would be seriously affected by a loss of voltage for ⅛ sec?

C.   What is the potential for expansion of the unit? How will it expand?

D.   What is the planned operating cycle between plant turnarounds? The electrical portion of the job must be built so as not to be the limiting factor.

E.   Do the electrical load blocks correspond to the process load blocks? One electrical fault should not affect more than one block of the process. No block of the process should be exposed to faults in more than one chain of electrical equipment.

F.   The electrical system should be isolated physically from the process as much as possible, minimizing exposure to fire, corrosion, and mechanical damage.

G.   Installation should allow for maintenance during operation or at shutdown, depending on the operating cycles between turnarounds of the unit.

H.   Equipment and material used in the design should be compatible with plant spares inventory as well as plant maintenance experience.

I.   Review each alarm indicated on the flow sheet for importance and list it either A or B. A alarms are those announcing approaching catastrophe. B alarms are for operating convenience.
    Different techniques are applied to the two by the design group but the importance should be established in the safety review.

TABLE 32.1. *(Continued)*

J.   Pilot lights should be reviewed as the alarms were. They too fall into two groups: (1) those whose failures to function properly may cause injury to people or major loss of money, and (2) those that are pure operating convenience.

    In this consideration, it is important to be sure that the listed condition for indication is the critical variable rather than a dependent variable.

K.   Interlocks fall into three levels of importance. Each interlock should be judged against the following criteria and given a rating:

    1.   Safety interlock—A safety interlock is one whose failure to operate would be likely to cause a major injury to people and/or a major dollar loss.

    2.   Interlock—An interlock is an arrangement of equipment to insure that certain specific conditions exist before an action can be taken or for action to continue. The failure of this interlock may cause an inconvenience and may cost money for repairs and clean-up.

    3.   Sequence control—These interlocks insure that the process is following a prescribed sequence or one causing a specific sequence to be followed. The failure of this interlock would stop the sequence and be an inconvenience. If the stoppage in some particular point in the sequence is critical, it may be backed up by a higher level of interlock.

L.   The electrical system should be simple in schematic and physical layout so that it can be operated in a straightforward manner. (This minimizes human error in switching for isolation load transfer.)

M.   The electrical system should be instrumented so that equipment operation can be monitored. The objective being to eliminate downtime due to equipment failures caused by unknown overloading.

N.   Overload and short circuit protective devices shall be provided.

    1.   These should be located in circuits for optimum isolation of faults.

    2.   Give the interrupting capacity.

    3.   Describe the coordination between the two.

    4.   Instruction must be furnished for field testing on installation and for later during life of equipment.

O.   Bonding and grounding should be provided to:

    1.   Protect against static build-up.

    2.   Provide lightning protection.

    3.   Provide for personnel protection from power system faults.

P.   Tankage grounding should be coordinated with cathodic protection.

Q.   Review the accessibility during mishaps of power disconnects, starters, etc.

R.   Describe the plant communication system (telephones, radios, signals, alarms, etc.)

S.   Provisions must be made to dissipate static electricity. (See Monsanto Standards E.2.)

T.   Structural steel must be grounded in accordance with Monsanto Standards E.2.

U.   Where sequence controllers are used, there should be an automatic check, together with alarms, at key steps after the controller has called for a change. There should also be a check together with alarms at key steps before the next sequence changes.

TABLE 32.1. *(Continued)*

V.   Review the effects of extremes of atmospheric humidity and temperature on instrumentation.

W.   Gauges, meters, and recorders should be designed and installed in such a manner that they can be read easily.

X.   Instruments which cannot be read casily should be replaced.

Y.   Verify that instrument packages are properly installed, grounded, and of the proper design for the environment.

Z.   Procedures must be established or in effect to test and prove instrument functions including interlocks and safe action in the event of power or instrument air failure.

AA.   Periodic testing to check performance and potential malfunction should be scheduled.

Section III—Fire Protection Review Checklist

Provide the following fire protection for all CED projects where the need is established by review with the S&PP Branch.

A.   Automatic sprinkler or automatic water spray (deluge) protection wherever there are combustible materials either in the construction or in the contents (See S&PP Design Guide #11).

B.   Drainage for spilled flammable liquids and fire protection water from sprinklers and hose streams.

C.   Arrange sprinkler waterflow alarms so that any flow of water through the sprinkler piping will give an audible local alarm and automatically transmit water flow signals to a central supervised location.

D.   Locate small hose stations inside buildings so that every square foot of floor area is within 20 ft of a hoze nozzle attached to not more than 75 ft of $1\frac{1}{2}$ in. woven jacketed rubber-lined hose. The nozzles should be the combination spray and solid stream with shutoff.

Small hose should preferably be attached to risers independent of the sprinkler system if hose streams are considered needed when sprinklers are not operating. If this cannot be arranged, small hose may be attached to $2\frac{1}{2}$ in. or larger pipe on a wet pipe system.

E.   Sufficient hydrants having two outlets to provide two $2\frac{1}{2}$ in. hose streams at any point in the property where fire may occur regardless of wind direction. This will require spacing hydrants 225–250 ft apart at plant with ordinary hazards. At plants having highly combustible occupancies, the spacing may need to be reduced to 100–150 ft. For plants with noncombustible buildings and nonhazardous occupancies, hydrant spacing may be extended to 250–300 ft.

For average conditions locate hydrants 50 ft from the building or equipment protected.

F.   Extension of undergound fire mains to feed proposed fire protection for project. A loop system with sectional valves is recommended because it assures greater reliability by providing water flow from two or more directions. The mains should be of ample size to carry the maximum fire water demand for sprinklers, hose streams, and monitors. See S&PP Design Guide #10.

TABLE 32.1. *(Continued)*

G.   Sectional control valves in the underground fire mains to divide the grid system into sections and thus limit the area subject to a single impairment. Not over a total combination of six sprinkler risers and hydrants should be out of commission during any change or repair. With any one section shut off at least one of the water supplies should be available on the remainder of the system.

H.   Fire protection monitors are not a substitute for automatic sprinkler protection or hose streams. They should be used to protect equipment containing flammable liquids that is not in a building or structure protected by automatic sprinklers. Their principal advantage is to provide a quick stream of water while hose lines are being laid, particularly where work force is a problem.

I.   Fireproofing of all main load-bearing structural members that support either process piping or equipment within hazardous areas. See CED Fireproofing Design Guide A4.10 Std. 1.

J.   Fire wall, fire partition, or barricades to separate: (1) important property damage values; (2) units important to continuity of production; (3) high hazard operations. See S&PP Design Guide #14.

K.   Adequate spacing, diking and fire protection for tanks storing flammable and combustible liquids. See S&PP Design Guides #1 and #13.

L.   Adequate spacing and fire protection for LP gas storage tanks. See S&PP Design Guide #6.

M.   Fire protection for indoor and outdoor storage of flammable liquids in drums. See S&PP Design Guide #2.

N.   Fire protection for warehouses. See S&PP Design Guide #4.

O.   Fire protection for cooling towers. See S&PP Design Guide #3.

P.   Special hazard protection where needed. This would include fixed carbon dioxide, foam, dry chemical, and explosion suppression systems.

Q.   Fire detectors where automatic sprinklers cannot be justified due to limited combustibles or compatible with hazard to be protected.

R.   Fire extinguishing agents must be compatible with process materials.

Section IV—Personnel Safety Review Checklist

A.   *Project Site Location*

    1. Consider exposure to or from the neighborhood from fire, explosion, noise, air and stream pollution, and exposure to or from existing operations. (See S&PP Design Guide #16 for fire. Consultation with appropriate CED or corporate medical department necessary in other areas.) (See Chaps. 3,9,11,12 and 34.)

    2. Provide a minimum of two means of access for the site to facilitate entrance by emergency vehicles and not more than one of these shall be subject to uncontrolled blockage, i.e., railroads, highways.

    3. Review the access to highways which handled heavy traffic. Where congestion is anticipated, such as at shift changes, consider traffic lights.

    4. Provide right-of-way, clearance, etc., for future heavy traffic such as four-lane divided highways and limited access highways.

    5. Prepare contingency plans for access roads to the sites which are exposed to railroad blockage or highway congestion and other hazards. Identify railroad crossings by warning signs.

TABLE 32.1. (*Continued*)

6. Provide proper banking, barricades and warning signs where sharp curves are a necessity in access roads. (In areas where fog conditions may obscure visibility, provide rumble strips to warn of an approaching curve.)

B.   *Building and Structures*

1. Follow CED Design Std. A9.2 Std. 1 and the appropriate referenced Std. in the design of stairways, platforms, ramps and fixed ladders. (Monsanto Standard A2.5, Std. 1, A4.2, Std. 2, A4.3, Stds. 9, 21 and 25.)

2. Provide sufficient general exit facilities and escape routes including alternate means of escape from roofs in accordance with S&PP Design Guide #12.

3. Furnish lighting to meet or exceed the recommended lighting intensities of Monsanto Std. E5.1, Std. 5.

4. Provide emergency lighting for escape or emergency shutdown operation in the event of power failure.

5. Hang doors and windows to avoid projecting into or blocking walkways and exits. Provide a safety railing where it is necessary that a door open into an aisle or a road for vehicles.

6. Provide a utility room on each floor level for storage of cleaning equipment (sink, sweeper, scrubber, broom, etc.).

C.   *Operating Areas*

1. Install and arrange equipment, steam, water, air and electrical outlets to keep aisles and operating floor areas clear of hoses and cables.

2. List materials in the operation which are highly toxic. (Source: *Dangerous Properties of Industrial Materials* by Sax.) (See also references in Chapters 14 and 15, and in Appendix, this volume.)

3. Provide adequate ventilation, to keep concentrations below TLV or known permissible levels of hazardous fumes, vapors, dust, and excessive heat (see Chapters 14 and 15).

4. Provide space for the temporary storage of raw materials at process points and for finished projects in the manufacturing area. This will permit continued operation during abnormal situations in continuous operations and alleviate safety and operational problems in batch operations.

5. Use laminated safety glass or wire glass for windows of control rooms exposed to potential fire and explosion hazards. (Size 8 in. × 10 in.)

6. Provide adequate platforms and sufficient clearance around equipment for conducting maintenance operations safely.

7. Size and locate nozzles and manholes with consideration for safe cleanout, maintenance operations, and emergency removal of people from vessels. (Minimum of 20 in. diameter) (see Chapters 25 and 36).

8. Provide personnel protection against contact with surfaces which may reach or exceed 139°F.

9. Provide guarding in accordance with Monsanto Standard D11.8, Std. 5 for exposed moving parts of power-driven equipment before operation start-up. Give special attention to "guarded by location" requirements of the Standard.

10. Interlock covers and guards with operating mechanism particularly on high-speed equipment and on equipment where high personnel or property hazard exists.

TABLE 32.1. (*Continued*)

    11. Manually operated valves, switches, and other controls shall be readily accessible to the operator from a safe location.

    12. Do not install free-swinging material hoist.

    13. Material hoists must meet the requirements of the appropriate material hoist or elevator code.

    14. Furnish elevator shaftway door interlocks and gate contact in accordance with the ANSI Safety Code for Elevators (A17.1) and the applicable local codes. Provide biparting doors with safety astragals. Do not install manually operated biparting doors.

    15. Describe the material handling operations. Avoid manual handling in excess of 50 lb.

    16. Provide Monsanto standard emergency showers and eye baths. The number and location of these units will vary depending on the type of operation. (See Monsanto Standards P6.3, Stds. 1, 6, 7 and 16.)

    17. *Consult the CED noise control engineer* where there is any question of noise levels in the hazardous range. Measures should be taken to reduce the noise level to a safe range (see Chapter 27).

    18. Do not locate offices inside hazardous (i.e., fire, explosion, toxic potential) manufacturing facilities.

    19. Install positive electrical power disconnects which provide actual physical isolation for purposes of lockout.

D.   *Yard*

    1. Lay out roadways with consideration for the safe movement of pedestrians, vehicles, and emergency equipment.

    2. Concerning railroad car puller installations:

       a. Protect the control stations against broken cable whiplash.

       b. Protect the operator from being caught between cable or rope, and capstan or cable drum. (See Monsanto Standards A8.4, Stds. 15 & 17.)

       c. Install an audible warning device to alert personnel to the operation.

    3. Provide static grounding on flammable liquid tank cars and tank truck loading and unloading docks. (See Monsanto Standards E2.3, Stds. 11 & 12.) (see Chapter 19).

    4. Provide approved loading platforms, including handrails, for access to work areas of tank cars, hopper cars, and trucks. (See Monsanto Standard A8.4, Stds. 10 & 13; S&PP TecFact 6.)

    5. Provide safe access to tops of storage tanks on which persons go for contents measurement and vent maintenance. (Frequency of operation determines design.)

Section V—Process Safety Review Checklist

Note: Consider the checklist in terms not only of steady state operation, but also start-up, shut-down, and upsets of all conceivable types.

A.   *Materials*

    1. What provisions are being made to control process materials which are unstable, pyrophoric, or spontaneously ignitable?

    2. What data are available on the amount and rate of heat evolution during decomposition of all materials in the process?

TABLE 32.1. *(Continued)*

3. Discuss precautions which are necessary for flammable materials including storage and piping systems. (See Chap. 19.) (Reference: S&PP Design Guide No. 8)

4. Define flammable dust hazards which may exist. (Reference: S&PP Brief No. 44.) What precautions are needed?

5. Describe what has been done to assure that materials of construction are compatible with the chemical process materials that are involved.

6. What maintenance control is necessary to assure replacement with proper materials, e.g., to avoid excessive corrosion, to avoid producing hazardous compounds with reactants?

7. What changes have occurred in composition of raw materials and resulting changes in process?

8. Define and review procedures established to control raw material identification and quality.

9. What hazards can be created by failure of supply of one or more raw materials or improper sequence of material additions?

10. Provide reliable inert gas supply for purging, blanketing, or inerting. (Reference: NFPA 69T.) What provisions are there for true sweep-through purge for start-up and shut-down?

11. What precautions need to be considered relative to stability of all materials in storage?

B. *Reactions*

1. Define potentially hazardous reactions. How are they isolated? Prevented? (See Chaps. 4, 5, and 16.)

2. Define process variables which could, or do, approach limiting conditions for hazard. What safeguards are provided against such variables?

3. What unwanted hazardous reactions can be developed through unlikely flow or process conditions or through contamination?

4. What combustible mixtures can occur within equipment?

5. What precautions are taken for processes operating near or within the flammable limits? (Reference: S&PP Design Guide No. 8.) (See Chap. 19.)

6. What are process margins of safety for all reactants and intermediates in the process?

7. List known reaction rate data on the normal and possible abnormal reactions.

8. How much heat must be removed for normal, or abnormally possible, exothermic reactions? (see Chapters 7, 17 and 18).

9. How thoroughly is the chemistry of the process including desired and undesired reactions known? (See NFPA 491 M "Manual of Hazardous Chemical Reactions.")

10. What provision is made for rapid disposal of reactants if required by emergency?

11. What provisions are made for handling impending runaways and for short-stopping an existing runaway?

12. Discuss the hazardous reactions which could develop as a result of mechanical equipment (pump, agitator, etc.) failure.

13. Describe the hazardous process conditions that can result from gradual or sudden blockage in equipment including lines.

TABLE 32.1. (*Continued*)

    14. Review provisions for blockage removal or prevention.

    15. What raw materials or process materials or process conditions can be adversely affected by extreme weather conditions? Protect against such conditions.

    16. Describe the process changes including plant operation that have been made since the previous process safety review.

C. *Equipment*

    1. In view of process changes since the last process safety review, how was adequate size of equipment assured?

    2. What procedure is there for assuring adequate liquid level in liquid seals?

    3. What is the potential for external fire which may create hazardous internal process conditions? (see Chapter 7).

    4. Is explosion suppression equipment needed to stop an explosion once started? (In any high hazard operation.)

    5. Do not use equipment made of glass or other fragile material. If a more durable material cannot be used, the fragile material must be adequately protected to minimize breakage., e.g., use Corgard with flex joints. In the event of breakage discuss the resulting hazard and describe provisions that will be made to handle the situation.

    6. When was pertinent equipment, especially process vessels, last checked for pressure capability?

    7. What provisions are needed for complete drainage of equipment for safety in maintenance?

    8. Review the requirements for concrete bulkheads or barricades to isolate highly sensitive equipment and protect adjacent areas from disruption of operations.

D. *Instrumentation Control* (see Chapter 6).

    1. What hazards will develop if all types of motive power used in instrumentation should fail nearly simultaneously?

    2. In the event that all instruments fail simultaneously, will the collective operation fail-safe?

    3. Provide for process safety when an instrument, instrumental in process safety as well as in process control, is taken out of service for maintenance. When such an instrument goes through a dead time period for standardization or when, for some other reason, the instrument reading is not available, provisions must be made to maintain operational safety.

    4. Minimize response time lag in instruments directly or indirectly significant to process safety.

    5. How has the process safety function of instrumentation been considered integrally with the process control function throughout plant design?

    6. Evaluate procedure and frequency of checking safety instruments for functionality.

    7. How are highly exothermic reactions protected by dual, independent instrumentation including alarms and interlocks?

E. *Operations*

    1. Review and revise the written operating procedure when process or operational changes are made. Review the operating procedure for revision at least annually.

TABLE 32.1. *(Continued)*

2. Train new operating personnel on initial operation and keep experienced operating personnel up to date on plant operating procedures, especially for start-up, shut-down, upsets, and emergencies.

3. What special clean-up requirements are there before start-up and how are these checked?

4. Provide easy-to-reach emergency valves and switches.

5. Observe safety precautions in loading liquids into or withdrawing them from tanks. Control the possibility of static electricity.

6. Review process hazards introduced by routine maintenance procedures and revise procedures to eliminate the hazards on an annual basis.

7. What evaluation is made of the hazards of sewered materials during normal and abnormal operation?

8. Inerting gas supplies must be dependable. Review the potential for interruption and emergency supply.

9. Review safety margins which have been narrowed by revisions of design or construction or process conditions in efforts to debottleneck operations, reduce cost, increase capacity, or improve quality.

10. What provisions does the operating manual have for coverage of start-up, shut-down, upsets, and emergencies?

11. What economic evaluation has dictated the choice on process between a batch process and a continuous one?

F. *Malfunctions*

1. Discuss the hazards created by the loss of each feed, and by simultaneous loss of two or more feeds.

2. Discuss the hazards which may result from loss of each utility, and from simultaneous loss of two or more utilities.

3. Describe the severest credible incident, i.e., the worst conceivable combination of reasonable malfunctions, which can occur.

4. Discuss the potential for spills and the hazards that would result from them. (see Chapters 28 and 29.)

G. *Venting*

1. Flame arresters shall not normally be installed on discharge of relief valves or rupture disks on pressurized vessels.

2. Design must include provisions for removal, inspection, and replacement of relief valves and rupture disks. What is the schedule for this? How well is it adhered to?

3. Discuss the basis for sizing emergency relief devices. (Breather vents, relief valves, rupture disks, flame arresters and liquid seals. Reference: S&PP Briefs #42 and #100. What is the needed *trim size* of relief valves? (See Chap. 7.)

4. Where rupture disks are used to prevent explosion damage, they must be sized relative to vessel capacity and design. (Reference S&PP Brief #100 and API-RP520, Part I.)

5. Where rupture disks have delivery lines to or from the disks, adequate line size relative to desired relieving dynamics must be assured. Prevent whipping of discharge end of line. How is liquid blow to the atmosphere prevented? Should it be?

6. Avoid manifolding venting systems, e.g., deposits in lines.

TABLE 32.1. (*Continued*)

7. Where manifold venting is necessary describe the system and the potential hazards and the hazard prevention means.

8. Equipment vents shall terminate outside buildings and structures and be so located as to prevent injury to personnel or damage to property in the event of discharge.

9. Describe equipment, operating under pressure, or capable of having internal pressures developed by process malfunction which is not protected by relief devices.

10. Support discharge piping independently of relief valves. Make piping as short as possible and with minimum changes in direction. Support properly to prevent bending or whipping upon relief discharge.

11. Provide drain connections or weep holes in discharge piping of relief valves and relief valves where condensate, rain, or snow could collect. Weep holes should not be permitted to discharge flammable gases or liquids in such a way as to expose tank shells to fire in the event of ignition during operation of relief valves. Provide low-inertia weather covers at end; provide liquid disengagement where necessary.

12. Provide relief valves on discharge side of positive displacement pumps, between positive displacement compressor and block valves, between back-pressure turbine exhaust flanges and block valves, and on any equipment where liquid can be blocked in and later warmed.

13. Where rupture disks are in series with relief valves to prevent corrosion on valve or leakage of toxic material, install rupture disk next to the vessel and monitor the section of pipe between disk and relief valve with pressure gauge and pressure bleed-off line. No valves in lines to the tee, to the pressure gauge, and to the bleed-off. Flow checks are required in bleed-off lines.

14. Make provisions for keeping piping to relief valves and vacuum breakers at proper temperature to prevent accumulation of solids from interfering with action of safety device.

15. Assure that the service temperature at which the rupture disk bursting pressure is stated by the supplier is actually the rupture disk operating temperature; this is often different from the process temperature.

H. *Sight Glasses*

1. Where sight glasses are necessary on process equipment which is subject to hazardous conditions (flammables, toxic, high pressure, temperature extremes), pressure equipment should be used. (Refer to S&PP TecFact #16 for details.)

I. *Control Rooms*

1. Adhere to applicable provisions of S&PP Design Guide #7, revised July 21, 1967, in the construction of process control houses and process control rooms. (See Chap. 23.)

TABLE 32.2. Environmental Control Checklist

This guide for conducting an environmental control review is primarily a checklist of items which should be considered (see Chapters 28 and 29).

The goal of the environmental control review is to insure that all plant effluents are adequately and economically handled to minimize pollution potential and the possibility of unfavorable publicity and payment of penalties. Special effort should be directed toward recycling by-products, controlling raw material and product losses, and reducing waste volumes as an economical approach to pollution control, especially as compared to potential collection and treatment costs.

I. *Environmental Control Audit*

An audit of all plant effluents must be made to provide a basis for evaluating pollution potential and determining the requirements for environmental control facilities. *Each* discharge to the environment should be reviewed and the following data assembled, both the normal or average, and the range for each component.

A. Atmospheric Discharges
Chemical composition
Quantity—lbs/hr
Specific volume—ft³/lb
Temperature of the discharge
Particulate matter included in the vent or stack
Chemical composition
Quantities
Particle size (if available)
Height of discharge (elevation above surroundings)
Toxicity (if available)

B. Liquid Discharges
Chemical composition
Flow or quantity
pH
Acidity or alkalinity
Specific gravity
Total solids
Total dissolved solids
Total suspended solids
Total settleable solids
BOD (biochemical oxygen demand)
COD (chemical oxygen demand, i.e., ultimate BOD)
Quantities of nutrients
Nitrogen (nitrates, nitrites, ammonia)
Phosphorus (phosphates)
Temperature
Color
Toxicity (if available)

C. Solid Discharges
Chemical composition
Quantity—lb/hr
Toxicity (if available)
Collection method

TABLE 32.2. (*Continued*)

II. *Questions*

The following questions should be considered when making an environmental control audit.

A. General

1. *Flow Measurement*

   Can the flow or quantities be measured? If so, are such facilities necessary? Are they required by law?

   Sampling

   Are sampling provisions necessary? Are they required by law? If so, have they been provided and are they adequate, especially those for gaseous discharges where particulates are present? Can sampling locations be reached safely or are access platforms necessary?

2. *Monitoring*

   Is special instrumentation necessary or justified to detect potential upsets which would result in unwanted (due to the value of lost materials or to potential for pollution) discharges to the environment, especially on sewer connections which may otherwise go undetected?

3. *Special Considerations*

   Special consideration should be given any potential discharge of the following materials to the environment. (See Chaps. 14, 15, 25, and 28.)

   a. Acid Gases and Mists

      That is, HCl, NO_x, SO_x, H_2S, HF

   b. Heavy Metals and Metal Oxides

      That is, arsenic, barium, beryllium, cadmium, chromium (hexavalent), copper, iron, lead, mercury, nickel, phosphorus, selenium, silver, zinc.

   c. Miscellaneous Gases

      Ammonia, bromine, carbon disulfide, carbon monoxide, chlorine, fluorides, hydrogen cyanide.

   d. Organic Compounds

      Acetone (and other solvents)

      Acrylonitrile

      Benzene

      Carbamate compounds

      Chlorinated biphenyls

      DDT

      Fuel oil

      Gasoline

      Highly chlorinated organic insecticides

      Naphtha

      Naphthalene

      Organophosphorus compounds

      Phenolic compounds

      Tetraethyl lead

      Tetramethyl lead

   e. Hydrocarbons emissions (atmospheric)

      Refer to the National Air Pollution Control Administration, Publication No. AP-64, March, 1970 entitled "Air Quality Criteria for Hydrocarbons."

TABLE 32.2. *(Continued)*

      f.   Radioactive materials (see Chapter 37).

NOTE: The above chemicals have appeared in various existing and/or proposed regulations; thus their current status versus local regulations should be checked. In addition, they should be considered as likely candidates for future regulations.

      g.   Others

          The U.S. Public Health Service Drinking Water Standards may also be referred to as a likely source of future regulations.

          Are chemicals which are objectionable because of their odors or lachrymatory properties properly contained or treated to avoid pollution and/or citations based on "nuisance" regulations? This includes potential problems due to discharge to city sewer systems. (Note regulation that all spills or discharges of hazardous materials should be reported to National Response Center on toll-free 800-424-8802.)

B. Atmospheric

1. Is dispersion intended as the prime means of meeting acceptable concentrations in the ambient air? If so, are there other sources of similar discharges in the vicinity which should be considered, and is this method reasonable based on current and anticipated regulations?

2. Will the discharge of the gas in question combine with other materials in the atmosphere to form an undesirable effect, i.e., ammonia and HCl to form an ammonium chloride "haze"?

3. Are undesired liquids (mists) or particulates entrained in the vent or relief stream or will they form after discharge? Particulates must not only meet local regulations but must not create a poor public image or nuisance.

4. Will the stream be colored, i.e., will it draw attention to itself, such as NO_2?

5. Are large quantities of water vapor being discharged and if so, is the discharge located so as to avoid visibility and icing problems, especially outside of our plant boundaries?

6. What are the potentials for paint damage from either regular or emergency discharges? Are special considerations justified?

7. Will "breathing" of process equipment or storage tanks carburet unwanted vapors into the environment?

C. Combustion Process, Including Flares and Incinerators

1. Have all fuels (and wastes to be flared or incinerated) been analyzed? What will be the products of combustion? Are acid gases, i.e., HCl, SO_3, NO_x, formed?

2. Flue gases should be reviewed closely as possible pollution sources, including flares.

3. If flares are to be used will a "smokeless" type be required? Is the flare sized for emergency conditions (this is particularly important for toxic gases to insure complete destruction). Have drawback problems (due to condensation of gases) been considered when handling hot gases? Are continuous purges required? Molecular seals? Knockout drums?

4. If incinerators are to be used, are scrubbers necessary? If so, is the scrubber effluent a source of pollution? Is an auxiliary fuel required, special pilot

TABLE 32.2. (*Continued*)

lights? Are materials present in the waste stream or auxiliary fuel which will affect refractory life, i.e., calcium or sodium? (See Chap. 29.)

    5. Has heat recovery been considered?

  D. Liquid Discharges

    1. Can a local city or area treatment plant handle liquid wastes? What are economics? What limits in quantitites or waste composition would be imposed? i.e., pH, phenols, solids, chromates or heavy metals, toxicity, inorganic salts (see Chapters 28 and 29).

    2. *By-Products*

Are there recoverable products in the plant waste streams? Can process recycle by-products? Can special plant waste streams (solvents, alcohols, etc.) be sold for reprocessing?

    3. *Spill and/or Fire Water Control*

What is the effect on waste treatment in the event a sprinkler or deluge system is operated? Are special considerations justified?

Are special provisions necessary for controlling possible spills, especially of liquids with high pollution potentials? Examples:

    Pump seal leakage

    Dikes for storage tanks

    Selective collection of storm water

    Provision for early detection

Are environmental control provisions compatible with safety requirements?

Are spills required to be reported to a pollution control agency? i.e., effect on potable water supplies. Does the plant make a spill report to the plant manager? (See note above in II 3 g regarding National Contingency Center report.)

    4. *Sewer Systems*

Are segregated sewers necessary for:

    Storm water

    Cooling water

    Sanitary wastes

    Concentrated or general process wastes

Are process area floor drains connected to process sewers? Are sewer grades compatible with existing and future sewers?

Are special materials of construction required for:

    Acid or caustic wastes

    Solvents

    Plasticizers

    Temperature

    Solids or scaling

Are in-battery facilities (API separators, surface condensers, settling basins, etc.) required for:

    Oil—many waste treatment systems will not accept oil

    Flammable materials

    Solids

    Concentrated organics

    Odors or lachrymators

TABLE 32.2. *(Continued)*

pH adjustment

Temperature reduction

5. *Solvent Spill Control*

Are facilities provided or necessary to prevent discharge of solvents, etc., to the sewer which might result in explosive vapors in the sewers? (see Chapter 19).

6. *Toxicity*

Does the discharge contain toxic concentrations of any materials (use the 96-hr mean tolerance limit (MTL) as a guide)? If so, what special precautions are necessary and/or provided to protect personnel who might be required to enter sewers? Will the ultimate plant discharge exceed this 96-hr MTL toxic limit and if so, how does this compare to existing or anticipated regulations? Also, is there any potential of fish kills as a result of this discharge, keeping in mind other potential sources and the possibility that similar chemicals may be present upstream of the plant. (See Chaps. 14 and 15.)

7. *Storm Water Contamination*

Has possible contamination of storm water runoff been properly considered?—i.e., curbing potential spill areas for drainage to process sewers. If so, has the waste treatment system been designed to accept wet weather surge? Or, if plant effluent is tied into a local city or area sewer system for treatment will they accept contaminated storm water?

8. *Barometric versus Surface Condensers*

Are surface condensers justified over barometric condensers in the light of environmental control requirements and economics?

9. *Decanting*

Decanting operations where a water layer is to be drained to the sewer should be considered as a potential source of high losses. This should be reviewed carefully and automatic monitoring considered to minimize losses from these areas.

10. *Temperature*

Effluent temperatures should be considered carefully versus existing and potential regulations

E. Solid Discharges

1. If solid wastes are to be landfilled, is the landfill properly designed and approved?

Can toxic materials be properly handled? (This question must also be considered in the case where our wastes are landfilled by others on a contract basis.)

Should selective landfills be employed to provide for possible future reclaiming operations?

2. Open burning should not be considered as a means of disposal unless specifically allowed by local regulations. (See Chaps. 28 and 29.)

III. *Regulations and Permits*

A. As a general guide the lowest governmental subdivision issuing regulations are those which must be met. The local plant can best verify the status of regulations affecting their area.

TABLE 32.2.   (*Continued*)

    B.   Have *all* potential discharges been checked against existing discharges and existing permits? Will new or revised permits be required?

    C.   Are there tax credits or tax depreciation allowances which apply to the cost of waste treatment?

    D.   Should federal or state grant money be sought?

IV.   *Receiving Streams*

    All new plant sites should consider carefully the details of the surrounding environment and the potential effect of the plant on this environment. Air quality, potentials for inversions and stream parameters such as type, minimum flow, history and use all must be considered (see Chapters 3 and 28).

V.   *Pollution Control Facilities*

    All major pollution control facilities should be listed, including their design capacity (specify limiting design parameter) and the expected loading.

---

this review should seek full understanding of recognized and existing hazards and of the procedures for avoiding or controlling them. It should detect details which have been overlooked, and concentrate on the adequacy of plans to cope with operating emergencies that might arise. The manufacturing superintendent conducting the review will arrange for participation by research, engineering and plant safety personnel and others necessary to ensure that operation supervisors are fully advised of hazard potentials and of safety provisions which have been engineered.

PERIODIC PLANT LOSS PREVENTION REVIEW (MANUFACTURING)

It is important that a loss prevention review be conducted at scheduled intervals to:

1.   Keep operating personnel alerted to the hazards.

2.   Determine whether operating procedures require revision.

3.   Carefully screen the operation for changes which may have introduced new hazards, or changes which should be made to reduce existing hazards.

4.   Reevaluate property and business interruption loss exposures.

5.   Uncover potential hazards not previously recognized, especially in the light of experience or new information.

The manufacturing superintendent is responsible for holding such periodic reviews. He or she should find the tentative process design manual and division operating manual helpful supplements to the "Loss Prevention Review Checklist," Table 32.1 and the "Environmental Control Checklist," Table 32.2. The review should look particularly for changes that have been introduced with or without the benefit of process amendments or alteration projects. Minor changes in operating details, equipment, piping connections, temperatures, pressures, operating rates, and raw material inevitably occur in an operation, making these periodic reviews essential. Sometimes it is not a single change which creates a hazard, but a combination of minor changes may substantially increase the risks involved. The periodic loss prevention review should also ferret out recurring operating abnormalities and see how they should be handled safely.

A review of maintenance checklists dealing with the inspection, cleaning and testing of critical control equipment should be included.

REVIEW REPORTS

Reports on loss prevention reviews should be sent to the corporate or plant safety group to enable the company to:

1. Include the improvements in the techniques of loss prevention reviews, developed at each location, in a compilation of suggestions for improving loss prevention reviews elsewhere.
2. Build up a backlog of safety information in the company to aid engineers when designing new plants.

Organization of Loss Prevention Reviews

TEAM APPROACH

The analysis of loss occurrence probability can best be accomplished by a review team approach utilizing the knowledge and experience of individuals competent in the various aspects of the hazards concerned. The technical backgrounds of those comprising the team will also vary depending on the type of loss prevention review being conducted. For new project development reviews, for example, the team in a small plant or company may include the same individuals who would cover reviews of existing operations. In a multiplant organization, the new project review team would more likely comprise specialists from the corporate loss prevention group with representatives from the plant operating department and safety department. To secure an objective review, an organic chemist or chemical engineer from a unit of the organization which is not concerned with the success or failure of the project should be included in the review team.

Company size limitations may affect ability to form a team of technically qualified people. Under these circumstances, the services of consultants may be acquired or the aid of insurance company specialists requested.

COMPOSITION OF REVIEW TEAM

The review team should normally be composed of a central group which participates in reviews of all units in the plant, and other groups appointed for the particular review at hand. A representative from the plant safety department should always be a member of the central group. For large plants, the central group usually consists of three people. For very small plants, the central group may consist of the safety department representative and a technical specialist. The contributions of medical and industrial hygiene specialists should not be overlooked. (See Chapters 9, 10 and 11.)

A group appointed for a particular review should include a representative selected from each of the functions concerned with the operation to be audited. These will normally include manufacturing, maintenance, plant technical service, and, depending on the operation to be reviewed, possibly research, engineering, or other groups. Each person should be selected on the basis of ability to contribute the most from the particular function. This usually means that the busiest people are the ones to be appointed for the audit.

Special Value of Central Group

Although the central group will be the most experienced, all groups should work as one team. The central group should neither direct nor dominate the review meetings; it should not ask all the questions, but should see that the proper questions are asked. The central group should participate in the deliberations, but only on a basis of equality with the other review team members.

The central group will bring to the reviews a wealth of experience derived from prior reviews. It will know of internal and external sources of information. It will also have a broad knowledge of interdependency of hazards among the various operations. Further, because of its experience, it can help in the framing of recommendations.

Scheduling of Loss Prevention Reviews

Loss prevention reviews for the different operations in a plant should be fairly evenly scheduled over the period concerned. This can help the achievement of adequate preparation for, and sufficient depth of, the review. A crash program of reviews will tend to lead to shallowness. A review should cover not only process operations, but also other parts of the plant such as utilities operation, maintenance shops, etc. (see Chapters 12 and 13).

Loss Prevention Review Meetings

A loss prevention review team should meet regularly until its job is completed. A suggested frequency, recommended for a number of reasons, is once or twice per week. The length of meetings should be sufficient to permit real accomplishment. A suggested minimum time is four hours but some teams find six or even eight hours to be desirable. If the meeting is too brief, too little will be accomplished per workhour.

During review meetings the team members should not be called away for other than the very direst emergencies. Attendance of team members at the meetings should definitely take precedence over any other phase of their jobs for the designated times of the meetings. All team members should be present at all meetings. This will result in the maximum benefit of the review for the fewest total workhours spent.

REVIEW IN DEPTH

The loss prevention review should be in depth rather than in petty detail. It should include a comprehensive step-by-step examination of the process to uncover unsuspected hazards that could lead to serious injury and property damage. It should always include review of:

1. Process chemistry.
2. Hazardous properties of all materials.
3. Physical operations, including flow charts and plant layout.

Desirably, the team should accomplish most of its work by free discussion guided by some logical sequence of operation coverage. It should preferably reserve reference to, or use of, a checklist to the end of its deliberations. Faithful following of a checklist throughout all its discussions, or use of a checklist as a discussion guide, will throttle the thinking that is so necessary for objective evaluation of an operation for hazard analysis. Checklists do serve a useful function when the audit team uses them to help prevent overlooking points which should be covered. Table 32.1, "Loss Prevention Review Checklist," and Table 32.2, "Environmental Control Checklist," can be used for this purpose.

RECOMMENDATIONS BY TEAM

The review team should develop recommendations for attacking recognized hazards, but it should not try to engineer the solutions. Again, the function of the team is to analyze in depth, not to design. Rather, it should assign to a particular department the responsibility for execution of each recommendation. Sometimes the audit team can suggest desirable approaches.

REVIEW TEAM REPORT

The activity of the review team should culminate in a concise, definite report. This should have two primary sections or areas of coverage. The first is a listing of loss exposures, both property and business interruption (B/I or U & O). The second is a listing of hazards, recommendations, assignments of responsibility for actions, and target dates for completion. In this latter section should also be dates for a check on action completion and an assignment for doing this, as well as provision for closeout when action is satisfactorily completed.

Each report should be dated and should be addressed to a specific person. The plant should also decide upon the list of people who should receive carbon copies of the report. Specifically, it should be determined whether the plant manager wants a copy; in most plants the plant manager is the person to whom the report should be addressed.

REFERENCES

1. Arthur Spiegelman, "Risk Evaluation of Chemical Plants," paper published by American Institute of Chemical Engineers, New York, 1969.
2. "Guidelines for Risk Evaluation and Loss Prevention in Chemical Plants," Chemical Manufacturing Association, Washington, D.C., 1970.
3. "Dow's Safety and Loss Prevention Guide Hazard Classification and Protection," a CEP Technical Manual, American Institute of Chemical Engineers, New York, 1973.

33

Tools and Techniques For Chemical Safety Training

F. Owen Kubias

Many studies have been made of industrial accident causes. In each of these studies the central theme is that by far the vast majority of industrial accidents are caused by the acts of individuals involved in the incident—either by omission of something that should have been done or commission of something that was not authorized. The physical plant and its hardware generally are not the prime cause of the accident. Thus, the key to safe operation in the chemical industry, or any other industry, is adequate initial employee training and the maintenance of that training level.

Before discussing the tools and techniques that are successfully used in the chemical industry for employee training, it would be well to review a few basic concepts of human behavior as it relates to industrial safety, and to the learning process.

First, we must recognize that there is no such thing as the average person. No two individuals are exactly alike. People differ in their intelligence, their knowledge, their abilities to learn, their attitudes, their beliefs, their physical abilities, their physical skills—in all personal characteristics that come to mind. However, in spite of individual differences, certain general patterns are found common to all people. All of us are continually striving to satisfy our needs and our wants. Our behavior is directed toward satisfying these needs and wants. In today's society, the basic needs of food, clothing, and shelter are for the most part satisfied, although we still continually strive to improve our quality of life with respect to these basic needs. We also have a long list of material wants which continually change with our age and with our family situations. But beyond these, psychologists have found that there are several psychological wants common to all of us, and when the basic needs and the material wants are reasonably well satisfied, the psychological needs and wants are perhaps our strongest motivators.

Of these half a dozen or so psychological wants, perhaps three are most important to supervisors in achieving adequate levels of training. They are a recognition for our efforts and achievements, being accepted, and maintaining our self-respect and self-approval. In our training activities, and in the supervision of our people, we must work towards fulfilling these wants and providing satisfaction to our people.

We must always be aware of effort and achievement and be ready to recognize it even if it is as simple as a word of praise or an expression of your appreciation for some extra effort. Normally our employees want to be part of the work group, be accepted as a person, be involved whenever and wherever possible in the decisions and activities of the work effort.

Similarly, we all have a self-image of ourselves and we strive to maintain this image. As supervisors we want to be sure that we understand this, have an assessment of what our employees' self-images are, and work to reinforce them rather than tear them down. In short, we must direct our training activities in such a manner as to satisfy wherever possible basic psychological needs; at the same time, tailoring their application to each individual, based on differing personal characteristics.

Most of the difference between people, and for that matter the similarities, are the results of learning. Through learning people have developed certain attitudes and behavioral patterns. To change attitudes one must substitute new learning for old concepts and ideas. To change a behavioral pattern we must teach better means to achieve our wants and desires. In short, new learning must be substituted for the old. Through the years some dozen laws of learning have been developed. When our training procedures incorporate these laws of learning, they become more efficient and more thorough. In applying our training tools and techniques three of the laws of learning are of most import. These are primacy, recency, and frequency.

We tend to remember better and to accept more readily ideas and concepts that seem of prime importance to the person with whom we are in contact. If operational safety is conveyed as one of the items of prime importance in our company, we will help insure a positive response to our safety activities. Perhaps the most important use of the learning law of primacy is an initial contact with a new employee. When an employee comes to work he or she immediately begins to learn and to form attitudes about the job and the company. Therefore, it is vitally important that we convey the importance of operational safety to the employee at initial contact with the company. Primacy is also important in a routine training activity. We must be certain that the safe method is the first method that is conveyed to the employee. Only by starting safety training in an organized, orderly manner from the beginning of employment can we develop positive safety attitudes and the motivation that we need.

We all recognize that we do best those things that we do most frequently. Thus, frequency is an important part of the learning process and certainly an important part of our safety program. It is not enough to give a new worker a copy of the safety rules and regulations and review them with him or her on the first day on the job. We must be certain that safety rules, regulations, and safe practices are repeated during job training and on other regular occasions throughout his or her work career. Continual reminders and regular use of safe methods will create safe habit patterns that will eventually be followed almost automatically. The law of frequency will be utilized in several of the training techniques.

Recency might be considered a corollary to frequency in the laws of learning. The process of forgetting goes on along with that of learning. Forgetting curves indicate that much is lost by forgetting almost immediately after learning has taken place. Thus, that which is learned last can usually be most easily recalled. Therefore, we cannot depend upon instructions and training demonstrations given some time ago to insure the safety of our people and our operations today. Our training tools and techniques must be designed to provide constant reminders of rules, regulations and safe practices.

Some years ago, the National Safety Council undertook a research study of mass communications to increase the effectiveness of information and persuasion campaigns in traffic safety. One of the theories developed in this study might be applied to the question often asked in the chemical industry about how much information should be communicated to the employee about the potential hazards of the individual chemicals. Paraphrasing the theory as it might apply to the chemical industry, if a situation is potentially hazardous and if the employee recognizes the extent of the hazard, then accidents will be reduced. Communications and training activities should be directed towards objectively assessing the potential risks involved. Once workers are aware of the potential risks, they will act accordingly. This has been said many times—"An informed worker is a safe worker."

We have long said that all chemicals can be handled safely with proper procedures and equipment. We can only assure this if our individual operator fully understands and respects the need for proper procedures and equipment. Thus, in our industry, as in any industry, our employees must be aware of the potential hazards of the work they are doing and the material they are handling.

Further information on human behavior and industrial safety may be found in Chapter 7 of the National Safety Council's *Accident Prevention Manual for Industrial Operations*[1] and in Normax Publication's *Accident Prevention Fundamentals.*[2]

33.1. TRAINING TOOLS

Standard Operating Procedures

The keystone of chemical process safety is the standard operating procedure (SOP) or master process. A complete standard operating procedure must contain sufficient information for plant management and the chemical operator to produce safely a quality product at standard cost. Such procedure is not only necessary to derive a quality product at standard cost but is also necessary to properly train operating personnel and to assure maintenance of safe practices by trained operating personnel. It is a valuable reference for all personnel during process operations and should be readily available to everyone in the work area. It should not be locked in a supervisor's desk drawer. Typical contents of a standard operating procedure are listed in Table 33.1.

Material safety information may be inserted into the standard operating procedure in the form of a material safety data sheet, as outlined later, or in an abridged version with reference to the material safety data sheet. The process safety information should contain major potential hazards, how they may be recognized, and process provisions for their control either mechanical or procedural. Special process safety controls should be listed along with their test procedures to assure proper functioning.

Critical operation steps from a safety standpoint, as well as from a yield or quality standpoint, should be made part of the operating report completed for each batch or shift. Check off should be required on the part of the operator to verify that the steps were performed. With the procedures of a particularly critical nature it may be desirable to have it checked off and initialed by two separate operators.

The standard operating procedure should provide a detailed description of the required procedures. Each sequence must be clearly defined. Protective equipment requirements should be listed prior to the step in which they are required as well as in the separate protective equipment requirement section. Steps in the process that are particularly critical from a process safety standpoint should be indicated and practices to assure safe operation outlined at this point.

Emergency procedures must be a part of the standard operating procedures. Considerable attention should be given on the part of the process engineer to the development of these procedures. The procedures should be checked out in actual operation in a dry run drill prior to their need in any emergency situation.

In order that a standard operating procedure be made of value it must be kept up-to-date continuously. Procedures must be established for providing process amendments to the standard operating procedure. In the process of amendment approval, review should be made of the safety aspects of the process change from a procedural and an equipment standpoint to assure that adequate safe practices and process hardware have been provided.

TABLE 33.1. Typical Standard Operating Procedure Contents

Raw material quantities and yields
Equipment and utility requirements
Chemical reactions
Process flowchart
Standard quantities and labor requirements
Operating reports
Material safety information
Process safety information
Protective equipment requirements
Process operating procedure
Shut-down procedures—standard and emergency
Emergency procedures
By-product and waste disposal

Process changes or experimentation without a formal amendment review must be prohibited.

Chemical Safety Data Sheet

As noted earlier, an informed worker is a safe worker. It is most important that our process operators understand the potential hazards of the materials that they are handling. The best procedure is to provide the operating personnel with chemical safety data sheets outlining the material's potential hazards, proper safe practices, and engineering control methods to prevent a hazardous condition from occurring and other information vital to safe handling of the item. Such information should be provided in terms readily understandable by the operating personnel. While technical data such as flash point, flammable range, and threshold limit value may be of interest to the process supervision and to the process engineer and may be included in a chemical safety data sheet, it should also be interpreted in terms common to the operating personnel as well as through use of standardized terminology such as flammable, reactive, corrosive, etc. First-aid procedures should be limited to those to be applied immediately in the field before transfer to the first-aid room or dispensary. Detailed first-aid procedures are best provided for the first-aid room or dispensary where they may be applied by trained personnel. (see Chaps. 9 and 11). These immediate emergency-aid procedures are quite frequently listed by type of potential exposure. Table 33.2 lists the typical contents of a chemical safety data sheet.

New materials must not enter the plant for use before chemical safety data sheets are available. Procedures for development of chemical safety data sheets must include control of the release of the new raw materials. Handling precautions, safe practices, and engineering control procedures as well as protective equipment and field first-aid requirements must be reviewed with all personnel who will handle new materials before the first pound is intro-

TABLE 33.2.   Contents of Typical Chemical Safety Data Sheet

Written safe handling information shall be developed and employees informed for materials classified as potentially hazardous. Such information shall contain but not be limited to:
1. Name and description of material (chemical name, synonym, commercial name).
2. Nature of the hazard (flammable, reactive, corrosive, toxic, etc.)
3. Immediate emergency aid procedure for personnel exposure.
4. Proper protective equipment.
5. Handling precautions.
6. Immediate spill control.

Where appropriate, the data shall also contain information on:
7. Engineering control method.
8. Fire extinguishment agent.
9. Incompatibility with other chemicals.
10. Unusual fire and explosion hazards.
11. Simple description of critical symptoms relevant to the type of exposure likely to be encountered.

duced into the process. Chemical safety data sheets must remain readily available to the operating floor. Chemical safety data sheets for materials currently in use in the plant must be reviewed periodically with operating personnel to assure all employees remain current on the potential hazards, proper practices, and engineering controls of the materials that they handle.

Table 33.3 outlines a typical set of criteria for identifying materials for development of chemical safety data sheets.

Sources of information for development of chemical safety data sheets can be found with the National Safety Council data sheets,[2] the Chemical Manufacturers Association chemical information sheets,[3] the "National Fire Protection Association Guide on Hazardous Materials,"[4] and from individual chemical manufacturers' product data sheets.

Plant and Area Safety Rules

Safety and fire protection rules are necessary to provide guides for controlling potential hazards in chemical operations. These rules should be written and distributed to all employees. For smaller operations they can be in the form of a one or two-page handout. Larger locations may provide plant safety rules in the form of a bound booklet. Wherever possible the safety rule format should be pocket-sized so that the rule book or handout may be carried on the employee's person for ready reference. Safety rules can be made more interesting by providing cartoons or pictures applicable to the particular location and rule. Table 33.4 provides a list of subjects usually covered by plant and area safety rules.

Certain basic precepts are necessary in preparation of safety rules. Only logical and enforceable rules should be selected. A rule that cannot be en-

TABLE 33.3. Scheme for Identifying Materials for Development of Chemical Safety Data Sheets

The following system can be used to identify materials with a higher degree of flammability, toxicity or reactivity.

Health

1. Skin absorbable chemicals
2. Dusts with TLVs at or below 1 mg/m^3
3. Vapors and gases with TLVs at or below 1000 ppm
4. Corrosives and primary irritants
5. Others as judged necessary by the industrial hygienist

Flammability

All materials with a flash point below 100°F closed cup (essentially NFPA 704M signals 3 & 4)

Reactivity

1. Materials readily capable of detonation themselves or of explosive decomposition or reaction at normal temperatures and pressures.
2. Materials capable of detonation themselves or explosive reaction but require a strong initiating source, or which must be heated under confinement for initiation, materials that react explosively with water.
3. Normally unstable materials that readily undergo violent chemical change but do not detonate, materials that react violently with water or which may form potentially explosive mixtures with water.

forced will impair the effectiveness of the other rules. In selecting rules the conditions existing in a work place and the equipment being used should be kept constantly in mind.

Rules should be prepared jointly by the first line supervisor, the plant safety coordinator, and higher plant management. Assistance of the personnel manager is also of value in selecting rules which can be enforced. When preparing rules know the people for whom the publication is intended. Rules must be presented in terms of their knowledge and comprehension. Positive statements are more effective than negative ones. Whenever meaning permits, rules should state what is to be done rather than what is prohibited. General plant rules and specific instructions for certain jobs or departments should be presented under separate sections.

Safety rule books or booklets are designed to explain to newly hired employees, and to recall for existing employees, the company's accident prevention policies, practices, and conditions of employment. To fully benefit from this material one cannot simply hand out a rule book and tell the employee to read it. For the new employee the safety rules should be re-

TABLE 33.4.   Subjects Included in Plant Safety Rules

The policy of the company regarding accident prevention
Reporting accidents
Emergency first aid
Information for laboratory workers
Handling chemicals
Unloading chemicals from drums, trucks, and cars
Housekeeping
Safety equipment available
Waste disposal
Use of tools
Fork truck operation
Eye protection
Fire prevention
Smoking regulations
Welding rules
Use of fire extinguishers and fire alarm
Sprinkler systems
Flammable liquids
Health hazards
Inspection of equipment
Shutting down processing equipment
Safety organization
Specific instructions
General safety information

viewed with him at time of employment. Techniques for new employee orientation are discussed in New and Transferred Employee Training. Departmental safety rules should not only be reviewed with the new employee but also with the older employee that is newly transferred into that particular department. In addition to reviews with workers new on the job, plant and department safety rules should be reviewed regularly with existing employees. This can be done in the form of group safety meetings or individual contact. To fully benefit from the materials they must be utilized regularly.

Hazardous Job Procedures

Certain operations have a higher potential for severe injury or property damage if deviation is made from standard operating procedures. To insure safe conditions for these operations, written detailed procedures must be developed and be available for employee training, and for reference and review by trained personnel. Frequently these procedures involve maintenance or repair operations where two or more different supervisory responsibilities are involved. In these cases, permits may be a part of the hazardous job procedure. The operating procedures should clearly point out the hazards involved and set forth safe practices to control the hazards. These procedures should be laid out in steps. In some cases, these procedures may involve repeated operations carried out in many parts of the plant. Hazardous job procedures provide a standard means of approaching these jobs and eliminate the need to provide this detail in each standard operating procedure.

Examples of tasks requiring hazardous job procedures are cutting and welding operations, entering tank cars, tank trucks and small enclosures, acid handling operations, caustic handling operations, specialized high-speed process equipment (centrifuges, high-speed mixers, etc.), equipment under pressure, gas-fired equipment, and lock out procedures for maintenance work on normally moving or electrically energized equipment.

A discussion of hazardous job procedures requiring work permits will be found in Chapter 36.

Job Safety Analysis—JSA

Job safety analysis is a tool that has been used by safety professionals for many years; it is just as effective today in job hazard identification as it was when originated. Job safety analysis is a technique for identifying potential hazards that accompany each step of a task and for providing an orderly procedure to eliminate the potential hazards by changes in procedures, equipment, or tools.

The first step in performing a job safety analysis is selection of the job or task to be analyzed. With a myriad of jobs available for analysis and with the limitations on supervisory time, we must be selective in choosing a task for job safety analysis. By being selective, we get the quickest possible return and the maximum benefit for our efforts. Things to be considered when selecting a task for job safety analysis are:

1. Has the job had a repeated history of accidents? The greater the number of accidents associated with a job, the greater should be its priority.
2. Has the task produced more disabling injuries? If this is the case, existing procedures certainly have not proven adequate and a careful look should be taken of current practices.
3. Some jobs may not have a history of accidents but they have a high potential for severe injury or death. These jobs are good candidates for the job safety analysis technique.
4. Lastly, when we take on a new job, we have no accident history to guide us toward satisfactory procedures. To fully appreciate the accident potential of a new job, a job safety analysis is an excellent technique. Utilizing supervisory personnel who are familiar with tasks of similar nature can bring a great deal of experience to the new job procedure and provide optimum safety for the new task.

Job safety analyses are usually recorded on a three-column form. The first column contains the list of steps to complete the task, the middle column lists potential accidents and hazards, and the right-hand column lists the developed solutions to control hazards and potential accidents.

Once the job has been selected, the next step in job safety analysis is to break it down into its successive work steps. Care should be taken that the breakdown is not too general so that we fail to recognize important job steps that may have accident potential. At the same time, we must be certain that we do not have too fine a breakdown of the job so we have an excessively large number of job steps. Job steps should be listed in their normal sequence and should usually start with an action word such as remove, carry, place, etc.

After the steps have been identified and listed, the next step in job safety analysis is to identify hazards or potential accidents associated with each step. This calls for a brainstorming type of approach by the participants. No attempt should be made at this time to pass judgment on the probability or validity of the identified hazards and potential accidents. Also, at this time, no attempt should be made to develop solutions to control the hazards and accidents identified. One technique to assist in spotting potential hazards and accidents is to utilize the accident type categories used in injury analysis and look for exposures that can cause struck by accidents, struck against accidents, falls, etc.

When hazards and potential accidents for each step have been identified, the next step is to develop corrective action to prevent the potential hazards from producing an accident. Frequently, a simple change in the job procedure will provide adequate protection. Perhaps a change in equipment or tools will be necessary or improved guarding or interlocks required. In providing solutions for potential accidents, one should never overlook substitution of an entirely new approach to completion of the task.

Once the job safety analysis has been completed and revised based on the solutions developed for the hazards and potential accidents, we have a training tool without peer. The job safety analysis now provides us with a step by step procedure for accomplishing the task, identification of potential accident hazards and an outline of procedures and equipment that have been provided to prevent potential hazards from producing accidents. This is the best training aid possible for teaching a new employee to do the job. It can be readily utilized with experienced employees to refresh themselves on the important points of the job. The job safety analysis is also valuable when developing a program of employee safety contacts, or when making unsafe act observations as part of your inspection program. Job safety analyses on tasks that are infrequently or irregularly performed are a valuable tool for reviewing, with the work crew, the proper steps of the task and their potential accident hazards before the work commences. One final benefit from a job safety analysis is that frequently as the analysis is being done, ideas are developed for labor and material savings that had not been previously discovered for that particular task. Quite often, a job safety analysis may pay for the time and effort in material and labor savings alone.

There are many methods for doing a job safety analysis including actual plant observation of the work as the analysis form is completed to simply sitting in an office and attempting to recall the procedures and the potential accidents associated with each step. The exact method to be used is dependent upon the circumstances. Perhaps as important as method is who participates in the job safety analysis. A job safety analysis should not be done simply by the safety coordinator for the location or by the industrial engineering staff. The most qualified personnel for doing a job safety analysis is the line supervisor of that job and the work crew. By utilizing this group to produce a job safety analysis, you bring the greatest experience to bear on solutions of hazards and potential accidents as well as provide the best talent for identifying potential accidents. Quite often in the course of a job safety analysis with the experienced work crew, previously unknown examples of regular malfunctions of equipment are identified. Doing the job safety analysis with the work crew also provides a subtle retraining of that crew in the job's safe practices and the identification of potential hazards associated with the work.

Detailed information on conducting job safety analyses may be found in references 1 and 2.

Job Instruction Training—JIT

At the beginning of World War II, our nation was faced not only with the tremendous task of training large numbers of new workers in sophisticated tasks in the armed services but also training a large civilian work force that had not had previous industrial experience. Work done at Ohio State University at the beginning of the war resulted in what is known today as "job instruction training," a quick, simple and effective way to teach a new job to the worker. Today, over 35 years later, job instruction training is still effectively utilized in industry and our armed forces. The most notable and consistent use of job instruction training has been in the field of pilot training.

Studies have shown that when individuals learn a new job they retain the job information best if they follow oral and visual instructions by doing the job themselves and recalling the important points of the job as they proceed (See Table 33.5). Job instruction training combines these two techniques in the training activity.

Job instruction training is done in two parts: a preparation period and an instruction period. While the ultimate aim of job instruction training is the instruction period, the preparation period can be equally important. Here, the instructor establishes the game plan: how fast the teaching will be, how much skill the learner is expected to obtain, and how soon the learner is expected to have it. In addition, the instructor must have a job procedure available, either through job safety analysis or through use of the standard

TABLE 33.5. Information Retention

Method of Instruction	Three-Day Information Recall
Telling only	10%
Showing only	20%
Combination of telling and showing	65%
People Generally Remember:	
10% of what they *read*	
20% of what they *hear*	
30% of what they *see*	
50% of what they *hear* and *see*	
70% of what they *say*	
90% of what they *say as they do a task*	

operating procedure. Equipment must be made ready. Materials and supplies need to be on hand. Nothing will defeat the entire purpose of instruction more quickly than the interruption of apologies and confusion caused by looking for necessary supplies, equipment, and tools. It is important that the work place be properly arranged just as the employee will be expected to keep it. It is equally important that during the instruction period the employees be placed in the work position while observing the instruction so they will see the task as they would normally view it as if they were doing it. Do not make the mistake of placing the employers opposite you so that they watch your demonstrations from the backwards position.

The instruction period is comprised of four steps: preparing the worker for the job and advising what it involves and in a manner which will erase anxieties over the new task. Next, presenting the operation by illustrating each step, explaining the what and why of the operation and encouraging questions. Third, allowing the trainee to try out what has been demonstrated. After repeating the task several times, you are ready for the fourth and final step where the new operator is on his or her own but advised to seek the instructor for assistance if need be. The supervisor or the instructor must check back frequently during the follow up step until satisfied that the new worker can proceed independently.

The following summary from the Chemical Manufacturers Association Safety Guide No. 15, "Training of Process Operators," describes how job instruction training can be utilized in the chemical industry.[3]

Have a Time Table.   How much skill do you expect him to have and by what date? Decide upon plan for back-checking at regular intervals.

Analyze the Job

1. Check existing standard operating procedure step by step.
2. What are the health problems? Are provisions adequate?
3. Any personal protective equipment required?
4. Are the material handling methods safe?

5. What opportunities (if any) for operator to err in using proper sequence for adding materials?

6. What equipment hazards are present and what protective provisions have been taken?

7. What steps must be taken to counteract any fire or explosion hazards?

8. Define any potential emergency situations. What steps can be arranged for their control?

9. Has emergency shut-down procedure been defined?

Have Everything Ready.   Have process equipment checked to make sure that it has been properly maintained and ready to operate. Make sure that necessary supplies and materials are on hand. Arrange your schedule so that operator gets off to a proper start. If the instructor will be someone other than the foreman, make sure that the instructur has his instructions beforehand. Emphasize points that he will be expected to stress.

Have the Work Properly Arranged.   Make sure that everything has been arranged just the way you will want it to be kept. No greater opportunity for building good habits will ever be present than during the training period. Impressions gained at this time are likely to be the most lasting.

Giving Work Instructions

STEP 1—PREPARE THE WORKER.   Describe the job. Tell what it will involve. Discuss only the more important points such as strict adherence to standard operating procedures, safety rules, reporting accidents, suggestions, etc. Tell why, when you say "don't." Remember to keep this a two-way street. Try to find out what he or she already knows about the job. Correct any false notions. Try to build up interest in the job. Explain its importance in the scheme of things. Tour the work area. Explain equipment operation and purpose. Let him or her watch the operation in progress.

STEP 2—PRESENT THE OPERATION.   Refer to the standard operating procedure frequently as the instructor illustrates each step of the operation. Encourage questions. Ask questions—explain what operator is doing and why it should be done. Stress the key points. Care must be taken to be sure that instructions are clear, that they do not conflict with standard operating procedures and that they are understood. Instructions should be complete, but not faster than can be absorbed.

STEP 3—TRY OUT PERFORMANCE.   Check what trainee has learned. Have the trainee tell you how the process is operated and describe what goes on at each step. Be sure to correct any apparent misconceptions. Let the trainee operate the process with instructor watching—correcting, explaining, etc. Have the trainee stress each key point during the entire procedure. Have the trainee repeat the cycle until you know that he or she knows what he or she is doing and why.

STEP 4—FOLLOW-UP.   Put operator-candidate on his or her own. Tell the operator-candidate to whom to go when help is needed. Try to impress upon the operator-candidate the concept that "if you are not sure—don't do it"; that it is always better to ask questions than to explain a mistake; that suggestions for improvement are always welcome, but procedures must not be altered unless first approved officially. The supervisor must back-check frequently until sure the operator has learned to do the operation correctly, and must follow-up periodically but continuously by an occasional session on what to do in case some abnormal situation should occur. If the worker has not learned, the instructor has not taught.

Programmed Learning

One of the newer training techniques now available for safety training is programmed instruction. It may be used as a substitute for classroom and textbook training or it may simply supplement other training techniques. Programmed learning is self-contained teaching material. It allows the trainee to set an individual pace for learning and to absorb knowledge in easy to take bits.

The trainee receives a carefully arranged sequence of information, questions, and answers. As the trainee reads the text, answers questions, and checks the answers he moves from simple to more complex concepts. Each step builds on the step before. In programmed instruction the learning process is reinforced by requiring the trainee to answer questions and correct errors before progressing with the material.

This principle of reinforcement, in which the trainee's responses to questions of gradually increasing difficulty are confirmed immediately gives reward and encouragement for continuing the learning cycle until a predetermined level of achievement is reached. Questions and bits of information are arranged in a careful sequence that lead to rapid, efficient mastery of the concepts. Slow learners learn just as thoroughly as fast learners.

Studies have shown that trainees using programmed instruction spend 25% less time learning the subject and achieve 13% higher scores on the final exam than those who received the conventional instruction. These trainees were retested 18 months later and the programmed instruction group had much greater learning retention.

Programmed instruction can be given to one person at a time eliminating the need to delay training until a class group can be assembled. The training can be given at any time on any shift. Proceeding at the trainee's own pace, the fast trainee learns faster without being bored, while the slower learner learns without being embarrassed or left behind.

Programmed learning materials are available for safety training on such topics as lifting, ladder safety, fire-fighting techniques, slips and falls, and fork truck operation.

One series of programmed learning materials of particular interest to the chemical industry is that produced by the Applied Technology Division of the E. I. du Pont Company.[5] The material was developed for internal use by the du Pont Company and later released through their Applied Technology Division. Hence, it is particularly suited for chemical operations. The initial two lessons in the series develop an awareness on the part of the chemical operator for safe operations by teaching some basic safety principles and safety planning techniques. After these are mastered, programmed learning training texts are available for various activities peculiar to the chemical industry such as handling flammable liquids, static electricity, tank entry, and personal protective equipment.

33.2. TRAINING TECHNIQUES

No matter how well safety is engineered into a plant, process, or piece of machinery, much of the safety of employees depends upon their own actions. Employees' actions in turn depend on how well they have been taught and trained regarding potential hazards, necessary safeguards, and safe work practices associated with their jobs. Accident prevention depends basically upon training and motivation of employees to work safely. To achieve this, certain basic training programs are necessary.

New and Transferred Employee Training

As discussed previously, one of the important laws of learning is that of primacy. Certainly, its application can be no more important than when it comes to the training of the new employee. At least a portion of the first day on the job for the new employee is spent in the Personnel Department, obtaining basic information about the company, the plant, hours of work, and employee benefits. This is the time to provide the new employee with the plant safety rule booklet and to take time during this period to discuss the general plant safety rules. This general safety material must be presented with care and attention to ensure that the employee recognizes the importance of safety to our operations from the very start. Topics covered in this initial safety orientation should include information on the location's safety program and its safety performance, procedures for reporting accidents and injuries and obtaining medical attention, emergency procedures, information on obtaining protective equipment, the plant's fire alarm and evacuation procedures, exit facilities, security program, etc.

After the employee has received the basic information, he or she usually reports to the new supervisor to begin the first day's work. At this initial contact with the supervisor, it is important that departmental safety rules and the safety requirements of the new job are reviewed carefully. A brief review of the previously covered general plant rules might also be in order. During this period, the supervisor has the opportunity to assess the employee's experience and to become better acquainted. Topics usually covered by the supervisor include department safety rules, hazards that may exist in the work area, fire prevention for the department or area, instruction on use of personal protective equipment required for the job, location of emergency showers and eye wash units, housekeeping, and the department's safety program.

To assure that all items are covered for each new employee and to assist the personnel and supervisory staff in conducting new employee indoctrination, a safety orientation checklist is recommended. Topics covered by the personnel staff and the supervisor can be checked and the form filed in the employee's record to attest that the training has taken place.

It is important that we recognize that even the veteran employee is a new employee when moved to a new department or a new job, and will require the same departmental safety information provided by the supervisor to the newly hired employee. The same checklist approach can be used for assuring complete information is provided to this employee. The veteran employee in the new position or new department probably has not reviewed the plant safety rules recently and so it would be beneficial for the supervisor to at least cover them in an abridged fashion.

On-the-Job Instruction

After the new employee or the veteran employee new on the job has received initial orientation information, specific job instruction is in order. This is the time to apply the tools of job safety analysis, standard operating procedures, and job instruction training. Initial job instruction may be given by the employee's immediate supervisor or a key employee under the supervisor's general guidance. It is important that when a key employee or trainer is used that the supervisor periodically review the instruction material with the key employee or trainer to assure that correct and safe procedures are being instructed. This periodic review also provides for updating of job procedures where changes have been or need to be made.

Group Safety Meetings

The learning laws of frequency and recency mandate that regular safety input be provided to the work force. Group safety meetings are an effective method of getting safety information to a number of people in a minimum of time. For this reason, they are a preferred contact program for many plants. Topics selected for the group meetings should be associated with the department's work situation rather than topics selected at random. Department, plant, and division injury experience should also be utilized in group safety meetings. An off-the-job safety topic appropriate to the season or group can be selected occasionally for a change of pace. Periodically, on a regular frequency, the group safety meeting should review the plant emergency program. Similarly, the group meeting should be used to periodically review the use of plant fire-extinguishing equipment and procedures for sounding emergency alarms and summoning emergency help.

Group safety meetings are particularly effective for demonstrations, the use of visual aids, and for subjects provoking lively discussions and information exchange. Group safety meeting format should be varied from meeting to meeting to maintain interest and to encourage participation by the chemical operators.

Frequency and length of time for group safety meetings vary widely and must be to some extent tailored to the work method and shift rotation procedures of the location. Experiences show, however, that shorter meetings held more frequently are better than longer meetings held at great intervals. Weekly meetings of short duration tend to be the most effective. This simply is a reemphasis of the learning laws of frequency and recency. People tend to remember best those things they have heard most recently. The basic format for group safety meetings should provide five to ten minutes of topic presentation and five to ten minutes of discussion and comment. The meeting should involve a single topic and discussion should be limited to the subject of the meeting. Employees with questions or comments not related to the meeting topic should be instructed to contact the supervisor at a later time. It is important that the meeting stay on the topic assigned for everyone to obtain full value from the meeting.

Personal Contacts

Planned safety contacts are a powerful tool for the development of favorable employee attitudes toward safety and loss prevention. Such a program is also a visible demonstration of the management's commitment to safe operations. In addition to initial safety orientation and on-the-job instruction, employees require some kind of regular safety training or instruction. Conducted on a regular basis, such contacts reinforce employee awareness of hazards, strengthen their knowledge of safe job procedures, and generally enhance employee safety-mindedness. Regular planned safety contacts are a key to reducing injury frequency and maintaining safe operations.

Personal safety contacts can be used to supplement on-the-job instruction and group safety meetings. Personal contacts are one-to-one situations where the safety message can be tailored personally to the employee. The employee feels important that your attention and interest are directed solely at him or her. Personal contact also assures better understanding and acceptance of your safety message. The employee is more likely to express opinions to you on a one-to-one basis. Personal contacts also communicate concern for the welfare of the workers as nothing else can. The personal contact should be on a topic suited to the task being performed by the worker.

Pre-Job Instruction

While most jobs in our operations are repetitive and safety information can be reinforced readily after the initial on-the-job instruction through group safety meetings and personal contacts, there are tasks in the chemical industry that are infrequently done. Many fall in the maintenance department operations. Quite often, such work is not done by the same employee who did the job the last time. Pre-job safety instruction is especially appropriate for this type of potentially hazardous, infrequently done, one-time job. This type of work is also a good candidate for a job safety analysis, and the completed job safety analysis will prove an effective tool in the safe execution of this type of work. In pre-job safety instruction, the supervisor should include information on potentially serious hazards that could occur if safe practices are not followed. Required protective equipment, where appropriate, should also be reviewed. Short-cuts and unsafe practices that have led to accidents and injuries are also to be discussed. Of particular interest to maintenance operations are other allied precautions—protecting workers who may be in the vicinity of the job or who might move into the path of the work. Pre-job instruction is actual instruction on how to do the job before the crew sets out for the work site. It is not enough to simply tell the workers to be careful and send them into the field.

Further information on training techniques can be found in the booklet, "Practical Safety Communications," published by the International Safety Academy[6] and from reference 2.

33.3. ACTIVITY REPORTS

The tools and techniques discussed in this chapter, when properly applied, will resort in a decreasing work injury experience. Injury statistics themselves can be a measure of the application of these techniques and tools, but they are after-the-fact measurements and might even be considered measurements of failures in the accident-prevention effort. A more current measure of safety performance must be aimed at what we are doing to prevent accidents and injuries. An accident control report, much like a quality control report or production control report, provides a measure of the work activity directed at loss prevention. Administrative control of the loss prevention effort can be obtained by the use of activity reports. An activity report is a record filed weekly or monthly by each supervisor of the various activities completed during the report period involving the job safety analysis development or updating, completion of area safety rules or hazardous job procedures, conducting group safety meetings or personal safety contacts, prejob instruction or new employee orientation. In the simplest form, the supervisor simply reports the number of times the particular activity was conducted during the report period.

More sophisticated safety activity reports establish goals or quotas for the report period, and activity level percentages are developed based on activity reported versus goals. A more refined system can provide value factors for each activity in recognition that some tools and techniques may be more effective than others in promoting safe work performance. With value factors, the raw percentages are adjusted based on the importance of the activity as well as the supervisor's performance as compared to the goal. The individual supervisor's reports can be combined into department safety activity reports and those further assembled into a plant safety activity report.

Such management control of the safety activities are important as changes in safety activity levels can be detected before their results show in the work injury statistics. With timely evidence of a slackening of safety activity, management corrective action can be applied before the slack in activity affects work injury performance.

REFERENCES

1. Accident Prevention Manual for Industrial Operations, 7th ed., National Safety Council, Chicago, Ill., 1980.
2. M.U. Eninger, *Operation Zero--Accident Prevention Fundamentals*, Normax Publications, Pittsburgh, 1976.
3. Chemical Manufacturers Association, 2501 M Street, N.W., Washington, D.C. 20037.
4. National Fire Protection Association, Batterymarch Square, Quincy, Mass. 02269.
5. Department of Applied Technology, E. I. du Pont, Wilmington, Delaware 19898.
6. "Practical Safety Communications," International Safety Academy, Houston, Texas.

BIBLIOGRAPHY

DeReamer, R., *Modern Safety and Health Technology*, Wiley, New York, 1980.
Hammer, W., *Occupational Safety Management and Engineering*, Prentice-Hall, Englewood Cliffs, N.J. 1976.
Leslie, J. H. and Adams, S. K., "Programmed Safety Through Programmed Learning," *Human Factors J.*, **15** (3) 223–236 (1973).
Wilson, C., "Supervisor's Book of Safety Meetings," National Foremen's Institute, Waterford, Conn., 1982 (monthly).

34

Developing a Community-Preparedness Capability for Sudden Emergencies Involving Hazardous Materials

Kathleen J. Tierney

This discussion of community preparedness for hazardous materials emergencies offers a new prespective on the problem in two ways. First, it applies the concepts and analytical strategies of social science to a problem which is usually thought of as primarily technical nature—dealing on the community level with the threat of life, property, and community life posed by hazardous chemicals. Second, it applies knowledge about community planning for natural disasters to the problems of preparation for sudden chemical emergencies.

Some readers may ask what contribution a social scientific perspective can make to preparedness for sudden chemical emergencies. They may reason that advice about how to respond to hazardous materials threats should

This chapter is condensed from a longer monograph by the author entitled *A Primer for Preparedness for Acute Chemical Emergencies,* Disaster Research Center Book and Monograph Series #14, Disaster Research Center, Ohio State University, Columbus, Ohio. That work was supported by grant #PFR-771445, from the Division of Problem-Focused Research, National Science Foundation, to the Disaster Research Center, Ohio State University, E. L. Quarantelli, Principal Investigator. Any findings, conclusions and recommendations expressed are those of the author and the Disaster Research Center and do not necessarily reflect those of the National Science Foundation.

759

come from experts such as chemical engineers, chemical corporation health and safety personnel, toxicologists, and fire service personnel. After all, is not the important information on dealing with chemical hazards primarily technical in nature? The answer is yes—*and* no. The production and safe handling of chemicals are areas of great technical complexity, requiring training, skill, and expertise. Yet, the problems encountered in organizing a response to chemical threats are also social in nature. A person would not consider handling hazardous chemicals without the appropriate background, equipment, and knowledge about how that chemical might react; similarly, one should try to base planning and response for chemical emergencies on knowledge of how individuals and organizations react under pressure. In short, technical and social problems are intermingled in the disaster setting.

This chapter is based on the premise that many problems in disaster planning and response are ''people'' problems; the consequences of unresolved people problems can be just as serious as those resulting from unsolved technical problems. For example, a special resource such as foam may not be mobilized in a chemical fire because it does not exist in or near the stricken community. On the other hand, it may be there, but may not be used simply because community emergency personnel are unaware of its existence or because it is not known who has the responsibility for *authorizing* its use. The consequences are the same in either case, but the second case is a ''people'' problem, the kind of problem social scientific research focuses on.

34.1. THE LOCAL COMMUNITY AS A SETTING FOR HAZARDOUS MATERIALS EMERGENCY PREPAREDNESS

Local emergency organizations should be concerned with chemical emergency planning for several reasons. First, the initial consequences of a hazardous chemical episode are invariably borne first by *some local community,* regardless of how much outside aid is eventually sent to deal with the emergency. The fact that groups from other places may make their services available to a local community in certain situations does not relieve emergency planners and responders of their ongoing responsibilities for community safety. Moreover, although certain tasks relating to chemical agents themselves—for example, suppression, neutralization, and disposal—can perhaps best be handled by trained specialists, other tasks—evacuation, for example—almost always will have to be planned and carried out by knowledgable local emergency personnel.

Second, local planning is important because the initial response in the first few minutes of a chemical incident can be critical to the way the incident later proceeds. Many chemical agents are relatively unstable: substances treated improperly with water can burst into flame, producing a fire hazard; they can give off toxic or lethal fumes. Two hazardous agents released at the same time can combine to create a third unless the proper steps are taken. Some agents present very different hazards to human beings, on the one hand, and the natural environment, on the other.

Local personnel who have not planned and received training on the proper response to chemical agents—and who fail to respond appropriately in urgent situations—could unknowingly increase the threat to life and property such substances pose. Therefore, some type of chemical hazard planning and training for local personnel is essential.

Finally, the local community is the logical and appropriate setting for carrying out chemical disaster preparedness activities because the local community is precisely the place where planning can make a difference. In addition to facilitating a good response, preparedness measures for hazardous materials emergencies can actually reduce the likelihood that a chemical incident will occur. A systematic assessment of traffic patterns and of the volumes and types of hazardous materials that are transported through a community, for example, can lead to the designation of special hazardous materials routes and to a subsequent reduction of accident potential.

Many of the most effective safety measures—including activities such as risk assessment, training, and public education—are most appropriately carried out in the local community. Local emergency personnel are in the best position to know about the hazards present in their own community. They have access to detailed and specific information on threats facing the community and on community emergency resources and are thus in a position to reduce both the probability and the potential severity of incidents involving hazardous chemicals.

Should chemical emergency planning be coordinated with planning for natural disasters and other emergencies or should it be separate? Probably the most cost-effective way for local communities to plan for hazardous materials incidents is to integrate tasks relevant to chemical hazards within the activities of established emergency-response organizations. While the threats posed by disaster agents frequently necessitate specialized emergency resources, this does not mean the community should adopt the unnecessarily duplicative strategy of setting up new organizational structures and plans for each type of disaster agent. The assumption will be made in this discussion that traditional community emergency organizations—the police and fire departments, the civil defense office, and hospitals, for example—will be working along with chemical producers and transporters in disaster preparedness and response.

34.2. WHAT IS DISASTER PREPAREDNESS?

In this discussion, the term disaster preparedness refers to *all documents, activities, formal and informal agreements, and social arrangements which, over the long or short term are intended to reduce the probability of disaster and/or the severity of the community disruption occasioned by its occurrence.*

Some examples of preparedness activities are:

Convening meetings for the purpose of sharing knowledge on disaster planning.

Holding disaster drills, rehearsals, and simulations.

Developing techniques for training, information transfer, and hazard assessment.

Formulating memoranda of understanding and mutual aid agreements.

Public education.

Drawing up community and organizational disaster plans.

It should be noted that, while formal disaster plans are an important element in disaster preparedness, they are only one of several sets of activities devised to improve the efficiency and effectiveness of a community disaster response.

Common Misconceptions

In addition to defining disaster preparedness, it may also be worthwhile to state what it is *not*. Among the more common misconceptions about preparedness are the following:

1. *Preparedness is the same as the development of formal disaster plans.* People tend to believe that, once a written document is produced outlining resources, lines of responsibility, and disaster-related tasks for an organization or community, the planning task has been accomplished. This may be one of the reasons behind the tendency for local community officials to seek and use "model" plans; it seems quicker, easier, and more economical to devise a local play by copying or adapting one from another community than to "start from scratch" in the local community. In reality, however, these kinds of short cuts do not produce the desired results. Despite the fact that a formal plan is an essential element in the planning process, community preparedness cannot be achieved merely by drafting plans.

To be useful, a local disaster plan must rest upon a strong foundation, consisting both of accurate facts and the proper social and political supports. Good preparedness begins with the recognition of the need for hazard assessment, resource assessment, the cultivation of a hospitable social climate (e.g., supportive laws and community attitudes), and social networks that are conducive to getting things done. When these elements are present, they enhance the probability that official documents such as disaster plans will be used and used well. Disaster plans are important, but they stand in the same relation to community preparedness as a blueprint does to a building. Much more is actually involved in constructing a building than is shown on a blueprint.

To be useful, plans must actually be *used* in disaster. The probability that plans will actually be used is higher if they are factually accurate, relevant, widely understood, and perceived as legitimate by emergency organizations in the community. There is no substitute for the experience that is to be gained in going through all the steps involved in disaster planning—the meetings, discussions, debates, rehearsals, training sessions and related preparedness activities.

2. *Once developed, plans never go out of date.* It would, of course, be very convenient if a state of preparedness could be achieved once and for all, but the nature of community settings, as well as the nature of disaster agents, dictate that preparedness be an ongoing process. Preparedness is affected by changes in threats as well as by community and organizational changes in resources or capability; examples of the latter include budget or equipment cuts or the provision of new hardware or facilities. Even if material resources remain more or less constant, new people are continually entering emergency organizations, and they must be brought into the planning process. On the "demand" side, any number of factors can affect the nature and degree of threat: population shifts can alter the number of people at risk from different disaster agents; changes in land use can occur; the addition of new manufacturing concerns into the community can entail new risks to residents. Because the local scene is not static, preparedness can never be accomplished once and for all.

3. *Disaster response is like everyday emergency and safety operations, only more so.* The notion is frequently expressed by chemical industry personnel that preparedness for serious, acute toxic releases, chemical explosions, and other mishaps is but an extension of everyday corporate health and safety measures. They are not alone in this kind of thinking; it is also common for public safety personnel to believe disaster operations differ from daily activities only *quantitatively*.

This type of thinking is far more common among communities and organizations that have never experienced a serious disaster than among those that have. Familiarity with the functioning of communities in actual disasters or mass casualty situations leads to an awareness of crucial *qualitative* differences between these situations and the ongoing, everyday activities of community emergency organizations. For example, because large-scale emergencies place increased demands on many organizations and because community resources may at the same time be depleted, community organizations must depend upon one another to a much greater degree in disaster situations than during normal times. In this situation of increased interdependence, everyday boundaries (e.g., among political jurisdictions or between organizations) may not be maintained. There may be more sharing of personnel, tasks, and equipment than would be possible during normal times. Thus, community organizations must understand one another's functions and capabilities and must be prepared to work together smoothly because this is what will *have* to happen in a disaster.

Similarly, performance standards for some organizations may change drastically in a disaster. In the chemical hazards area, there are differences in standards of action between everyday and emergency operations for many responding groups. Swift response is an absolute necessity for the fire service, when responding to structural fires on an everyday basis. Dealing with unidentified chemical substances or materials whose properties are not thoroughly understood may require a very *different* response on the part of the fire fighter; this situation may require *delaying* the response until more information is received. People who do not recognize that some emergencies call for different types of performance are liable to make poor decisions in crisis situations.

In a disaster situation, organizations may be faced with a whole new set of challenges: taking on new personnel, tasks, and responsibilities; working within a different chain of command; being judged by standards different from those which are normally applied. For these reasons, it seems ill-advised for organizations to think of disaster-related demands as simply "more of the same" in comparison with everyday activities.

4. *Disaster preparedness must be costly.* Adequate disaster preparedness need not necessarily carry a high price tag. There are several reasons why attaining an adequate state of preparedness need not be expensive. First, many local and outside groups are probably *already* performing activities that could contribute to community preparedness. People in the local community need to recognize and take advantage of opportunities to upgrade preparedness by working with these groups. Hospitals, for example, must conduct disaster drills twice yearly. It would be very instructive, and not very expensive, for a number of other community organizations—fire departments, chemical companies, the local civil defense, and others—to participate in hospital drills, when they occur, on a community-wide or a regional basis. In another example, industry or community safety personnel who have received specialized training could, in turn, pass this training on to members of other organizations or to the general public, thus educating more people for just about the same amount of money.

Second, in any community, some resources exist which are overlooked. When vigorous efforts are made to identify and link local resources which are not salient, it may be discovered that it is not necessary to bring new ones into the community. Finally, before dimissing disaster preparedness as too costly, it might be a good idea for public and industry officials to ponder for a moment the potential costs of *not* planning. A chemical facility, for example, stands to lose a great deal if a fire or rupture is not contained or is handled badly. It is a sobering fact that a large proportion of those killed and injured in hazardous materials incidents are members of local emergency-response organizations. If better disaster planning could save a valuable chemical facility or help reduce response-related casualties, does this not justify the expenditure of time, money, and effort by local companies and agencies?

The idea that disaster preparedness almost pays for itself in the long run becomes even more plausible once two points are recognized: "an ounce of prevention" now may possibly avert a catastrophe in the future; and individuals and organizations which are better prepared to deal with major emergencies may also perform better during minor emergencies and everyday operations.

Good Disaster Preparedness

Preparedness is more than plans—it should be viewed as an ongoing process. It is based on the idea that disaster response is not just an extension of everyday operations, and it need not be costly. The following are also among the characteristics of good disaster preparedness:

1. *Preparedness seeks to insure appropriate actions by responders.* In the urgency and confusion of the emergency period, the pressure on responders to engage in action—*any* action—may be almost overwhelming. Sometimes, however, particularly in hazardous materials incidents, the best action to take may be *no* immediate action. Good planning reduces the understandable tendency to act impulsively in a crisis situation and emphasizes the payoffs which result when measures that are known to be correct and effective are undertaken judiciously.

2. *Good preparedness is based on the way people typically behave.* Those individuals who are responsible for community preparedness should avoid adopting measures which require people to drastically change their usual ways of doing things in the event of disaster. Rather than expecting people to change their behavior in order to conform to disaster plans, planning measures should be tailored to the behavior of people. Directions to emergency organizations should be expressed simply and in a straightforward manner. Elaborate systems of passwords and authorization should be bypassed in favor of simple badges and color-coded clothing, so as to make mutual identification simpler for responders. The natural tendency for members of the victim population to converge on a disaster site or to inundate the telephone system with requests for information about loved ones should be taken into consideration. An awareness of the fact that, in disaster, people are going to behave in ways that are natural to them, and not according to scripts devised with the ideal response in mind, can help planners avoid costly mistakes.

3. *Preparedness focuses on general principles.* It is impossible to plan in detail for every contingency that may arise in the course of a natural disaster or hazardous materials incident. Thus, there is much to be said for a very *general* plan, which clearly and explicitly outlines tasks, responsibilities, lines of authority, and places to go for resources, but which does *not* spell out in exhaustive detail *every* aspect of an anticipated response. Potential users of disaster plans are generally not willing to plow through a multivolume document comprised of several hundred pages. To make sure that

the plan will be read and used, it should be relatively short and simple, perhaps with accompanying appendixes which describe the disaster responsibilities of specific agencies in more detail and which relate the activities of individual organizations to the overall community response.

4. *Preparedness efforts must overcome resistance.* Planning for disasters is not always met with enthusiasm; indeed, it is almost always resisted—if not actively, then passively. Measures which seem necessary and self-evident to those charged with responding to disasters may seem frivolous to others whose participation is essential. Thus, those who are responsible for community emergency preparedness must also be ready to "sell" others on the idea that preparedness is necessary.

34.3.   PARTIES IN THE PLANNING PROCESS

Local Organizations

LOCAL INDUSTRY

Historically, chemical manufacturers have tended to focus primarily on in-plant safety and the health and welfare of plant employees in their emergency plans. In communities which have many chemical facilities in close proximity to one another, chemical companies have also formed either official or informal mutual aid systems for the primary purpose of assisting one another in case of a serious in-plant emergency. Some mutual aid organizations have a long history of continuous existence and have formulated written plans and highly developed communications networks. Chemical manufacturing or processing firms are a major potential source of expertise and resources which might be utilized in developing emergency preparedness. For this potential to be realized, however, the safety efforts of private chemical concerns must be linked with those of public sector safety organizations. This means that chemical companies must be willing to expand their traditional definition of emergency planning beyond their own gates, to encompass the larger community as well. It also means that local government officials, that is, the mayor, mayor's representative, the county commissioners, or key personnel in public safety organizations, must be willing to initiate contact and coordinate with representatives of the chemical industry on matters relating to the safety of the entire community.

The following are among the kinds of activities and working agreements that can be established between government and private industry in the chemical emergency planning area:

Meetings, drills, and joint training sessions between members of the local fire service and plant safety personnel.

Discussions about how public safety organizations such as the local police might provide aid to local industry in emergency.

The provision of tours of chemical facilities for members of local government to increase awareness of in-plant safety procedures.

Joint problem-solving or planning sessions between industry or mutual aid representatives and local police, fire or civil defense representatives, focusing on common emergency-related problems such as citizen warning, public information, or evacuation.

Formal arrangements signifying the intention to coordinate emergency related tasks. Examples of such arrangements between the public and private sector might include writing chemical company mutual aid plans and community disaster plans which are consistent or mutually reinforcing; making sure local industry has a representative and a designated role to play at the community emergency operations center (EOC); and enlisting the membership of representatives of local industry on the community disaster council, or conversely, including one or more representatives of local government at mutual aid meetings.

LOCAL GOVERNMENT AND PUBLIC SAFETY ORGANIZATIONS

The Local Executive.   In the event of a major catastrophe, the chief executive of the local jurisdiction is the official who by law is responsible for disaster operations in the community. This is one reason why the mayor or city manager, or a representative of the executive office, should be directly involved in preparations for community emergencies. Another reason is that the support of the office of the mayor or city manager lends legitimacy to disaster-preparedness activities. Thus, while the chief executive may have no personal disaster planning duties to perform, there are many things the mayor's office can do to facilitate planning. Examples include making disaster preparedness an explicit priority for the local community, and drafting executive orders to governmental units requesting their cooperation in disaster-preparedness efforts such as writing plans, participating in drills, and conducting risk-assessment surveys.

The Local Civil Defense Office.   The local civil defense office can and should play a major role in community-based preparedness for chemical emergencies, for several reasons. First, civil defense organizations are in a position to obtain Federal technical and financial assistance for disaster preparedness efforts which is beneficial. Second, such offices are likely to contain persons who have knowledge and experience with a number of different kinds of emergencies because civil defense offices typically have a major role in planning and response for natural disasters and other community emergencies. Third, involving civil defense personnel in chemical disaster preparedness may help reduce duplication, inconsistency, and planning gaps.

The Local Fire Department.   The fire department should play an important role in local emergency preparedness efforts, due to its specialized knowledge and expertise and its strong ongoing links with private sector manufacturers and transporters. Besides being the logical choice for on-site coordinator in any incident involving hazardous materials, the fire department can also play a major role in preparedness tasks such as hazard assessment, training, and mobilization of specialized equipment and resources for use in chemical emergencies.[5a,b]

Local government officials should see to it that the emergency preparedness efforts of the fire department are recognized and supported. At the same time, however, an effort should be made to make sure emergency planning for dangerous chemical incidents is not viewed *solely* as a fire department responsibility. Fire departments typically perform their specialized functions with great speed and competence in standard fire disaster situations, and they can certainly be expected to play a major role in the response in any chemical emergency. However, because other essential tasks (evacuation, assuming overall responsibility for the entire response) will probably be performed by other organizations such as the police or the local civil defense organization, it is important that *all* potential emergency responders meet together for planning purposes and maintain close contact during nondisaster times.

Local Law Enforcement.   In recent years, there has been a tendency for some law enforcement to plan on their own for certain types of emergencies, considered to be part of "police work"—civil disturbances and hostage episodes, for example. This approach tends to be accompanied by a comparative lack of interest and involvement in other types of community emergencies. It is important from a preparedness standpoint, that law enforcement agencies coordinate with other organizations in planning for the *entire range* of community crises.

Law enforcement organizations have a clear role in preparedness for chemical emergencies. For example, police representatives can meet with local chemical industry personnel to preplan routes and procedures for evacuation. Police department personnel are also among those who might be involved in assessing hazards associated with the transportation of hazardous materials by road, rail, and other methods.

PRIVATE RELIEF ORGANIZATIONS

Local units or chapters of private-sector, disaster-relevant agencies, such as the Red Cross and the Salvation Army, have, as part of their organizational mandates, a responsibility to respond at times of community emergencies. Often these agencies are among the few community groups which have relatively immediate access to large quantities of food, cots, blankets, first-aid materials, and other supplies necessary to support a large-scale evacuation. Because of past experience in mass emergencies, such groups are able to find emergency shelters and housing through their own channels. They can also be a source of volunteer personnel as well as certain kinds of facilities, for example, radio communication networks, which can be quickly mobilized when needed.

HOSPITALS AND EMERGENCY MEDICAL SERVICES ORGANIZATIONS

Hospitals, ambulance services and other emergency medical service (EMS) organizations frequently engage in disaster planning and in drills among themselves, but too often this is done in isolation from other community sectors. It is important to include EMS organizations in community-wide planning for chemical emergencies for two reasons. First, as the organizations which transport and provide emergency medical care to disaster victims, they have a key role to play. Second, individuals in these kinds of organizations can be a potential source of valuable information and assistance where the nature and effects of hazardous materials are concerned. Community planners may wish to identify members of the medical community who possess a high degree of expertise and experience with toxic chemicals and arrange for their participation in disaster preparedness and response efforts.

OTHER COMMUNITY ORGANIZATIONS

Any number of other organizations might conceivably provide needed services in the event of a natural or chemical disaster. Examples of such organizations and resources are the following:

Mass media (radio, television)	Warning, evacuation messages, public information
Schools and churches	Shelter, supplies, food, volunteers
Labor unions	Specialized skills, information
Pharmacies	Prescription replacements, medical supplies
Military organizations	Specialized equipment, supplies, food

LOCAL RESIDENTS

Public education is often a neglected aspect of disaster preparedness. Sometimes hazards are downplayed for public relations reasons; or officials, as a result of a well-intentioned desire to avoid alarm in the community, keep the public "in the dark" about certain hazards. This is counterproductive because people tend to react in more positive ways in emergencies if they have some idea of what to do at the outset. Community residents will comply more readily if they have knowledge of measures they can take in the event of an evacuation order. They should be informed in advance which routes to use, what items to take with them, how to obtain additional information or assistance, and who is likely to be in charge at such a time, for example. The general principle is that the more local citizens understand what to do in a disaster, the fewer unfamiliar, complicated instructions will have to be conveyed at the actual time of the emergency, and the greater the probability of citizen cooperation.

It is also a cardinal principle of disaster planning of any kind that the potential victim population should see whatever is done by local organizations is being done *for* them and not *to* them. Apart from being consistent with the values of a democratic society, citizens are much more likely to be cooperative if they know ahead of time what might be expected of them in a crisis rather than to be suddenly faced with demands to do something unusual and unexpected as a result of a strange and unfamiliar danger—which is usually how a chemical threat is perceived. As disaster preparedness plans are developed, area residents should be kept informed via stories in the press, talks to local groups, and contacts with key community leaders.

State Organizations

State-level offices are involved in natural and technological disasters in a variety of ways. Agencies such as the state office of emergency services or emergency management offer assistance with disaster preparedness and may furnish aid in local emergencies which are so serious they exhaust local capabilities. These offices assist local planners on a day-to-day basis, coordinate aid to local governments, and become operational at the local level following a formal request for aid through an official declaration of emergency.

There are also other state-level agencies that respond to emergencies involving chemical agents with or without an official emergency declaration because these types of hazards are part of their organizational responsibilities. A state EPA typically sends representatives to the scene of hazardous materials episodes. Offices such as the State Department of Natural Resources, the Fish and Wildlife Department, or the DOT may also have mandates to respond to or make a report of hazardous materials incidents. Additionally, state regulations often require local jurisdictions to report hazardous materials incidents to these and other state agencies.

In some states—Tennessee, Illinois, and Michigan, for example—special units within state emergency or public safety agencies are given authority to respond directly and to take control of hazardous materials incidents. In many others, the State Police or Highway Patrol has special authority for the management of emergency situations involving dangerous chemicals on state routes and highways.

It is important that local communities seeking to develop or extend their chemical disaster response capabilities establish planning contacts with state-level organizations for three reasons. First, these organizations may possess resources which could serve as back-up for local responders. Second, they may be capable of handling particular, exotic chemicals for which specialized equipment or personnel are needed and not available at the local level. Third, since in many cases state organizations may respond on their own to these kinds of emergencies without a formal request by local officials, a disaster response can be expected to proceed much more smoothly if all parties know what to expect from one another. *Adding outside resources means greater potential effectiveness in combating chemical emergencies, but it also means that coordination of the responding groups presents an increasing challenge.* State disaster response organizations have their own disaster plans, and efforts should be made to insure that state and local responders will be working within a common framework, rather than at cross-purposes, in the event of a major disaster.

Regional and Federal Entities

Like branches of state government, some regional and Federal organizations offer technical assistance and funds to planners and may provide resources in major catastrophes, some monitor chemical incidents, and some respond directly to particular types of chemical emergencies occurring within the federal government's jurisdiction.

In the first category are the government agencies such as the Federal Emergency Management Administration (FEMA), the new Federal agency which combines the functions of several, formerly separate disaster preparedness, relief and recovery bodies such as the Defense Civil Preparedness Agency (DCPA) and the Federal Disaster Assistance Administration (FDAA). FEMA is in the process of restructuring and reorganizing the whole Federal posture toward mass emergencies, and serious consideration is being given to its assuming a major role in large-scale chemical emergencies. Whatever the ultimate decisions, final structure, and disaster responsibilities assumed, it is almost certain FEMA will provide advice and resources that will be helpful to all those interested in upgrading their response capability in the chemical hazards area.

Most prominent among the second group of Federal organizations is the NTSB (National Transportation Safety Board). It often sends investigators to look into transportation emergencies involving hazardous materials. Local emergency groups concerned with transportation accidents should familiarize themselves with the board's work and its publications (see Chapter 31).

The passage of new Federal legislation in the early 1970s and the subsequent development of the National Oil and Hazardous Substances Pollution Contingency Plan (NCP) have resulted in the creation of Federal units or teams, organized on a regional basis, which respond directly to certain types of pollution emergencies and threats. Termed Regional Response Teams (RRTs), these units consist of both primary response agencies such as the U.S. Coast Guard and the Federal EPA and advisory agencies. Which Federal agency has overall responsibility in an incident depends upon the site on which it occurs—on waterways or inland. The NCP outlines the responsibilities of the RRTs to include both the development of predisaster plans and the provision of an official to coordinate the on-site Federal response to the oil or hazardous materials spill. The intention of the Federal planning effort is to develop an emergency response capability, both across agencies at the Federal level and among groups from the Federal, state, and local levels. To assure consistency and to avoid duplication of effort, emergency planners on the local level (and on the state level as well) should attempt to obtain as much information as possible on the status and the emergency plans of the group of Federal agencies comprising the RRT for their particular jurisdiction. (For more information on the development and responsibilities of RRTs, see reference 1, although the National Weather Service (NWS) does not respond directly at the site of a chemical emergency, it is a Federal agency which can play a role in emergency planning and response. The NWS is a source of information on wind direction and velocity, precipitation, and temperature changes which can influence the spread and activity of chemical emissions, and, consequently, affect the response to a hazardous materials incident.

Other Federal agencies which can become involved in planning for or responding to hazardous materials emergencies are the Departments of Defense and the Interior; the Department of Energy (re: radioactive material); the Department of Agriculture (re: pesticides), and the Department of Transportation (DOT). The DOT includes the Materials Transportation Bureau which has prime responsibility for the implementation of the Hazardous Materials Transportation Act of 1972. This involves two main tasks: (1) safety regulation of hazardous materials by all modes of transportation; and (2) regulation and enforcement of all safety aspects of pipeline transportation under DOT jurisdiction. Of particular relevance to emergency planners is the DOT handbook, *Emergency Action Guide for Selected Hazardous Materials,* which is issued annually and is prepared to help emergency service personnel during the first 30 minutes of an incident involving a spill of a volatile, toxic, gaseous and/or flammable material that is shipped in bulk.

Private Response Organizations and Information Clearinghouses

Private sector groups are organized in a variety of ways. One type is the *industry-sponsored team* which becomes active in incidents involving a specific type of chemical agent. The Chlorine Institute is an example of an agent-specific response group sponsored by various segments of the industry. Another example is the HCN system, made up of the producers of hydrogen cyanide. Also included in this category are response teams funded and directed by specific chemical manufacturers, for example, Union Carbide, to neutralize hazards posed by their own products.

Another type of private sector response organization responds to incidents involving particular modes of *transportation*. Examples of organizations in this category are the teams dispatched by railroads to handle incidents arising in the shipping of hazardous materials, and groups such as the Bureau of Explosives, operating under the direction of an industry-wide organization, the Association of American Railroads.

There are also private organizations around the country, not attached to the manufacture or transportation of chemicals, which specialize in the neutralization, removal, and clean-up of hazardous materials. These groups typically possess trained personnel and specialized equipment for dealing with a range of hazardous substances and are set up to travel to the site of a chemical emergency on short notice.

Finally, the Chemical Manufacturers Association (formerly the Manufacturing Chemists Association) maintains CHEMTREC, a 24-hour emergency hotline which provides information to persons needing technical assistance in handling a hazardous materials incident resulting from a transportation emergency. By calling the toll-free CHEMTREC number (800-424-9300), local responders can receive assistance with identifying an unknown chemical, advice on ways to handle the chemical initially, and information on resources available for neutralization of the substance. During the first 10 years of its operation, the Center received over 155,000 telephone inquiries of which 27,400 involved emergencies. Of these, CHEMTREC was able to provide information and/or assistance to more than 21,700 of the cases. (See Chapter 31.)

Training and Educational Groups

A number of the groups mentioned above are also involved in training and educating local emergency personnel to prepare for acute chemical emergencies. Some produce publications; for example, the Bureau of Explosives of the Association of American Railroads has a well-known book, *Emergency Handling of Hazardous Materials in Surface Transportation.* Other groups offer seminars and classes.

Often, several organizations collaborate in an educational effort. For example, an interindustry Task Force on Rail Transportation of Hazardous Materials was formed in 1978 with representatives from the Chemical Manufacturers Association, The Association of American Railroads, the Chlorine Institute, the Compressed Gas Associates, the National Liquified Petroleum Gas Association, the Fertilizer Institute, and rail car manufacturers. Among other things, this task force developed a training program for volunteer and other smaller emergency response forces that might be called to the scene of a transportation chemical emergency.

There are still other groups and associations interested in the training of personnel to respond to acute chemical emergencies. For example, NFPA has developed through a contract with DOT, a training course entitled, *Handling Hazardous Materials Transportation Emergencies*.

Outside and Local Resources: The Need for Balance

Depending upon the nature and variety of local hazards, as well as the extensiveness of local emergency resources, communities will have to rely to a greater or lesser degree on extracommunity sources of aid. Some localities may face danger from relatively few chemical threats and may be relatively self-sufficient with regard to local response capability. Others may face multiple or exotic threats or may possess little in the way of equipment or expertise and thus may have a greater need to rely on outside resources. All communities could conceivably require involvement by outside agencies at some time. Therefore, planners should make an effort to identify chemical and transportation industry sources of aid and specialized emergency response organizations which could assist with hazardous materials emergencies *before* these groups are urgently needed. Additionally, steps should be taken to insure that information about whom to contact and how is recorded in ways that make it accessible to those likely to be making decisions during the emergency period. Rather than residing "in someone's head," or even within a single organization, lists of possible outside resources should be included in the community disaster plan—perhaps in a special appendix —or at the community Emergency Operations Center (EOC) so that they are available when needed.

The stress placed here on outside sources of aid is not intended to minimize the importance of local first responder capability in chemical incidents. Local personnel are the first line of defense against catastrophe because the emergency measures which are taken during the first few moments of an acute chemical episode can often make or break the entire response. While it may be that the majority of communities will still need some outside resources to deal definitively with hazardous substances, it takes time to obtain this type of aid, and the local community is still the place where early action can make a tremendous difference in the management of these kinds of emergencies.

34.4.   DIRECTIVES FOR ACHIEVING COMMUNITY PREPAREDNESS

This section outlines the planning process itself, indicating steps communities can take to insure improved readiness for chemical emergencies. The directives are presented as if there were *no* existing community-wide disaster plan; however, most potential users of this information will probably use it to broaden or update existing plans.

Threat, Risk Reduction, and Vulnerability

Vulnerability, or the likelihood that the hazards a community faces will produce casualties or loss of property, is a result of both the threats which the environment poses and the efforts made to reduce those threats:

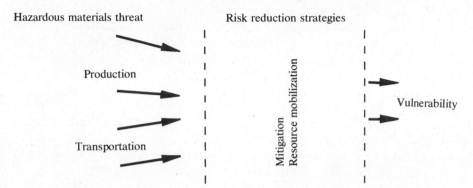

Hazardous materials threat · Risk reduction strategies · Production · Transportation · Mitigation · Resource mobilization · Vulnerability

When risk reduction or disaster preparedness strategies are employed, vulnerability is considerably lower than the objective threat posed by hazardous chemicals.

Preparedness can be viewed as a state in which response *capabilities* offset threats, risks, or potential *demands* on the response system. Local disaster planning efforts can be visualized as *those activities which either reduce the risk side of the equation, or increase the resource side.* Community preparedness for chemical incidents can be viewed as consisting of three major phases: risk assessment, resource assessment, and the reduction of risk through hazard mitigation and resource mobilization.

Risk Assessment

In this discussion, the term "risk" means the objective hazard to life and property posed by the manufacture or transportation of dangerous chemicals. *Risk* can be distinguished from *vulnerability,* at least conceptually, and it is important not to confuse these two terms. Vulnerability is the "net hazard" that remains after preparedness efforts have reduced risks. Because emergency preparedness efforts can offset environmental hazards, the vulnerability of a community—the probability of a chemical incident—may not be as high as the actual risk the community faces.

Risk analysis attempts to rate or scale the danger posed by some agent or agents to some specific unit. Risk analysis can focus on different time periods and a variety of units. Pre, trans, and postemergency analyses can be conducted for buildings, neighborhoods, or entire communities. The interest here is in one form of analysis: preemergency risk assessment, particularly the type that focuses on the community as the unit of analysis.

Few approaches attempt to take estimates of preemergency risk for large geographic or political units. Guides and outlines do exist in the area, however, Zajic and Himmelman[3] have developed a method for this type of assessment which includes indicators, measures, checklists and directives. Gabor and Griffith[4] advocate a slightly less elaborate and more community-oriented approach, which takes into consideration (1) the density of chemical manufacturing and storage in the jurisdictional unit (city, county, region, etc.); (2) the proximity of chemical facilities to residential and commercial areas; (3) the hazard associated with the transportation of hazardous materials through the community; and (4) the variety of hazards to which the geographic unit is exposed.

Local disaster preparedness personnel may wish to use one or both these approaches, they may adopt some other risk assessment formula, or they may decide to develop their own schema. The following are some suggestions on how to develop standards for measuring risk in the local community, using the four dimensions identified by Gabor and Griffith.

STATIONARY THREATS

Density—Chemical Industry Strength in the Community.   The relative proportion of chemical manufacturing and processing personnel in the labor force is one possible indicator of chemical industry strength in a city or county. *The proportion of county retail sales accounted for by the chemical industry is another.* Both kinds of information can be obtained from U.S. Census documents.

*Proximity—Nearness of Chemical Facilities to Populated Areas.   As*sessing hazards in terms of their proximity to populated areas is important because it gives an indication of problems that could arise with warning, evacuation, or care of the injured in a chemical emergency. It is possible for a community to be high in chemical manufacturing density and at the same time relatively low in proximity. The proportion of households in the community located within a 2000 ft radius of a facility handling large volumes of explosives, flammables, or highly volatile materials is one possible index of proximity.

Toxic clouds or plumes pose another hazard to community residents. Some measure of the degree of risk should be calculated when facilities which process, use, or store large quantities of chlorine are located near populated sections of the community. A relatively easy way of estimating the size of the population at risk from a toxic emission might be to calculate *the population density per square mile for the community* and to determine its rank, relative to others in the state or to similar-sized communities. Densely population communities containing facilities from which toxic clouds could be released would rank ''high'' on the proximity dimension.

TRANSPORTATION THREATS

When evaluating risks, it is important to consider the volume of chemicals *passing through* an area, in addition to those produced or stored there. A first step in assessing transportation hazards is to determine the number of transportation *modes* present. Is the community a rail center? A major port? Do large trucking concerns route loads through the area? Cities and towns which are centers for multiple transportation modes probably have an above-average potential for chemical mishaps.

The measurement of transportation hazards can be further refined. Sampling and the use of trained observers on highways or at railroad yards can help planners estimate the volume of hazardous materials traffic through the community. With some cooperation by shippers, it would also be possible to arrive at estimates of the amount of materials (e.g., chlorine, anhydrous ammonia) which pass through or near the community, posing a hazard to large numbers of people.

TYPES OF HAZARDS

Risk assessment should include not only the different *sources* of chemical threats (manufacturing and transportation) but also their *variety*. Different hazards entail different containment and neutralization strategies and, therefore, present a more complex challenge to planners and potential responders. Gabor and Griffith[4] suggest a simple five-point scale, in which communities are given one point for the presence of each of the following:

Chemicals which are hazardous due to their flammability.
Explosives.
Chemicals which can form toxic clouds.
Water pollutants.
Chemicals which produce acute corrosion.

The extent to which these threats are present in a community can be determined once something is learned of the nature of the hazardous materials produced in and shipped through the community. Reports available from the NTSB can also yield information about the kinds of chemicals involved in reported accidents in a given locality—another indicator of the variety of hazards to which an area is subject.

A thorough, careful risk assessment should provide planners with information on questions such as:

1. What potentially hazardous chemicals exist in the community?
2. Where are the highest concentrations of chemicals found?
3. What types of carriers handle the most hazardous materials?

4. What routes or locations present the greatest threat?

5. Are there specific localities where hazards are markedly higher than in other areas?

6. What are the types of threats posed by hazardous materials in the community (explosion, fire, toxic clouds, etc.)?

7. How many people (or what proportion of the population) are at risk from a failure to adequately contain dangerous chemicals?

8. Are there particular groups within the community (e.g., the elderly) which may require special attention in the event of a serious chemical incident?

Assessment of Resources

It is not the intent of this discussion to offer precise technical information on exactly which specialized resources are needed for containing and neutralizing particular chemicals. This kind of information is available elsewhere.[2,5-8] Similarly, this section will not attempt to discuss exactly what types of radio equipment is best for use in disasters or exactly how many pieces of fire apparatus are needed in different communities (see Froebe[9] for a list of equipment and reference literature which are recommended for starting a local hazardous materials response team). This discussion will, however, highlight crucial resource dimensions and suggest relatively economical ways in which an assessment of the resources can be made.

One way to gain a rough estimate of the quantity and distribution of community resources and of the extensiveness of existing disaster preparedness networks is to use a checklist. The checklist at the end of this chapter takes into consideration both human and material resources. It attempts to provide a picture of overall response capability by focusing on five areas:

1. The current community disaster plan.

2. The extent to which different organizations have developed emergency procedures specifically for chemical incidents.

3. Disaster drills and training exercises.

4. Disaster-relevant equipment, information, and facilities.

5. The number of types of disaster planning networks present in the community.

The checklist notes whether particular resources exist in the community and which organizations control those resources. This is important because certain resources may be present in the community, but not formally *linked* via plans and agreements to those community organizations that are charged with major responsibilities in chemical disasters.

The same methods can be used to assess resources as are used to assess risks; along certain lines, in fact, locating and counting resources may be somewhat easier than assessing risks since the latter almost invariably involves sampling, estimation, and individual judgment. Responsibility for completing the checklist could be given to an official or task force in a particular organization such as the civil defense office or the fire department. Information on resources can be obtained through an analysis of documents (disaster and emergency plans), combined with either a mail questionnaire or a telephone survey to responsible individuals in key organizations. Calling a meeting of emergency agency and industrial personnel for the purpose of assessing resources is another way of approaching the resource assessment question.

When data on the resources present in the community have been gathered, knowledgable people in the community will be in a position to begin making judgments about whether local response capabilities are appropriate, adequate, and properly distributed. They can then decide upon appropriate strategies for risk reduction.

Risk Reduction

This section will focus on two major forms of risk reduction: mitigation and resource mobilization. The former category includes measures enacted to *prevent* a hazardous materials incident; the latter term refers to measures taken to insure that injuries, damage, and disruption will be *minimized* in the event an incident does in fact occur.

MITIGATION

One very good way to reduce losses from hazardous materials emergencies is to prevent their occurrence. Several mechanisms exist at the community level for change threats. Discussed below are some mitigation strategies local officials may wish to pursue.

Land-Use Management. In recent times, industry has become more sensitive to safety and environmental-quality issues. Nowadays, potential hazards to residents are likely to be a factor in the decision to build a facility on a given site. Where feasible, officials at the city and county level should take the opportunity to work with corporate personnel to insure that new development is consistent with emergency preparedness needs: away from population concentrations, away from fire hazards, in an area where prevailing winds would not carry a toxic release directly over densely populated areas. Safety should also be a concern in highway planning. Bypasses and outerbelts make it possible for drivers carrying dangerous loads to avoid congested areas.

Enforcement of Existing Codes and Regulations. "Cracking down" on enforcing of Federal, state, and local laws and ordinances is another way local personnel can detect and reduce hazards. The range of interventions available to local agencies is quite broad and includes:

Enforcing highway speed limits.

Enforcing placarding regulations for hazardous materials carriers.

Citing carriers for use of inappropriate transportation routes.

Inspecting for violations of shipping regulations.

Notifying proper agencies (DOT, EPA) of illegal toxic releases.

Inspecting facilities for building code violations on a stepped-up basis and taking swift action with violations.

Some forms of mitigations are long range; others are not. Some are very expensive; others are not. No easy prescription exists for telling communities which mitigation approaches to use. This is something each community must determine based on a consideration of community needs, available resources, and political realities.

RESOURCE MOBILIZATION

A second dimension of risk reduction consists of efforts to enhance disaster response resources. This process occurs in three basic ways:

The new resources a community needs are obtained.

Existing resources are upgraded.

Resources are linked to one another to increase efficiency.

A resource checklist like Table 34.1 can be the point of departure for efforts to enhance response capability. The checklist indicates organizations, tasks, specialized resources and facilities which ought to be taken into account in planning for chemical emergencies. The focus in this is on those organizations on the local level which are highly likely to be involved in an emergency response to a serious chemical incident—the fire department, the police, the hospitals, and so on. Different items in the checklist aim at determining the quality and type of resources and the degree of expertise possessed by personnel in these local organizations; an attempt is also made to find out the extent to which these organizations are part of a comprehensive community disaster plan. The tasks listed encompass the range of demands the emergency response system will have to meet in a serious chemical emergency, beginning with warning and carrying through to the recovery period. Ideally, the local disaster plan should address these task areas and assign clear responsibility for each task to some organization or group. Like the tasks, the facilities and supplies listed—particularly the emergency operations center—are resources which every community should be capable of mobilizing.

In addressing more specialized resources, the checklist becomes more

TABLE 34.1. Community Checklist of Resources and Disaster Preparedness Activities

I. Community Disaster Plan

a. Is the community disaster plan_____ Written?_____ Unwritten?

If plan is written, which organizations are included in the plan?

_____ Fire Department
_____ Police Department
_____ Local Civil Defense or Emergency Preparedness Office
_____ Local Executive Office
_____ Local Industry
_____ Hospitals and Emergency Medical Service Sector
_____ Social Service Organization
_____ County Public Safety
_____ State Emergency Preparedness
_____ Outside Chemical Emergency Response Organizations
_____ Other (List) _____

b. What disaster-related tasks are addressed in the plan? What organizations are given responsibility for their performance?

Tasks	Organization responsible
1. Pre disaster overall community emergency planning	_____
2. Warning	_____
3. Stockpiling emergency supplies and equipment	_____
4. Search and rescue	_____
5. Evacuation	_____
6. Compiling lists of missing persons	_____
7. Care of the dead	_____
8. Maintenance of community order	_____
9. Housing victims	_____
10. Providing food and clothing to victims	_____
11. Establishing a pass system	_____
12. Overall coordination of disaster response	_____
13. Handling of radioactive material	_____
14. Identifying substances as toxic or chemically dangerous	_____
15. Handling or neutralizing toxic or chemically dangerous substances	_____

TABLE 34.1. (*Continued*)

II. Internal Emergency Procedures for Hazardous Materials Incidents

Which of the organizations listed below have developed written emergency plans for use in hazardous materials emergencies?

_____ Fire Department
_____ Police Department
_____ Local Civil Defense or Emergency Preparedness Office
_____ Local Executive Office
_____ Local Industry
_____ Hospitals and EMS Sector
_____ Social Service Organization
_____ County Public Safety

III. Drills and Training

Which organizations have participated in a disaster drill with some other organization during the past year? Which organizations have held their own internal rehearsals of disaster operations? Which organizations have at least one person who was given training in some aspect of hazardous materials response during the last year?

	Drills				Training	
	Joint		Individual			
	No	Yes[a]	No	Yes[a]	No	Yes
1. Fire Department	_____	_____	_____	_____	_____	_____
2. Police Department	_____	_____	_____	_____	_____	_____
3. Local Civil Defense	_____	_____	_____	_____	_____	_____
4. Local Executive	_____	_____	_____	_____	_____	_____
5. Local Industry	_____	_____	_____	_____	_____	_____
6. Hospitals and EMS Sector	_____	_____	_____	_____	_____	_____
7. Social Service Organizations	_____	_____	_____	_____	_____	_____
8. County Public Safety	_____	_____	_____	_____	_____	_____

TABLE 34.1. (*Continued*)

IV. Inventory of Material, Human and Informational Resources

a. What special materials or equipment are present in the community (use county as area for assessment) for containing or neutralizing chemical threats? What organization(s) control these materials? (Materials may include special foams, heavy equipment, etc.)

Materials	*Controlling Organization*
_____	_____
_____	_____
_____	_____
_____	_____
_____	_____
_____	_____

b. What special materials or equipment are present in the community for protecting primary responders (masks, acid suits, etc.)? What organization(s) control them?

Materials	*Controlling Organization*
_____	_____
_____	_____
_____	_____
_____	_____
_____	_____

c. What sources or expert advice and specialized information exist in the community? In what organization(s) can they be found? (Include materials such as handbooks, Chemcards, and also individuals if giving advice is part of their job responsibilities).

Informational Resources	*Organization*
_____	_____
_____	_____
_____	_____

d. Which of the following facilities and equipment exist in the community for use in a disaster? Which organizations possess or control these resources?

_____ Community-wide Emergency Operations Center (EOC) _____

_____ Mobile Radio Communication Equipment _____

_____ Alternative Sources for Power Evacuation Center Site and Supplies _____

_____ First Aid Equipment _____

TABLE 34.1. *(Continued)*

e. Of the organizations listed below, which ones: 1) have recorded the telephone number and notification process for exchanging information with CHEMTREC, the State Environmental Protection Agency, and the most relevant response teams; and, 2) have informed personnel of proper notification procedures?

	CHEMTREC	State EPA	Response Teams
Local Fire Department	_____	_____	_____
Local Police Department	_____	_____	_____
Local Civil Defense	_____	_____	_____
Local Executive or Other City Government Branch	_____	_____	_____
Local Industry	_____	_____	_____
Hospitals and EMS Sector	_____	_____	_____
County Public Safety	_____	_____	_____

V. Disaster Preparedness and Response Networks

Which of the following networks, groups, or systems exist in the community (county as unit of analysis)?

_____ Industrial Mutual Aid or Mutual Assistance Pacts
_____ Fire Department Mutual Aid
_____ Volunteer Organizations (REACT, Red Cross)
_____ Disaster Planning Councils (either governmental or governmental/private)
_____ Hospital/Ambulance/Emergency Medical Service System

*ª*If yes, for joint drills: 1) list, by number (1-8), other organizations participating; 2) indicate whether field exercise (F), or telephone or paper drill (T) was carried out. For individual drills, indicate F or T.

open-ended and less explicit about which resources should be present in the community. This is because the local need for specialized materials and equipment depends in large measure on the types and the severity of the chemical hazards that are present. The level of protection needed will vary, depending upon the level of the risk faced. Once the risks faced by the community have been documented and local chemical-specific resources have been assessed, the adequacy of the latter can be judged by health and safety professionals. If, on the basis of their knowledge, authorities in the field (fire service personnel, authors of technical manuals, chemical plant safety personnel) judge the specialized resources identified by the assessment as inadequate for dealing with existing threats, efforts to obtain materials providing an acceptable level of protection should begin. Outright purchase, leasing, contracting, or joint purchase with other communities are ways which might be employed to obtain equipment which is not present anywhere in the community.

Weak disaster planning is often not so much a matter of poor *resources* as a matter of poor *organization*. For this reason, besides having a purely quantitative aspect—recording specific items and amounts of equipment—the checklist is written so as to yield information on the *quality* of the local preparedness network—that is, on the degree of *resource integration* that is present. Noting which organizations possess vital information on chemical hazards and crucial material resources and *then* noting whether these organizations are part of a comprehensive plan gives an idea of how easily available such resources would be in the event of a community-wide emergency. Lack of clear preemergency guidelines for the delivery of resources could result in lost time when an emergency occurs; thus, to be most effective, resources must be identified *and* linked in an overall response system.

Generally, this need for linkage is best addressed by the development of a general community disaster plan. *The disaster plan is a document which links a range of community organizations for the purpose of accomplishing disaster-related tasks through the applicable of appropriate human and material resources.* This definition contains virtually all the elements which need to be addressed in upgrading community preparedness: organizations, linkages, tasks, and resources. An adequate plan contains references to linkages: authority for overall operations as well as for specific subareas of responsibility is clearly spelled out. (This is why disaster plans often contain charts and diagrams.) Lines of communication (e.g., from disaster site to the EOC or the command post) and notification (of potential victims, outside responders, and the like) have been specified. The necessity for performing the full range of disaster tasks—from warning to long-term recovery—has been addressed, and the specific organizations responsible for each task are aware of their roles. Clear, written understandings exist regarding the quantity, quality, types, and location of resources various organizations will contribute to the response, and the steps for utilizing them have been detailed.

What other properties characterize an adequate disaster plan? As indicated earlier, rather than requiring individuals to perform in unaccustomed ways, a good plan is based on *realistic* expectations. Similarly, a good plan is brief and *concise:* personnel in participating organizations are unlikely to adhere to disaster plans that are too voluminous to read even once. (Detailed directives and emergency procedures for individual organizations can be attached to the plan as appendices.) A good disaster plan is one which details a response that can be expanded by stages, calling up resources as needed and avoiding the potentially disruptive effects of overresponse and convergence at the site. Finally, a good plan is one which possesses an *official stamp of authority*. Government and private industry officials must endorse the plan and show a willingness to implement it.

Human resources are as important as equipment and facilities insofar as disaster preparedness is concerned; and here, too, it is frequently quality, rather than sheer quantity of personnel that is a key issue. A principle long known in the field of drama and often pointed out by students of human behavior is that people do better at carrying out their roles when they have had an opportunity to rehearse. Community disaster plans and organizational emergency procedures are only of value to the extent they are understood and complied with by emergency personnel. Since prior rehearsal improves both the understanding of emergency operations and the probability of compliance, risk reduction also involves continuous training and periodic drills of disaster operations. Moreover, to be most effective, training and drills must closely resemble actual disaster operations. From an operational standpoint, adequate performance in disaster involves learning not only how to anticipate and resolve problems with the tasks of one's own organization, but also how to anticipate and resolve problems that arise when working with other individuals, groups, and organizations. The very best disaster drills are those which are (1) *realistic*—that is, performed in the field, and lasting as long as they would in an actual situation, and (2) *interorganizational*. This is why the resource assessment checklist attempts to determine which organizations hold their own drills in contrast to which organizations conduct joint drills, and why it seeks information on the frequency of file exercises.

34.5 SUMMARY AND CONCLUSION

This chapter opened by presenting a definition of disaster preparedness and discussing the essential features of community preparedness. The public and private organizations that frequently play a role in planning for and responding to hazardous materials emergencies were then described. Next, strategies for reducing the occurrence and negative consequences of these kinds of incidents were outlined.

The probability of an emergency involving hazardous materials is a function of threat, on the other hand, and preparedness, or risk-reduction strategies, on the other. Once local emergency personnel know the types and magnitude of chemical threats, they can reduce these threats through whatever combination of mitigation and resource mobilization strategies seems most likely to succeed. Several steps in the disaster preparedness process were discussed; it might be useful to reiterate the major points here.

1. *Assess Risks.* Using agreed upon methods and the most thorough means of data collection feasible, obtain information on the density, proximity, transportation threat, and variety of hazards present. Methods need not be elaborate and mathematically sophisticated, but they should be capable of yielding information on the types of chemical threats the community faces, the size of the population at risk, and possible trouble spots.

2. *Assess Human and Material Resources.* Determine the extent to which formal disaster preparedness and response networks exist in the community and specific organizations have been assigned essential disaster tasks. Note whether general disaster-relevant resources

are present. Determine the extensiveness of knowledge of disaster operations (both general and specific to chemical incidents) in key emergency organizations. Note existence or absence of realistic interorganizational drills in the community. Use experts to gauge adequacy of specialized materials and equipment.

3. *Acquire, Link, and Upgrade Resources to Minimize Threat.* Obtain essential resources which are not present in the community; link needed resources by means of a comprehensive disaster plan; institute training procedures to close knowledge gaps.

REFERENCES

1. C. R. Corbett, "A dynamic regional response team," *Control of Hazardous Materials Spills,* U. S. Environmental Protection Agency, Rockville, 1978.
2. U.S. Department of Transportation, "Emergency Action Guide for Selected Hazardous Materials." National Highway Traffic Safety Administration and Materials Transportation Bureau, Washington, D.C., 1980.
3. J. E. Zajic and W. A. Himmelman, *Highly Hazardous Materials and Emergency Planning.* Dekker, New York, 1978.
4. T. Gabor and T. K. Griffith, "The assessment of community vulnerability to acute hazardous materials incidents." *J. Hazard. Mater.* **4**, 343–355 (1981).
5a. C. W. Bahme, *Fire Officer's Guide to Dangerous Chemicals,* National Fire Protection Association, Quincy, Mass. 02269. 1972.
5b. International Asso. of Fire Chiefs, *Disaster Planning Guidelines for Fire Chiefs,* Prepared under FEMA Contract DCPA 01-79-C-0303, Final Report, July 1980, Federal Emergency Management Agency, Washington, D.C. 20476.
6. International Association of Chiefs of Police, "Evaluating Hazardous Materials Emergencies," Training Key Vol. 9. International Association of Chiefs of Police, Gaithersburg, MD, 1977.
7. Missouri Pacific Railroad Company, "Safety First—Emergency Handling of Hazardous Materials in Railroad Cars, Missouri Pacific Railroad Company, St. Louis, 1977.
8. National Fire Protection Association, "Fire Protection Guide on Hazardous Materials" National Fire Protection Association, Quincy, Mass. 1979.
9. L. R. Froebe, "The organization of a local environmental response team." *Proceedings of the 1976 National Conference on Control of Hazardous Materials Spills,* New Orleans, La., April, 1976.

BIBLIOGRAPHY

Gann, D. S., Chairman, *Medical Control in Emergency Medical Services Systems,* Subcommittee on Medical Control, Committee on Emergency Medical Services, Division of Medical Sciences, Assembly of Life Sciences, National Research Council, National Academy Press, 1981. Maclean, A.D., "Chemical Incidents Handled by the United Kingdom Public Fire Service in 1980, *J. Hazardous Mater.* 5(1-2), 3–40 (1981). Special Issue on Social Aspects of Acute Chemical Emergencies, *J. Hazardous Mater.*, 4 (4), 309–394 (March 1981) seven papers.

35

Practical Application of Fault Tree Analysis

Richard W. Prugh

35.1. DEVELOPMENT

The fault tree analysis concept was originated by H. A. Watson of Bell Laboratories in 1961 to evaluate the safety of the Minuteman Launch Control System. A team of analysts at the Boeing Company modified the fault tree technique to facilitate quantitative analyses using a digital computer. During the next ten years, great strides were made in refining and applying the technique, particularly in the aerospace industry.[1-7]

One large chemical company found that prior to 1970 fault tree analysis was suited to batch systems but that much more development was required for general application to chemical process safety. Early in 1970, equations appropriate for continuous processing were devised and use of fault tree analysis progressed rapidly. Powers and Tompkins[8] presented a procedure for automatically generating fault trees and applied them to chemical processing. Powers and Lapp[9] described computer-aided fault tree synthesis.

Considerable technical effort is required to construct and evaluate fault trees; hence procedures were devised that would minimize the effort without reducing the benefits of intensive investigation. In this chapter are some guiding principles which can assist fault tree analysts in constructing trees for chemical processing.

35.2. EFFECT-TO-CAUSE PROCEDURE FOR CONTROL SYSTEMS

The first example (Fig. 35.1) shows the standard symbols: rectangles, for failure events, including the top event at the top of the tree or branch;

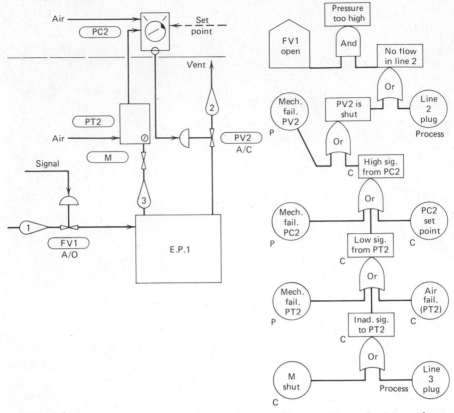

Figure 35.1. Effect-to-cause (control system) primary/command/(secondary) procedure.

circles, for basic causes; "house" symbols for high-probability normal events; and symbols for the AND and OR gates.

Construction of a fault tree is a deductive process, and the analyst works backward from effects toward causes. A good example of this method is found in the procedure for constructing branches for control systems. One starts at the component which acts on the process (usually a control valve) and works backward along the signal path toward the source (usually a sensor). Each component along the path has a primary (mechanical or internal) failure mode, and many components also can be commanded to fail by applying an undesirable control signal. For example, the pressure-control valve PV2 can be commanded to close by applying a high-pressure air signal to this air-to-close valve. Similarly, the controller can be commanded to produce an unsafe high output signal by reducing the signal from the pressure transmitter PT2. Secondary (external) causes are those which can cause simultaneous failures of several components, as the result of an earthquake, nearby explosion, etc.

Note that the branch below "PV2 is shut" consists only of OR gates, and this branch would be qualitatively and quantitatively the same with one OR gate and seven inputs. The reason for proceeding stepwise backward along the signal path is to assure that all of the important causes are included.

35.3. EFFECT-TO-CAUSE PROCEDURE FOR PROCESSES

Occasionally, the method involving backward progression, to develop branches, can be applied to processes. This is particularly true if the top event is at the end of the process. In the example shown (Fig. 35.2), the analyst is concerned with venting a toxic material to the atmosphere. The analyst starts backward into the process to develop causes for scrubber failure if the material enters the scrubber and also determines why the material would enter the scrubber. The path leads backward through valves, the

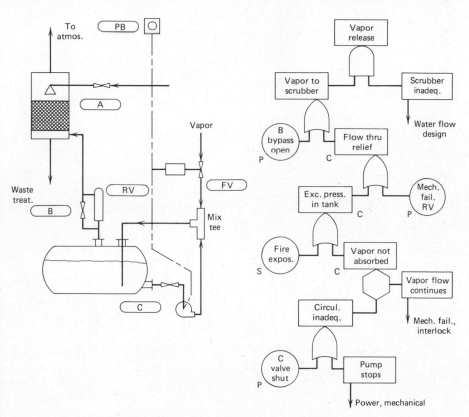

Figure 35.2. Effect-to-cause (process) primary/command/secondary procedure.

storage vessel, the mixing circuit, and into the control systems, to develop the required branches of the tree.

35.4. SOURCE-OF-HAZARD ANALYSIS

Vessel rupture is a frequent cause for concern for safety analysts, because personnel may be injured and property loss may be severe. The example shown (Fig. 35.3) illustrates many of the common causes of vessel rupture, including mechanical (primary) failure and rupture from excessive internal pressure. At each of the precursor subevents, branches would extend downward to show how the subevent could occur. For example, at "heating/cooling" for a distillation column, there would be a branch showing excessive boilup without compensating increased condensing capability OR failure of condensing capability without shutoff of boil-up.

It is likely that a relief device provided on the vessel of interest would perform differently for each cause of overpressure; for example, a relief

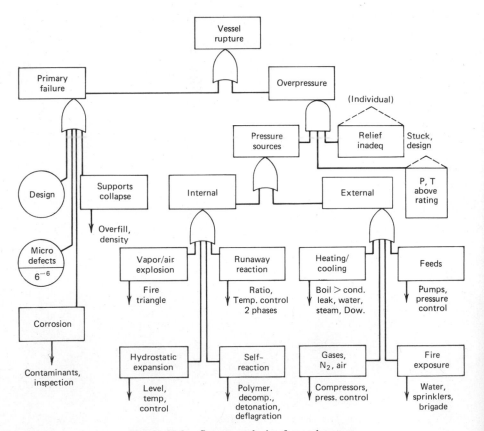

Figure 35.3. Source analysis of vessel rupture.

valve may safely vent only 1% of vapor–air explosions, but may safely vent 95% of runaway reactions, and 99% of excessive nitrogen-purge rates. Therefore, it may be necessary to construct a tree which shows the effectiveness of relief for each cause; thus, the tree would be considerably more complicated than the one shown here.

35.5.   FIRE TRIANGLE

The causes of hazardous fire or explosion (Fig. 35.4) can be described in a fault tree in a manner analogous to the fire triangle. Not only does fuel have to be present, but the fuel concentration must be within the flammable range. Whether or not the concentration is above the lower limit and below the upper limit will depend on temperature, pressure, and composition of the vapor source. Presence of an oxidizer may involve in-leakage of oxidizer and

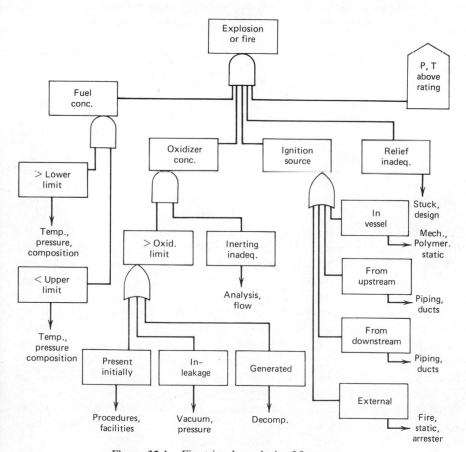

Figure 35.4.   Fire triangle analysis of fire cause.

failure of an inerting system. Several types of ignition sources may occur in a process, and the causes for each type should be determined and placed on the tree.

Of course, an explosion may not be hazardous if it can be safely contained or relieved. Thus, the probability of the vessel withstanding an explosion (or fire) and the causes of overpressure-relief (or fire-protection) failure should be developed.

35.6.   COMBINATIONS

In many well-designed processes, there are several safeguards to prevent hazardous loss of control (Fig. 35.5). These include temperature, pressure, and ratio controls, alarms, overrides or excursion limits, interlocks, emergency-cooling facilities, automatic vent valves, relief valves, and rupture disks. Failure of all of them most probably would cause a serious incident, while failure of only one would be inconsequential. However, failure of three or four in various combinations may also lead to an undesired incident. The tree structure shown covers a batch process where one combination of three failures, two combinations of four failures, and one combination of five failures would cause the undesired "excessive pressure." In this case, seven safety features were identified, and from the equation for combinations, there were 127 combinations which had to be evaluated to arrive at the four "critical" combinations shown in the Tree.

35.7.   ACTION POINTS

One approach to constructing a fault tree for a process is to determine the temperature or pressure levels at which automatic systems or the operator should act to control a process. In the example shown (Fig. 35.6), four important temperatures were identified: the normal operating temperature (185°C), the temperature alarm point (190°C), the temperature at which the interlock system should shut down the process (195°C), and the temperature at which a building-evacuation alarm sounds (205°C). To estimate the probability of explosion and injury, it would be necessary to evaluate the probability of the operator (or other persons) being present, the frequencies of control, interlock, and evacuation-alarm failures, and the probability of the unstable process material exploding at each temperature of interest.

To estimate the probability of explosion, it would be necessary to test the material and determine the temperature at which it would autodecompose and explode. As is probably typical, there is not one autodecomposition temperature but, instead, a range of temperatures. The data for tests on 25 batches

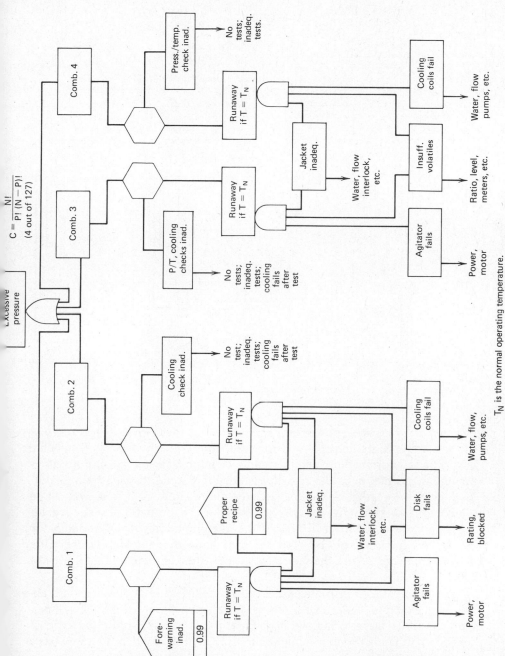

Figure 35.5. Combinations of causes of misadventure in batch operations.

T_N is the normal operating temperature.

$$C = \frac{N!}{P!\,(N-P)!}$$
(4 out of 127)

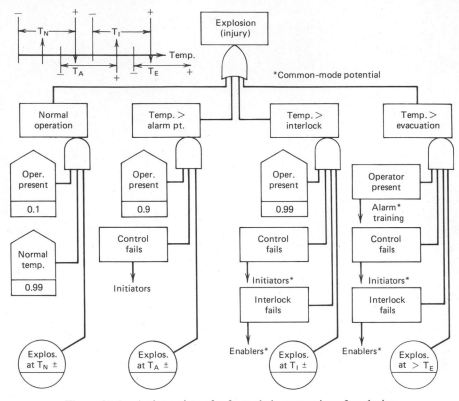

Figure 35.6. Action points of safeguards in prevention of explosion.

of the starting ingredient plotted on a probability graph (Fig. 35.7) indicate that there is a normal distribution of autodecomposition temperatures. If this apparently normal behavior can be assumed to extend to temperatures corresponding to normal operation, and the alarm, interlock, and evacuation temperatures, then it is possible to estimate the probabilities needed for evaluation of the fault tree. For example, the graph indicates that one batch in each two million could be expected to explode after the high-temperature alarm sounds but before the interlock system could actuate to shut down the process.

This example also introduces the problem caused by common-mode failure. For example, failure of a temperature sensor may cause the temperature control system to fail and also cause the interlock circuit and evacuation alarm to fail. Common-mode failures are identified when the same basic cause appears in two or more branches which meet at an AND gate. The tree must be restructured to show the common-mode failure properly, or the subsequent qualitative and quantitative evaluations will be seriously incorrect and in an unsafe direction. A simple procedure for common-mode fail-

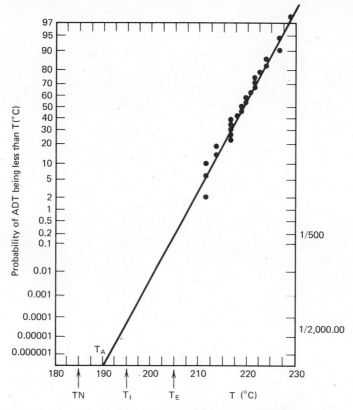

Figure 35.7. Normal distribution of autodecomposition temperatures.

ure resolution (Fig. 35.8) involves a stepwise replacement of events by inputs to their gates, following the rule: AND gates—inputs placed horizontally in the table; OR gates—inputs inserted vertically in the table.

35.8. REACTANT RATIOS

An incorrect ratio of reactants can lead to runaway reactions or other incidents in many processes. The tree structure that applies (Fig. 35.9) involves three situations which should be evaluated. Situation 1 is where the flow of one reactant (A) is too high and the flow of the other reactant (B) fails to increase proportionately, to maintain the required safe ratio. Situation 2 is the reverse of the first, where the flow of one reactant (B) is too low and the flow of the other (A) fails to decrease proportionately—or stop altogether—to maintain a safe ratio. The third situation is where the reactant flows deviate unsafely in opposite directions, at the same time; in most

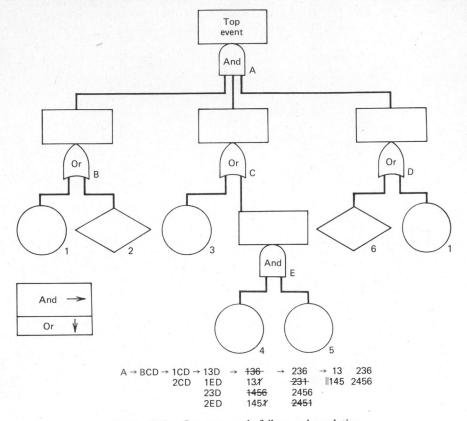

Figure 35.8.   Common-mode failure and resolution.

cases, the probability of this occurring is much less than either of the first two situations and does not need to be shown on the Tree.

35.9.   INITIATORS, ENABLERS, AND INTERVENTION

The concept of initiating failure events and enabling failure events can aid in constructing a fault tree (Fig. 35.10).

Initiating events cause unsafe changes in a process, enabling events allow these initiating events to become more serious that is allow upward progression of failure events in a fault tree. Typically, enabling failures (such as stuck relief valves) are not detected when the failure occurs, but only during deliberate tests or as the result of a "demand" by the process; thus, rather long failure durations (detection plus repair times) are associated with some enabling failures. On the other hand, the duration of an initiating event (such

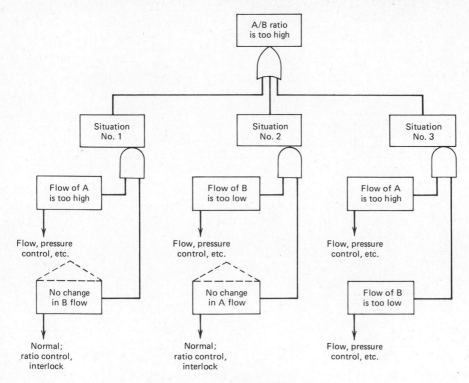

Figure 35.9. Reactant ratios analysis.

as failure of a steam pressure regulator, or a lightning stroke) usually is short enough to be inconsequential.

For most processes, the human operator is an important element of process control. Indicators, alarms, and emergency-shutdown devices are provided by the process designer to enable the operator to intervene in case the automatic controls fail, and then bring the process to a safe condition. The important aspects of human intervention (presence, capability, and operating aids) are shown on the tree for later quantitative evaluation.

35.10. "DAISY CHAIN"

Some hazard incidents occur in a stepwise manner (Fig. 35.11), such that there is a sequential relationship between the causes. This applies particularly to vehicle accidents and probably to many types of batch processes, also. As shown in a fault tree for a ship accident, the analyst determines how frequently a collision course occurs (while entering a harbor, for example); then estimates the probability that the ship would fail to change course to

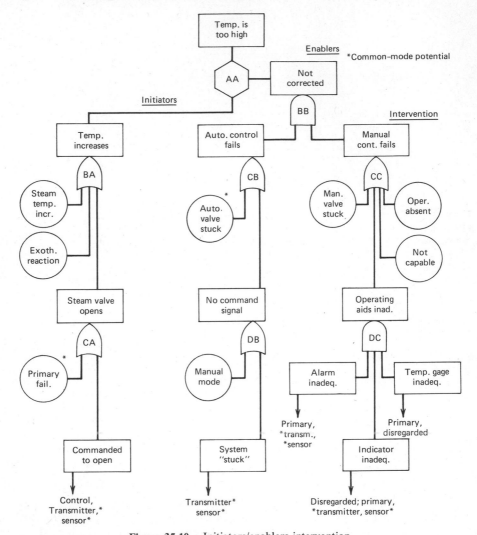

Figure 35.10.   Initiators/enablers intervention.

avoid collision. In this combination of failures, a collision would occur unless the ship which is about to be struck does not move out of the way. Other aspects of accidents are evaluated similarly to estimate the frequency of cargo release.

35.11.   COMBINING BATCH AND CONTINUOUS OPERATIONS

In many processes, the hazards involved in starting up or shutting down a process are more serious than those which could occur during the continu-

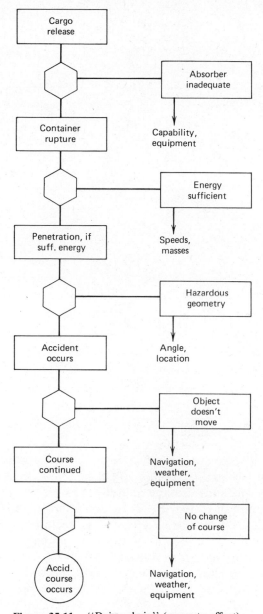

Figure 35.11. ''Daisy-chain'' (cause-to-effect).

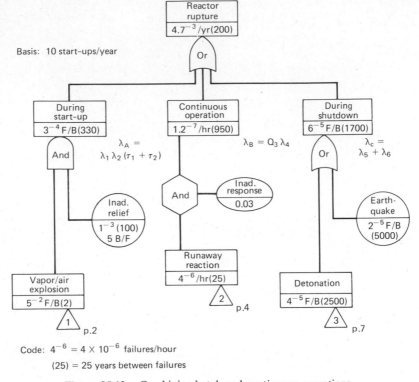

Basis: 10 start-ups/year

Code: $4^{-6} = 4 \times 10^{-6}$ failures/hour

(25) = 25 years between failures

Figure 35.12. Combining batch and continuous operations.

ous phase of operation. This situation might arise if it is necessary to bypass interlock systems during start-up or if it is necessary to pass through a hazardous ratio of reactants during some types of shutdowns. To combine these different phases of operation in a fault tree, it is necessary to combine batch operations (the start-ups and shutdowns) with the continuous operation. This is accomplished (Fig. 35.12) by determining the per-year frequencies of occurrence and then adding them. The reciprocal of the sum is the average interval of top event occurrences (or the probability of the top event occurring in terms of chances per year).

This example also shows the mathematical method for combining frequencies and durations at AND, INHIBIT, and OR gates. [The equations are presented in Table 35.1]

35.12. FUTURE STRUCTURE-DEVELOPMENT PROCEDURES

The fault tree structures presented here cannot be assumed to cover all aspects of all processes, but may serve as starting points or models for fault tree construction. As fault tree analysis continues to be applied to chemical

TABLE 35.1 Fault Tree Equations[a]

		Logic Gate Formulas		
		2 Inputs	3 Inputs	i Inputs
AND	λ	$\lambda_1\lambda_2(\tau_1 + \tau_2)$	$\lambda_1\lambda_2\lambda_3(\tau_2\tau_3 + \tau_1\tau_3 + \tau_1\tau_2)$	$\Sigma(1/\tau_i) \times \Pi Q_i$
	τ	$\dfrac{\tau_1\tau_2}{\tau_1 + \tau_2}$	$\dfrac{\tau_1\tau_2\tau_3}{\tau_2\tau_3 + \tau_1\tau_3 + \tau_1\tau_2}$	$\dfrac{1}{\Sigma(1/\tau_i)}$
OR	λ	$\lambda_1 + \lambda_2$	$\lambda_1 + \lambda_2 + \lambda_3$	$\Sigma\lambda_i$
	τ	$\dfrac{\lambda_1\tau_1 + \lambda_2\tau_2}{\lambda_1 + \lambda_2}$	$\dfrac{\lambda_1\tau_1 + \lambda_2\tau_2 + \lambda_3\tau_3}{\lambda_1 + \lambda_2 + \lambda_3}$	$\dfrac{\Sigma Q_i}{\Sigma\lambda_i}$
INHIBIT	λ	$\lambda_1 Q_2$	$\lambda_1 Q_2 Q_3$	—
	τ	τ_1	τ_1	—

[a]The above relationships apply only where the product of λ and τ (which equals Q) is substantially less than 1.0.

processes, it is likely that other useful models will be developed. For example, the primary–command–secondary concept may be applicable to the *chemistry* of a process: "primary" (internal) failures may involve instability of ingredients or reactions, "command" failures may involve mixing, catalysis, etc., and "secondary" (external) failures would involve fire exposure, loss of utilities, etc.

35.13.   QUANTITATIVE RESULTS

One of the important results of fault tree analysis is the calculated frequency of top event occurrence. For example, a fault tree may indicate that the process of interest would explode at average intervals of ten years. One of the advantages of fault tree analysis is that the process can be improved by the analyst—with subsequent restructuring of the tree and recalculation of the top event frequency—to any desired extent. Perhaps the frequency of process explosion could be reduced to one chance in 1000/year (an average interval of 1000 years) by adding an interlock system to the process controls. Several questions then arise: Would the original 10 years have been "acceptable"? If not, is the 1000-year interval obtained with the interlock system acceptable? At what points (frequency or interval) would proposed process improvements be considered "required" or "gold-plating"? The answers to these questions involve evaluating the exposure of operators, other personnel, and the public to process hazards (in relation to other day-to-day exposures to accidents and "Act of God" hazards) and the extent of property damage and business-interruption losses. Imperial Chemical Industries, Limited, and the United Kingdom Health and Safety Executive have recently taken important steps in establishing criteria for acceptability of process risks. S. B. Gibson presents some of the considerations and criteria in his paper on this subject.[10]

REFERENCES

1. D. B. Brown, "Systems Analysis and Design for Safety," *Fault Tree Analysis*, Prentice-Hall, Englewood Cliffs, 1976, p. 152.

2. J. B. Fussell, "Fault Tree Analysis—Concepts and Techniques," *"Generic Techniques in Systems Reliability Assessment,"* E. J. Henley and J. W. Lynn, Noordhoff International Publishing, 1976. Alphen aan den Rijn, Netherlands.

3. S. B. Gibson, "The Design of New Chemical Plants Using Hazard Analysis," *Process Industry Hazards, Institute of Chemical Engineers Symposium* Series No. 47, 1976, p. 135.

4. S. W. Malasky, "System Safety," *Fault Tree Analysis*, Spartan Books, 1974, p. 142.

5. N. Rasmussen, et al., "Reactor Safety Study," *Fault Tree Methodology*, USAEC, WASH-1400, Appendix II, Atomic Energy Commission, Washington, D.C., August 1974.

6. T. W. Yellman, "Comments on 'Fault Trees—A State of the Art Discussion'," *IEEE Trans. on Reliab.*, **R-24**(5), 344 (December 1975).

7. J. Young, "Using the Fault Tree Analysis Technique," *Reliability and Fault Tree Analysis,* R. E. Barlow et al., Society for Industrial & Applied Mathematics 1975, p. 827.

8. G. J. Powers and F. C. Tompkins, "Fault Tree Synthesis for Chemical Processes," *AIChE J.*, **20**(2), 376 (March 1974).

9. G. J. Powers and S. A. Lapp, "Computer-Aided Fault Tree Synthesis," *Chem. Eng. Prog.*, **72**(4) 89–92 (April 1976).

10. S. B. Gibson, "Hazard Analysis and Numerical Risk Criteria," AIChE Loss Prevention, **14**, p. 11 (1980).

36

Hazards in Chemical System Maintenance: Permits

Trevor A. Kletz

36.1. BASIC PRINCIPLES OF A PERMIT SYSTEM

Many of the raw materials, intermediates, and products handled by the chemical industry are hazardous. They may be flammable or explosive, toxic or corrosive; they may be handled hot or under pressure; they may possess all these properties. Therefore, before any equipment is opened up for inspection, repair, or modification, all hazardous materials should, if possible, be removed. If this is not possible then the workers carrying out the inspection, repair, or modification must be told of the hazards that remain and told what precautions they should take.

These objectives are usually achieved by issuing a permit-to-work or clearance certificate. It

1. Provides a checklist for the worker preparing the equipment and this reduces the chance that any part of the procedure will be missed.
2. Informs the workers carrying out the repair, modification, or inspection of the hazards that are present and the precautions that should be taken.

Many accidents have occurred because the permit system was not satisfactory or because the system, though satisfactory, was not followed. Errors in the preparation of equipment for maintenance are one of the commonest causes of serious accidents in the chemical and allied industries.

Isolation

Any piece of equipment on which work is to be done must be isolated from equipment containing hazardous materials. This is best done by the use of blinds (also known as spades or slip plates) or by physical disconnection. Valves are liable to leak and may be opened in error and therefore should not be used as the means of isolation unless the job to be done is so quick that fitting blinds would take longer than the main job and be as hazardous. The four types of isolation are sketched in Fig. 36.1.

The blinds should be made to the same engineering standard as the pipelines. Tables are available giving the thicknesses required for various line sizes and design pressures.

Certain proprietary blinds may allow leakage past them if not fully tight. These should not be used to isolate equipment for maintenance (except in circumstances in which valves would be accepted). Blinds used for isolation for maintenance should be of types which, if they leak, will leak into the surroundings.

Where valves are used as the means of isolation they should be locked shut using a padlock and chain or similar device. This rule should apply to valves which have to be isolated so that blinds can be inserted (or disconnections made).

When physical disconnection is used as a method of isolation, the lines leading to the rest of the plant should be blanked.

When a whole plant or section of a plant is shut-down and freed from hazardous materials it is not necessary for each item of equipment to be isolated individually. It is sufficient to make sure that the whole plant or section is isolated as described above. Care is needed to make sure that process materials have been removed from "dead-ends".

Separate permits should be issued for:

1. The initial isolation by fitting blinds or disconnecting.
2. The main job.
3. Removing the blinds or reconnecting the disconnected pipework.

Permit 2 should not be issued until job 1 is complete, permit 1 has been handed back and the person issuing the permits has checked that job 1 is complete and signed off. Permit 3 should not be issued until job 2 is similarly complete.

Additional precautions are necessary when equipment has to be prepared for hot work or entry. See section 36.2.

Special care is needed when isolating relief valve or vent lines. Whenever possible these should be disconnected rather than blinded. If the arrangement of lines is such that disconnection is impossible, then the relief valve or vent line should be blinded last and deblinded first.

When electricity is connected to a piece of equipment the supply is best

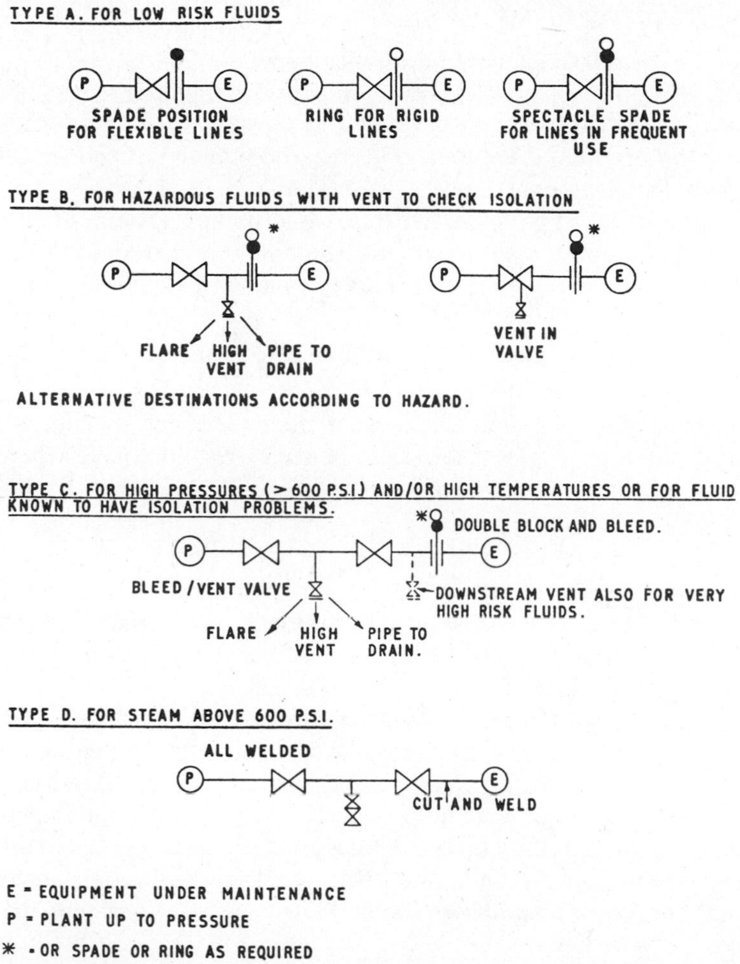

TYPE A. FOR LOW RISK FLUIDS

| SPADE POSITION FOR FLEXIBLE LINES | RING FOR RIGID LINES | SPECTACLE SPADE FOR LINES IN FREQUENT USE |

TYPE B. FOR HAZARDOUS FLUIDS WITH VENT TO CHECK ISOLATION

FLARE   HIGH   PIPE TO
        VENT   DRAIN

VENT IN
VALVE

ALTERNATIVE DESTINATIONS ACCORDING TO HAZARD.

TYPE C. FOR HIGH PRESSURES (> 600 P.S.I) AND/OR HIGH TEMPERATURES OR FOR FLUID KNOWN TO HAVE ISOLATION PROBLEMS.

DOUBLE BLOCK AND BLEED.

BLEED / VENT VALVE

DOWNSTREAM VENT ALSO FOR VERY HIGH RISK FLUIDS.

FLARE   HIGH   PIPE TO
        VENT   DRAIN.

TYPE D. FOR STEAM ABOVE 600 P.S.I.

ALL WELDED

CUT AND WELD

E = EQUIPMENT UNDER MAINTENANCE
P = PLANT UP TO PRESSURE
✳ = OR SPADE OR RING AS REQUIRED

Figure 36.1.   Four approaches to isolation of hazardous equipment before maintenance.

isolated by removal of fuses after first checking that the circuit is dead. An attempt should be made to start the equipment before work is started in order to check that the correct fuses have been withdrawn.

Some companies lock off isolators before working on equipment. However, experience has shown that this method of isolation can prove inadequate. On certain designs of starters, faults can occur that bypass the isolator, making equipment "live." Consequently this method of isolation is not recommended when the "work to be done" is on the electric circuitry.

The rules recommended in this section cannot always be followed. Exceptions may be necessary. They should be authorized in writing by a senior level of management, for example, the third level of professionally qualified management, and reviewed regularly.

Sweeping Out (purging) and Testing

Before equipment is opened up for repair, inspection, or modification it must be freed, so far as is possible, from hazardous materials, and tested to confirm that they are absent.

The methods adopted for removing hazardous materials depend on their nature and the degree of hazard. Flammable or toxic gases at pressure should be blown off, if possible, to a flare or scrubbing system, or discharged from a vent stack at a safe height and the equipment then swept out with inert gas or steam. Sweeping with inert gas can be carried out either by:

1. Raising the pressure in the equipment to that of the inert gas supply and then blowing off.
2. Passing inert gas in at one side of the equipment and out at the far side. Method 1 is best for equipment of irregular shape where dead-ends might otherwise remain unswept. Method 2 is best for long, thin equipment such as pipelines.

Hazardous liquids should be drained or blown out by water, steam or inert gas, preferably into a blow-down tank, another part of the plant, or into a closed drainage system. Water-soluble liquids are best removed with water. Air can be used to remove nonflammable liquids.

The need for maintenance should be foreseen during the design of the plant and suitable connections provided for adding inert gas, sweeping to flare, and so on. It sometimes happens that maintenance has to be carried out in circumstances that were not foreseen and equipment can then be freed of hazardous materials only by breaking a joint and allowing gas to discharge to atmosphere or liquid to drain onto the floor. In these circumstances steam jets, water sprays, or water flows should be used as appropriate to dilute and remove the gas or liquid.

Special precautions are necessary with materials such as benzene and other substances which have a low threshold limit value. All pumps and other equipment on which maintenance is foreseen should have a closed connection to the drainage system. Sweeping out normally takes place after valves have been closed and locked but before blinds are inserted (or disconnections made).

After the cleaning operation has been carried out it may be necessary to check that no hazardous materials are present. This is particularly important if a whole plant, or section of a plant, is being prepared for maintenance, and individual items of equipment are not being isolated individually, as discussed under Isolation. The tests will depend on the materials present. With flammable gases a test with a combustible gas detector is adequate if the atmosphere being tested contains some oxygen. With toxic gases a test for the specific gas may be necessary. With liquids, no testing may be necessary.

Special attention must be paid to testing when equipment is being prepared for hot work or entry. See Section 36.2.

If water or steam are used to sweep out equipment it may be necessary to make sure that no pockets of water remain, as sudden vaporization of such residual pockets while a plant is warming up when coming back on-line can cause considerable damage. Water should not be used to sweep out stainless steel equipment unless the water is chloride-free, since stress corrosion cracking may occur.

Identification

Many accidents have occurred because the wrong item of equipment was opened up. The worker doing a job must be left in no doubt which pipeline, pump, etc. is to be worked on. Experience shows that describing the equipment, pointing it out, or even showing it to the worker is not sufficient. By the time the correct tools have been collected, the worker may have forgotten and may open up the next pipeline or pump.

If equipment is provided with a permanent number, this number should be stated on the permit. If the equipment does not have a permanent number then a numbered tag should be fixed to the equipment and the number written on the permit or a duplicate tag given to the worker who will do the job. If a pipeline is to be broken, the tag should be fixed to the pipeline at the point at which it is to be broken.

Numbering systems for equipment should be logical and consistent. Accidents will occur if pump 5 is not located between pumps 4 and 6 (Fig. 36.2.) At each site equipment should be numbered from, for example, north to south and east to west.

If all the lines leading into a vessel are to be blinded or broken, then a tag

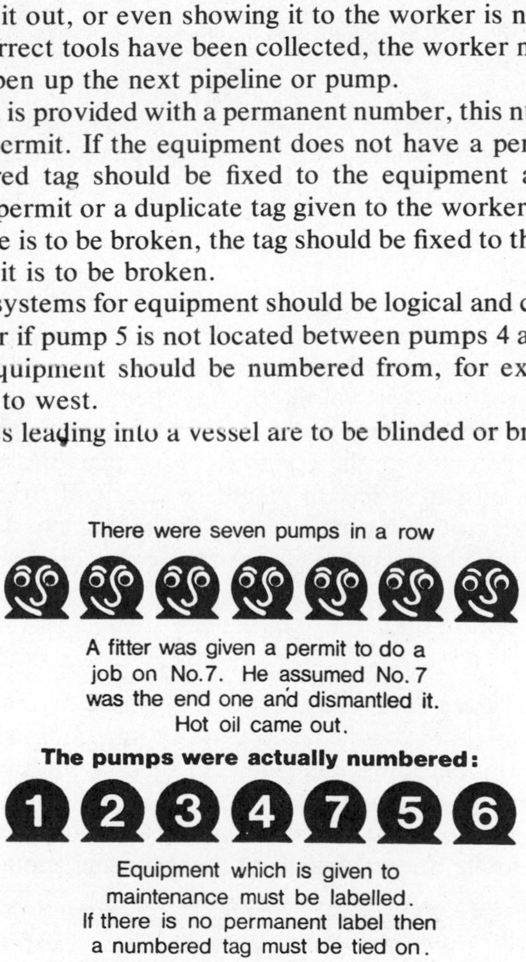

There were seven pumps in a row

A fitter was given a permit to do a
job on No.7. He assumed No. 7
was the end one and dismantled it.
Hot oil came out.

The pumps were actually numbered:

Equipment which is given to
maintenance must be labelled.
If there is no permanent label then
a numbered tag must be tied on.

Figure 36.2.   Proper numbering of equipment.

**They make the job exact and neat...
Please return when job complete.**

Figure 36.3.   Joint tagged for blind or disconnect.

should be fixed to each joint to be broken and the tag numbers written on the permit, Fig. 36.3. It is not sufficient to say "Blind all lines on tank no....."

Multiple Jobs

Sometimes several jobs have to be carried out at the same time on one piece of equipment. For example, while a pump is being repaired mechanically the electrical department may wish to work on the motor and the instrument department may wish to work on the bearing thermocouple.

Only one permit should be in force at a time for one item of equipment. Provision should be made on the permit for more than one trade or craft to accept it. In the case described it would be accepted by the mechanical department, the electrical department, and the instrument department.

When the various jobs are complete each trade should sign off the permit.

Changes of Intention

A common cause of accidents is a change of intention by the workers doing the job. Instead of working on the bearings of a pump, for example, they may decide to dismantle the pump. Unfortunately the pump has not been prepared for dismantling and contains a hazardous liquid.

It is therefore essential that if there is any significant change in the nature of the work to be done, the original permit is cancelled and a new one taken out.

The job for which the permit is issued should be stated as unambiguously as possible.

Issuing and Receiving Authorities

Permits are usually issued by a lead operator or other person in a supervisory position. The supervisor's knowledge and commitment are more important than the position in the hierarchy. On a small unit operated by only one worker he may be responsible for issuing permits. On each plant or part of a plant only one person at a time should be permitted to issue permits.

Whoever issues permits must understand the purpose and working of the permit system and have some knowledge of incidents that have occurred because the system was not followed.

Permits should be accepted by the worker who is going to do the job. When several workers will be doing a job the permit should be accepted by the one in charge. In some companies it is the practice for a supervisor to accept all permits. If this system is followed the permit should be shown to the worker or workers who will do the job. A convenient way of doing this is to display the permit on the job in a plastic bag.

Most plants in the chemical industry are operated on shifts, though repair work may be carried out in daytime. When a permit is issued the person issuing it commits colleagues on other shifts. Difficulties sometimes arise because different shifts place different interpretations on the words used or on the plant rules. When many permits have to be issued, for example, at a shutdown lasting several days, or if new construction is taking place alongside operating plant, then a day process supervisor should be appointed with special responsibility for issuing permits and for liaison with maintenance or construction.

All permits should cease to be valid after a stated period, usually one shift or one day but never more than one week (one shift when hot work or entry is involved), and must then be handed back or renewed. The period of validity should be stated on the permit. Permits should be issued only when the maintenance team is ready to start the job. It is bad practice to issue permits for jobs which will be done "If we have some spare time."

Testing and Disarming of Protective Systems

A permit should always be issued before any protective system such as a relief valve, alarm, trip, water sprinkler system, or other protective or firefighting system is tested or disarmed (i.e., made inoperative) for testing, modification, or repair. This should apply even if the test does not involve breaking into process lines.

Issuing a permit ensures that:

1. The operating team knows what is going on, and is fully informed.
2. The workers doing the job know precisely what they may do, any precautions they must take, any limitation on timing, and so on.

A permit should also be issued before any change is made to the setting of a protective system, for example, a change in the set point of an alarm or trip. Authorization for the change should be given only by higher management.

Communications

Many accidents have occurred despite issue of a permit because of failures in understanding and communication—words meant different things to different people. For example, when a worker says a job is complete, it means that the job that worker thought was to be done, is complete. This may not be the same as the job that the person issuing the permit wanted done.

Issue of a permit should always be preceded by a discussion between the issuer and the worker doing the job or the supervisor. It is not good practice to issue a permit and then leave it on the table for another worker to accept.

When a job is complete, the issuer should inspect the job carefully before accepting the permit back.

A Typical Permit

A good design of permit is illustrated. Some of the supervisors who use it were involved in its development and are involved in regular reviews of its application. See Fig. 36.4.

Part A is concerned with preparation.

Part A.1 is a checklist of hazards to be covered. Those not relevant are crossed out.

Part A.2 states whether or not the equipment is isolated and, if so, by what means (see Isolation).

If a fire permit (for hot work) or an entry certificate is required this is recorded in A.4 and A.5. The fire permit or entry certificate should be attached.

Part A.6 covers radioactive sources, A.7 electrical isolation and A.8 records that preparation is complete.

Note that Section A is a checklist as well as a record.

Part B lists the jobs to be done, the protective clothing required and any other precautions necessary.

Part C1 is completed by the supervisor preparing the equipment and Part C2 by the worker doing the job. Note that there is space for four independent groups of workers to work on the equipment (see Issuing and Receiving Authorities).

The rest of Part C is completed when the permit is handed back.

Training

We live in a society when instructions will not be followed merely because we are told to follow them. We need to be convinced that the instructions are reasonable and necessary. Therefore, while the recommendations made in this chapter should be incorporated in a plant instruction, this alone is insufficient. They must be backed up by an ongoing program of training.

When the plant instructions are changed the changes should be explained and discussed with the individuals who will have to operate the system. Those who issue permits and those who accept them should be trained together. This may bring to light ways in which the instructions can be improved or perhaps even show that they are impracticable.

The plant instructions should be explained to new workers, or to those taking up new responsibilities, in clear and simple terms. Tape-recorded descriptions of the permit form have been found useful. In particular, accidents which have occurred because the plant instructions were not followed should be described. Those described later may be used but accidents which have actually occurred on the plant are better. If possible, participants should be allowed to draw their own conclusions from the incidents and work out for themselves the need for the procedures described in the plant instructions.

Checking

Even though we have convinced those concerned of the need for the permit system, it is necessary to check from time to time that it is operating satisfactorily. Workers doing routine tasks are often tempted to become slack or take shortcuts.

When an accident occurs because the permit procedure has not been followed it is unlikely that it was the first time that the procedure was not followed. It is probable that the procedure has not been followed for a long time and that regular checking would have detected this.

All first-line managers should therefore check a number of permits every week. The check should include a visit to the job to make sure that blinds are in position, valves locked, protective clothing in use, as well as a check of the paperwork. More senior managers should check from time to time. Finally, there should be occasional checks by someone from outside, for example from company headquarters.

If checks are not made, then those who operate the system assume that managers do not regard the system as important and they operate it less conscientiously, if at all. *The first step down the road to a serious accident occurs when a manager turns a blind eye to a missing blind!*

Design

Plants must be designed so that the procedures which have been described, particularly in isolation, can be followed. Valve handles must be of a type that can be locked off. Where there is equipment which may have to be isolated for maintenance with the rest of the plant on line, the pipework

PLANT   SECTION OR EQUIPMENT:

PART A   PREPARATION

A.1   WHILE DOING THE JOBS DETAILED IN B.1 THE FOLLOWING HAZARDS MAY BE MET

| Strike out those for which there will be no hazard specific to the job when preparation is complete | Gas, fume
Corrosive hot and other liquids
Gas or liquid under pressure
Fire and explosion
Hot metal
Steam
Noise | Electricity
Moving machinery
Overhead cranes
Traffic (road and rail)
Radioactive substances
Asbestos
Other jobs nearby |

Add other
hazards

A.2   PHYSICAL ISOLATION (other than electrical)

Strike out   Physical isolation is NOT APPLICABLE
   TWO   The place, equipment, etc. is NOT ISOLATED
Statements   The place, equipment, etc. is ISOLATED

                    METHOD OF ISOLATION - delete as appropriate:

                    SINGLE/DOUBLE ISOLATION VALVE CLOSED AND LOCKED OFF
                    LINES SLIP PLATED
                    PHYSICAL DISCONNECTION - OPEN END BLANKED OFF
                    VENT, DRAIN OR BLOW OFF OPEN
                    Other Methods:

A.3   PRECAUTIONS ALREADY TAKEN

A.4   FIRE PERMIT

Strike out   A fire permit is NOT NECESSARY
   ONE   A fire permit is NECESSARY   Permit issued No.------------------
Statement   Further Permits can be noted on the back of the Clearance.

A.5   FACTORIES ACT 1961, Section 30, and CHEMICAL WORKS REGULATION 7

Strike out   DO NOT APPLY
   ONE   DO APPLY - I have seen the Signature of the responsible person
Statement   on Certificate No.---

                    Further Entry Certificates can be noted on the back of the
                    Clearance.

A.6   INSTALLED RADIOACTIVE SOURCES

Strike out   There is NO INSTALLED RADIOACTIVE SOURCE
   as   There is AN INSTALLED RADIOACTIVE SOURCE and the
Necessary   Safety Instruction for the installation
                    DOES NOT REQUIRE
                    DOES REQUIRE
                    the authority of the qualified person before work is done.

                    I have made the installation safe for the duration of the
                    Clearance Certificate by:

                    Signature of qualified person------------------------------

---

A.7   ELECTRICAL ISOLATION

Strike out   Electrical isolation is NOT APPLICABLE
   TWO   Electrical equipment is NOT ISOLATED
Statements   Electrical equipment is ISOLATED and I have tested the isolation.

                    Method of isolation:

High Voltage   --
Permit to   --
Work Card   --
No.   --
------------   Signature---

A.8   PREPARATION COMPLETE
                    Signature--

Figure 36.4. Wilton Works clearance certificate.

PART B   OPERATION

B.1   JOBS TO BE DONE | Tag Nos.

            Equipment Rotation Check: NECESSARY/NOT NECESSARY
            (To be carried out in accordance with Plant instructions)

B.2   PRECAUTIONS TO BE TAKEN AND WHY
   (Strike out those printed precautions not required, enter others in the space
   provided)

                    PROTECTIVE CLOTHING - EQUIPMENT

Wear P.V.C. Suit   Wear full face shield
Wear P.V.C. Gloves   Wear goggles
Wear ear protection (State type)   Beware Trapped Pressure - Wear Gloves
                                           and Goggles for first break in

----

         OTHER PRECAUTIONS THAT RELATE TO THE HAZARDS IN A.1
      Beware trapped pressure   Beware hot metal

817

PART C   TRANSFER AND ACCEPTANCE OF RESPONSIBILITY

THIS CERTIFICATE IS ONLY VALID WHEN ALL SECTIONS (A.1 to C.1
INCLUSIVE) HAVE BEEN COMPLETED

C.1   HANDED   Name (Printed)   Signed or Initialed
       OVER

       By------------------------   -------------------------------

       To------------------------   -------------------------------

C.2   ENDORSEMENT BY ENGINEERING SERVICES (WILTON) OR OTHER TRADES OR
       CONTRACTORS

                (1)   (2)   (3)

Signing On

Trade   ----------------------   ----------------------   ----------------------

Name   ----------------------   ----------------------   ----------------------
(Printed)

Signed or   ----------------------   ----------------------   ----------------------
initialed

Tel. No.   ----------------------   ----------------------   ----------------------

Date   ----------------------   ----------------------   ----------------------

       (Where daily endorsements required use the back of this Clearance)

Signing Off

Name   ----------------------   ----------------------   ----------------------
(Printed)

Signed or   ----------------------   ----------------------   ----------------------
initialed

Date   ----------------------   ----------------------   ----------------------

C.3   COMPLETION OF JOB

Strike out   THE JOB IS COMPLETE
    ONE   THE JOB IS INCOMPLETE and a further Clearance is required
Statement   for the following work:

HANDED   Signatures   By--
BACK   To--

C.4   INSTALLED RADIOACTIVE SOURCES (delete if not applicable)

       I have recommissioned the installation in accordance with the SI No. 808

       Signature of qualified person--

C.5   ELECTRICITY MADE AVAILABLE (Delete if not applicable)

       Method--

       Signature---

C.6   EQUIPMENT ROTATION CHECK (Delete if not applicable)

Strike out   DIRECTION IS INCORRECT
    ONE
Statement   DIRECTION IS CORRECT

                    Signed---

should have sufficient flexibility for blinds to be inserted, or, if this is not possible, slip-rings or spectacle plates should be provided.

The hazards of inserting a slip-plate should also be considered. For relatively safe materials at moderate temperatures and pressures it is sufficient to close a valve and break a flanged joint. For more hazardous materials more elaborate systems of isolation, as indicated in Fig. 36.1, are necessary. Note that when double block and bleed valves are shown, their purpose is to ensure that blinds can be inserted safely; they do not provide adequate isolation for long repair jobs (see Isolation).

Modifications

Many accidents have occurred because the results of plant modifications were not foreseen. In each plant there should be a system for controlling plant modifications. Before they are authorized they should be considered in a systematic way in order to make sure, so far as is possible, that their consequences are foreseen.[1]

In particular, care is needed that plants are not modified during repair unless a modification has been authorized. Maintenance teams are often tempted to put the plant back in what seems a better way. They should be encouraged to *suggest* modifications, but should put the plant back exactly as it was unless a modification has been authorized, Fig. 36.5.

It is a good idea to print on the permit: "Is this a modification? if so, has it been authorized?"

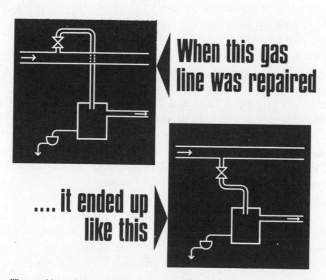

When this gas line was repaired

.... it ended up like this

When repairing a plant, put it back as it was unless a modification has been authorised

Figure 36.5. Unauthorized modification.

36.2. SPECIAL PERMITS

Special precautions are necessary for certain operations and these should be authorized on special forms or on a special part of the permit form.

Entry into Vessels and Other Confined Spaces

Many workers have been killed or injured while working inside vessels or other confined spaces. Special care is therefore necessary before entry is allowed.

Isolation should always be by blinds or physical disconnection, and the blinds should be inserted or the disconnections made close to the vessel. There should be no valves or other restrictions between the blind or disconnection and the vessel as process material may remain between the restriction and the blind. Valve isolations are not satisfactory even for quick jobs.

Any electrical equipment or machinery inside the vessel should be disconnected from the electrical supply. Fuse withdrawal is not sufficient.

Special care is needed in sweeping out and testing. Vessels which have contained oils should be steamed for several hours and then filled with water (provided the vessel and supports can take the weight).

If entry is permitted without breathing apparatus then the atmosphere at several points well inside the vessel, not just near the manhole (utility hole), should be tested for oxygen content, flammable gas, and materials which were previously inside the vessel. Entry without breathing apparatus should not be permitted if the oxygen content is below 20% or if the concentration of any toxic gas or vapor is above the threshold limit value. Continuous monitoring of the oxygen content should be considered, using, for example, pocket-size oxygen alarms.

Entry should not be permitted at all, even with breathing apparatus, if the oxygen content is above 22% or if the concentration of flammable gas or vapor is more than 5% of the lower flammable limit, as there may be pockets of higher concentration. (See Chap. 25.)

Entry permits should be authorized by a supervisor of greater seniority than those who authorize normal permits. Permits should be valid for not more than 8 hours and all tests should be repeated before renewal. The supervisor authorizing entry should *personally* inspect all connections before issuing the initial permit, and at each renewal, to make sure that they are blinded or disconnected and that any other special precautions necessary, such as forced ventilation, are in operation. The supervisor should also make sure that rescue facilities are available in case a worker becomes injured or ill while inside the vessel. It may be necessary to issue a permit for "entry for inspection" before a permit is issued for entry for work.

It is convenient to classify entry permits into one of three types.

Type A permits are issued when the atmosphere is fit to breath indefinitely. No special precautions are necessary other than those already described. However, another worker must remain present at the entrance.

Before issuing a type A permit remember that disturbing sludge or scale, burning, welding, or painting may change the atmosphere inside the vessel. Special care is necessary if oxygen or compressed gases are introduced into a vessel so that welding can take place.

Type B permits are issued when the atmosphere in the vessel contains some toxic or unpleasant gas or vapor, though not enough to produce an immediate danger to life. Breathing apparatus must be worn and a lifeline should normally be worn. A worker qualified in the use of breathing apparatus must be continuously on duty at the entrance to the vessel and additional breathing apparatus and resuscitation equipment must be available. The observer should also be able to summon assistance quickly and rescue plans must be practiced so that all concerned know what they have to do. See Chapters 25 and 27.

Type C permits are issued when the atmosphere in the vessel is irrespirable or contains so much toxic material that there is immediate danger to life. If possible, situations of this type should not be allowed to develop. On some plants, however, it may be necessary occasionally to enter a vessel which contains, for example, a catalyst which must not be exposed to air. For a type C permit, in addition to the precautions for a type B, two workers trained in rescue should be on duty at the entrance to the vessel, they should keep the one inside continuously in view, and should be in radio contact with the medical and rescue services. Again, rescue plans must be practiced.

Some other points which should be noted in connection with entry permits are:

1. It is not normally possible for one worker to lift another out of a vessel with a rope, hence lifting gear and/or additional manpower may have to be provided.

2. Manhole sizes (utility holes) should not be less than 18 in. and, if possible, not less than 24 in. internal diameter.

3. Particular care is necessary with vessels containing baffles or other obstructions. Unless the whole of a vessel can be seen to be clean it should be assumed that hazards are present, Fig. 36.6. Rescue may be complicated and a rescue plan should be agreed beforehand.

4. A group of vessels may be treated as a single vessel and blinded or disconnected as a group if the connecting lines are large, free from valves or other obstructions, and clean, for example, a distillation column, overhead condenser, and reflux drum may be isolated as a unit.

5. Excavations should be treated as confined spaces and entry permits issued if they are more than 1-m deep, unless the width at the widest point is more than twice the depth.

6. Entry permits should be issued for all new vessels under construction as soon as they become more than 1-m deep.

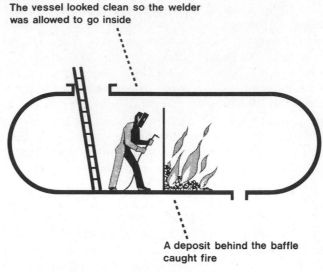

The vessel looked clean so the welder was allowed to go inside

A deposit behind the baffle caught fire

If you cannot see the whole of a vessel, assume it contains hazardous materials

Figure 36.6.   Hidden hazard.

7. Entry for rescue should never take place without breathing apparatus. Many workers have been killed while trying to rescue others.

8. Rust formation may reduce the oxygen content of an enclosed mild steel vessel, resulting in an oxygen-deficient atmosphere.

Hot Work

Where flammable materials are handled a special permit (usually known as a fire permit or hot work permit) is necessary before any operations involving sources of ignition are allowed. Such operations include welding, burning, use of nonclassified electrical equipment, introduction of vehicles and chipping of concrete.

Before a hot work permit is issued two distinct hazards have to be considered:

Hazards due to the presence of flammable gas in the atmosphere.

Hazards due to the presence of flammable materials in the equipment being worked on.

Permits for hot work should not be issued if abnormal plant conditions make a leak more likely than usual. The atmosphere should be tested with a combustible gas detector and no permit should be issued if more than 20% of the lower flammable limit is present. See Chapter 19.

Whenever welding or burning is carried out or cranes or other vehicles are operating in the plant area, portable combustible gas detector alarms should be placed nearby and the workers concerned warned to shut off their welding equipment or engines if an alarm sounds. It is good practice to place several alarms around the place of work. Care should be taken that workplace noise does not mask the alarm.

As a general rule, particularly for welding and burning, a stand-by worker should be in attendance with fire-extinguishers, hoses, or whatever fire equipment is considered appropriate. See Chapters 20 and 21.

On many plants gasoline or diesel engine vehicles are allowed on certain designated roads without special permission but a hot work permit is required before they can leave these designated roads. There is no justification for treating diesel engined vehicles as less dangerous than gasoline engined vehicles; diesel engines can ignite flammable gases unless they have been protected.

Special care is necessary before welding or burning are allowed near pools of water. Leaks of gasoline or similar liquids hundreds of meters away may spread on top of the water and be set alight. Care is also needed that flammable vapor cannot come out of the drains; they may have to be covered with blankets of wet sand. This source of vapor is often overlooked.

Before welding or burning is allowed on equipment it must be freed from flammable liquids, usually by steaming, or from flammable gases by steaming or by sweeping with inert gas, and then the inside tested to make sure that any gas present amounts to less than 20% of the lower flammable limit.

Note that a combustible gas detector will not detect flammable gas in a vessel filled with an inert gas. When no air is present, the sample will have to be mixed with air.

Whenever possible the equipment should be moved to a safe area before welding or burning is allowed. Welding should not normally be allowed on the *outside* of a vessel when workers are inside.

Special care is needed before welding is allowed on equipment which has contained heavy oils or materials which polymerize or are solid at ordinary temperature. The polymers and the heavy oil residues cannot be completely removed by steaming or any other cleaning methods and cannot be detected by combustible gas detectors. Often traces are hidden behind the lap-welded plates of storage tanks or behind rust. When welding is started the residues or polymers are vaporized and ignited. Many explosions have been caused in this way.[2]

Before welding is allowed on vessels which may have contained residues or polymers, the vessels should be filled with water or inert gas. Before welding is allowed on pipelines which may contain residues or polymers, there should be a number of open ends for fumes to escape and the welder should be provided with a clear and easy escape route. Fire-fighting equipment should be available.

Service lines should always be tested for the presence of flammable vapor before welding or burning is allowed because of the ease with which they can become contaminated with process material. See Chapter 19.

Excavations and the Like

A permit is required for an excavation, for grading or for driving stakes into the ground. Before issuing the permit the site plan must be checked by a competent person who will either:

1. Certify that no cables or pipelines are present.
2. If cables or pipelines are present within 1 m of the place of work, specify that excavation may take place only with hand tools, taking care not to disturb the cable or pipe. If possible, cables should be isolated.

The limits of an excavation may have to be marked on the ground.

Radiation

A permit should be obtained before portable radiation-producing equipment, such as radiographic equipment for welds, is introduced into a plant. The permit should specify the precautions necessary, for example, fencing of the surrounding area, and should not be signed until the fencing is complete.

If there are any sealed radioactive sources, such as radioactive level controllers, in a plant they may have to be removed by a competent person before a permit is issued for work on the item of equipment to which they are attached. Local instructions should define the work that can be done without removal of the sealed sources and space should be provided on the permit form for the competent person to certify that the installation has been made safe.

If there are any unsealed radioactive sources on a plant, such as radioactive tracers or naturally radioactive catalysts, then no permit to work should be issued until a competent person is satisfied with the precautions taken and has signed the permit accordingly.

The dust collected on air filtration systems may contain sufficient radioactive material to make it "legally radioactive" in some countries. Thus special precautions may have to be taken before filter elements are removed from such systems; see Chapter 37.

Interplant Pipelines

Special precautions are necessary if work has to be done on pipelines connecting two units which are under the control of different supervisors. One supervisor should complete a certificate stating the condition of the pipeline, for example, "The acid it contained has been blown out but some may remain, particularly in low-lying sections, and the line has been isolated by a blind." This certificate is *not* a permit and does not authorize any work. This certificate is then attached to the permit issued for work on the pipeline. When work is complete the certificate is returned.

Similar problems arise when work has to be done on a pipeline *passing through* another area of the factory. The maintenance workers are exposed to two distinct sets of hazards: those due to the surroundings and those due to the contents of the pipeline. The permits should be signed by two supervisors: one responsible for the pipeline and one responsible for the area in which the work will be done.

Equipment Sent Outside the Plant

Before equipment is sent outside a plant for repair, modification, or scrapping, all traces of hazardous materials should, if possible, be removed. The organization to which the equipment is sent is probably not familiar with hazardous materials and the precautions necessary.

When equipment cannot be made completely clean without damaging it beyond repair or excessive expense, then a certificate should be attached to the equipment listing the residual hazards and the precautions necessary, for example, "Traces of residual oils are present which may vaporize and explode on exposure to heat." If necessary someone should visit the receiving organization to explain the hazards or even supervise the repair or modification work.

Hand-Over of New Equipment by Contractors

A permit should be required to install or construct new equipment and the normal hot work procedure and the normal entry permit procedure should apply in all cases (see Entry into Vessels and Other Confined Spaces).

Before new equipment is connected to the rest of the plant it should be formally handed over in writing by the contractor. Thereafter the full permit procedure should apply for all engineering work.

Where construction takes place in an area which can be separated from the rest of a plant by a fence, this area can be designated a construction area, and, as such, some of the above requirements may be rescinded. However, contractors should be made aware of the limits set on the work they can carry out and of the action to be taken in the event of abnormality, such as a leak, on adjacent plant. In particular, they should realize that new equipment should not be connected up to existing plant without a special permit. Such connections are best made by plant personnel.

Opening-Up of Equipment by Operators

Normally equipment is prepared by one person who then issues a permit to another person who breaks into the equipment. Special problems arise when the same worker prepares and opens up equipment under pressure, for example, when a process operator opens up a filter, autoclave or batch reactor.

In these circumstances it is sooner or later inevitable, through oversight or neglect, that someone will open the equipment before the pressure has been released. It is therefore necessary to do either or both of the following:

1. Install a two-stage opening mechanism in which an initial movement releases the door a few millimeters. The mechanism should be capable of holding the full pressure and, if the pressure has not been blown-off, this is immediately apparent and the door can be closed again or the pressure allowed to blow off through the gap. A separate movement is then required to release the door fully.

2. Install an interlock to prevent the door being opened until a vent is open. This second option alone is not satisfactory for materials which may cause choking of the vent.

When lines are broken by operators to clear chokes or insert blinds the operator should first complete a written checklist, in effect issue a "self-permit."

The Law

In some countries special requirements are laid down by law, for example, concerning radioactive materials or entry to vessels. It is necessary to make sure that the permit procedure complies with these local requirements.

36.3. HAZARDS TO WHICH PERSONNEL ARE EXPOSED

Flammable and Explosive Materials

These constitute by far the biggest hazard in the chemical industry; most of the materials handled form flammable mixtures with air. It is therefore necessary to ensure that flammable gases or vapors are removed before maintenance work takes place. Even though everything possible is done to remove known sources of ingition, experience shows that flammable mixtures of gas and air may ignite, and workers should never be placed in such a position that they will be injured if ignition occurs. For example, as already stated, no-one should be allowed to enter a vessel if the amount of gas present exceeds 20% of the lower flammable limit. See Entry into Vessels and Other Confined Spaces and Chapter 25.

It is sometimes desirable to tighten up leaking joints and thus avoid a shutdown. Before this is permitted the size of the flammable gas cloud formed by the leak should be ascertained using a combustible gas detector. If the gas is blowing away from the leak a worker may be able to stand on the far side and tighten up the joint. If the leak is small, the worker may be permitted to place hands, protected by suitable gloves, inside the flammable zone. In no circumstances should the worker's body be immersed in flammable vapor. Leaks of gas can sometimes be moved to one side by water spray or dispersed by steam jets.

When removing flammable gas from process equipment by sweeping with nitrogen or other inert gas, sweeping should be continued until less than 2% flammable gas is present. Note that, as already pointed out, a combustible gas detector will not detect flammable gases in the absence of oxygen. A detector that mixes air with the sample is available.

After maintenance work is complete air must be removed from equipment with inert gas before flammable gas or liquid is introduced. The concentration of oxygen must be reduced to less than 5% (2% if hydrogen is to be introduced).

When equipment contains flammable liquids that are below their flash points no special precautions are necessary unless hot work has to be done (see Section 36.2). If equipment contains flammable liquids that are above their flash points then flammable vapor will be present and the precautions described above will be necessary.

Special precautions are necessary with materials such as peroxides, acetylene, or high-pressure ethylene, which are liable to decompose explosively without the presence of air or oxygen. Special measures may have to be taken to ensure that all traces are removed.

Inert Gases

Inert gases, mainly nitrogen, are widely used to prevent the formation of explosive mixtures, either in normal operation or when preparing equipment for maintenance. Unfortunately many workers have been killed by inert gas, as it gives little or no warning and its hazards are not always understood, The name "inert gas" suggests a harmless material, rather than one which can cause death by asphyxiation in a few minutes, without any warning or feeling of distress.

Those who work with inerted equipment should be aware that workers have been overcome when working close to a leaking joint or when standing over an open manhole (utility hole) on a vessel full of inert gas. (See Chap. 25).

Toxic Gases

When working on equipment which has contained toxic gases the aim should be to ensure, whenever possible, that the concentrations of toxic materials are below the threshold limit values. If this cannot be achieved by sweeping out and testing then breathing apparatus must be worn. However, it should be remembered that the TLV is the concentration that can be breathed by most workers without harm for a working lifetime. Occasional exposures to somewhat higher concentrations may be authorized by a competent person in accordance with the short term limits published by the American Conference of Government Industrial Hygienists. (See Chaps. 14, 15, and 25.)

Corrosive Materials

When working on equipment which has contained acids, alkalis, phenolic compounds, and other materials which are corrosive or are absorbed through the skin, goggles as well as gloves of impermeable and resistant material should always be worn. If there is liable to be more than a trace of the corrosive material left in the equipment, a suit of impermeable and resistant material known to be impervious to the substances at hand should be worn and possibly a transparent helmet covering the whole head. The specific protective clothing required should be specified on the permit. (See Chaps. 26 and 27.)

Trapped Energy

When equipment has been operating at pressure it is possible, as the result of chokes, that pressure is trapped in some part of the equipment. Joints should therefore be broken cautiously. The bolts away from the worker breaking the joint should be loosened first and any pressure allowed to blow off. Even though no pressure blows off, it is possible that a choke will clear later and that gas or liquid will be blown out.

Particular care is needed when dismantling equipment that has been out of service for some time. The nature of any chemicals present may not be known, acids may have reacted with the material of construction, causing chokes, or air may have been used up by rust formation forming an oxygen-deficient atmosphere.

Moving equipment, such as inclined conveyor bands, should be in a state of minimum potential energy.

A Note on Protective Clothing

Supervisors issuing permits are often tempted to err on the safe side by asking for too much protective clothing; it is not unusual to find every permit on a plant marked "Wear gloves and goggles."

Such tendencies should be resisted. Unnecessary protective clothing makes the job of the repair workers difficult. Furthermore, they soon realize that the protective clothing is unnecessary and cease to wear it. Accidents then occur when the protective clothing should really have been worn.

Ask only for the protective clothing that is really necessary and then insist that it is worn. Different clothing may be necessary for the initial blinding (or disconnection), and the removal of the blind (or reconnection) as compared with the main job.

General

The procedures described in this chapter are onerous but practicable. Experience shows that they can be operated without undue expense or interference with production. Experience also shows that without these or similar procedures accidents will occur. Some examples of such accidents are described in the next section.

In many plants there are traditional methods of working. Change should not be made for its own sake. Existing methods should be followed if they are satisfactòry. But they should be checked against the principles described above and modified if necessary. If this is not done the accidents described will take place as surely as the sun will rise tomorrow morning.

36.4. SOME ACCIDENTS WHICH WOULD NOT HAVE OCCURRED IF THE PRINCIPLES DESCRIBED ABOVE HAD BEEN FOLLOWED

The sections are in the same order as those in Sections 36.1. and 36.2.

Isolation

1.   A serious fire occurred on a distillation plant. Three workers were killed, one was seriously injured, and the plant was extensively damaged. Maintenance workers were working on a pump and decided to dismantle it. When they removed the cover, hot oil, above its autoignition temperature, came out and caught fire as the suction valve had not been shut.

2.   A fatal accident occurred when a process operator, in error, opened the suction valve on an ammonia pump, which was under maintenance and partially dismantled, instead of opening the suction valve on the spare pump. He was killed by escaping ammonia.

3.   An explosion in a reactor killed two workers. The reactor had been swept out for maintenance but, as no welding was to be done and entry was not required, it was decided not to blind the vessel but to rely on valve isolations. Some flammable vapor leaked in, was ignited by a cutter-grinder and exploded.

4.   A lost-time accident occurred while a fitter was working on a pipeline separated from a plant at pressure by a single isolation valve. He was injured by a sudden spurt of pressure. It is believed that someone may have opened the isolation valve.

5.   A fitter narrowly escaped injury when corrosive liquids came out of a pipeline on which he was working. Although the line had been blinded, a branch which was covered in lagging had been overlooked.

Identification

1.   A man was seriously injured by a corrosive chemical when it came out, under pressure, from a pump on which he was working. The pump had been isolated by blinds but these had been put in the wrong position so they did not effectively isolate it. The permit asked for the pump to be isolated but did not specify precisely which joints should be blinded.

2.   A fitter was asked to repair a leaking joint on a water line on a pipebridge. Staging was erected but because of the difficulty of access the process supervisor pointed out the joint to the maintenance supervisor from the ground. The maintenance supervisor, in turn, pointed it out to the fitter from the ground. The fitter broke a joint in a carbon monoxide line, was gassed, and was lucky to escape with his life. If the process supervisor had fixed a tag onto the joint concerned he would have realized that the joint concerned was not, in fact, on a water line.

3.   A fitter was asked to clear a choke in a caustic soda line leading to a storage tank. He was not shown or told which joint to break first and therefore decided to start on the tank side of the last isolation valve. When a movement was made into the tank corrosive liquid came out of the broken joint, fortunately without injuring anyone.

4.   A plumber broke into a line full of kerosene instead of an empty line which had been prepared for welding. The correct line had been pointed out to him.

5.   A plumber was stopped in time while he was hacksawing his way through a live line. The correct line had been marked with chalk and the chalk had been washed off by rain.

6.   A fitter was shown a steam valve by his supervisor from a pipebridge and told to remove the bonnet. He went down a flight of stairs to reach the valve, mistook the valve he had been shown and unbolted the bonnet from a compressed air valve. It flew off grazing his face.

7.   A fitter took a blank off the end of what he understood to be a nitrogen line. In fact it was a fuel gas line and fuel gas escaped under pressure just as welders were starting work nearby.

8.   A company hot tapped and drilled what was thought to be a nitrogen main but turned out afterwards to be an ethylene main. In this case the line was tagged but a process supervisor put the tag on the wrong line. He did not walk the line back to a point where it could be identified with certainty.

Changes of Intention

The first incident described under Isolation involved a change of intention. The original permit asked the fitters to attend to the pump bearings but did not authorize them to dismantle it. If they had asked for a fresh permit when they decided that it was necessary to dismantle it then it is possible that the process supervisor would have checked the isolation valves again and discovered that one was open.

Communication

A permit was issued to remove a pump for overhaul; the pump motor was defused, the pump removed, and the open ends blanked. Next morning the maintenance supervisor signed the permit to show that the job—removing the pump—was complete.

The morning shift lead operator glanced at the permit, and, seeing that the job was complete, asked the electrician to replace the fuses. The electrician replaced them and signed the permit to show that he had done so. By this time the afternoon shift lead operator had come on duty. He went out to check the pump and found that it was not there.

The job on the permit was to remove the pump for overhaul. Permits are often issued to remove a pump, overhaul it, and replace it, but in this particular case the permit was for removal only. When the maintenance supervisor signed the permit to show that the job was complete, he meant that the job of removal was complete. The lead operator, however, did not read the permit thoroughly and assumed the overhaul was complete.

The message is clear: read permits carefully—do not just glance at them.

Modifications

Thirty years ago a special network of compressed air lines was installed for use with breathing apparatus only. A special branch was taken off the top of the compressed air main as it entered a plant.

For 30 years this system was used without any complaint. Then one day a man got a faceful of water while wearing a face mask inside a vessel. Fortunately, he was able to signal to the stand-by worker that something was wrong and he was rescued before he suffered any harm (see Fig. 36.5).

Entry to Vessels and Other Confined Spaces

1.   An analyst sampled the atmosphere in a tank and reported that it was safe for entry. Fortunately the supervisor suspected that something was wrong. He had further samples taken and it was found that the atmosphere was toxic. The first sample had been taken too near an open manhole (utility hole).

2.   Tests with a portable oxygen analyzer showed that the oxygen content of the air in a tank was normal. Fortunately a supervisor suspected that something was wrong. The tests were repeated using another instrument and this showed that the oxygen content was low. There was a choke in the first instrument and the sample was not reaching the sensitive element.

3.   As the atmosphere in a vessel was a bit smelly a man connected up a hose to what he thought was a compressed air line and put the other end in a vessel. It was actually a nitrogen line. Fortunately the mistake was discovered before anyone was asphyxiated.

4.   The normal entry procedure was omitted because a tank contained water only and was not connected to any other equipment. Unfortunately

rusting had used up some of the oxygen and three men were overcome. Two recovered but one died.

5.   A man stood on a ladder above a drain manhole (utility hole) before putting on his lifeline and breathing apparatus. He was overcome by fumes rising out of the drain and fell into it. His body was recovered at the outfall a mile away.

6.   An excavation in an old factory was sweetened with compressed air and the atmosphere tested. Later a man felt unwell. It was then found that acid had seeped out of the ground into the excavation and reacted with limestone in the soil to produce carbon dioxide.

7.   A supervisor went down a ladder into a reactor to remove a piece of rubbish. He did not ask for another man to stand by as it would take him only a minute and all the other men were having a meal break. He fell and was knocked unconscious. His tongue blocked his throat and he suffocated.

8.   Instrument personnel were working inside a series of new tanks, installing and adjusting the instruments. About eight weeks earlier a nitrogen manifold to the tanks had been installed and pressure tested; the pressure was then blown off and the nitrogen isolated by a valve at the plant boundary.

The day before the accident the nitrogen line was put back up to pressure as the nitrogen was required on some of the other tanks. On the day of the accident an instrument artificer entered a 2 m³ tank to adjust the instruments. There was no written entry permit as the people concerned believed, mistakenly, that they were not required on new plant until water or process fluids had been in them. Although the tank was only 2-m tall and had an open manhole (utility hole) at the top, the instrument man collapsed. An engineer arrived at the vessel about 5 min later to see how the job was getting on. He saw the first man lying on the bottom, climbed in to rescue him and was overcome as soon as he bent down.

Another engineer arrived after another 5 or 10 min. He summoned the process supervisor and then entered the vessel. He also collapsed. The supervisor called the works fire service. Before they arrived the third man recovered sufficiently to be able to climb out of the vessel. The second man was rescued and recovered but the first man died.

It is believed that an hour or two before the incident somebody opened the nitrogen valve leading to the vessel and then closed it.

This incident shows that if someone is overcome inside a vessel or pit we should never attempt a rescue without breathing apparatus. We must curb our natural human tendency to rush to aid or two people will require rescue instead of one (see Chap. 25).

Hot Work

Many explosions have occurred during the repair or demolition of tanks which have contained heavy oils or polymers.

1. Repairs had to be carried out on the roof of a storage tank which had contained heavy oil. The tank was cleaned out as far as possible and two welders started work. They noticed smoke was coming out of the vent and flames coming out of the hole which they had cut. They started to leave, but before they could do so the tank erupted audibly and a flame 80-ft long leapt out. One of the men was killed and the other was badly burned. The residue in the tank continued to burn for 10–15 min.

2. In another incident the roof was blown off an empty tank while a welder was repairing it.

In both these incidents the tanks had been cleaned and freed from oil as far as possible, but traces of heavy oils were trapped between the plates or behind rust, or were stuck to the sides. The welder's torches vaporized these heavy oils and ignited them. See Chapter 19.

3. One of the best known incidents of this type occurred at Dungeon's Wharf in London in 1969.[3] A tank containing a gummy deposit on the walls and roof had to be demolished. The deposit was unaffected by steaming but gave off flammable vapor when a welder's torch was applied to the outside of the tank. The tank blew up, killing six firemen.

It is almost impossible to clean completely a tank which has contained heavy oils, residues or polymers, or any material which is solid at ordinary temperatures, particularly if the tank is corroded, so that oil can get between the plates where there is a defect in the welding (some old tanks are welded along the outside edge of the lap only). Tanks which have contained heavy oils are more dangerous than tanks which have contained lighter fractions such as gasoline, as gasoline can be completely removed and it is possible to detect any traces that are left with a combustible gas detector.

4. A similar incident to those just described occurred while some old pipelines were being demolished. They were cleaned as far as possible and then tested with a combustible gas detector. No flammable gas or vapor was detected and so a burner was given permission to cut them up. While he was doing so, sitting on the pipes 12 ft above the ground, a tarry substance seeped from one of the pipes and caught fire. The fire spread to the burner's clothing and he ended up in hospital with burns to his legs and face.

The tarry deposit in the pipe caught fire when it was heated by the burner's torch. The deposit was not flammable when it was cold, so it could not be detected by the combustible gas detector.

5. A permit was issued for the removal of a blind from a 12-in. naphtha line which was located in a pipe trench. While this was being done, several gallons of naphtha ran out and was ignited by a welder working 65 ft away who was constructing a new pipeline. One of the men working on the blind was killed.

The vapor from a small spillage of naphtha will not normally spread for 65 ft. However, the pipe trench was flooded after heavy rain and the oil spread across the surface of the water. This had not been realized by the man who issued the permit.

The incident was due, in part, to the fact that the pipe trench was over

1640 ft. or 500 m from an operating unit and the supervisors concerned were primarily responsible for operating the unit. For this reason no one visited the scene before issuing the permit for removal of the blind, though it had been visited earlier in the day when a permit was issued for the welding. Had the scene been visited before the second permit was issued, the significance of the flooded pipe trench might have been noticed.

After the incident a special supervisor was appointed to supervise the construction of the new pipeline.

Excavations and the Like

A contractor asked for and received a permit to "level and scrape the ground." As no excavation was requested the process supervisor did not consult the electricians. The contractor used a mechanical shovel, removed several feet from the ground, and cut through a live electric cable.

Equipment Sent Outside the Plant

A large shell and tube heat exchanger was sent to a workshop for retubing. The tubes contained traces of process material. They were first cleaned by steaming except for certain tubes which were plugged and could not be cleaned.

The contractor removed most of the tubes and then asked some men to go into the shell to grind out the remaining tubes. They were affected by fume and ended up in hospital.

The contractor asked the plant if it was safe for men to enter the shell. He was told, correctly, that the shell side of the exchanger was clean. He did not say why he wanted men to go into the shell and no one asked him.

Hand-Over of New Equipment by Contractors

1.   While a welder was finishing off the connecting pipeline to a new methanol tank an explosion occurred and the roof landed some distance away.

It was then found that the contractors had connected up a nitrogen line to the tank. They would not, they said, have connected up the process line but they thought it would be safe to connect up the nitrogen line.

Although the nitrogen valve was closed it was leaking and the nitrogen was contaminated.

2.   A contractor was asked to leak test a new tank designed for use at atmospheric pressure. He misunderstood his instructions and connected up

a compressed air supply. The pressure in the tank reached 25 psig before the roof blew off.

Opening-Up of Equipment by Operators

Every day equipment which has been under pressure is opened for repair, but this is normally done under clearance—one worker prepares the equipment and issues a clearance to another worker who opens the equipment. Both incidents described below occurred when equipment was opened as part of a process operation.

1. A suspended catalyst was removed from a process stream in a pressure filter. After filtration was complete, steam was used to blow the remaining liquid out of the filter. The pressure in the filter was blown off through a vent valve and the fall in pressure observed on a pressure gauge. The operator then opened the filter for cleaning. The filter door was held closed by eight radial bars which fitted into U bolts on the filter body. The bars were withdrawn from the U bolts by turning a large wheel fixed to the door. The door could then be withdrawn.

One day an operator started to open the door before blowing off the pressure. He was standing in front of it and was crushed between the door and part of the structure and was killed instantly.

The plant operators were surprised that the pressure in the filter, about 30 psig, could cause so much injury. They did not realize that 30 psig over 10 ft^2, the area of the door, produces a *force* of *20 tons*.

2. Plastic pellets were blown out of a road tanker by compressed air. When the tanker was empty the driver opened a manhole (utility hole) cover on top of the tanker to check that the tanker was empty. One day a driver, not the regular man, started to open the manhole (utility hole) before releasing the pressure. When he had opened two of the quick release couplings the cover was blown off. The driver was blown off the tanker and killed by the fall.

Either the driver forgot to vent the tanker or thought that it would be safe to let the pressure (10 psig) blow off through the manhole (utility hole).

Many of those concerned were surprised that 10 psig could cause so much injury—10 psig is not a small pressure. The damage at Flixborough was caused by a pressure of about 10 psig.[4]

ACKNOWLEDGMENTS

Thanks are due to the many colleagues who supplied ideas for this chapter or commented on the draft, and to the companies where the incidents occurred for allowing their descriptions to be published.

REFERENCES

1. T. A. Kletz, *Chem. Eng. Prog.*, **72**, 48 (1976).
2. T. A. Kletz, *J. Hazard. Mater.*, **1**, 165 (1976).
3. *Public Enquiry into a Fire at Dudgeon's Wharf on 17 July 1969*, Her Majesty's Stationery Office, London, 1970.
4. Report of the Court of Inquiry, *The Flixborough Disaster*, Her Majesty's Stationery Office, London, 1974.

37

Radiation Hazards: Their Prevention and Control

Joseph K. Kielman and
Howard H. Fawcett

The term radiation generally refers to the process of emission, transmission, reflection, or absorption of energy. Considered in this way, radiation can take several forms, such as heat, sound, or electromagnetic energy. Electromagnetic radiation is a special type of radiant energy that does not require a medium for propagation and comprises a continuous spectrum of energy, the effects and properties of which are determined by its frequency and wavelength. Some radiation within that spectrum, such as x and gamma rays, is not obvious to or perceived by us under normal circumstances and is called ionizing radiation because of its ability to physically interact with matter to release electrons. Other radiation, such as radiowaves and television waves as well as visible light rays, is part of our daily life and, because of its low energy content, is considered to be nonionizing. The breadth of the electromagnetic spectrum is noted in Table 37.1.

The properties and physiological effects of radiofrequency and microwave radiation have been well summarized elsewhere.[1-4] Those of visible light and of ultraviolet and infrared radiation are also well documented.[5-7] Laser beams, for example, present unique problems because of the coherent monochromatic nature of the visible light emitted.[8-11] The purpose of this chapter is to survey briefly the physical and physiological properties of the short-wavelength radiation and high-energy particles that constitute one portion of the electromagnetic spectrum usually referred to as ionizing radiation and to supply references for a more complete study.

Ionizing radiation has been part of the earth's environment since its formation (see Table 37.2),[12] but only recently has society learned of its potential for beneficial use. At present, the practical uses of radiation are limited

TABLE 37.1.   The Electromagnetic Spectrum[a]

Type of Wave		Wavelength
Radio, standard AM and international bands		100 km–10 m
Shortwave, TV channels 2–13, and FM		10 m–1 m
Ultrashortwave and TV channels 14–83		1 m–10cm
Microwave (radar)		10 cm–0.5 mm
Infrared		0.5 mm–770 nm
Far	500–10 μm	
Intermediate	10–1.3 μm	
Near	1.3 μm–770 nm	
Visible light		770–390 nm
Red	770–630 nm	
Orange	630–590 nm	
Yellow	690–550 nm	
Green	550–490 nm	
Blue	490–450 nm	
Violet	450–390 nm	
Ultraviolet		390–10 nm
Near	390–300 nm	
Middle	300–200 nm	
Extreme	200–10 nm	
X rays		10–0.1 nm
Superficial		
Therapy	5–0.9 nm	
Diagnostic	0.9–0.1 nm	
Therapeutic	0.1–0.3 nm	
Gamma rays		1–0.001 nm
Cosmic rays		0.1–0.001 nm

Source.   M. S. Litwin and D. H. Glew, "The Biological Effects of Laser Radiation," *J. Am. Med. Assoc.*, **187**(11), 846 (March 14, 1964).

[a]Band separation is not clearcut and overlap occurs.

to techniques in which its unique properties make those techniques more efficient or refined than other methods. Understanding and appreciating the benefits of radiation,[13] from the mundane (as in locating leaks in underground piping) to the exotic (as in nuclear power sources to supply energy to space vehicles), will doubtlessly encourage wider interests and applications. Atomic power from nuclear reactors is already economically attractive; power from thermonuclear fusion has been demonstrated publicly.[14]

37.1.   SOURCES AND PROPERTIES OF IONIZING RADIATION

Ionizing radiation can take the form of either waves or particles of matter. X and gamma rays are electromagnetic waves. Ionizing particles include such varied atomic and nuclear constituents as the electron (or beta particle); the

TABLE 37.2. Body Tissue Dose Rates Due to External and Internal Irradiation from Natural Sources of Radiation in "Normal Regions" [a,b]

	Dose rates in mrem/yr		
Source of Irradiation	Gonad	Haversian canal	Bone marrow
External irradiation			
Cosmic rays (including neutrons)	50	50	50
Terrestrial radiation (including air)	50	50	50
Internal irradiation:			
^{40}K	20	15	15
^{226}Ra and decay products (35% equilibrium)	0.5	5.4	0.6
^{228}Ra and decay products equilibrium)	0.8	8.6	1.0
^{210}Pb and decay products (50% equilibrium)	0.3	3.6	0.4
^{14}C	0.7	1.6	1.6
^{22}Rn (absorbed into bloodstream)	3	3	3
Total	125	137	122

Source. M. Eisenbud, *Environmental Radioactivity,* McGraw-Hill, New York, p. 170.

[a]UN (1962).

[b]^{210}Pb in excess of that expected from ^{226}Ra and decay products in 35% equilibrium.

proton; the neutron; the deuteron, which consists of a proton and a neutron; and the alpha particle, which consists of two protons and two neutrons. As is shown later, some of the mass of the heavier deuterons and alpha particles is transformed into the energy that binds their constituents. All of these can be produced in a variety of ways.

Ionizing radiation sources may be broadly classed as either nuclear or electronic (or machine). The chart below lists the principal sources of ionizing radiation.[15]

Nuclear Sources	Electronic Sources
Nuclear reactor	High-voltage transformer
Gaseous fission products	Impulse generator (capacitron)
Spent fuel rods	Van de Graaff generator
Liquid reactor fuel	Resonant transformer
Fission products (solutions or separated, such as ^{137}Cs and ^{90}Sr)	Linear accelerator
	betatron
	cyclotron
Natural radiation sources (such as Ra, Th, U)	Bevatron
	other particle accelerators
Manmade radioactive isotopes (such as ^{60}Co)	

The range of uses for and equipment producing various types of ionizing radiation is wide. It includes high-voltage electron tubes, such as the magnetrons used to produce radio- and microwaves, cathode ray tubes, and electron guns; gas and aerosol detectors, such as smoke alarms; such laboratory devices as electron microscopes; such industrial components as static eliminators; and common as well as recently developed medical equipment, such as chest or dental x-ray machines, fluoroscopes, whole-body scanners, and therapeutic radiation equipment. Other uses are gauging liquid levels, measuring the thickness of sheet materials and films, monitoring flow through pipes, determining wear rates of metal parts, sterilizing food and medical supplies, and powering cardiac pacemakers. In deciding whether to use machine-produced or radioactively released radiation, such factors as convenience, cost, ease and absoluteness of control, and precision of manipulation are often considered.

The most widely used sources of radiation are radioactive isotopes. Isotopes are forms of an element that have identical chemical properties, but because of different numbers of neutrons in their nuclei have different atomic weights. Eighty of the 103 known elements have at least one stable isotope; some have as many as nine. A total of 280 stable isotopes has been discovered. Sixty-six naturally occurring radioactive isotopes (or species) have been identified,[16] and over 1000 known radioactive isotopes (also called radionuclides) have been produced artificially. Of the latter, nearly 100 are commercially available from several sources in the United States and Canada (see Table 37.3).

Radioactive (or unstable) isotopes are distinguished from other so-called stable isotopes by the fact that their nuclei decay; that is, they emit high-energy ionizing radiation, such as beta or alpha particles or gamma radiation, during spontaneous disintegration or nuclear transformation. Radioactive isotopes (or radionuclides) tend to become stable as they give off radiation, much as a mechanical clock tends to run down as the energy in the spring is expended. The time required for half of the atoms to decay is called the half-life. Each radioactive isotope has a unique half-life, which may last from a fraction of a second to billions of years (Fig. 37.1). During this decay, radioactive isotopes emit one or more forms of energy. These include:

Alpha particles (mass number 4, charge +2) are helium nuclei evolved from elements of high atomic number, such as thorium, radium, polonium, and plutonium. As decay occurs, the parent alpha-emitting element acquires an atomic number of two less and a mass number of four less than the parent. Because of their relatively large mass, alpha particles have relatively little penetrating power. Paper will stop alpha particles, and even the most energetic alphas are attenuated or dissipated by less than 3 in. air or 0.002 cm aluminum. Alpha emitters are not without potential hazard. Although they pose little hazard when external to the body, they do present serious problems as internal radiation sources, and this, together with the long half-life and intrinsic toxicity of many of the alpha-emitting metals, makes it mandatory to prevent their possible entry into the body by ingestion, inhalation, or other routes.

Beta rays (or particles) are electrons. They are the only negatively charged radioactive particles of direct interest to the study of radiation effects. Since their mass is very small, they can penetrate deeper than alpha particles. When liberated from the nucleus, the beta particle acts as an ionizing particle, and the electrons or electromagnetic radiation released by its interaction with other atoms is sufficiently energetic to produce further ionization. The energy of the primary electron determines the amount of secondary ionization that occurs. Since ionizing particles have almost two-thirds of their energy carried away and distributed throughout the environment by these secondary electrons, that process greatly contributes to the total ionization or damage occurring in any cell.

Beams of high-energy radiation, such as fast electrons from accelerators, have been used for modifying polymers to improve such physical properties as tensile strength, food preservation, and drug sterilization. Typical electron accelerators include the Van de Graaffs, producing electrons of 1 to 6 MeV (million electron volts; an electron volt is a measure of the energy of a particle or wave); resonant transformers producing 1- to 3.5-MeV electrons, betatrons producing 6- to 30-MeV electrons, and linear accelerators yielding 3- to 25-MeV, or higher, electrons. Both low-voltage electrons, also called cathode rays, and high-energy electrons may be emitted by a heated filament (cathode) in a vacuum tube.

The relative energy of the electron has a significant effect on the shielding required to protect personnel who may be exposed to the particles. Glass, plastic, and aluminum are often used for shielding from electrons, but glass and many plastics discolor and become brittle in high-energy or dense radiation fields. With low-energy electrons, iron, copper, concrete, or concrete blocks are used as shields.

Gamma and x rays are electromagnetic radiation analogous to light but of shorter wavelength. They differ from each other fundamentally only in their frequency (wavelength) and by the fact that gamma rays are emitted from nuclei whereas x rays are released from the electron shell during transitions to the lower energy levels. X rays are frequently called Roentgen rays in honor of their discoverer, W. Roentgen. Both x rays (usually supplied by a machine employing a tube, such as the Coolidge tube) and gamma rays (usually emitted from a radioactive isotope or radionuclide, such as ^{60}Co) may be used for clinical radiotherapy in humans, for radiography in the inspection of welds or flaws in pipes, vessels, and other metallic forms, or for many other uses.

Neutrons are not emitted spontaneously from radioactive nuclei; rather, their release depends on the interaction of alpha, gamma, or other neutron radiation with the nuclei of certain target materials such as uranium, thorium, and plutonium. Neutrons are also essential to the production of most radioactive isotopes. Neutrons are electrically neutral; thus, they ex-

TABLE 37.3. Available Radioisotopes Listed According to Half-Life[a,b]

Half-Life[c]	Radioisotopes[d]	Radiation	Half-Life[c]	Radioisotopes[d]	Radiation
24.2 s	Silver-110	β, γ	43 d	Cadmium-115m	β, γ
30 s	Rhodium-106 (Ru^{106})	β, γ	44.3 d	Iron-59	β, γ
72 s	Indium-114	$EC, \beta^-, \beta^+, \gamma$	44.6 d	Hafnium-181	β, γ
2.6 m	Barium-137m (Cs^{137})	IT, γ	45.4 d	Mercury-203	β, γ
17.5 m	Praseodymium-144	β, γ	50 d	Indium-114m	IT, γ
12.47 h	Potassium-42	β, γ	50.5 d	Strontium-89	β
12.5 h	Iodine-130	β, γ	57.4 d	Iodine-125	γ
12.82 h	Copper-64	$EC, \beta^-, \beta^+, \gamma$	59.1 d	Yttrium-91	β, γ
13.6 h	Palladium-109	β, γ	60.9 d	Antimony-124	β, γ
14.2 h	Gallium-72	β, γ	64 d	Strontium-85	γ
15.05 h	Sodium-24	β, γ	65 d	Zirconium-95	β, γ
19.0 h	Iridium-194	β, γ	71.3 d	Cobalt-58	EC, β^+, γ
19.2 h	Praseodymium-142	β, γ	74.2 d	Iridium-192	β
24.0 h	Mercury-197m	EC, IT, γ	75.8 d	Tungsten-185	EC, β, γ
24.0 h	Tungsten-187^2	β, γ	84.2 d	Scandium-46	β, γ
26.8 h	Arsenic-76	β, γ	89 d	Sulfur-35	β
35.55 h	Bromine-82	β, γ	115.1 d	Tantalum-182	β, γ
38.7 h	Arsenic-77	β, γ	119 d	Tin-113	EC, γ
40.3 h	Lanthanum-140	β, γ	119.9 d	Selenium-75	EC, γ
46.8 h	Samarium-153	β, γ	129 d	Thulium-170	β, γ
53.5 h	Cadmium-115	β, γ	165.1 d	Calcium-45	β
64.4 h	Yttrium-90	β	246.4 d	Zinc-65	EC, β^+, γ
64.8 h	Gold-198	β, γ	249 d	Silver-110m	IT, β, γ
65 h	Mercury-197	EC, γ	285 d	Cerium-144	β, γ
66 h	Antimony-122	β, γ	371 y	Ruthenium-106	β

Half-life	Radioisotope	Decay	Half-life	Radioisotope	Decay
67 h	Molybdenum-99	β, γ	1.3 y	Cadmium-109	EC, γ
75.4 h	Gold-199	β, γ	2.07 y	Cesium-134	β, γ
88.9 h	Rhenium-186	EC, β, γ	2.5 y	Promethium-147	β, γ
4.53 d	Calcium-47	β, γ	2.78 y	Antimony-125	β
5.00 d	Bismuth-210	β	2.94 y	Iron-55	EC
5.27 d	Xenon-133	β, γ	3.57 y	Thallium-204	EC, β
7.5 d	Silver-111	β, γ	5.24 y	Cobalt-60	β, γ
8.05 d	Iodine-131	β, γ	10.27 y	Krypton-85	β, γ
11.06 d	Neodymium-147	β, γ	10.7 y	Barium-133	EC, γ
12.0 d	Barium-131	EC, γ	12.46 y	Hydrogen-3	β
12.8 d	Barium-140	β, γ	12.7 y	Europium-152	EC, β, γ
13.7 d	Praseodymium-143	β	16 y	Europium-154	β, γ
14.3 d	Phosphorus-32	β	28 y	Strontium-90	β
16 d	Osmium-191	β, γ	30 y	Cesium-137	β
18.68 d	Rubidium-86	β, γ	125 y	Nickel-63	β
27.8 d	Chromium-51	EC, γ	5.57×10^3 y	Carbon-14	β
32.5 d	Cerium-141	β, γ	2.12×10^5 y	Technecium-99	β
34.3 d	Argon-37	EC	3.08×10^5 y	Chlorine-36	β
35 d	Niobium-95	β, γ	1.56×10^7 y	Iodine-129	β, γ
39.7 d	Ruthenium-103	β, γ			

[a]Suppliers of radioisotopes include: Oak Ridge National Laboratory, Isotopes Division, P.O. Box X, Oak Ridge, Tennessee 37830. General Electric Co., Vallecitos Atomic Laboratory, P.O. Box 460, Pleasantown, Calif. 94566. Union Carbide Corp., P.O. Box 324, Tuxedo, New York, N.Y. 10987. University of Missouri, Research Reactor Facility, Columbia, MO 65211.

[b]Decay allowance for isotopes with half-lives of less than 15 days is shown by brackets in half-life column. Allowance calculated from 8 a.m. on day of shipment; 1—one half-life; 2—one day; 3—two days; 4—four days.

[c]s, second; m, minute; h, hour; d, day; y, year.

[d]The radioisotope in parenthesis is the one under which the listed radioisotope is cataloged.

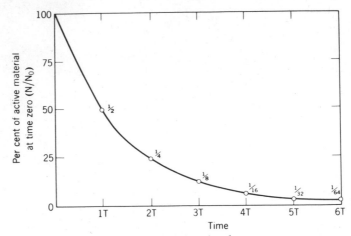

Figure 37.1. Radioactive decay.

hibit no electric interactions and have very high penetrating power. By their ability to convert or transmute other elements by entering their nuclei (a process known as neutron capture), neutrons produce from stable atoms most of the manufactured radioactive isotopes (or radionuclides). Nuclear reactors are the most common source of neutrons. The flux inside one reactor has been estimated to be a mixture of the following:

Thermal neutrons	$1.1 \times 10^{12}/cm^2 \cdot s$
Neutrons ($\geqslant 0.1$ MeV)	1.4×10^{11}
Neutrons ($\geqslant 0.5$ MeV)	6.7×10^{10}
Fast neutrons ($\geqslant 1$ MeV)	4.2×10^{10}
Gamma photons (avg 1 MeV)	5×10^{11}

The General Electric test reactor has a peak flux of:

Core position

3×10^{14} nv thermal and 6×10^4 nv fast

Pool position

1×10^{14} nv thermal and 3×10^{13} nv fast

The properties of the various forms of ionizing radiation are summarized in the table below. As is shown later, the relative effect of these different types of radiation on biological and other materials does not depend solely on their energies:

Radiation or Particle	Relative Mass	Charge	Typical Energy
X ray	0		0.4 keV–0.4 MeV
Gamma ray	0		10 keV–10 MeV
Beta ray (electron)	1	–	3 keV–4 MeV
Proton	1836	+	—
Neutron	1839	0	
Fast			>0.1 MeV
Intermediate			1 eV–0.1 MeV
Slow			<0.1 eV
Thermal			0.025 eV
Deuteron	3671	+	—
Alpha particle	7296	+ +	4 MeV–8 MeV

37.2. RADIATION MEASUREMENTS

Units of Radiation

The fundamental unit of radioactivity is the curie. One curie (Ci) is that amount of radioactive material in which 3.7×10^{10} nuclear disintegrations occur per second. The millicurie (mCi) is one one-thousandth curie; the kilocurie (kCi) is one thousand curies. Many laboratories experiments or operations are performed using less than 1 mCi of a radioactive isotope. Sources of one to several kilocuries are used routinely for medical therapy, whereas food sterilization may require several hundred kilocuries.

Since the effect of radiation exposure depends upon the type and quantity of radiation, the type and amount of interaction of the radiation with matter, as well as biological differences, a unit of exposure, the roentgen (R), has been established to define the ionizing effect. One roentgen is the quantity of x or gamma radiation such that the associated corpuscular emission per cubic centimeter or 0.001293 g of air produces ions carrying one electrostatic unit (esu) of electricity of either sign. One roentgen of x or gamma radiation is equivalent to the absorption of about 87 ergs energy per gram of air. (The associated corpuscular emission referred to in the definition is electrons produced by the interaction of x or gamma rays with air. All ion pairs produced are considered if we wish to account for the entire amount of energy.) Because it is defined in terms of air and refers only to x or gamma radiation, the roentgen must be converted to other units to be practical as a measure of exposure. For x rays used in radiation therapy and for soft tissue, 1 R yields an energy transfer of about 93 ergs/g of soft tissue. The energy transfer for other types of radiation and for other materials differs significantly.

Since it is the transfer of energy to and its interaction with tissue or materials that produce effects, the rad was developed as a unit of absorbed dose. The rad represents the absorption of 100 ergs of nuclear (or ionizing) radiation per gram of absorbing material or tissue. Its use is valid for radiation of any type or energy and for any material with which the radiation interacts.

Equal absorbed doses produced by different types of ionizing radiation or by radiation of different energies produce different biological effects. In addition to the quality of the radiation, its intensity or rate of delivery and continuous versus fractionated administration have been identified as important variables that determine its subsequent effect on matter. These differences have led to the concept of relative biological effectiveness (RBE), in which various ratios have been established to define the relative effectiveness of various types of radiation:

Radiation	RBE
X rays, gamma rays, electrons, beta rays, and nuclear weapon neutrons	1
Fast neutrons and protons up to 10 MeV	10
Thermal neutrons	4–5
Heavy recoil nuclei	20
Naturally occurring alpha particles (from Ra, Po, Th, U)	10–20

As the chart indicates, the alpha particle is ten times as effective and the neutron is five times as effective as beta, x, or gamma rays in producing biological effects.

By incorporating these RBE ratios into the previous definition of a rad, a unit known as the rem has been developed to indicate the extent of the biological injury (of a specific type) that can result from absorption of nuclear radiation. It is calculated by the formula:

$$\text{Dose in rems} = \text{RBE} \times \text{dose in rads}$$

The rem is a dose equivalent unit of *biological effect,* whereas the rad is a unit of *absorbed energy dose* and the roentgen (for x and gamma rays) is a unit of *exposure* or total dose. Metric system (or SI) equivalents have been proposed for several of the base units of radioactivity. These include the becquerel (Bq) for activity, gray (Gy) for absorbed dose, and sievert (Sv) for dose equivalent:

$$1 \text{ Bq} = 1 \text{ s}^{-1} = 2.7 \times 10^{-11} \text{ Ci}$$
$$1 \text{ Gy} = 1 \text{ J/kg} = 1 \times 10^2 \text{ rad}$$
$$1 \text{ Sv} = 1 \text{ J/kg} = 1 \times 10^2 \text{ rem}$$

Another term used to describe the deposition of energy in tissue, and, consequently, its site of action, is linear energy transfer (LET), which represents the energy released by the radiation per unit length of the absorbing tissue. Alpha particles and neutrons are high-LET radiation, whereas electrons and x and gamma rays are low-LET radiation. Thus, in comparing equally energetic particles, the latter will penetrate farther, whereas the former will deposit most of their energy closer to the surface. Nevertheless, high-LET particles are generally more damaging than low-LET particles.

Measuring Radiation

Measurement is a prime prerequisite of radiation control,[17-19] and awareness of why radiation must be measured is essential for its safe use. Important reasons for making radiation measurements and keeping records of exposures include:

1. To maintain a cumulative record of exposure to radiation of every individual exposed. (This has become increasingly important, both legally, in the disposition of suits and claims by employees against employers who exposed them to potentially harmful levels of radiation, and because of the interest in the so-called cradle-to-grave documentation of the medical history of each individual.)

2. To measure over a short period the exposure of individuals carrying out specific operations during which the radiation level is abnormally high. For example, if a source is manually moved or physically transferred from a shielded storage location to a work station or if a radioactive isotope solution is removed from its shipping container and checked for identity and activity, exposure cannot be practically prevented, but it can usually be kept within recognized limits. Measurements are essential to control such exposures.

3. To survey the flux or radiation level in all parts of the facility to detect areas of unnecessarily high radiation levels. For example, a bottle or drum of liquid radioactive waste may have been temporarily stored with other chemicals under circumstances in violation of standard operations and procedures. Prompt discovery of this error by survey aids in restoring proper control.

4. To locate radioactive contamination on workbenches, floors, and other surfaces or, when necessary, in the air. Spills will occasionally occur, even in meticulous operations, and, unless careful surveys and clean-up (often referred to as decontamination) are instituted promptly and effectively, radiation can be spread over a wide area and eventually result in unnecessary exposure of other personnel.

5. To monitor contamination of hands, feet, hair, skin, or clothing of personnel who are working with or have contact with radioactive

materials. In large operations where this is a part of normal daily routine, special instrumentation may be made available in a locker or washroom area for self-monitoring of feet, hands, and body; in smaller operations, readily available survey meter may suffice. For removal of radioactive material from the skin, special cleaners that combine chelating agents with surface-active detergents can be used. Shower facilities or other means of thoroughly washing the whole body, including the hair, are useful for emergency decontamination and for medical control of chemicals and radioactive materials, even if showers are not used routinely.

6.   To establish legal protection for employers and employees, in compliance with local, state, and Federal codes and other laws controlling use of radioactive materials and exposure to radiation.

Radiation surveys and services are not glamorous occupations; however, they do require the same intelligence, patience, time, initiative, and dedication that other accident prevention work requires. The seriousness, the moral and legal responsibility, and the accountability of the position must be fully appreciated if the job is to be done properly on a routine basis.

Radiation Instrumentation

FILM BADGES

Photographic film of the proper type will respond to radiation much as it does to visible light, and this fundamental fact is the basis for the film badge dosimeter—the device of choice by many individuals for measuring and recording accumulated radiation exposure. Low cost, convenient size, capability for integrating doses over relatively long periods of time, and permanency of the record are the unique advantages of film badges. Film badges measure exposures from a minimum of hundredths to a maximum of hundreds of roentgens and detect radiation from low-energy x and beta rays to very high-energy gamma and neutron radiation. Film badges will not respond to alpha particles, since these are stopped by the paper necessary to shield the film from exposure to light.

The badges consist of small sheets of sensitive photographic or x ray-type film supported by a plastic or metal frame holder. The holder may contain small disks or pieces of metals of different density, positioned in front of the film to intercept the impinging radiation: aluminum, copper, silver, cadmium, and lead are frequently used to help discriminate different energy levels and types of radiation. Normally, the badge and its holder are worn on the lapel or shirt pocket to indicate total body exposure. Where fingers and hands may be more directly exposed, as in the use of x-ray diffraction equipment, smaller film badges may be worn on the fingers to measure hand

and finger exposures. If the forearm and hands are likely to receive exposures, a film badge may be worn on the wrist or lower arm.

Where exposure to neutrons may be encountered, additional sensitive film is placed behind the beta-gamma film and a special shield is used. A shield of cadmium, which has a high cross section for the absorption of slow neutrons, can be used to discriminate among three types of neutrons. Only fast neutrons penetrate the cadmium shield, producing recoil protons which leave tracks in the film. Thermal neutrons leave proton tracks in the areas of the film not shielded by cadmium. Slow neutrons do not leave tracks. On developed film, the tracks are counted by microscope. A new approach to analysis of minute concentrations of uranium uses tracks from fission particles in solid state media.[21]

Badges are usually worn for 7–10 days and are returned to the service supplier for processing and reporting of results. Since a few days normally elapse between return of the badge and receipt of the exposure report, emergency processing and reporting (on a 24-hr basis) is necessary to help determine whether an unusual exposure has occurred. This is important both clinically and legally, since high exposures must be immediately reported to medical and regulatory agencies. In addition to regular reports listing all badges used during the period, separate reports showing totals for for the previous period, for the past quarter year (or 13 weeks), and for the total accumulated exposure since the inception of the badge service may also be obtained for each individual.[20] Although badges do not give immediate information and are not self-reading, they represent an excellent radiation control device if they are supplied and processed by a reputable service. Quality control in processing and reporting is therefore essential to ensure accurate results and records.

IONIZATION CHAMBERS

Ionization of air or other gases by radiation provides a method of measurement based on the movement of ions from one part of a chamber to an electrically charged plate or post. Since a single alpha particle from radium, for example, produces 137,000 positive ions and an equal number of electrons when it is completely absorbed in air, the consequent electric current can be measured with the aid of an amplifier. The ions formed in a gas (ions of nitrogen and oxygen molecules in the case of air) are usually exceedingly reactive; they often undergo chemical change during the time required to collect them at an electrode. The total electric charge, however, is not altered. For this and numerous other reasons, the ionization method of detection has proved remarkably trustworthy. Ionization chambers are widely employed to count energetic particles and to determine their individual energies. If the instrument is properly designed, with thin walls and high sensitivity, beta as well as x and gamma rays may be measured.

The ionization effect also serves as the basis for another method of personal monitoring using small, pocket ionization chambers, which are devices about the size and shape of a fountain pen. Initially, an auxiliary charger is used to bring the device, which is essentially a capacitor, to a known potential, for example, 150 V. The change in the capacitance due to discharge by ionization is then noted by reading the residual charge on a reader device. Another type, not unlike the first in external appearance, is a combined electrostatic ionization chamber and fiber electroscope. The quartz hair-like fiber is charged with an external battery to produce a scale reading of zero. As the radiation produces ionization, partial discharge occurs and the fiber moves across the scale. Its location may be immediately observed by viewing the scale through the lens built into one end and observing the shadow against a light.

Both types discharge as the dose accumulates, and they may be read as frequently as desired. One style, designed for civil defense purposes, may be used to indicate the rate of exposure. In this application the device is exposed for a specified number of minutes and the dose per unit time is interpreted as a rate (such as roentgens per hour). One disadvantage of pocket ionization chambers is their relatively high sensitivity to shock, which may be incorrectly attributed to radiation exposure, if the device is accidentally dropped. For this reason, pocket ionization chambers are often worn in pairs.

For higher levels of radiation, other types of dosimeters are available. Phosphate glass dosimeters measure doses in the range of 10–600 R, which is a critical range of determining the probable survival of any individual exposed to high levels of radiation. (Single doses in the range of 25–100 rems over the whole body will produce only blood changes; does of 100–200 rems will result in illness but will rarely be fatal; for doses between 200 and 1000 rems, the probability of survival is good at the lower end of the range but poor at the upper end.) (See the next section.) The special glass incorporates traces of silver and other metals as activators and, after exposure to radiation, flucresces under ultraviolet irradiation.[22] Another type, the chemical dosimeter, utilizes the principle of color change. In this device an acid, such as hydrochloric, is liberated from chloroform or other chlorinated hydrocarbons in the presence of an organic dye indicator, such as methyl red or methyl orange. Indicators of this type may be dissolved in wax or a paraffin and are commonly used to measure penetration of the radiation by the color change at various depths. The Fricke dosimeter is based on another chemical reaction, the oxidation of ferrous sulfate dissolved in sulfuric acid, to give an accurate measure of absorbed energy.

The principle of ionization, when coupled with appropriate electronic circuitry, serves as a basis for the dose-rate meter. Available in a wide range of types, incorporating several scales to accommodate various intensities, from less than 1 mR/h to 5 R/h, and equipped with a movable shield to permit discrimination between beta and gamma radiation, ionization dose-rate meters are widely used for surveys.

GEIGER–MUELLER COUNTER

The Geiger-Mueller counter is a count-rate meter in which a thin-walled tube reacts with radiation to produce electronic pulses that are then amplified so they can be displayed on a meter and heard through a speaker or earphones. Some counters rely on a high voltage to further break down the gas once a passing particle has supplied the initial requirement of ions; such a device can yield a burst of current strong enough to be observed without amplification. Depending on the type of tube and how it is housed and shielded, the Geiger counter can pinpoint local radiation areas on contaminated floors or benches, as well as clothing or skin, especially if a sensitive end-window detector tube is used. Although useful for counting large numbers of ionizing particles, the Geiger counter cannot determine their energies. Some counter scales are calibrated in counts or pulses per minute, while others read milliroentgens per hour.

SCINTILLATION COUNTERS

Another method of detection rests on the emission of light by molecules. When an atom is excited to a higher electronic energy state, it loses its excitation energy and returns to its lowest energy state by emitting one or more photons. Since few atoms exist in isolation, however, the excitation energy is usually dissipated by collisions with other atoms or molecules and converted to other forms of energy such as heat. Special substances called scintillators are available in which that excitation energy is not degraded and light is emitted under irradiation. Inorganic phosphors, such as zinc sulfide, and organic substances (usually aromatic polycyclic compounds) serve as scintillating media, and the burst of light is detected by a sensitive photocell and amplified into an electrical signal. A portable device operating similarly to scintillation detectors is the Low-Intensity X ray Imaging Scope, originally developed by NASA at the Goddard Space Flight Center. The Lixiscope uses a phosphor screen and a photoconverter to produce a visible image of an x or gamma ray emitting radioactive material.

ALPHA DETECTORS

Where alpha sources are used, an alpha-detecting instrument is required for check and control. Except for the scintillation counter, the instruments described previously will not perform this service. Typical alpha detectors have a very thin plastic-covered window that permits the alpha particles to enter the electric field of a relatively large probe. The resulting perturbation is translated directly into an electronic pulse or, through ionization of a gas such as propane (as in the gas proportional counter), the particle produces secondary electrons that are then amplified and indicated on a meter. Such meters read in counts per minute over a wide range, which can be chosen using several scale multipliers, and must be calibrated for the particular alpha energy to be surveyed.

37.3.   EXPOSURE TO RADIATION

Biological Effects of Radiation

The biological effects of radiation, which include genetic and cellular damage to both the reproductive and somatic tissues of the body, have been widely publicized in connection with nuclear weapons, and because of the high emotional content of much that has been written, it is difficult to approach the subject in a scientific manner. Recent attempts to place radiation into proper perspective will not succeed until knowledge replaces fear in the minds of many.

As stated above, ionizing radiation represents a part of the natural environment of every person, regardless of his or her occupation.[23] Exposure of the general populace results from continuous contact with the natural background radioactivity as well as occasional contact with medical equipment, such as diagnostic x-ray and cancer treatment machines. The levels of exposure expected from such sources are listed below.[27]

Source	Level of Exposure	
Background Radiation		
Cosmic radiation		
At sea level	30	mrem/year
At 5000 ft	70	mrem/year
Natural radioactivity		
In the open	50	mrem/year
In buildings	80	mrem/year
Estimated dose due to fallout from atomic testing	3	mrem/year
Medical Equipment		
Average dose of medical X rays	100	mrem/year
Dose from one chest x ray		
To whole body	50	mrem
To gonads	0.4	mrem
Dose from pelvimetry during pregnancy		
To mother	1.3	rem
To child	2.7	rem
Dose from one set of dental x rays		
To face	300	mrem
To gonads	5	mrem (male)
	1	mrem (female)

Source	Level of Exposure	
Dose from foot x rays		
To feet	25	rem
To gonads	75	mrem (male)
	15	mrem (female)
Local dose in cancer treatment	3	rem
Dose from wristwatch dial (skin)	600	mrem/year

Terrestrial sources of radiation include uranium, actium, thorium, ^{14}C, and ^{3}H. Radon and its decay products (or daughters) contribute the greatest amount of radioactivity to the natural background that is inhaled or ingested[28]; for example, the concentrations of ^{222}Rn and ^{220}Rn have been measured to be 0.2–0.5 pCi/liter at heights of 1 m above the ground.[29] A 1980 report from the U.S. EPA stated that none of its 67 monitoring stations located throughout the United States reported levels of radioactivity in air particulate samples above normal fluctuations. It should be noted that exposures to radon in tightly insulated buildings (or homes) can reach as high as 225–1500 mrem/year,[28] and that water supplies contain usually 30–300 pCi/liter and occasionally up to 3000 or 240,000 pCi/liter of radon.[29]

These figures have been provided as a basis for making judgments about the relative hazard of occupational exposures to ionizing radiation, since these would obviously add to the burden induced by the background. The following doses have been accepted as thresholds for producing effects in humans[27]:

Mean lethal dose (whole body)	500 rem
Dose for radiation burn	300–1000 rem
Dose for radiation sickness	150 rem
Estimated dose to double spontaneous mutation rate	5–150 rem
Life-shortening rate	5–15 days/rem

It is true that the body can, within certain limits, repair the damage from radiation, whatever its source. At what point exposure to radiation becomes harmful depends on many factors. For example, different types of ionizing radiation differ greatly in their ability to penetrate the skin and body and in their relative biological effectiveness. The effects of ionization may be direct (localized within the cells directly affected, as a skin exposure) or indirect (since the action on one organ or group of cells, such as the bone marrow, may be transmitted to other cells remote from the bone by the blood). Effects may be immediate or delayed, depending on intensity, type, and other factors characteristic of the radiation involved. Finally, effects may be reversible (repairable by the body through normal repair and replacement processes) or irreversible.

The somatic effects of ionizing radiation can be either acute or chronic. Short-term exposures to radiation are followed by any or all of the following acute effects: erythema or burns, nausea and vomiting, loss of appetite, loss of hair, diarrhea, bleeding (external and internal), anemia and leukopenia, infection, and death.[30] In many cases of high exposure (single exposure to 300 rads or above), the symptoms progress in the order given above and death ensues within a few days to 2 weeks. As the preceding list indicates, tissues whose cells undergo rapid turnover appear to be affected by ionizing radiation first. Certain body cells have been found to be more sensitive to radiation than others.[24] These are listed below in decreasing order of sensitivity to radiation:

1. Lymphocytes, or white blood cells formed by the tissues of the spleen and lymph nodes. Because of their sensitivity, complete blood counts at frequent intervals following a suspected excessive exposure to a large area of the body may reflect any serious damage and pinpoint the stage in the recovery cycle. Reliable and accurate preexposure counts are essential for comparison purposes.

2. Granulocytes, or white blood cells formed in bone marrow. These are needed in proper proportion to combat bacterial infection.

3. Basal cells, from which the complex and specialized cells of the gonads, bone marrow, skin, and the alimentary canal are generated.

4. Alveolar cells of the lungs, which are essential in oxygen-carbon dioxide exchange of the body.

5. Bile duct cells.

6. Cells of the tubules of the kidneys.

7. Endothelial cells, which line the closed cavities of the body, such as the heart and the blood vessels.

8. Connective tissue cells, which are structural cells of the tissues supporting the organs and other specialized tissues of the body.

9. Muscle cells.

10. Bone cells.

11. Nerve cells.

The relatively short life span of blood cells may explain their radiosensitivity. The life spans for typical blood cells are:

Cells	Source	Life Span
Erythrocytes	Bone marrow	17 weeks
Granulocytes	Bone marrow	3 days
Lymphocytes	Lymphatic tissue	8–24 hr
Platelets	Bone marrow	3–6 days

In considering internal radiation exposures, in which various radioactive materials, either naturally occurring or synthetic, are inhaled, ingested, or otherwise gain entry into the body, it is interesting to note that certain radionuclides or isotopes tend to concentrate in a few vital organs, especially the lungs, kidneys, liver, and bone. This property, coupled with the known half-life of the isotope and the biological excretion rate (at which the material is known to be eliminated from the body), makes monitoring of exhaled breath, urine, and other body wastes a precise method for indicating the body burden, or the total amount of the radioactive substance actually in the body. Whole body counters of great sensitivity have also been used to measure the radiation emitted from the body by internally deposited radioactive materials (Fig. 37.2). Some isotopes known to concentrate in certain body organs include[16]:

Lungs	Kidneys	Liver
Nickel-63	Chromium-51	Manganese-56
Radon-222	Manganese-56	Nickel-59
Polonium-210	Germanium-71	Cobalt-60
Uranium-238	Arsenic-76	Copper-64
Plutonium-239	Rhodium-105, 106	Silver-105
	Ruthenium-106	Cadmium-109
	Technecium-127	Silver-109, 111
	Tellurium-129	
	Iridium-190, 192	
	Gold-198	
	Uranium-238	

Bone	Bone
Beryllium-7	Cerium-144
Carbon-14	Promethium-147
Fluorine-18	Samarium-151
Phosphorus-32	Europium-154
Calcium-45	Holmium-166
Vanadium-48	Thulium-170
Zinc-65	Lutecium-177
Gallium-72	Tungsten-185
Strontium-89, 90	Lead-203
Yttrium-90, 91	Radium-226
Niobium-95	Uranium-233
Molybdenum-99	Thorium-234
Tin-113	Plutonium-239

Bone	Bone
Barium-140	Americium-241
Lanthanum-140	Curium-242
Praseodymium-143, 144	

The major chronic effect of radiation exposure to the somatic cells of the body is considered to be carcinogenesis; genetic effects, such as mutations and chromosome aberrations, are known to occur in animals but have not been unequivocally established in humans.[30] Consequently, the mutagenic and carcinogenic potential of ionizing radiation to humans has been the object of intense study by various agencies of the United States government, such as the National Academy of Sciences Committee on the Biological Effects of Atomic Radiation (BEAR Report of 1956) and the Committee on the Biological Effects of Ionizing Radiations (BEIR). A 1980 report[30,31] from the BEIR Committee as well as a 1979 report[32] from the Interagency Task Force on the Health Effects of Ionizing Radiation addressed two areas of special concern: low-level exposures and carcinogenesis, respectively. The former report has been, as its two predecessors were, an object of careful scrutiny and widespread controversy, since several methodologies are available for estimating cancer or risk potential at low levels of exposure and none has been unequivocally established as accurate. The question of the hazard of long-term exposure to low-level doses (below 50 rems) of radiation remains unresolved at present.[30] The latter report[32] also dealt with this problem and pointed out the lack of sufficient data to make reliable predic-

Figure 37.2. Whole body counter for counting internal radiation.

tions of the risk of exposure. Current estimates of the production of radiogenic cancers indicate that the increases in the incidence of various cancers due to radiation exposure are as follows:

Type of Cancer	Lifetime Incidence/10^6 Person/rem
Breast, thyroid	100
Lung	25–50
Leukemia (acute and chronic granulocytic)	20–50
Stomach, liver, colon	10–15
Bone, esophagus, small intestine, pancreas, rectum, lymphatic tissue	2–5

These figures are of more concern to occupational exposure than to background exposure situations. Although cancers have been reported in uranium miners and workers in nuclear facilities, the measured dose levels were low (in some cases well below allowable limits) and the interpretation of the results is unclear. Presumably, as the use of radioactivity increases, exposure will be more widespread, and the incidence of radiogenic cancers will increase. Current occupational exposure limits are discussed later in this chapter.

Control of Exposure to Radiation

Regardless of the source (nuclear or electronic) or type of ionizing radiation, a few fundamental control measures are appropriate. No unnecessary exposure should be permitted, and the risk of exposure should always be balanced against the importance of the results to be obtained. Stronger sources require meticulous care or complex precautions to prevent excessive exposures of the operators or patients. Preplanning of a facility, careful selection and use of equipment, standardized procedures, and adequate education and training of personnel are important aspects of a radiation program, especially where large sources are involved. Every operation or activity involving ionizing radiation should be preplanned and, if possible, rehearsed with a dry run to ensure that the desired operation can be accomplished in the minimum amount of time with the least exposure. This does not mean that exposure should be avoided at all times but, rather, that exposures should be controlled and measured so they can be kept within the limits that have been established. Obviously, amounts of radioactivity (in the case of nuclear sources) or voltages and times of exposures (in the case of x-ray and other electronic sources) should be kept to a workable minimum. From a practical viewpoint, most exposures are controlled by the application of three principles: time, distance, and shielding.

TIME

Radiation dose is directly related to the time of exposure—this elementary fact is one of the keys to radiation control. For this reason, each action or activity in the radiation field or area should be preplanned so that delays or other complications do not result in unnecessarily long exposure. The formula

$$\text{exposure dose} = \text{rate} \times \text{time}$$

is fundamental, since time is the easiest factor for an individual to control. Stopwatches, clocks, or calendars thus become important instruments in the radiation control program, and the importance of time should not be minimized. Keeping all other factors equal, the same total dose is less likely to cause biological damage the longer the period of time over which it is accumulated.

DISTANCE

Next to time, distance is probably the second most fundamental concept in radiation control. Radiation follows the same inverse square law that light does, namely, the intensity decreases as the inverse square of the distance from the source. This means that, if the radiation flux or level is 1 at a given distance, it will be only $\frac{1}{4}$ at twice the distance, since $(\frac{1}{2})^2 = \frac{1}{4}$. To illustrate further: if a source has an exposure rate of 100 mR/h at a distance of 1 ft, the corresponding rate for other distances may be tabulated as follows:

Distance (ft)	Ratio	Exposure Rate (mR/h)
1	1	100
2	$\frac{1}{4}$	25
5	$\frac{1}{25}$	4
10	$\frac{1}{100}$	1

In practice, distance may be used to advantage in decreasing radiation exposures by remote-handling aids, such as tongs, long-handled wrenches, cranes, and remotely operated mechanical hands. Viewing aids, such as mirrors, telescopes, closed-circuit television, and windows made of leaded glass and installed so they may be filled with water solutions of heavy-metal salts, permit visibility from a distance.

SHIELDING OR FILTERS

Any matter or mass placed between a radiation source or beam and the point of exposure will screen or decrease the exposure to at least some extent. As noted previously, even air has some shielding effect. Whether or not shielding of a particular type is adequate for a given situation depends on the specific exposure conditions, and each type of radiation is most effectively controlled by a particular shielding material. The type and energy of radiation must be considered in choosing the kind of shield required to cut levels of radiation down to acceptable limits. Lead and other heavy metals are excellent shields for radiation, but their cost is high. Concrete mixed with iron ore or heavy mineral aggregate (to increase the density) is effective but expensive and bulky. Poured walls, precast concrete made with iron ore or heavy aggregates in walls and ceilings, or solid concrete blocks laid into wall form are the most common shielding materials.

The concrete thickness estimated to be necessary to reduce radiation levels to established limits for personnel near a 1-MV electron beam generator is 2.5 ft; for a generator of 2 MV, 4 ft. These numbers are estimates, and the exact thickness will depend on the voltage and wattage of the electron-beam generator, size of the room, room geometry, and occupancy of adjacent areas. An alternate scheme is to construct two walls of timber and planking, filling the space between with poured sand. Such a barrier is easily erected and dismantled, and the construction materials may be reclaimed. However, the lower density of the sand as a barrier requires that its thickness be about 70% greater than the equivalent concrete shielding, thus requiring a greater overall floor area.

A frequently used approach is to locate a machine or source in a pit or depression underground, so that the earth acts as a natural shield on the sides and bottom. Deep pools of water are also used as shielding material if the material is not adversely affected by water. In this connection, it has been reported that the total radiation exposure of personnel aboard an atomic-powered submarine when submerged is less than that to a person on land near the same area, because of the shielding of the cosmic and other background radiation by the water.

It can be noted that safe work practices are as important to limiting exposures to radioactive materials and ionizing radiation as engineering controls. Uranium mining and milling, for example, has been studied for its impact on worker health and radiogenic cancer.[33] The U.S. DOT has established guidelines for the safe transport of radioactive materials,[34a,b] and the Nuclear Regulatory Commission and the National Research Council have instituted methods for safely transporting and handling special nuclear materials and wastes.[35,36] Such methods are an invaluable source of information on the potential hazards of radionuclides and how exposures to these substances can be controlled.

Limits of Exposure

Any attempt to establish exposure limits must be predicated on the belief that exposure to radiation should be kept at the lowest levels practical, always balancing the benefits to be gained from the use of radiation with the potential hazards of exposure and recognizing the established limits. As in many other phases of life, the gain should be weighed against the risk to achieve an understanding of the necessity for the risk.

EXTERNAL DOSE LIMITS

The permissible limits for external radiation exposure are:

Site of exposure	Dose for 13 consecutive weeks (rem)	Accumulated dose
Whole body, head, and trunk, blood-forming organs, gonads, lens of the eye, and other organs	3 rem	5 (*N*-18) rem[a]
Skin of the whole body	10 rem	30 rem/year
Hands and forearms, feet and ankles	25 rem	75 rem/year

[a]*N* is the age in years, greater than 18.

INTERNAL DOSE

Since the main route of entry of radioactive materials is the respiratory tract, the primary method for controlling exposure is to limit the average rates at which materials are breathed. Permissible concentrations have also been recognized for drinking water and foodstuffs. Handbook No. 69, "Maximum Permissible Body Burdens and Maximum Permissible Concentrations of Radionuclides in Air and in Water for Occupational Exposure," published by the National Bureau of Standards, United States Department of Commerce, Washington, D.C., provides further information on this subject.

PERMISSIBLE LIMITS FOR NONOCCUPATIONAL EXPOSURE

For external radiation exposure, the maximum dose for nonoccupational exposure to the whole body, head, and trunk, active blood-forming organs, gonads, or lens of the eye should not exceed 0.5 rem in any year. Handbook No. 69 also presents internal exposure limits.

37.4. SUMMARY

The material presented in this chapter is intended to serve as an introduction to an understanding of radiation as a controllable hazard. Perhaps more than any other field of scientific knowledge, even within the scientific community itself, radiation is inadequately understood and even feared. The plea of Madame Curie, one of the discoverers of radium and a pioneer in radiation research, still applies:

> Nothing in life is to be feared;
> It is only to be understood.

The attached references furnish a key to this understanding, and the reader is urged to consult them in detail.[25,26]

REFERENCES

1. *Medical Physics*, The Year Book Publishers, Chicago, 1944, pp. 1145–1164.
2. W. W. Mumford, Some Technical Aspects of Microwave Radiation Hazards, *Proc. IRE*, **49**, 427–447 (February 1961).
3. M. F. Payton, *Proceedings of the Fourth Annual Tri-Service Conference on the Biological Effects of Microwave Radiation*, Vol. 1, Plenum Press, New York, 1961.
4. G. H. Mickey, "Electromagnetism and Its Effect on the Organism," *N.Y. State J. Med.*, **63**, 1935–1942 (July 18, 1963).
5. L. Koller, *Ultraviolet Radiation*, Wiley, New York, 1952.
6. P. W. Kruse, L. D. McGlauchlin, and R. B. McQuistan, *Elements of Infrared Technology: Generation, Transmission, and Detection*, Wiley, New York, 1962.
7. S. Duke-Elder, *System of Ophthalmology*, Vol. 7, C. V. Mosby Co., St. Louis, 1962, pp. 759–763.
8. M. Brotherton, *Masers and Lasers: How They Work, What They Do*, McGraw-Hill, New York, (1964).
9. H. E. Tebrock, W. N. Young, and W. Machle, "Laser-Medical and Industrial Hygiene Controls," *J. Occup. Med.*, **5**, 564–567 (December 1963).
10. M. S. Litwin and D. H. Glew, "The Biological Effects of Laser Radiation," *J. Am. Med. Assoc.*, **187**,(11) 842–847 (March 14, 1964).
11. "Laser Seen as a Potentially Useful Tool, but Scientists Hear Warnings of Hazards," *J. Am. Med. Assoc.* **188**, 35 (June 15, 1964).
12. M. Eisenbud, *Environmental Radioactivity*, McGraw-Hill, New York, 1963.
13. W. W. Schultz and R. S. Rocklin, *Radioisotopes in Industry*, Reinhold Pilot Book, New York, 1959.
14. "Research Laboratory Bulletin," Special Issue on Nuclear Fusion, GP-0288A, Research Laboratory, General Electric Co., Schenectady, Summer 1964.
15. A. Charlesby, *Atomic Radiation and Polymers, Typical Radiation Sources and Their Intensity*, 1960, p. 53.
16. D. T. Goldman, "Chart of the Nuclides," APH 66F, Educational Relations, Department MWH, General Electric Co., Schenectady, (December 1962).
17. W. J. Price, *Nuclear Radiation Detection*, McGraw-Hill, New York, 1962. Hanson Blatz, *Radiation Hygiene Handbook*, McGraw-Hill, New York, 1962. D. C. Fleckenstein, *Radiation and Radiation Protection*, Z-3531, General Electric Co., Schenectady, 1962.
18. "Instrumentation and Monitoring Methods for Radiation Protection," NCRP Report No. 57, and "A Handbook of Radioactivity Measurements Procedures," NCRP Report No. 58, National Council on Radiation Protection and Measurements, Washington, D.C., 1978.
19. "Radiological Monitoring Methods and Instruments," Handbook 51, National Bureau of Standards, Washington, D.C., 1952.
20. *"Question,"* periodic publication on radiation control, R. S. Landauer, Jr. and Co., Matteson, Ill., 1968.
21. R. L. Fleischer, P. B. Price, and R. M. Walker, "Track Registration in Various Solid-State Nuclear Track Detectors," *Phys. Rev.*, **133**(5A) A1443–A1449 (March 2, 1964).
22. S. J. Malsky, B. Roswit, C. G. Amato, H. M. Jones, B. Reid, and H. Patterson, "Measurement of Radiation Dosage," *J. Am. Med. Assoc.* **187**(11), 839–841 (March 14, 1964).
23. *Pinhead and Planets*, courtesy Atomic Power Equipment Department, General Electric Co., San Jose, Calif., 1964.
24. *Atomic Radiation*, R.C.A. Service Co., Camden, N.J., 1959, pp. 53–59.
25. *Nuclear Terms, a Brief Glossary*, U.S.AEC, Oak Ridge, 1964.

26. D. F. Janes, "Ionizing Radiation—The Safety Engineer's Most Technical Challenge," *J. Am. Soc. Safety Eng.*, **LX**(7) 9–14 (July 1964).

27. "U.S. Congressional Hearings on Nature of Radiation Fallout and Its Effects on Man" (1956). Report of the United Nations Scientific Committee on Atomic Radiation, New York, 1958.

28. "The Radon Question," *Consumer's Research Magazine,* **63**(11), 12–14 (November 1980).

29. *McGraw-Hill Encyclopedia of Science and Technology,* Vol. 11, McGraw-Hill Book Company, New York, 1977.

30. George M. Wilkening, "Ionizing Radiation," *Patty's Industrial Hygiene and Toxicology,* Vol. 1. 3rd. rev. ed., G. D. and F. E. Clayton eds., Wiley-Interscience, New York, 1978, pp. 441–512.

31. The Effects on Populations of Exposure to Low Levels of Ionizing Radiation," Committee on the Biological Effects of Ionizing Radiation, National Academy of Sciences of the United States of America, National Academy Press, Washington, D.C., 1980.

32. "Report of the Working Group on Science," Interagency Task Force on the Health Effects of Ionizing Radiations, U.S. Department of Health and Human Services, June 1979.

33. Michael H. Momeni, "Pathways of Exposure from Uranium Mining and Milling Operations," American Chemical Society Meeting, Washington, D.C., September 9–14, 1979.

34. "Proposed Rules," *Fed. Reg.*, **44**(5) 4910-60-M, 1852–1885 (January 8, 1979).

35. Final Environmental Statement on the Transportation of Radioactive Material by Air and Other Modes (40 FR 23768), NUREG-0170, Office of Standards Development, U.S. Nuclear Regulatory Commission, December 1977.

36. "The Geological Criteria for Suitable Sites of High-Level Radioactive Waste Repositories," Committee on Radioactive Waste Management, National Research Council, National Academy of Sciences of the United States of America, 1978.

BIBLIOGRAPHY

Biological Effects of Radiation

Andrews, H. L., *Radiation Biophysics,* Prentice-Hall, Englewood Cliffs, N.J., 1961.

Biological Effects of Radiation, Vol. 11, Proceedings of the International Conference on the Peaceful Uses of Atomic Energy, Geneva, 1955, United Nations Publications, Sales No. 1956 IX.1, 11, New York, 1956.

Biological Effects of Radiation, Vol. 22, Proceedings of the Second United Nations International Conference on the Peaceful Uses of Atomic Energy, Geneva, 1958, United Nations, Geneva, Sales No. 58 IX.2, 22, 1958.

Browning, E. *Harmful Effects of Ionising Radiations,* Elsevier, New York, 1959.

Elson, L. A., *Radiation and Radiomimetic Chemicals: Comparative Physiological Effects,* Cancer Monograph Series, Butterworth, Washington, D.C., 1963.

Frigerio, N. A., "Your Body and Radiation," U.S. Atomic Energy Commission Office of Information Services (now Department of Energy), Library of Congress Catalogue Card No. 67-60927, 1967.

Glasstone, S., *The Effects of Nuclear Weapons,* prepared by the U.S. Department of Defense, U.S. Atomic Energy Commission, April, 1962, U.S. Government Printing Office, Washington, D.C. (Nuclear bomb effects computer insert is sold separately.)

Harris, R. J. C. *The Initial Effects of Ionizing Radiations on Cells,* Academic Press, New York, 1961.

Hollaender, A., *Radiation Protection and Recovery,* Pergamon Press, New York, 1960.

Novick, Sheldon *The Careless Atom,* Houghton Mifflin, Boston, 1969.

Radiological Health Handbook, rev. ed., U.S. Dept. of Health, Education, and Welfare, Bureau of Radiological Health, Rockville, Md., January 1970.

Shapiro, J., *Radiation Protection: A Guide for Scientists and Physicians,* Harvard University Press, Cambridge, Mass., 1978.

Simmons, G. H., "A Training Manual for Nuclear Medicine Technologists," BRH/DMRE 70-3, U.S. Dept. of Health, Education, and Welfare, Public Health Service, Bureau of Radiological Health, Rockville, Md. October 1970.

Umbarger, C. J., "New Small Devices for Radiation Detection: the Wee Pocket Chirper and the Portable Multichannel Analyzer," LASL Mini-reviewer, LASL 80-35, Los Alamos Scientific Laboratory, Los Alamos, Mexico 87545, August 1980.

Wang, C. H. and Willis, D. L., *Radiotracer Methodology in Biological Science,* Prentice-Hall, Englewood Cliffs, N.J., 1965.

Isotopes

"A" is for Atom, 16 mm sound color movie, 15 minutes, produced in 1953 by General Electric Co., Schenectady, available on loan from Division of Public Information, U.S. Atomic Energy Commission (now U.S. Department of Energy), Washington, D.C.

"Radioisotopes," a reprint with revisions from *Atomic Energy Facts,* Item No. 220, U.S. Atomic Energy Commission (now U.S. Department of Energy), Division of Isotopes Development, Washington, D.C.

"Radioisotopes in Science and Industry," A Special Report of the U.S. Atomic Energy Commission (now U.S. Department of Energy), U.S. Government Printing Office, Washington, D.C. January 1960.

Rothman, M. A. *The Short-Lived Radioactive Isotopes,* reprinted from *Foote Prints,* 29(1) (1957). [Available from Foote Mineral Co., 18 West Chelten Ave., Philadelphia, Pa.]

"Special Sources of Information on Isotopes in Industry, Agriculture, Medicine, and Research (a bibliography)," T.I.D. 4563 (3rd rev ed.), U.S. Atomic Energy Commission, Division of Isotopes Development, Washington, D.C., January 1962.

Laws and Standards

American Nuclear Standards are available from American National Standards Institute, 1430 Broadway, New York, N.Y. 10018.

Federal Law 10 CFR 20, Title 10, Atomic Energy, Chapter 1, AEC, Part 20, Available from Division of Licensing and Regulation (now U.S. Department of Energy and Nuclear Regulatory Agency), Washington, D.C.

Ionizing Radiation, The Sanitary Code, Chapter XVI, New York State Department of Health, Albany, New York.

National Committee on Radiation Protection and Measurement: handbooks containing these recommendations are published by the National Bureau of Standards, U.S. Department of Commerce, Washington, D.C. Twenty-one handbooks on specific phases of radiation protection are currently available from U.S. Government Printing Office, Washington, D.C.

National Safety Council

"Sources of Information on Nuclear Energy," Data Sheet 446, "X-Rays in Industry," Data Sheet 475; "Ionizing Radiation." See also *Ionizing Radiation, Accident Prevention Manual for Industrial Operations*, 5th ed., National Safety Council, Chicago, Chap. 42, pp. 42–41 to 42–51

State, city, county, town, and other local regulations, laws, and codes should be consulted.

Periodicals

American Journal of Roentgenology, Radium Therapy, and Nuclear Medicine, monthly publication, Charles C. Thomas, 301 E. Lawrence Ave., Springfield, Ill.

Atomics, bimonthly publication, Technical Publishing Co., 308 East James St., Barrington, Ill.

Bertell, R., "The Nuclear Worker and Ionizing Radiation," *Am. Ind. Hyg. Assoc. J.* **40**(5), 395–401 (1979).

Health Physics, Official Journal of the Health Physics Society, published bimonthly by Pergamon Press, New York.

The International Journal of Applied Radiation and Isotopes, a monthly publication, Pergamon Press, New York.

Nucleonics, monthly publication, McGraw-Hill, New York.

Organic Substances with Radiation

Samuel, A. H., "Radiation Chemistry," *Ind. Res.,* **6**(3), 42–47 (March 1964).

Charlesby, A., *Atomic Radiation and Polymers,* Vol. I, *International Series of Monographs on Radiation Effects in Materials,* Pergamon Press, New York, 1960.

Bolt, R. O. and Carroll, J. G., *Radiation Effects on Organic Materials,* Academic Press, New York, 1963.

Bovey, F.A., *The Effects of Ionizing Radiation on Natural and Synthetic High Polymers,* Interscience, New York, 1958.

"Report of the United Nations Scientific Committee on the Effects of Atomic Radiation," General Assembly, Official Records: Seventeenth Session, Supplement No. 16 (A/5216), United Nations, New York 1962.

Selected Bibliography of Radiation Protection Organizations, Committee on Ionizing Radiation, American Conference of Governmental Industrial Hygienists, Cincinnati, Ohio 1963.

Oerlein, K.F., *Radiological Emergency Procedures for the Nonspecialist, Prepared for the Interagency Committee on Radiological Assistance, Training Division, Headquarters, Defense Atomic Support Agency, Washington, D.C., May 1964.*

Finkel, A. J. et al., "Radiobiological Parameters in Human Cancers Attributable to Long-Term Radium Deposition," IAEA-SM-118/7, International Atomic Energy Agency, Vienna, Austria 1969 (Reprint from *Radiation-Induced Cancer*).

Finkel, A. J. et al., "Radium-Induced Malignant Tumors in Man," (reprinted from: Delayed Effects of Bone-Seeking Radionuclides, C. W. Mays et al., University of Utah Press, Salt Lake, Utah 1969.)

Proceedings, National Conference on X-Ray Technician Training, held at Center of Adult Education, Univ. of Maryland, College Park, Md. September 7–9, 1966, Division of Radiological Health, Public Health Service, U.S. Dept. of Health, Education and Welfare, Rockville, Md. 1966.

Holthusen, H., Meyer, H. and Molineus, W., *Ehrenbuch der Rontgenologen and Radiologen aller Nationen, Zweite Auflage,* Verlag von Urban & Schwarzenberg, Munchen und Berlin, 1959 (in German) (sketches of 359 persons who died from excessive exposure to radiation up to 1959).

Mathews, J., FDA, Answering Medfly Threat, Eases Produce Irradiation Bans, Washington Post, p. A-10, Dec. 12, 1981.

Appendix **1**

Data on Selected Hazardous Chemicals†

The following table lists important physical chemical properties, as well as information for safe storage and handling, of many typical chemical compounds commonly encountered in laboratories. The data are condensed from various references as noted. Codes used in presenting data and sources of data are as follows:

1. *Physical State.*   G—gas, L—liquid, S—solid.
2. *Water Solubility.*   (Solubility in water at room temperature.)
   M—Miscible (infinity soluble)
   V —Very soluble (50 g/100 ml H_2O)
   S —Soluble (5–50 g/100 ml H_2O)
   F —Fairly soluble (5 g/100 ml H_2O)
   I —Insoluble
   D—Decomposes
   Source: *Handbook of Laboratory Safety*, 2nd Ed., CRC Press, Boca Raton, FL, 1971.
3. *Specific Gravity.*   Ratio of weight of substances to the weight of an equal volume of water, rounded off to nearest tenth. Source: NFPA Code No. 325M, 1977.
4. *Vapor Density.*   Ratio of weight of a vapor or gas (without air pressure) to the weight of an equal volume of air, rounded off to the nearest tenth. Source: NFPA Code No. 325M, 1977.
5. *Boiling Point (°C).*   Temperature of liquid at which its vapor pressure equals atmospheric pressure.

---

† For substances not listed, refer to Index and to references given in chapters in text. For on-line interactive data for chemical, toxicological, and environmental effects, refer to National Library Medicine *Toxicology Data Bank*, operated by Union Carbide, Information Center Complex, Oak Ridge, TN 37830. Phone contact is (615) 574-7587. For relation of materials and processes to hazard control, refer to *Recognition of Health Hazards in Industry*, by W. A. Burgess, Wiley-Interscience, N.Y., 1981.

D—Decomposes

Source: NFPA Code No. 325, 1977.

6. *Flash Point (°C).*   Temperature at which liquid gives off sufficient vapor to form an ignitable mixture with air near the surface of the liquid. Flash point data represent closed cup tests, except where open cup test is designated by an asterisk (*). Source: NFPA Code No. 325M, 1977.

7. *Ignition Temperature (°C).*   Minimum temperature required to initiate self-sustained combustion of the substance. Source: NFPA Code No. 325M, 1977.

8. *Flammability Limits (% by volume in air).*   Lower Limit—Minimum concentration of vapor in air below which burning or explosion cannot occur.

9. *Flammability Limits.*   Upper Limit—Maximum concentration of vapor in air above which burning or explosion cannot occur).

Source: NFPA Code No. 325M, 1977.

10. *Class of Flammable or Combustible Liquid, NFPA Code No. 30, 1977.*

11. *Type of Hazard*

    1—Flammable material.

    2—Oxidizing material; contact with other combustible may cause fire.

    3—Gas or vapor rapidly toxic or extremely irritating on exposure for short time or to low concentrations.

    4—Gas or vapor harmful or irritating on prolonged or repeated exposure, or exposure to high concentrations.

    5—Gas or vapor physiologically inert, but displaces oxygen available for breathing.

    6—Dust hazardous when inhaled or touched.

    7—Irritant, sensitizer, corrosive; causes skin irritation or burn.

    8—Toxic through skin absorption.

    9—Impact or shock sensitive, spontaneously ignites, decomposes or polymerizes.

    Source: Manual L-1, Chemical Manufacturers Association, Inc.

12. *Hazard Ratings.*   Health (type of possible injury):

    0—On exposure under fire conditions, material is no more hazardous than is an ordinary combustible (e.g., paper or rags).

    1—Exposure causes irritation, but only minor residual injury, even if untreated.

    2—Intense, continued exposure could cause temporary incapacitation or possible residual injury, unless promptly treated.

    3—Short exposure could cause serious temporary or residual injury even if promptly treated.

4—Very short exposure could cause death or major residual injury, even if promptly treated.

13. *Hazard Rating.* Flammability (susceptibility of material to burning:

0—Will not burn.

1—Must be preheated before ignition can occur.

2—Must be moderately heated or exposed to relatively high temperature environment before ignition can occur.

3—Can be ignited at almost all temperatures.

4—Will rapidly or completely vaporize at atmospheric pressure and normal temperatures, or will readily disperse in air and burn easily.

14. *Hazard Rating.* Reactivity (susceptibility of materials to release energy either by themselves or in combination with other materials).

See also NFPA Code No. 491-M, 1977.

0—Normally unstable; even under fire exposure conditions.

1—Normally stable except in combination with certain other materials or at elevated temperatures and pressures.

2—Normally unstable; readily undergoes violent chemical change at elevated temperatures and pressures.

3—Can detonate or explode under a strong initiating force or after heating under confinement.

4—Readily detonates or explodes at normal temperatures and pressures.

Source: NFPA Codes No. 49, 1975 and 325M, 1977.

15. *Life Hazards.* Skin: Local action on normal skin from brief contact with the undiluted material. Source: "Occupational Diseases," DHEW(NIOSH)77-181, 1977.

1—Relatively harmless.

2—Sensitizer, can cause allergic reactions.

3—Irritant, can cause inflammation.

4—Primary skin irritant, can cause severe eruptions and burns.

16. *Life Hazard.* Oral: Amount to produce death when swallowed by an average (150 lb) man. Source: Gleason, Gosselin, and Hodge, "Clinical Toxicology of Commercial Products", 2nd ed., Williams and Wilkins, Baltimore, 1963.

1—Practically nontoxic, takes more than one quart (2 lb).

2—Slightly toxic, takes one pint to one quart.

3—Moderately toxic, takes one ounce (30 g) to one pint.

4—Very toxic, takes one teaspoon (4cc) to one ounce (30 g).

5—Extremely toxic, takes 7 drops to one teaspoon (4 cc).

6—Super toxic, takes just a taste (7 drops or less).

17. *Life Hazard.* Respiratory (threshold limit value, ppm). The

threshold limit values refer to air-borne concentrations of substances and represent conditions under which it is believed that nearly all workers may be repeatedly exposed, day after day, without adverse effect. Threshold limits should be used as guides in the control of health hazards and should not be regarded as fine lines between sale and dangerous concentrations. Source: "Threshold Limit Values", American Conference of Governmental Industrial Hygienists, revised annually. All values are expressed in parts per million parts of air (ppm), except where units of milligrams per cubic meter of air (mg/M^3) is designated by an asterisk (*). Additional listings are:

Ala—Human carcinogen with assigned TLV
Alb—Human carcinogen without assigned TLV
A2 —Suspect of carcinogenic potential for humans
F   —Simple asphyxiant
sk —Skin

18. *Precautions to Take.*

1—Keep away from heat, sparks, open flame.
2—Avoid spilling, contacting skin, eyes, clothing. May require gloves, goggles, apron, etc.
3—Use adequate ventilation. Avoid breathing dust, fumes, mists, gases or vapors. Personal respiratory protection may be required.
4—Avoid contact with acids, combustibles, moisture.
5—Do not handle until safety precautions outlined in References (Column 19) are understood.

19. *Fire Extinguishing Method.*

0—No water.
1—Water spray.
2—Carbon dioxide.
3—Dry powder.
4—Special extinguishing agent.
5—Do not attempt to fight fire.
6—Stop flow of gas.

Source: NFPA Code No. 325M (1977).

20. *References.* Data sheets on individual compounds are available from organization as follows:

1—American National Standards Institute (ANSI) (Formerly American Standards Association).
2—American Industrial Hygiene Association (AIHA).
3—National Safety Council (NSC).
4—American Insurance Association (AIA).
5—Compressed Gas Association (CGA).

Hazardous Materials Data Table (page 869)

NAME OF COMPOUND	PHYSICAL STATE	WATER SOLUBILITY	SPECIFIC GRAVITY	VAPOR DENSITY	BOILING POINT (°C)	FLASH POINT (°C)	IGNITION TEMPERATURE (°C)	FLAMMABILITY LIMITS Lower (vol.% in air)	FLAMMABILITY LIMITS Upper	NFPA CLASS	TYPE OF HAZARD	HAZARD RATING Health	HAZARD RATING Flammability	HAZARD RATING Reactivity	LIFE HAZARDS Skin	LIFE HAZARDS Oral	LIFE HAZARDS Respiratory (TLV - ppm)	PRECAUTIONS TO TAKE	FIRE EXTINGUISHING METHODS	REFERENCES
COLUMN NO.	1	2	3	4	5	6	7	8	9	10	11	12	13	14	15	16	17	18	19	20
Acetaldehyde	L	M	0.8	1.5	21	-39	175	4.0	60	IA	1,4,7	2	4	2	2,4	3	100	1,2,3	1,2,3	2
Acetic Acid, Glacial	L	M	1.0	2.1	118	39	465	5.4	20	II	7,4,1	2	4	2	2	2	10	2,3	0,2,3	2,3,4
Acetic Anhydride	L	D	1.1	3.5	140	49	390	2.7	10	II	7,3,1	2	2	1*	2	2	5	2,3,4	0,2,3	3
Acetone	L	M	0.8	2.0	56	-20	465	2.6	13	IB	1	1	3	0	3	3	1000	1,3	1,2,3	2,3,4
Acetonitrile	L	M	0.8	1.4	82	6	524	3.0	16	IB	3,8,1	2	3	1	2	2	40	1,2,3	2,3	2
Acetyl Chloride	L	D	1.1	2.7	51	4	390	5.0	100	IB	7,1,3	3	3	2*	4	2	-	1,2,3	0,2,3	3,5
Acetylene	G	V	0.9	0.9	-83		305	2.5			1,5	3	4	3	1	-	-	1,3	6	2,4
Acrylonitrile	L	V	0.8	1.8	77	0	481	3.0	17	IB	3,8,1	4	3	1	4	4	A1a	3,2,1	2,3	3
Aluminum Dust	S	I	2.7								7				1	-	10*	1	0,4	-
Aluminum Chloride	S	V	2.4								7	3			3	3	2*	2,4	-	3
Aluminum Stearate	S	I	-								1				-	-	-	-	-	-
2-Aminoethane	L	M	1.0	2.1	172	85	410	15	28	IIIA	1,4,7	2	2		3	3	3	1,2,3	1,2,3	1,2,3,5
Ammonia, anhydrous	G	V	-	0.6	-33		651				3,1,7	3	1	0	4	3	25	3,1,2	6,1	2,3
Ammonium bifluoride	S	V	1.2								6,7	3			4	4	-	3,2,5	-	2,3
Ammonium Fluoride	S	V	1.3								6,7	3			4	4	-	3,2,5	-	2,3
Ammonium hydroxide (28%)	L	S	0.9		302d						7,3	0	0	0	4	-	25	2,3	-	3,5
Ammonium nitrate	S	S	0.8								2,9	0	0		1	3	-	5,4	-	-
Ammonium perchlorate	S	V	0.8								2,9			3	1	2	-	4	1	-
Ammonium persulfate	S	S	0.9								2,9			4	1	1	-	4	-	-
Ammonium vanadate	S	S	0.9								6,7,8				-	-	-	2,3,4	-	-
Amyl acetate	L	F	0.9	4.5	149	16	360	1.0	7.1	IB	1,4,7	1	3	0	3	3	100	1,2,3	1,2,3	2,3
Amyl alcohol, normal	L	F	0.8	3.0	138	33	300	1.4	10	IC	1	1	3	0	3	2	-	1,3	1,2,3	2,3
Amyl alcohol, iso	L	F	0.8	3.0	118	34	435	1.4	10	IC	1	1	3	0	3	2	-	1,3	1,2,3	-
Amyl ether, normal	L	V	0.9	5.5	190	57	170	0.7		II	1,4	1	2	0	1	3	-	1,2,3	1,2,3	-
Amyl nitrite	L	F	0.9	4.0	104		210	1.0		II	9,1,4	1	0	2	-	-	-	2,3	2,3	-
Aniline	L	F	1.3	3.2	184	70	615	1.2	8.3	IIIA	1,8,3	3	2	0	2,4	3	2sk	2,3,5	2,3	2,3
Anthracene	S	I	1.24	6.15	340	121	540	0.65		IIIB	7,8	0	1	0	-	0	-	2,3	2,3	2,3
Antimony, dust	S	I									7,6				-	-	2*	3,2	-	1,2
Antimony trichloride	S	V									7,4				-	-	2*	3,2	-	2,3
Arsenic	S	I									6				2	-	0.2*	2,3	-	2,3
Arsenic trioxide	S	F									4,7,6	3			2,4	5	A1a	3,2,5	-	2,3
Arsine	G	F		1.5							3,8	3			-	-	0.05	2,3	-	2,3
Atropine sulfate	S	V									6,7,8				2	4	-	3,2	-	-
Aziridine (Ethyleneimine)	L	M	0.8	1.5	56	-11	320	3.6	46	IIB	1,3,7,8	3	3	3	2,3	-	-	1,2,3	1,2,3	-
Barium sulfate	S	I										3	3		3	-	-	3,2	-	-

Name of Compound	Physical State	Water Solubility	Specific Gravity	Vapor Density	Boiling Point (°C)	Flash Point (°C)	Ignition Temperature (°C)	Flammability Limits — Lower (vol.% in air)	Flammability Limits — Upper (vol.% in air)	NFPA Class	Type of Hazard	Health	Flammability	Reactivity	Skin	Oral	Respiratory (TLV – ppm)	Precautions to Take	Fire Extinguishing Methods	References
Column No.	1	2	2	4	5	6	7	8	9	10	11	12	13	14	15	16	17	18	19	20
Benzaldehyde	L	F	1.1	3.7	179	63	192			IIIA	1,4,8 / 8,6	2	2	0	2	4	10,A2	1,3	1,2,3	-
1,2 Benzanthracene	S	I													3	3		1,3	1,2,3	1,2,3,4
Benzene	L	F	0.9	2.8	80	-11	560	1.3	7.9	IB	1,4,8 / 4,7,8	2	3	0	4	3		1,3,5 / 2,3	1,2,3	-
Benzenesulfonic acid	S	S									7,8				4	-		2,3	1,2,3	1,2,3,4
Benzidine	S	S									7,8	2			-		A1b	2,3,5	1,2,3	2,3,4
Benzoyl chloride	L	D								IIIA	1,4,7 / 9,2,1,7	3	2	1	3	3	5*	1,4,3	0,2,3	-
Benzoyl peroxide	S	F	1+								7,4,1	2	4	4	4	3	5*	5,1,4	5	-
Benzyl alcohol	L	F			206		436			IIIA	7,4,1	2	2	0	3	2	1	2,3	1,2,3	-
Benzyl chloride	L	I	1.1	4.4	179	93	585	1.2			6,7	2	1	1	4	2	1	2,3	1,2,3	-
Beryllium	S	-									6,7	4	1	0	2,4	3	0.002*	5,3	0,4	2,3,4
Biphenyl	S	I	1.2		254	113	540	0.7	5.8	IIIB	1,4,7,8 / 3,3	2	1	0	3	3	0.2	1,2,3 / 5,1,3	1,2,3 / 0,2,3	2
Boron hydrides (decaborane)	S	D	0.9		213	80		0.2		IIIA	3,7	4	0	1	4	2	0.05*	5,1,3	0,2,3	2,3,4
Boron trifluoride	G	F									7,4,2	4	0	0	4		0.1	5,3	1	2,3,4
Bromine	L	F									1,3	2	0	2	4		1000	1,3	6,2,3	2
Butadiene	G	I	1.9		-4		420	2.0	12		1,3	2	4	2	1					
Butane	G	V	2.0	2.0	-1		405	1.8	8.4		1,6,7	1	4	0	1		600	1,2,3	1,2,3,6	2
Butyl alcohol, tert.	L	S	0.8	2.6	83	11	478	2.4	8.0	IB	7,4,1	1	3	0	4	3	100	3,2,1	1,2,3	2
Butylamine	L	M	0.8	2.5	78	-12	380	1.7	8.9	IB	1,4,8	2	3	0	2,4	4	5sk	1,2,3	1,2,3	2
Butyl cellosolve	L	-						1.1	11		1,7	3	2		2,4	3	50sk	1,2,3	1,2,3	
Butyllithium	L	-			-192		245				1,7,9	3	4	2	3,4	3		2,3	4	
Butyl mercaptan, normal	L	F	0.8	3.1	98	2				IB	1,4,7 / 9,1,4	2	3	0	3	3	0.5	1,2,3 / 4,1	1,2,3	
Butyl nitrite	L	I									6				3	3	0.05*	3	5	3,4
Cadmium compounds	S	-										1	4	2	2	4	0.05*	2,3,4	1	2
Calcium carbide	S	D									6,7	1	0	2	3			2,3,4	0	
Calcium oxide	S	V									6,7	1	0	1	3		2*	2,3,4	-	
Calcium stearate	S	I	1.0	5.6	196	82		1.3	50	IIIA	1	3	2	0	4	3		1,2,3	1,2,3	1,2,3
Caprylyl chloride	L	D	1.3	4.2	46	-30	90	12.5	74	IB	1	2	4	0	4		10	5,1,3,2	0,2,3	1,2,3
Carbon disulfide	L	F	1.0	2.6	-192		609				1,4,8	2	4	0	1	3	50	3,1	1,6	1,2,3
Carbon monoxide	G	F		1.0							4,1	2	4	0	1		10sk	3,2,5		1,2,3,4
Carbon tetrachloride	L	F									3,8	3	0	0	5	4				
Chlorine	G	S	1.1	3.9	132	29	640	1.4	9.6	IC	3,2,7	3	0	0	4	3	75	5,3,4	3,2,1	3,4
Chlorobenzene	L	I									3,7,1	3	3	0	4	3	0.5*	2,3	-	2,4
Chlorodiphenyl (54% Cl)	L	I									4,7	1	0	0	3	2	2	2,3	-	-
Chloroform	L	F	1.1	2.6	45	-32	485	2.9	11.1	IB	4,7	3	0	1	5	3	10,A2	2,3	-	2,4
3-Chloropropene (Allyl chloride)	L	I	0.9	2.6	45	-32	485	2.9	11.1	IB	1,3,7	3	3	1	4	3	1	1,2,3	2,3	2,4

Chemical Hazards Reference Table

Name of Compound	Physical State (1)	Water Solubility (2)	Specific Gravity (3)	Vapor Density (4)	Boiling Point (°C) (5)	Flash Point (°C) (6)	Ignition Temp (°C) (7)	Flam. Limit Lower (8)	Flam. Limit Upper (9)	NFPA Class (10)	Type of Hazard (11)	Health (12)	Flammability (13)	Reactivity (14)	Skin (15)	Oral (16)	Respiratory TLV-ppm (17)	Precautions to Take (18)	Fire Extinguishing Methods (19)	References (20)
Chromic acid	S	V	–	–	–	–	–	–	–	–	2,6,1	3	0	1	4	4	0.05*	4,2,3	–	1,2
Copper nitrate	S	V	–	–	–	–	–	–	–	–	8,7,4	3	0	0	–	4	–	3	–	–
Cresol, ortho	L	F	1.1	3.7	191	81	599	1.1	–	IIIA	1,3,7,8	3	2	0	4	4	5sk	2,3	1,2,3	2
Crotonaldehyde	L	V	0.9	2.4	102	13	232	2.1	16	IB	1,3,7,8	3	3	2	4	3	–	1,2,3	1,2,3	2
Cumene	L	I	0.9	4.1	152	35	425	0.88	6.5	IC	1,4,7	2	3	0	3	2	50sk	1,2,3	1,2,3	2
Cyanamide	S	V	1.07	1.45	260d	141	–	–	–	IIIB	3,7,8	4	1	3	4	–	2*	1,2,3,5	–	2
Cyanogen	G	S	–	1.8	-21	–	–	6.6	32	–	3,6,7,8	4	4	3	–	–	10	1,2,3,5	5,6	2
Cyanogen bromide	S	S	–	–	61	–	–	–	–	–	3,6,7,8	4	0	2	3	–	–	1,2,3,5	–	2
Cyanogen chloride	L	S	–	–	13	–	–	–	–	–	3,6,7,8	3	0	1	3	3	0.3	1,3	6	2
Cyclohexane	L	I	0.8	2.9	82	-20	245	1.3	7.8	IB	1,4,7	1	3	0	3	3	300	1,3	1,2,3	2
Cyclohexylamine	L	S	0.9	3.4	134	31	293	–	–	IC	1,4,7	2	3	0	3	3	10sk	1,2,3	1,2,3	2
Cyclopentanone	L	I	0.9	2.9	131	26	–	–	–	IC	1,4,7	2	3	0	–	–	–	1,2,3	1,2,3	–
Cyclopropane	G	I	–	1.5	-34	–	500	2.4	10.4	–	1,4	1	4	0	–	–	–	1,3	6	–
Decalin (decahydronaphthalene)	L	I	0.9	4.8	194	58	250	0.74	4.9	II	1,4,7	1	2	0	3	3	–	2,3	–	–
Deuterium	G	I	–	–	–	–	–	4.9	75	–	1,5	0	4	0	–	–	–	1	1,2,3,6	–
Diazomethane	G	D	–	–	-23	–	–	–	–	–	1,3,7	3	4	3	–	–	0.2	1,2,3	1,2,3,6	2
Diborane	G	D	–	1.0	-93	–	38	0.8	88	–	1,3,7	3	4	3W	–	–	0.1	1,2,3	0,2,3,6	2
Dichloroacetic acid	L	M	1.6	4.5	194	66	–	–	–	IIIA	4,7,1	3	1	0	3,4	4	–	2,3	–	–
Dichlorobenzene, ortho	L	I	1.3	5.1	180	57	648	2.2	9.2	IIIA	3,7,1	2	2	0	4	3	50	3,2,1	1,2,3	2,4
1,1 Dichloroethane	L	F	1.2	3.4	57	-17	–	5.6	–	IB	3,7,1	2	3	0	3	3	200	1,2,3	1,2,3	–
1,2 Dichloroethane	L	F	1.3	3.4	84	13	413	6.2	16	IB	1,4,7	2	3	1	3	3	50	1,2,3	1,2,3	2
1,1 Dichloroethylene	L	I	1.3	3.4	37	-18	570	7.3	16	IA	1,4,7	2	4	2	3	3	–	1,2,3	1,2,3	2
1,2 Dichloroethylene	L	I	1.3	3.4	61	-6	460	9.7	12.8	IB	1,4,7	2	3	2	3	3	200	1,2,3	1,2,3	2
Dichloromethane	L	S	1.3	2.9	40	–	556	12	19	–	4,7	2	0	0	4	3	100	2,3	2,3	2
Diethylamine	L	V	0.7	2.5	57	-23	312	1.8	10	IB	1,3,7	2	3	0	4	3	25	2,3	1,2,3	–
Diethyl carbonate	L	I	1.0	4.1	126	25	–	–	–	IC	1,7	2	3	1	3	3	–	1,2	1,2,3	2
Diethylene glycol	L	S	1.1	–	244	124	224	–	–	IIIB	4,7	1	1	0	3,4	3	–	1,2,3	1,2,3	2
Diethyl sulfate	L	D	1.2	–	d	104	436	–	–	IIIB	3,7,8	3	1	1	4	4	–	2,3	1,2,3	2
Diglycidyl ether	L	–	–	–	–	–	–	–	–	II	1,7,8	3	2	–	3	3	0.5	1,2,3	1,2,3	2
Diisobutyl ketone	L	I	0.8	4.9	168	49	396	0.79	6.2	II	1,7,8	1	2	0	3	3	25	1,2,3	1,2,3	2
1,1 Dimethyl hydrazine	L	V	0.8	2.0	63	-15	249	2.0	95	IB	1,4,7	3	3	1	3	3	0.5sk	1,2,3	1,2,3	2
Dimethyl sulfate	L	S	1.3	3.4	188	83	188	–	–	IIIA	1,4,8	4	2	0	3	4	0.1,A2	5,2,3	1,2,3	4
Dimethyl sulfoxide	S	S	1.1	–	189	95	215	2.6	42	IIIB	4,1	1	1	0	1	2	1.5sk*	3,2,1	5	2,3
Dinitrotoluene	S	I	1.5	6.3	300	207	–	–	–	IIIB	9,8,6,1	2	1	3	2	4	1.5sk*	3,2,1	5	5
Dioxane	L	M	1.0+	3.0	101	12	265	2.0	22	IB	1,4,8	2	3	1	3	3	50sk	1,2,3	1,2,3	2

NAME OF COMPOUND	1 PHYSICAL STATE	2 WATER SOLUBILITY	3 SPECIFIC GRAVITY	4 VAPOR DENSITY	5 BOILING POINT (°C)	6 FLASH POINT (°C)	7 IGNITION TEMPERATURE (°C)	8 FLAMMABILITY LIMITS Lower	9 FLAMMABILITY LIMITS Upper	10 NFPA CLASS	11 TYPE OF HAZARD	12 Health	13 Flammability	14 Reactivity	15 Skin	16 Oral	17 Respiratory (TLV - ppm)	18 PRECAUTIONS TO TAKE	19 FIRE EXTINGUISHING METHODS	20 REFERENCES
Epichlorhydrin	L	I	1.2	3.2	115	32	411	3.8	21	IC	3,7,8	3	2	2	4	4	2sk	2,3,5	2,3	2
Ethanethiol	L	F	0.8	2.1	35	<-18	300	2.8	18	IA	1,4,7	2	4	0	4	-	0.5	1,2,3	1,2,3	-
Ether, ethyl	L	S	0.7	2.6	35	-45	160	1.9	36	IA	1,4	2	4	1	3	3	400	1,3	0,2,3	2,3,4
Ethyl acetate	L	S	0.9	3.0	77	-4	426	2.2	11	IB	1,4	1	3	0	3	3	400	1,3	1,2,3	2
Ethyl alcohol, anhydrous	L	M	0.8	1.6	78	13	365	3.3	19	IB	1,4	0	3	0	3	2	1000	1,3	1,2,3	2,3
Ethyl benzene	L	I	0.9	3.7	136	15	430	1.0	6.7	IB	1,4	2	3	0	4	2	100	1,3	1,2,3	2
Ethyl bromide	L	F	1.4	3.8	38	-	511	6.8	8.0	IA	1,4	2	4	1	3	-	200	1,3	1,2,3	2
Ethyl chloride	L	F	0.9	2.2	12	-50	519	3.8	15.4	IA	1,5	1	4	0	3	-	1000	1,3	1,2,3	2
Ethylene	G	I	-	1.0	-104	-	490	2.7	36		1,5	1	4	2	-	-	F	1,2,3	1,2,3,6	2
Ethylenediamine	L	S	0.9	2.1	116	34	385	4.2	14.4	IC	7,4,1	3	2	0	2	2	10	2,3,1	1,2,3	2,3
Ethylene dichloride	L	F	1.3	3.4	84	13	413	6.2	16	IB	1,4,7	2	3	1	4	3	10	1,2,3	1,2,3,6	2,3,4
Ethylene oxide	G	I	0.9	1.5	11	<-18	429	3.6	100	IA	1,4,7	2	4	3	3	-	50	1,2,3	1,2,3,6	2
n-Ethylmorpholine	L	F	1.1	3.1	88	10	-	3.0	50	IB	9,1,4	2	3	0	3	-	20	4,1	5	-
Ethyl nitrite	L	S	0.9	4.0	138	32	-	-	-	IC	1,4,7	2	3	0	4	3	-	1,2,3	-	2,4
Fluorine	G	S	-	3.5	-188	-	-	-	-	IB	2,3,7	4	0	3W	4	1	1	5,4,3	0	2,4
Formaldehyde sol'n(37%)	L	S	1.0	1.0	101	85	424	7.0	73	IIIA	7,4,1	2	2	0	2,4	3	2	3,2	1,2,3	1,2,3
Formic acid	L	M	1.2	1.6	101	69	539	18	57	IIIA	7,4,1	3	2	0	4	2	5	3,2,1	1,2,3	2
Furfural	L	S	1.1	3.3	161	60	316	2.1	19.3	IIIA	8,4,1	2	2	1	4	3	5sk	3,2,1	0,2,3	2
Gasoline	L	I	0.8	3-4	30	-45	440	1.3	7.1	IB	1,4	1	3	0	3	-	-	1,3	1,2,3	2
Heptane	L	I	0.7	3.5	98	-4	215	1.05	6.7	IB	1,4	1	3	0	3	1	400	2,3	1,2,3	2
Hexane	L	I	0.7	3.0	69	-7	225	1.2	7.4	IB	1,4	1	3	0	3	1	100	1,3	1,2,3	2
Hydrazine	L	V	1.0+	1.1	113	38	-	4.7	100	II	1,3,7	3	3	2	3	-	0.1,A2	1,2,3,4	1,2,3	2,4
Hydrazoic acid	L	M	1.1	1.4	37	-	-	-	-		1,3,7	4	3	3	3	-	-	1,2,3	1,2,3	3,4
Hydrogen	G	F	0.1	0.1	-252	-	400	4.0	75		1,5	0	4	0	-	-	-	2,3	1,2,3	5
Hydrogen bromide, anhyd.	L/G	S	-	2.7	-67	-	-	-	-		7,4	3	0	0	4	4	3		-	2
Hydrogen chloride, anhyd.	L/G	V	1.2	1.2	-85	-	-	-	-		7,4	3	0	0	4	4	5	2,3,5	-	2,4
Hydrogen fluoride, anhyd.	L/G	V	0.7	0.7	19.5	-	-	-	-		3,7	4	0	0	4	4	3	5,2	-	3,4
Hydrogen iodide, anhyd.	L/G	M	-	-	-35	-	-	-	-		7,4	3	0	0	4	4	-	2,3	-	2
Hydrogen cyanide	L	M	0.7	0.9	26	-18	538	5.6	40	IA	3,1,8	4	4	2	4	6	10sk	3,1,2,5	2,1	2
Hydrogen peroxide (27%)	L	M	-	-	-	-	-	-	-		9,2,7	2	0	1	4	3	1	4,2	1	2
Hydrogen selenide	G	F	-	-	-	-	260	4.0	44		1,2,7	4	4	0	-	-	0.05	1,2,3	5,6	2
Hydrogen sulfide	G	F	-	1.2	-60	-	516				3,1	3	4	0	3	-	10	5,3,1	6	1,2,3,4
Hydroquinone	S	S	1.3	3.8	286	165				IIIB	1,4,7	2	1	0	4	5	2*	2,3	1,2,3	2
Iodine	S	S	4.9	-	-	-	-	-	-		3,6,7	3	0	0	3	5	0.1	3,2	-	2,3
Isoprene	L	I	0.7	2.4	34	-54	395	1.5	8.9	IA	1,4,7	2	4	2	3	-	-	1,2,3	1,2,3	-

Hazardous Chemical Data (continued)

NAME OF COMPOUND	PHYSICAL STATE	WATER SOLUBILITY	SPECIFIC GRAVITY	VAPOR DENSITY	BOILING POINT (°C)	FLASH POINT (°C)	IGNITION TEMPERATURE (°C)	FLAMMABILITY LIMITS Lower	FLAMMABILITY LIMITS Upper	NFPA CLASS	TYPES OF HAZARD	HAZARD RATING Health	HAZARD RATING Flammability	HAZARD RATING Reactivity	LIFE HAZARDS Skin	LIFE HAZARDS Oral	LIFE HAZARDS Respiratory (TLV – ppm)	PRECAUTIONS TO TAKE	FIRE EXTINGUISHING METHODS	REFERENCES
COLUMN NO.	1	2	3	4	5	6	7	8	9	10	11	12	13	14	15	16	17	18	19	20
Isopropyl alcohol	L	M	0.8	2.1	83	12	399	2.2	12.7	IB	1,4	1	3	0	3	3	400sk	1,3	1,2,3	-
Isopropyl ether	L	F	0.7	3.5	69	-28	443	1.4	7.9	IB	1,4	2	3	1	3	3	250	1,3	0,2,3	-
Kerosene	L	I	0.7	-	180	43	210	0.7	5	II	1,4	0	2	0	3	2	-	1,3	1,2,3	-
Ketene	G	D	-	1.4	-50	-	-	-	-	-	1,3,9	4	1	4	3,4	4	0.5	2,3,4	6,1,2,3	2
Lithium aluminum hydride	S	D	1.7	-	300	-	-	-	-	-	4,8	3	1	2W	3,4	-	-	2,3,4	0,4	2
Lithium hydride	S	D	0.8	-	-	-	-	-	-	-	1,3,9	3	4	2W	3,4	-	0.025*	1,2,3	0,4	2
Magnesium dust	S	I	1.7	-	-	-	-	-	-	-	1,7	0	1	1	-	3	-	1,5	0,4	2
Maleic anhydride	S	I	1.5	3.4	202	102	477	1.4	7.1	IIIB	7,4	3	1	1	4	3	0.25	3,5	0,2,3	4
Mercury	L	I	13.6	7	357	-	-	-	-	-	4	2	0	0	3	6	0.05	3,5	-	2,3
Mercury alkyls	L	I	2.9	-	-	-	-	-	-	-	4,8	2	0	0	4	5	0.01sk	3,2,5	-	1,2,3,4
Methyl alcohol	L	M	0.8	1.1	64	11	385	6.7	36	IB	1,4,7	1	3	0	3	3	200	1,3,5	1,2,3	1,2,3,4
Methylamine	G	V	-	1.1	-6	-	430	4.2	20.7	-	1,7,4	3	4	0	4	-	10	2,3	6	-
Methyl bromide	G	F	1.7	3.3	4	-	537	10	15	-	1,4	3	1	0	-	4	15sk	2,3	6,1,2,3	2
Methyl cellosolve acetate	L	S	1.0	4.1	144	44	393	1.7	8.2	II	1,4,7	2	2	0	3	-	25sk	2,3	1,2,3	2
Methyl chloride	G	V	-	1.8	-24	-	632	7	17	-	1,4,7	2	4	0	4	-	100	1,3,2	6	2
Methyl ethyl ketone	L	S	0.8	2.5	80	-9	404	1.9	10	IB	1,4,7	1	3	0	3	3	200	1,2,3	1,2,3	2,4
Methyl isobutyl ketone	L	F	0.8	3.5	118	18	448	1.2	8	IB	1,4,7	2	3	0	3	3	100	1,2,3	1,2,3	-
Methyl methacrylate	L	F	0.9	3.6	100	10	-	1.7	8.2	IB	9,1,4	2	3	2	3	1	100	2,3	2,3	-
Methyl salicylate	L	S	1.2	5.2	222	96	454	-	-	IIIB	1,4,7	1	1	0	3	-	-	2,3	1,2,3	-
Methyl sulfide	L	I	0.8	1.7	37	<-20	195	2.2	19.7	IA	1,3,7	2	4	0	3	-	-	1,2,3	1,2,3	-
Monoethylamine	L	M	0.7	1.6	17	-18	385	3.5	14.0	IA	1,4,7	3	4	0	4	2	10	1,2,3	1,2,3	-
Morpholine	L	M	1.0	3.0	128	38	310	2.0	11.2	-	1,4,7	2	3	0	3	-	10sk	1,2,3	2,3	-
2-Naphthylamine	S	I	1.1	4.0	306	79	526	0.88	5.9	IIIA	6,8	2	1	0	2,4	1	A1b	2,3	1,2,3	-
Nickel carbonyl	L	I	1.3	5.9	43	<-24	-	2	-	IB	3,8	4	3	3	3	2	0.05	3,2,1	1,2,3	2,4
Nickel nitrate	S	S	-	-	-	-	-	-	-	-	6,7	0	0	0	-	-	0.01*	2,3	-	-
Nicotine	L	V	1.0	5.6	246	-	244	0.75	4.0	-	3,8	4	1	0	3	4	0.5sk*	2,3	5	2,4
Nitric acid	L	S	1.5	-	86	-	-	-	-	-	2,3,7	3	0	0	4	-	2	2,4,3	1	-
Nitroaniline, para	S	M	1.4	4.8	336	199	418	1.8	-	IIIB	8,3,1	3	1	1	2,4	4	1sk	2,3,1	1,2,3	2,4
Nitrobenzene	L	F	1.2	4.3	211	88	482	-	-	IIIA	8,3,1	3	2	0	2,4	5	1sk	2,3,1	1,2,3	2,4
Nitrogen dioxide	G	S	-	-	21	-	-	-	-	-	2,3,7	3	0	0	4	-	5	5,4,3	-	1,2,3,4
Nitromethane	L	S	1.1	2.1	101	35	418	7.3	-	IC	9,1,4,7	1	3	4	4	3	100	5,1,3	1	2
Nitrotoluene, para	S	I	1.2	4.5	238	106	-	-	-	IIIB	9,7,6	3	1	1	4	4	5sk	3,2,1	1	-
2-Octanol (capryl alcohol)	L	F	0.8	4.5	184	88	-	-	-	IIIA	1,4	1	2	0	3	-	-	1,3	1,2,3	-
Octane, iso (2,2,4 Trimethylpentane)	L	I	0.7	3.9	99	-12	415	0.95	6.0	IB	1,4	0	3	0	3	-	300	1,3	1,2,3	-

Table header (columns):

Column No.	1	2	3	4	5	6	7	8	9	10	11	12	13	14	15	16	17	18	19	20
	PHYSICAL STATE	WATER SOLUBILITY	SPECIFIC GRAVITY	VAPOR DENSITY	BOILING POINT (°C)	FLASH POINT (°C)	IGNITION TEMPERATURE (°C)	FLAMMABILITY LIMITS (vol.% in air) Lower	Upper	NPFA CLASS	TYPE OF HAZARD	HAZARD RATING Health	Flammability	Reactivity	LIFE HAZARDS Skin	Oral	Respiratory (TVL – ppm)	PRECAUTIONS TO TAKE	FIRE EXTINGUISHER METHODS	REFERENCES
Osmium tetroxide	S	S		4.2	149d						6,7	1	–	–	4	1	0.0002*	2,3	–	–
Oxalic acid	S	S									6,7	1	–	0	4	4	1*	2,3	1,–	3
Paraformaldehyde	L	D	0.6	2.2	60	70	300	7.0	73	IIIA	7,4,1	3	2	0	4	4		2,3	1,2,3	3
Pentaborane	L	D	0.6	2.2	60	30	–	0.42		IC	1,3,7,8	3	3	2	4	2	0.005	2,3	1,2,3	–
2,4 Pentanedione (Acetylacetone)	L	S	1.0	3.5	140	34	340	1.7		IC	1,4	2	3	0	3	3		1,2,3	1,2,3	–
Pentane, iso	L	I	0.6	2.5	28	<-50	420	1.4	7.8	IA	1,4	1	4	0	3	3	600	1,2,3	1,2,3	2
Perchloric acid	L	V	1.6		200						9,2,7	3	0	3	4	2		4,2,5	1	–
Petroleum ether	L	I	0.6	2.5		<-18	288	1.1	5.9	IB	8,7,4,1	1	4	0	3	4	500	1,2,3	1,2,3	3,4
Phenol	L	S	1.1	3.2	181	79	715	1.8		IIIA	9,8,4,1	3	2	0	4	4	5sk	2,3,5	1,2,3	–
Phenyl hydrazine	L	F	1.1	3.7	d	88				IIIA		3	2	0	2	3	5sk	2,3,1	1,2,3	2,3
Phenyl isocyanate	L	D	1.1		165	–					9,3,8	3	2	1	4	6		3,2,5	0	4
Phosgene	G	D		3.4	7						3,7,8	4	0	1	–	–	0.1	2,3,5	5,6	2
Phosphine	G	F		1.17	-88		100				3,7,8	3	4	1	–	3	0.3	1,–	5,6	2
Phosphorus, red	S	I					260					0	1	1	1	6		1,–	1,4	–
Phosphorus, white/yellow	S	I					30				6,1,7	3	3	3	4	6	0.1*	5,2,3	1,4	3,4
Phosphorus oxychloride	L	D	1.5		107						7,3	3	0	2W	4	4		2,3,5	0,2,3	–
Phosphorus trichloride	L	D									7,3,1	3	1	2W	4	1		5,4,3	0,2,3	–
Phthalic anahydride	L	F	1.5		284	152	570	1.2	9.2	IIIB	7,6	2	2	2	2	1	0.5	2,1	1,2,3	2,4
4-Picoline	L		1.0-	3.2	144	57				II	1,3,7	2	4	0	2,3	5		1,2,3	5	3
Picric acid	S	F	1.8	7.9	>300	150	<300	1.4		IIIB	9,1,2,6	2	4	4	2,4	5	0.1sk*	1,4,3		
a-Pinene	L	F	0.9	4.7	156	33	255			IC	1,7	1	3	0	3	2		1,2,3	1,2,3	–
Piperidine	L	M	0.9	3.0	106	16				IB	1,3,7	2	3	0	2,4	3		1,2,4	0,4	–
Potassium, metal	S	D	0.8								9,1,7	3	1	2W	4	4		1,2,4	1	3
Potassium chlorate	S	S	2.3		d						6,7	0	0	3	4	4		4,2,1	1	3
Potassium cyanate	S	S	2.0									2	0	0	4	6		5,3,2	1	4
Potassium cyanide	S	V	1.5								6,7	3	0	0	4	6	5sk*	5,3,2	1	4
Potassium hydroxide	S	S	2.0								7,6	3	0	1	4	4	2	4,2,1	1	–
Potassium iodate	S	F	3.9								2,7,6	0	0	0	4	4		4,2	1	–
Potassium nitrate	S	S	2.1		d							1	0	2	4	2		4,2,1	1	–
Potassium perchlorate	S	S	2.5		d						9,2,7,6	0	0	2	4	4		4,2,1	1	3
Potassium persulfate	S	S	2.5		d	–					2,7,6	1	0	0	4	2		4,2,1	6,1,2,3	–
Propane	G	I	0.5		-42		450	2.1	9.5		1,5	1	4	0	1	1		1,2,3	1,2,3	–
1,2 Propanediol	L	M	1.0-2.8		188	99	410	2.5	12.5	IIIB	4	0	1	0	3	1		2,3	1	–
Propionic acid	L	M	1.0-2.5		147	52	465	2.9	12.1	II	1,7	2	2	0	4	–	10	1,2	1,2,3	–
Propionic anhydride	L	D	1.0-4.5		169	63	285	1.3	9.5	IIIA	1,7,4	2	2	1	4	–		4,2,1	0,2,3	–

NAME OF COMPOUND	PHYSICAL STATE	WATER SOLUBILITY	SPECIFIC GRAVITY	VAPOR DENSITY	BOILING POINT (°C)	FLASH POINT (°C)	IGNITION TEMPERATURE (°C)	FLAMMABILITY LIMITS (vol. in air) Lower	Upper	NFPA CLASS	TYPE OF HAZARD	HAZARD RATING Health	Flammability	Reactivity	LIFE HAZARDS Skin	Oral	Respiratory (TLV - ppm)	PRECAUTIONS TO TAKE	FIRE EXTINGUISHING METHODS	REFERENCES
COLUMN NO.	1	2	3	4	5	6	7	8	9	10	11	12	13	14	15	16	17	18	19	20
Propionic chloride	L	D	1.1	3.2	80	12	-	-	-	1B	1,7,4	3	3	1	4	4	-	4,2,1	0,2,3	-
Propylene	G	I	0.9	1.5	-47	-	460	2.4	11	IA	1.5	1	4	1	4	2	100	1,2,3	6,1,2,3	-
Propylene oxide	L	V	1.0-	2.0	35	-37	449	2.8	37	IB	9,1,4	2	4	2	4	2	100	1,4,3	1,2,3	2
Pyridine	L	M	1.0-	2.7	115	20	482	1.8	12	1B	1,3,7	2	3	0	2,3	3	5	1,2,3	1,2,3	2,3
Quinoline	L	L	0.9	4.5	238	-	480	1.0	-		1,4	0	1	0	3	3	-	4,2,1	1,2,3	-
Silver nitrate	S	S	4.5		d						7,6,2	0	0	0	4	2	0.01*	2,4	-	-
Sodium, metal	S	D	1		d						1,7	3	0	2#	4	6	-	5,4,1	0,4	3
Sodium arsenate	S	V	1.8								6,7	0	1	0	4	6	-	3,2	1	2,3
Sodium chlorate	S	V	2.5		d						9,2,7,6	0	0	2	2,4	6	-	5,4,2	1	2,3
Sodium cyanide	S	S	1.9								6,7	3	0	0	-	6	5sk	5,3,2	2	2,3
Sodium dichromate	S	V	2.5								6,7,2	1	0	1	4	4	0.05*	3,2,5	-	-
Sodium fluoride	S	F	2.6		d						6,7	2	0	0	1	2	2.5*	3,2,5	1	2,3
Sodium hydrosulfite	S	V	2.1								9,1,6	2	1	2	1	2	2*	1,2,3	0,2,3	3
Sodium hydroxide	S	S	2.3		d						7,6	3	0	1	4	3	-	2,3	1	3
Sodium nitrate	S	V									9,2	0	0	0	-	2	-	4,2	-	-
Sodium peroxide	S	S	2.8		d						9,2,6	3	0	2#	4	3	-	4,3,2	0,3	2,4,5
Sodium sulfide	S	S	1.9		-17						1,6	1	1	0	1	2	-	1,3	1	-
Stibine	G	S		4.3	14		76				3,7	2	0	0	4	-	0.1	5	-	2
Sulfur dioxide	G	M		2.3							3	2	0	0	4	-	2	3,5	-	2,4,5
Sulfuric acid	L	M	1.8	-	327		537				7,3,2	3	1	2#	4	3	1*	3,4	-	2,3,4
Tetraethyl lead	L	I	1.6	8.6	110d	93	110d	1.8	11.8	IIIA	1,3,8	3	2	3	3	4	0.1sk*	1,2,3,5	1,2,3	-
Tetrahydrofuran	L	I	0.9	2.5	66	-14	321	2.0	7.1	IB	1,7	3	3	0	3	4	200	1,3	1,2,3	3
Toluene	L	F	0.9	3.1	111	4	480	1.2	7.1	IB	1,4,7	2	3	0	4	4	100	1,2,3	1,2,3	1,2,3,4
Trichloroacetic acid	S	I	1.6	5.6	197	-	-				4,7	2	1	1	3,4	-	1*	2	-	-
1,1,1 Trichloroethane	L	I	1.3	4.5	74	-	537	8.0	10.5		4,7,8	2	1	0	-	-	350	2,3	-	2,3
1,1,2 Trichloroethane	L	I	1.3	4.5	113	-	76	11	41	IC	4,7,8	3	1	0	3	4	10sk	2,3	-	2,3
Trichloroethylene	L	I	1.5	4.5	87	32	420	12	40	IC	4,7	1	1	0	3	2	100	3,2	1,2,3	1,2,3,4
Xylene, ortho	L	I	0.9	3.7	144	32	465	1.1	6.4		1,4,7	2	3	1	1	4	100	1,3,2	1,2,3	1,2,3,4
Zinc, dust	S	I									1,6	0	1	1	-	3	-	1,3,5	3	2,3

Appendix 2

The Carcinogen Assessment Group's List of Carcinogens

July 14, 1980

(Confirmed April 6, 1982.)

CHEMICALS HAVING EVIDENCE OF CARCINOGENICITY

In response to requests from several EPA offices, the Carcinogen Assessment Group (CAG), Office of Health and Environmental Assessment in EPA's Research and Development Office has prepared a list of chemical substances for which substantial or strong evidence exists showing that exposure to these chemicals, under certain conditions, causes cancer in humans, or can cause cancer in animal species which in turn, makes them potentially carcinogenic in humans.

The list was initially prepared in response to the needs of the Office of Pesticides and Toxic Substances (OPTS) to develop labeling regulations under Section 6 of TSCA and the Office of Solid Waste (OSW) to develop hazardous waste regulations under Section 3001 of RCRA. It is anticipated that it will serve other purposes within the agency according to the needs of the program offices.

The sources of information used in selecting agents as candidates for the list are of two types: chemicals which CAG previously has evaluated and has determined pose a potential human cancer risk; and chemicals, the carcinogenicity of which the CAG reviewed because one or more of three organizations—the International Agency for Research on Cancer (IARC), the National Cancer Institute Bioassay Program which has been reorganized into the National Toxicology Program (NTP), and the FDA of the U.S. Department of Health and Human Services—had concluded that these chemicals are potentially human carcinogens. (Chemicals regulated as carcinogens by OSHA and CPSC are also on this list but are not noted as such since they have been evaluated as being carcinogens by one of the other organizations previously mentioned). CAG evaluated the studies upon which IARC, NTP, or FDA relied and agreed with all the NTP and FDA evaluations that the chemicals presented a potential human cancer risk. CAG agreed with most of IARC's evaluations. There are inconsistencies between the CAG and IARC evaluations for a few chemicals because the CAG considered information not available to or not otherwise used by IARC, and because there are differences in the criteria used in making the qualitative evaluations.

The list is not a comprehensive listing of *all* chemicals having substantial or strong evidence of carcinogenicity. As the CAG continues to analyze chemicals for carcinogenicity, chemicals which do not now appear on the list will be added. A continuing review of evaluations by organizations such as IARC, NTP, FDA, OSHA, and CPSC may result in periodic revisions to the present list.

The CAG evaluates substances for possible carcinogenicity according to the procedures outlined in the agency's Interim Guidelines for Carcinogen Risk Assessment found in Interim Procedures and Guidelines for Health Risk and Economic Impact Assessments of Suspected Carcinogens (41 *Fed. Reg.* 21402, May 25, 1976). These guidelines are consistent with the Inter-agency Regulatory Liaison Group's Scientific Basis for Identification of Potential Carcinogens and Estimation of Risks [*Journal of the National Cancer Institute*, **63** (1), 243–268 (1979), 44 *Fed. Reg.* 39858 (July 6, 1979)], and the Regulatory Council Statement on Regulation of Chemical Carcinogens [44 *Fed. Reg.* 760037 (October 17, 1979)].

Evidence concerning the carcinogenicity of chemical substances is of three types: (1) epidemiologic evidence derived from studies of exposed human populations; (2) experimental evidence derived from long-term bioassays on animals; and (3) supportive or suggestive evidence derived from studies of chemical-structure or from short-term mutagenicity, cell transformation, or other tests that are believed to correlate with carcinogenic activity.

CAG evaluates all available evidence on the carcinogenicity of a chemical before reaching a conclusion based on the "weight of the evidence," about the chemical's human carcinogenic potential. Conclusions about the overall weight of evidence involve a consideration of the quality and adequacy of the data and the kinds of responses induced by the suspect carcinogen. The best evidence that an agent is a human carcinogen comes from epidemiologic studies in conjunction with confirmatory animal tests. Substantial evidence is provided by animal tests that demonstrate the induction of malignant tumors in one or more species or of benign tumors that are generally recognized as early stages of malignancies. Suggestive evidence includes indirect tests of tumorigenic activity, such as mutagenicity, in vitro cell transformation, and initiation-promotion skin tests in mice. Ancillary data that bear on judgments about carcinogenic potential, for example, evidence from systematic studies that relate chemical structure to carcinogenicity, are also considered.

Substances were placed on the CAG list only if they had been demonstrated to induce malignant tumors in one or more animal species or to induce benign tumors that are generally recognized as early stages of malignancies, and/or if positive epidemiologic studies indicated they were carcinogenic. Although the CAG has determined that there is substantial evidence of carcinogenicity for each chemical substance on the list, the data varies to some extent with respect to the scope and quality of the studies.

Not uncommonly, CAG reports are updated because new evidence becomes available. Because of this, it is important that the most recent CAG evaluation be consulted.

Some of the reports prepared by CAG are subject to confidentiality claims. Because of these claims (primarily under the Federal Insecticide, Fungicide, and Rodenticide Act) some reports may not be released. Therefore, all requests for CAG reports and related documentation must be submitted through EPA's Freedom of Information Office (A-101), Washington, D.C. 20460, and should be marked CAG/LOC.

2-Acetylaminoflourene
Acrylonitrile (CAG, IARC)
Aflatoxins (IARC)*
Aldrin (CAG, NCI)
4-Aminobiphenyl (IARC)
Amitrole (IARC)
Aramite (IARC)
Arsenic and arsenic compounds (CAG, IARC)
Asbestos (CAG, IARC)
Auramine and the manufacture of auramine (IARC)
Azaserine (IARC)†
Benz(c)acridine (IARC)‡
Benz(a)anthacene (IARC)
Benzene (CAG, IARC)
Benzidine (CAG, IARC)

This is not a comprehensive list of *all* chemicals having substantial evidence of carcinogenicity. Other chemicals will be added. No attempt has been made to select chemicals based upon appropriateness for regulation by EPA. The list is intended to be a basis for selection by the various program offices according to their specific needs.
*Fungal toxin, not an industrially manufactured product.
†Used as a drug.
‡Evaluated by IARC as not having sufficient evidence of carcinogenicity.

Benzo(a)pyrene (IARC)

Benzo(b)fluoranthene (IARC)

Benzo(j)fluoranthene (IARC)‡

Beryllium and Beryllium Compounds (CAG, IARC)

N,N-Bis(2-Chloroethyl)-2-Napthylamine (Chlornaphazine) (IARC)†

Cadmium and Cadmium Compounds (CAG, IARC)

Carbon Tetrachloride (CAG, IARC)

Chlorambucil (IARC)†

Chloroalkyl Ethers
  Bis(2-chloroethyl)ether (BCEE) (CAG), (IARC)‡
  Bis(chloromethyl)ether (BCME) (CAG, IARC)
  Chloromethyl methyl ether (CMME), technical grade (IARC)

Chlordane (CAG, NCI)

Chlorinated Ethanes
  1,2-Dichloroethane [Ethylene Chloride, Ethylene Dichloride (EDC)]
(CAG, IARC, NCI)
Hexachloroethane (CAG)
  1,1,2,2-Tetrachloroethane (CAG)
  1,1,2-Trichloroethane (CAG, NCI, IARC)†

Chlorobenzilate (CAG)

Chloroform (CAG, IARC)

Chromium Compounds, Hexavalent (CAG, IARC)

Chrysene (IARC)‡

Citrus Red No. 2 (IARC)

Coal Tar and Soot (CAG, included in IARC's soots, tars, and oils designation)

Coke Oven Emissions [Polycyclic Organic Matter (POM)] (CAG)

Creosote (CAG)

Cycasin (IARC)

Cyclophosphamide (IARC)†

Daunomycin (IARC)†

DDT (dichlorodiphenyltrichloroethane) (CAG)

Diallate (CAG) (IARC)‡

Dibenz(a,h)acridine (IARC)

Dibenz(a,j)acridine (IARC)

Dibenz(a,h)anthracene (IARC)

7H-Dibenzo(c,g)carbazone (IARC)

Dibenzo(a,e)pyrene (IARC)

Dibenzo(a,h)pyrene (IARC)

†Used as a drug.
‡Evaluated by IARC as not having sufficient evidence of carcinogenicity.

Dibenzo(a,j)pyrene (IARC)

1,2-Dibromo-3-chloropropane (DBCP) (CAG, IARC, NCI)

1,2-Dibromoethane [Ethylene Bromide, Ethylene Dibromide (EDB)] (NCI, CAG, IARC)

3,3'-Dichlorobenzidine (DCB) (CAG, IARC)

Dieldrin (CAG)

Diepoxybutane (IARC)

1,2-Diethylhydrazine (IARC)

Diethylstilbestrol (DES) (IARC)†

Dihydrosafrole (IARC)

3,3'-Dimethoxybenzidine (o-Dianisidine) (IARC)

p-Dimethylaminoazobenzene (IARC)

7,12-Dimethylbenz(a)anthracene (see Appendix 9)

3,3'-Dimethylbenzidine (o-Tolidine) (IARC)

Dimethylcarbamoyl Chloride (IARC)

1,1-Dimethylhydrazine (IARC)

1,2-Dimethylhydrazine (IARC)

Dimethyl Sulfate (IARC)

2,4-Dinitrotoluene (CAG, NCI)

1,4-Dioxane (NCI)

1,2-Diphenylhydrazine (CAG)

Epichlorohydrin (CAG)

Ethylene Bis Dithiocarbamate (EBDC) (CAG)

Ethyleneimine (Aziridine) (IARC)‡

Ethylene Oxide (CAG, IARC)

Ethylenethiourea (CAG, IARC)

Ethyl Methanesulfonate (IARC)

Formaldehyde (CAG)

Glycidaldehyde (IARC)

Heptachlor (CAG, NCI)

Hexachlorobenzene (CAG, IARC)

Hexachlorobutadiene (CAG)

Hexachlorocyclohexane (HCH)
  α HCH (CAG)
  β HCH (CAG)
  γ HCH (Lindane) (CAG)
  Technical HCH (CAG)

Hydrazine (IARC)

†Used as a drug.
‡Evaluated by IARC as not having sufficient evidence of carcinogenicity.

Indeno (1,2,3-cd)pyrene (IARC)

Iron Dextran (IARC)†,‡

Isosafrole (IARC)

Kepone (Chlordecone) (CAG, NCI)

Lasiocarpine (IARC, NCI)

Melphalan (IARC)†

Methapyrilene (FDA)†

3-Methylcholanthrene

4,4'-Methylenebis(2-Chloroaniline) (MOCA) (IARC)

Methyl Iodide (CAG, IARC)

Methyl Methanesulfonate (IARC)

N-Methyl-N'-nitro-N-nitrosoguanidine (IARC)

Methylthiouracil (IARC)†

Mitomycin C (IARC)†

Mustard Gas (IARC)

1-Naphthylamine, technical grade (CAG)

2-Naphthylamine (IARC)

Nickel and Nickel Compounds (CAG, IARC)

Nitrogen Mustard and its hydrochloride (IARC)

Nitrogen Mustard N-oxide and its hydrochloride (IARC)

5-Nitro-o-toluidine (NCI)

4-Nitroquinoline-1-oxide

Nitrosamines

   N-Nitrosodiethanolamine (IARC)

   N-Nitrosodiethylamine (DENA) (CAG, IARC)

   N-Nitrosodimethylamine (DMNA) (CAG, IARC)

   N-Nitrosodi-n-butylamine (IARC)

   N-Nitrosodi-n-propylamine (IARC)

   N-Nitrosomethylethylamine (IARC)

   N-Nitrosomethylvinylamine (IARC)

   N-Nitroso-N-Ethylurea (NEU) (CAG, IARC)

   N-Nitroso-N-Methylurea (NMU) (CAG, IARC)

   N-Nitroso-N-methylurethane (IARC)

   N-Nitrosomorpholine (IARC)

   N-Nitrosonornicotine (IARC)

   N-Nitrosopiperidine (IARC)

   N-Nitrosopyrrolidine (IARC)

   N-Nitrososarcosine (IARC)

Pentachloronitrobenzene (PCNB) (CAG)

†Used as a drug.
‡Evaluated by IARC as not having sufficient evidence of carcinogenicity.

Phenacetin (IARC)†

Polychlorinated Biphenyls (PCBs) (CAG, IARC)

Pronamide (CAG)

1,3-Propane Sultone (IARC)

β-Propiolactone (IARC)

Propylthiouracil (IARC)†

Reserpine (NCI)†

Saccharin (FDA)§

Safrole (CAG, IARC)§

Selenium Sulfide (NCI)

Streptozotocin (IARC)†

2,3,7,8-Tetrachlorodibenzo-p-dioxin (TCDD) (CAG)

Tetrachloroethylene (Perchloroethylene) (CAG, NCI)

Thioacetamide (IARC)

Thiourea (IARC)

o-Toluidine Hydrochloride (NCI)

Toxaphene (CAG, IARC, NCI)

Trichloroethylene (CAG, NCI)

2,4,6-Trichlorophenol (NCI)

Tris)1-aziridinyl)phosphine sulfide (Thio-TEPA) (IARC, NCI)†

Tris(2,3-dibromopropyl)phosphate (IARC, NCI)

Trypan Blue, commercial grade (IARC)

Uracil Mustard (IARC)†

Urethane (IARC) (Ethyl carbamate; ethyl ester of carbamic acid)

Vinyl Chloride (CAG, IARC)

Vinylidene Chloride (CAG)

†Used as a drug.
§Used as a food.

Appendix 3

List of IARC Monographs

The following is a list of IARC monographs on the evaluation of the carcinogenic risk of chemicals to humans, published in Lyon, France.

Vol. 1.—*Some Inorganic Substances, Chlorinated Hydrocarbons, Aromatic Amines, N-Nitroso Compounds, and Natural Products,* 1972.

Vol. 2.—*Some Inorganic and Organometallic Compounds,* 1973.

Vol. 3.—*Certain Polycyclic Aromatic Hydrocarbons and Heterocyclic Compounds,* 1973.

Vol. 4.—*Some Aromatic Amines, Hydrazine and Related Substances, N-Nitroso Compounds and Miscellaneous Alkylating Agents,* 1974.

Vol. 5.—*Some Organochlorine Pesticides,* 1974.

Vol. 6.—*Sex Hormones,* 1974.

Vol. 7.—*Some Anti-Thyroid and Related Substances, Nitrofurans and Industrial Chemicals,* 1974.

Vol. 8.—*Some Aromatic Azo Compounds,* 1975.

Vol. 9.—*Some Aziridines, N-, S- and O-Mustards and Selenium,* 1975.

Vol. 10.—*Some Naturally Occurring Substances,* 1976.

Vol. 11.—*Cadmium, Nickel, Some Epoxides, Miscellaneous Industrial Chemicals and General Considerations on Volatile Anaesthetics,* 1976.

Vol. 12.—*Some Carbamates, Thiocarbamates and Carbazides,* 1976.

Vol. 13.—*Some Miscellaneous Pharmaceutical Substances,* 1977.

Vol. 14.—*Asbestos,* 1977.

Vol. 15.—*Some Fumicants, the Herbicides 2,4-D and 2,4,5-T, Chlorinated Dibenzodioxins and Miscellaneous Industrial Chemicals,* 1977.

Vol. 16.—*Some Aromatic Amines and Related Nitro Compounds—Hair Dyes, Colouring Agents and Miscellaneous Industrial Chemicals, 1978.*

Vol. 17.—*Some N-Nitroso Compounds, 1978.*

Vol. 18.—*Polychlorinated Biphenyls and Polybrominated Biphenyls, 1978.*

Vol. 19.—*Some Monomers, Plastics and Synthetic Elastomers, and Acrolein, 1979.*

Vol. 20.—*Some Halogenated Hydrocarbons, 1979.*

Supplement 1.—*Chemicals and Industrial Processes Associated with Cancer in Humans, 1979.*

Vol. 21.—*Sex Hormones, 1979.*

Vol. 22.—*Non-Nutritive Sweetning Agents, 1980.*

Vol. 23.—*Some Metals and Metallic Compounds, 1980.*

Vol. 24.—*Some Pharmaceutical Drugs, 1980.*

Vol. 25.—*Wood, Leather, and Some Associated Industries, 1981.*

Vol. 26.—*Anti-Neoplastic and Immunosuppressive Agents, 1981.*

Vol. 27.—*Aromatic Amines, Anthraquinones, Nitroso Compounds, and Inorganic Fluorides Used in Drinking Water and Preparations, 1982.*

Vol. 28.—*Exposures in the Rubber Manufacturing Industry, 1982.*

Compounds Evaluated by IARC But Not as Having Sufficient Evidence of Carcinogenicity

Compound	IARC Monograph
Benz(c)acridine	**3**, 241
Benzo(j)fluoranthene	**3**, 82
Bis(2-chloroethyl) ether	**9**, 117
Chrysene	**3**, 159
Diallate	**12**, 69
Ethyleneimine (aziridine)	**9**, 37
Iron detran	**2**, 161
1,1,2-Trichloroethane	**20**, 533

Appendix **5**

Sources of Information

1 NATIONAL ORGANIZATIONS

Air Pollution Control Association.   211 S. Dithridge, Pittsburgh, PA 15213.

The American Academy of Industrial Hygiene.   475 Wolf Ledges Parkway, Akron, OH 44311.

The American Academy of Occupational Medicine.   150 North Wacker Drive, Chicago, IL 60606.

American Association of Occupational Health Nurses, Inc.   (AAOHN).   3500 Piedmont Ave, N.E., Atlanta, GA 30324.

American Chemical Society.   1155 Sixteenth Street, N.W., Washington, D.C. 20036. (Division of Chemical Health & Safety)

American Conference of Governmental Industrial Hygienists, Inc.   Executive Secretary, Mr. W. D. Kelley. 2205 South Road, Cincinnati, OH 45238.

American Industrial Hygiene Association.   William E. McCormick, Managing Director. 475 Wolf Ledges Parkway, Akron, OH 44311.

American Institute of Chemical Engineers.   345 E. 47th St., New York, N.Y. 10017. (Division of Safety & Health.)

American Medical Association.   535 North Dearborn Street, Chicago, IL 60610.

American National Standards Institute.   1430 Broadway, New York, NY 10018.

American Occupational Medical Association.   150 North Wacker Drive, Chicago, IL 60606.

American Public Health Association.   1015 15th St., N.W., Washington, D.C. 20005.

American Society for Testing and Materials.   1916 Race Street, Philadelphia, PA 19103.

889

American Society of Safety Engineers.   850 Busse Highway, Park Ridge, IL 60068.

The British Occupational Hygiene Society.   Nuffield Dept. of Ind. Health, Medical School, The University, Newcastle-upon-Tyne NEI 7RU.

British Safety Council.   62-64 Chancellors Rd. London W6 9RS.

Chemical Industry Institute of Toxicology.   Research Triangle Park, NC 27709.

Industrial Health Foundation, Inc.   5231 Centre Avenue, Pittsburgh, PA 15232.

National Safety Council.   444 N. Michigan Avenue, Chicago, IL 60611.

Society of Toxicology.   475 Wolf Ledges Parkway, Akron, OH 44311.

Health Physics Society.   4720 Montgomery Lane, S.

2   INDUSTRIAL AND TRADE ASSOCIATIONS

American Iron and Steel Institute.   1000 16th Street, N.W., Washington, D.C. 20036.

The Aluminum Association.   818 Connecticut Avenue, N.W., Washington, D.C. 20006.

American Foundrymen's Society.   Golf and Wolf Roads, Des Plaines, IL 60016.

American Petroleum Institute.   2101 L Street, N.W., Washington, D.C. 20037.

American Society for Metals.   Metals Park, OH 44073.

American Welding Society.   2501 Northwest Seventh Street, Miami, FL 33125.

Association of American Railroads.   1920 L St., N.W., Washington, D.C. 20036.

Chemical Manufacturers' Association.   2501 M Street, N.W., Washington, D.C. 20037.

Chemical Specialties Manu. Assoc.   1001 Conn. Ave., N.W., Washington, D.C. 20036.

Chlorine Institute.   342 Madison Ave., New York, N.Y. 10017.

Compressed Gas Association, Inc.   500 Fifth Avenue, New York, NY 10110.

The Fertilizer Institute.   1015 18th Street, N.W., Washington, D.C. 20036.

Forging Industry Association.   1121 Illuminating Building, Cleveland, OH 44113.

Investment Casting Institute.   8521 Clover Meadow, Dallas, TX 75243.

Metal Finishers Supplies Association.   1025 E. Maple Road, Birmingham, MI 48011.

Metal Powder Industries Federation. 105 College Rd., E., P.O. Box 2054, Princeton, NJ 08540.

Metal Treating Institute. 1300 Executive Center, Suite 115, Tallahassee, FL 32301.

National Association of Metal Finishers. 111 E. Wacker Drive, Chicago, IL 60601.

National Tank Truck Carriers, Inc. 1616 P St., N.W. Washington, D.C. 20036.

Society of Manufacturing Engineers. P. O. Box 930, Dearborn, MI 48128.

The Society of the Plastics Industry. 355 Lexington Avenue, New York, NY 10017.

3 FEDERAL AGENCIES

Department of Commerce
National Bureau of Standards. Washington, D.C. 20234.

National Technical Information Service (NTIS). 5285 Port Royal Road, Springfield, VA 22151.

Department of Defense
Explosives Safety Board, 2461 Eisenhower Ave., Alexandria, VA 22331.

Department of Health and Human Services
Public Health Service
Center for Disease Control, Atlanta, GA 30333.

National Institute for Occupational Safety and Health (NIOSH). Parklawn Building, 5600 Fishers Lane, Rockville, MD 20857.

Bureau of Radiological Health. 12720 Twinbrook Parkway, Rockville, MD 20852.

National Clearinghouse for Poison Control Centers. 5401 Westbard Avenue, Bethesda, MD 20816.

National Institutes of Health
National Cancer Institute. Bethesda, MD 20205.

National Heart and Lung Institute. Bethesda, MD 20205.

National Institute of Environmental Health Sciences (NIEHS). P. O. Box 12233, Research Triangle Park, NC 27709.

National Library of Medicine. 8600 Rockville Pike, Bethesda, MD 20814.

Department of Labor
Bureau of Labor Statistics. 200 Constitution Avenue, N.W., Washington, D.C. 20210.

Mine Safety and Health Administration. 4015 Wilson Bvld., Arlington, VA 22203.

Occupational Safety and Health Administration.   200 Constitution Avenue, N.W., Washington, D.C. 20210.

Environmental Protection Agency

Air and Water Programs Office. 401 M Street, S.W., Washington, D.C. 20460.

Pesticides Programs Office. 401 M Street, S.W., Washington, D.C. 20460.

Radiation Programs Office. Rockville, MD 20850.

Water Program Operations Office. 401 M Street, S. W., Washington, D.C. 20460.

Air Pollution Technical Information Center (APTIC), Office of Technical Information and Publications (OTIP), Air Pollution Control Office (APCO). P. O. Box 12055, Research Triangle Park, NC 27709.

National Environmental Research Center. Research Triangle Park, NC 27711 (also located in Cincinnati, OH 45268; Corvallis, OR 97330; and Las Vegas, NV 89114).

Western Environmental Research Laboratory. P. O. Box 15027, Las Vegas, NV 89114.

National Council on Radiation Protection and Measurements (NCRP). 7910 Woodmont Avenue, Suite 1016, Bethesda, MD 20814.

Nuclear Regulatory Commission. 1717 H Street, N.W., Washington, D.C. 20555.

U.S. Government Printing Office, Washington, D.C. 20402.

4   UNIONS

AFL-CIO Dept. of Health and Safety
AFL-CIO Building, 815 16th St., N.W.
Room 507
Washington, D.C. 20006.

Amalgamated Clothing & Textile Workers Union
770 Broadway
New York, New York 10003

Boilermakers, International Brotherhood of—
  Local 802
P.O. Box 618
Chester, Pennsylvania 19016

Chemical Workers Union, International
1655 W. Market Street
Akron, Ohio 44313

IUE—Local 201
Health and Safety Committee
100 Bennett Street
Lynn, Massachusetts 01905

Molders and Allied Workers Union
1216 E. McMillan Street
Suite 302
Cincinnati, Ohio 45206

Oil, Chemical and Atomic Workers
1636 Champa
P.O. Box 2812
Denver, Colorado 80201

Rubber, Cork, Linoleum and Plastic Workers
    of America, United
URWA Building
87 S. High Street
Akron, Ohio 44308

Steelworkers of America, United
Five Gateway Center
Pittsburgh, Pennsylvania 15222

United Auto Workers
Solidarity House
8000 East Jefferson Avenue
Detroit, Michigan 48214

Teamsters, International Brotherhood of
25 Louisiana Avenue, N.W.
Washington, D.C. 20001

Author Index

Subject Index